Designing cdma2000® Systems

John Wiley & Sons, Ltd.
The Atrium, Southern Gate
Chichester, PO19 8SQ, England
+44(0)1243-770277 (work)
+44(0)7718-577081 (mobile)
   mhammond@wiley.co.uk
http://www.wileyeurope.com/go/commstech

Designing cdma2000® Systems

Leonhard Korowajczuk
CelPlan Technologies, Inc.

Bruno de Souza Abreu Xavier
CelTec Technologia de Telecomunicações, Ltda.

Arlindo Moreira Fartes Filho
CelTec Technologia de Telecomunicações, Ltda.

Leila Zurba Ribeiro
CelPlan Technologies, Inc.

Cristine Korowajczuk
CelPlan Technologies, Inc.

Luiz A. Da Silva
Virginia Tech

JOHN WILEY & SONS, LTD

Email (for orders and customer service enquiries): cs-books@wiley.co.uk
Visit our Home Page on www.wileyeurope.com or www.wiley.com

Reprinted July 2004

This publication is designed to provide accurate and authoritative information in regard to the subject matter
covered. It is sold on the understanding that the Publisher is not engaged in rendering professional services. If
professional advice or other expert assistance is required, the services of a competent professional should be
sought.

Other Wiley Editorial Offices

John Wiley & Sons Inc., 111 River Street, Hoboken, NJ 07030, USA

Jossey-Bass, 989 Market Street, San Francisco, CA 94103-1741, USA

Wiley-VCH Verlag GmbH, Boschstr. 12, D-69469 Weinheim, Germany

John Wiley & Sons Australia Ltd, 33 Park Road, Milton, Queensland 4064, Australia

John Wiley & Sons (Asia) Pte Ltd, 2 Clementi Loop #02-01, Jin Xing Distripark, Singapore 129809

John Wiley & Sons Canada Ltd, 22 Worcester Road, Etobicoke, Ontario, Canada M9W 1L1

Wiley also publishes its books in a variety of electronic formats. Some of the content that appears in
print may not be available in electronic books.

British Library Cataloguing in Publication Data

A catalogue record for this book is available from the British Library

0-470-85399-9

Typeset in 10/12pt Times by Thomson Press (India) Limited, New Delhi
Printed and bound in Great Britain by TJ International, Padstow, Cornwall
This book is printed on acid-free paper responsibly manufactured from sustainable forestry
in which at least two trees are planted for each one used for paper production.

This book is dedicated to

Our families for their support and love.

Trademarks

cdma2000® is a registered trademark of the Telecommunications Industry Association (TIA–USA)

CelPlan® is a registered trademark of CelPlan Technologies, Inc. (CelPlan–USA)

CelPlanner™, CelOptima™, CelTools™, CelEnhancer™, CelLink™, CelPerformance™, CelData™ are trademarks of CelPlan Technologies, Inc. (CelPlan–USA)

Contents

Preface

Wireless technology had a growth as explosive as the number of mobile subscribers in the past few years. It evolved from simple analogue radio technology to sophisticated adaptive technologies, employing advanced hardware, modulation and coding techniques and processing solutions not imaginable a decade ago.

Several of the new techniques are too complex to be predicted by us, and trial and error solutions have been used to maintain even the simplest CDMA networks. These new techniques have increased network capacity well beyond our initial expectations but there is still plenty of room to evolve before we reach Shannon's predicted capacity limits.

The design of new networks is becoming extremely complex, mainly when based on CDMA techniques, particularly in the case of advanced cdma2000 networks. Existing network designs do not optimise resource usage and are usually below their achievable capacity.

The objective of this book is to provide engineers with the background required to understand the cdma2000 technology, describing design techniques and available algorithms.

cdma2000 is one of the technologies selected to be a part of the ITU-R IMT-2000 solutions for the evolution of wireless communications. From the selected technologies it is certainly the one most widely deployed worldwide. cdma2000 is a highly standardised technology, fully compatible with its previous generation, the IS-95. These standards are very extensive and built upon previous documents; therefore reading them becomes a cumbersome and difficult chore.

This book started as a CDMA training class provided by CelPlan Technologies, Inc. for its customers who were adopting the new technology. With the technology evolution several customers requested that the contents of the class be expanded to cover the new technologies and the advances in network design. We felt that the developed material could be beneficial to others, so the idea of writing a book about the subject was born.

This book is an effort of many hands, with the contributions of different authors for its chapters. Even though some topics may be repeated in part between chapters, we felt that this would provide a diversity of views and keep each chapter more independent of the others. The idea is to allow readers to understand each chapter by just having a general knowledge about the subject.

The authors encourage reader comments and suggestions through CelPlan Technologies, Inc. website at www.celplan.com, where a link is available for readers to send messages to the authors.

The first part of this book consists of Chapters 1–8 and covers the main standards involved in the technology specification. The idea is not to replace the standards but to explain them in a more didactical manner focusing mainly on the aspects required for the design of CDMA systems.

Chapter 1 provides general information about spread-spectrum systems, their basic concepts and their use in the CDMA technology.

Chapter 2 covers the evolution of commercial CDMA systems from the initial IS-54 to the different flavours of cdma2000, such as 1X RTT, EVDO and EVDV. It provides also a brief explanation about the logical channels and their functionality.

Chapter 3 addresses spreading codes used in CDMA covering Maximum Length Sequences (MLS), Walsh codes and quasi-orthogonal functions, explaining specific functionalities of each sequence and its application in CDMA technology.

Chapters 4 and 5 describe, respectively, the forward and reverse link channels used in IS-95 and cdma2000 presenting their structure and functionality.

Chapter 6 covers call processing in CDMA systems describing call states, processes and main messages. It describes also the security and identification process performed during the authentication process.

Chapter 7 describes the resource management, covering handoff, authentication and power control processes. It analyses how the network can benefit from and efficient resource management that optimises code allocation and explores fast power control capabilities to perform sophisticated handoffs.

Chapter 8 presents the EVDO and EVDV standards.

The second part of this book consists of Chapters 9–11 and covers practical network design aspects.

Chapter 9 provides the fundamentals of CDMA radio network engineering covering the topics required by engineers that design or maintain CDMA networks.

Chapter 10 provides insights into the network design process, suggesting a design sequence and highlighting design procedures to be taken during the design.

Chapter 11 covers traffic issues related to existing and the new packetised voice and data networks.

Leonhard Korowajczuk

CEO and CTO of CelPlan Technologies, Inc.

Acknowledgements

The authors would like to acknowledge CelPlan Technologies, Inc. for its support in writing this book and the disclosure of some proprietary algorithms and procedures.

The authors would like to acknowledge that this book started from works done by Mr. Samuel Rocha Lauretti and Mr. Mario Penas Marcos in the preparation of CelPlan's training manuals and some of the contents are based on their initial work.

The authors have a lot of gratitude for the work performed by Ms. Cristine Korowajczuk, who revised and edited every chapter, standardising the overall presentation and contents.

We will be ever indebted to Dr. Leonard Miller who provided comments and technical guidance on the different topics of the book, and his infinite patience in discussing some topics with the authors.

The manuscript was commented in different phases of its preparation by Mr. Gustavo Nader and Mr. Antonio Vivaldi Rodrigues.

Finally we must thank Mr. Mark Hammond publisher from John Wiley & Sons, Ltd. and his staff, Ms. Sophie Evans and Daniel Gill, for their ever-continuing support and understanding.

About the Authors

Leonhard Korowajczuk – *CelPlan Technologies, Inc.*

He graduated from the Universidade Federal do Rio de Janeiro in 1969, and has 34 years experience in the telecommunications R&D field. Previous experience include ITT telecommunication engineering departments (Brazil, England, Belgium and Spain). He was one of the founders of the Brazilian Telecommunications Research Center (Telebras-CPqD), Switching department manager and wireless division director at Elebra Telecomunicações/Alcatel. CEO and CTO of CelPlan Technologies, Inc., and also one of its founders in 1992. Was responsible for the design and deployment of wireless networks as early as 1988 and headed the team that developed an advanced MTSO and base station for Comsat-Plexsys from 1993 to 1996. Since then he has led the development of the family of planning and design tools provided by CelPlan Technologies, Inc. Holds several patents in the telecom field and is a member of IEEE.

Bruno de Souza Abreu Xavier – *CelTec Tecnologia de Telecomunicações, Ltda.*

Telecommunications and RF engineer; he graduated as electrical engineer from the Federal University of Minas Gerais (UFMG) in 1996 and did MS from the University of Campinas (UNICAMP), researching on land-satellite CDMA performance analysis. He is responsible for the design of several wireless networks in South America and for training classes on planning, deployment and comissioning of wireless equipment infrastructure.

Arlindo Moreira Fartes Filho – *CelTec Tecnologia de Telecomunicações, Ltda.*

Senior systems engineer; graduated as an electrical engineer from the University of Campinas (UNICAMP) in 1975 and completed MS in 1978. He has several years experience in the telecommunications R&D field, covering system specifications and software. He participated in the design and development of several telecom equipments (harware and software). He planned and supervised the design of nationwide telecommunication networks, covering transmission, switching and wireless elements.

Dr. Leila Zurba Ribeiro – *CelPlan Technologies, Inc.*

Director of Systems Engineering at CelPlan Technologies, Inc.; she received her Ph.D. in Electrical Engineering from Virginia Tech in May, 2003. She joined CelPlan Technologies, Inc., in 1993, and her work experience includes the design, traffic dimensioning and optimisation of wireless communication systems in several countries. She has also taught graduate courses as an adjunct professor at the Electrical and Computer Engineering Department of George Mason University. She is a member of IEEE.

Cristine Korowajczuk – *CelPlan Technologies, Inc.*

She graduated in 1999 in computer engineering from PUCC (Pontifícia Universidade Católica de Campinas), and has more than 10 years experience in the wireless field with CelTec and CelPlan. She has designed and optimised several wireless networks covering all main technologies. She is responsible for customer support, documentation and training at CelPlan Technologies, Inc.

Dr. Luiz A. DaSilva – *Virginia Tech*

He joined Virginia Tech as an assistant professor at the Bradley Department of Electrical and Computer Engineering in 1998, after receiving his Ph.D. in electrical engineering at the University of Kansas. He has previously worked for IBM for six years. His research interests currently focus on performance and resource management in wireless mobile networks and Quality of Service (QoS) issues. He is a member of the Center for Wireless Communications (CWT) and associated faculty at the Mobile and Personal Radio Group (MPRG) at Virginia Tech. He is a senior member of IEEE and a member of ASEE.

1

Introduction to Spread Spectrum Systems

BRUNO DE SOUZA ABREU XAVIER and
ARLINDO MOREIRA FARTES FILHO

This chapter provides a broad overview of spread spectrum systems while introducing many of the concepts that are covered in depth later, including frequency-hopping and direct-sequence spread spectrum.

1.1 MULTIPLE ACCESS TECHNIQUES

Because of the finite amount of radio spectrum allocated for wireless systems, the telecommunications industry developed multiple access techniques to allow multiple users to share the available communication channels efficiently. The most common multiple access techniques are the following (see Figure 1.1).

- Frequency Division Multiple Access (FDMA): allocates a discrete amount of bandwidth to each user.

- Time Division Multiple Access (TDMA): allocates unique time slots to each user, that is, each user has a specific set of time intervals to transmit information (data/voice).

- Code Division Multiple Access (CDMA): all users share the same frequency all the time. A unique code assigned to each user allows it to be distinguished from other users.

CDMA was created to provide secure communications and navigation systems for military applications, thus promoting the development of Spread Spectrum (SS) technology, whose main objectives included the following.

- Multiple access over a single carrier frequency.

- Interference reduction.

Designing CDMA 2000 Systems   L. Korowajczuk et al.
© 2004 John Wiley & Sons, Ltd   ISBN: 0-470-85399-9

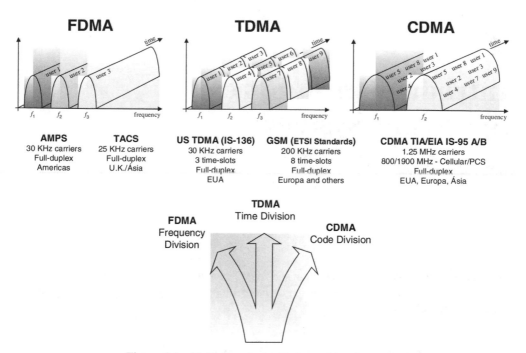

Figure 1.1   Multiple access techniques (see glossary).

Multipath: reduction of the undesired effects of delayed versions of the same signal arriving at the receiver through different paths, thus causing self-interference and mutual interference.

Intra-system: minimisation of interference caused by a different base station or mobile terminal belonging to the same system.

External: reduction of network operation disturbances caused by external agents (e.g. jammers).

- Privacy.

Between users: information transmitted in the network can only be understood by the intended recipient.

Interception: unauthorised receivers cannot intercept waveforms and information transmitted/received in the network.

1.2   THE SPREAD SPECTRUM CONCEPT

Transmission systems have two basic characteristics related to frequency spectrum: centre frequency (or transmitted carrier signal) and bandwidth (see Figure 1.2). While observation of signals in the time domain is very common (e.g. oscilloscopes), it is often more useful to observe signals in the frequency domain using spectrum analysers. There are mathematical tools that allow the conversion of time to frequency domain functions and vice versa, such as Fourier and Laplace transforms.

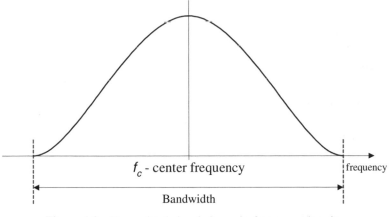

Figure 1.2 Transmitted signal shown in frequency domain.

The bandwidth occupied by a conventional radio transmission system is directly related to that of the original (baseband) information waveform and to the carrier modulation scheme. If Amplitude Modulation (AM) is used, a bandwidth of twice the size of the baseband signal is required, whereas Single-Side Band Amplitude Modulation (AM-SSB) and Frequency Modulation (FM) require a transmission bandwidth comparable to the bandwidth of the baseband signal. The FM bandwidth depends on the frequency deviation applied to the carrier and the information signal and is calculated using Carson's rule

$$B_{RF} = 2(\Delta f + f_m) \tag{1.1}$$

where

B_{RF} bandwidth of modulated carrier
Δ_f peak frequency deviation of the modulated carrier
f_m maximum frequency of the information signal

Digital modulation schemes with higher spectral efficiency can be used to minimise the bandwidth of the signal, for data transmission at a given bit rate in bits per second (bps). Phase modulations frequently used, for example, are Binary Phase Shift Keying (BPSK) and Quadrature Phase Shift Keying (QPSK), together with their variants. These modulations allow a 1:1 relation between the data rate and the carrier bandwidth in Hz. For instance, 100 kHz of bandwidth is required to modulate a QPSK carrier with a data rate of 100 kbps. In spread spectrum systems, the transmitted signal is "spread" using a bandwidth much larger than that required by mixing the data signal with a spreading code signal.

Two of the most important characteristics of spread spectrum systems are the following.

- Transmission bandwidth much larger than the bandwidth (or rate) of the baseband data signal.

- Transmitted bandwidth dependent on the rate of the code employed for spreading.

Spread spectrum systems are based on the Hartley-Shannon law that calculates the asymptotic transmission capacity in a channel disturbed by Additive White Gaussian Noise (AWGN)

$$C_{Sh} = B_{RF} \times \log_2(1 + \text{SNR}) \tag{1.2}$$

where

C_{Sh}	asymptotic channel transmission capacity in bps
B_{RF}	channel bandwidth in Hz
SNR	signal-to-noise ratio

According to this law, systems employing a wideband spectrum require an SNR lower than that of narrowband systems with the same transmission capacity in bps. Table 1.1 presents modulation schemes, carrier bandwidth, communication quality criteria and desired SNR for different cellular technologies.

cdma2000 systems following 3GPP2 standards have three traffic channel types: (1) the same used in IS-95A and IS-95B systems, to maintain compatibility among them; (2) Multi-Carriers (MC), that is, three 1.25 MHz carriers in the forward link and one 3.75 MHz carrier in the reverse link; and (3) one 3.75 MHz carrier in both links, almost the same bandwidth used in the multi-carrier case. The latter type has not been deployed because of the transmission rates and efficiency already achieved by the other types.

Each system has its own quality evaluation criteria according to its transmission characteristics. Analogue systems, such as AMPS, usually are based on SNR. Digital systems, however, are based on Bit Error Rate (BER) or Frame Error Rate (FER).

Although these criteria seem to be completely disconnected from one another, the relationship between them is quite close. For example, for GSM systems to operate with an adequate BER, it is important to have a high C/I (Carrier signal-to-Interference ratio, equivalent to SNR)[1] for the signal received at the base station. Therefore, regardless of the system, quality criteria are somehow related to the SNR, directly or indirectly.

Table 1.1 Comparison of cellular systems (see glossary)

Technology	Modulation scheme	RF carrier bandwidth	Quality criteria	Signal-to-Noise Ratio (SNR)	Bit energy to noise density ratio (E_b/N_0)
AMPS (Advanced Mobile Phone Service)	FM	30 KHz	C/I	17 dB	—
N-AMPS (Narrowband AMPS, also referred as IS-88)	FM	10 KHz	C/I	17 dB	—
D-AMPS (Digital AMPS, also referred as IS-54)	DQPSK	30 KHz	BER	17 dB	—
GSM (Global System for Mobile Communications)	GMSK	200 KHz	BER	6 to 9 dB	—
CDMA IS-95 (Code Division Multiple Access)	QPSK/ O-QPSK	1.25 MHz	FER	—	−14 dB
cdma2000 (Code Division Multiple Access)	QPSK/ O-QPSK	1.25/3.75 MHz	BER/FER	—	$(*)^a$

[a] Quality criteria for cdma2000, this value may vary according to the specific characteristics of each channel (e.g. data and/or voice transmission) at each time interval.

[1] The C/I corresponds to the ratio of the communication channel signal strength to the interference, which is commonly used to evaluate quality of analogue systems.

1.3   SPREAD SPECTRUM TECHNIQUES

There are four main types of spread spectrum techniques: Direct-Sequence (DS), Frequency-Hopping (FH), Time-Hopping (TH) and Multi-Carrier CDMA (MC-CDMA). These techniques can be combined or used separately. The first two types, DS and FH, are the most commonly employed and are illustrated in Figure 1.3.

Direct Sequence

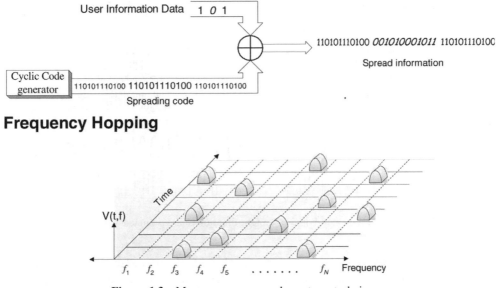

Frequency Hopping

Figure 1.3   Most common spread spectrum techniques.

1.3.1   Frequency-Hopping Spread Spectrum

In this technique, the carrier frequency changes at regular time intervals. Frequencies are selected from a pre-determined group within the available spectrum and they change in an order defined by a pseudo-random sequence, with characteristics similar to thermal noise; thus the name, PN (Pseudo-Noise) sequence.

Each bit in a PN sequence is called a *chip*. The idea of this name is to differentiate between data, expressed in bits, and codes (e.g. spreading sequences) that do not carry any information.

The PN sequence controls a frequency synthesiser used for the generation of the carrier. Figure 1.4 shows a simplified block diagram of a frequency-hopping system and its associated waveforms.

For the system of Figure 1.4, during transmission, information bits are baseband modulated.[2] PN sequences define the hopping sequence $\{f_1, f_2, f_3, \ldots, f_n\}$ to be generated

---

[2] There are also other ways of implementing frequency hopping, such as upbanding a conventional IF signal using a PN-controlled synthesiser.

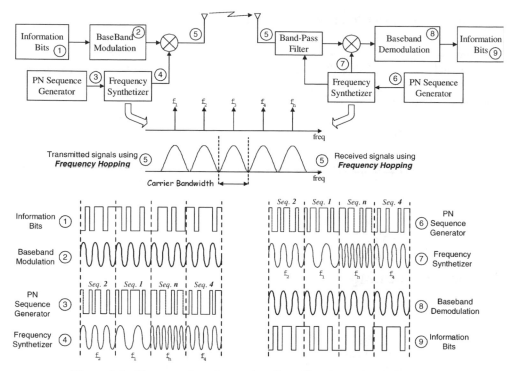

Figure 1.4   Frequency-hopping system diagram and associated waveforms.

by a frequency synthesiser for transmission through the air interface (see Figure 1.4). On the reception side, the PN sequence generator defines the centre frequency of the bandpass filter and the frequency for the de-modulation process. Modulation, transmission, reception and de-modulation only succeed if both PN sequence generators are compatible and synchronised. The number of frequencies used varies from a few to several thousands. There are other possible implementation schemes for FH systems, however. FH systems are usually classified in two categories, depending on the hopping rate:

- *Fast Frequency Hopping (FFH)*, the hopping rate is equal to or greater than the bit rate of the baseband signal;

- *Slow Frequency Hopping (SFH)*, the hopping rate is less than the bit rate of the baseband signal.

The greater the desired privacy and interference protection, the higher the hopping rate and the number of frequencies required.

1.3.2   Direct-Sequence Spectrum Spreading

The DS spread spectrum technique is the most used in spread spectrum systems due to its ease of implementation. It does not require a high-speed frequency synthesiser circuit as FH systems do. As the name indicates, spectrum spreading is obtained from direct modulation of the carrier carrying the PN sequence. This technique occupies the whole available frequency band continuously, unlike FH. Modulation can be AM, FM or any form of angular modulation. In practice, the most employed modulation schemes are phase modulation, such as BPSK, QPSK and their variants. Distinct and more efficient modulation schemes (in terms of number of bits per symbol), such as 8-PSK, 16-QAM and 64-QAM, are being evaluated for further evolutions of CDMA standards to achieve higher transmission capacities.

Figure 1.5 shows the diagram of a direct sequence spread spectrum system using BPSK modulation. The associated waveforms are shown at the bottom.

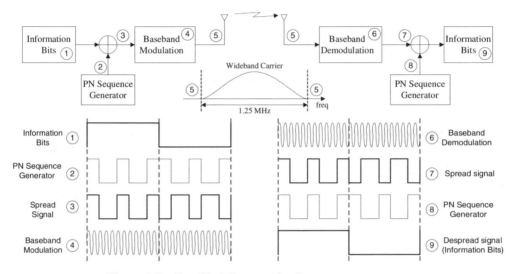

Figure 1.5 Simplified diagram of a direct-sequence system.

Information bits, usually a narrowband digital signal, are spread into a wideband signal by the PN sequence and transmitted after being modulated. At the receiver, the signal is demodulated and de-spread by the PN sequence locally generated. Similar to frequency-hopping systems, transmitter and receiver must have circuits to guarantee synchronisation between PN sequences.

The information signal (bits), for example digitised voice, is summed to the PN sequence used for the spread spectrum process. This is a modulo-2 sum, that is, an exclusive-OR logic function (XOR). Figure 1.6 shows diagrams representing the logical function XOR and the analogue product.

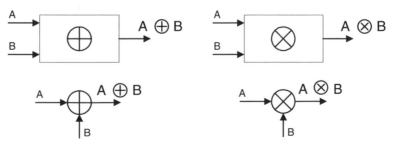

Figure 1.6 Logical function XOR and analogue product.

Table 1.2 Equivalence between logical function XOR and analogue product

A	B	$A \oplus B$
0	0	0
0	1	1
1	0	1
1	1	0
A	B	$A \otimes B$
+1	+1	+1
+1	−1	−1
−1	+1	−1
−1	−1	+1

Table 1.2 shows the equivalence between the logical function XOR and the analogue product by mapping the analogue values '−1' and '+1', respectively, to the binary values '1' and '0'. This equivalence is important because analogue values can carry information such as gain.

Figure 1.7 shows a diagram for the direct-sequence spectrum spreading process using a particular PN code, designated c_1, or PN sequence.

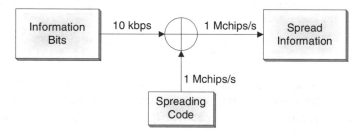

Figure 1.7 Direct-sequence spreading functional diagram.

The PN sequence rate is usually much higher than the information rate. For example, for an information rate of 10 Kbps, a code rate of 1 Mcps (Mega chips per second) is applied, producing a spread signal rate of 1 Mcps. In this case, the exclusive-OR function uses 100 consecutive chips of the c_1 code to process each bit of information, '0' or '1'.

1.4 PROCESSING GAIN P_G

In spread spectrum systems, the transmitted signal is de-spread in the receiver, returning to its original bandwidth. This is done through correlation, between received signal and locally generated PN sequences. To de-spread and recover the signal, the receiver must identify the locally generated PN sequence that matches the transmitted PN sequence, that is, both PN sequences have to be identical and synchronised. Other signals present at the receiver input, noise and interference, remain spread. The energy of these signals is reduced in the narrow-band filter after the correlation process.

The de-spreading process strengthens the desired signal in relation to other signals. The processing gain allows quantisation of the gain due to de-spreading. It is defined as the ratio of the SNR at the output to the SNR at the input of the signal processor (see Figure 1.8).

Figure 1.8   Processing gain calculation (linear).

If values are expressed in dB, the processing gain P_G is defined by the following expression

$$P_G(\text{dB}) = \text{SNR}_{out}(\text{dB}) - \text{SNR}_{in}(\text{dB}) \tag{1.3}$$

According to the previous equation, a signal processor with a SNR_{out} of 10 dB and SNR_{in} of 4 dB has a processing gain of 6 dB.

The relation between total transmitted bandwidth (B_{RF}) and bit rate of the information signal (R_{Info}) can also express processing gain in a system employing contiguous frequency bands

$$P_G = B_{RF}/R_{Info} \tag{1.4}$$

1.4.1   Processing Gain in Frequency-Hopping Systems

Equation (1.4) does not apply to a frequency-hopping system because this technique uses a different narrowband/wideband carrier at each time interval, which makes the use of W_{RF} (total transmitted bandwidth) not possible in the processing gain calculation.

The processing gain for frequency-hopping systems is, in a simplistic way, defined as the number of frequencies available for hopping

$$P_G = N_{freq} \tag{1.5}$$

or, in dB

$$P_G = 10\log_{10}(N_{freq}) \tag{1.6}$$

According to the previous equations, processing gain increases with the number of frequencies available. The availability of more frequencies for hopping also lowers the interference in each specific frequency, that is, it reduces spectral power density, assuming that the interference power is spread over the total bandwidth.

1.4.2   Processing Gain for Direct-Sequence Systems

A direct-sequence system uses a single carrier frequency, therefore eqn (1.4) can be directly applied to calculate processing gain. The bandwidth required for transmitting the information is identical to the information rate, thanks to spectrally efficient modulation schemes; therefore processing gain can be defined by the following expression

$$P_G = R_{PN}/R_{Info} \tag{1.7}$$

where

R_{PN} chip rate of the spread information (equal to the PN sequence rate),
R_{Info} information signal bit rate.

Considering processing gain in dB

$$P_G = 10\log_{10}(R_{PN}/R_{Info}) \tag{1.8}$$

According to eqn (1.8), for a spread information rate (R_{PN}) of 1 Mcps and information rate (R_{Info}) of 10 kbps, the processing gain is 100, or 20 dB (see Figure 1.9).

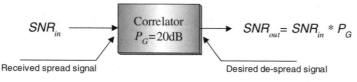

Figure 1.9   Correlator with processing gain.

1.5   SPREADING AND DE-SPREADING OF THE DS-CDMA SIGNAL

Recall that direct-sequence CDMA systems do not change the centre frequency over time as hopping systems do. A specific coded modulation with a high transmission rate is applied to the low data rate information of each user. At the modulator output, or exclusive-OR logic circuit, the data has the same transmission rate as the code. This is the spreading part of the process (see Figure 1.10.)

Figure 1.10 illustrates the concept of a DS-CDMA transmission. All users simultaneously share the same frequency band for communication, thus it is not possible to distinguish them from each other by filtering in the frequency or time domain, as in FDMA or TDMA, respectively.

In CDMA cellular systems, codes are assigned to each user to allow distinguishing them from each other. Walsh codes are used in the forward link whereas PN sequences with user specific phase offset are used in the reverse link. A careful choice of codes is capable of minimising multi-user interference.

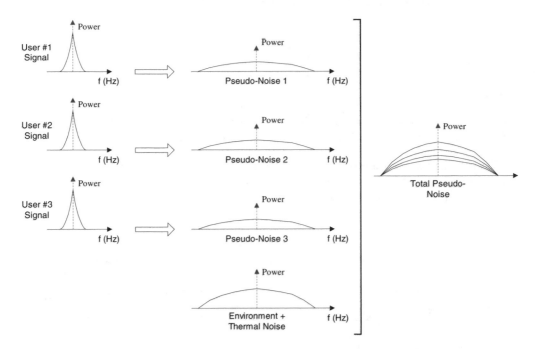

Figure 1.10   Spectrum spreading in DS-CDMA systems.

Figures 1.10 and 1.11 show that each transmitted signal $s_i(t)$ is modulated by a code $c_i(t)$. All coded signals, which correspond to modulation symbols showing amplitude and gain information, are combined at baseband. The result of this operation modulates the carrier, as illustrated by eqn (1.9)

$$S_{TX}(t) = [s_1(t)c_1(t) + s_2(t)c_2(t) + \cdots + s_n(t)c_n(t)]A\cos(w_c t) \qquad (1.9)$$

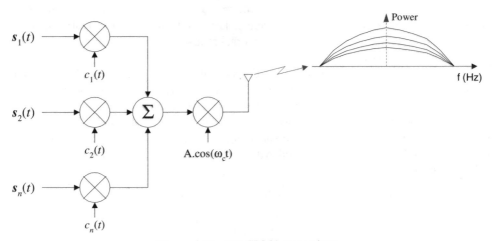

Figure 1.11   DS-CDMA transmitter.

or

$$S_{TX}(t) = \left[\sum_{i=1}^{n} s_i(t)c_i(t)\right] A\cos(w_c t) \tag{1.10}$$

where

$s_i(t)$ user number i signal (e.g. digital voice bearing amplitude information)
$c_i(t)$ code number i (Walsh code or PN sequence associated to user i)

The antenna transmits a series of stacked channels, modulating the same RF carrier. The receive antenna captures the same signal that, even though attenuated, still carries all the stacked channels, as shown in Figure 1.12.

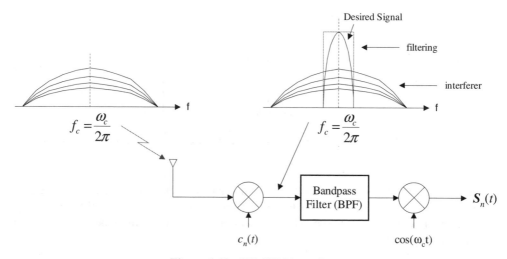

Figure 1.12   DS-CDMA receiver.

Figure 1.12 shows the de-spreading and de-modulation processes used to recover the signal of a single user's communication channel. The simultaneous de-modulation of several channels can be represented by repeating this diagram several times, using as many codes $c_n(t)$ as the number of active users.

The signal received always consists of the sum of all communication channels in use within the same RF carrier. In the frequency domain, or even in the time domain, CDMA transmission channels appear in an overlapped or stacked manner.

To de-spread a given channel, the two codes $c_i(t)$ generated by the transmitter and by the receiver must be synchronised. The de-spreading is done through the multiplication of the received signal by the code associated with the desired channel, as illustrated from eqn (1.11) to eqn (1.15)

$$S_{RX}(t)c_1(t) = [s_1(t)c_1(t) + s_2(t)c_2(t) + \cdots + s_n(t)c_n(t)]A'\cos(w_c t)c_1(t) \tag{1.11}$$

$$S_{RX}(t)c_1(t) = \left\{\left[\sum_{i=2}^{n} s_i(t)c_i(t)\right] + s_1(t)c_1(t)\right\}A'\cos(w_c t)c_1(t) \tag{1.12}$$

Therefore

$$S_{RX}(t)c_1(t) = \left[\sum_{i=2}^{n} s_i(t)c_i(t)c_1(t)\right]A'\cos(w_c t) + [s_1(t)c_1(t)c_1(t)]A'\cos(w_c t) \qquad (1.13)$$

But

$$c_1(t)c_1(t) = 1 \quad \text{and} \quad c_i(t)c_1(t) \neq 1 \qquad (1.14)$$

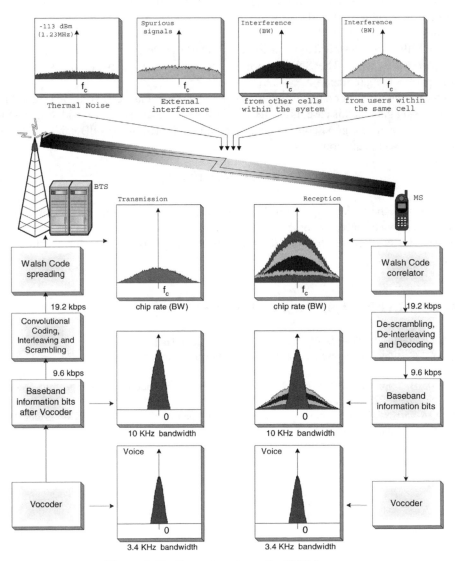

Figure 1.13 Processing steps of a CDMA system.

when i is different from 1. Thus

$$S_{RX}(t)c_1(t) = \left[\sum_{i=2}^{n} s_i(t)c_i(t)c_1(t)\right]A'\cos(w_c t) + s_1(t)A'\cos(w_c t) \qquad (1.15)$$

The expression in brackets represents the sum of all spread non-desired user signals, which cause interference to user 1. The last part of the equation $[s_1(t)A'\cos(w_c t)]$ represents the desired signal, i.e. the de-spread signal of user 1.

If the codes generated by the transmitter and receiver are synchronised, the de-spreading of the desired signal (including amplitude information) is accomplished and the signal is demodulated to its original bandwidth. Other signals using different codes remain spread in the spectrum. The bandpass filter, shown in Figure 1.12, rejects most of the energy from interfering signals.

Channels that remain spread are considered as interferers to user 1 and determine the SNR at the correlator output. The SNR determines the channel capacity of the system when specific parameters are fixed.

As an example, consider a maximum data rate of 9.6 kbps and a code rate of 1.2288 Mcps. The processing gain, in this case, is 21 dB, or 128 times in linear scale. Considering a SNR_{out} of 7 dB as adequate for good communication quality, a SNR_{in} of -14 dB can be calculated using the following equation

$$\text{SNR}_{out}(\text{dB}) = \text{SNR}_{in}(\text{dB}) + P_G(\text{dB}) \qquad (1.16)$$

This means that the desired signal level can be 14 dB below the interference caused by the sum of interference caused by other users and noise.

Figure 1.13 summarises the spreading and modulating processes performed in CDMA systems.

BIBLIOGRAPHY AND REFERENCES

1. CelTec/CelPlan, "CDMA IS-95 and cdma2000 Systems - Training Course."
2. C.S0002-C, "Physical Layer Standard for cdma2000 Spread Spectrum Systems," Release C, 3GPP2, May 2002.
3. C.S0005-B, "Upper Layer (Layer 3) Signaling Standard for cdma2000 Spread Spectrum Systems," Release B, 3GPP2, April 2002.
4. TIA/EIA-95-B, "Mobile Station – Base Station Compatibility Standard for Dual-Mode Spread Spectrum Cellular System," November 1998.

2

CDMA Evolution

BRUNO DE SOUZA ABREU XAVIER and
ARLINDO MOREIRA FARTES FILHO

This chapter starts by briefly tracing the historical development of CDMA technology. The evolution of the technology is discussed, presenting the main requirement and characteristics of each implementation stage of IS-95 and cdma2000 standards. The second half of the chapter presents an overview of the channel structure of IS-95 and cdma2000 systems. A full description of each channel is provided in the following chapters.

2.1 CDMA STANDARDS AND EVOLUTION

Cellular systems that use analogue technology, such as Advanced Mobile Phone System (AMPS), Nordic Mobile System (NMS) and Personal Digital Communications (PDC) are known as the First Generation (1G) of cellular systems. The desire of achieving higher quality standards for voice communication and more capacity forced these systems to evolve.

At the beginning, the main idea of the evolution process was to give more mobility to subscribers, allowing them to move freely across the network service area. However, as the need for higher capacity increased, new technologies arose, starting the Second Generation (2G) of cellular systems. The systems in this group are digital systems, using distinct transmission technologies, such as CDMA and TDMA in North America, and GSM in Europe. Even though 2G systems considered some data transmission at low data rates, voice was still the focus of the transmission.

In September 1988, the CTIA (Cellular Telecommunications and Internet Association) released the User Performance Requirements (UPR), specifying cellular operational requirements for a second generation of cellular systems (2G) using digital technology. The main requirements listed in this document are the following.

- Ten-fold increase in relation to the existing analogue system (AMPS).

- Privacy for voice and data users.

- Ability to introduce new services.

Designing CDMA 2000 Systems   L. Korowajczuk
© 2004 John Wiley & Sons, Ltd   ISBN: 0-470-85399-9

- Ease of transition from legacy to new systems.

- Use of dual-mode terminals; that is, mobiles capable of simultaneously operating in analogue and digital systems.

- Compatibility with the frequency spectrum of the existing analogue system.

- Reasonable infrastructure and mobile terminal costs.

Telecommunications Industry Association (TIA) committees were created to propose different solutions to meet CTIA requirements.

- TR 45.3 (sub-committee on digital cellular systems) presented, in 1989, the TIA/EIA/ IS-54 (dual-mode subscriber equipment–network equipment compatibility specification). IS-54 adopted Time Division Multiple Access (TDMA) as the air interface access scheme for the communication channel.

- TR 45.5 (sub-committee on wideband spread spectrum digital technology) presented, in 1993, the TIA/EIA/IS-95 (mobile station–base station compatibility standard for dual-mode wideband spread spectrum cellular system).

The TR 45.5 subcommittee for the IS-95 standard based its work on a CDMA cellular system developed by Qualcomm with the support of the following companies.

- Infrastructure equipment manufacturers, including Lucent, Motorola and Nortel.

- Mobile terminal manufacturers, including Motorola, Oki Telecom, Clarion, Sony, Alps Electric, Nokia and Matsushita-Panasonic.

- Cellular operators, including PacTel Cellular, Ameritech Mobile, GTE Mobile Communications, Bell Atlantic Mobile Systems, Nynex Mobile, US West New Vector Group and Bell Cellular Canada.

In July 1993, the first CDMA interim standard IS-95 was published. The document was based on the specifications of the TR 45.5. IS-95 and defines compatibility requirements for 800 MHz AMPS and CDMA systems.

These sub-committees also specified the reference network architecture of 2G cellular systems as shown in Figure 2.1.

Here

AUC   Authentication Center

HLR   Home Location Register

VLR   Visitor Location Register

EIR   Equipment Identity Register

ISDN   Integrated Service Digital Network

PSPDN   Packed-Switched Public Digital Network

PSTN   Public Switched Telephone Network

PLMN   Public Land Mobile Network

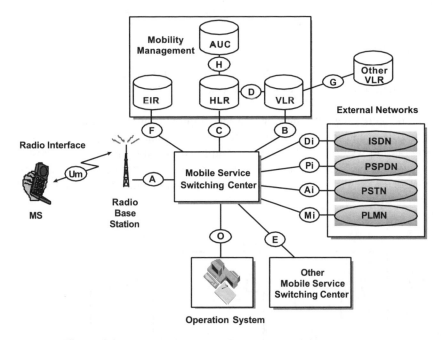

Figure 2.1 Basic architecture of American cellular systems [1].

MS   Mobile Station

. . . .   Interface

The inter-operation protocol for cellular systems, IS-41, regulates B, C, D, E, G and H interfaces. The air interface, UM or radio interface, is found between the MS and the Radio Base Station (RBS), known as Base Transceiver Station (BTS) in some systems. The definition of this interface is different for each cellular system, such as AMPS, TDMA IS-136 and CDMA IS-95. The A interface, between the BTS and Mobile Switching Center (MSC), was standardised to conform to IS-634 in the beginning of 1996.

The CDMA standard IS-95A was published in 1995. The operational features of the system described in IS-95A include the following.

- Channel bandwidth of 1.25 MHz (wideband).

- Chip rate of 1.2288 Mcps (chips per second).

- Power control.

- Call processing and handoffs.

Also in 1995, two documents, ANSI-J-STD-008 and TSB-74, were published, defining compatibility standards and interaction between CDMA and Personal Communication Systems (PCS), in the range of 1800–2000 MHz and specifying an optional circuit-switched data rate of 14.4 Kbps. IS-95B was then released, combining both of these documents with the previous IS95-A specifications. This new revision is also known as TIA/EIA-95. Some of the features of this revision are the following.

- Backward compatibility with IS-95A.

- Packet-switched data.

- Traffic channel speed of up to 115.2 kbps by combining Walsh codes (this also means setting some traffic channels to the same user).

The idea of increasing data transmission and improving users' connectivity led to more research and development, aiming to create new communication systems belonging to a Third Generation (3G). Figure 2.2 shows a simplified evolution path for the most common cellular systems. The concept of an intermediate generation, referred to as 2.5G, was created to accommodate systems with characteristics of 2G technologies, but providing higher data throughput.

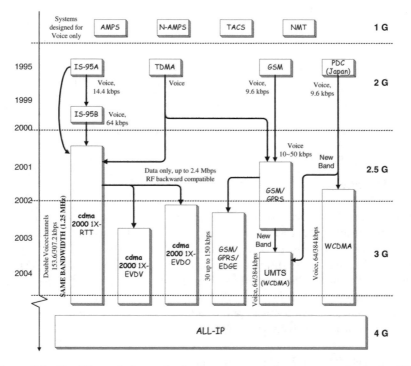

Figure 2.2   Possible evolution paths for the current wireless access technologies [2].

Trying to meet all industry requirements, the main features CDMA IS-95 networks offer are the following.

- High traffic capacity: Increase of system traffic capacity up to ten times when compared to former analogue systems.

- High Quality of Service (QoS) standards, by using fading reduction techniques, such as rake receivers.

- Paging services, data and fax traffic achieving data rates greater than provided by analogue systems (9.6 kbps).

- High communication privacy standards.

- Inter-operation with existing analogue systems through dual-mode terminals (AMPS/CDMA) operating on the same frequency band (e.g. A and B bands, from 824 to 849 MHz and 869 to 894 MHz).

- Three types of diversity techniques: frequency diversity (wideband carriers), time diversity (convolutional and forward error correction codes and interleaving), and space diversity (multiple reception antennas, rake receivers and soft handoff).

- Variable rate vocoders, to take advantage of users' voice activity factor.

- Fast and efficient power control schemes over forward and reverse link channels, increasing system capacity.

- Frequency re-use factor (K) equal to 1, i.e. all sectors and/or BTSs re-use the same carriers simultaneously.

- Soft capacity ability, allowing balancing increase on system traffic with performance degradations.

- Possibility to increase available communication channels (traffic capacity) per carrier in BTSs with low-traffic neighbors.

IS-95 standards for CDMA, also known as cdmaOne^{TM}, trademark of the CDMA Development Group (CDG), represent the second generation of CDMA based systems. These standards evolved into a new set of specifications called cdma2000, targeting the development of third generation technologies based on CDMA.

cdma2000 represents a family of technologies which includes the following.

- cdma2000-1XRTT, where 1X represents use of a single carrier and RTT stands for Radio Transmission Technology.

- cdma2000-1XEVDO, where EV stands for Evolution and DO for Data Optimised (or Data Only).

- cdma2000-1XEVDV, where DV stands for Data and Voice.

- cdma2000-3XRTT, where 3X represents use of multiple carriers.

The main requirements of the cdma2000 family include the following.

- Compatibility with IS-95B.

- Provision for IP-based packet switching.

- Greater efficiency through use of QPSK (*Quadrature Phase Shift Keying*) modulation for spreading. Different information is carried in I and Q branches.

- Complex spreading (IS-95 uses balanced and dual spreading).

- Coverage and total mobility for voice and data with bit rates of up to 144 kbps, which is equivalent to *Integrated Services Digital Network* (ISDN) channels 2B + D, in first phase of implementation (1XRTT).

- 1XRTT implementation provides speed of 307k, whereas 3XRTT and 1XEV provide higher rates.

- Limited coverage and mobility for bit rates of up to 2 Mbps, which is equivalent to ISDN channel H12. These rates are supported only by 1XEV and 3XRTT implementations.

- Improvement in spectral efficiency.

- Coherent de-modulation in the reverse link due to the introduction of a user pilot channel (reverse pilot channel).

- Auxiliary pilot channel in the forward link to enable use of smart antennas.

- Fast power control in the forward link.

- Capability to provide service for multiple information rates.

- Use of Multi-User Detection (MUD), consequently obtaining better efficiency when treating interference.

- High flexibility to introduce new services.

- Chip rate of 1.2288 Mcps for the one-carrier implementations (1XRTT, 1XEVDO, 1XEVDV). 3XRTT uses three carriers of 1.2288 Mcps in the forward link and 3.6864 Mcps in the reverse link.

- *Quick paging channels* to provide an increase of up to 15 times in mobile terminals' battery life.

Another option of third generation technology that has spread in Europe as the natural evolution for existing GSM systems is called UMTS (Universal Mobile Telecommunications System). UMTS can theoretically achieve 2 Mbps, whereas cdma2000-1X is only capable of reaching 614.4 kbps. Equipment manufacturers, however, will initially support only speeds of up to 153.6 kbps in both cases.

A great advantage of cdma2000-1X is the full compatibility with cdmaOne^{TM} systems, what allows the technologies to share all the transmission equipment and most of the infrastructure of existing networks, providing ease of technology migration. They also share the same frequency band, thus cdma2000-1X does not require new licenses.

Despite the existence of the cdma2000-3X implementation, it seems that the most promising paths in the IS-2000 evolution are the cdma2000-1XEV implementations (EVDO, EVDV), which maintain 1.25 MHz carriers.

cdma2000-1XEVDO adopts an exclusive carrier for data that can reach 2457.6 kbps (same capacity as 16 supplemental channels of 153.6 kbps) in the forward link and 307.2 kbps (two supplemental channels) in the reverse link. This technology is derived from the HDR (High Data Rate) version proposed and implemented by Qualcomm. This implementation of cdma2000 is only capable of data transmission and requires additional network equipment. This system was created for non-real-time services at high data rates. It can be represented as a wireless pipeline, where all traffic resources are shared between users, allowing base stations to increase transmission power compensating the additional E_b/N_0 required for high data rate transmission. This system foresees the use of 8-PSK (Phase Shift Keying) modulation and 16-QAM (Quadrature Amplitude Modulation) schemes in the forward link.

cdma2000-1XEVDV standards are still being discussed. The Nokia–Motorola association proposed a new technology denominated 1XTREME, which provides rates of up to 5 Mbps. This standard returns to simultaneous voice and data transmission and is foreseen to be available in 2004. This implementation is considered as an evolution of cdma2000-1X hardware and is intended for higher data rate and real-time services. It uses modulation schemes introduced in EVDO (8-PSK and 16-QAM).

Figure 2.3 illustrates the relation among the several CDMA standards, evolution path and trends.

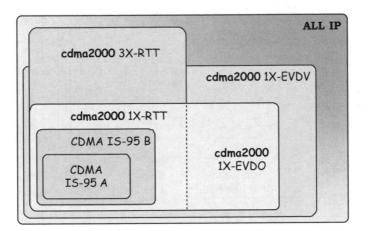

Figure 2.3 Evolution for CDMA systems.

2.1.1 Third Generation Systems Structure

As in the second generation, regulatory agencies also specified reference network architectures for 3G systems. Figure 2.4 shows the architecture suggested by 3GPP2 (Third Generation Partnership Project 2) for cdma2000 networks.

Each of the squares, triangles and rectangles in Figure 2.4 represents a Network Entity (NE) while circles represent reference points. NEs may be a complete physical device, a part of it or even distributed over a number of physical devices. Reference points divide two groups of functions and do not represent an interface [3].

AAA – *Authentication, Authorisation and Accounting*: Provides Internet Protocol functionality to support authentication, authorisation and accounting functions.

AC – *Authentication Center*: Manages authentication information related to the MS. It may be located within and impossible to differentiate from an HLR. The AC may serve multiple HLRs.

BS – *Base Station*: Consists of a BSC and a BTS and provides the means for MSs to access network services via radio.

BSC – *Base Station Controller*: Provides control and management for one or more BTSs, exchanging messages with both the BTS and the MSC. Traffic and signalling related to call control, mobility and MS management pass transparently through the BSC.

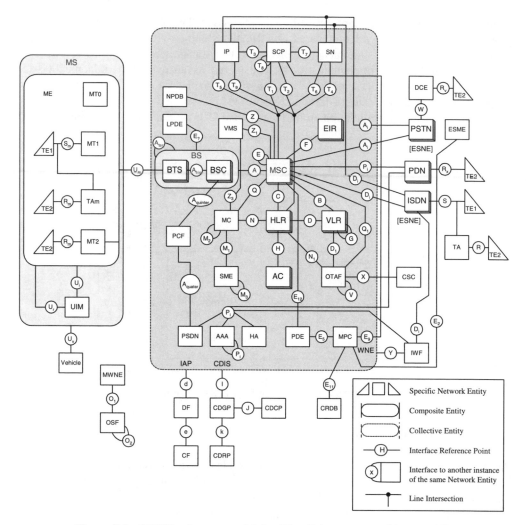

Figure 2.4   3GPP2 reference model for 3G cellular system architecture [3].

BTS – *Base Transceiver Station*: Consists of radio devices, antenna and equipment. Provides transmission capabilities on the air interface between the BS and MS.

CDCP – *Call Data Collection Point*: Collects call details in ANSI-124 format.

CDGP – *Call Data Generation Point*: Provides call details in ANSI-124 format to the CDCP. If call details use a proprietary format, the CDGP concerts the information into ANSI-124 before sending it.

CDIS – *Call Data Information Source*: Source of call details, may use proprietary format.

CDRP – *Call Data Rating Point*: Charges and taxes un-rated call details (in ANSI-124 format).

CF – *Collection Function*: Collects intercepted communications for authorised law enforcement agencies. Some of the functions include reception and processing of calls for each intercept subject and reception and processing information (e.g. call associated or non-call associated) from the delivery function regarding each intercept subject.

CRDB – *Coordinate Routing Database*: Translates positions expressed as latitude and longitude into a string of digits.

CSC – *Customer Service Centre*: Location where service provider representatives receive telephone calls from customers willing to subscribe to initial wireless service or request changes in their existing service; also known as Call Centres (CCs). Performs network and MS related changes required to complete service-provisioning request.

DCE – *Data Circuit Equipment*: Termination that provides a non-ISDN user–network interface.

DF – *Delivery Function*: Responsible for delivering intercepted communications to one or more collection functions. Some of the delivery functions include acceptance of information from multiple data channels, combining it into a single data flow for each intercept subject, filtering and delivering it to a collection function using one or more channels as authorised for each law enforcement agency. It can optionally detect audio in-band DTMF digits for translation before delivery to collection functions. It is able to duplicate and deliver information on the intercept subject to multiple collection functions and to provide security to restrict access.

ESME – *Emergency Service Message Entity*: Routes and processes out-of-band messages related to emergency calls. The entity can be incorporated into selective routers (also known as routing, bridging and transfer switches), public safety answering ports, emergency response agencies and Automatic Location Information (ALI) database engines.

ESNE – *Emergency Service Network Entity*: Composed by selective routers, public safety answering points and emergency agencies, it routes and processes voice in emergency calls.

EIR – *Equipment Identity Register*: User equipment identity is assigned to this register for record purposes.

HA – *Home Agent*: Authenticates mobile IP registrations; re-directs packets to and from foreign agent components of the PDSN; establishes, maintains, and terminates secure communications with the PDSN; receives provisioning data from the AAA; and assigns dynamic home IP addresses.

HLR – *Home Location Register*: Stores subscriber information, such as Electronic Serial Number (ESN), Mobile Directory Number (MDN), profile information, current location and authorisation period.

ISDN – *Integrated Services Digital Network*: Defined in accordance with ANSI T1 standards.

IP – *Intelligent Peripheral*: Performs specialised resource functions such as playing announcements, collecting digits, speech-to-text or text-to-speech conversion, recording and storing voice messages, facsimile services, data services, etc.

IAP – *Intercept Access Point*: Provides access to the communications to or from equipment, facilities or services of an intercept subject.

IWF – *Inter-working Function*: Provides information conversion for one or more WNEs.

LPDE – *Local Position Determining Entity*: Facilitates location of a MS, through one or more location methods.

MWNE – *Managed Wireless Network Entity*: Any entity that requires OS wireless management, including another OS.

MC – *Message Centre*: Stores and forwards short messages, providing supplementary services for Short Message Services (SMS).

ME – *Mobile Equipment*: Consists of the MS without a UIM, being only capable of accessing the network through local service configuration, e.g. emergency services, service centre, etc.

MPC – *Mobile Position Center*: Selects a PDE (or LPDE) to determine the position of a mobile station. This centre restricts access to location information by requiring, for example, the MS to be engaged in an emergency call. It may also restrict access by disclosing location information only to authorised NEs.

MS – *Mobile Station*: Consists of an ME with a programmed UIM. MSs are the equipments used by subscribers to access the wireless network.

MSC – *Mobile Switching Centre*: Connected to one or more BSs, it switches circuit mode traffic originated or terminated by the MS. It may also connect to other public networks and to other MSCs.

MT0 – *Mobile Terminal 0*: Data capable ME that does not support an external interface.

MT1 – *Mobile Terminal 1*: ME with an ISDN user–network interface.

MT2 – *Mobile Terminal 2*: ME with a non-ISDN user–network interface.

NPDB – *Number Portability Database*: Provides portability information for portable directory numbers.

OSF – *Operations Systems Function*: Defined by the Telecommunications Management Network (TMN) OSF in ITU M.3100. OS functions include element, network, service and business management layer.

OTAF – *Over-The-Air Service Provisioning Function*: Interfaces proprietarily to CSCs to support service-provisioning activities, interfacing with the MSC to send orders required to complete requests.

PCF – *Packet Control Function*: Manages relay of packets between the BSC and the PDSN.

PDN – *Packet Data Network*: Network, such as the Internet, that performs packet data transmission between processing NEs.

PDSN – *Packet Data Serving Node*: Routes packet data traffic originated or terminated by the MS, establishing, maintaining and terminating link layer sessions. It may interface multiple BSs and PDNs.

PDE – *Position Determining Entity*: Facilitates location of an MS through one or more location techniques. Multiple PDEs serve the coverage area of an MPC. When residing at the BS, these are known as LPDEs.

PSTN – *Public Switched Telephone Network*: Defined according to ANSI T1 standards.

SCP – *Service Control Point*: Behaves as a real-time database and transaction processing system, providing service control and service data functionality.

SN – *Service Node*: Provides service control, service data, specialised resources and call control functions to support bearer-related services.

SME – *Short Message Entity*: Composes and decomposes short messages. It may be located within and impossible to differentiate from an HLR, MC, VLR, MS or MSC.

TA/TAm – *Terminal Adapter*: Converts signalling and user data from a non-ISDN to an ISDN interface.

TE1 – *Terminal Equipment 1*: Data terminal providing an ISDN user–network interface.

TE2 – *Terminal Equipment 2*: Data terminal providing a non-ISDN user–network interface.

UIM – *User Identity Module*: Contains subscription information such as NAM. May also contain subscription information. It can be removable or integrated into an ME.

Vehicle – Location where the MS may be installed. Provides power, audio and antenna connections to the MS along with control and user data gateway to vehicle-based networks.

VLR – *Visitor Location Register*: Location register used by the MSC to handle calls to or from a visiting subscriber. May interact with multiple MSCs. It may be located within and impossible to differentiate from an MSC.

VMS – *Voice Message System*: Stores voice and data messages. May also support messages managing function such as retrieval of previously stored messages, notification of stored messages and of change in the number of messages.

WNE – *Wireless Network Entity*: A network entity in a wireless network.

Besides the reference model for network structure, regulatory agencies felt the need to understand the subscriber perspective of the network and started analysing the types of services that are used with 3G systems and their requirements. The IMT2000 proposed a set of Quality of Service (QoS) classes that was widely accepted by the wireless community as a guideline for evolution efforts of the new technologies. Four different classes are listed in the proposal,

- conversational,
- streaming,
- interactive,
- background.

Table 2.1 QoS classes proposed for the 3G wireless systems [4]

Traffic class	Real time classes		Best effort classes	
	Conversation	Streaming	Interactive	Background
Main features	Preserve time relation between information entities of the stream Conversational pattern (stringent and low delay)	Preserve time relation between information entities of the stream	Request response pattern Preserve payload content	Destination is not expecting the data within a certain time Preserve payload content
Main applications	Voice	Streaming video	Web browsing	E-mails download

The main characteristic distinguishing the classes is their delay sensitivity, that is, how the class performance is affected by delay. Conversational and streaming classes are the most delay sensitive classes because of their need to preserve the time relation between information entities of the stream.

Interactive and background classes are less delay sensitive because their channel coding and re-transmission characteristics provide a better error rate. However, the interactive applications still have a higher priority in traffic scheduling than the background class. Table 2.1 presents the QoS classes suggested for 3G systems.

2.1.1.1 Conversational Class

This class is mostly used in speech telephony, but new applications such as VoIP (Voice over IP) and video-conferencing are also included here. The main characteristic of these services is real-time conversation, which implies in the communication of two (or more) users. The requirements of this class are defined to meet human perception needs.

The transfer time of this scheme has to be low because of the conversational nature of this class and to preserve the time variation between information entities of the transmitted data. The class is highly delay sensitive; that is, the limit for traffic delay is very strict because of the human perception of video and audio. Not meeting delay requirements results in unacceptable lack of quality in the application.

2.1.1.2 Streaming Class

The streaming class represents one-way transport applications aiming at a live destination (user). The most common example of this class is real-time video and audio. As in the conversational class, the time variation between information entities within the data flow must be preserved, however, this class does not have requirements on low transfer delay.

Transfer delay still has to be kept within certain limits to preserve time relation in the flow, but, because the stream is usually time aligned at the receiving end (user equipment), the acceptable delay variation is higher than in the conversational class, which is limited because of human perception.

2.1.1.3   Interactive Class

This class represents users requesting data from remote equipment (e.g. a server). Applications in this class include web browsing, database retrieval and server access. Interactive traffic is characterised by the request response pattern of users and by the preservation of payload content.

The request response pattern indicates that the end user is expecting a response within a certain time, what gives this class some restrictions in terms of round trip delay time. Preservation of the payload content implies in transmission of packets with a low BER.

2.1.1.4   Background Class

Even though background and interactive traffic are very similar, the first is not as delay sensitive as the latter. In background applications, such as E-mail, SMS and download of databases, end users are not expecting the data within a certain time, that is, this class can be considered as delivery time insensitive. The payload content, however, still needs to be preserved, requiring a low BER during data transfer.

2.2   CDMA TIMELINE

CDMA origins start with spread spectrum technology studies for military and navigation applications. The following timeline presents the main events in the creation of CDMA.

[1935] Kotowski and Dannehl, working for Telefunke in Germany, combine voice to a broadband noise signal produced by a rotator generator. This invention is the starting point in the development of the Direct-Sequence Spread Spectrum (DS-SS) communication systems.

[1949] John Pierce writes a technical memorandum contemplating the spreading of non-synchronised coded signals through a time-multiplexed system (*Time Hopping Spread Spectrum*–TH-CDMA). In this process, the original signal is stored according to its original bit rate (baseband), it is then sampled with a much higher bit rate (extended band), and later transmitted in bursts, spaced according to codes associated to each user.

[1949] Claude Shannon and Robert Pierce discuss ideas of direct-sequence coding (CDMA) and analyse the effects of white noise interference in the transmission of information.

[1950] Claude Shannon announces the theorem establishing that the information transmission capacity (bps) in an environment disturbed by white noise is proportional to the product of the transmission band (Hz) and $\log_2 (1 + \text{SNR})$, where SNR stands for Signal-to-Noise Ratio.

[1950] Louis De Rosa and Mortimer Rogoff propose a system that implements the idea of modulating information using a coded sequence with a high transition rate (CHIP–*Code Hopping Information Piece*) and a behavior similar to white noise (pseudo-random). They called it DS-SS. According to this concept, the environment noise is multiplexed and, when the original signal is reconstructed, a processing gain is observed.

[1956] Price and Green file for patenting the 'rake'. The rake is a receiver that searches for information signals coming from different propagation paths and combine their contents, obtaining multi-path gain.

[1961] Magnuski analyses the effects of a strong interference signal on the processing gain of a weak spread-spectrum signal. This phenomenon is known as 'near-far problem'. This problem is later corrected using power control techniques.

[1978] G. R. Cooper and R. W. Nettleton propose the use of spread-spectrum technology in cellular telephony.

[1984] A hybrid technique (CDMA/FDMA) is suggested for multiple access in GSM systems in 1984, due to the promising results shown by DS-CDMA studies. The TDMA technique with eight timeslots is only presented in 1987.

[1986] S. Verdu proposes the Multiple-User Detection (MUD) technique as a way of minimising error probability in an asynchronous Gaussian signals multiple access system. MUD tries to detect the desired signal without considering other signals as interference, because CDMA capacity is interference limited.

[1988] Europe begins the selection of a third generation system with the RACE I/II programs (*Research of Advanced Communication Technologies in Europe*) resulting in the development of a CDMA based system called CODIT (*Code Division Testbed*) between 1992 and 1995.

[1988] The CTIA (*Cellular Telecommunications Industry Association*) releases, in September, a document called User Performance Requirements (UPR), specifying cellular operators' requirements for the second generation of cellular systems.

[1989] Qualcomm makes the first field test with components researched and developed in the 80's. The test system is deployed in November in San Diego.

[1990] Qualcomm shows, in September, the first version of a common air interface.

[1990] Wideband CDMA (WCDMA) studies begin. The technology is proposed as an evolution of GSM systems. It later results in the UMTS specification (1999), which provides transmission rates of 2 Mbps.

[1991] In December 1991, field test results are presented to CTIA on a presentation called '*Presentation of the Results of the Next Generation Cellular Field Trials*'. This presentation motivated the creation of the TR 45.5 sub-committee.

[1992] In January, TIA begins the CDMA standardisation process and in March the sub-committee TR 45.5 is created to develop the IS-95.

[1993] The IS-95 specification is finished in July 1993 and is called IS-95A. The standard provides compatibility to AMPS systems.

[1994] The CDG (CDMA Development Group) is established with the mission of promoting the IS-95.

[1995] The first revision of the IS-95A is finished.

[1995] The ACTS (Advanced Communication Technologies and Services) program is created in Europe, culminating in the FRAMES (Future RAdio wideband Multiple accEss System) project involving Nokia, Siemens, Ericsson, France Télécom and CSEM/PRO Telecom together with European Universities. Two projects are created: FMA1 (FRAMES Multiple Access based in wideband TDMA) and the FMA2 (FRAMES Multiple Access based in wideband CDMA).

[1996] Commercial operation of CDMA IS-95 systems starts.

[1996] Initial studies targeting high bit rate transmissions.

[1998] The IS-95B standard is approved, offering the ability to concatenate up to eight Walsh codes, what provides a higher bit rate transmission capacity (115.2 kbps). The IS-95B standard was implemented in Korea (SKT), Japan and Peru.

[1995–1998] Several efforts around the world aiming to develop 3G systems.

- In Europe, FMA2 evolves with the contributions of Japanese manufacturers Fujitsu, NEC and Panasonic, originating the WCDMA standard. WCDMA is later denominated UMTS. The standard features asynchronous base stations, unlike IS-95, which uses GPS (Global Positioning System) for synchronisation.

- In US, the TR45.5 group is now in charge of specifying the third generation cellular system based in CDMA technology. The IS-2000 standard (originally called IS-95C) is proposed, beginning the *cdma2000* family.

- Two systems called TTA1 (Telecommunication Technology Association 1) and TTA2 are evaluated in Korea. TTA1 is an asynchronous wideband CDMA system, like UMTS, whereas TTA2 is a synchronous wideband CDMA system, similar to cdma2000.

[1998] A group called 3GPP (*Third Generation Partnership Project*) is created to produce a common standard for asynchronous wideband systems. The group brings together several standard organisations such as ARIB and TTC (Japan), ETSI (Europe), T1P1 (USA), TTA (Korea) and CWTS (China). They establish a standard that is backwards compatible with the existing GSM/GPRS network structure. Later, the 3GPP is also assigned the task of creating specifications for GSM and its derived standards (GPRS/EDGE). Their work resulted in the Releases 1999 and 2000.

[1999] The 3GPP2 group is created to harmonise the use of the multi-carrier cdma2000. The group is initially composed by the TR45.5 (TIA-USA) and TTA (Korea). ARIB, TTC (Japan) and CWTS (China) join the group later.

[1999] The harmonisation process is concluded and generates three operating concepts.

- (DS–Direct Sequence) direct sequence typically used in UMTS (Europe).

- (MC–Multi-Carrier) multiple carriers typically used in cdma2000 (USA).

- (TDD–Time Division Duplex) duplex operation in time, UMTS option currently preferred by China.

[2000] South-Korean Telecom (SKT) launches, in October, a commercial cdma2000-1X
network. 300000 users subscribe by June 2001, operating with typical rates of 70–
90 kbps.

[2001] NTT DoCoMo deploys a commercial UTMS network in Japan. In the same month,
Korean operators, LG Telecom (LGT) and Korea Telecom Freetel (KTF), and some
North-American operators, Verizon and Sprint PCS, began commercial operation of
cdma2000-1X systems.

2.3   EVOLUTION OF CDMA STANDARDS

2.3.1   IS-95

CDMA IS-95 networks belong to the second generation of cellular systems and represent the
transition from analogue to digital technologies. The following sections describe IS-95
systems structure and forward and reverse link characteristics.

2.3.1.1   CDMA IS-95 System Structure

CDMA networks usually employ full-duplex access methods, such as Frequency Division
Duplex (FDD).

In these networks, one carrier transmits forward link channels (or downlink) from Base
Transceiver Stations (BTSs) to Mobile Stations (MSs), whereas another frequency is
allocated for the reverse link (or uplink), transmitting from MSs to their server BTSs.

A BTS can utilise more than one pair of forward and reverse carrier frequencies to
increase capacity in its cell (Figure 2.5). Carriers have 1.23 MHz of bandwidth (1.2288
Msps) but the separation between carriers is 1.25 MHz, thus it is common to see the CDMA
carrier width declared both as 1.23 MHz and 1.25 MHz. Separation between uplink and
downlink carriers is, usually, 45 MHz for the 850 MHz band and 80 MHz for the 1.9 GHz
band.

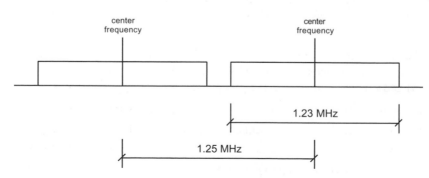

Figure 2.5   Carrier width in CDMA IS-95 systems.

Carriers represent the physical transmission interface, i.e. frequencies used to transmit data, whereas logical channels are used to transmit and organise transmitted data, identifying each data type, such as traffic, control and broadcast data.

Every CDMA channel in any BTS is identified by two parameters: a RF carrier (radio frequency) and a code. The first parameter defines the 1.25 MHz carrier centre frequency, whereas the second consists of a Walsh code or an offset mask for PNLC sequences.

On the forward link, every channel is identified by one of the 64 Walsh codes available (64 chips long). On the reverse link, channels are identified by long code offset masks and by the carrier in use. The channel assignment process, i.e. carrier and code definition for each MS, is performed by the BTS serving in the area where the MS is located.

2.3.1.1.1 Forward Link in CDMA IS-95 Systems

CDMA IS-95 systems present four logical channel types in the forward link: pilot, synchronism, paging and traffic. Figure 2.6 presents the CDMA IS-95 forward link channel structure.

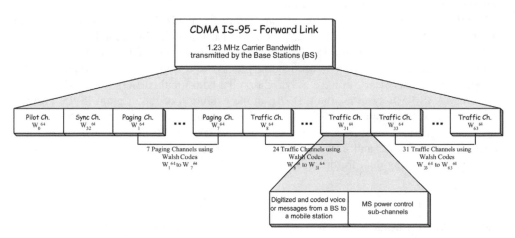

Figure 2.6 Forward link channel structure for CDMA IS-95 systems.

Every CDMA IS-95 carrier has a Forward Pilot Channel (FPiCh) that uses Walsh Code zero (W_0^{64}). This channel only transmits bits '0', orthogonally modulated by PN-I and PN-Q sequences (short PN sequences) using a specific PN phase offset, which uniquely identifies each BTS within the network. The MSs acquire the forward pilot channel, transmitted on the default forward link carrier if the BTS uses more than one frequency, to synchronise their internal PN sequence generator to the sequences transmitted over the air interface. Because Walsh Codes are synchronised with short PN sequences, the MS also obtains Walsh code synchronism.

Every BTS on a CDMA system transmits on one or two default carrier frequencies, also known as primary and secondary carriers, which have to contain the forward pilot channel (FPiCh) and the synchronisation channel (SyncCh). After acquisition and successful processing of the SyncCh, the MSs are able to identify their server BTS.

The SyncCh uses Walsh Code 32 (W_{32}^{64}) and transmits messages containing data about the BTS. This information allows MSs in the service area of this BTS to synchronise their long code sequence generators, and identify the BTS offset (Pilot_PN).

The identification of a BTS consists of a unique offset number, ranging from 0 to 511. Each offset represents steps of 64 chips on short PN sequences, i.e. PN-I and PN-Q sequences. The beginning of the Walsh Codes and of these sequences is coincident, both of them using the same transmission rate of 1.2288 Mcps. All forward channels transmitted by a BTS are synchronised and processed by short PN sequences after application of the BTS offset (Pilot_PN).

The default carrier must also transmit the Primary Forward Paging Channel (FPCh), which uses Walsh Code 1 (W_1^{64}). The FPCh is used by the BTSs to transmit system data, such as additional paging channels and identification of CDMA carriers to be used by the MSs on the forward and reverse links. The data transmitted through the FPChs gives the MSs all the information required to access the CDMA network. Each CDMA carrier may contain up to seven FPChs, using Walsh Codes 1–7. Each MS must be associated to one of the FPChs. The MSs in the service area of any BTS are pseudo-randomly distributed over the available FPChs, using a hash algorithm.

All other Walsh Codes are used for Forward Traffic Channels (FTChs), which are the logic channels used during a call. Table 2.2 shows Walsh Codes assignment for primary and secondary carriers (setup) in a BTS. Table 2.3 shows assignment in additional carriers of the same BTS. Because the system acquisition process is exclusively performed on setup carriers, it is not necessary to have a SyncCh on the additional carriers.

In CDMA IS-95 B systems, the first step in the evolution of the original IS-95, Forward Supplemental Code Channels (FSCChs) were implemented to improve data traffic capabilities. These channels use the same Walsh codes available for traffic channels in previous systems (see Tables 2.2 and 2.3). Up to seven FSCChs and one FTCh may be assigned to a subscriber, increasing data traffic capacity to 115.2 kbps.

Table 2.2 Walsh codes assignment for primary and secondary setup carriers in CDMA IS-95 systems

Channel type	Number of channels	Walsh codes
Traffic channel (FTCh)	Up to 55	W_8^{64}–W_{63}^{64}, except W_{32}^{64}
Paging channel (FPCh)	From 1 to 7	W_1^{64}–W_7^{64}
Synchronisation channel (SyncCh)	Only 1	W_{32}^{64}
Pilot channel (FPiCh)	Only 1	W_0^{64}

Table 2.3 Walsh codes assignment for additional carriers, on the same BTS, in CDMA IS-95 systems

Channel type	Number of channels	Walsh codes
Traffic channel (FTCh)	From 55 to 62	W_2^{64}–W_{63}^{64} except W_{32}^{64}
Paging channel (FPCh)	From 0 to 7	W_1^{64}–W_7^{64}
Pilot channel (FPiCh)	Only 1	W_0^{64}

2.3.1.1.2 Reverse Link in CDMA IS-95 Systems

In the reverse link of CDMA IS-95 systems there are only two logical channel types: the Reverse Access Channel (RACh) and the Reverse Traffic Channel (RTCh). Reverse link channels are identified by long code offset masks (Figure 2.7).

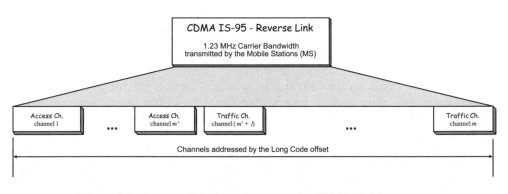

Figure 2.7 Reverse link channel structure for CDMA IS-95 systems.

Mobile stations use Reverse Access Channels (RACh) to access the system. This access may be the origination of a call or the transmission of a response to a specific message, or order, sent by the server BTS. Each RACh is uniquely associated to an FPCh, totalising 32 RAChs per FPCh.

In conversation, i.e. during a call, the MS uses a Reverse Traffic Channel (RTCh). Traffic channels assigned to a BTS, on both forward and reverse links, are defined by the system. This information is transmitted to the MSs over the associated FPCh.

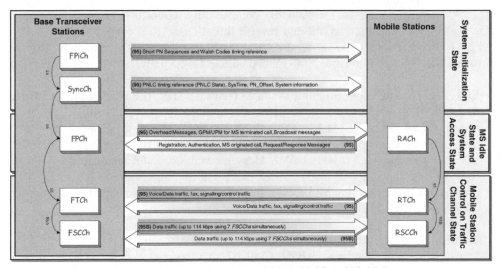

Figure 2.8 Logical channels used on CDMA IS-95 and IS-95 B systems.

As in the forward link, the evolution to CDMA IS-95B also included new logical channels, the Reverse Supplemental Code Channels (RSCCHs), with the same functionalities as in the forward link.

Figure 2.8 presents, graphically, the relationships among the logical channels of CDMA IS-95 A and B systems.

2.3.2   cdma2000

CDMA systems based on cdma2000 standards are capable of using distinct carrier configurations, employing carriers with bandwidths of 1.25 MHz (backward compatibility to CDMA IS-95 systems) and 3.75 MHz (to achieve data transmission rates compatible to 3G requirements).

The need of increasing system capacity (voice and data traffic channels) has motivated the implementation of additional logical channel types. To allow smooth system backwards compatibility, some of the original logical channels were kept. The use of cdma2000 channels, however, depends on new characteristics, such as Radio Configuration (RC) and Spreading Rate (SR). There are ten RCs for the forward link and six RCs for the reverse link.

cdma2000 logical channels use two spreading rate configurations:

- SR1: Used with 1.25 MHz bandwidth carriers, for compatibility with cdmaOne systems, i.e. networks based on IS-95A and IS-95B standards, with transmission rate of 1.2288 Mcps. Systems using this spreading rate are referred to as 1X, due to the use of one 1.25 MHz carrier per communication channel, such as cdma2000 1X-RTT (also called cdma2000-1X).

Some sources treat CDMA IS-95 systems as part of cdma2000-1X systems, because of the backward compatibility that assures that CDMA IS-95 mobile terminals may also be used in cdma2000-1X networks. To maintain the compatibility, cdma2000-1X systems employ RCs 1 and 2 on both the forward and reverse links. These RCs correspond to 9.6 kbps and 14.4 kbps vocoders (voice coders) defined for IS-95 systems. The remaining RCs are considered as belonging to cdma2000-1X and cdma2000-3X only.

- SR3: There are two configurations available for using SR3. The first, called SR3 Direct Sequence (DS), employs 3.75 MHz carriers in both the forward and reverse links, using transmission rates of 3.6864 Mcps (3×1.2288 Mcps), three times greater than what is used in SR1 systems. The second, called SR3 Multi-Carrier (MC), employs three distinct 1.25 MHz carriers (each with 1.2288 Mcps transmission rate) simultaneously on the forward link and one 3.75 MHz carrier (with 3.6864 Mcps transmission rate) on the reverse link. Systems using SR3 are referred to as 3X, such as cdma2000-3X systems (or cdma2000 3X-RTT).

The trend, however, seems to be the adoption of the MC version only. Reflecting this tendency, cdma2000 standards 3GPP2 C.S0002 that presented both configurations in Release 0 only describe the SR3 MC option in Release C.

A Walsh Code, with length varying from 4 to 128 chips, identifies each channel on the forward link, depending on the SR and RC being used by the mobile station. Long code offset masks, Walsh codes, and the carrier in use identify channels on the reverse link. The channel assignment process, i.e. carrier and code definition for each MS, is performed by the BTS (Figure 2.9).

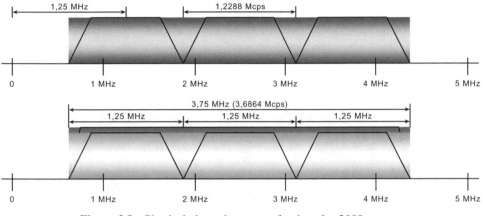

Figure 2.9 Physical channel structure for the *cdma2000* systems.

2.3.2.1 *Forward Link in cdma2000 Systems*

There are ten main types of logic channels in the forward link of cdma2000 systems. Figure 2.10 shows these channels and their subdivision into sub-channels. The highlighted blocks indicate channels inherited from IS-95 systems. Channel configuration depends on the radio configuration and spreading rate being used by the mobile station.

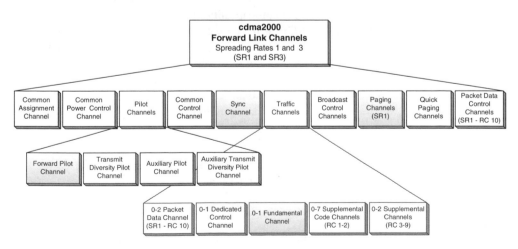

Figure 2.10 Forward link channel structure for cdma2000 systems [5].

Table 2.4 Radio configurations available in the forward link of cdma2000 systems [2]

RC	Associated SR	Data rate (Kbps)	FEC convolutional encoder (R)	Modulation schemes	Diversity
1	1	1.2; 2.4; 4.8; **9.6**	1/2	BPSK pre-spreading	
2	1	1.8; 3.6; 7.2; **14.4**	1/2	BPSK pre-spreading	
3	1	1.2; 1.35; 1.5; 2.4; 2.7; 4.8; **9.6**; 19.2; 38.4; 76.8; 153.6	1/4	QPSK pre-spreading	TD (OTD and/ or STS)
4	1	1.2; 1.35; 1.5; 2.4; 2.7; 4.8; **9.6**; 19.2; 38.4; 76.8; 153.6; 307.2	1/2	QPSK pre-spreading	TD (OTD and/ or STS)
5	1	1.8; 3.6; 7.2; **14.4**; 28.8; 57.6; 115.2; 230.4	1/4	QPSK pre-spreading	TD (OTD and/ or STS)
6	3	1.2; 1.35; 1.5; 2.4; 2.7; 4.8; **9.6**; 19.2; 38.4; 76.8; 153.6; 307.2	1/6	QPSK pre-spreading	
7	3	1.2; 1.35; 1.5; 2.4; 2.7; 4.8; **9.6**; 19.2; 38.4; 76.8; 153.6; 307.2	1/2	QPSK pre-spreading	
8	3	1.8; 3.6; 7.2; **14.4**; 28.8; 57.6; 115.2; 230.4; 460.8	1/4 (20 ms) 1/3 (5 ms)	QPSK pre-spreading	
9	3	1.8; 3.6; 7.2; **14.4**; 28.8; 57.6; 115.2; 230.4; 259.2; 460.8; 518.4; 1036.8	1/2 (20 ms) 1/3 (5 ms)	QPSK pre-spreading	
10	1	81.6; 158.4; 163.2; 312.0; 316.8; 326.4; 465.6; 619.2; 624.0; 633.6; 772.8; 931.2; 1238.4; 1248.0; 1545.6; 1862.4; 2476.8; 3091.2	1/5	QPSK, 8-PSK, 16-QAM	

Table 2.4 shows all radio configurations available in the forward link of cdma2000 networks. The table provides the main characteristics of each RC and associates it with a spreading rate.

All radio configurations employ at least one of the two basic Rate Sets (RS), 9.6 Kbps and 14.4 Kbps. This feature assures backward compatibility because these RSs are also used in CDMA IS-95 systems (RCs 1 and 2). RC10 is the only exception to this rule and is specially assigned to cdma2000-1XEV systems, representing the general evolution pattern for the cdma2000. Basic data rates are shown in bold in the table and represent maximum vocoder rates.

Not all RCs are supported by cdma2000 systems, but a minimum set of radio configurations is required. However, this set may vary according to the technology. The following list presents the requirements of each network:

- cdma2000-1X: RC1 and RC3,

- cdma2000-1X evolutions: RC1, RC3 and RC10,

- cdma2000-3X: RC1, RC3 and RC7.

All remaining RCs presented in Table 2.4 (RCs 2, 4, 5, 6, 8 and 9) are optional but some rules apply to their usage:

- all BTSs supporting RC2 must also support RC1,

- in cdma2000-1X networks, BTSs supporting RC4 and RC5 must also support RC3,

- in cdma2000-3X networks, BTSs supporting RC6, RC8 and RC9 must also support RC7.

Some RCs use a diversity technique, solely in the forward link, known as Transmit Diversity (TD). Two types of TDs are implemented: Orthogonal Transmit Diversity (OTD) and Space Time Spreading (STS). The OTD technique consists of a transmission method that distributes forward link channel symbols among multiple antennas and spreads them using the Walsh codes or Quasi-Orthogonal Functions (QOF) associated with each antenna. The STS technique also distributes forward link channel symbols among multiple antennas but it spreads them using complementary Walsh codes or QOFs.

The modulation efficiency provided by CDMA IS-95 systems and cdma2000 systems is also presented in Table 2.4. CDMA IS-95 systems are said to use BPSK (or half-efficient QPSK) because of how the information bit is modulated for transmission. One coded bit is used per modulation symbol. The modulator obtains the QPSK modulated symbol by duplicating this bit and applying a distinct orthogonal modulation (PN sequences) to each of the duplicated bits prior to transmission.

Conversely, on cdma2000 systems, two coded information bits are used to generate a real QPSK modulated symbol prior to transmission. This allows improvement of the data transmission rate or use of further coding and protection schemes, improving system security and reliability.

Table 2.5 relates the quantity and availability of each type of logical channel in the forward link of cdma2000 systems to SRs and RCs.

Table 2.5 Forward link channel availability according to supported SR and RC [2]

Forward logical channels	Maximum number allowed		Availability	
	SR1	SR3	SR1	SR3
Forward pilot channel	1	1	Y	Y
Transmit diversity pilot channel	1	1	Y	Y
Auxiliary pilot channel	Not defined	Not defined	Y	Y
Auxiliary transmit diversity pilot channel	Not defined	Not defined	Y	Y
Synchronisation channel	1	1	Y	Y
Forward paging channel	7		Y	N
Broadcast control channel	8	8	RC3–RC5	Y
Quick paging channel	3	3	RC3–RC5	Y
Common power control channel	15	4	RC3–RC5	Y
Common assignment channel	7	7	RC3–RC5	Y
Forward common control channel	7	7	RC3–RC5	Y
Forward dedicated control channel	1^a	1^a	RC3–RC5	Y
Forward fundamental channel	1^a	1^a	Y	Y
Forward supplemental code channel	7^a		RC1 and RC2	N
Forward supplemental channel	2^a	2^a	RC3–RC5	Y
Forward packet data control channel	2		RC10 only	N
Forward packet data channel	2		RC10 only	N

^{a}Quantity of logic sub-channels per traffic channel.

Channels available in cdma2000 networks are more specialised than channels available in CDMA IS-95 systems. For example, the Forward Pilot Channel (FPiCh) and Synchronisation Channel (SyncCh) used in CDMA IS-95 systems were kept in cdma2000, but specialised channels were added to aid on acquisition and synchronisation, the Auxiliary Pilot Channels (FAPiChs) and the Transmit Diversity Pilot Channels (FTDPiChs).

cdma2000 networks also distributed IS-95 Forward Paging Channel (FPCh) functions among specialised channels when using RC3 or higher. These specialised channels are: Forward Common Control Channels (FCCChs), Forward Broadcast Control Channels (FBCChs) and Common Assignment Channels (CAChs). The FCCCh performs all functions assigned to the previous FPCh, such as transmission of broadcast messages, support for call establishment or system access and traffic channel assignment to the MSs. The FBCCh aids the FCCCh in the transmission of some of the broadcast messages and/or SMS. The CAChs perform faster channel assignments relieving the FCCChs workload.

Another addition of cdma2000 is the Forward Quick Paging Channel (FQPCh or QPCh), which allows an increase of the MSs' batteries lifetime.

The Forward Common Power Control Channels (FCPCChs) execute power control on cdma2000 reverse logic channels while the MS is accessing or trying to access the network.

Another example of specialisation is the set of CDMA IS-95 Forward/Reverse Traffic Channels (FTChs and RTChs), used to transmit voice and data traffic, fax and signalling/control during a call. In cdma2000 these channels evolved into Forward Fundamental Channels (FFChs), Forward Dedicated Control Channels (FDCChs) and Forward Supplemental Channels (FSChs).

FFChs is similar to the original FTChs, but are used preferably for voice traffic and low data rates transmissions. Signalling and control traffic are transmitted to the mobile stations by FDCChs, whereas high data rate traffic uses the FSChs.

The Forward Supplemental Code Channels (FSCChs) implemented in CDMA IS-95B networks to transmit high data rate traffic is used only for MSs working with RCs 1 and/or 2. FSChs, however, provide a much higher data transmission capacity when compared to the FSCChs.

Forward Packet Data Channels (FPDChs) and Forward Packet Data Control Channels (FPDCChs) are implemented only in 1XEVDO networks using RC10.

The variety of channels in cdma200 networks caused the industry to divide them into two groups: common channels and dedicated channels. The term 'common channels' refers to channels interacting with several MSs simultaneously. They consist of the FPChs, FCCChs, QPChs, FBCChs, FCPCChs and CAChs. Dedicated Channels, conversely, are the channels interacting with only a single MS at a time, especially during communication, such as the traffic channels.

As in CDMA IS-95 systems, each BTS is identified by the short PN sequence offset. There are 512 64-chip long offsets available (numbered from 0 to 511).

2.3.2.2   Reverse Link in cdma2000 Systems

There are five main types of logic channels in the reverse link of cdma2000 systems. Figure 2.11 shows these channels and their subdivision into sub-channels. The highlighted blocks indicate channels inherited from IS-95 systems. As in the forward link, channel

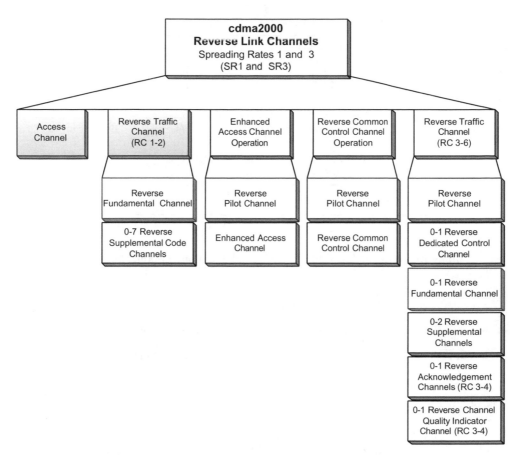

Figure 2.11 Reverse link channel structure for cdma2000 systems [2].

configuration depends on radio configuration and spreading rate being used by the mobile station.

Table 2.4 shows all radio configurations available in the reverse link of cdma2000 networks, presenting their main characteristics and associating them with spreading rates.

As in the forward link, all RCs employ at least one of the two basic Rate Sets (RS), 9.6 Kbps and 14.4 Kbps. This feature assures backward compatibility because these RSs are also used in CDMA IS-95 systems (RCs 1 and 2). Basic data rates are presented in bold in the table.

The minimum operating configuration of the mobile stations must support a minimum set of radio configurations. This set, however, may vary according to the technology. The following list presents the requirements for each type of network:

- cdma2000-1X: RC1 and RC3,

- cdma2000-1X evolutions: RC1 and RC3,

- cdma2000-3X: RC1, RC3 and RC5.

Table 2.6 Available Radio Configurations (RC) for the cdma2000 reverse link [2]

RC	Associated SR	Data rate (kbps)	FEC–Conv. Encoder (R)	Modulation
1	1	1.2; 2.4; 4.8; **9.6**	1/3	Orthogonal–order 64
2	1	1.8; 3.6; 7.2; **14.4**	1/2	Orthogonal–order 64
3	1	1.2; 1.35; 1.5; 2.4; 2.7; 4.8; **9.6**; 19.2; 38.4; 76.8; 153.6; 307.2	1/4 1/2	BPSK with RPiCh
4	1	1.8; 3.6; 7.2; **14.4**; 28.8; 57.6; 115.2; 230.4	1/4	BPSK with RPiCh
5	3	1.2; 1.35; 1.5; 2.4; 2.7; 4.8; **9.6**; 19.2; 38.4; 76.8; 153.6;307.2; 614.4	1/4 1/3	BPSK with RPiCh
6	3	1.8; 3.6; 7.2; **14.4**; 28.8; 57.6; 115.2; 230.4; 460.8; 1036.8	1/4 1/2	BPSK with RPiCh

All remaining RCs presented in Table 2.6 (RCs 2, 4 and 6) are optional but some rules apply to their usage:

- MSs supporting RC2 must also support RC1,

- in cdma2000-1X networks, MSs supporting RC4 must also support RC3,

- in cdma2000-3X networks, MSs supporting RC6 must also support RC5.

Table 2.7 relates the quantity and availability of each type of logical channel in the reverse link of cdma2000 systems to SRs and RCs.

When working with RCs 3 and higher, the Enhanced Access Channels (EAChs) and Reverse Common Control Channels (RCCChs) execute tasks that are performed by RACh in IS-95 systems, such as system access. These channels also perform additional functions such as transmission of users and system information (signalling and control).

A great innovation introduced by cdma2000 networks is the use of a Reverse Pilot Channel (RPiCh). The MSs use this channel before and during calls. Before the call it serves

Table 2.7 Reverse link channel availability according to supported SR and RC [2]

	Maximum number allowed		Availability	
Reverse logical channels	SR1	SR3	SR1	SR3
Reverse pilot channel	1	1	RC3 and RC4	Y
Reverse access channel	1		Y	N
Enhanced access channel	1	1	RC3 and RC4	Y
Reverse common control channel	1	1	RC3 and RC4	Y
Reverse acknowledgement channel	1		RC3 and RC4	N
Reverse channel quality indicator channel	1		RC3 and RC4	N
Reverse fundamental channel	1	1	RC3 and RC4	Y
Reverse dedicated control channel	1	1	RC3 and RC4	Y
Reverse supplemental code channel	7		RC1 and RC2	N
Reverse supplemental channel	2	2	RC3 and RC4	Y

as a preamble and is used for timing and power reference. During a call, it still performs these functions but also carries power control sub-channels to control forward link channels.

The Reverse Channel Quality Indicator Channels (RCQIChs) and Reverse Acknowledgement Channels (RAckChs) are available only for RCs 3 and 4 and are compatible to cdma2000-1X EVDO systems. These channels provide acknowledgement and FPDChs acquisition messages to the BTS and also inform about the reception quality of these channels.

Figure 2.12 presents, graphically, the relationships among the logical channels of CDMA IS-95 and cdma2000 systems.

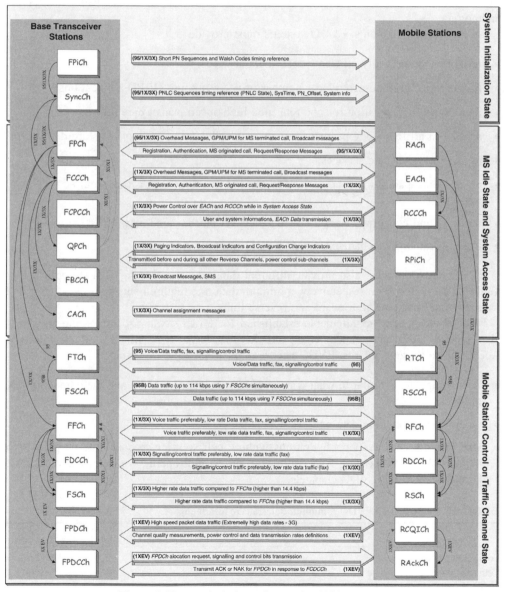

Figure 2.12 Logical channels on cdma2000 systems.

2.3.3   cdma2000-1X EVDO Systems

2.3.3.1   Requirements (IS-856)

The original evolutionary path for cdma2000 considered 1XRTT coming after IS-95 system and 3XRTT coming after that. However, a more realistic approach, for non-real-time services, is the evolution of 1XRTT into 1XEVDO (also known as EVDO, where EV stands for Evolution and DO stands for Data Optimised or Data Only).

EVDO systems are regulated by the IS-856 standard, High Rate Packet Data (HRPD). These standards establish a set or requirements to be met by networks using the EVDO technology.

As general requirements, EVDO systems must provide

- reliable and spectrally efficient packet data services;

- traffic capacity to absorb a large number of users running HRPD applications;

- data optimised operation for non-real time high-speed packet data services;

- service in all frequency bands in areas where cellular and PCS systems are deployed;

- asymmetric data rate services;

- always-on capability, that is, no need to use dial-in and no user action required for PDSN log-on;

- multiple and concurrent packet sessions support on the same network;

- traffic balancing mechanisms for all EVDO frequency channels.

EVDO networks must also provide different data rates suitable for HRPD applications. Tables 2.8 and 2.9 show, respectively, minimum data rate requirements per type of user and available data rates in the forward and reverse links.

Besides the general requirements established for EVDO systems, the IS-856 standards also specify radio environment characteristics that have to be met by this technology. According to these standards, EVDO networks must have the following.

- A coverage area similar to what is provided by the CDMA IS-95 and the cdma2000-1X networks.

- Operate on a separate data-optimised channel.

- Use existing cell/sector configurations without cell splitting, because the purpose is to deploy EVDO on existing BTS equipment, which will operate a mix of CDMA IS-95, cdma2000-1X and EVDO channels.

Table 2.8   cdma2000-1XEVDO minimum data rate requirements

	Forward link (to user) (Mbps)	Reverse link (from user) (Kbps)
Vehicular peak data rate	1.25	144
Vehicular average data rate	0.60	144
Pedestrian/fixed peak data rate	2.00	144

Table 2.9 cdma2000-1XEVDO (HPRD) available data rates

Forward link	Reverse link (Kbps)
38.4 Kbps	9.6
76.8 Kbps	19.2
153.6 Kbps	38.4
307.2 Kbps	76.8
614.4 Kbps	153.6
1.2288 Mbps	
1.8432 Mbps	
2.4572 Mbps	

- Provide a smooth transition from CDMA IS-95 A/B and cdma2000-1X networks to minimise the impact on mobile terminals and on infrastructure.

- Conform to cdma2000-1X systems requirements for out-of-band emissions.

- Support both mobile and fixed users.

- Provide bandwidth efficiency by dynamically allocating resources (power, code space and time slots) and adapting to the time varying nature of shadowing and fast fading.

- Fully exploit the capability of maximising spectral efficiency through the use of Adaptive Modulation and Coding (AMC). The AMC includes the selection of the Modulation and Coding Scheme (MCS) that better matches the channel environment for the required data rate. It also includes a collection of techniques referred to as Link Adaptation (LA), which consists of fast feedback channel state information, adaptive modulation, incremental redundancy, repetition coding, time diversity adaptation, hybrid ARQ, selection diversity and multi-user diversity.

As per inter-operability support, the standards require EVDO networks to enable the following.

- An air interface operating with a Radio Access Network (RAN) designed according to the inter-operability specification [A.S0001].

- Inter-operability (including handoff) with GSM-MAP systems and with cdma2000-1X channels (for packet services).

- An embedded transparent system method to re-direct user invoked voice and real-time services to cdma2000-1X.

The EVDO radio access network must also comply with the following authentication support requirements. Note that an Access Terminal (AT) in EVDO networks is the same as a Mobile Station (MS) in other systems.

- Use of an authenticated AT identifier for hybrid EVDO/cdma2000 devices. This identifier must be compatible with a mobile station identifier that is also authenticated by IS-2000 for the same hybrid device (e.g. IMSI).

- Use of authenticated AT to permit the cdma2000 RAN and EVDO RAN to coordinate the operation of hybrid devices.

- Possibility for the EVDO RAN to deny an AT access to any dedicated RAN resource (i.e. resources supporting transfer of user data to or from the PDSN, not including power and rate control) until after the MS identity has been authenticated.

- Possibility to preclude any end-user IP data traffic from being exchanged until after the MS identity has been authenticated.

- Possibility to minimise the time required to authenticate a hybrid device on the cdma2000-1X system to minimise the consumption of system resources by an invalid device.

- Reduction of the total transmission power required for authenticating a hybrid device on the EVDO system to minimise interference imposed by an invalid device.

2.3.3.2   EVDO Channel Structure

As in previous versions of CDMA based networks, regulatory agencies also suggested a reference model for EVDO network architecture (Figure 2.13).

Here

Pilot channel. Channel constantly radiating but without transmitting any information (string of 1'0s'). Used to evaluate power level of the current server and neighbor cells, to evaluate BER, to help in coherent detection, to keep the connected route by triggering route messages and to monitor cell with control channel.

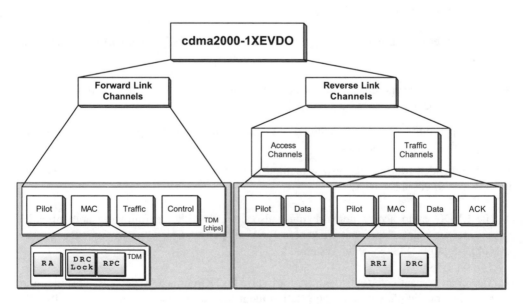

Figure 2.13   cdma2000-1X EVDO channel structure.

MAC channel. Dedicated to media access control activities, divided into three sub-channels.

- Reverse link.

  Reverse link data rate index (RRI): indicates the data rate of the data channel being transmitted on the reverse traffic channel.

  Data rate control (DRC): indicates the data rate at which the MS can receive the forward traffic channel and the sector from which the MS wishes to receive it.

- Forward link.

  Reverse link activity (RA): indicates the activity state of the slots on the reverse channel.

  DRCLock: indicates if the BTS can receive the DRC sent by the MS.

  Reverse link power control (RPC): used by the BTS to control the reverse channel power of a mobile station.

Traffic channel. Used by BTSs and MSs to transmit user-specific traffic or signalling information.

Control channel. Used by MSs to broadcast messages or signalling information.

Data channel. Part of the access channel or the traffic channel containing MAC layer packets for control or for traffic data on the reverse link.

ACK channel. Used by MSs to inform if the physical layer packet transmitter on the forward traffic channel was received successfully.

Access channel. Used by MSs to initiate communication with BTSs or to respond to any BTS with a directed message.

2.3.4 cdma2000-1X EVDV Systems

2.3.4.1 Requirements (R.S0026)

Even though EVDO is a more realistic evolution for 1XRTT systems, this technology lacks voice capacity. Therefore a new standard, based on EVDO principles but integrating voice capacity was proposed. This technology is known as cdma2000-1X EVDV (EVolution for Data and Voice), and can reach data transmission rates even higher than EVDO systems. This technology, however, does not seem to be as widely accepted in the market as EVDO was, thus it may never even be deployed.

EVDV networks are regulated under the R.S0026 specifications (3GPP2). The main requirements and expectations for this technology are the following:

- high-speed data capability to support existing and future Internet based services;

- packet data transmission schemes to support significant amount of bursty data;

- increased spectral efficiency allowing data and voice transmissions on the same carrier (1.25 MHz bandwidth), considering different busy periods for distinct applications on the same air interface;

- packet scheduling and traffic balancing mechanisms for all frequency channels and underlying technologies (CDMA IS-95, cdma2000).

Because of implementation schedule requirements in 1.25 MHz carriers, the evolution from cdma2000-1X to cdma2000-1X EVDV is divided in two, distinct phases.

2.3.4.2   Phase 1

Phase 1 considers system optimisation to allow high-speed packet data services for non-real-time applications.

In this phase, a specific carrier is separated for the initial high-speed packet data services. If users require voice or another real-time service, the EVDV network uses cdma2000-1X resources to provide these services, making this solution transparent to users.

The requirements for phase one include the following.

- Incorporation of all cdma2000 requirements, behaving as an extension of the existing cdma2000-1X features, functions, applications and services. This means that EVDV specifications maintain all voice and packet data capabilities of cdma2000-1X and EVDO networks.

- Inter-operability with cdma2000-1X channels for packet services, including handoff.

- High-speed packet data capability for non-real-time processes.

- Improved signalling and packet data throughput.

- Evolution from cdma2000-1X by maximising re-use of existing cdma2000-1X family standards, including support of existing cdma2000 vocoders.

- Supersede EVDO data services with real-time capability delivering traffic in three distinct modes:

  Real-time;

  Non-real-time;

  Mixed real-time/non-real-time on the same carrier.

- Use of multiple and concurrent packet sessions supported on the same network.

- Support to voice over IP (VoIP).

- QoS control and adaptive handoff thresholds.

- Deployment in all frequency bands in areas where cellular and PCS systems are deployed.

- Operation under asymmetric data rates conditions in order to serve the majority of Internet applications.

- Implementation of 16 Class of Services (CoS) for packet data transmissions to achieve the best system performance and maximise data throughput:

  class 1: real-time services and high-speed packet data (phase 2 only);

class 2: non-real-time high-speed packet;

class 3: scheduled delivery;

classes 4–16: reserved for future CoS (phase 2 only).

- Fast connection time compared to existing IS-95 standards.

- Always-on configuration for data services.

- Backward compatibility to CDMA IS-95 A/B and cdma2000 systems, keeping the same antennas but with the possibility of adding active, smart and directional antennas for specific applications.

- Simple user device activation using over-the-air activation.

- Smooth evolution from CDMA IS-95 A/B and cdma2000 systems and to phase 2 to minimise the impact on terminals and infrastructure.

2.3.4.3 Phase 2

Phase 2 of EVDV implementation focus on the support to concurrent high-speed packet data and real-time services. The main idea is to integrate all services of phase 1 on the same carrier while retaining the ability to maintain packet data services on a separate carrier.

The requirements for the second phase of implementation for EVDV networks are the following.

- Backward compatibility to CDMA IS-95 A/B and cdma2000 systems, using the same antennas and including cdma2000-1X channels interoperability for packet and voice services.

- Multiple and concurrent packet sessions supported on the same network.

- Voice and packet service options and capability to add future, globally harmonised audio and video CODECS.

- Same Classes of Service (CoS 1-16) defined for Phase 1 but operating for voice and packet services.

- Higher voice quality standards when in the highest quality mode.

- Double of voice service spectral efficiency when compared to other cdma2000-1X networks.

- Quality of Service (QOS) parameters to be followed:

  priority;

  minimum acceptable data rate;

  maximum permitted data loss rate (expressed as a percentage of either packet loss or frame loss per data message, or as a BER);

  latency or delay.

- Power consumption reduction by one-half, extending active transmission time and increasing standby time by 100%.

- Asymmetric and symmetric data rate services.

- Use of traffic balancing mechanisms for all EVDV frequency channels.

- Network adaptation of data transmission speeds and other operating parameters to maximise system capacity, while satisfying QoS constraints, for a given mix of RF conditions and system workload.

- Compatibility to any combination of traffic types with varying QoS constraints, including voice, video and data on a single radio channel.

- Segregation of traffic types with varying QoS constraints, including voice, video and data onto separate radio channels.

- Handoff of voice and data services between EVDV radio channels and other radio channels operating according to cdma2000 standards. The EVDV specification does not mandate that a mobile station that supports EVDV has also to support EVDO. However, if a mobile station does support both EVDV and EVDO, then it must support handoff of commonly supported data services.

- Voice quality that meets or exceeds the quality provided by other cdma2000 systems.

- Support GSM full rate and AMR vocoders.

EVDV networks must provide different data rates suitable for data package only services. Tables 2.10 and 2.11 show, respectively, minimum data rate requirements per type of user and available data rates in the forward and reverse links [6–42].

When EVDV networks are operating in a radio channel that only supports packet data services that are not QoS restricted, they must be capable of providing a peak rate and a

Table 2.10 cdma2000 1X EVDV minimum data rate requirements

	Forward link (to user) (Mbps)	Reverse link (from user) (Mbps)
Vehicular peak data rate	2.40	1.25
Vehicular average data rate	0.60	0.60
Pedestrian/fixed peak data rate	2.40	2.40

Table 2.11 cdma2000 1X EVDV available data rates

Forward link (Kbps)	Reverse link (Kbps)
$76.8 \times 1N$	$9.6 \times 2^P/N$
$115.4 \times 1N$	$460.8/N$
$76.8 \times 2N$	$P = 0\text{–}6$
$115.4 \times 2N$	$N = x\text{–}y$
$76.8 \times 4N$	
$115.4 \times 4N$	
$N = 1\text{–}14$	

system-wide average data rate in a fully loaded system that is at least the same as in EVDO specifications.

According to R.S0026 standards, EVDV networks must comply with the same radio environment specifications proposed for EVDO systems in the IS-856 standards.

R.S0026 standards introduced the concept of backward compatible class, which represents mobile terminal devices that support the set of standards and specifications C.S0001 through C.S0006, its precursors (e.g. TIA/EIA-95-B) and all ancillary standards and specifications. This concept is essential in the understanding of the compatibility requirements for EVDV networks. According to this requirements, EVDV networks must provide the following.

- A radio channel to provide service to mobile stations belonging to the backward compatible class and a radio channel to provide service to mobile stations that conform to EVDO. Here the version of specification of the mobile and the base station differ, the services provided are consistent with the level of quality associated with the limiting element.

- Compatibility with cdma2000 chip rate and band plan, supporting existing frame lengths. This requirement, however, does not preclude the addition of new frame lengths.

As per inter-operability support, the standards require EVDV networks to enable

- an air interface operating with a Radio Access Network (RAN) designed according to the inter-operability specification [A.S0001];

- inter-operability (including handoff) with GSM-MAP systems and with cdma2000-1X channels (for packet services).

BIBLIOGRAPHY AND REFERENCES

1. CelTec/CelPlan, CDMA IS-95 and cdma2000 Systems – Training Course.
2. Irwin Mark Jacobs, QUALCOMM, 3D World Congress Presentation, June 11, 2002.
3. C.S0005-B, Upper Layer (Layer 3) Signaling Standard for cdma2000 Spread Spectrum Systems, Release B, 3GPP2, April 2002.
4. Recommendation ITU-R M.1079-1, Performance And Quality of Service Requirements for International Mobile Telecommunications-2000 (IMT-2000).
5. C.S0002-C, Physical Layer Standard for cdma2000 Spread Spectrum Systems, Release C, 3GPP2, May 2002.
6. CDG, cdma2000 Overview.
7. The Shosteck Group, *Third Generation Wireless (3G): The Continuing Saga.* Maryland: Wheaton, 2001, pp. 263–264.
8. The Shosteck Group, CDG, GSM or CDMA: The Commercial and Technology Challenges for TDMA Operators, June, 2001.
9. TIA/EIA-95-B, Mobile Station – Base Station Compatibility Standard for Dual-Mode Spread Spectrum Cellular System, November 1998.
10. CDG, Detailed Info: Standard Requirements for the Evolution of cdma2000. May 30, 2000.
11. TIA/EIA-41-D (ANSI-41), Cellular Radio Telecommunications Intersystem Operations, December, 1997.

12. TIA/EIA/95 (ANSI-95), Mobile Station – Base Station Compatibility Standard for Dual-Mode Wideband Spread Spectrum Cellular Systems, Telecommunications Industry Association; May 1995.
13. TSB74, Support for 14.4 kbps Data Rates and PCS Interaction for Wideband Spread Spectrum Cellular Systems, December, 1995.
14. TIA/EIA/124-C, Wireless Radio Telecommunications Intersystem Non-Signaling Data Communications (DMH) Data Message Handler, September 2000.
15. TIA/EIA/136B, TDMA Third Generation Wireless Revision B, March 2000.
16. EIA/TIA/553, Mobile Station – Land Station Compatibility Specification, November 1999.
17. TIA/EIA/634-B, MSC – BS Interface for Public Wireless Communications Systems, April 1999.
18. TIA/EIA/IS-658, Data Services Interworking Function Interface for Wideband Spread Spectrum Systems, July 1996.
19. TIA/EIA/IS-658-1, Data Services Interworking Function Interface for Wideband Spread Spectrum Systems – Addendum 1, April 1999.
20. TIA/EIA/IS-683-A, Over-The-Air Service Provisioning of Mobile Stations in Spread Spectrum Systems, June 1998.
21. TIA/EIA/IS-725-A, Cellular Radio Telecommunications Intersystem Operations – Over-The-Air Service Provisioning (OTASP) & Parameter Administration (OTAPA), July 1999.
22. TIA/EIA/IS-728, Intersystem Link Protocol, April 1998.
23. TIA/EIA/IS-737, IS-41-C Enhancements to Support Circuit Mode Services, May 1998.
24. TIA/EIA/IS-756-A and TIA/EIA41-D, Enhancements for Wireless Number Portability Phase II, December 1998.
25. TIA/EIA/IS-771, Wireless Intelligent Network, July 1999.
26. TIA/EIA/IS-788, Connector Specification for the Portable Phone Interface, June 1999.
27. TIA/EIA/IS-789-A, Electrical Specification for the Portable Phone to Vehicle Interface, April 2000.
28. TIA/EIA/IS-816, IDB Message Set Definitions for the Electrical Interface Between Portable Phone and Vehicle, 2000.
29. TIA/EIA/IS-820, Removable User Identity Module (R-UIM) for TIA/EIA Spread Spectrum Standards, March 2000.
30. TIA/EIA/IS-823, Wireless Intelligent Network Capabilities for Pre-Paid Charging, August 2000.
31. TIA/EIA/IS-835, CDMA Wireless IP Network Standard, December 2000.
32. TIA/EIA/IS-841, Network Based Enhancements for the User Identity Module (UIM), August 2000.
33. TIA/EIA/IS-2000.1-A, Introduction for cdma2000 Spread Spectrum Systems, March 2000.
34. TIA/EIA/IS-2000.2-A, Physical Layer Standard for cdma2000 Spread Spectrum System, March 2000.
35. TIA/EIA/IS-2000.3-A, Medium Access Control (MAC) Standard for Spread Spectrum Systems, March 2000.
36. TIA/EIA/IS-2000.4-A, Link Access Control (LAC) Standard for Spread Spectrum Systems.
37. TIA/EIA/IS-2000.5-A, Upper Layer (Layer 3) Signaling Standard for Spread Spectrum Systems (S.R005-B).
38. TIA/EIA/IS-2000.6-A, Analog Signaling Standard for Spread Spectrum Systems.
39. TIA/EIA/IS-2001, Access Network Interfaces Interoperability Specification (IOS), December 2000.
40. ITU M.3100, Generic Network Information Model, July 1995.
41. TIA/EIA/J-STD-025, Lawfully Authorised Electronic Surveillance, 2000.
42. TIA/EIA/J-STD-036, Wireless Enhanced Emergency Services, 2000.

3

Codes and Sequences

BRUNO DE SOUZA ABREU XAVIER and
ARLINDO MOREIRA FARTES FILHO

CDMA systems employ several types of codes, each with a specific characteristic, to allow multiple users to communicate simultaneously. The codes uniquely identify users so that they are able to transmit information while limiting interference to each other. cdmaOne systems use codes, such as maximal length sequences and Walsh codes, both in the forward and reverse links. cdma2000 systems implement an additional code type, known as quasi-orthogonal functions. This new code allows cdma2000 to have more users and to provide a higher Quality of Service (QoS), the main requirements for third generation systems. This chapter addresses the main characteristics of these codes and functions, focusing on the concept behind each code, while avoiding a deep mathematical approach, to allow an easier understanding of the overall process.

3.1 INTRODUCTION

Base Transceiver Stations (BTS) and Mobile Stations (MS) generate specific codes and sequences to allow encrypting, multiplexing, spreading and de-spreading transmitted signals.

An appropriate choice of spreading codes avoids excessive interference among users. The codes contribute to the inherent privacy of CDMA systems and define their capacity in terms of simultaneous users. There are several types of spreading codes, each with specific characteristics, which makes some codes more suitable to particular applications than others. CDMA IS-95 systems use Maximal Length Sequences (MLS) and Walsh codes. cdma2000 systems also implement Quasi-Orthogonal Functions (QOFs), aiming to improve privacy and reduce potential capacity limitations.

Pseudo-random-Noise (PN) sequences or MLS are used to multiplex and spread the data to be transmitted, depending on the different usage approach for each link direction (forward and reverse). Mobile stations also use these sequences to identify BTSs within the network.

PN sequences can be used to spread the signal, giving it characteristics similar to thermal noise, while remaining totally predictable, i.e. deterministic. These codes are configured and synchronised in a way to assure perfect alignment between sequences transmitted by the BTS and locally generated at the MS and vice versa. The fact that their properties directly

Designing CDMA 2000 Systems   L. Korowajczuk
© 2004 John Wiley & Sons, Ltd   ISBN: 0-470-85399-9

influence synchronisation performance between transmitters and receivers makes the choice of codes extremely important.

The synchronisation has a fundamental role in the de-spreading and de-multiplexing processes, because it allows correlated sequences to be recovered whereas uncorrelated sequences remain spread, behaving like noise. The synchronism between PN sequences involves retrieving the original information by using two XOR (exclusive-OR) functions, one at the transmitter and one at the receiver. If the sequences are different or not synchronised between BTS and MS, the information retrieved presents errors.

The following sections in this chapter describe each type of sequences and codes. The following concepts, however, are essential for a better understanding of the topic.

- Bit: binary digit, that is, an element that can assume one of two values, e.g. 0 or 1, +1 or −1. The original information (before any coding or spreading) is usually referred to as 'bits'.

- Chip: type of bit. The name is applied to represent bits in sequences or codes used for spreading and in the coded or spread information.

- Coded bit or coded symbol: information bits after being processed by any Forward Error Correction (FEC) encoder type, such as convolutional or turbo.

- Modulation symbol: element of a finite alphabet that is the output of a modulator carrying analogue signal characteristics, such as amplitude. The alphabet may be binary or have a larger dimensionality. Bits enter the modulator and generate symbols as outputs. The information rate is always the same, but the baud rate is usually different between input and output.

3.2   MAXIMAL LENGTH SEQUENCES

By definition, MLSs (also known as PN Sequences) are the longest sequences that can be generated by a certain arrangement of shift registers or similar delay elements. This section describes the process when using shift registers.

PN sequences, also known as Pseudo-random-Noise sequences, are binary sequences with noise-like random characteristics. The term pseudo-noise is used because these sequences are neither completely deterministic nor truly random. The first case would allow anybody to understand information scrambled by the sequence whereas the second would avoid even the intended receiver to understand anything. The idea of PN sequences is to look random to unintended receivers while being known to both transmitter and intended receiver.

The most common MLS generator consists of a shift register working according to a specific logic that provides a feedback path, which combines the states of two or more stages of the shift register. The length of the MLS, generated by a Linear Feedback Shift-Register (LFSR) circuit, is given by the following expression

$$L = 2^N - 1 \text{ chips} \tag{3.1}$$

where

N number of shift registers used in the generator circuit

A polynomial expression can be associated with each set of shift registers. These expressions are primitive, that is, their only roots are '1' and itself

$$P(x) = a_n x^n + a_{n-1} x^{n-1} + \cdots + a_2 x^2 + a_1 x^1 + 1 \qquad (3.2)$$

where

a_n existence of feedback path (when there is a feedback path from x_i output, a_i equals 1, otherwise, a_i equals 0)

x_n shift register output

There are two basic feedback shift-register configurations for generating PN sequences using N-stage shift registers: the Simple Shift-Register Generator (SSRG) and the Modular Shift-Register Generator (MSGR). Figure 3.1 shows a 3-stage shift-register LFSR PN generator, which generates a sequence of seven chips in length. Note that if the shift registers are not connected correctly, the generated sequence will not be an MLS, i.e. it will not be ($2^3 - 1$ chips) long.

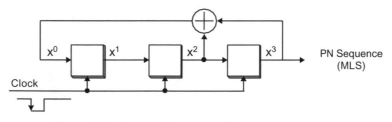

Figure 3.1   3-stage shift-register PN sequence generator.

Based on eqn (3.2), the 3-stage shift-register sequence generator presented in Figure 3.1 is represented by the following expression

$$P(x) = x^3 + x^2 + 1 \qquad (3.3)$$

Table 3.1 lists the number of MLSs that can be obtained with an N-stage shift-register circuit.

Table 3.1   Maximum number of MLSs generated by a sequence generator

Number of shift registers (N)	Number of possible distinct MLSs
3	2
5	6
10	60
15	1800
20	24 000

The main properties of an MLS are the following.

- Thermal noise-like behavior, even though deterministic.

- The amount of '1s' (Q_{one} chips) in the MLS exceeds the amount of '0s' (Q_{zero} chips) by one, i.e. $Q_{one} = 2^{N-1}$ and $Q_{zero} = 2^{N-1} - 1$ chips.

- The XORing of two different shifts of the sequence produces a third shift of the same sequence (closure property).

- The distribution of consecutively repeated '1s' and '0s' over the sequence is well known and defined.

For N-stage shift-register circuits, the probability of a bit (1 or 0) being repeated a given number of times consecutively ("run" of length M) is given by 2^{-M}, as in Figure 3.2. For example, the probability of three consecutive "1s" or "0s" happening in a sequence is of 2^{-3}, or 12.5% (1/8).

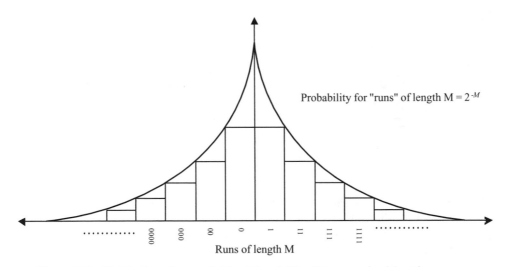

Figure 3.2 Distribution of runs of chips '0' and '1' within a maximal length sequence.

This concept also reinforces the pseudo-random-noise behaviour because higher probabilities are estimated for the non-consecutive repetition of symbols.

The MLS generator continuously produces output chips whose sequence repeats every L chips, thus requiring a convention to define the beginning of an MLS period.

Because the probability of consecutive repeated numbers being part of the sequence is low, it is defined that the MLS period starts on the first chip '1' after a set of $N - 1$ chips '0', where N is the number of shift registers in the generator circuit.

- Possibility of generating known code-phase offsets by applying bit masks (offset masks). Figure 3.3 shows the effect of a mask applied to the output of shift registers in a sequence generator of 15-bit long MLSs. Each of the 15 different masks generates one of the 15 possible shifts of sequence.

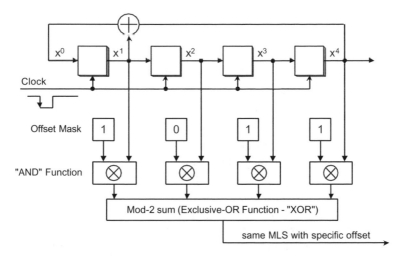

Figure 3.3 4-stage shift-register MLS generator with offset mask and mod-2 sum.

The offset mask is applied to the MLS through a mod-2 sum, resulting in an MLS that is a shifted version of the original sequence, i.e. it has a distinct phase offset. The mask used in the circuit has to be a shifted version of the original MLS. The first two columns of Table 3.2 illustrate the concept of phase offset.

3.2.1 Correlation Characteristics of Spread Spectrum Systems

The similarity between a function $F(t)$ and itself at a different time (phase offset) is known as auto-correlation and is quantified by the *Auto-Correlation Function*

$$\text{ACF} = \int_{-\infty}^{+\infty} F(t)F(\tau - t)dt \qquad (3.4)$$

Table 3.2 Auto-correlation function for an MLS of length $L = 7$ and $N = 3$

Phase offset (chips)	Sequence phase	Number of Coinciding Chips (CC)	Number of Non-Coinciding Chips (NCC)	Auto-Correlation Function (ACF)
0	*1011100*	7	*0*	7
1	0111001	3	4	-1
2	1110010	3	4	-1
3	1100101	3	4	-1
4	1001011	3	4	-1
5	0010111	3	4	-1
6	0101110	3	4	-1
7	*1011100*	7	*0*	7

The ACF for a PN sequence can be understood as the chip-by-chip comparison of a sequence (length L) with shifted versions of itself (offsets from 1 to L), as shown in eqn (3.5)

$$ACF = CC - NCC \qquad (3.5)$$

where

CC   number of Coinciding Chips between sequences
NCC   number of Non-Coinciding Chips between sequences

Table 3.2 shows ACF results when comparing an MLS of length $L = 7$ and itself with different phase offsets. For non-zero offset, the number of non-coinciding chips always exceeds the number of coinciding chips by one, regardless of the offset. The maximum value (auto-correlation spikes) of the ACF is obtained only when both sequences being compared have the same phase. Table 3.2 uses the sequence '1011100' obtained from eqn (3.3) as the reference sequence (offset 0) and compares it to each of its other phases.

From Table 3.2, the comparison of phase offsets 0 and 6 is the following.

Offset 0 $\Rightarrow$ 1 0 1 1̲ 1̲ 0 0
Offset 6 $\Rightarrow$ 0 1 0 1̲ 1̲ 1 0̲
Coincident Chips (CC) $= 3$
Non-Coincident Chips (NCC) $= 4$
Therefore, ACF $=$ CC $-$ NCC $= 3 - 4 = -1$

The maximum values for the ACF are known as auto-correlation spikes. Figure 3.4 presents auto-correlation function results for a 7-chip period MLS, showing the spikes every time both the sequences compared have the same phase offset. The spikes' magnitude is $2^N - 1$, where N is the number of shift registers in the MLS generator.

The auto-correlation plays an important role in the synchronisation between sequences generated at the BTS and at the MS.

Figure 3.5 shows ACF results for a 4-chip long non-MLS generated by a 3-stage shift-register circuit $(L = 7)$. The sequence '1100' is considered as the reference sequence (offset 0).

In this case, the spikes do not represent the maximum value expected of $2^N - 1$. This makes it harder (if not impossible) to establish synchronism between sequences generated at the BTS and at the MS. Figures 3.5 and 3.6 illustrate this behaviour.

Figure 3.6 presents the normalised ACF result of a 31-chip non-MLS sequence, generated by

$$P(x) = x^5 + x^4 + x^3 + x^2 + x + 1 \qquad (3.6)$$

As stated before, the MLS's ACF has a very important role in the synchronisation of two distinct circuits. The synchronisation process can be divided in two steps: acquisition and tracking.

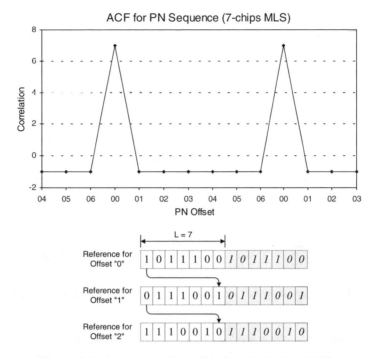

Figure 3.4 Auto-correlation spikes for a 7-chip long MLS.

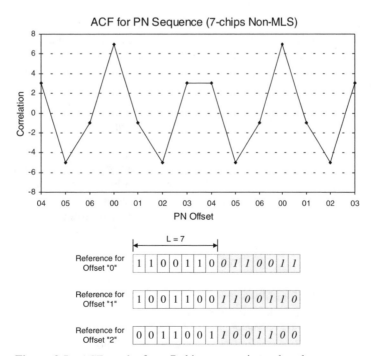

Figure 3.5 ACF results for a 7-chip non-maximum length sequence.

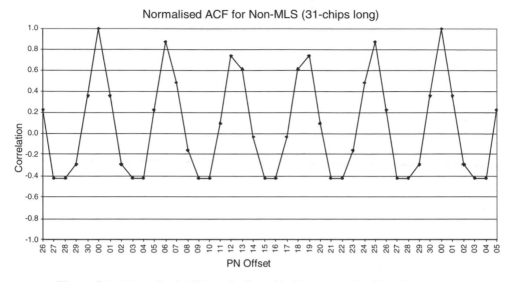

Figure 3.6 Normalised ACF results for a 31-chip non-maximal length sequence.

In the acquisition phase, the reception circuit tries to achieve code alignment by shifting the time alignment of the locally generated sequence and comparing it with the one received. A correlation threshold (similarity) defines the transition from acquisition to tracking, based on the expected ACF spike, as shown in Figure 3.4. Tracking maintains the correct code rate and alignment.

Whereas the ACF represents the comparison between a sequence and itself on a different phase, the Cross-Correlation Function (CCF) represents the comparison between two sequences of same length, as expressed in eqn (3.7). The CCF becomes equal to the ACF if $G(t)$ equals $F(t)$. Cross-correlation functions are also relevant in the understanding of spreading codes

$$\text{FCC} = \int_{-\infty}^{+\infty} F(t)G(\tau - t)dt \tag{3.7}$$

The CCF represents the similarity between two time-variant signals (or functions) taken with a time difference τ between them. Thus considering two binary sequences of same length and distinct phase offsets, the CCF can be defined as the result of a chip-by-chip comparison between sequences $F(t)$ and $G(t)$, for all possible phase offsets (from 1 to L, where L is the sequence length)

$$\text{CCF} = \text{CC} - \text{NCC} \tag{3.8}$$

Figure 3.7 presents CCF results when comparing two 31-chip periodic MLSs. CCFs only present partial correlation spikes, different from $(2^N - 1)$. The figure indicates expected ACF spikes if comparing the same sequence. Exceeding the valid correlation threshold for synchronism means that the two sequences are equal and almost synchronised. In this case, the phase error between the sequences would typically be 1 or ½ chip.

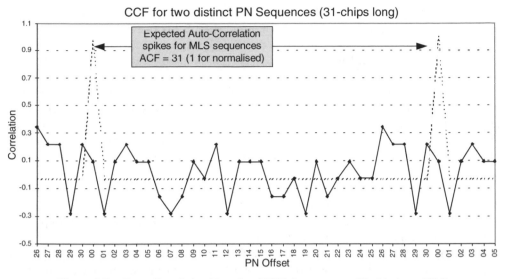

Figure 3.7 Cross-Correlation Function (CCF) between two 31-chip long MLSs.

Each CDMA carrier can accommodate several logical channels modulated and multiplexed (on the reverse link) by PN sequences. In CDMA IS-95 systems, three PN sequences perform these functions: two short PN sequences and one long PN sequence.

All pseudo-random-noise sequences used in CDMA systems (PN-I, PN-Q and long code (PNLC)) are synchronised to a reference time defined as the beginning of the CDMA IS-95 systems: January 6th, 1980, at 00:00:00 hours.

3.2.2 Short PN Sequences (PN-I and PN-Q)

CDMA IS-95 short PN sequences are based on two distinct MLSs generated by a circuit with 15 shift-register stages (i.e. one shift register with 15 stages) arranged to produce the following polynomials

$$\text{PN-I}: \quad P(x) = x^{15} + x^{13} + x^9 + x^8 + x^7 + x^5 + 1 \tag{3.9}$$

$$\text{PN-Q}: \quad P(x) = x^{15} + x^{12} + x^{11} + x^{10} + x^6 + x^5 + x^4 + x^3 + 1 \tag{3.10}$$

Each of the circuits generates 32767-chip long sequences ($2^N - 1$), composed of 16384 '1s' and 16383 '0s'. However, an external circuit inserts an additional chip '0' into each sequence after reading a set of 14 consecutive 0s, which happens only once on a complete set of 32767 chips.

This extra chip intends to balance the number of '1s' and '0s' in the MLS, forcing the sequence to be of even length (32768 chips), and reinforcing the starting point, which is represented by the first '1' after a set of 15 consecutive '0s'.

Considering that the transmission rate of these sequences is 1.2288 Mcps and that the sequence is 32768-chip long, the sequence has a repetition rate of 37.5 times per second.

PN-I and PN-Q sequences modulate all logical channels in the forward link, therefore serving as time reference for synchronisation acquisition between BTS and MS.

All BTSs always use the same PN-I and PN-Q sequences. Because all BTSs are synchronised in the forward link, these sequences can be used to identify them within the network. The set of 32768 chips is divided into 512 smaller sets of 64 chips, each set representing one distinct phase offset, known as PN offset. A phase offset (from 0 to 511) is assigned to each BTS to allow the MSs to distinguish the transmissions from different sites. Sequence offsets can be mapped into distance, when considering signal propagation time, with one chip offset representing a distance of 244.1m. The reverse link does not use PN offset but the locally generated short PN sequences are still synchronised to the initial CDMA reference time.

cdma2000-1x standards, also known as SR1 (Spreading Rate 1), apply the same IS-95 short PN sequences concept for the forward and reverse links, given by eqns (3.9) and (3.10). cdma2000-3x (SR3) systems with Multiple Carriers (MC), however, apply this concept only on the forward link, for each 1.25 MHz carrier.

According to 3GPP2 C.S0002 Rev. C standard, the reverse link of cdma2000-3x systems uses carriers with 3.75 MHz of bandwidth (3.6864 Mcps). Considering that the rate of these systems is three times larger than IS-95 systems (3.6864 Mcps $= 3 \times 1.2288$ Mcps), the number of chips in the PN sequence must also be three times larger. To meet this requirement for cdma2000-3x (SR3 operation mode), the PN-I and PN-Q sequences are generated by truncating an MLS generated by a 20-stage shift-register circuit, described in eqn (3.11) by the following polynomial

$$P_{(x)} = x^{20} + x^9 + x^5 + x^3 + 1 \qquad (3.11)$$

This polynomial generates a 1048575-chip long MLS ($2^{20} - 1$ chips), which is more than enough to cover the required number of chips; therefore, this sequence is truncated after 3×2^{15} chips to generate the desired PN sequences. Because the MLS is not entirely used, the starting point of the PN-I and PN-Q sequences cannot be referenced to a sequence of consecutive '0s' as in IS-95 systems. The first 20 chips of the PN-I sequence are '1000 0000 0001 0001 0100'. The start point of the PN-Q sequence is the same but delayed by 2^{19} chips,

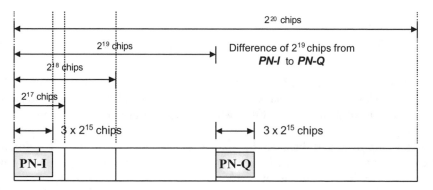

Figure 3.8 Representation of PN-I and PN-Q for the polynomial generator used in SR3CDMA systems.

thus the first 20 chips are '1001 0000 0100 0100 0101'. This structure allows one generator to produce two distinct sequences, as in Figure 3.8.

3.2.3 Long Code or PNLC (Long PN Sequence)

The long code is a 4 398 046 511 103-chip long MLS ($2^{42} - 1$ chips) generated by a 42-stage shift-register circuit described by the following polynomial

$$P(x) = x^{42} + x^{35} + x^{33} + x^{31} + x^{27} + x^{26} + x^{25} + x^{22} + x^{21} + x^{19}$$

$$+ x^{18} + x^{17} + x^{16} + x^{10} + x^{7} + x^{6} + x^{5} + x^{3} + x^{2} + x^{1} + 1 \qquad (3.12)$$

The transmission rate of this sequence is 1.2288 Mcps, resulting in one repetition of the sequence every 42 days, 10 hours, 12 minutes and 19.4 seconds approximately. The starting point of the sequence is the first '1' after all 41 consecutive '0s'.

All mobile stations and BTS channel elements have a long code generator. After the MS initialisation state, the MS and its best server BTS have their long codes synchronised. A unique PNLC is used within the entire network, however, masks of bits are applied to the sequence to reflect user-specific characteristics, as in Figure 3.9.

The masks vary according to the function of the long code. Long code sequences can be used, for example, for privacy (message scrambling) on forward and reverse links and for identification of mobiles and access channels on the reverse link. Figure 3.9 illustrates the PNLC generator and the offset mask application. There are two types of Long Code (LC) masks defined by CDMA standards.

- Public Long Code Offset Mask: The configuration of this mask varies according to the application of the long code. The mask structure is defined by CDMA IS-95 standards and is presented in the next chapters of this book.

- Private Long Code Offset Mask: The structure of this mask is described in the Annex A of the IS-95 Standard, whose distribution is controlled by TIA and supervised by the U.S. International Traffic and Arms Regulation (ITAR) and the Export Administration Regulations. The PNLC with phase offset using the private long code mask provides full privacy to users (cryptography).

As an example, the mask applied to each traffic channel is a public mask combining a reserved sequence of 10 bits and a permutation of the Electronic Serial Number (ESN) of the mobile, which is unique within the system. The ESN is permuted to avoid the close correlation of two mobiles that may have consecutive ESNs. The use of the 32 ESN bits allows the generation of up to 4 294 967 296 (2^{32}) distinct masks. Mobiles requiring more privacy can use the private long code mask instead.

In the reverse link, MSs use the PNLC for privacy, multiplexing and spreading purposes and a different mask is defined for each user. Figure 3.10 illustrates the signal flow in the reverse link, showing two MSs communicating with the same BTS but with distinct propagation delays. This example shows that the reverse link is asynchronous in terms of user signal reception, leading to the adoption of specific and distinct user long code offset masks for multiplexing purposes.

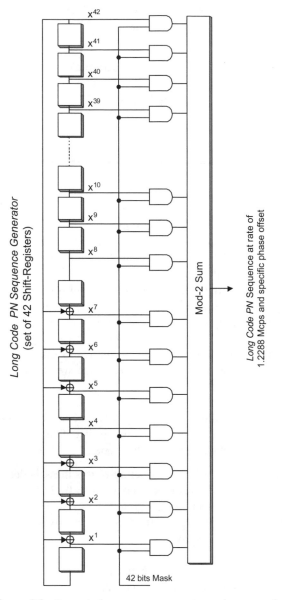

Figure 3.9   Long code sequence generator and user mask.

CDMA IS-95 and cdma2000-1x (SR1) operating with RC 1-2 systems employ the long code sequence described by eqn (3.12). cdma2000 systems using SR1 and RC 3-5 use two long code sequences, which are generated by the same polynomial and have the same rate (1.2288 Mcps) as IS-95 systems, as in Figure 3.9.

For SR3 systems however, even though the same MLS generator polynomial is used, the chip rate is three times larger when compared to CDMA IS-95 systems. This is achieved with a different configuration of the PNLC generator circuit, as depicted by Figure 3.11.

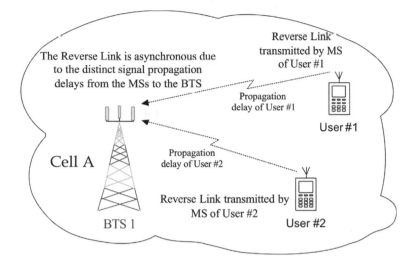

Figure 3.10 Distinct propagation delays observed on reverse link channels.

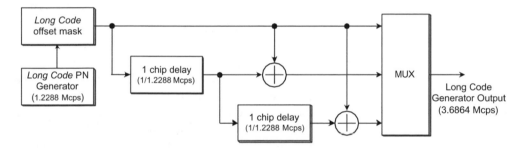

Figure 3.11 Long code sequence generator for the cdma2000 SR3 configuration – at rate of 3.6864 Mcps.

3.3 WALSH SEQUENCES

Walsh sequences are often referred to as Walsh codes or Hadamard codes. The Walsh matrix (M_{Walsh}) is a square matrix (equal number of elements in rows and columns) of binary elements, '0s' and '1s'. The matrix order (dimension) is always a power of 2, i.e. it is only possible to form Walsh code matrices of orders 2^x.

By definition, the first-order Walsh matrix (one row and one column, i.e. only one element) is represented as '0'. All other Walsh matrices are created using a recursive pattern that follows the following rule

$$M_{2N} = \begin{bmatrix} M_N & M_N \\ M_N & \overline{M_N} \end{bmatrix} \tag{3.13}$$

where

M_N Walsh matrix of order N, i.e. N rows and N columns
$\overline{M_N}$ binary complement of the M_N matrix
M_{2N} Walsh matrix of order $2N$

Figure 3.12 presents Walsh matrices of orders 2, 4 and 8 created according to eqn (3.13).

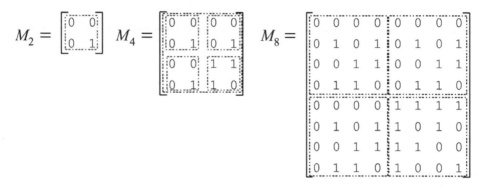

Figure 3.12   Walsh matrices of orders 2, 4 and 8.

The first row and column are always composed of '0s'. The nomenclature used to describe the rows follows the format W_i^m, where m indicates the matrix length and i the row number. Rows are numbered from top to bottom, from 0 to $m - 1$.

Following this concept, the Walsh sequence, or code, indicated by W_1^8 represents the second row (row number 1) of a matrix of order 8. This Walsh code, whose elements are [0 1 0 1 0 1 0 1], may also be represented in terms of signal amplitude, with bits '0' mapped as a signal of amplitude '+1' and bits '1' as '−1'. Therefore, Walsh code W_1^8 represented in terms of signal amplitude is [+ − + − + − + −].

A Walsh matrix is generated in a systematic and recursive way. The code formation pattern can also be represented as a tree, as in Figure 3.13, which shows codes hierarchy and allows determination of any Walsh code without the need of building the complete Walsh matrix. Figure 3.13 shows the formation of 64-chip codes (W^{64}), the number of leaves at the tree determines the length of the code.

The sequence of bits at the bottom of Figure 3.13 represents the Walsh code index (in decimal and in binary) in the matrix and depicts the branch represented by each code. The decimal value for the Walsh code index also corresponds to the row number in the Walsh matrix.

The Walsh matrix root (W^1) is '0' by definition, and this is the starting point for generating all other codes. The sequence that determines the code number is also read from bottom-up, that is, from the leaves to the root of the tree. The code formation, however, follows the tree natural direction, starting at the root (Figure 3.14).

One of the main characteristics of Walsh codes is that rows (codes) are orthogonal to each other within the same matrix, for the same phase offset, i.e. the result of the CCF is constant and equal to '0'. On a matrix of order N, the number of coinciding and non-coinciding bits

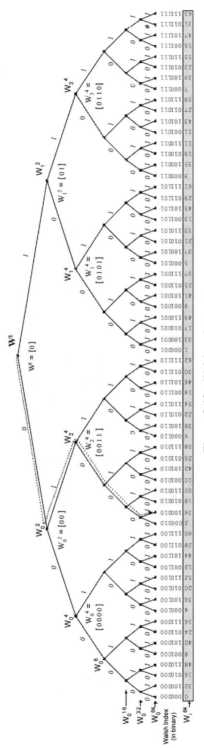

Figure 3.13 Walsh code tree.

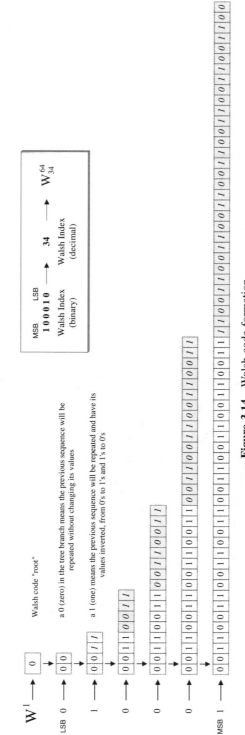

Walsh code "root"

a 0 (zero) in the tree branch means the previous sequence will be repeated without changing its values

a 1 (one) means the previous sequence will be repeated and have its values inverted, from 0's to 1's and 1's to 0's

MSB LSB
100010 $\longrightarrow$ **34** $\longrightarrow$ W_{34}^{64}
Walsh Index   Walsh Index
(binary)   (decimal)

Figure 3.14 Walsh code formation.

Table 3.3 64-bit long Walsh codes used by CDMA systems (IS-95, IS-95B and cdma2000)

Code	Sequences
0	0000 0000 0000 0000 0000 0000 0000 0000 0000 0000 0000 0000 0000 0000 0000 0000
1	0101 0101 0101 0101 0101 0101 0101 0101 0101 0101 0101 0101 0101 0101 0101 0101
2	0011 0011 0011 0011 0011 0011 0011 0011 0011 0011 0011 0011 0011 0011 0011 0011
3	0110 0110 0110 0110 0110 0110 0110 0110 0110 0110 0110 0110 0110 0110 0110 0110
4	0000 1111 0000 1111 0000 1111 0000 1111 0000 1111 0000 1111 0000 1111 0000 1111
5	0101 1010 0101 1010 0101 1010 0101 1010 0101 1010 0101 1010 0101 1010 0101 1010
6	0011 1100 0011 1100 0011 1100 0011 1100 0011 1100 0011 1100 0011 1100 0011 1100
7	0110 1001 0110 1001 0110 1001 0110 1001 0110 1001 0110 1001 0110 1001 0110 1001
8	0000 0000 1111 1111 0000 0000 1111 1111 0000 0000 1111 1111 0000 0000 1111 1111
9	0101 0101 1010 1010 0101 0101 1010 1010 0101 0101 1010 1010 0101 0101 1010 1010
10	0011 0011 1100 1100 0011 0011 1100 1100 0011 0011 1100 1100 0011 0011 1100 1100
11	0110 0110 1001 1001 0110 0110 1001 1001 0110 0110 1001 1001 0110 0110 1001 1001
12	0000 1111 1111 0000 0000 1111 1111 0000 0000 1111 1111 0000 0000 1111 1111 0000
13	0101 1010 1010 0101 0101 1010 1010 0101 0101 1010 1010 0101 0101 1010 1010 0101
14	0011 1100 1100 0011 0011 1100 1100 0011 0011 1100 1100 0011 0011 1100 1100 0011
15	0110 1001 1001 0110 0110 1001 1001 0110 0110 1001 1001 0110 0110 1001 1001 0110
16	0000 0000 0000 0000 1111 1111 1111 1111 0000 0000 0000 0000 1111 1111 1111 1111
17	0101 0101 0101 0101 1010 1010 1010 1010 0101 0101 0101 0101 1010 1010 1010 1010
18	0011 0011 0011 0011 1100 1100 1100 1100 0011 0011 0011 0011 1100 1100 1100 1100
19	0110 0110 0110 0110 1001 1001 1001 1001 0110 0110 0110 0110 1001 1001 1001 1001
20	0000 1111 0000 1111 1111 0000 1111 0000 0000 1111 0000 1111 1111 0000 1111 0000
21	0101 1010 0101 1010 1010 0101 1010 0101 0101 1010 0101 1010 1010 0101 1010 0101
22	0011 1100 0011 1100 1100 0011 1100 0011 0011 1100 0011 1100 1100 0011 1100 0011
23	0110 1001 0110 1001 1001 0110 1001 0110 0110 1001 0110 1001 1001 0110 1001 0110
24	0000 0000 1111 1111 1111 1111 0000 0000 0000 0000 1111 1111 1111 1111 0000 0000
25	0101 0101 1010 1010 1010 1010 0101 0101 0101 0101 1010 1010 1010 1010 0101 0101
26	0011 0011 1100 1100 1100 1100 0011 0011 0011 0011 1100 1100 1100 1100 0011 0011
27	0110 0110 1001 1001 1001 1001 0110 0110 0110 0110 1001 1001 1001 1001 0110 0110
28	0000 1111 1111 0000 1111 0000 0000 1111 0000 1111 1111 0000 1111 0000 0000 1111
29	0101 1010 1010 0101 1010 0101 0101 1010 0101 1010 1010 0101 1010 0101 0101 1010
30	0011 1100 1100 0011 1100 0011 0011 1100 0011 1100 1100 0011 1100 0011 0011 1100
31	0110 1001 1001 0110 1001 0110 0110 1001 0110 1001 1001 0110 1001 0110 0110 1001
32	0000 0000 0000 0000 0000 0000 0000 0000 1111 1111 1111 1111 1111 1111 1111 1111
33	0101 0101 0101 0101 0101 0101 0101 0101 1010 1010 1010 1010 1010 1010 1010 1010
34	0011 0011 0011 0011 0011 0011 0011 0011 1100 1100 1100 1100 1100 1100 1100 1100
35	0110 0110 0110 0110 0110 0110 0110 0110 1001 1001 1001 1001 1001 1001 1001 1001
36	0000 1111 0000 1111 0000 1111 0000 1111 1111 0000 1111 0000 1111 0000 1111 0000
37	0101 1010 0101 1010 0101 1010 0101 1010 1010 0101 1010 0101 1010 0101 1010 0101
38	0011 1100 0011 1100 0011 1100 0011 1100 1100 0011 1100 0011 1100 0011 1100 0011
39	0110 1001 0110 1001 0110 1001 0110 1001 1001 0110 1001 0110 1001 0110 1001 0110
40	0000 0000 1111 1111 0000 0000 1111 1111 1111 1111 0000 0000 1111 1111 0000 0000
41	0101 0101 1010 1010 0101 0101 1010 1010 1010 1010 0101 0101 1010 1010 0101 0101
42	0011 0011 1100 1100 0011 0011 1100 1100 1100 1100 0011 0011 1100 1100 0011 0011
43	0110 0110 1001 1001 0110 0110 1001 1001 1001 1001 0110 0110 1001 1001 0110 0110
44	0000 1111 1111 0000 0000 1111 1111 0000 1111 0000 0000 1111 1111 0000 0000 1111
45	0101 1010 1010 0101 0101 1010 1010 0101 1010 0101 0101 1010 1010 0101 0101 1010
46	0011 1100 1100 0011 0011 1100 1100 0011 1100 0011 0011 1100 1100 0011 0011 1100
47	0110 1001 1001 0110 0110 1001 1001 0110 1001 0110 0110 1001 1001 0110 0110 1001
48	0000 0000 0000 0000 1111 1111 1111 1111 1111 1111 1111 1111 0000 0000 0000 0000
49	0101 0101 0101 0101 1010 1010 1010 1010 1010 1010 1010 1010 0101 0101 0101 0101
50	0011 0011 0011 0011 1100 1100 1100 1100 1100 1100 1100 1100 0011 0011 0011 0011
51	0110 0110 0110 0110 1001 1001 1001 1001 1001 1001 1001 1001 0110 0110 0110 0110
52	0000 1111 0000 1111 1111 0000 1111 0000 1111 0000 1111 0000 0000 1111 0000 1111
53	0101 1010 0101 1010 1010 0101 1010 0101 1010 0101 1010 0101 0101 1010 0101 1010
54	0011 1100 0011 1100 1100 0011 1100 0011 1100 0011 1100 0011 0011 1100 0011 1100
55	0110 1001 0110 1001 1001 0110 1001 0110 1001 0110 1001 0110 0110 1001 0110 1001
56	0000 0000 1111 1111 1111 1111 0000 0000 1111 1111 0000 0000 0000 0000 1111 1111
57	0101 0101 1010 1010 1010 1010 0101 0101 1010 1010 0101 0101 0101 0101 1010 1010
58	0011 0011 1100 1100 1100 1100 0011 0011 1100 1100 0011 0011 0011 0011 1100 1100
59	0110 0110 1001 1001 1001 1001 0110 0110 1001 1001 0110 0110 0110 0110 1001 1001
60	0000 1111 1111 0000 1111 0000 0000 1111 1111 0000 0000 1111 0000 1111 1111 0000
61	0101 1010 1010 0101 1010 0101 0101 1010 1010 0101 0101 1010 0101 1010 1010 0101
62	0011 1100 1100 0011 1100 0011 0011 1100 1100 0011 0011 1100 0011 1100 1100 0011
63	0110 1001 1001 0110 1001 0110 0110 1001 1001 0110 0110 1001 0110 1001 1001 0110

when comparing two codes is always $N/2$. For example, the comparison of code W_1^8 with W_5^8 shows four coinciding and four non-coinciding bits

$$W_1^8 \quad \text{Line 1 :} \quad \underline{0\,1\,0\,1}\,0\,1\,0\,1$$
$$W_5^8 \quad \text{Line 5 :} \quad \underline{0\,1\,0\,1}\,1\,0\,1\,0$$

Therefore, $\text{CCF} = \text{CC} - \text{NCC} = 4 - 4 = 0$. The CCF between two identical phase offsets of the same code also equals '0'.

Table 3.3 shows 64-bit long Walsh codes used by CDMA systems based on IS-95, IS-95B, and cdma2000 standards.

The orthogonality of Walsh codes allows them to be used in the multiplexing of communication channels (logical channels) in the forward link of CDMA systems. A Walsh code is assigned to each logical channel and the system informs user terminals of the code to employ during communication (call or signalling). In the process of multiplexing, the Walsh codes (other than W_0) also spread the data sequence for the assigned channel.

CDMA IS-95 systems consider Walsh codes of order 64, that is, 64-chip long, whereas cdma2000 systems use codes of variable lengths, ranging from 4 to 256 chips.

3.3.1   Walsh Codes in Multiplexing and Spectrum Spreading

This section illustrates the use of Walsh codes for multiplexing (for channellisation purposes) and spectrum spreading. The example considers three users employing 4-bit long Walsh codes, with one code per user (codes 0, 2 and 3).

Table 3.4 lists the four Walsh codes in a matrix of order 4. The codes are represented as chips, '0s' and '1s', and mapped into symbols of amplitude '+1' and '−1', respectively.

The information bit rate is smaller than the chip rate in such a way that an input bit interval corresponds to the whole Walsh code period. The mod-2 sum of the input data to the Walsh code (W_1^4) results in the spectrum spreading (Figure 3.15) and multiplexing processes. Input data bits '0' are represented as the Walsh code itself whereas bits '1' are the complement (inverse) of the Walsh code.

Table 3.4   Representation of 4-bit long Walsh codes

Code	Chips	Symbol Representation
Walsh 0	0 0 0 0	
Walsh 1	0 1 0 1	
Walsh 2	0 0 1 1	
Walsh 3	0 1 1 0	

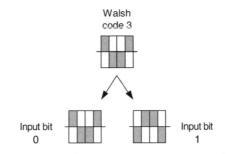

Figure 3.15 Spreading and coding using Walsh codes.

Considering that all forward link channels are synchronous, Figure 3.16 presents all possible combinations of data bits coming from three users and the spreading process using Walsh codes.

Figure 3.17 shows information bits (from users *A*, *B* and *C*) spread using Walsh codes and then added in baseband. The highlighted part depicts output RF carrier modulation symbols after the sum representing the input bits '1 0 1', which were generated, respectively, by users *A*, *B* and *C*.

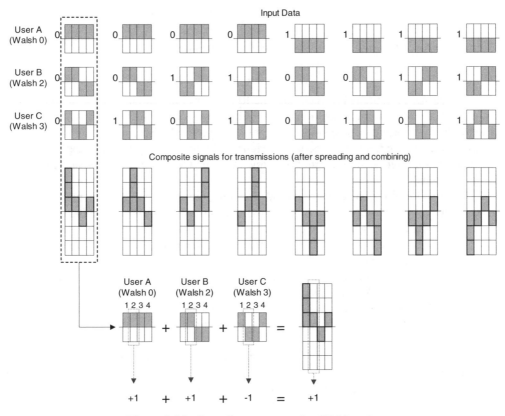

Figure 3.16 Spreading process using Walsh codes.

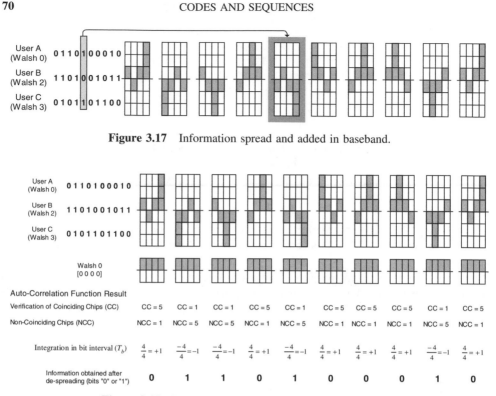

Figure 3.17 Information spread and added in baseband.

Figure 3.18 Recovered information transmitted by user *A*.

The resulting information bits modulate the RF carrier at the output of an analogue adder. The output signal is the superposition of information bits of all users after spreading and is synchronously transmitted. The system assigns a temporary Walsh code to each MS, which the MS uses to de-multiplex incoming signals. Figure 3.18 illustrates the recovered information for user *A* after de-multiplexing (and, in the process, de-spreading) of the incoming signal.

The correlation function recovers the information contained within the symbols using a comparison at analogue level (sum of all chips). This operation must provide the same result as if the comparison was done separately (counting coinciding and non-coinciding chips) and the final results were added up. In fact, the received set of data channels is demultiplexed by a process, which consists of a chip-by-chip multiplication with the user assigned Walsh code, in this example, Walsh 0. The result of this process (correlation) is divided by the number of chips in the Walsh sequence, presenting the desired transmitted data in analogue forms of '+1' (for a bit with value 0) and '−1' (for a bit with value 1). Only the data of the desired channel survives the correlation process, because Walsh codes are orthogonal time functions.

Figure 3.19 shows what happens if an MS applies the wrong code for de-spreading. In the example, the MS is applying Walsh code 1 to information spread using Walsh codes 0, 2 and 3. Because the auto-correlation function can not determine whether there are more coinciding chips than non-coinciding, the information can not be recovered. This can be also understood as if there is no channel data transmitted multiplexed by this Walsh code.

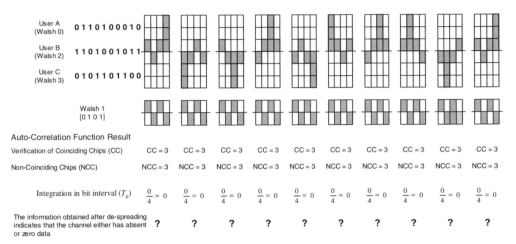

Figure 3.19 Demodulation process of a communication channel considering an unused Walsh code.

Figures 3.18 and 3.19 prove that the correlation between orthogonal codes, expressed as an integral taken for the entire bit interval (code interval), always results in zero, according to eqn (3.14)

$$\frac{1}{T_b} \int_0^{T_l} a(t)b(t)dt = 0 \qquad (3.14)$$

In eqn (3.14), if $a(t)$ and $b(t)$ are orthogonal to each other, they are represented by distinct Walsh codes.

3.3.2 Walsh Codes in IS-95 Systems

3.3.2.1 Forward Link

CDMA IS-95 systems use Walsh codes of length 64. The codes are used to identify, i.e. to enable isolation or selection of the logical channels (pilot, synchronism, paging and traffic) simultaneously present in the forward link (BTS → MS). Because the forward link is synchronous, i.e. all the logical channels are synchronously processed, the orthogonality of Walsh codes acts as a performance-improving factor. Figure 3.20 compares signal behaviour in synchronous and asynchronous systems.

Walsh codes are used in combination with PN codes, the former causing information bits to be spread into a 64-chip sequence. During de-modulation, the receiver compares the energy of the PN de-modulated signal multiplied by a known Walsh sequence to a single bit value, '0' or '1'. Because of transmission errors (caused by variations on the amplitude signal of the received chips), this comparison may not produce an exact match; the existence of Walsh codes, however, allows an FEC-like behaviour from the system that will select the closest match in the comparison process.

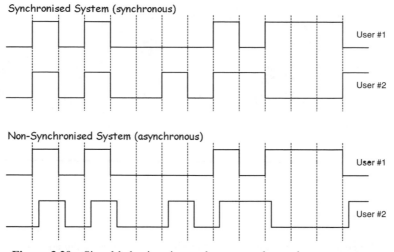

Figure 3.20   Signal behaviour in synchronous and asynchronous systems.

Synchronisation implies that all BTSs of a CDMA network use the same 1.2288 MHz clock to simultaneously multiplex and spread signals from all users in the forward link. From the point of view of a single MS, all logical channels received on the forward link, from one specific BTS, are synchronised and suffer the same attenuation factor and the same propagation delay, as in Figure 3.21. This is very convenient for the reception side, because it allows reduction of the interference generated within each cell, improving system capacity.

Theoretically, interference between forward link channels should be negligible, since Walsh codes are orthogonal between themselves. In reality, there is always a certain amount of interference generated by users covered by the same BTS, due to loss of orthogonality in urban environments, caused by multi-path signal components, and because orthogonality requires synchronisation.

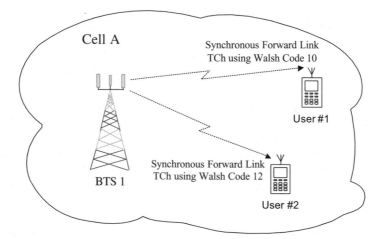

Figure 3.21   Synchronism of forward link logical channels as seen by the mobile station.

Table 3.5 Walsh code lengths in cdma2000 systems

Spreading Rate (SR)	Walsh code length	
	Min	Max
SR1	4	128
SR3	4	256

3.3.2.2 Reverse Link

The Walsh codes used in the reverse link of IS-95 systems are also of length 64. Walsh codes are used for spreading and 64-ary orthogonal modulation; however, unlike the forward link, these sequences are not used for channellisation (multiplexing) purposes.

At the reverse link of IS-95 CDMA systems, an orthogonal modulator, of order 64, maps every 6-bit set into one of the 64 Walsh codes of order 64, acting like a $(n, k) = (6, 64)$ block code. The reception at the BTS performs the opposite process, using the maximum likelihood concept to decide in case of errors in the data stream received; this process may be understood as an FEC scheme. The spreading and coding processes are combined in the reverse link. Every set of 6 bits of information is mapped into a set of 64 chips. Each of these 6-bit sequences is mapped into a unique row of the Walsh matrix, therefore defining a Walsh code, which is sent to the BTS. On the reception side, the BTS compares the 64-chip sequence received to the Walsh codes, selecting the closest match and recovering the 6-bit sequence that maps that code.

3.3.3 Walsh Codes in cdma2000 Systems

3.3.3.1 Forward Link

cdma2000 systems use Walsh codes of variable lengths, according to the Spreading Rate (SR), information bit rate for transmission and channel in use. Table 3.5 identifies minimum and maximum lengths of Walsh codes for the cdma2000 systems.

3.3.3.2 Reverse Link

The reverse link makes use of some Walsh codes of length in the range from 4 up to 64 as Walsh covers, but only for operation with radio configurations 3–4 (SR1) and 5–6 (SR3).

3.4 QUASI-ORTHOGONAL FUNCTION

IS-95 systems use Walsh codes of fixed length (64), whereas cdma2000 systems allow variable Walsh length; cdma2000-1X systems use Walsh codes up to length 128 and cdma2000-3X up to length 256. The limitation on Walsh code length may cause limitations to the system due to unavailability of Walsh codes, especially when dealing with high data rates. Therefore, cdma2000 standards propose the use of an additional function, referred to as Quasi Orthogonal Function (QOF), to enhance system capacity. The QOFs are specially employed when Transmit Diversity (TD) techniques are used, such as Space Time Spreading (STS).

QOFs consist of the multiplication of the Walsh codes by a QOF mask, which is a vector of binary symbols. The resulting codes are not fully orthogonal to the original Walsh codes, but the masks selected by the standard are optimal in terms of minimising the correlation (non-orthogonality).

QOF have a length of 256 chips, while Walsh codes can have variable lengths. This implies that, depending on the size of the Walsh code, a single QOF may be applied into more than one code sequence (Walsh code vector). For example, a single QOF mask multiplies, componentwise, four Walsh codes of length 64, that is, each chip of the mask multiplies the corresponding chip of the Walsh code vector.

There are four different QOF sets (numbered from 0 to 3), each one generated by a different QOF mask. Each QOF mask corresponds to a row in a Walsh matrix of size 256. The masks selected by cdma2000 standards are optimal in the sense that they minimise the correlation between QOFs generated and regular Walsh codes with the same length. Table 3.6 shows QOF masks in their hexadecimal form and their respective rows in a Walsh matrix.

Table 3.6 Quasi-orthogonal function masks

| QOF Set | Masking function | |
	QOF$_{sign}$ hexadecimal representation	Walsh$_{rot}$
0	00000000000000000000000000000000 00000000000000000000000000000000	W_0^{256}
1	7228d7724eebebb1eb4eb1ebd78d8d28 278282d81b41be1b411b1bbe7dd8277d	W_{130}^{256}
2	114b1e4444e14beeee4be144bbe1b4ee dd872d77882d78dd2287d277772d87dd	W_{173}^{256}
3	1724bd71b28118d48cbddb172b187eb2 e7d4b27ebd8ee82481b22be7dbe871bd	W_{47}^{256}

All QOFs within a set are orthogonal to each other whereas the cross-correlation between QOFs of two, distinct sets is constant. This makes QOF predictable and helps to overcome eventual problems that could be caused by the non-orthogonality.

QOFs are applied to the system in two steps: the first is known as QOF$_{sign}$, and the second as Walsh$_{rot}$. Figure 3.22 illustrates the implementation of quasi-orthogonal functions.

The QOF$_{sign}$ changes the polarity of the symbols: when the QOF$_{sign}$ chip is '1' the Walsh code symbols is inverted ('+1' becomes '−1' and vice versa); when the QOF$_{sign}$ chip is '0' the Walsh code symbol remains the same. Thus QOF set 0 corresponds to the original Walsh code vector.

The Walsh$_{rot}$ corresponds to a symbol phase modification, or rotation. Rotation is only enabled when the Walsh$_{rot}$ chip is '1'. To perform the rotation, the QOF multiplies the Walsh

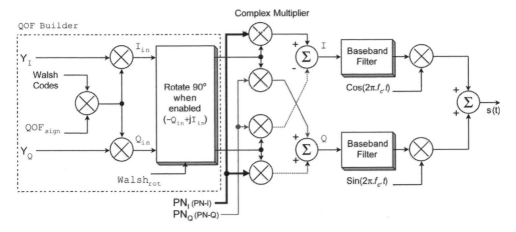

Figure 3.22 Quasi-orthogonal function implementation.

code vector by '*j*', where '*j*' is a complex variable that corresponds to a 90° phase rotation. If the Walsh$_{rot}$ chip is '0', the Walsh code symbol remains the same.

QOF usage is flexible and optional. According to cdma2000 standards, QOFs may be used only in the forward link of systems with both spreading rates, SR1 and SR3.

Carriers may choose to deploy QOF in a variety of ways, e.g. QOFs may be used only when all Walsh codes have been exhausted. In this case, QOFs would increase system capacity, eventually the system would be limited again due to interference problems caused by the non-orthogonality of QOFs and Walsh codes.

Another option is to use QOFs on the dedicated channels of low mobility users within a specific coverage area. This approach offers a low self-interference scenario considering that QOFs are orthogonal to each other within the same set; however if users move out of the planned coverage area, they will require increased signaling for code channel reassignment.

BIBLIOGRAPHY AND REFERENCES

1. 3GPP2, C.S0002-C, *Physical Layer Standard for cdma2000 Spread Spectrum Systems (Release C)*, May 2002.
2. 3GPP2, C.S0005-B, *Upper Layer (Layer 3) Signaling Standard for cdma2000 Spread Spectrum Systems (Release B)*, April 2002.
3. TIA/EIA-95-B, *Mobile Station – Base Station Compatibility Standard for Dual-Mode Spread Spectrum Cellular System*, November 1998.
4. CelTec/CelPlan, *CDMA IS-95 and cdma2000 Systems – Training Course*, December 2002.
5. Jhong Sam Lee and Leonard E. Miller, "CDMA Systems Engineering Handbook, 1998, Artech House, Inc.

4
Forward Link Channels

BRUNO DE SOUZA ABREU XAVIER

This chapter is concerned with the forward link channel structure of IS-95 and cdma2000 networks. The characteristics of each type of channel are surveyed. The behaviour and processing of each channel are examined in detail.

4.1 FORWARD LINK CHANNEL STRUCTURE IN IS-95 CDMA SYSTEMS

In IS-95 CDMA networks, each BTS has four different types of channel that are required for transmissions on the forward link. Each of these channels is described in detail in this chapter.

- Forward Pilot Channel (FPiCh).

- Forward Sync Channel (SyncCh).

- Forward Paging Channels (FPCh).

- Forward Traffic Channels (FTCh).

Figure 4.1 presents the structure of IS-95 forward link channels.

4.1.1 Phase, Quadrature and Carrier Modulation

Figure 4.2 illustrates the processing of forward channels within a CDMA IS-95 network. After orthogonal multiplexing and spreading using Walsh codes, the bits of each channel are converted into analogue format (signal point mapping). This process consists of converting bits '0' into signals with amplitude '+1' and bits '1' into amplitude '−1'. A channel gain, specific to each logic channel, is then applied to this signal. This gain [1–5] is variable per cell due to power control schemes and depends on the energy level required for the signal to be transmitted, considering the bit energy (E_b) available for transmission.

Designing CDMA 2000 Systems L. Korowajczuk
© 2004 John Wiley & Sons, Ltd ISBN: 0-470-85399-9

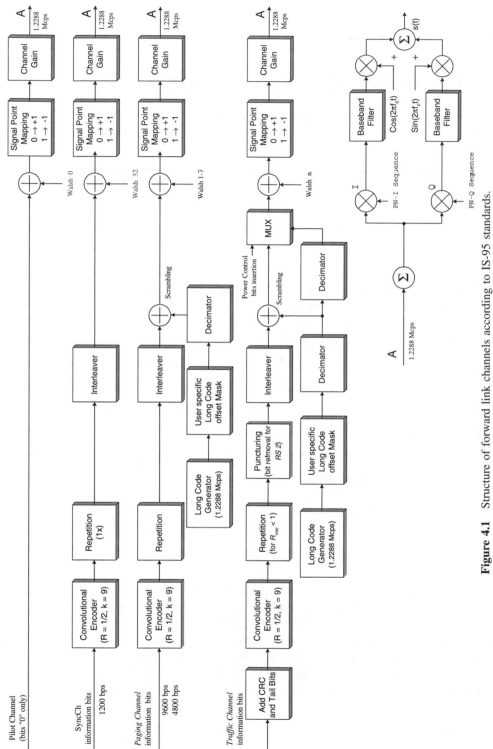

Figure 4.1 Structure of forward link channels according to IS-95 standards.

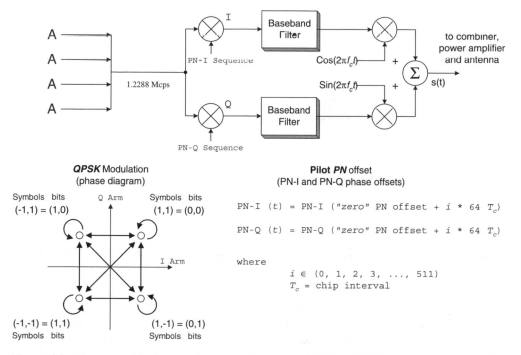

Figure 4.2   Carrier modulation, quadrature modulation and PN-I and PN-Q sequences phase offset assignment.

After gain is applied to the signal, each channel is phase and quadrature modulated by PN-I and PN-Q sequences, as in Figure 4.2, process that is also known as PN *orthogonal modulation*. The same PN-I and PN-Q sequences modulate the signal of a carrier within each BTS or BTS cell sector 2 using the same phase offset. Additional RF carriers in the same BTS do not require different phase offsets for PN sequences.

Logical channels can be combined after gains are applied, before PN application, or, in a more usual way, at the antenna after spread channels are individually weighted. A combination of these techniques can also be used. The final configuration depends on the modularisation adopted by the BTS vendor.

After the modulation with PN sequences, each channel passes through a digital waveshaping filter to control the spectrum of the modulating signal. Prior to transmission, all logical channels are combined, amplified and/or unhanded radiated, and this result modulates the RF carrier.

The carrier modulation performed in the forward link has a special characteristic regarding modulation efficiency. Figure 4.2 shows that at point *A*, the information is transmitted both through I and Q branches prior to the PN modulation (PN-I and PN-Q). This results in half efficiency of QPSK, that is, the modulation method acts like BPSK [5].

I and Q branches carry the communication channel amplitude (or power) that modulates $\cos(\omega_c t)$ and $\sin(\omega_c t)$ carrier components. This characteristic allows the sum in baseband and

Table 4.1   Carrier phase mapping in QPSK modulation

I	Q	Carrier phase
0	0	$\pi/4$
1	0	$3\pi/4$
1	1	$-3\pi/4$
0	1	$-\pi/4$

is illustrated by the following equations

$$I_{(composed)} = I_1 \cos(\omega_c t) + I_2 \cos(\omega_c t) + \cdots + I_n \cos(\omega_c t) = (I_1 + I_2 + \cdots + I_n) \cos(\omega_c t) \tag{4.1}$$

$$Q_{(composed)} = Q_1 \sin(\omega_c t) + Q_2 \sin(\omega_c t) + \cdots + Q_n \sin(\omega_c t) = (Q_1 + Q_2 + \cdots + Q_n) \sin(\omega_c t) \tag{4.2}$$

Figure 4.2 also shows a QPSK phase diagram, depicting the transitions and phase constellation of the resulting modulated signal at the transmitting antenna, and the PN-I and PN-Q sequences phase offset assignment used in the quadrature modulation.

Table 4.1 shows I (in phase) and Q (quadrature) mapping performed by forward link channels using QPSK modulation.

4.1.2   Forward Pilot Channel

The Forward Pilot Channel (FPiCh) is the simplest channel in terms of channel processing, but, at the same time, it is the most important logical channel in CDMA based networks. All carriers being used in the network continuously transmit this channel. Its main functionalities include providing timing reference for synchronisation and enabling coherent detection and power estimation and control. Figure 4.3 shows that only bits '0', inserted by Walsh code 0 (W_0^{64}), are transmitted by FPiChs.

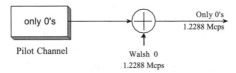

Figure 4.3   Forward pilot channel processing block diagram.

Pilot channels do not transmit any information other than PN-I and PN-Q sequences. The acquisition process, executed by the mobile terminals, is performed very quickly and efficiently, as the repetition rate of the short PN sequences is 37.5 times per second, considering that one complete sequence is transmitted every 26.667 ms.

During the initial synchronisation acquisition process, or *system initialisation state*, mobile terminals use the FPiCh as a phase and timing reference for coherent detection and de-modulation of received channels. The BTS transmits the FPiCh on the primary setup carrier (see Chapter 6 for the explanation of primary and secondary carriers). This acquisition process consists of the MSs synchronising their internally generated PN-I and PN-Q sequences with the ones transmitted by the BTS.

All BTSs in the network transmit the same synchronised PN sequences. Each BTS, however, is identified by a specific phase offset of these sequences. This identification method allows high re-use of CDMA IS-95 carrier frequencies, thus allowing the frequency re-use factor to be equal to one, that is, the same carrier frequency can be simultaneously used by all BTSs in the network.

There are 512 different phase offsets available in CDMA IS-95 systems, with a 64-chip interval between each two consecutive offsets. Therefore, the assignment of offsets 0 and 10 to two BTSs, for example, means that the second BTS is transmitting PN-I and PN-Q sequences with a delay of 640 chips (520.83 µs) in relation to the first one.

Offset planning must be done carefully because of re-use and propagation delay. A good offset assignment assures that a mobile terminal will not de-code a channel with data intended to a neighbouring MS that may be using the same Walsh code.

Walsh codes are 64-chip long and are transmitted at the same rate as the short PN codes (1.2288 Mcps). The beginning of Walsh code transmission is aligned to the PN-I and PN-Q phase offset, established for each BTS in the system.

This synchronisation of the mobile station with the BTS pilot channel assures that three of the codes employed by IS-95 standards are perfectly synchronised between the MS and the network:

- PN-I and PN-Q codes,

- Walsh codes.

FPiChs also aid mobile stations during handoff. In this process, mobiles are constantly monitoring and evaluating the signal level received from pilot channels transmitted by other BTSs, to keep their pilots lists (active, candidate, neighbour and remaining) always updated. Pilot channels are also used as power reference, for example, in system access procedures, and as timing reference for coherent demodulation of the signal received on the forward ink.

4.1.3 Synchronisation Channel

The SyncCh also plays a major role in the *system initialisation state*. Mobiles use the synchronisation channel to receive sync channel messages that allow them to synchronise locally generated codes to the network.

The SyncCh transmission data rate is 1200 bps. The channel is orthogonally multiplexed by mean of XOR combining with the 64-bit long Walsh code number 32 (W_{32}^{64}) and is organised in 80 ms super-frames. Each super-frame consists of 96 information (data) bits, organised in three frames of 26.667 ms (32 bits per frame) as shown in Figure 4.4.

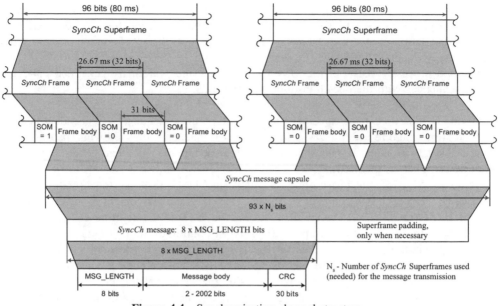

Figure 4.4 Synchronisation channel structure.

Each SyncCh frame is composed of two parts: the Start Of Message (SOM) bit and the frame body. The SOM indicates if the frame is starting a new message (SOM = 1) or if it is part of the current message (SOM = 0). Therefore, in every super-frame, there are 93 information bits and three SOM bits, i.e. 96 bits.

The sync channel message is divided into message capsules, which are always multiple of 93 bits. If the message itself does not satisfy this requirement, padding is added to the message. This is equivalent to state that the message capsules are $93N_S$-bit long, where N_S is the number of super-frames needed to transmit the entire message.

The sync channel message is structured in three parts.

- Message length (*MSG_LENGTH* parameter): Represents the total length of the message, including body, CRC and this field itself. The length is indicated in number of bytes and is coded in 8 bits.

- Message body: Main part of the message, from 2 to 2002 bits.

- Cyclic Redundancy Check (CRC): Message quality indicator, composed of 30 bits.

The frame duration (26.667ms) corresponds to the period of the PN-I and PN-Q sequences transmitted by the pilot channel, i.e. there are exactly 75 frames every 2 s, or 25 super-frames every 2 s. Therefore, the beginnings of SyncCh frames and short PN sequences coincide with each other, making it easier for mobile stations to acquire SyncCh data. In fact, MSs only have to apply Walsh code 32 to de-spread and de-code SyncCh messages, as in Figure 4.5.

Figure 4.5 shows time alignment of SyncChs in relation to the pilot channel PN-I and PN-Q sequence offsets.

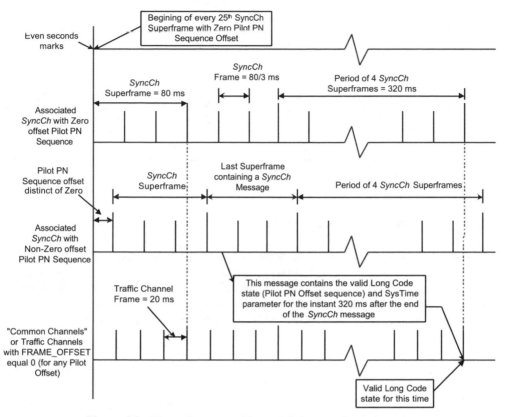

Figure 4.5   Time alignment of forward link transmitted channels.

During system acquisition, or initialisation, after synchronising PN sequences, MSs use the SyncCh to synchronise the long code generator and acquire system data. At call termination, mobile stations tune in the synchronisation channel again searching for system configuration data, executing a process similar to re-initialisation or re-synchronisation.

The SyncCh message includes system data information parameters such as

- state of the long code generator (LC_STATE) and system time (SYS_TIME), 320 ms after the end of the current super-frame, to allow full synchronisation of the MS internal codes;

- PN offset used by the BTS (PN_OFFSET);

- Current transmission rate of the forward paging channel (P_{RAT});

- System Identification (SID) and Network Identification (NID);

- protocol revision levels supported by the network (P_REV and MIN_P_REV);

- forward paging channel assigned to MSs on a specific CDMA RF carrier ($CDMA\_FREQ$).

Channel coding consists of processing the SyncCh message bits with a convolutional coder ($R = 1/2$), a coded symbol repetition block (1), and an interleaver. After channel coding, bits are channelised by Walsh code 32 (W_{32}^{64}), at 1.2288 Mcps, as shown in Figure 4.6.

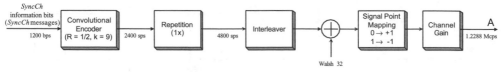

Figure 4.6 Channel processing block diagram for the SyncCh.

4.1.3.1 Convolutional Coding and Coded Bit Repetition

The convolutional coder circuit is composed of an 8-stage shift register, as illustrated in Figure 4.7. This process generates two output bits for each bit input into the circuit. The output of a sub-set of stages is modulo-2 summed ('exclusive OR') generating a resulting bit called c_0. A second sub-set of outputs is also modulo-2 summed generating c_1. This process generates two output coded symbols (c_0 and c_1) for each information bit that enters the circuit.

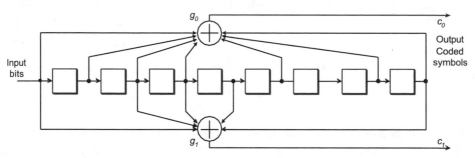

Figure 4.7 Convolutional coder ($R = 1/2, k = 9$) used in the forward link.

Each pair of coded symbols, taken from outputs c_0 and c_1, is based in the information of k contiguous data bits that enter the first register and pass through the $k-1$ outputs of the remaining shift registers.

The parameter k is called the constraint length and corresponds to the number of bits considered to determine coded bits c_0 and c_1. Each information bit passes through all shift-register stages participating in the determination of $2k$ contiguous coded bits at the convolutional encoder output.

In case there is a degradation of one coded symbol (c_0 or c_1) during transmission, the detection and correction of errors in the reception is easier. However, if a burst of $2k$ (or more) consecutive coded symbols is received with error, it is impossible to re-generate the original information.

After 1/2 rate convolutional encoding and bit repetition processes, the output rate becomes 4.8 kbps, corresponding to 26.66 ms frames with 128 bits each.

Because there is no insertion of encoder tail bits (8 bits with value '0') on the SyncCh frames, the coding of a frame depends on the previous frame. That means the last bits in a frame contribute to the convolutional coding of the first bits of the next frame, and the encoder is not reset after each frame processing.

4.1.3.2 Interleaving

The interleaving is intended to improve reliability of coded transmitted signals, re-arranging the bit transmission order. In case of burst errors during transmission, this avoids errors that damage many consecutive bits, which could affect detection and de-coding capabilities.

The advantage of this process is spreading of burst errors, which can be caused by fading, for example. With interleaving, errors are spread throughout several blocks and also within a single block, providing more efficiency in error detection and correction. This process is especially important because convolutional coding does not present a good performance in correcting sequential errors (burst).

During interleaving, bits at the convolutional encoder output are sequentially written in an interleaving matrix (memory buffer) with 16 lines and 8 columns (Figure 4.8).

Figure 4.8 Interleaving process for the SyncCh on the forward link.

The interleaving process does not modify the number of bits, only rearranges the transmission sequence.

Bits are written by columns, from left to right, as presented in Table 4.2. After the initial matrix is complete, it is internally interleaved generating a re-organised matrix, assembled

Table 4.2 Coded bits writing sequence (input) in the SyncCh interleaving matrix, after bit repetition

1	9	17	25	33	41	49	57
1	9	17	25	33	41	49	57
2	10	18	26	34	42	50	58
2	10	18	26	34	42	50	58
3	11	19	27	35	43	51	59
3	11	19	27	35	43	51	59
4	12	20	28	36	44	52	60
4	12	20	28	36	44	52	60
5	13	21	29	37	45	53	61
5	13	21	29	37	45	53	61
6	14	22	30	38	46	54	62
6	14	22	30	38	46	54	62
7	15	23	31	39	47	55	63
7	15	23	31	39	47	55	63
8	16	23	32	40	48	56	64
8	16	24	32	40	48	56	64

Table 4.3 Interleaved matrix of the SyncCh

1	3	2	4	1	3	2	4
33	35	34	36	33	35	34	36
17	19	18	20	17	19	18	20
49	51	50	52	49	51	50	52
9	11	10	12	9	11	10	12
41	43	42	44	41	43	42	44
25	27	26	28	25	27	26	28
57	59	58	60	57	59	58	60
5	7	6	8	5	7	6	8
37	39	38	40	37	39	38	40
21	23	22	24	21	23	22	24
53	55	54	56	53	55	54	56
13	15	14	16	13	15	14	16
45	47	46	48	45	47	46	48
29	31	30	32	29	31	30	32
61	63	62	64	61	63	62	64

with the output bits of the convolutional coder, which are read by columns – from left to right, as illustrated in Table 4.3.

4.1.3.3   Phase, Quadrature and Carrier Modulation

After interleaving, the SyncCh data is channelised or 'covered' using Walsh code 32 (W_{32}^{64}). None of the information (SyncCh *message*) transmitted by this channel is encrypted using the long code because mobiles, at this time, are still unaware of the accurate phase offset of this sequence, therefore, they are unable to decode data ciphered using it.

After channelisation, the coded bit sequence is modulated by the PN-I and PN-Q codes, as in Figure 4.2.

4.1.4   Forward Paging Channels

The Forward Paging Channels (FPChs) are used to transmit overhead messages (system parameter information) and specific messages to mobile stations, such as call setup procedures, registration orders and information about mobile terminated calls.

An RF CDMA carrier has up to seven FPChs that use 64-bit long Walsh codes 1–7 (W_{1-7}^{64}). At least one FPCh, called primary FPCh (or FPCh1), is always available in any RF carrier, using W_1^{64}.

The primary paging channel transmits system data (*overhead messages*) paging and access response messages, while the remaining FPChs transmit information related to paging and access responses.

The transmission data rate of the FPChs is either 4.8 or 9.6 kbps, and the choice of rate is communicated to the mobile via the P_{RAT} parameter, sent in the SyncCh message. All FPChs transmitted within the same network must have the same transmission rate and use the same System ID (*SID*).

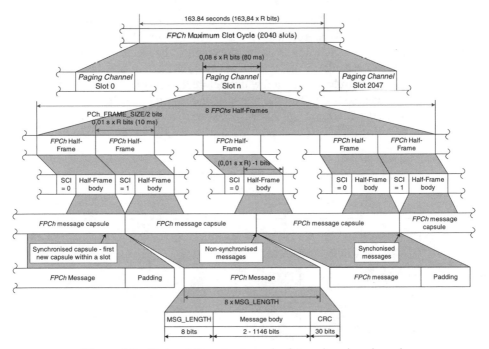

Figure 4.9   Slot and time structure of a forward paging channel.

Mobiles obtain identity and carrier assignment from the *CDMA_FREQ* parameter, sent in the SyncCh message by the BTS that also transmits FPCh1.

When more than one FPCh is used, the network tries to uniformly distribute MSs among existing channels. Hence the system applies a hashing algorithm, using the Electronic Serial Number (*ESN*) or the International Mobile Station Identity (*IMSI*), to choose which carrier and FPCh the mobile station will use.

Figure 4.9 presents the logical structure of the paging channel. Transmitted messages are divided into 80 ms slots. Each slot is divided into smaller periods of 20 ms, called frames. The frames are sub-divided into 10 ms half frames. FPCh slots have the same length as SyncCh super-frames, and are always time aligned to each other within the same BTS, as in Figure 4.5.

The paging message capsule comprises the information bits (actual message) and, if required, padding bits (sequence of bits '0' to complete a capsule). The FPCh message is divided in three parts (Figure 4.9):

- message length *(MSG_LENGTH)*: size of the message body, in bytes, encoded in 8 bits,

- message body: actual information, ranging from 2 to 1146 bits,

- Cyclic Redundancy Check (CRC): message capsule quality indicator, composed of 30 bits.

Each FPCh half frame transmits 32 bits, where the first bit represents the Synchronised Capsule Indicator (SCI) flag and the remaining 31 are data (or information bits).

If the BTS transmits FPCh message capsules that start at the second bit of a half frame, these capsules are called synchronised capsules. Otherwise, they are defined as non-synchronised capsules. The SCI flag is used to distinguish between synchronised ('1') and non-synchronised ('0') capsules.

If an FPCh message capsule finishes a few bits before the end of a half frame, the BTS can complete the remaining bits with '0s' (padding bits) and start the next message capsule at the second bit of the next half frame. In this case, the next half-frame SCI flag value is '1'. Otherwise, the SCI flag is '0'.

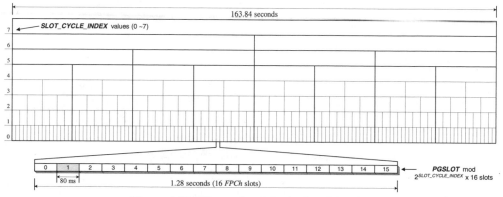

Figure 4.10   Slot cycle of the paging channels.

A set of 2048 consecutive slots, from 0 to 2047, with length of 163.84 s, is called maximum slot cycle of the FPCh, as in Figure 4.10. Usually, all messages sent by the BTS fit in one, or two, consecutive 80 ms slots.

Mobile stations in idle state can monitor the FPChs in two, distinct ways.

- *Non-slotted mode*: Mobile terminals are programmed to continuously monitor ('listen') the FPChs. The BTS transmits messages through the FPCh to any specific MS, associated to it, at any time, because terminals are always able to decode it.

- *Slotted mode:* Mobile stations are programmed by the system to monitor the associated FPCh at certain times. All other times, mobiles 'sleep', keeping only a few active circuits (transmitter and receiver off, thus not able to transmit or receive messages). This strategy results in substantial battery life saving. However, synchronism between the BTS and each MS is required, and messages to a specific MS must be sent at times when the terminal is activated, or 'awake'.

The frequency that each MS monitors its associated FPCh, i.e. 'listens' to its assigned FPCh slot, varies from once every 1.28 s up to a maximum 'dormant period' of once every 163.84 s, as in Figure 4.10. This is usually configured by the *SLOTTED_MODE* and *SLOT_CYCLE_INDEX* parameters, available in the following messages:

- registration message;
- origination message;
- page response message.

When working in the *slotted mode* operation, *SLOTTED_MODE* set to '1', mobile stations send their preferred slot cycle in the *SLOT_CYCLE_INDEX* parameter, whose valid values range from 0 to 7.

In this mode, the 2048 slots are grouped in cycles of ($16 \times 2^{SLOT\_CYCLE\_INDEX} \times 80$) ms slots. When the *SLOT_CYCLE_INDEX* parameter is '0', it represents the minimum slot cycle, composed of only 16 slots. This means that at every 16 slots, or 1.28 s, MSs awake and monitor the associated FPCh. Conversely, a *SLOT_CYCLE_INDEX* of 7 represents the maximum slot cycle, corresponding to 2048 consecutive slots, or 163.48 s. The beginning of a maximum slot cycle (slot zero) is set when the system time, informed to the MS by the SyncCh message, is a multiple of 163.48 s.

If operating in slotted mode, MSs must be able to define the slot to 'wake up', whereas BTSs must be able to identify these slots to transmit messages to MSs at the correct time. Mobile terminals and the network apply the Hash algorithm with the Mobile Identification Number (*MIN*) to choose a reference *PGSLOT* (paging slot) from the 2048 possible slots of the selected FPCh in an RF carrier.

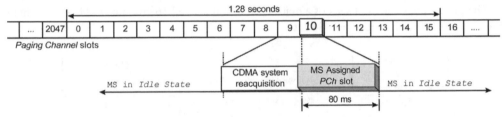

Figure 4.11 Operation of an MS in slotted mode.

Figure 4.11 shows an example of a mobile terminal configured with *SLOT_CYCLE_INDEX* '0' (slot cycle of 16) and *PGSLOT* '10'. The terminal 'wakes up' in slot 9, re-acquires the IS-95 network and, in slot 10, monitors the associated PCh in search of messages. If there is no message, the MS goes back to 'sleep' (idle state) in slot 11, remaining idle until slot 25 (9 + 16), when it 'wakes up' again and monitors slot 26. The process is indefinitely repeated at every 1.28 s, or 16 slots.

When an MS monitors a slot and finds a paging message addressed to another mobile, it de-codes the data to obtain the parameter *MORE_PAGES*, which indicates if there are additional paging messages in the same slot. If this parameter is '1', the terminal stays 'awake', monitoring the slot, otherwise it goes back to idle state. When paging messages can not fit in a single slot, mobile terminals stay 'awake', receiving and de-coding data sent in the next slot.

Figure 4.12 illustrates FPCh processing with a block diagram.

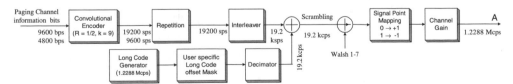

Figure 4.12 Forward paging channels processing block diagram.

4.1.4.1 *Convolutional Coding and Coded Symbol Repetition*

The convolutional encoder in Figure 4.7 is the same as that used for all other CDMA IS-95 forward link channels. The coding process is described in Section 4.1.3.1. As in the SyncCh, no encoder tail bits are used, i.e. the convolutional coder is not re-set at the end of each 20 ms frame.

The repetition is performed with the coded symbols at the convolutional coder output, but only when the FPCh transmission rate is 4.8 kbps (baud rate). Therefore, the coded symbol rate of 19.2 ksps is always achieved prior to the interleaving process, regardless of the FPCh rate (P_{RAT}).

4.1.4.2 *Interleaving*

Figure 4.13 illustrates the interleaving process indicating frame sizes at the matrix input and output.

The process basically consists of a matrix with 24 lines and 16 columns (384 cells), which means, capable of interleaving 384 bits.

Figure 4.13   Interleaving process for FPChs.

Bits are written in the matrix by columns, from left to right, until all cells are filled, as shown in Figure 4.14.

Table 4.4 presents an example of an interleaved matrix with a burst error that occurred during transmission causing erroneous reception of the highlighted coded bits.

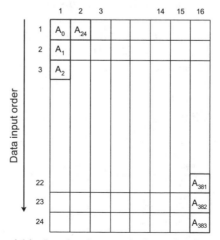

Figure 4.14   Interleaving matrix formation for FPChs.

Table 4.4 Interleaved matrix (with example of burst error)

0	8	4	12	2	10	6	14	1	9	5	13	3	11	7	15
64	72	68	76	66	74	70	78	65	73	69	77	67	75	71	79
128	135	132	140	130	138	134	142	129	137	133	141	131	139	135	143
192	200	196	204	194	202	198	206	193	201	197	205	195	203	199	207
256	264	260	268	258	266	262	270	257	265	261	269	259	267	263	271
320	328	324	332	322	330	326	334	321	329	325	333	323	331	327	335
32	40	36	44	34	42	38	46	33	41	37	45	35	43	39	47
96	104	100	108	98	106	102	110	97	105	101	109	99	107	103	111
160	168	164	172	162	170	166	174	161	169	165	173	163	171	167	175
224	232	228	236	226	234	230	238	225	233	229	237	227	235	231	239
288	296	292	300	290	298	294	302	289	297	293	301	291	299	295	303
352	360	356	364	354	362	358	366	353	361	357	365	355	363	359	367
16	24	20	28	18	26	22	30	17	25	21	29	19	27	23	31
80	88	84	92	82	90	86	94	81	89	85	93	83	91	87	95
144	152	148	156	146	154	150	158	145	153	149	157	147	155	151	159
208	216	212	220	210	218	214	222	209	217	213	221	211	219	215	223
272	280	276	284	274	282	278	286	273	281	277	285	275	283	279	287
336	344	340	348	338	346	342	350	337	345	341	349	339	347	343	351
48	56	52	60	50	58	54	62	49	57	53	61	51	59	55	63
112	120	116	124	114	122	118	126	113	121	117	125	115	123	119	127
176	184	180	188	178	186	182	190	177	185	181	189	179	187	183	191
240	248	244	252	242	250	246	254	241	249	245	253	243	251	247	255
304	312	308	316	306	314	310	318	305	313	309	317	307	315	311	319
368	376	372	380	370	378	374	382	369	377	373	381	371	379	375	383

Table 4.5 Spread errors after de-interleaving

0	24	48	72	96	120	144	168	192	216	240	264	288	312	336	360
1	25	49	73	97	121	145	169	193	217	241	265	289	313	337	361
2	26	50	74	98	122	146	170	194	218	242	266	290	314	338	362
3	27	51	75	99	123	147	171	195	219	243	267	291	315	339	363
4	28	52	76	100	124	148	172	196	220	244	268	292	316	340	364
5	29	53	77	101	125	149	173	197	221	245	269	293	317	341	365
6	30	54	78	102	126	150	174	198	222	246	270	294	318	342	366
7	31	55	79	103	127	151	175	199	223	247	271	295	319	343	367
8	32	56	80	104	128	152	176	200	224	248	272	296	320	344	368
9	33	57	81	105	129	153	177	201	225	249	273	297	321	345	369
10	34	58	82	106	130	154	178	202	226	250	274	298	322	346	370
11	35	59	83	107	131	155	179	203	227	251	275	299	323	347	371
12	36	60	84	108	132	156	180	204	228	252	276	300	324	348	372
13	37	61	85	109	133	157	181	205	229	253	277	301	325	349	373
14	38	62	86	110	134	158	182	206	230	254	278	302	326	350	374
15	39	63	87	111	135	159	183	207	231	255	279	303	327	351	375
16	40	64	88	112	136	160	184	208	232	256	280	304	328	352	376
17	41	65	89	113	137	161	185	209	233	257	281	305	329	353	377

Table 4.5 *(Continued)*

18	42	66	90	114	138	162	186	210	234	258	282	306	330	354	378
19	43	67	91	115	139	163	187	211	235	259	283	307	331	355	379
20	44	68	92	116	140	164	188	212	236	260	284	308	332	356	380
21	45	69	93	117	141	165	189	213	237	261	285	309	333	357	381
22	46	70	94	118	142	166	190	214	238	262	286	310	334	358	382
23	47	71	95	119	143	167	191	215	239	263	287	311	335	359	383

At the de-interleaving process, burst errors are spread throughout the matrix, as presented in Table 4.5.

4.1.4.3 Data Scrambling

Data scrambling happens after interleaving at a rate of 19.2 ksps. Scrambling is not used for 'spreading'; its main function is to improve security by encrypting the transmitted data.

Scrambling consists of a modulo-2 addition ('exclusive OR' circuit) of the interleaver output coded symbols (data) and a binary value determined by the long code generator after applying the 42-bit offset mask (one decimated LC chip out of every 64 chips), as in Figure 4.15. More specifically, scrambling consists of applying the data sequence through XOR function with a PN code (chip sequence decimated from the long code) at the same rate as the data.

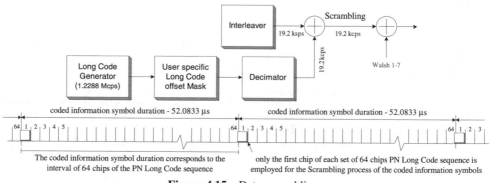

Figure 4.15 Data scrambling.

The chip sequence used for scrambling is the long code sequence (PNLC) with a user-specific phase offset, where only the first out of every 64 chips is used in the process.

Figure 4.16 shows the structure of the 42-bit mask used to obtain the phase of the long code for FPChs.

41 29	28 24	23 21	20 9	8 0
1100011001101	00000	PCN	000000000000	Pilot_PN

Figure 4.16 42-bit mask used by paging channels.

In the figure,

PCN   number of associated FPCh (1–7)
Pilot_PN   PN sequence offset used by the BTS transmitting the FPCh

4.1.4.4 Phase, Quadrature and Carrier Modulation

Prior to modulation, FPChs data must be channelised using Walsh codes. The primary FPCh always uses W_1^{64}. Even though the primary carrier always needs to reserve Walsh codes 1–7 to FPChs, any additional CDMA RF carrier that is not using all forward paging channels can use these Walsh codes to spread and channelise FTChs.

The quadrature PN modulation process is performed as in Figure 4.2.

4.1.5 Forward Traffic Channels

Forward Traffic Channels (FTChs) are responsible for transmitting voice, data (low rates) and signalling information bits during a call. Processing of TChs is more complex, because it also considers voice coding. Figure 4.17 presents all stages involved in the process, which is described in detail in the following sections.

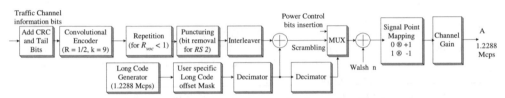

Figure 4.17   Processing block diagram for forward traffic channels.

4.1.5.1 Voice Coding

This section explains in a simplified manner the voice coding process in IS-95. The study will be developed in two parts:

- Pulse Code Modulation (PCM) coding,

- VOice CODER (vocoder).

4.1.5.1.1 PCM Coding

Pulse Code Modulation (PCM) is the most common method used to code analogue voice into digital bit flows. To digitise the voice, the signal is sampled, measuring amplitudes in well-defined time intervals. The voice sampling rate is determined by the uniform sampling criteria, or Nyquist's sampling theorem.

Nyquist states that the sampling rate has to be equal or greater than twice the highest voice frequency in the original signal to be transmitted, allowing the receiver to re-generate the original signal without significant errors when compared to the original voice.

The voice signal used in telephony is usually frequency limited in a bandwidth between 300 and 3400 Hz. Thus, the sampling frequency must be higher than 6800 Hz.

An amplitude value, representing the signal level, is assigned to each sample. The system takes positive and negative peaks of the voice signal and divides the interval [Peak$_{negative}$, Peak$_{positive}$] into smaller values. These values, taken as reference unit, can be constant or not, in a process called quantisation. Figure 4.18 shows an example of linear quantisation, in which all reference units are constant.

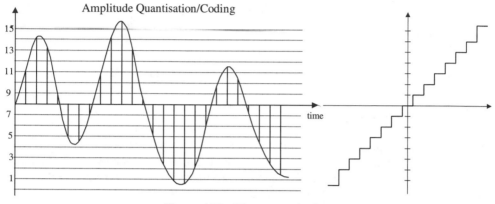

Figure 4.18   Linear quantisation.

The quantisation process is not very effective for voice signals with low amplitude because the smaller the sampled signal level, the greater the possibility of quantisation errors. This happens because a discrete value must be assigned to the sampled signal and this value may be above or below the real value.

To minimise quantisation errors for signals with low amplitude, a non-linear quantisation scale can be used. Thus lower amplitudes use smaller quantisation units, whereas higher amplitudes use bigger units, as in Figure 4.19. The formation law (scale) that is usually applied follows a logarithmic behaviour. This technique is called companding, meaning compression of high values and expansion of lower values.

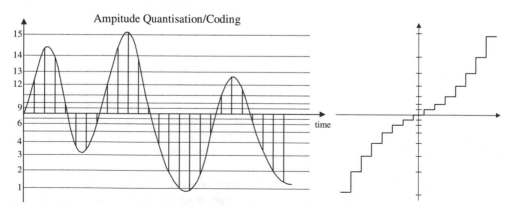

Figure 4.19   Non-linear quantisation.

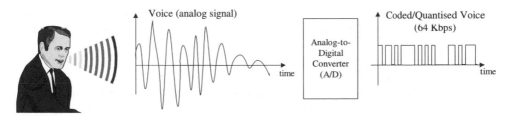

Figure 4.20   A/D converter.

A codeword is assigned to each quantised value, regardless of the formation law, achieving the signal conversion from analogue to digital, as in Figure 4.20.

Current commercial systems use non-linear quantisation methods. The two methods most commonly used are known as μ law and A law. The first was developed by Bell Laboratories and became an American standard. The latter, known as A law, was created by CCITT and became a European standard.

The formation law for each of the methods is the following

$$\mu \text{ law}: \quad y = \ln(1 + \mu|x|)\text{sgn}(x) \tag{4.3}$$

$$
\begin{aligned}
A \text{ law}: \quad & y = 0.18(A|x|)\text{sgn}(x) && \text{for } 0 \leq x < A - 1 \\
& y = 0.18[1 + \ln(A|x|)]\text{sgn}(x) && \text{for } A - 1 \leq x \leq 1
\end{aligned}
\tag{4.4}
$$

where

$\mu = 25$
$A = 87.6$
y companded signal level
x actual sampled signal level
$|x|$ modulus of sampled signal, or amplitude without polarity indication
$\text{sgn}(x)$ polarity of sampled signal

Any signal reconstructed from digitised samples always contains a quantisation error, due to resolution characteristics. The greater the number of bits used for quantising, the greater the number of discrete levels.

Systems that employ PCM coding usually have a sampling rate of 8000 times per second (one sample every 125 μs, or 125 microseconds). Because each sample is coded in 8 bits, the process provides a stream of 64 kbps (8000×8-bit samples).

4.1.5.1.2   Vocoder

When a call from the fixed network, or Public Switched Telephone Network (PSTN), arrives at CDMA IS-95 systems, the PCM signal is already coded as 8 bits using non-linear quantisation (usually μ law), with a bit rate of 64 kbps. Before this signal goes through the vocoder, it is transcoded into 13 bits using linear quantisation, achieving a data rate of 104 kbps at the transcoding circuit output. The amount of quantisation levels goes from 256 (2^8), using 8-bit coding, to 8192 (2^{13}) in the 13-bit coding. If the vocoder is used in the mobile station, however, because the input is analogue voice, an analogue-to-digital converter codes the information directly into 13 bits using linear quantisation.

The digitised voice signal with a constant rate of 104 kbps is divided into 20 ms frames before being sent through the vocoder. Each frame corresponds to 160 samples of the original voice signal. The vocoder processes the information and generates output data with a variable rate, according to the voice activity.

The voice coder is a key element in the network. Due to the human voice intermittent behaviour in a telephone conversation, during the total usage time of an RF channel, there are periods with transmission of digital voice data and periods with pauses between words and syllables.

The first CDMA IS-95 vocoder adopted had a rate of 8.6 kbps, and used Qualcomm's Code-Excited Linear Predictive (QCELP) coding algorithm. This vocoder allowed users to perfectly understand the conversation, but they were often unable to identify the other party, due to absence of voice timbre. To achieve higher voice communication quality standards, the industry subsequently developed a QCELP vocoder with a 13.3 kbps rate. However, the gain in speech quality because of the higher data rate is reflected in a capacity loss.

The industry, however, refers to these vocoders using their gross data rates, 9.6 Kbps and 14.4 Kbps, which include the nominal data rates, 8.6 Kbps and 13.3 Kbps, plus CRC and tail bits for cleaning the contents of the convolutional coder's memory in a way that aids in the de-coding.

Another high-quality speech vocoder with smaller data rates, 8.6 kbps, was developed by Bell Labs, Lucent Technologies' research and development team, using the Enhanced Variable Rate Coder (EVRC) technology, also called IS-127. The main goal of this project was to aid CDMA network capacity expansion without sacrificing voice quality, delivering high-quality speech, low-bit-rate operation and easy implementation.

EVRC technology is based on a Relaxed Code-Excited Linear Predictive (RCELP) coding algorithm, developed in 1994 by Bell Labs. RCELP consists of a CELP speech-coding algorithm generalisation, ideal for variable operations rate and robust CDMA networks.

The two basic types of vocoder output rates are called Rate Sets (RSs) 1 and 2, and have maximum net data rates of 8.6 and 13.3 kbps, respectively. The first vocoder available operated at a rate of 8.6 kbps, i.e. RS1. For this vocoder, possible net data rates (coded voice plus flags indicators) are 8.6 kbps, 4.0 kbps, 2.0 kbps and 800 bps, as presented in Figure 4.21. Adaptive thresholds, which vary according to the background noise level, determine the data rate to be used. This feature results in the suppression of background noise, providing good voice quality, even in noisy environments.

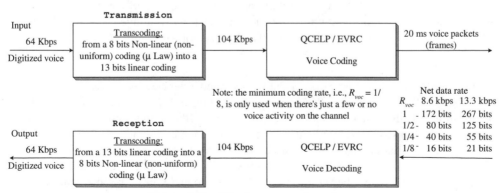

Figure 4.21   Vocoder rate configurations.

Higher data rates (8.6 and 4.0 kbps) are only activated when there is voice activity. The transmission power is lowered when using lower rates, reducing interference to other users and favouring increase of capacity in number of channels, therefore allowing more traffic per carrier.

Voice activity is determined by comparing the energy in a 20 ms frame with a set of three adaptive thresholds that change dynamically as a function of the existing background noise.

CDMA IS-95 systems have initially employed vocoders with 8.6 kbps of output data rate, which corresponds to 172 bits per 20 ms frame. When the frame is only transmitting coded voice data, these bits are divided into 171 bits for coded voice and one bit, set to '0', representing the Mixed Mode (MM) flag.

When the network needs to send commands to mobile terminals that are in conversational state, forward traffic channel frames use a signalling traffic structure. Signalling messages can have from 16 to 1160 bits. Therefore, vocoders with maximum rate of 9600 require more than one frame to transmit messages longer than 168 bits.

The first bit of the traffic signalling structure is always the Start Of Message (SOM) indicator, where '1' indicates beginning of a new message and '0' means the message of the previous frame is being continued. If the last frame used to transmit a message is not fully occupied, the system completes the frame with padding bits '0'.

The signalling and data message transmitted in the traffic channel is composed of the following.

- *Message length* (*MSG_LENGTH*) – message body length, in bytes, coded into 8 bits. Minimum and maximum values of this field are, respectively, 5 and 148, corresponding to lengths of 40 and 1184 bits.

- *Message body* – message data, from 16 to 1160 bits.

- *Cyclic Redundancy Check* (CRC) – message capsule quality indicator, 16 bits.

4.1.5.1.3   Characteristics of 8,6 kbps Vocoders (RS1)

Each FTCh frame consists of information bits (voice and/or data) and additional bits used to organise and structure the frame. The additional bits include flags (TT, MM and MT), CRC and encoder tail bits. Table 4.6 shows the number of bits added to each 20 ms frame that produce the gross rates of 9600, 4800, 2400 and 1200 bps.

The CRC field, or frame quality indicator, has two tasks in the reception:

- allow detection of frame errors,

- allow determination of the transmission data rate, R_{voc} (aided by other mechanisms).

Table 4.6 Gross data rates for the 8.6 kbps vocoder (gross rate of 9.6 kbps)

Data rate (R_{voc})	Bits/frame	Net rate (kbps)	CRC	Encoder tail bits	Total of bits/ frame	Gross rate (kbps)
1	172	8.6	12	8	192	9.6
1/2	80	4.0	8	8	96	4.8
1/4	40	2.0	0	8	48	2.4
1/8	16	0.8	0	8	24	1.2

Frames with data rate of '1' ($R_{voc} = 1$) use CRC generator circuits composed of a 12-stage shift register. These circuits are synchronised by a single clock generator and can be represented by the following polynomial

$$g(x) = x^{12} + x^{11} + x^{10} + x^9 + x^8 + x^4 + x + 1 \qquad (4.5)$$

The feedback in the circuit is implemented using 'exclusive OR' gates. Figure 4.22 shows a block diagram of the circuit.

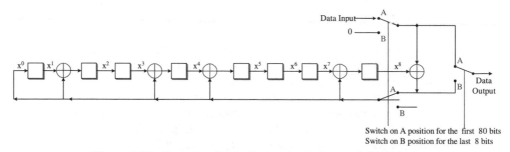

Figure 4.22 Frame quality indicator circuit for rate 1 (9600 bps).

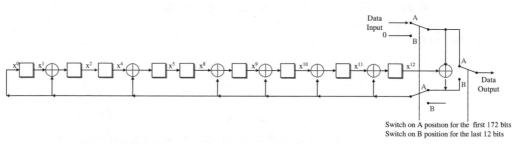

Figure 4.23 Frame quality indicator circuit for rate 1/2 (4800 bps).

Similarly, half-rate frames use a CRC generator circuit with an 8-stage shift register. Figure 4.23 shows the circuit and the polynomial representing the feedback connections

$$g(x) = x^8 + x^7 + x^4 + x^3 + x + 1 \qquad (4.6)$$

Frames with rates '1' and '1/2' have, respectively, CRC fields composed of 12 and 8 bits, obtained as follows:

- in the beginning of each 20 ms frame, all elements (stages) in the shift registers are set to '1' and all switches are set in position 'A';

- the circuits are loaded, respectively, with 172 and 80 bits, corresponding to the net data output of the vocoder for rates 1 and 1/2;

- the registers generate CRCs for 172 and 80 bits (as shown in Figures 4.22 and 4.23);

Table 4.7 Flags and information bits in 20 ms frames on the forward traffic channels for 8.6 kbps vocoders

Transmit rate (bps)	Format bits			Primary traffic Bits/frame	Signalling traffic Bits/frame	Secondary traffic Bits/frame
	Mixed Mode (MM)	Traffic Type (TT)	Traffic Mode (TM)			
	0	–	–	171	0	0
	1	0	00	80	88	0
	1	0	01	40	128	0
	1	0	10	16	152	0
9600	1	0	11	0	168	0
[a]	1	1	00	80	0	88
[a]	1	1	01	40	0	128
[a]	1	1	10	16	0	152
[a]	1	1	11	0	0	168
4800				80	0	0
2400				40	0	0
1200				16	0	0

[a]Structure is optional.

- the switches are set to position 'B', and the clock generator pulses 12 or 8 times, according to the data rate, sending the CRC bits to the output;

- the final frames have a total of 184 and 88 bits, which include the original data and CRC bits.

The receiver has a similar CRC generator circuit that recalculates the CRC for the 172 or 80 bits received and compares the results to the CRC sent with the data. If both CRCs match, the frame is considered to be correct. The system does not perform CRC calculation for output rates (R_{voc}) of 1/4 and 1/8.

Regardless of the data rate in use, however, all frames have eight encoder tail bits. These bits are a set of '0s' inserted after the information and CRC bits.

Each frame starts with flags that indicate the structure, or multiplexing option format, of the information contained in the frame. Table 4.7 shows the types of flags and their configuration in each type of channel.

Depending on the configuration a different number of information bits is sent per frame, with the maximum being 171 bits/frame when transmitting primary traffic only.

The following configurations are possible when using the $R_{voc} = 1$ data transmission rate:

- only primary traffic;

- only secondary traffic;

- only signalling traffic;

- primary and secondary traffic;

- primary and signalling traffic.

4.1.5.1.4   Characteristics of the 13.3 kbps Vocoder (RS2)

Similarly to 8.6 kbps vocoders, each FTCh frame of 13.3 kbps vocoders consists of information bits (voice and/or data), flags, CRC bits (frame quality indicators) and encoder tail bits. Table 4.8 shows the number of output bits per frame.

Table 4.8   Gross rates for 13.3 kbps vocoder

Data rate (R_{voc})	Bits/ frame	Net rate (kbps)	Reserved flag (R)	CRC	Encoder tail bits	Total of bits/ frame	Gross rate (kbps)
1	267	13.35	1	12	8	288	14.4
1/2	125	6.25	1	10	8	144	7.2
1/4	55	2.75	1	8	8	72	3.6
1/8	21	1.05	1	6	8	36	1.8

Table 4.9   Flags and information bits in 20 ms frames on forward traffic channels for 13.3 kbps vocoders

Transmit rate (bps)	Format bits		Primary traffic Bits/frame	Signalling traffic Bits/frame	Secondary traffic Bits/frame
	Mixed Mode (MM)	Format Mode (FM)			
14400	0		266	0	0
	1	0000	124	138	0
	1	0001	54	208	0
	1	0010	20	242	0
	1	0011	0	262	0
	1	0100	124	0	138
	1	0101	54	0	208
	1	0110	20	0	242
	1	0111	0	0	262
	1	1000	20	222	20
7200	0		124	0	0
	1	000	54	67	0
	1	001	20	101	0
	1	010	0	121	0
	1	011	54	0	67
	1	100	20	0	101
	1	101	0	0	121
	1	110	20	81	20
3600	0		54	0	0
	1	00	20	32	0
	1	01	0	52	0
	1	10	20	0	32
	1	11	0	0	52
1800	0		20	0	0
	1		0	0	20

Table 4.8 shows bits added to the 20 ms frame on the 13.3 kbps vocoder output. Unlike 8.6 kbps vocoders, 13.3 kbps vocoders calculate CRC for all frames, regardless of the output rate.

There are three types of flags in 13.3 kbps vocoders: Mixed Mode (MM), Format Mode (FM) and Reserved (R). The flags indicate the structure, or format, of the frame. Table 4.9 shows flag configurations in each type of frame.

Frames transmitted through traffic channels, on the forward and reverse link, may be aligned with the time reference received at the BTS by the GPS receiver. When they are aligned, their frame offset is zero. Otherwise, the frame offset is given by the parameter *FRAME_OFFSET* multiplied by 1.25 ms. Mobiles transmit this parameter in the traffic channel assignment message or in the handoff direction message delaying TCH transmission is important for the correct resource assignment of the MSC and the system's interconnection links.

4.1.5.2   Convolutional Coding, Bit Repetition and Puncturing

After coding the voice in 20 ms frames and inserting additional bits, the system applies convolutional coding to the vocoder output data, as shown in Figure 4.24. Convolutional coding consists of serially adding selected derivations of a shifted data sequence (Figure 4.7).

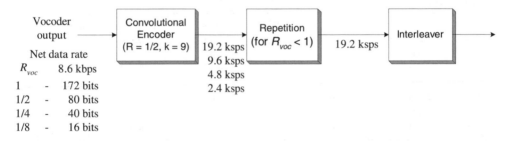

Figure 4.24 Convolutional coding $(R = 1/2)$, bit repetition, and interleaving for the 9.6 kbps vocoder.

CDMA IS-95 systems always use half-rate $(R = 1/2)$ convolutional encoders on the paging, SyncCh and traffic channels, that is, two output bits are generated for every input bit (Figure 4.7).

When the vocoder rate is less than 1 ($R_{voc} < 1$), the convolutional encoder is followed by a bit repetition procedure, as shown in Figure 4.25.

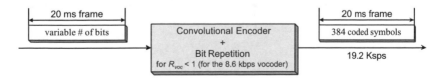

Figure 4.25 Convolutional coding and bit repetition in the FTCh for 9600 bps vocoders.

Table 4.10 Number of repetition per symbol as a function of vocoder output rates

Vocoder rate	Data (bps)	Convolutional coder output (sps)	Number of repetitions	Interleaving block (Ksps)
1	9600	19200	–	19.2
1/2	4800	9600	1	19.2
1/4	2400	4800	3	19.2
1/8	1200	2400	7	19.2

Table 4.10 shows the number of repetitions as a function of vocoder output rates. The final rate of 19.2 kbps, or 384 coded bits per 20 ms frame, is the input to the block interleaver.

Table 4.11 shows a summary of data rates at different stages in the circuit when using the 9.6 kbps vocoder.

Table 4.11 Summary of data rates for 9.6 kbps vocoders with half rate $R = 1/2$ for coding

Vocoder input			Vocoder output		Convolutional coder output		
PSTN PCM	Linear coding	Rate (R_{voc})	Net rate	Gross rate	Before repetition block	After repetition block	
1280	2080	1	172	192	384	384	Symbol/frame
64000	104000		8600	9600	19200	19200	sps
1280	2080	1/2	80	96	192	384	Symbol/frame
64000	104000		4000	4800	9600	19200	sps
1280	2080	1/4	40	48	96	384	Symbol/frame
64000	104000		2000	2400	4800	19200	sps
1280	2080	1/8	16	24	48	384	Symbol/frame
64000	104000		1000	1200	2400	19200	sps

Data processing when using the 14.4 kbps vocoders is different from processing for 9.6 kbps vocoders. Figure 4.26 shows a simplified block diagram for the 14.4 kbps vocoder.

For 14.4 kbps vocoders, the coded bit rate at the convolutional coder output can be 28 800, 14 400, 7200 or 3600 bps, depending on the transmission rate (R_{voc}), 1, 1/2, 1/4 or 1/8. For

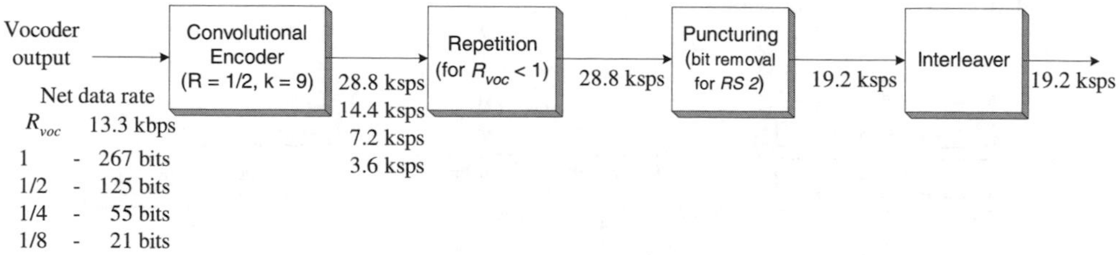

Figure 4.26 Convolutional coding ($R = 1/2$), bit repetition, puncturing and interleaving for 14.4 kbps vocoders.

Table 4.12 Summary of data rates for 14.4 kbps vocoders with half rate $R = 1/2$ for coding

Vocoder input			Vocoder output		Convolutional coder output		
PSTN PCM	Linear coding	Rate (R_{voc})	Net rate	Gross rate	Before repetition block	After repetition block and puncturing	
1280	2080	1	267	288	576	384	Symbol/frame
64 000	104 000		13 350	14 400	28 800	19 200	sps
1280	2080	1/2	125	144	288	384	Symbol/frame
64 000	104 000		6250	7200	14 400	19 200	sps
1280	2080	1/4	55	72	144	384	Symbol/frame
64 000	104 000		2750	3600	7200	19 200	sps
1280	2080	1/8	21	36	72	384	Symbol/frame
64 000	104 000		1050	1800	3600	19 200	sps

rates lower than 1, the coded bit flow goes through the repetition block. Bits are repeated one, three or seven times, according to the rate 14 400, 7200 or 3600 bps. Therefore, all 20 ms frames after the repetition block have 576 coded bits (28 800 bps).

To allow common processing of the channel, regardless of the vocoder in use, a bit removal block (puncturing) is applied between the repetition and interleaving processes. Puncturing removes two out of every six coded bits in the data flow causing the final data rate to be 19.2 kbps, which corresponds to 384 coded bits in a 20 ms frame, as in Figure 4.25.

Table 4.12 shows a summary of the data rates in several stages of frames processing, when using 14.4 kbps vocoders.

4.1.5.3 Interleaving

Interleaving of forward traffic channels works as interleaving for paging channels (FPChs). The process is fully described in Section 4.1.4.2.

Thus, the interleaving process is done in 20 ms frame time basis, i.e. the process is time aligned to the frame durations, where there are 384 coded bits being interleaved per 20 ms frame.

4.1.5.4 Data Scrambling

The scrambling process is performed with the output data from the interleaving block, at a rate of 19.2 kbps.

Scrambling consists on a modulo-2 addition ('exclusive OR' circuit) of the interleaver output coded bits and a binary value determined by the long code generator after the 42-bit mask application (one decimated LC chip out of every 64 chips), as in Figure 4.27.

The chip sequence used for scrambling is, in fact, the long code sequence (PNLC) itself but with a user-specific phase offset, as in Figure 4.28. However, only the first, out of every

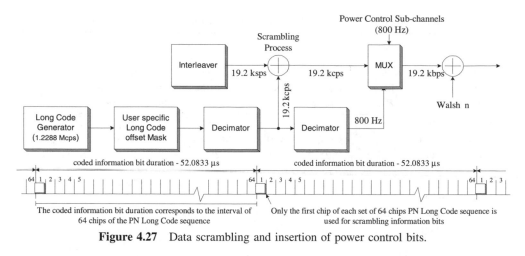

Figure 4.27 Data scrambling and insertion of power control bits.

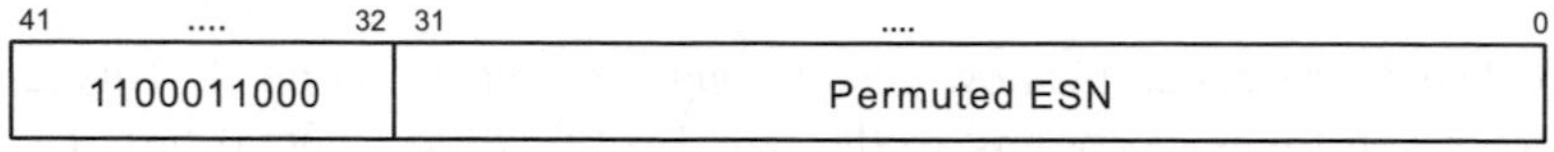

41	....	32	31	....	0
1100011000			Permuted ESN		

Figure 4.28 Mask used to generate the long code offset for the traffic channel.

64 chips in the sequence, is used for scrambling. The selection of only this first chip is called decimation.

In Figure 4.28, *ESN* represents the Electronic Serial Number, coded in 32 bits. The chapter on call processing and authentication procedures provides additional information about the *ESN*.

The *ESN* permutation avoids similar long code phases for mobile terminals with similar *ESN*s (for example, from the same production set). The existence of long code phases too close to each other within the same service area can cause errors in de-modulation and decoding processes.

The scrambling process also provides encryption and security of user data. Because each long code mask depends on the *ESN*, which is unique in the network, the long code sequence used for scrambling, after mask application, is different for each user.

4.1.5.5 Insertion of Power Control Sub-Channels (Power Control Bits)

The Power Control Sub-Channel (PCSChs) is transmitted every 1.25 ms in the forward traffic channel (800 bps – one bit per sub-channel). This sub-channel carries commands for the mobile to increase or decrease transmission power on the reverse traffic channel.

One bit '0' transmitted by the PCSCh indicates that the mobile needs to increase its transmission power, whereas, a bit '1' indicates that power reduction is required. The power control is executed in steps defined by the *PWR_CNTL_STEP* parameter, shown in Table 4.13. Power control techniques for CDMA systems are described in detail in Chapter 7.

Table 4.13   Power control steps for FTChs

PWR_CNTL_STEP	Power control steps (dB)
0	1.00
1	0.50
2	0.25

During the 1.25 ms period between the PCSChs transmission, the BTS estimates the level received in the reverse traffic channel. The BTS uses this information to determine the power control bit, '0' or '1', to be transmitted in the PCSCh assigned to each mobile, as presented in Figure 4.29. This process consists of closed loop power control scheme.

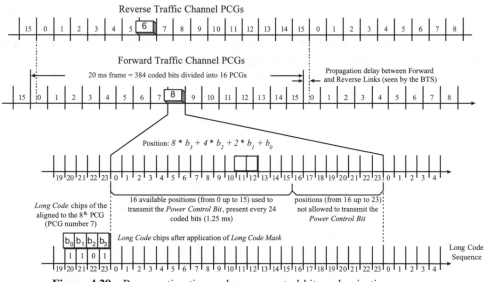

Figure 4.29   Power estimation and power control bit randomisation process.

Figure 4.29 shows that there are 384 coded bits per 20 ms frame, that is, a transmission rate of 19.2 kbps. Each 20 ms frame is composed of 16 Power Control Groups (PCG), from 0 to 15, each 1.25 ms long (800 Hz) and with 24 coded bits.

An estimated power control bit is only transmitted two PCGs after it was calculated, that is, in PCG $(i + 2)$. For example, if the bit is estimated in group 9 of the reverse link channel, the correspondent power control bit is sent in group $11 = (9 + 2)$ of the forward traffic channel. If the estimation happens in the last two groups of a frame, the power control bit is transmitted in a group of the next frame.

Because the power control bit length corresponds to two coded bits in the forward link each power control bit replaces two consecutive bits in the FTCh frame.

From the 24 coded bits in a power control group in the forward link, IS-95 standards assign 16 possible positions to insert power control bits in a PCG, corresponding to the first 16 coded bits (0–15).

The insertion position of the power control bits within its PCG is not fixed, and needs to be defined in a frame-by-frame basis, in a process called power control bit randomisation.

After applying the 42-bit mask and the decimator process on the long code, the system generates a sequence with pseudo-random behaviour at a rate of 19.2 kbps. At every 1.25 ms, this sequence provides 24 chips used for data scrambling. The last four chips of the decimated PN sequence, corresponding to the PCG immediately before the one being processed, determine the position of the power control bit, as in Figure 4.29.

Chip 23 is the most significant chip, and chip 20 is the least significant. Figure 4.29 shows an example where the scrambling bits b_{23}, b_{22}, b_{21} and b_{20} are '1011', or 11 in decimal notation. In this case, the power control bit is inserted in position 11, replacing data symbols 11 and 12, in power control group 8.

To understand the need of power control in CDMA IS-95 it is necessary to understand that these systems do not use C/I ratio as a quality criterion. They use, instead, the Frame Error Rate (FER), which statistically represents the number of wrong frames received in each group of 100 frames.

As the FER lowers (through an increase of channel power, for example), there is an improvement in the communication quality. Even though an increase in channel power decreases the FER, it also leads to an interference increment on the remaining channels sharing the same frequency spectrum. The interference reduces the amount of channels that can be stacked, yielding to a lower traffic capacity. Figure 4.30 illustrates this concept.

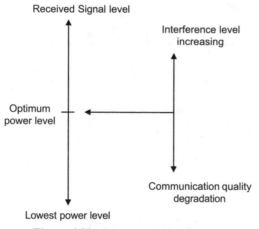

Figure 4.30 Power control objective.

The ideal transmission power of BTS and mobile terminal is achieved when the receiver is capable of de-modulating and de-coding the information received within the frame error rate limits specified by the system (MSC), while providing the lowest interference level to other users.

The system administrator may eventually decide to de-grade communication quality, that is, increase the FER, to add traffic capacity to the network, in the busiest hours, for example.

However, this strategy is only valid within specific FER limits that usually range from 0.2% to 3%, and when the BTS, or sector, has hardware channels available.

The ratio E_b/N_0 is the relation between the energy of a data bit and the power spectral density of the noise plus interfering signals. E_b is defined as

$$E_b = T_b \times S = S/R_b \tag{4.7}$$

where

S power assigned to the communication channel, in Watts
T_b interval of an information bit, in seconds
R_b information transmission bit rate (9.6 or 14.4 Kbps)

and

$$T_b = 1/R_b \tag{4.8}$$

The total power density of the thermal noise plus interfering signals (N_0), internally and externally generated, is given by

$$N_0 = N/W \tag{4.9}$$

where

N total power of the noise plus interferers, in Watts
W spectrum bandwidth, 1.23 MHz

From eqns (4.7) and (4.9)

$$\frac{E_b}{N_0} = \left(\frac{S}{N}\right) \Big/ \left(\frac{R_b}{W}\right) \Rightarrow \frac{E_b}{N_0} = \frac{S}{N} \times \frac{W}{R_b} \tag{4.10}$$

As the equation for Processing Gain (PG) is

$$P_G = \frac{W}{R_b} \tag{4.11}$$

Eqn (4.10) can be written as

$$\frac{E_b}{N_0} = P_G \times \frac{S}{N} \tag{4.12}$$

For 9.6 kbps vocoders, the processing gain is 128 (or 21 dB), whereas 14.4 kbps vocoders have a P_G of 85.3 (or 19.3 dB).

E_b/N_0 can also be expressed in dB, as follows

$$\frac{E_b}{N_0}(\text{dB}) = P_G(\text{dB}) + \frac{S}{N}(\text{dB}) \tag{4.13}$$

CDMA systems are designed in a way that all channels transmitted by several users (within a cell) reach the BTS with an adequate E_b/N_0. The adequate E_b/N_0 provides the desired FER and its value varies according to the mobile speed.

Figure 4.31 shows the E_b/N_0 required (in dB), as a function of the mobile station speed, to obtain a frame error rate of 1% (considered to provide good voice quality). The required E_b/N_0 ranges between 2.5 and 6.1 dB for lower speeds because the power control is highly effective to compensate fading in this situation. At higher speeds, where power control is not so effective, the required E_b/N_0 lowers because the benefits of bit interleaving are higher.

Power control, performed 800 times per second, is divided in two steps: outer loop control (only at BTSs) and inner loop control (simultaneously on BTSs and MSs). In the outer loop,

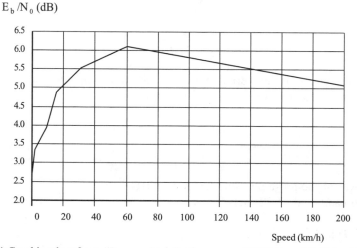

E_b/N_0 (dB)

Speed (km/h)

(*) Graphic taken from *"Reverse Link Performance of IS-95 Based Cellular Systems"*, *R. Padovani, IEEE Personal Communications, Third Quarter, 1994*

Figure 4.31   Required E_b/N_0 for a 1% FER, as a function of the mobile speed.

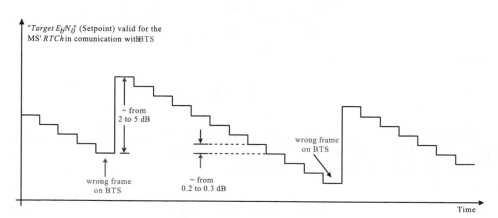

Figure 4.32   Obtaining the 'target E_b/N_0'.

the target E_b/N_0 is reduced, frame by frame, in small steps, typically 0.2 dB, until there is a frame received in error. When an error is detected, the target E_b/N_0 is increased to a value usually between 3 and 5 dB, and the frame-by-frame incremental power reduction process is restarted, as in Figure 4.32. In the inner loop, the BTS compares the E_b/N_0 of the reverse traffic channel with the target E_b/N_0 every 1.25 ms.

4.1.5.6 Phase, Quadrature and Carrier Modulation

Prior to modulation, FTChs must be channelised using Walsh codes 8–63 (except for Walsh code 32, which is used in the synchronisation channel), as described in Figure 4.2.

4.1.5.7 Power Control of FTChs as a Function of Vocoder Output Data Rate

The transmitted coded bit rate is constant (19.2 kbps) regardless of the vocoder type being used, 9.6 kbps (RS1) or 14.4 kbps (RS2) (Figure 4.33). This allows expressing energy per bit in terms of information bit length

$$E_b = T_b \times S = S/R_b \tag{4.14}$$

where

E_b energy per bit (Joule)
S transmission power assigned to the channel (Watt) = (Joule/s)
T_b information bit length (seconds)

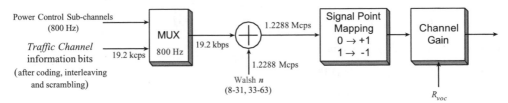

Figure 4.33   BTS transmission power control as a function of the vocoder data output rate.

To keep the energy per baud rate constant, keeping the same E_b/N_0, the transmission power is lowered when the symbol length increases, that is, when the information rate is lowered.

Coded symbols associated to lowest information rates are, thus, transmitted with a smaller amount of energy. Table 4.14 shows transmitted symbol energy as a function of the vocoder gross rate.

The energy of each information bit (E_b) at the convolutional encoder input is divided among coded bits at the repetition block output. For example, considering rate 1, for each

Table 4.14 Symbol energy transmitted as a function of vocoder data rates

Vocoder output gross data rate R_{voc} (bps)		Coded symbol rate (ksps)	Energy per coded symbol (E_S)
1	9600	19.2	$E_b/2$
1/2	4800	19.2	$E_b/4$
1/4	2400	19.2	$E_b/8$
1/8	1200	19.2	$E_b/16$
1	14400	19.2	$3E_b/4$
1/2	7200	19.2	$3E_b/8$
1/4	3600	19.2	$3E_b/16$
1/8	1800	19.2	$3E_b/32$

bit, there are two coded symbols taken from the convolutional coder, each with bit energy $E_b/2$. For rates 1/2, 1/4 and 1/8, coded bits must also be repeated (1, 3 and 7 times, respectively) after convolutional encoding, implying in a total of 2, 4 and 8 coded bits, respectively, with energies of $E_b/4$, $E_b/8$ and $E_b/16$.

As a direct consequence of smaller vocoder rates ($R_{voc} < 1$), the FTCh transmission power is also reduced, leading to less interference to other users, therefore providing an increase in system capacity.

4.2 FORWARD LINK CHANNEL STRUCTURE IN cdma2000 SYSTEMS

In cdma2000 networks, each BTS can be configured to offer the following logical channels:

- Forward Pilot Channels (FPiCh);

- Forward Sync Channel (SyncCh);

- Forward Paging Channels (FPCh);

- Broadcast Control Channel (BCCh);

- Quick Paging Channel (QPCh);

- Common Power Control Channel (CPCCh);

- Common Assignment Channel (CACh);

- Forward Common Control Channel (FCCCh);

- Forward Dedicated Control Channel (FDCCh);

- Forward Fundamental Channel (FFCh) – equivalent to the FTCh of IS-95;

- Forward Supplemental Channel (FSCh);

- Forward Supplemental Code Channel (FSCCh).

4.2.1 Forward Pilot Channels

Forward pilot channels (FPiChs) are non-modulated channels that do not carry information data. These channels transmit only bits '0', being channelised by Walsh sequences W_0^{64} for SR1 and SR3.

The main functions of FPiChs are to allow MSs to synchronise with the system, to provide timing and phase reference for coherent detection and to provide power reference for power control schemes.

Differently than in IS-95 systems, cdma2000 networks can transmit up to four distinct types of pilot channels in the forward link:

- Forward Pilot Channel (FPiCh) – same logical channel described for IS-95 systems (Section 4.1.2);

- Transmit Diversity Pilot Channel (TDPiCh);

- Auxiliary Pilot Channel (APiCh);

- Auxiliary Transmit Diversity Pilot Channel (ATDPiCh).

Figure 4.34 shows a block diagram representing part of the processing for pilot channels SR1 and SR3. Because FPiChs transmit only bits '0', they contain only short PN sequences (shown later). This allows MSs to synchronise with these sequences, as described in Section 4.1.1 for CDMA IS-95 systems. In this figure, the X_Q branch is set to zero (0), which means that no transmission is performed, i.e. all data in this branch is gated-off.

Figure 4.34   Pilot channel processing diagram block.

The forward pilot channel is continuously transmitted by BTSs in each CDMA RF carrier available in the system, unless the BTS or sector is a hopping pilot beacon.

Hopping pilot beacon transmissions follow timing characteristics described in Section 4.2.1.6. Transmission frequencies of hopping pilot beacons may be periodically modified to simulate several pilot channels transmitting in the system.

4.2.1.1   Transmission Diversity Techniques

If the Transmission Diversity (TD) option is supported by the BTS, a Transmit Diversity Pilot Channel (TDPiCh) is continuously transmitted for each physical channel available in the BTS, except for FPiCh. TDPiChs aid MSs to acquire and de-code data on logical channels transmitted if TD is employed at the BTS.

Even if the TDPiCh is transmitted, the BTS must continue to transmit the FPiCh with enough power to ensure that MSs can detect and estimate data in the forward link without needing the transmit diversity pilot channel. TDPiChs are transmitted with the same power or with power reduced by 3, 6 or 9 dB in relation to the forward pilot channel. TDPiChs are channelised using Walsh sequence W_{16}^{128}.

BTSs can transmit none, one or more Auxiliary Pilot Channels (APiCh) on the same CDMA carrier. BTSs must only transmit an ATDPiCh, with an auxiliary pilot channel, when the Orthogonal Transmit Diversity (OTD) mode is supported in the forward link. These channels are described in detail in Section 4.2.1.4.

All cdma2000 logical channels have an additional channel-processing block, called symbol de-multiplex (or DEMUX), because of specific diversity transmission configurations, such as OTD and Space Time Spreading (STS). Each logical channel has its own operation mode, employing some specific DEMUX processing.

OTD is a transmission method that distributes forward link channel symbols among multiple antennas and multiplexes them using Walsh codes or QOFs associated with each antenna, whereas STS technique multiplexes them using complementary Walsh Codes or QOF.

The main DEMUX functionality is to divide the data stream between I and Q branches. This procedure increases the data transmission rate, because different data is processed in each branch of the orthogonal modulator (PN-I and PN-Q modulation). Unlike IS-95 systems, this is a real QPSK modulation. In IS-95, the same information data generates both I and Q modulated data, that is, the modulation efficiency is half of what is achieved in cdma2000. cdma2000 systems have a modulation efficiency of 2 bits per modulation symbol. Logical channels may operate in distinct diversity configurations, such as non-TD, OTD or STS.

4.2.1.2   De-multiplexing Processes for Pilot Channels

cdma2000 forward pilot channels can only use the non-TD configuration for the de-multiplexing, i.e. serial-to-parallel block operation (DEMUX).

The DEMUX operation for SR1 in non-TD mode (Figure 4.35) is basically the transmission of bits to the binary sequence spreading and multiplication process block, as shown in Figure 4.37.

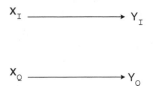

Figure 4.35 DEMUX operation in non-TD mode.

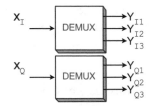

Figure 4.36 DEMUX operation for SR3.

As depicted in Figure 4.34, no pilot channel bits are transmitted in the X_Q branch, therefore there are no bits in the Y_Q outputs shown in Figures 4.35 and 4.36.

The DEMUX operation for SR3 (only for forward pilot and auxiliary pilot channels) consists in the transmission of the first symbol from the frame, in the X_I branch, to the Y_{I1}

output. The next symbols are transmitted to outputs Y_{I2} and Y_{I3}, respectively. Then the serial-to-parallel process starts again in Y_{I1}.

There is no data at input X_Q, because no bits are processed in branch Q.

4.2.1.3 Channelisation by Walsh Sequences and/or QOF and Complex Multiplier Spreading by PN Sequences

After de-multiplexing, the channel data is channelised using Walsh codes, complex spread by PN-I and PN-Q sequences (complex multiplier spreading), and carrier modulated with QPSK, as presented in Figure 4.37. Indices i (1, 2 and 3), in this figure, represent each of the 1.25 MHz carriers used in SR3 systems.

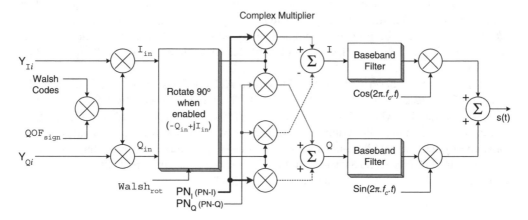

Figure 4.37 Spreading and complex multiplier process for forward pilot channels – SR1 and SR3.

A comparison between the carrier modulators presented in Figure 4.3, for CDMA IS-95, and in Figure 4.37, for cdma2000, reveals some differences that, at first, might suggest that the systems are incompatible. The fact that there are no bits in the Y_{Qi} branch, however, makes both configurations compatible.

Figure 4.37 represents I and Q branches processing with solid lines, whereas the complex multiplier is represented with dotted lines. Because there is no information on the Q_{in} branch, bits on the I_{in} branch are processed and then orthogonal modulated by PN-I and PN-Q sequences at the complex multiplier.

4.2.1.4 Auxiliary Pilot Channels and Auxiliary Transmit Diversity Pilot Channels

The association of Walsh code (W_i^m) sequences or of distinct quasi-orthogonal functions to each auxiliary pilot creates the auxiliary pilot channels.

Extended Walsh sequences, 128-bit long for SR1 and 256-bit long for SR3, are concatenated, creating another Walsh matrix of order $N \times m$ (matrix $N \times W_i^m$), where m is the order of the initial Walsh matrix, as shown in Table 4.15.

Table 4.15 Walsh Codes and QOFs for auxiliary pilot channels

SR	Number of concatenation (N_w)	Maximum length ($N = N_w \times m$)	Used code	Sequence index
SR1	1, 2 or 4	512	W_i^{128}	$1 \leq i \leq 128$
SR3	1 or 2	512	W_i^{256}	$1 \leq i \leq 256$

Polarity-inverted sequences, for the sequential Walsh code concatenation process, must be carefully selected to generate additional Walsh sequences of order $N_W \times m$, for $N = 4$. As an example, the sequence could be $W_i^m \overline{W_i^m} \overline{W_i^m} W_i^m$, with the second and the third Walsh sequence with their polarity inverted, as logically complementary. One or more Walsh sequences can be used in a complementary way. Some Walsh codes, however, can not be combined, such as codes W_0^m and W_{64}^{128} for SR1, or codes W_{64}^{256}, W_{128}^{256} and W_{192}^{256} for SR3.

The same features described for APiCh apply to ATDPiCh associated to them, except for spreading and channel multiplexing. Sequence W_i^N is assigned to APiChs whereas sequence $W_{i+N/2}^N$ is assigned to ATDPiChs, where $(1 \leq i \leq (N/2) - 1)$ and $N \leq 512$ for SR1 and SR3.

Summarising, cdma2000 systems operating in non-TD mode have to transmit an FPiCh for each physical channel but may also transmit one or more auxiliary pilot channels. When operating in TD mode, BTSs transmit FPiChs and APiChs, but may also transmit one TDPiCh and one ATDPiCh per auxiliary pilot channel.

4.2.1.5 Pilot Channel Offset Index (PN_OFFSET)

The system's PN sequence offset (for PN-I and PN-Q only) identifies pilot channels of each BTS. There are 512 offsets available (from 0 to 511).

Each offset index represents the delay related to a point of reference called zero offset. This delay is expressed in multiples of 64 chips (for SR1 and SR3). The zero offset reference is set in a way that short PN sequences (PN-I and PN-Q) begin at each even second.

The same pilot PN offset must be used for all carriers available within a BTS. Offsets can be re-used in other BTSs of the system as long as they are assigned so as to avoid reception problems (PN conflicts).

4.2.1.6 Hopping Pilot Beacon Timing

If BTSs support this function, the hopping pilot beacon must be periodically transmitted. It is transmitted as any FPiCh of the server BTS, using, for example, the same *PN_OFFSET* and transmission power. The beacon however also presents some distinct characteristics such as discontinuous transmission.

Beacons are usually transmitted by BTSs located at network boundaries, in systems where more than one CDMA RF carrier is available for example, allowing handoff between carriers.

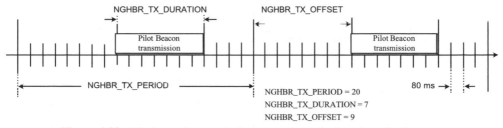

Figure 4.38 Timing and transmission parameters of a hopping pilot beacon.

The transmission timing configuration for hopping pilot beacons is defined by three parameters, each of them defined in 80 ms intervals, as in Figure 4.38:

- *NGHBR_TX_PERIOD*;

- *NGHBR_TX_OFFSET*;

- *NGHBR_TX_DURATION*.

The parameter *NGHBR_TX_PERIOD* represents the period of time between successive transmissions of the hopping pilot beacon. The transmission period begins aligned with the system time beginning.

The parameter *NGHBR_TX_OFFSET* represents the delay between the beginning of the transmission period assigned and the pilot's transmission. The parameter *NGHBR_TX_DURATION* specifies the transmission time of a hopping pilot beacon, in 80 ms intervals.

4.2.2 Synchronisation Channel

The Synchronisation Channel (SyncCh) for SR1 and SR3 presents the same characteristics as in IS-95 and IS-95B standards, using the same Walsh code, W_{32}^{64}.

SyncCh messages can have from 40 to 2040 bits and are organised in structures called message capsules, as described in Figure 4.4.

The maximum transmission rate is 1200 bps, and the channel is divided in 32-bit frames, 26.667 ms each. Groups of three frames correspond to 80 ms super-frame structures, as defined in IS-95 standards and represented in Figures 4.4 and 4.5. Frames are composed of 32 bits whereas message capsules in the SyncCh have 31 bits. The additional bit required to complete a frame is the Start Of Message (SOM) bit. Frames are repeated 75 times every 2 seconds and start aligned with the beginning of PN sequences, as in Figure 4.5.

Section 4.1.3 provides a detailed explanation of SyncCh functionalities in CDMA systems.

In Figure 4.39, 'common channels' include the following: broadcast control, paging, quick paging, common power control, common assignment and forward common control channels.

Figure 4.40 shows the SyncCh processing block diagram for SR1 and SR3. The X_Q branch input is set to zero (0) because, as explained for FPiChs, there is no transmission through this branch, i.e. this branch is gated-off.

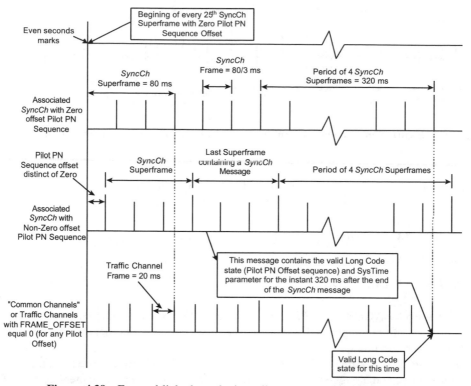

Figure 4.39   Forward link channels time alignment on cdma2000 systems.

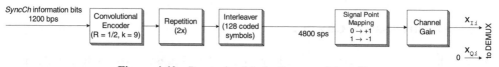

Figure 4.40   Processing block diagram of SyncChs.

Table 4.16 shows bit rates obtained during SyncCh processing.

4.2.2.1 Convolutional Coding and Coded Symbol Repetition

To maintain backward compatibility, the convolutional encoder (Figure 4.41) used in the SyncCh of cdma2000 systems is the same employed in IS-95 systems, which is fully described in Section 4.1.3.1.

Table 4.16   Bit rates during SyncCh frame processing

Spreading Rate (SR)	Bits/frame	Convolutional encoder	Repetition	Interleaver block	Coded symbol rate after interleaving
SR1 and SR3	32	$R = 1/2, k = 9$	1	128	4.8 Ksps

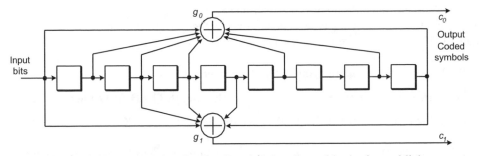

Figure 4.41 Convolutional coder ($R = 1/2, k = 9$) used in the forward link.

After convolutional coding ($R = 1/2$) and coded symbol repetition (1), the coded symbol rate is 4.8 kbps, allowing bits to be split into 26.66 ms frames with 128 bits each. Because encoder tail bits are not used, the convolutional coder is not reset between consecutive frames.

4.2.2.2 Interleaving

As in Figure 4.40, after the convolutional coder and coded symbol repetition stages, each frame goes through interleaving.

Table 4.17 Parameters m and J, used for interleaving, listed according to the number of bits per frame

Block of N bits/frame	m	J
48	4	3
192	6	3
768	6	12
3072	6	48
12 288	7	96
288	5	9
1152	6	18
4608	7	36
18 432	8	72
128	7	1
96	5	3
384	6	6
1536	6	24
6144	7	48
144	4	9
576	5	18
2304	6	36
9216	7	72
36 864	8	144

cdma2000 has three different types of frame interleaving for forward link channels. The selection of an interleaving type depends on the SR, on the bit rate, and on the logical channel. The following are the types available:

- Bit-Reversal Order Interleaving (SR1 BRO Interleaving),

- Forward-Backwards Bit-Reversal Order Interleaving (SR1 FB BRO Interleaving),

- Complex Cyclic-Shift Bit-Reversal Order Interleaving (SR3 CCS BRO Interleaving).

Only cdma2000 channels (such as forward common control, broadcast control, and common assignment channels) use forward-backwards and complex cyclic-shift BRO.

The interleaving process uses two parameters (m and J), presented in Table 4.17, which vary according to the logical channel and to the channel bit rate per frame.

The SyncCh only uses the BRO Interleaver, which is described next. Other possible interleaving methods are discussed later in this chapter.

4.2.2.2.1   BRO Interleaving

Figure 4.42 shows the interleaver block diagram for systems operating in SR1, whereas Figure 4.43 illustrates the BRO Interleaving interleaving process.

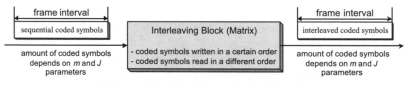

Figure 4.42   Interleaver block diagram.

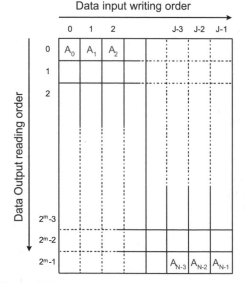

Figure 4.43   Bit reversal order interleaving for SR1.

According to the parameters (m and J) presented in Table 4.17, to process a frame of N coded bits, the interleaver creates a matrix with 2^m lines (0 to $2^m - 1$) and J columns (0 to $J - 1$), as in Figure 4.43.

Bits (denoted A_i) are written in the matrix according to (4.15)

$$A_i = 2^m(i \bmod J) + BRO_m(\lfloor i/J \rfloor) \tag{4.15}$$

where

$i = 0, 1, 2, 3, \ldots, N - 1$	index of each bit in the interleaving input
$BRO_m(x)$	or m Bit-Reversal Order – returns the decimal value of the m-length reversed binary sequence of x
$\lfloor x \rfloor$	truncates x to the lowest integer number (e.g. returns 3 for $x = 3.6$)

In eqn (4.15), the term 2^m ($i \bmod J$) represents the column where the input bit will be written, whereas the term $BRO_m(\lfloor i/J \rfloor)$ calculates the line in the matrix.

Coded bits are written (input) in the interleaving matrix by rows, as in Figure 4.43, the sequence of rows is given by eqn (4.15).

Table 4.18 shows an example of BRO interleaving for $m = 4$. Bits are read (output) by columns, starting from the first column on the left and reading consecutive columns to the right.

For the specific case of the SyncCh, which has 128 coded bits to be interleaved, from Table 4.17 it is possible to evaluate that the parameters will have values $m = 7$ and $J = 1$, i.e. only one column with 128 cells (rows).

Table 4.18 Example of the Bit Reversal Order (BRO) function

X	Binary sequence ($m = 4$)	Reversed binary sequence	Value of the reversed binary sequence ($BRO_m(x)$)
0	0000	0000	0
1	0001	1000	8
2	0010	0100	4
3	0011	1100	12
4	0100	0010	2
5	0101	1010	10
6	0110	0110	6
7	0111	1110	14
8	1000	0001	1
9	1001	1001	9
10	1010	0101	5
11	1011	1101	13
12	1100	0011	3
13	1101	1011	11
14	1110	0111	7
15	1111	1111	15

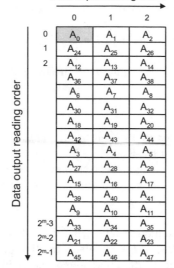

Figure 4.44 Example of bit positioning in the BRO Interleaver matrix for SR1.

For compatibility, the interleaving used in cdma2000 SyncChs is equivalent to the method described in Section 4.1.3.2, for CDMA IS-95 SyncChs.

Figure 4.44 shows an interleaving example with a matrix of order $N = 48$, $m = 4$ and $J = 3$.

4.2.2.3   De-multiplexing (Serial-to-Parallel Conversion)

After interleaving, SynchCh I and Q branches are de-multiplexed to be later spread and channelised with Walsh sequences and/or QOF functions.

In the SyncCh, the DEMUX only operates in the non-TD mode, as in Figure 4.45, consisting only in sending coded bits to Walsh spreading and short PN modulation processes.

In the DEMUX operation for SR3, the first bit at input X_I is transmitted to the Y_{I1} output. The following symbols are transmitted to outputs Y_{I2} and Y_{I3}. The serial-to-parallel process starts again in Y_{I1}. The same happens to symbols from the X_Q branch, transmitted to outputs Y_{Q1}, Y_{Q2} and Y_{Q3}, as in Figure 4.46.

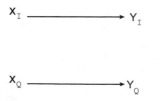

Figure 4.45 DEMUX operation for SR1 in non-TD mode.

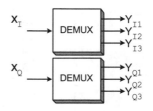

Figure 4.46 DEMUX operation for SR3 in non-OTD mode.

Even though the DEMUX block is present in SyncCh processing, when operating with SR3 there is no real de-multiplexing, because the SyncCh is only transmitted over one carrier.

4.2.2.4 Channelisation by Walsh codes and Complex Multiplier Spreading by PN sequences

Table 4.19 shows the Walsh code and spreading rate calculation in the SyncCh.

The number of Walsh repetitions per coded bit indicates that, prior to transmission, Walsh code W_{32}^{64} is applied four times for each coded symbol at the interleaving output, i.e. each coded symbol results in 256 chips (4×64). The effective baud rate, however, is still 4.8 kbps.

Table 4.19 Walsh sequence and spreading rate for the SyncCh

Spreading rate	Bit rate after interleaving (kbps)	Walsh code	Number of Walsh repetitions per symbol	Coded symbol rate after spreading (Mcps)
SR1 and SR3	4.8	W_{32}^{64}	4	1.2288

In a different approach, it can also be understood that Walsh code W_{32}^{64} consists of a 32-chip set of 0's and 32-chip set of 1's. Mapping the 0's into '+1' and 1's into '−1', it can also be seen a square waveform of frequency 1.2288 Mcps/64 chips = 9.6 KHz. Thus, the Walsh code perform a change in coded symbol polarisations in a rate twice the waveform frequency, i.e. 1.2288 Mcps/32 chips (1's or 0's) = 19.2 KHz. Due to this, it is also possible to understand that, due to the pseudo-random noise-like behaviour of the PN sequences, they are responsible for the 'information spreading' of the channel data, although the number of chips after the Walsh code application is already 1.2288 Mcps.

Figure 4.47 shows processing steps after the SynchCh data is converted from serial to parallel. These steps consist of channelising and spreading the information using 64-bit long

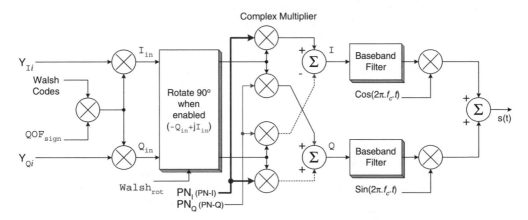

Figure 4.47 Channelising and spreading with Walsh codes and/or QOFs, complex multiplier spreading by PN sequences and transmission processing for SR1 and SR3.

Walsh code 32 (W_{32}^{64}), complex spreading of the data using PN-I and PN-Q sequences, and finally QPSK modulating the carrier prior to transmission.

For the SyncCh in SR3 mode, as in Figure 4.47, indexes $i = 1$, 2 and 3 represent each of the 1.25 MHz carriers used for signal transmission.

4.2.3 Forward Paging Channels

Forward Paging Channels (FPChs) only use SR1 (Figure 4.48) so that there is backward compatibility with IS-95 systems. For the same reason, the channel structure is very similar to IS-95 systems (Section 4.1.4). FPChs use Walsh codes 1–7 (W_{1-7}^{64}) and are used to transmit system information and messages to mobile stations, such as call origination messages, responses to MS requests, and registration orders.

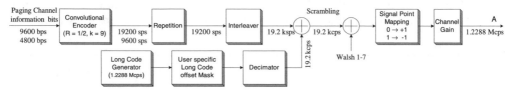

Figure 4.48 Forward paging channel processing diagram block for SR1 operation.

The primary FPCh (W_1^{64}) transmits system data, paging messages and access response messages. Other FPChs transmit information related to paging and access response messages.

Table 4.20 shows bit rates during FPCh channel processing.

Because encoder tail bits are not used, the convolutional coder is not reset between consecutive frames, that is, bits from the previous frame influence bits from the frame being processed.

Figure 4.9 shows that FPCh message capsules can have from 40 to 1182 information bits. The smallest data transmission structure of FPChs is the half frame, which is 10 ms long.

A maximum slot cycle of 2048 FPCh slots (163.84 s long) allows paging and control over MSs operating in two distinct modes, slotted and non-slotted, as in Figure 4.10. Terminals are configured to 'listen' to FPChs according to the parameters *PGSLOT* and *SLOT_ CYCLE_INDEX*.

Table 4.20 Bit rates in paging channel processing

Bit rate (kbps)	Bits/frame (20 ms)	Convolutional encoder	Repetition	Interleaver block (bits)	Coded symbol rate after interleaving (ksps)
4.8	96	$R = 1/2, k = 9$	1	384	19.2
9.6	192	$R = 1/2, k = 9$	–	384	19.2

4.2.3.1   Convolutional Coding and Coded Bit Repetition Processes

For backward compatibility, the convolutional encoder used on cdma2000 FPChs is the same as the one adopted by IS-95 systems, illustrated by Figure 4.7 and described in Section 4.1.3.1.

Bit repetition is required, after convolutional coding, if the transmission rate is set to $P_{RAT} = 4800$ bps. This repetition provides only one rate for all channels (384 coded bits per 20 ms frame) at the interleaving stage, i.e. 19.2 kbps.

4.2.3.2   Interleaving

The interleaving process starts aligned to the beginning of FPCh frames. Figure 4.49 presents the interleaving block diagram for FPChs, which consists of a BRO interleaving process, as for the SyncCh (Figure 4.43). Interleaving parameters for FPChs are $m = 6$ and $J = 6$ to obtain a matrix with 384 cells according to Table 4.17.

Figure 4.49   Interleaving process for FPChs.

BRO interleaving, with parameters $m = 6$ and $J = 6$, works as the process described in Section 4.1.4.2 for FPChs on CDMA IS-95 systems.

4.2.3.3   Data Scrambling

After interleaving, FPCh bits go through a scrambling process. There are two distinct data scrambling processes for cdma2000 forward link channels. The difference between them is the number of long code chips used for scrambling. The first process is the same used in CDMA IS-95 systems, where scrambling uses only one long code chip, whereas the second process, developed for cdma2000 logical channels, uses two distinct chips per every 64 long code chips.

For compatibility, FPChs employ the first scrambling method. This process is described in Section 0. This process consists of a modulo-2 addition of the interleaving output coded bit and a valid chip from the decimator block output (taken from the long code PN sequence after applying the offset mask).

The chip used for data scrambling starts aligned to the FPCh coded bit, as in Figure 4.50. For SR1, the decimator only uses the first chip for each group of 64 chips, which corresponds to the transmission period of 1 coded bit (at 19200 bps).

Logical channels use the long code generator with rate 1.2288 Mcps when operating with SR1 and 3.6864 Mcps for SR3. Forward paging channels use the following long code offset mask for scrambling.

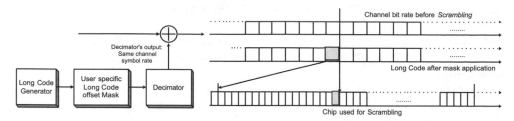

Figure 4.50   Data scrambling for FPChs.

41 29	28 24	23 21	20 9	8 0
1100011001101	00000	PCN	000000000000	Pilot_PN

Figure 4.51   Long code mask for forward paging channels.

In Figure 4.51, PCN represents the FPCh being used and *Pilot_PN* is the pilot channel index offset.

Only logical channels created specifically for cdma2000 systems (such as the broadcast control or common control channels) use the second type of scrambling, described in Section 4.2.4.2.

4.2.3.4   Symbol De-multiplexing (Serial-to-Parallel Conversion)

After scrambling, bits are mapped, de-multiplexed and prepared for transmission through the air interface. Because of compatibility requirements, PChs can only use SR1 and operate in non-TD mode (Figure 4.52), as the SyncCh. The DEMUX process is the same shown in Figure 4.45.

$X_I \longrightarrow Y_I$

$X_Q \longrightarrow Y_Q$

Figure 4.52   DEMUX operation for SR1 in non-TD mode.

4.2.3.5   Channelisation and Spreading by Walsh Sequences and Complex Multiplier Spreading by PN Sequences

Table 4.21 shows Walsh codes and rates on the spreading process for FPChs.

The channelling and spreading process using Walsh codes and PN sequences, and the preparation for the transmission channel are the same used for the SyncCh, described on Section 4.1.3.3.

Table 4.21 Walsh code used in paging channels spreading and channelisation

Coded symbol rate after interleaving (ksps)	Walsh code	Sequence index	Total chip rate after spreading (Mcps)
19.2	W_i^{64}	1–7	1.2288

4.2.4 Forward Common Control Channel

The Forward Common Control Channel (FCCCh) is designed for cdma2000 systems and its main function is to perform the same tasks of FPChs, but operating at both spreading rates, SR1 and SR3.

Each RF carrier can have up to seven FCCChs, which are used by BTSs to send control messages to MSs within their coverage area. FCCChs are organised in 5, 10 and 20 ms frames and use bit rates that vary according to the frame length. Frames are organised in 80 ms slots, as in FPChs.

The frame structure of FCCChs is always the same, regardless of the bit rate and frame length, as in Figure 4.53 (F represents frame quality indicator bits (CRC) and T, the encoder tail bits (8 bits '0').

Figure 4.53 Frame structure of a forward common control channel.

The operation of FCCChs and FPChs is very similar, because their slots are organised in a maximum slot reading cycle with a total of 2048 slots, corresponding to 163.84 s. Figure 4.54 shows the structure of FCCChs.

Note that, in addition to every frame having its own CRC and encoder tail bits (T), the FCCCh message, as a whole, also has distinct CRC, as in Figure 4.54, and may also have additional padding to complete the message structure.

The parameters in Figure 4.54 are described next:

- *FCCCH_FRAME_SIZE*: information bits in a FCCCh frame;

- LAC PDU fragment: fragment (part or body) of a protocol message (Protocol Data Unit) sent by the Layer Access Control (LAC) layer;

- SCI: synchronised capsules indicator;

- *B*: bits in each message fragment, where
  $B = 92$ for *FCCCH_FRAME_SIZE* = 172;
  $B = 94$ for *FCCCH_FRAME_SIZE* = 360;
  $B = 95$ for *FCCCH_FRAME_SIZE* = 744;

- T: message capsule tail bits, which are distinct from the encoder tail bits;

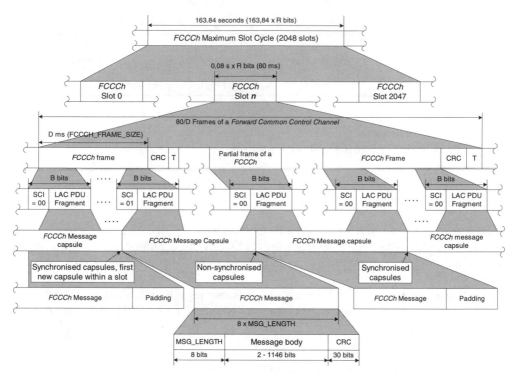

Figure 4.54 Structure of the FCCChs (frames and slots).

- CRC: message quality indicator (Cyclic Redundancy Check), 30 bits;
- R: FCCCh transmission rate (9600, 19200 or 38400 bps);
- D: frame duration (5, 10 or 20 ms).

FCCChs use two bits per frame as Synchronised Capsules Indicators (SCI). SCI set as '00' means non-synchronised capsules, whereas SCI set as '01' represents synchronised capsules. Figure 4.55 shows a block diagram describing FCCChs frame processing.

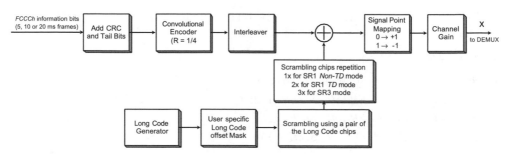

Figure 4.55 Forward common control channel frame processing block diagram.

Table 4.22 Bit rate during FCCCh frame processing for SR1 and convolutional coder $R = 1/2$

Frame (ms)	Information bits per frame	CRC	Tail bits	Bits/ frame	Bits (kbps)	Convolutional coder	Interleaving block (bits)	Coded symbol rate after interleaving (ksps)
5	172	12	8	192	38.4	$R = 1/2, k = 9$	384	76.8
10	172	12	8	192	19.2	$R = 1/2, k = 9$	384	38.4
10	360	16	8	384	38.4	$R = 1/2, k = 9$	768	76.8
20	172	12	8	192	9.6	$R = 1/2, k = 9$	384	19.2
20	360	16	8	384	19.2	$R = 1/2, k = 9$	768	38.4
20	744	16	8	768	38.4	$R = 1/2, k = 9$	1536	76.8

FCCChs use long code generators with rate 1.2288 Mcps for SR1 and 3.6864 Mcps for SR3. Tables 4.22, 4.23 and 4.24 show bit rates achieved during processing for SR1 ($R = 1/2$ and $R = 1/4$) and SR3 (for $R = 1/3$).

The convolutional coder ($R = 1/2$) used to obtain the rates in Table 4.22 is the same presented in Figure 4.7.

Table 4.23 Bit rate during FCCCh frame processing for SR1 and convolutional coder ($R = 1/4$)

Frame (ms)	Information bits per frame	CRC	Tail bits	Bits/ frame	Bits (kbps)	Convolutional coder	Interleaving block (bits)	Coded symbol rate after interleaving (ksps)
5	172	12	8	192	38.4	$R = 1/4, k = 9$	768	153.6
10	172	12	8	192	19.2	$R = 1/4, k = 9$	768	76.8
10	360	16	8	384	38.4	$R = 1/4, k = 9$	1536	153.6
20	172	12	8	192	9.6	$R = 1/4, k = 9$	768	38.4
20	360	16	8	384	19.2	$R = 1/4, k = 9$	1536	76.8
20	744	16	8	768	38.4	$R = 1/4, k = 9$	3072	153.6

Table 4.24 Bit rate during FCCCh frame processing for SR3 and convolutional coder ($R = 1/3$)

Frame (ms)	Information bits per frame	CRC	Tail bits	Bits/ frame	Bits (kbps)	Convolutional coder	Interleaving block (bits)	Coded symbol rate after interleaving (ksps)
5	172	12	8	192	38.4	$R = 1/3, k = 9$	576	115.2
10	172	12	8	192	19.2	$R = 1/3, k = 9$	576	57.6
10	360	16	8	384	38.4	$R = 1/3, k = 9$	1152	115.2
20	172	12	8	192	9.6	$R = 1/3, k = 9$	576	28.8
20	360	16	8	384	19.2	$R = 1/3, k = 9$	1152	57.6
20	744	16	8	768	38.4	$R = 1/3, k = 9$	2304	115.2

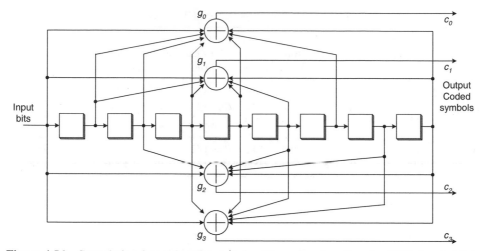

Figure 4.56 Convolutional encoder ($R = 1/4, k = 9$) employed on the cdma2000 forward link.

Figure 4.56 shows another type of convolutional encoder for SR1, with $R = 1/4$ and $k = 9$, and Table 4.23 shows coded symbol rates obtained with this encoder configuration.

Figure 4.57 shows the convolutional encoder for FCCCh operation in SR3, with $R = 1/3$. Table 4.24 shows coded symbol rates obtained with this encoder.

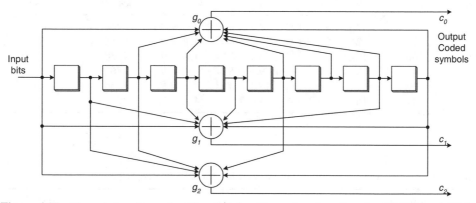

Figure 4.57 Convolutional encoder ($R = 1/3, k = 9$) employed on the cdma2000 forward link.

4.2.4.1 Interleaving

cdma2000 FCCChs do not use the interleaving method employed in CDMA IS-95 systems (BRO interleaving, described in Section 4.2.2.2). When working with SR1, FCCChs use Forward-Backwards Bit Reversal Order Interleaving (FB BRO interleaving, explained next)

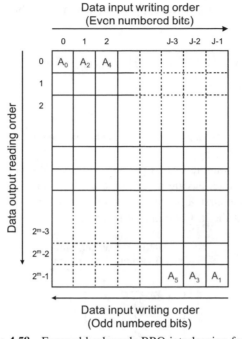

Figure 4.58 Forward-backwards BRO interleaving for SR1.

in Figure 4.58. For SR3, the method used is the Complex Cyclic-Shift BRO Interleaving (CCS BRO Interleaving), in Figure 4.60.

4.2.4.1.1 Forward-Backwards BRO Interleaving

Figure 4.58 illustrates the forward-backwards bit-reversal order interleaving process, used in channels operating with SR1. The process is performed in two steps: bit writing (data input in the interleaver) and data reading (data output).

Table 4.18 shows parameters m and J according to processing block sizes, defining a matrix with 2^m lines (0 to $2^m - 1$), and J columns (0 to $J - 1$).

Bits with even index are written in the interleaving matrix by rows, from left to right, according to the following equation

$$A_i = 2^m \left[\left(\frac{i}{2} \right) \bmod J \right] + \mathrm{BRO}_m \left[\left\lfloor \left(\frac{i}{2} \right) \middle/ J \right\rfloor \right] \tag{4.16}$$

where

$i = 0, 1, 2, 3,\ldots, N - 2$	index of each bit in the interleaving input
$BRO_m(x)$	or m Bit-Reversal Order – returns the decimal value of the m-length reversed binary sequence of x
$\lfloor x \rfloor$	truncates x to the lowest integer number (e.g. returns 3 for $x = 3.6$)

Bits with odd index are also written by rows, but from right to left, according to the following equation

$$A_i = 2^m \left[\left(N - \frac{(i+1)}{2} \right) \bmod J \right] + \mathrm{BRO}_m \left[\left[\left(N - \frac{(i+1)}{2} \right) \Big/ J \right] \right] \qquad (4.17)$$

where

$i = 0, 1, 2, 3, \ldots, N - 1$ index of each bit in the interleaving input

Figure 4.59 shows an example of this process, with $N = 48$, $m = 4$ and $J = 3$.

Bits are read from the interleaving matrix by columns, as in the example in Figure 4.59. Bits reading starts on the first left column, going from top to bottom, and successively reading all columns until the last column on the right.

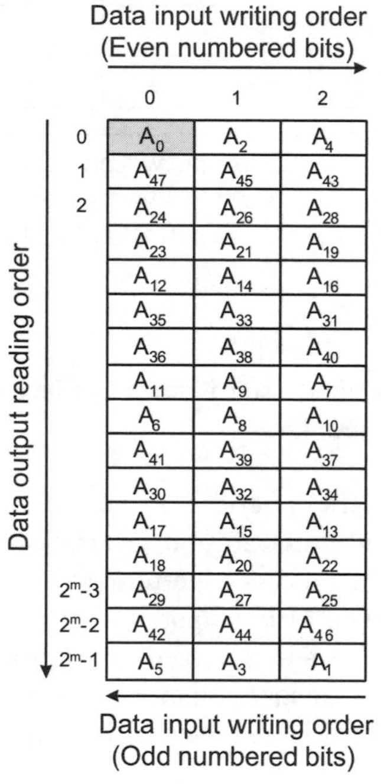

Figure 4.59 Example of bit positioning in a forward-backwards BRO interleaver matrix.

4.2.4.1.2 Complex Cyclic-Shift BRO Interleaving

Another interleaving method is the complex cyclic-shift BRO interleaving, used for system operating with SR3. Figure 4.60 presents the general block diagram for this type of interleaving.

The CCS BRO interleaving first de-multiplexes coded symbols into smaller interleaving matrices, each with $n = N/3$ bits. They are sequentially written in the DEMUX output, from top to bottom, in each of the k blocks, with k varying from 0 to 2, as in Figure 4.60.

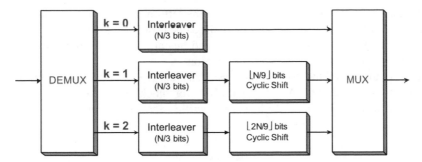

Figure 4.60 Complex cyclic-shift BRO interleaving for SR3.

Parameters m and J for interleaving in SR3, in Table 4.18, are set according to the number of bits in each block with $N/3$ bits. This also means that even with a total of N bits to be interleaved, parameters for each k block are defined according to an $n = N/3$ matrix.

Coded bits at the input of each k block (length $n = N/3$) are stored in n-bit matrices created according to parameters m and J, that is, 2^m lines (0 to $2^m - 1$) and J columns (0 to $J - 1$). Bits are stored in the matrices according to eqn (4.18)

$$A_i = 2^m \left[\left(i + \left\lfloor \frac{kN}{9} \right\rfloor \right) \bmod J \right] + \mathrm{BRO}_m \left\{ \left\lfloor \left[\left(i + \left\lfloor \frac{kN}{9} \right\rfloor \right) \bmod \frac{N}{3} \right] \middle/ J \right\rfloor \right\} \qquad (4.18)$$

where

$i = 0, 1, 2, 3, \ldots, N/3 - 1$ index of each bit in the input of each k block (length $N/3$)

Figure 4.61 describes the coded bits writing sequence in each of the $N/3$-bit long interleaving matrices, without considering the cyclic-shift process. In the figure, the first

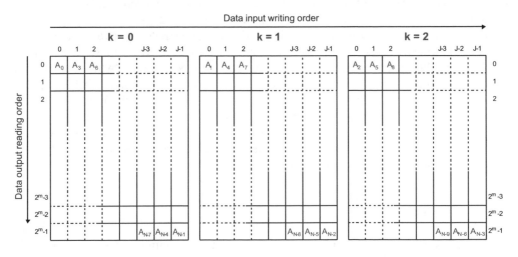

Figure 4.61 Writing sequence of an N-bit frame in the three blocks (length $N/3$), prior to the cyclic-shift process.

coded bit (A_0) is written in the first column on the left on the uppermost position, in block $k = 0$. The second coded bit (A_1) is written in the same position, but in block $k = 1$. The third bit (A_2) is written in block $k = 2$. The next coded bits are written by rows, from the left to right, on the column defined by the BRO function.

Interleaver outputs in blocks $k = 1$ and $k = 2$ also undergo another type of bit processing, the cyclic shift, with orders $\lfloor N/9 \rfloor$ and $\lfloor 2N/9 \rfloor$ for blocks $k = 1$ and $k = 2$, respectively. Figure 4.62 illustrates the final result, i.e. the writing sequence in the matrices and cyclic shift within blocks $k = 1$ and $k = 2$.

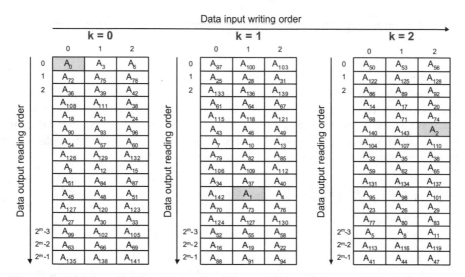

Figure 4.62 Bit positioning after cyclic shift, considering bit input order in the DEMUX.

Equation (4.18) describes the operation of a BRO interleaver considering the bit shift within the matrix. The interleaver for SR3 can also be understood as three BRO interleavers in parallel followed by a cyclic shift on blocks $k = 1$ and $k = 2$.

Figure 4.62 shows the final bit positioning during frame interleaving, in an example with $N = 144$, showing the indexes related to the bit input order in the DEMUX, in a sequential order.

Bits are read from the matrices by columns, one bit per block, starting by the first bit from the leftmost column and, then, repeating the process up to the last column. That is, the first bit from the leftmost column of block $k = 0$ is read; then the first bit from the leftmost column of block $k = 1$ and then the first bit from the same position of block $k = 2$. The process then is repeated for the second bit of the leftmost column of each block, until all coded bits are read from all three blocks. After the reading process is completed, bits are multiplexed, i.e. converted from parallel to serial.

4.2.4.2   Scrambling

After being convolutionally encoded and interleaved, bits are scrambled using two long code chips, as in Figure 4.63. Scrambling is performed in groups of $2M$ coded bits, where $M = 1$ for non-TD in SR1, $M = 2$ for SR1 in OTD or STS modes and $M = 3$ for SR3.

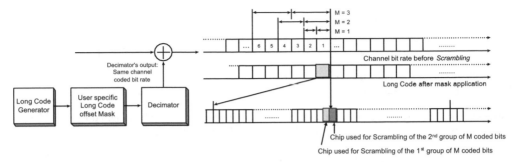

Figure 4.63 Scrambling process considering two chips.

This process consists of a modulo-2 addition of each $2M$ bits set at the interleaving output with a valid pair of chips from the decimator block output (taken from the long code PN sequence after applying the offset mask). The modulo-2 addition has the same rate as the interleaving block output.

For the first group of M-coded bits, the chip used for scrambling starts at the same time as the first coded bit is read from the interleaving matrix, as in Figure 4.63. For the second group, the chip used is the last valid chip of the long code sequence before the $2M$ bits begin. For example, considering $M = 3$, the first decimated chip is used for scrambling bits 1, 2 and 3. The second chip scrambles bits 4, 5 and 6.

Figure 4.63 shows how long code chips are selected for scrambling forward common control channel bits.

Logical channels use the long code generator with rate 1.2288 Mcps when operating with SR1 and 3.6864 Mcps for SR3. Figure 4.64 shows the forward common control channels long code offset mask.

41	....	29 28	....	24 23	....	21 20	....	9 8	....	0
1100011001101		01000		000		000000000000		000000000		

Figure 4.64 Long code offset mask for forward common control channels.

Figure 4.55 depicts an additional process, called scrambling bit repetition. This process consists of repeating the decimated chip, which is then used to scramble the programmed number of coded bits. For example if $M = 3$, each decimated chip must be repeated three times to scramble three coded bits.

4.2.4.3 De-multiplexing (Serial-to Parallel Conversion)

FCCChs may operate in three distinct modes: non-TD (Figure 4.65), STS and OTD (Figure 4.66) for SR1 or SR3 (Figure 4.67). Figure 4.55 also shows that the symbols are sent to a de-multiplexing process prior to transmission.

The de-multiplexing process aims to match the number of coded bits to the number required for transmission in each branch. Section explains this process further when describing Walsh coded lengths used for spreading.

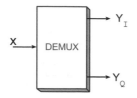

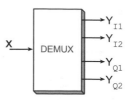

Figure 4.65 DEMUX operation for SR1 in non-TD mode.

Figure 4.66 DEMUX operation for SR1 in OTD or STS mode.

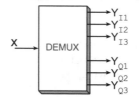

Figure 4.67 DEMUX operation for SR3.

4.2.4.4 Channelisation and Spreading by Walsh Sequences and Complex Multiplier Spreading by PN Sequences

Walsh sequences and QOFs are used to channelise and spread the coded symbols, which are then complex multiplied by PN sequences through a modulo-2 addition to be transmitted through the air interface.

FCCChs use Walsh codes whose length varies according to the convolutional encoder and transmission data rate, as in Table 4.25.

Table 4.25 presents the chip rates after spreading without considering the de-multiplexing process. These values explain the need of de-multiplexing to match these rates to the maximum supported for transmission by the I and Q branches (1.2288 Mcps).

To exemplify this need, consider a system operating in OTD mode with SR1. The information rate is 38 400 bps, coded by a convolutional encoder with $R = 1/4$, as in

Table 4.25 Walsh code length and chip rates used by forward common control channels

SR	Data rate (bps)	Convolutional encoder (R)	Walsh code	Sequence index	Chip rate after spreading (Mcps)
SR1	38 400		W_i^{32}	$1 \leq i \leq 31$	2.4576
	19 200	1/2	W_i^{64}	$1 \leq i \leq 63$	2.4576
	9600		W_i^{128}	$1 \leq i \leq 127$	2.4576
SR1	38 400	1/4	W_i^{16}	$1 \leq i \leq 15$	2.4576
	19 200		W_i^{32}	$1 \leq i \leq 31$	2.4576
	9600		W_i^{64}	$1 \leq i \leq 63$	2.4576
	38 400		W_i^{64}	$1 \leq i \leq 63$	7.3728
SR3	19 200	1/3	W_i^{128}	$1 \leq i \leq 127$	7.3728
	9600		W_i^{256}	$1 \leq i \leq 255$	7.3728

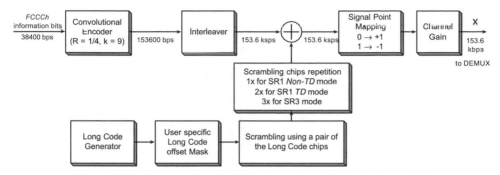

Figure 4.68 Number of coded bits prior to de-multiplexing (example).

Figure 4.68. Prior to de-multiplexing, there are 153.6 kbps to be spread and channelised using a 16-bit long Walsh code. After channelisation, there are 2.4576 Mcps to be transmitted, a condition not supported by the maximum of 1.2288 Mcps. The de-multiplexing process makes the number of coded bits compatible to transmission requirements of I and Q branches.

If the DEMUX in Figure 4.66 is connected to the output of Figure 4.68 (X input), the total amount of coded bits (153.6 kbps) is split into four distinct outputs, Y_{I1}, Y_{I2}, Y_{Q1} and Y_{Q2}, each with 38.4 kbps.

Depending on its configuration, each BTS may transmit in OTD or STS mode, as in Figures 4.69 and 4.70, respectively.

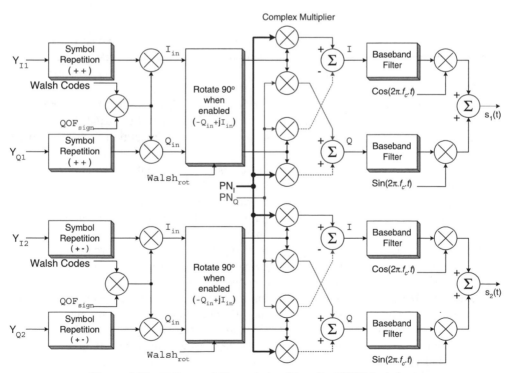

Figure 4.69 Orthogonal Transmission Diversity (OTD) technique.

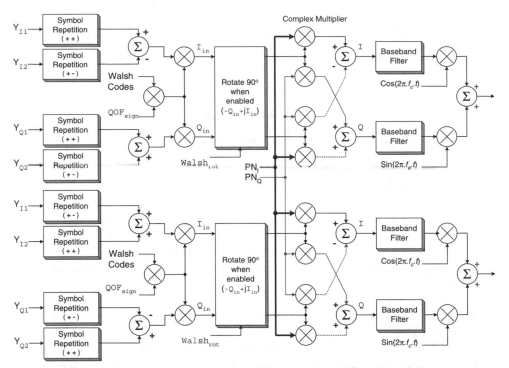

Figure 4.70 Space Time Spreading (STS) transmission diversity technique.

In the OTD transmission method, the forward link channel symbols are distributed among multiple antennas and are spread and channelised using Walsh codes or QOFs associated with each antenna, whereas the STS technique channelises them using complementary Walsh codes or QOF.

Figure 4.69 represents the OTD mode. Symbols at Y_{I1}, Y_{I2}, Y_{Q1} and Y_{Q2} are repeated, generating a pair of symbols for each symbol at the DEMUX input. In Y_{I2} and Y_{Q2} branches, the second symbol always has inverse polarity. The symbol repetition aims to increase system diversity.

Furthermore, the total rate of modulation symbols at each output ($s_1(t)$ and $s_2(t)$) will be always equal to 1.2288 Msps.

In the procedure shown in Figure 4.70 which represents operation in STS mode, notice that each of the symbols at Y_{I1}, Y_{I2}, Y_{Q1} and Y_{Q2} points are repeated, generating a pair of symbols for each symbol at the DEMUX input.

In both cases, each I and Q branches receive distinct parts of the information data. The final set of data employed for the carrier modulation will lead to a QPSK modulation with its maximum spectral efficiency, i.e. 2 bits per modulation symbol.

Figure 4.71 shows a block diagram representing the spreading process, complex multiplier and carrier modulation for SR1 (non-TD) and SR3 operation modes. In the figure, indices $i = 1$, 2 and 3 represent each of the 1.25 MHz carriers used for signal transmission.

The length of Walsh codes is selected to obtain an adequate chip/symbol rate after spreading and channelisation, regardless of the spreading rate, convolutional encoder and information bit rate.

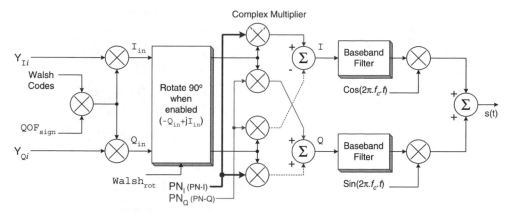

Figure 4.71 Walsh code and/or QOF spreading and channelling, PN sequences complex multiplier and transmission processing for SR1 (non-TD) and SR3 modes.

4.2.5   Forward Broadcast Control Channel (BCCh or FBCCh)

Broadcast control channels transmit broadcast messages, such as SMS, to all users compatible with cdma2000 configurations within each BTS coverage area, that is, all users compatible with Radio Configurations 3–9 (RC3–RC9) on the forward link.

BCChs transmit data at rates 19200, 9600 or 4800 bps. They are organised in slots with lengths of 40, 80 or 160 ms. Slots are divided into 40 ms frames. Each frame consists of 768 bits, of which 744 are information bits, 16 are CRC (F) and 8 are encoder tail bits (T), as in Figures 4.72 and 4.73.

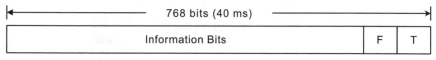

Figure 4.72   Broadcast control channel frame structure.

Although all BCCh frames have 768 bits, FBCChs messages may be larger, and the system will use as many frames as necessary to transmit the entire message.

For control purposes, frames are sub-divided into eight smaller structures called sub-slots, each with 93 bits, as in Figure 4.73, and with their first bit representing the Synchronised Capsule Indicator (SCI). SCI = '1' indicates that a new message starts in the current sub-slot, whereas SCI = '0' indicates that the current sub-slot is the continuation of a message initiated in a previous sub-slot.

If the number of information bits of a message is greater than the number of consecutive sub-slots (92 $N_{\text{sub-slots}}$), padding bits must be added to complete the last sub-slot structure.

The CDMA BTSs must support discontinuous transmission of this channel, and this decision must be made considering the slot period.

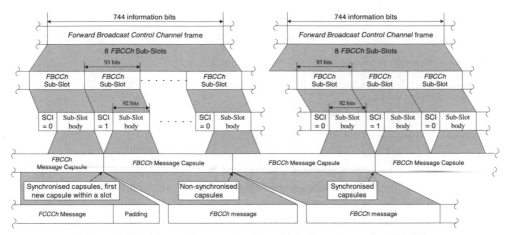

Figure 4.73   Slots and messages chronological structure of an FBCCh.

Figure 4.74 shows a diagram block for the FBCCh frame processing in cdma2000 systems (SR1 and SR3), but only for RCs 3–9, whereas Table 4.26 shows bit rates per frame during this processing.

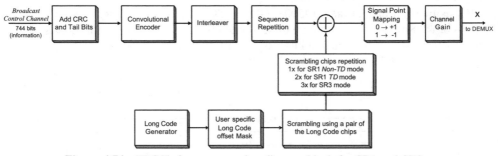

Figure 4.74   FBCCh frame processing diagram block for SR1 and SR3.

As explained previously, slots are divided in 40 ms frames; for slots more than one frame long, however, additional sequential frames are a copy of the first one. That is, for a 160 ms slot, the second, third and fourth frames are a copy of the first frame.

A convolutional encoder (Figure 4.41 for $R = 1/2$, Figure 4.56 for $R = 1/4$ and Figure 4.57 for $R = 1/3$) processes each FBCCh frame according to the Spreading Rate (SR) and frame length. Afterwards the block interleaver also processes the frames.

4.2.5.1   Interleaving

For SR1, the forward-backwards BRO interleaving is used, as in Figure 4.58. For SR3, the interleaving process is the Complex Cyclic-Shift BRO (CCS BRO) interleaving in

Table 4.26 Bit rates during frame processing of the broadcast control channels

SR	Slots (ms)	Bits/ frame	Bits (kbps)	Convolutional coder	Interleaver Block	Sequence repetition	Coded symbol rate after interleaving (ksps)
SR1	40	768	19.2	$R = 1/2, k = 9$	1536	1	38.4
	80	768	9.6	$R = 1/2, k = 9$	1536	2	38.4
	160	768	4.8	$R = 1/2, k = 9$	1536	4	38.4
SR1	40	768	19.2	$R = 1/4, k = 9$	3072	1	76.8
	80	768	9.6	$R = 1/4, k = 9$	3072	2	76.8
	160	768	4.8	$R = 1/4, k = 9$	3072	4	76.8
SR3	40	768	19.2	$R = 1/3, k = 9$	2304	1	57.6
	80	768	9.6	$R = 1/3, k = 9$	2304	2	57.6
	160	768	4.8	$R = 1/3, k = 9$	2304	4	57.6

Figure 4.60. Both processes use matrices defined by the parameters m and J, given in Table 4.17.

The long code generator rate depends on the spreading rate of the system. For SR1, the rate is 1.2288 Mcps and for SR3, it is 3.6864 Mcps. Figure 4.75 shows the offset mask used in this channel. In this figure, BCN represents the number of broadcast channels in use (from 1 to 8) and Pilot_PN is the FPiCh offset transmitted by the BTS.

41	....	29	28	....	24	23	....	21	20	....	9	8	....	0
1100011001101			00100			BCN			000000000000			Pilot_PN		

Figure 4.75 FBCCh long code offset mask.

4.2.5.2 Scrambling

After interleaving, coded bits are repeated according to the channel bit rate, and then scrambled as in Figure 4.74. Two long code chips scramble sets of $2M$-coded bits, as in Figure 4.63. This process is described in Section 4.2.4.2.

After scrambling, symbols are mapped, de-multiplexed and prepared for transmission through the air interface.

4.2.5.3 De-multiplexing (Serial-to-Parallel Conversion)

For systems operating with SR1 in non-TD mode (Figure 4.76), the first coded bit of each frame is always put in the Y_I branch. The second bit goes through Y_Q, the third goes through Y_I and so on.

For SR1 operation with Transmission Diversity (TD), OTD or STS (Figure 4.77), the first coded bit at the DEMUX input is sent to output Y_{I1}. The following bits are sent trough outputs Y_{I2}, Y_{Q1} and Y_{Q2}. The serial-to-parallel process starts again at Y_{I1}.

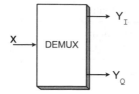

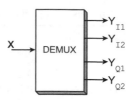

Figure 4.76 DEMUX operation for SR1 non-TD mode.

Figure 4.77 DEMUX operation for SR1 TD mode.

The DEMUX process for operation in SR3 (Figure 4.78) consists in sending the first bit through output Y_{I1}. The following coded bits are sequentially sent through Y_{I2}, Y_{I3}, Y_{Q1}, Y_{Q2}, Y_{Q3} and Y_{I1}. The process starts again at Y_{I1}.

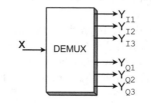

Figure 4.78 DEMUX operation for SR3.

4.2.5.4 Channelisation and Spreading by Walsh Sequences and Complex Multiplier Spreading by PN Sequences

Multiplexing (combining logical channels into one waveform) of the information bits uses Walsh sequences with length variable according to the convolutional coder and spreading rates, as in Table 4.27.

Because FBCChs are restricted to cdma2000 operation modes only (RC 3–9), TD schemes may be employed, allowing four transmission operation modes: SR1 non-TD, SR3 through air interface (Figure 4.37), SR1 OTD (Figure 4.69) and STS (Figure 4.70).

Table 4.27 Walsh code length used for spreading and channelising FBCChs

SR	Convolutional coder	Coded bit rates after interleaving (kbps)	Walsh codes	Sequence index	Total symbol rate after spreading (Msps)
SR1	$R = 1/2, k = 9$	38.4	W_i^{64}	$1 \leq i \leq 63$	2.4576
SR1	$R = 1/4, k = 9$	76.8	W_i^{32}	$1 \leq i \leq 31$	2.4576
SR3	$R = 1/3, k = 9$	57.6	W_i^{128}	$1 \leq i \leq 127$	7.3728

4.2.6 Quick Paging Channel

Quick Paging Channels (QPChs) are designed for cdma2000 systems and have the main purpose of reducing MS system monitoring time while in *mobile station idle state* (explained in detail in the chapter on call processing), consequently increasing battery life.

BTSs use QPChs to inform mobiles of messages in the next slot of a Forward Common Control Channel (FCCCh) or a Forward Paging Channel (FPCh). Each cdma2000 carrier may have from one to three QPChs, the first channel is defined as the primary QPCh.

FPCh and FCCCh messages are sent to MSs, in idle state, operating in slotted mode, within the coverage area. Because of the quick paging channel, in spite of monitoring an entire FPCh or FCCCh 80 ms slot, MSs only need to monitor their associated indicator intervals within QPChs, which are much smaller than a complete 80 ms frame.

QPCh slots are also structured in 80 ms slots and transmit three types of indicators (one bit/indicator): paging indicators, broadcast indicators and configuration change indicators. Because of indicator characteristics, quick paging channels are non-coded, even though they are spread using Walsh codes, and are On-Off Keying (OOK) modulated at the signal point mapping processing block.

Figure 4.79 presents the QPChs processing diagram block for SR1 and SR3.

Figure 4.79 Processing diagram block of quick paging channels for SR1 and SR3.

Paging Indicators must indicate that the BTS is about to transmit a paging message directed to an MS on the associated FPCh or FCCCh. Broadcast indicators inform all MSs that a broadcast message is going to start at the next FPCh or FCCCh slot. Configuration change indicators inform all MSs that the system configuration parameters have changed and the changes are going to be transmitted on the next FPCh or FCCCh messages, starting at the next slot. The last two indicator types, broadcast and configuration change indicators, are only transmitted on the primary QPCh.

The 80 ms slots are organised in 20 ms intervals, numbered 1–4, as in Figure 4.80. The slots must be aligned in a way that the first interval starts 20 ms before FPCh and/or FCCCh slots start. Because of this requirement, the structure of QPChs slots also has a maximum slot cycle of 2048 slots (representing a maximum of 163.84 s), which corresponds to the same configuration of FPChs and FCCChs.

The read cycle assignment for QPChs is similar to FPChs and FCCChs when MSs are operating in slotted mode. That is, the *SLOT_CYCLE_INDEX* defines the QPCh reading time according to the following expressions

$$T = 2^{SLOT\_CYCLE\_INDEX} \times 16 \text{ slots} \qquad \text{for } 0 \leq SLOT\_CYCLE\_INDEX \leq 7$$

$$T = 2^{SLOT\_CYCLE\_INDEX} \times 16 \times 80 \text{ ms} \qquad \text{for } 0 \leq SLOT\_CYCLE\_INDEX \leq 7$$

The data bit rate of a QPCh can be 4800 bps (384 bits per 80 ms slot) or 2400 bps (192 bits per 80 ms slot). Each information bit is used to generate an indicator. Because there are two

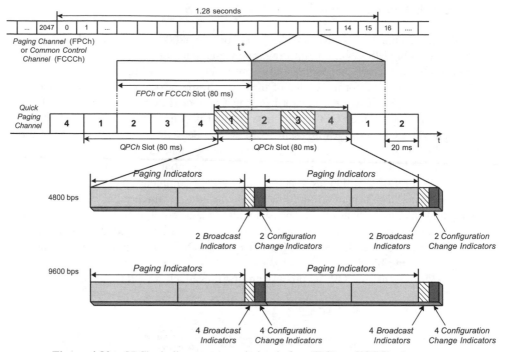

Figure 4.80 QPChs indicators transmission before FPCh or FCCCh slots start.

indicators per MS, and the second indicator is a copy of the first, the number of indicators transmitted is twice the number of data bits (blocks 3 and 4 in Figure 4.80 are copies of blocks 1 and 2). That is, if the information data rate is 4800 bps or 2400 bps, corresponding indicator transmission rates are 9600 bps and 4800 bps, respectively. Table 4.28 shows the available QPCh data and indicator rates.

Each BTS enables two paging indicators for each MS that is about to receive messages through an FCCCh or FPCh, which begins 20 ms after the end of the current QPCh slot. BTSs transmit each paging indicator in a distinct QPCh 20 ms slot (Figure 4.80). BTSs must

Table 4.28 Bit rate (and indicators rate) for quick paging channels

Spreading Rate (SR)	Data rate (kbps)	Indicators/ information bit	Indicators rate (kbps)	Indicators/ slot (80 ms)	Repetition of indicators	Symbol rate prior to DEMUX (ksps)
SR1	2.4	2	4.8	384	4	19.2
	4.8	2	9.6	768	2	19.2
SR3	2.4	2	4.8	384	6	28.8
	4.8	2	9.6	768	3	28.8

'turn off' the transmission signal of *paging indicators* not assigned to any MS, i.e. for MSs that do not monitor any quick paging channel. This is equivalent to the '0' at the symbol point mapping block when no indicator is assigned, in Figure 4.79.

Indicators are transmitted either in the first and third quarter or in the second and fourth quarter of the 80 ms slot, as in Figure 4.81. Blocks 1, 2, 3 and 4 in Figure 4.80 represent these quarters. Broadcast and configuration change indicators are only transmitted at the end of the second and fourth 20 ms intervals of a slot, as in Figures 4.80 and 4.81.

If the QPCh data rate is 2400 bps, the last four indicators of the first and second half-slots are reserved to be two *broadcast indicators* and two *configuration change indicators*, as in Figure 4.81.

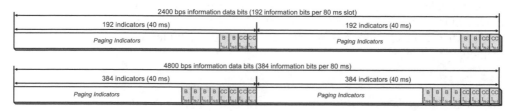

Figure 4.81 Structure of QPChs indicators for the two halves of one slot period.

Similarly, if the QPCh information rate is 4800 bps, the last eight indicators of each half slot are reserved to be broadcast indicators and configuration change indicators.

BTSs must enable configuration change indicators for all slots of their primary QPChs during a certain period of time after changing system configuration parameters. Usually, configuration change indicators are enabled during an interval determined by the current *SLOT_CYCLE_INDEX* in use ($2^{SLOT\_CYCLE\_INDEX} \times 16 \times 80$ ms). This guarantees that mobiles are notified of changes and are updating internally stored system parameters transmitted by the BTSs on overhead messages.

After repetition, indicators go through symbol mapping, spreading and preparation for transmission. Because this channel transmits general-purpose messages to MSs, there is no convolutional encoder frame processing, interleaving or scrambling, as in other cdma2000 channels (Figure 4.79).

4.2.6.1 QPCh Bit Repetition and De-multiplexing

Figure 4.79 depicts the data rate before and after bit repetition. Systems using SR1 have a final data rate of 19 200 bps obtained by repeating bits twice if the input data rate is 4.8 kbps or four times if the input rate is 2.4 kbps. Systems using SR3 have a final data rate of 28 800 bps obtained by repeating bits three times if the input rate is 4.8 kbps or six times if the input rate is 2.4 kbps. Table 4.28 presents data rates throughout the repetition process.

After bit repetition, two processes are responsible to match the indicator power levels to the transmission requirements: symbol point mapping and channel gain. The parameters *QPCH_POWER_LEVEL_PAGE*, *QPCH_POWER_LEVEL_CONFIG* and *QPCH_POWER_ LEVEL_BCAST* define the power level, respectively, for paging, configuration change and broadcast indicators.

Table 4.29 Walsh sequences used to channelise QPChs

Spreading Rate (SR)	Symbol rate after mapping (ksps)	Walsh code	Sequence index	Symbol rate after spreading (Msps)
SR1	19.2	W_i^{128}	80, 48 and 112	2.4576
SR3	28.8	W_i^{256}	$1 \leq i \leq 255$	7.3728

De-multiplexing of QPChs has the same characteristics of de-multiplexing of FBCChs, described in Section 4.2.5.3. QPChs associated to FPChs are only de-multiplexed in non-TD mode, whereas other quick paging channels can be de-multiplexed by any other DEMUX method.

4.2.6.2 Channelisation and Spreading by Walsh Sequences and Complex Multiplier Spreading by PN Sequences

In cdma2000 systems using SR1, Walsh sequences W_{80}^{128}, W_{40}^{128}, and W_{112}^{128} channelise and combine (multiplex) quick paging channels. Walsh code 80 is always assigned to the primary QPCh. Table 4.29 lists Walsh sequences and data rate before and after spreading for systems operating with SR1 and SR3.

Figure 4.37 shows Walsh code spreading and channelisation, PN complex multiplier spreading and QPSK carrier modulation for QPChs operating on SR1 non-TD mode or on SR3 mode for transmission through the air interface. Figure 4.69 shows these processes for systems operating on SR1 OTD mode, whereas Figure 4.70 shows the same for systems employing SR1 STS.

4.2.7 Common Power Control Channel (CPCCh)

cdma2000 BTSs use CPCChs to transmit power control sub-channels that perform power control over multiple logical channels on the reverse link. These sub-channels consist of only one bit per control period. Their purpose is to increase power control efficiency especially when MSs are on system access state (for RC 3–6 on the reverse link) or while allocating high-speed traffic channels.

MSs can receive information transmitted by CPCChs in two ways.

- *Reservation access mode*: MSs adjust the transmission power of the Reverse Common Control Channel (RCCCh) using parameters sent by the power control sub-channel on the associated CPCCh.

- *Power controlled access mode*: MSs adjust the transmission power of the Enhanced Access Channel (EACh), Reverse Acknowledgement Channel (RAckCh) and Reverse Channel Quality Indicator Channel (RCQICh). For the latter, power control is only adjusted when the channel is associated to a Forward Packet Data Channel (FDPCh) without assignment of a Forward Fundamental Channel (FFCh) or Forward Dedicated Control Channel (FDCCh).

BTSs support from 1 to 15 FCPCChs when operating on SR1 and from 1 to 4 when operating on SR3. Figure 4.82 shows a block diagram representing FCPCChs processing.

For FCPCCh processing, each multiplexer input (MUX) is also called an 'initial offset', as in Figure 4.82. Power Control Bits (PCBs) are multiplexed in two distinct data sets that are transmitted separately on FCPCChs I and Q branches, each with output data rate of 9600 bps.

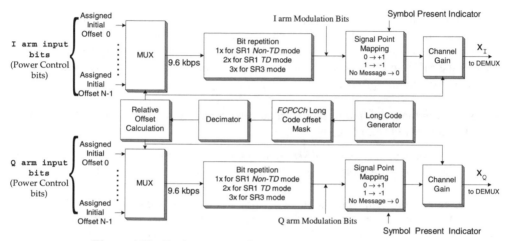

Figure 4.82   Basic structure of common power control channels.

FCPCChs are structured in 20 ms frames, divided into 16 PCGs of 1.25 ms each. Twelve power control bits are transmitted in every PCG, resulting in 192 bits per 20 ms frame.

Each FCPCCh is organised to allow three power control update rates: 800 bps, 400 bps or 200 bps. These rates correspond, respectively, to one power control bit per user every 1.25 ms, 2.5 ms and 5 ms. The power control rate also affects the number of simultaneous users that can be controlled: 12, 24 or 48, respectively.

Each PCG has $2N$ power control sub-channels, 0 to $(2N - 1)$, where N is the number of power controlled MSs. The first half of sub-channels is used in the I branch, inputs numbered 0 to $(N - 1)$, and the second half in the Q branch, N to $(2N - 1)$, as in Table 4.30.

Table 4.30   Number of power control sub-channels per FCPCCh

Rate (bps)	Duration (ms)	Number of power controlled MSs	Number (N) of sub-channels per I and Q branch	Total number ($2N$) of power control sub-channels
800	1.25	12	12	24
400	2.50	24	24	48
200	5.00	48	48	96

A power control bit assigned to one specific MS will be simultaneously inserted in one of the MUX inputs of both the I and Q branches, i.e. the same PCB is copied in the I and Q multiplexers, at the same MUX input number of I and Q branches, as in Figure 4.82.

To increase the probability of MSs receiving power control bits and to avoid interference problems during transmission, the bits transmission sequence within each PCG is pseudo-randomised according to the power control update rate (1.25, 2.5 or 5 ms).

4.2.7.1   Pseudo-Randomisation of the Power Control Bits Position

The PCBs transmission sequence is pseudo-randomised using '*initial offset*' values, each of them corresponding to a physical MUX input. The randomisation process consists of adding a '*relative offset*' to the '*initial offset*', modulo N, determining the '*effective offset*' that is used by the MUX to organise the transmission sequence. The following rules determine the assignment of '*initial offsets*':

- for the I branch, the '*initial offset*' value corresponds to the power control sub-channel index, numbered from 0 to $(N - 1)$;

- For the Q branch, the '*initial offset*' zero of the I branch corresponds to the sub-channel N, whereas the 'initial offset' $(N - 1)$ corresponds to the MUX input $(2N - 1)$.

Power control bits remain in the same position in I and Q branches (Figure 4.83) after the modulo-N operation, i.e. '*initial offset*' values are not changed, only the transmission sequence. If a power control sub-channel is not being used, the MUX must gate-off the transmission power associated to this bit position.

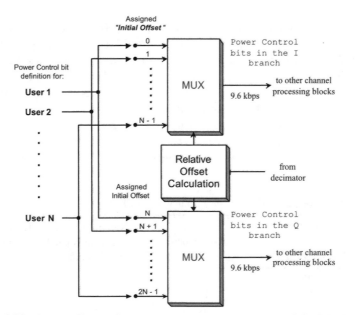

Figure 4.83   Initial offset assignment and multiplexing process of the FCPCCh bits.

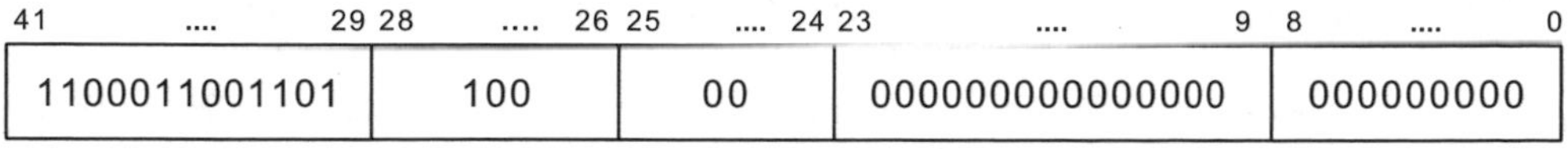

41 29	28 26	25 24	23 9	8 0
1100011001101	100	00	000000000000000	000000000

Figure 4.84 Long code mask used to pseudo-randomise power control bits transmission sequence.

FCPCChs use the decimator block output bit to pseudo-randomise PCBs, as in Figure 4.82. Figure 4.84 shows the mask that the decimator block uses to create the long code sequence offset.

For systems operating with SR1, the PCBs transmission sequence pseudo-randomisation process uses the first out of 128 long code chips (PNLC with rate of 1.2288 Mcps). For SR3, the process uses the first out of 384 long code chips (PNLC with rate 3.6864 Mcps). This produces a fixed output rate of 9600 chips per second (cps) at the decimator output, which corresponds to the same power control bit rate transmitted by FCPCChs. The long code sequence is not used to scramble power control bits, only to change the transmission sequence.

The '*relative offset*' value for PCBs is defined by part of the N chips per PCG interval at the decimator output (0 to $N - 1$), as in Figure 4.85. Thus, L chips out of the N previous PCG (1.25 ms) interval are used to calculate the '*relative offset*', as shown in Table 4.31.

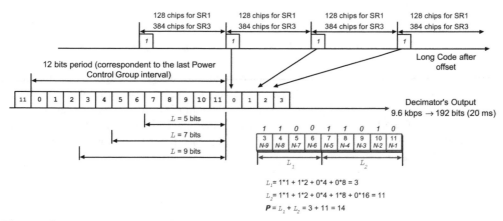

Figure 4.85 Relative offset used for power control bit position randomisation (this example takes $N = 12$).

Table 4.31 Relative offset for the PCB transmission sequence pseudo-randomisation process

PCG (ms)	Number of offset chips (L)	Number of chips in the first offset block (L_1)	Number of chips in the second offset block (L_2)	Calculated relative offset ($P = L_1 + L_2$)
1.25	5	2	3	0–10
2.50	7	3	4	0–22
5.00	9	4	5	0–46

Offset chips (*L*) are split into two blocks, L_1 and L_2. Block L_1 is transmitted first, as in Figure 4.85.

The '*relative offset*' P is the sum of positive integer numbers obtained from decimal conversion of chips in blocks L_1 and L_2, as in Figure 4.85. The result is a number within the interval from 0 to $N - 2$, as in Table 4.31. The first bit of each block (L_1 and L_2) is the Least Significant Bit (LSB).

The pseudo-randomisation process example in Figure 4.86 presents the PCB transmission sequence for user power control update with rates of 800 bps (maximum of 12 users per FCPCCh) and 400 bps (maximum of 24 user per FCPCCh).

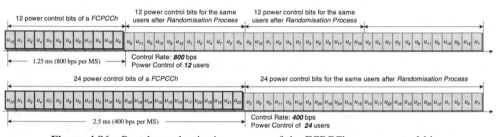

Figure 4.86   Pseudo-randomisation process of the FCPCCh power control bits.

4.2.7.2   FCPCCh Bit Repetition Processes

After the PCBs transmission sequence is determined, the bits go through the bit repetition process. For systems operating on SR1 non-TD mode, there is no repetition of power control bits, only one bit is transmitted per sub-channel in the I and Q branches. For systems on SR1 TD mode (OTD or STS), bits are repeated once, producing 2 bits per sub-channel. Systems using SR3 repeat the bits twice, resulting in three bits per power control sub-channel. Figure 4.82 illustrates this process.

4.2.7.3   De-multiplexing (Serial-to-Parallel Conversion)

After repetition, power control bits are mapped and go through the DEMUX to prepare for transmission through the air interface, according to the supported operating mode.

For FCPCChs operating on SR1 non-TD mode, symbols are just relayed to the following processes, as in Figure 4.87.

When operating with SR1 in TD mode (OTD or STS), as Figure 4.88, the first symbol per PCG from the X_I branch is sent to output Y_{I1}. The following bit is sent to output Y_{I2}. The serial-to-parallel process starts again with the following coded symbol at Y_{I1}.

The same process is applied to symbols on the X_Q branch that go to Y_{Q1} and Y_{Q2} outputs.

The DEMUX process for systems operating with SR3 is similar to the process described for systems with SR1 in TD mode. The only difference between both cases is that for SR3 the DEMUX has three outputs for each branch, as in Figure 4.89.

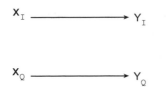

Figure 4.87 DEMUX for SR1 non-TD operation mode.

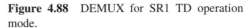

Figure 4.88 DEMUX for SR1 TD operation mode.

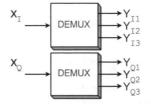

Figure 4.89 DEMUX operation for SR3.

4.2.7.4 Channelisation and Spreading by Walsh Sequences and Complex Multiplier Spreading by PN Sequences

Walsh Codes and/or QOF functions spread and channelise FCPCCHs according to the Spreading Rate (SR) and transmission diversity type supported by the system, as in Table 4.32.

When operating in cdma2000 radio configurations (RC 3–9), FCPCCHs can operate in any of the TD schemes. Figures 4.37, 4.69 and 4.70 illustrate Walsh code multiplexing, PN complex multiplier spreading and QPSK carrier modulation for FCPCCHs in systems operating, respectively, with SR1 non-TD or SR3, SR1 OTD and SR1 STS.

4.2.8 Common Assignment Channel (CACh)

CAChs are designed specifically for cdma2000 and provide fast responses to configuration and assignment commands for reverse link channels to support transmission of random access packets. Common assignment channel basic functions include the following.

Table 4.32 Walsh sequences used for multiplexing forward common power control channels

Spreading Rate (SR)	Repetition factor	Symbol rate after mapping (Ksps)	Transmission diversity technique	Walsh Code	Sequence index	Symbol rate after spreading (Msps)
SR1	1	9.6	Non-TD	W_i^{128}	$1 \le i \le 127$	2.4576
SR1	2	19.2	TD (OTD or STS)	W_i^{64}	$1 \le i \le 63$	2.4576
SR3	3	28.8	Multi-Carrier (MC)	W_i^{128}	$1 \le i \le 127$	7.3728

- Control of Reverse Common Control Channels (RCCChs) and associated power control sub-channels when in *reservation access mode* (in relation to the CPCCh).

- Providing acknowledgement (ACK) messages when in *power controlled access mode* (in relation to the CPCCh), especially when assigned to transmit random access packets on the reverse link.

- Implementing assignment and traffic control when there is system congestion.

BTSs can choose to support CAChs or not on a frame-by-frame basis. When CAChs transmission is enabled, BTSs must also support discontinuous transmission of these channels. BTSs communicate the decision of transmitting CAChs and supporting discontinuous transmission to MSs through Broadcast Control Channels (BCChs).

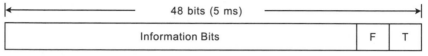

Figure 4.90   Frame structure of common assignment channels.

Common assignment channels are organised in 5 ms frames and transmit at a fixed rate of 9600 bps. The CACh frame structure consists of 48 bits: 32 data bits, 8 frame quality indicators (CRC) and 8 padding bits ('0s'), as in Figure 4.90. In the figure, F represents frame quality indicator bits and T, encoder tail bits.

Figure 4.91 presents a block diagram of the CACh processing and Table 4.33 shows bit rates obtained throughout the process.

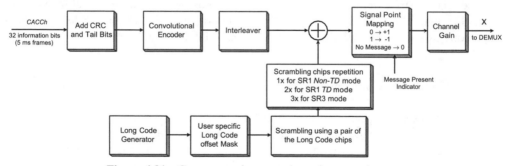

Figure 4.91   Common assignment channels processing.

Table 4.33   Bit rates during common assignment channel processing

Spreading Rate (SR)	Bits/frame (5 ms)	Convolutional encoder	Interleaving (bits)	Coded symbol rate after interleaving (ksps)
SR1	48	$R = 1/2, k = 9$	96	19.2
SR1	48	$R = 1/4, k = 9$	192	38.4
SR3	48	$R = 1/3, k = 9$	144	28.8

4.2.8.1 Convolutional Encoders and Interleaving

Figures 4.41 and 4.56 show convolutional encoders, with rates $R = 1/2$ and $R = 1/4$, used for CACHs when operating in SR1, whereas Figure 4.57 shows the coder used with SR3 systems ($R = 1/3$).

After convolutional coding, coded bits are interleaved with the forward-backwards BRO interleaving method (Figure 4.58) if operating with SR1, or with the complex cyclic-shift BRO interleaving method (Figure 4.60) if operating with SR3. Both methods are described in Section 4.2.4.1.

Interleaving is performed in a 5 ms frame basis, which corresponds to CACHs frame duration, and Table 4.17 provides the parameters that define interleaving matrices size.

4.2.8.2 Scrambling

Scrambling happens after frames are convolutional coded and interleaved (as in Figure 4.91). Because CACHs are cdma2000 exclusive channels they use two long code chips for scrambling groups of $2M$ coded bits, each M-bit set by one chip. This process is fully described in Section 4.2.4.2.

Figure 4.92 shows the long code offset mask used for common assignment channels scrambling. In the figure, CACN is the identifier of CACHs, and *Pilot_PN* is the pilot offset used by the BTS.

41	....	29 28	....	24 23	....	21 20	....	9 8	....	0
1100011001101		01100		CACN		000000000000		Pilot_PN		

Figure 4.92 Long code offset mask of common assignment channels.

4.2.8.3 De-multiplexing (Serial-to-Parallel Conversion)

De-multiplexing of CACHs has same characteristics as in FBCCHs, and this process is described in Section 4.2.5.3.

4.2.8.4 Channelisation and Spreading by Walsh Sequences and Complex Multiplier Spreading by PN Sequences

Table 4.34 shows the length of Walsh codes used to spread and channelise CACHs. The length varies according to the convolutional coder used by the system.

Table 4.34 Walsh codes used by common assignment channels

SR	Convolutional coder (R)	Coded bit rate after interleaving (kbps)	Walsh code	Sequence index	Symbol rate after spreading (Msps)
SR1	1/2	19.2	W_i^{128}	$1 \leq i \leq 127$	2.4576
SR1	1/4	38.4	W_i^{64}	$1 \leq i \leq 63$	2.4576
SR3	1/3	28.8	W_i^{256}	$1 \leq i \leq 255$	7.3728

Because CACHs are channels designed for cdma2000 Radio Configurations (RC 3–9), they can operate in any of the TD schemes. Figures 4.37, 4.69 and 4.70 illustrate Walsh code multiplexing, PN complex multiplier spreading and QPSK carrier modulation for FCPCCHs in systems operating, respectively, with SR1 non-TD or SR3, SR1 OTD and SR1 STS.

4.2.9 Forward Fundamental Channel

In CDMA IS-95A based systems, there is only one type of logical channel responsible for voice, signalling (control) and data traffic (at low rates) during a call. This channel is called Forward Traffic Channel (FTCh). One of the main characteristics of cdma2000 systems is the specialisation of logical channels. Therefore, cdma2000 implements specialised logical channels to perform tasks previously executed by FTChs. These channels include Forward Fundamental Channels (FFChs), Forward Dedicated Control Channels (FDCCHs), and Forward Supplemental Channels (FSChs).

FFChs have the same basic functionalities as IS-95 FTChs, but operating in all radio configurations, RC1–RC9, and focused on voice and low data rates traffic. Forward Dedicated Control Channels (FDCCHs) have the main task of transmitting signalling and control data traffic. Forward Supplemental Channels (FSChs) only operate for RCs 3–9, and have data transmission capabilities for higher data rate.

Figure 4.93 Frame structure of forward fundamental channels.

Figure 4.93 shows the frame structure of FFChs. In the figure, R/F represents a reserved bit or flag (EIB/QIB); F, the Frame quality indicator (CRC) and T, the padding bits (encoder tail bits, 8 bits '0').

Table 4.35 presents the bit rates during FFChs frame processing. Data bits are organised in 5 (RC 3–9) and 20 ms frames (RC 1–9). It also shows that, for IS-95 radio configurations (RC1 and RC2), bit rates and frame structures are the same presented for IS-95 FTChs, in Section 4.1.4.1, for 8.6 kbps (RS1) and 13.3 kbps (RS2) vocoders.

The decision on the frames length is taken on a frame-by-frame basis. For voice communications, the symbol transmission rate is always kept constant by applying bit repetition when the rate is lower than the maximum allowed ($R_{voc} < 1$).

Figure 4.94 shows the frame processing diagram block for FFChs, common for all radio configurations. Some of these processes depend on the radio configuration being used (e.g. encoder rate, addition of reserved bit); the sequence of the processes, however, is the same for all RCs.

Figure 4.95 shows the continuation of FFCh frames processing (Figure 4.94) when operating with radio configurations 1 or 2.

Processes in Figure 4.95 are equivalent to processes described in Section 4.1.5 for Forward Traffic Channels (FTChs) using RC1 and RC2 in non-TD operation mode. For compatibility with CDMA IS-95 systems, non-TD de-multiplexing follows the description

Table 4.35 FFCh frame structure for different RCs

Radio Configuration (RC)	Data rate (kbps)	Frame duration (ms)	Reserved bit	Information bits/frame	CRC (F)	Tail bits (T)	Total of bits/frame
1	9.6	20	0	172	12	8	192
	4.8	20	0	80	8	8	96
	2.4	20	0	40	0	8	48
	1.2	20	0	16	0	8	24
2	14.4	20	1	267	12	8	288
	7.2	20	1	125	10	8	144
	3.6	20	1	55	8	8	72
	1.8	20	1	21	6	8	36
3, 4, 6 and 7	9.6	5	0	24	16	8	48
	9.6	20	0	172	12	8	192
	4.8	5/20	0	80	8	8	96
	2.4	5/20	0	40	6	8	54
	1.2	5/20	0	16	6	8	30
5, 8 and 9	9.6	5/20	0	24	16	8	48
	14.4	5/20	1	267	12	8	288
	7.2	5/20	1	125	10	8	144
	3.6	5/20	1	55	8	8	72
	1.8	5/20	1	21	6	8	36

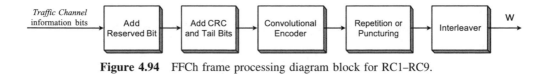

Figure 4.94 FFCh frame processing diagram block for RC1–RC9.

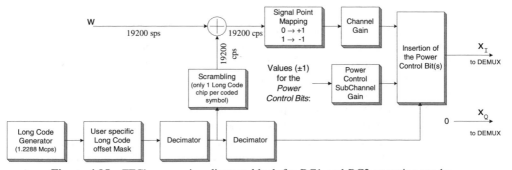

Figure 4.95 FFCh processing diagram block for RC1 and RC2 operation modes.

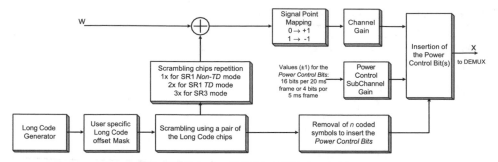

Figure 4.96   FFCh processing for RC 3–9 operation modes.

of Figure 4.35 (Section 4.2.1.2), and modulation and spreading processes are equivalent to Figure 4.37 (in Section 4.2.1.3).

Frame processing for RC3–RC9 operation uses most of the cdma2000 exclusive features developed to improve transmission rates and system reliability, as in Figure 4.96.

When operating in RC 3–9, BTSs can discontinue transmission of FFChs for up to three consecutive 5 ms periods within a 20 ms frame.

Tables 4.36–4.44 present bit rates during FFChs processing for the different radio configurations available in cdma2000.

4.2.9.1   Convolutional Coding and Bit Repetition

Tables 4.36–4.44 also show that convolutional encoders used for FFChs depend on the radio configuration.

Some of the convolutional encoders presented in these tables are described in previous sections:

- encoder with $R = 1/2$ for RC1, RC2, RC4 and RC9 – Figure 4.41,

- encoder with $R = 1/4$ for RC3, RC5 and RC8 – Figure 4.56,

- encoder with $R = 1/3$ for RC7 and RC8 – Figure 4.57.

RC6 uses a different convolutional encoder type, with $R = 1/6$ and $k = 9$, as in Figure 4.97.

After convolutional coding, the system performs a bit repetition process to equalise the number of coded bits for interleaving. This process also considers the RC in use because the number of repetitions depends on the number of information bits per frame, as in Tables 4.36–4.44.

An important feature of the coding process is the energy distribution among symbols being transmitted. The bits transmitted at lower rates are also transmitted with lower energy. Equation (4.1) determines the energy per modulated symbol at rates lower than nominal. In this equation, E_{max} represents the maximum energy per symbol, R_b is the information bit rate used by the channel and R_{max} is the highest permitted rate for the associated RC. The

Table 4.36 FFCh bit rates for RC1 operation

Data bits	Frame (ms)	Bits/frame	Total (kbps)	Convolutional coder	Repetition	Symbol puncturing	Block interleaving	Coded symbols after interleaving (ksps)
16	20	24	1.2	$R = 1/2, k = 9$	7	–	384	19.2
40	20	48	2.4	$R = 1/2, k = 9$	3	–	384	19.2
80	20	96	4.8	$R = 1/2, k = 9$	1	–	384	19.2
172	20	192	9.6	$R = 1/2, k = 9$	–	–	384	19.2

Table 4.37 FFCh bit rates for RC2 operation

Information bits	Frame (ms)	Bits/frame	Total (kbps)	Convolutional coder	Repetition	Symbol puncturing	Interleaving block	Coded symbols after interleaving (ksps)
21	20	36	1.8	$R = 1/2, k = 9$	7	2/6	384	19.2
55	20	72	3.6	$R = 1/2, k = 9$	3	2/6	384	19.2
125	20	144	7.2	$R = 1/2, k = 9$	1	2/6	384	19.2
267	20	288	14.4	$R = 1/2, k = 9$	–	2/6	384	19.2

Table 4.38 FFCh bit rates for RC3 operation

Information bits	Frame (ms)	Bits/frame	Total (kbps)	Convolutional coder	Repetition	Symbol puncturing	Interleaving block	Coded symbols after interleaving (ksps)
24	5	48	9.6	$R = 1/4, k = 9$	–	–	192	38.4
16	20	30	1.5	$R = 1/4, k = 9$	7	1/5	768	38.4
40	20	54	2.7	$R = 1/4, k = 9$	3	1/9	768	38.4
80	20	96	4.8	$R = 1/4, k = 9$	1	–	768	38.4
172	20	192	9.6	$R = 1/4, k = 9$	–	–	768	38.4

Table 4.39 FFCh bit rates for RC4 operation

Information bits	Frame (ms)	Bits/frame	Total (kbps)	Convolutional coder	Repetition	Symbol puncturing	Interleaving block	Coded symbols after interleaving (kbps)
24	5	48	9.6	$R = 1/2, k = 9$	–	–	96	19.2
16	20	30	1.5	$R = 1/2, k = 9$	7	1/5	384	19.2
40	20	54	2.7	$R = 1/2, k = 9$	3	1/9	384	19.2
80	20	96	4.8	$R = 1/2, k = 9$	1	–	384	19.2
172	20	192	9.6	$R = 1/2, k = 9$	–	–	384	19.2

Table 4.40 FFCh bit rates for RC5 operation

Information bits	Frame (ms)	Bits/frame	Total (kbps)	Convolutional coder	Repetition	Symbol puncturing	Interleaving block	Coded symbols after interleaving (ksps)
24	5	48	9.6	$R = 1/4, k = 9$	–	–	192	38.4
21	20	36	1.8	$R = 1/4, k = 9$	7	4/12	768	38.4
55	20	72	3.6	$R = 1/4, k = 9$	3	4/12	768	38.4
125	20	144	7.2	$R = 1/4, k = 9$	1	4/12	768	38.4
267	20	288	14.4	$R = 1/4, k = 9$	–	4/12	768	38.4

Table 4.41 FFCh bit rates for RC6 operation

Information bits	Frame (ms)	Bits/frame	Total (kbps)	Convolutional coder	Bit Repetition	Bit puncturing	Interleaving block	Coded symbols after interleaving (ksps)
24	5	48	9.6	$R = 1/6, k = 9$	–	–	288	57.6
16	20	30	1.5	$R = 1/6, k = 9$	7	1/5	1152	57.6
40	20	54	2.7	$R = 1/6, k = 9$	3	1/9	1152	57.6
80	20	96	4.8	$R = 1/6, k = 9$	1	–	1152	57.6
172	20	192	9.6	$R = 1/6, k = 9$	–	–	1152	57.6

Table 4.42 FFCh bit rates for RC7 operation

Information bits	Frame (ms)	Bits/frame	Total (kbps)	Convolutional coder	Bit Repetition	Bit puncturing	Interleaving block	Coded symbols after interleaving (ksps)
24	5	48	9.6	$R = 1/3, k = 9$	1	–	144	28.8
16	20	30	1.5	$R = 1/3, k = 9$	8	1/5	576	28.8
40	20	54	2.7	$R = 1/3, k = 9$	4	1/9	576	28.8
80	20	96	4.8	$R = 1/3, k = 9$	2	–	576	28.8
172	20	192	9.6	$R = 1/3, k = 9$	1	–	576	28.8

Table 4.43 FFCh bit rates for RC8 operation

Information bits	Frame (ms)	Bits/frame	Total (kbps)	Convolutional coder	Repetition	Symbol puncturing	Interleaving block	Coded symbols after interleaving (ksps)
24	5	48	9.6	$R = 1/3, k = 9$	1	–	288	57.6
21	20	30	1.8	$R = 1/4, k = 9$	7	–	1152	57.6
55	20	72	3.6	$R = 1/4, k = 9$	3	–	1152	57.6
125	20	144	7.2	$R = 1/4, k = 9$	1	–	1152	57.6
267	20	288	14.4	$R = 1/4, k = 9$	–	–	1152	57.6

Table 4.44 FFCh bit rates for RC9 operation

Information bits	Frame (ms)	Bits/frame	Total (kbps)	Convolutional coder	Repetition	Symbol puncturing	Interleaving block	Coded symbols after interleaving (ksps)
24	5	48	9.6	$R = 1/2, k = 9$	–	–	144	28.8
21	20	36	1.8	$R = 1/2, k = 9$	7	–	576	28.8
55	20	72	3.6	$R = 1/2, k = 9$	3	–	576	28.8
125	20	144	7.2	$R = 1/2, k = 9$	1	–	576	28.8
267	20	288	14.4	$R = 1/2, k = 9$	–	–	576	28.8

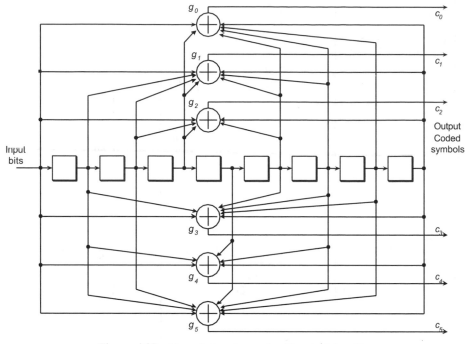

Figure 4.97 Convolutional encoder ($R = 1/6, k = 9$).

maximum energy per symbol occurs when the transmission rate is the maximum for the FFCh associated to the Radio Configuration (RC).

$$E_s = E_{\max}(R_b/R_{\max}) \tag{4.1}$$

Table 4.47 presents transmission energy values for the highest rates. For example, for RC1 transmitting at 4800 bps, symbols must have half of the power of symbols transmitted at 9600 bps. This lowers the transmitted power over the air interface and reduces interference within the system.

4.2.9.2 Convolutional Code Puncturing

There are two types of puncturing applied on the forward link of cdma2000 systems. The choice of puncturing type depends on the type of information (voice or data), transmission rate and encoder type being used by the traffic logical channel. Because of the latter parameter, the processes are called convolutional code puncturing and turbo code puncturing, respectively using a convolutional or turbo encoder.

As FFChs use convolutional encoders, the selected puncturing process for these channels is the convolutional code puncturing. For each FFCh frame, after convolutional coding and repetition, some bits are removed from the stream according to a puncturing pattern that depends on the radio configuration and the encoder rate, as in Table 4.45.

Table 4.45 Convolutional code puncturing patterns

Convolutional coder rate	Puncturing rate	Puncturing pattern	Associated Radio Configuration (RC)
1/2	2/6	'110101'	2
1/2	1/5	'11110'	4
1/2	1/9	'111111110'	4
1/2	2/18	'111011111111111110'	9
1/3	1/5	'11110'	7
1/3	1/9	'111111110'	7
1/4	4/12	'110110011011'	5
1/4	1/5	'11110'	3
1/4	1/9	'111111110'	3
1/6	1/5	'11110'	6
1/6	1/9	'111111110'	6

In the puncturing patterns, a '0' indicates that the associated coded bit must be removed (punctured) whereas a '1' means that the bit stays. If the frame is bigger than the length of the associated pattern, the pattern is repeated until the frame length is completed.

4.2.9.3 Interleaving

After puncturing, the coded bits are interleaved. The three interleaving methods are used for SR1 and SR3, depending on the radio configuration.

- Systems operating on SR1 with RC1 and RC2 employ the BRO interleaver (Figure 4.43), which is also used for FTChs and the SyncCh.

- Systems operating on SR1 with RC 3–9 employ the forward-backwards BRO interleaver (Figure 4.58).

- Systems operating on SR3, regardless of the RC (6–9), employ the complex cyclic-shift BRO interleaver (Figure 4.60).

4.2.9.4 Scrambling

Since RC1 and RC2 are compatible with CDMA IS-95 standards, scrambling for these radio configurations follows the description in Sections 4.1.5.4 and 4.2.3.3. This scrambling process uses one chip from the long code sequence, after the application of the offset mask, as in Figure 4.50.

Radio configurations 3–9 employ a scrambling process that uses two long code chips to process groups of $2M$ bits, as described in Section 4.2.4.2 (Figure 4.63).

In both cases, two types of offset masks can be used: private and public. Figure 4.98 presents the public long code offset mask that employs the mobile terminal *ESN*. The chapter on call processing provides more details about private and public long code offset masks.

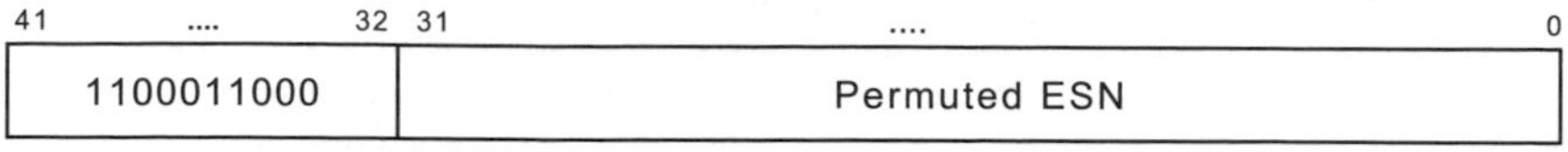

41	....	32	31	....	0
1100011000			Permuted ESN		

Figure 4.98　Public long code offset mask for forward fundamental channels.

4.2.9.5　*Insertion of Power Control Sub-channels*

If the parameter *FPC_PRI_CHANNEL* = '0' enables power control sub-channel transmission, BTSs must continuously transmit Power Control Bits (PCBs) in the forward fundamental channel associated to the target MS, as in Figure 4.95 for RC1 and RC2 and in Figure 4.96 for other RCs (3–9). If the parameter *FPC_PRI_CHANNEL* is set to '1', however, the forward dedicated control channels transmit Power Control Sub-channels (PCSChs).

PCBs are always transmitted at a rate of 800 bps, i.e. one bit every 1.25 ms, or one bit per PCG. If a bit '0' is transmitted, the mobile has to increase its average transmission power. Otherwise, the mobile has to decrease its transmission power.

Power control may be executed with three distinct power steps, defined by the *PWR_CNTL_STEP* parameter (Table 4.46). The power control process is explained in details in the chapter on power control.

Table 4.46　Power control steps for FTChs

PWR_CNTL_STEP	Power control step size (dB)
0	1.00
1	0.50
2	0.25

For RC1 and RC2, the forward link transmits the PCB in the second PCG after receiving the correspondent PCG from the reverse link channel, as in Figure 4.99. This process is equivalent to description in Section 4.1.5.5 for CDMA IS-95 systems.

For RC 3–9, the transmission of PCSChs depends on the gating rate applied to the Reverse Pilot Channel (RPiCh), as in Figure 4.100.

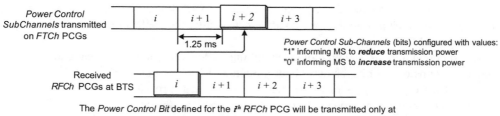

Figure 4.99　Transmission of a PCB related to a PCG received from the reverse link.

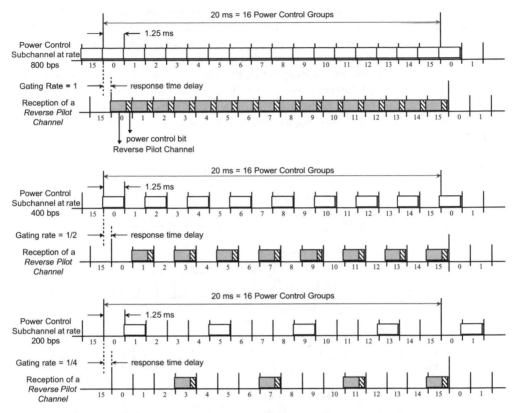

Figure 4.100   Structure of power control groups for different gating rates.

Table 4.47   Transmission energy and number of power control bits

Radio Configuration (RC)	Coded bits per frame (and per 1.25 ms PCG)	Punctured coded bits per PCG (n_{PCB})	Minimum power control bit energy	Initial PCB position	Decimator chips (MSB → LSB)
1	384 (24)	2	E_b	0, 1, …, 15	23, 22, 21, 20
2	384 (24)	2	$3E_b/4$	0, 1, …, 15	23, 22, 21, 20
3 (non-TD)	768 (48)	4	E_b	0, 2, …, 30	47, 46, 45, 44
3 (TD)	768 (48)	4	E_b	0, 4, …, 28	47, 46, 45
4 (non-TD)	384 (24)	2	E_b	0, 2, …, 14	23, 22, 21
4 (TD)	384 (24)	2	E_b	0, 2, …, 14	23, 22, 21
5 (non-TD)	768 (48)	4	E_b	0, 2, …, 30	47, 46, 45, 44
5 (OTD)	768 (48)	4	E_b	0, 4, …, 28	47, 46, 45
6	1152 (72)	6	E_b	0, 6, …, 42	71, 70, 69
7	576 (36)	3	E_b	0, 3, …, 21	35, 34, 33
8	1152 (72)	6	E_b	0, 6, …, 42	71, 70, 69
9	576 (36)	3	E_b	0, 3, …, 21	35, 34, 33

BTSs must estimate the signal level received from MSs, and use this value to determine the PCB ('0' or '1'). The number of power control bits per PCG (n_{PCB}) and their transmission energy on forward traffic channels depend on the RC in use for the channel and on the data rate being processed, as in Table 4.47.

In Table 4.47, the position for inserting the PCBs is given by the last n_{LC} decimated long code chips in a period that corresponds to the PCG (1.25 ms) immediately before the one where the power control bits are being inserted, as in Figure 4.101. Depending on the RC, a certain number of coded data bits (n_{PCB}), within the current PCG, is punctured allowing PCBs to be inserted.

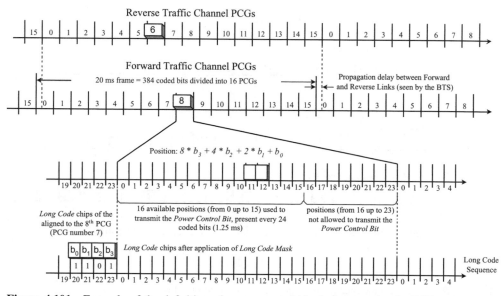

Figure 4.101   Example of the definition of power control bits insertion point for RC1 and RC2.

As an example, for RC5 in non-TD operation mode, from the 48 bits per PCG, four of them ($n_{PCB} = 4$) are punctured to insert four copies of the PCB, each with minimal transmission energy equal to E_b, as in Table 4.47. The channel gain block assigns the energy for the power control bit, in Figure 4.95 for RC1 and RC2 and in Figure 4.96 for other RCs (3–9).

4.2.9.6   De-multiplexing (Serial-to-Parallel Conversion)

No de-multiplexing process is performed for RC1 and RC2 operation (as illustrated in Figure 4.95) due to backward compatibility. Conversely, for RCs 3–9 operation mode (as in Figure 4.96), FFChs employ the same de-multiplexing processes as used for FBCChs, described in Section 4.2.6.3 (Figures 4.76, 4.77 and 4.78).

Table 4.48 Length of Walsh sequences used for FFChs

SR	RC	Coded symbol rate after interleaving (kbps)	Code	Sequence index	Symbol rate after spreading (Msps)
SR1	1 and 2	19.2	W_i^{64}	$1 \le i \le 63$	1.2288
SR1	3 and 5	38.4	W_i^{64}	$1 \le i \le 63$	2.4576
SR1	4	19.2	W_i^{128}	$1 \le i \le 127$	2.4576
SR3	6 and 8	57.6	W_i^{128}	$1 \le i \le 127$	7.3728
SR3	7 and 9	28.8	W_i^{256}	$1 \le i \le 255$	7.3728

4.2.9.7 Channelisation and Spreading by Walsh Sequences and Complex Multiplier Spreading by PN Sequences

Table 4.48 presents Walsh codes used for multiplexing FFChs.

Because forward fundamental channels are also designed for cdma2000 radio configurations (RC3–RC9), they can operate in any of the TD schemes. Figures 4.37, 4.69 and 4.70 illustrate Walsh code spreading, complex PN modulation and QPSK carrier modulation for FFChs in systems operating, respectively, with SR1 non-TD or SR3, SR1 OTD and STS.

4.2.10 Forward Dedicated Control Channel

Forward Dedicated Control Channels (FDCChs) transmit signalling and control data, including user information and signalling to MSs during a call (e.g. handoff procedures and service negotiation). BTSs can have one active FDCCh (operating with RC3–RC9) for each MS in *mobile control on the traffic channel state* (i.e. during a call).

Figure 4.102 shows the frame structure for FDCChs, which can have lengths of 5 and 20 ms. In the figure, R represents a reserved bit (EIB/QIB); F, the frame quality indicator (CRC) and T, the padding bits (encoder tail bits, 8 bits '0').

Figure 4.102 Frame structure of forward dedicated control channels.

Table 4.49 presents the bit rates and frames configuration during FDCChs processing.

FDCCh frames are processed according to Figures 4.103 and 4.104, in a way very similar to that presented for FFChs when operating with RC3–RC9.

After frames go through convolutional coding, interleaving and bit repetition/puncturing processes, as in Figure 4.103, they are scrambled and mapped, as in Figure 4.104.

Table 4.50 shows the data rate at the end of each processing block before scrambling. The rate varies according to the Radio Configuration (RC) and Spreading Rate (SR).

BTSs must support discontinuous transmission of the FDCCh, and this decision of transmitting or not must be taken frame by frame, that is, according to the frame duration.

Table 4.49 Bit rates and frame structure for FDCCHs, according to the RC in use

Radio Configuration (RC)	Frame (ms)	Information rate (bps)	Total of bits/frame	Reserved bit	Information bits/frame	CRC (F)	Tail bits (T)
3, 4, 6 and 7	20	9600	192	0	172	12	8
5, 8 and 9	20	14400	288	1	267	12	8
3–9	5	9600	48	0	24	16	8

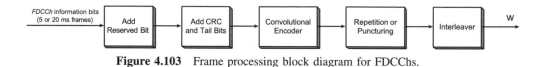

Figure 4.103 Frame processing block diagram for FDCCHs.

Table 4.50 Bit rates during FDCCh frame processing for distinct RC and SR

SR	RC	Frame (ms)	Information bits/frame	Convolutional encoder	Repetition/ puncturing	Interleaving	Coded symbols rate after interleaving (Ksps)
1	3	5	48	$R = 1/4, k = 9$	–	192	38.4
		20	192	$R = 1/4, k = 9$	–	768	38.4
1	4	5	48	$R = 1/2, k = 9$	–	96	19.2
		20	192	$R = 1/2, k = 9$	–	384	19.2
1	5	5	48	$R = 1/4, k = 9$	–	192	38.4
		20	288	$R = 1/4, k = 9$	4/12	768	38.4
3	6	5	48	$R = 1/6, k = 9$	–	288	57.6
		20	192	$R = 1/6, k = 9$	–	1152	57.6
3	7	5	48	$R = 1/3, k = 9$	–	144	28.8
		20	192	$R = 1/3, k = 9$	–	576	28.8
3	8	5	48	$R = 1/3, k = 9$	2	288	57.6
		20	288	$R = 1/4, k = 9$	–	1152	57.6
3	9	5	48	$R = 1/3, k = 9$	–	144	28.8
		20	288	$R = 1/2, k = 9$	–	576	28.8

BTSs must also support frame transmission offsets of multiples of 1.25 ms, defined by parameter *FRAME_OFFSET*, sent in overhead messages. 20 and 5 ms frames with offset zero start when the system time is a multiple of 20 or 5 ms, respectively. For offsets other than zero, 20 and 5 ms frames start at $1.25 \times FRAME\_OFFSET$ and $1.25 \times (FRAME\_OFFSET \bmod 4)$, respectively. Because of transmission offsets, the interleaving block must start aligned to the beginning of FDCCHs frames.

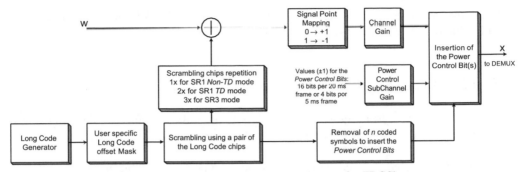

Figure 4.104   Frame processing block diagram for FDCChs.

The use of frame offsets (*FRAME_OFFSET*) is a CDMA resource related to traffic channels processing time. The system receives the information (voice and data) from multiple channels transmitted by MSs, but can process only one at a time. The information is kept in a buffer in the MSC, which decides if the information can be released or if a *FRAME_OFFSET* delay is required.

Before interleaving, FDCCh bits may go through symbol repetition or puncturing, as in Figure 4.103 and Table 4.50.

4.2.10.1   Convolutional Coding, Bit Repetition and/or Puncturing

FDCChs can use all types of convolutional encoders available for cdma2000 systems, depending on the spreading rate and radio configuration, as in Table 4.50. All the convolutional encoders are described in previous sections:

- encoder with $R = 1/2$ – Figure 4.41;

- encoder with $R = 1/3$ – Figure 4.57;

- encoder with $R = 1/4$ – Figure 4.56;

- encoder with $R = 1/6$ – Figure 4.97.

Puncturing performed for RC5 is equivalent to the process described for forward fundamental channels in Section 4.2.9.2.

4.2.10.2   Interleaving

The interleaving block starts aligned to the beginning of FDCChs frames. After convolutional coding, the coded bits are interleaved using forward-backwards BRO interleaving (Figure 4.58) for SR1 and complex cyclic-shift BRO interleaving (Figure 4.60) for SR3.

Both types of interleaving are described in Section 4.2.4.1. Table 4.17 presents parameters m and J that define the interleaving matrix size.

4.2.10.3   Scrambling

Scrambling of FDCChs is equivalent to the process described for forward common control channels (Figure 4.63), where two long code chips are decimated for scrambling groups of $2M$ bits.

The *public* long code offset mask used to scramble FDCCh coded bits is the same used for all traffic channels, and depends on the permuted *ESN* value of each MS, as in Figure 4.105.

FDCChs, as any traffic channel, can also be scrambled by chips decimated from the PNLC after application of the *private*, as described in the chapter on call processing.

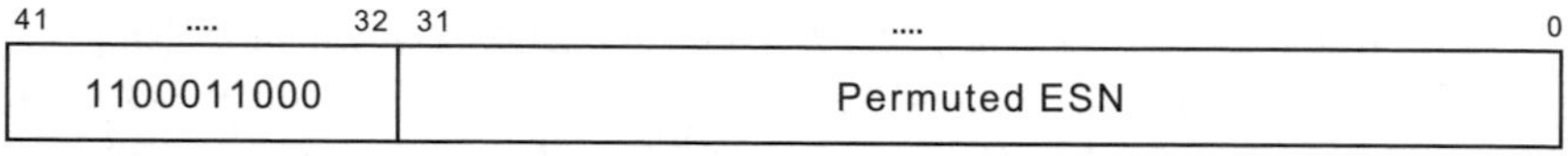

41 32	31 0
1100011000	Permuted ESN

Figure 4.105   Public long code offset mask of forward dedicated control channels.

4.2.10.4   Insertion of Power Control Sub-channels

Figure 4.104 shows the insertion of the power control bits process, which will only occur for the FDCCh in case the parameter *FPC_PRI_CHANNEL* = '1'. In this case, BTSs must continuously transmit Power Control Bits (PCBs) in the MS associated forward dedicated control channel. The power control bit insertion process is the same as that described for FFChs in Figure 4.101.

MSs can operate in two different modes to perform power control using forward dedicated control channels:

- non-gated transmission mode: the power control sub-channel is transmitted at the rate of 1 bit ('0' or '1') every 1.25 ms (that is, at 800 bps);

- gated transmission mode: the power control sub-channel is transmitted at 400 bps or 200 bps, when the gating rate of the control groups is 1/2 and 1/4, respectively.

4.2.10.5   Channelisation and Spreading by Walsh Sequences and Complex Multiplier Spreading by PN Sequences

Table 4.51 presents Walsh codes used for multiplexing FDCChs.

Because forward fundamental channels are designed for cdma2000 radio configurations (RC3–RC9), they can operate in any of the TD schemes. Figures 4.37, 4.69 and 4.70 illustrate Walsh code spreading and multiplexing, PN complex spreading and QPSK carrier

Table 4.51 Length of Walsh sequences used for FDCChs

SR	RC	Walsh code	Sequence index	Symbol rate after spreading (Msps)
SR1	3 and 5	W_i^{64}	$1 \le i \le 63$	2.4576
	4	W_i^{128}	$1 \le i \le 127$	2.4576
SR3	6 and 8	W_i^{128}	$1 \le i \le 127$	7.3728
	7 and 9	W_i^{256}	$1 \le i \le 255$	7.3728

modulation for FCPCChs in systems operating, respectively, with SR1 non-TD or SR3, SR1 OTD and SR1 STS.

4.2.11 Forward Supplemental Channel

Forward Supplemental Channels (FSChs) are only available for RCs 3–9 and transmit data traffic with data rates higher than the data transmitted by FFChs. Each traffic channel assigned to an MS can have up to two FSChs.

FSChs are organised in 20, 40 or 80 ms frames, and BTSs decide on supporting discontinuous transmission of these channels on a frame-by-frame basis, depending on the radio resource management performed by the system.

| R | Information Bits | F | R/T |

Figure 4.106 Frame structure of forward supplemental channels.

Figure 4.106 shows the frame structure for FSChs. In the figure, R represents a reserved bit; F, the frame quality indicator (CRC) and R/T, the padding bits (encoder tail bits). If the FSCh uses a convolutional encoder, R/T corresponds to eight encoder tail bits whose values are set to '0'. If it uses a turbo encoder, the first two, out of the eight bits, are reserved (value '0') and the remaining six bits are added during the turbo coding process.

FSCh frames are processed according to the steps in Figures 4.107 and 4.108.

Figure 4.108 presents steps between interleaving and de-multiplexing.

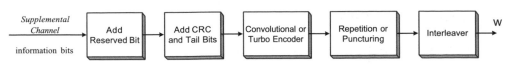

Figure 4.107 FSCh frame processing block diagram.

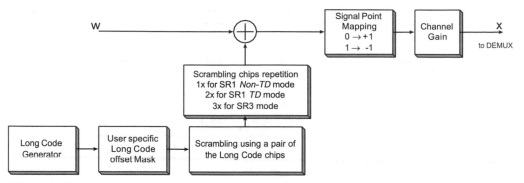

Figure 4.108 FSCh frame processing diagram block (between interleaving and de-multiplexing).

BTSs must also support frame transmission offsets of multiples of 1.25 ms. The parameter *FRAME_OFFSET*, sent in overhead messages, defines the delay. The delay can also be a multiple of 20 ms when transmitting 40 and 80 ms frames. In this case, the parameter *FOR_SCH_FRAME_OFFSET*[i] defines the delay ($i = 1$ stands for the primary FSCh; $i = 2$ for the secondary FSCh).

The following tables present bit rates obtained during frame processing with all radio configurations used by FSChs. In the tables, n_f represents the length of FSCh frames in multiples of 20 ms.

In Tables 4.52 and 4.53.

- for frames with up to 40 bits, n_f can be set to 1 or 2;

- for frames with more than 40 bits, n_f is 1, 2 or 4;

- for frames with 360 bits or more, the turbo encoder can be used or not, otherwise, the convolutional coder should be used.

In Table 4.54,

- for frames with up to 55 bits, n_f can be 1 or 2;

- for frames with more than 55 bits, n_f is 1, 2 or 4;

- for frames with 552 bits or more, the turbo encoder can be used, otherwise, the convolutional coder specified in the table should be used.

In Table 4.55, for frames with up to 40 bits, n_f can be 1 or 2. For all other cases, n_f is 1, 2 or 4.

In Table 4.56, related to RC7,

- for frames with up to 40 bits, n_f can be 1 or 2;

- for frames with more than 40 bits, n_f is 1, 2 or 4;

Table 4.52 FSCh bit rates for RC3 operation

Information bits	Frame (ms)	Reserved bit	CRC (F)	Reserved/tail	Total (kbps)	Convolutional/turbo encoder	Repetition	Symbol puncturing	Interleaving block	Coded symbols (ksps)
16	20	0	6	8	1.5	$R = 1/4, k = 9$	8	1/5	768	38.4
40	$20n_f$	0	6	8	$2.7/n_f$	$R = 1/4, k = 9$	4	1/9	768	$38.4/n_f$
80	$20n_f$	0	8	8	$4.8/n_f$	$R = 1/4, k = 9$	2	–	768	$38.4/n_f$
172	$20n_f$	0	12	8	$9.6/n_f$	$R = 1/4, k = 9$	1	–	768	$38.4/n_f$
360	$20n_f$	0	16	8	$19.2/n_f$	$R = 1/4, k = 9$	1	–	1536	$76.8/n_f$
744	$20n_f$	0	16	8	$38.4/n_f$	$R = 1/4, k = 9$	1	–	3072	$153.6/n_f$
1512	$20n_f$	0	16	8	$76.8/n_f$	$R = 1/4, k = 9$	1	–	6144	$307.2/n_f$
3048	$20n_f$	0	16	8	$153.6/n_f$	$R = 1/4, k = 9$	1	–	12 288	$614.4/n_f$

Table 4.53 FSCh bit rates for RC4 operation

Information bits	Frame (ms)	Reserved bit	CRC (F)	Reserved/tail	Total (kbps)	Convolutional/turbo encoder	Repetition	Symbol puncturing	Interleaving block	Coded symbols (ksps)
16	20	0	6	8	1.5	$R = 1/2, k = 9$	8	1/5	384	19.2
40	$20n_f$	0	6	8	$2.7/n_f$	$R = 1/2, k = 9$	4	1/9	384	$19.2/n_f$
80	$20n_f$	0	8	8	$4.8/n_f$	$R = 1/2, k = 9$	2	–	384	$19.2/n_f$
172	$20n_f$	0	12	8	$9.6/n_f$	$R = 1/2, k = 9$	1	–	384	$19.2/n_f$
360	$20n_f$	0	16	8	$19.2/n_f$	$R = 1/2, k = 9$	1	–	768	$38.4/n_f$
744	$20n_f$	0	16	8	$38.4/n_f$	$R = 1/2, k = 9$	1	–	1536	$76.8/n_f$
1512	$20n_f$	0	16	8	$76.8/n_f$	$R = 1/2, k = 9$	1	–	3072	$153.6/n_f$
3048	$20n_f$	0	16	8	$153.6/n_f$	$R = 1/2, k = 9$	1	–	6144	$307.2/n_f$
6120	$20n_f$	0	16	8	$307.2/n_f$	$R = 1/2, k = 9$	1	–	12 288	$614.4/n_f$

Table 4.54 FSCh bit rates for RC5 operation

Information bits	Frame (ms)	Reserved bit	CRC (F)	Reserved/ tail	Total (kbps)	Convolutional/turbo encoder	Repetition	Symbol puncturing	Interleaving block	Coded symbols (ksps)
21	20	1	6	8	1.8	$R = 1/4, k = 9$	8	4/12	768	38.4
55	$20n_f$	1	8	8	$3.6/n_f$	$R = 1/4, k = 9$	4	4/12	768	$38.4/n_f$
125	$20n_f$	1	10	8	$7.2/n_f$	$R = 1/4, k = 9$	2	4/12	768	$38.4/n_f$
267	$20n_f$	1	12	8	$14.4/n_f$	$R = 1/4, k = 9$	1	4/12	768	$38.4/n_f$
552	$20n_f$	0	16	8	$28.8/n_f$	$R = 1/4, k = 9$	1	4/12	1536	$76.8/n_f$
1128	$20n_f$	0	16	8	$57.6/n_f$	$R = 1/4, k = 9$	1	4/12	3072	$153.6/n_f$
2280	$20n_f$	0	16	8	$115.2/n_f$	$R = 1/4, k = 9$	1	4/12	6144	$307.2/n_f$
4584	$20n_f$	0	16	8	$230.4/n_f$	$R = 1/4, k = 9$	1	4/12	12 288	$614.4/n_f$

Table 4.55 FSCh bit rates for RC6 operation

Information bits	Frame (ms)	Reserved bit	CRC (F)	Reserved/ tail	Total (kbps)	Convolutional/turbo encoder	Repetition	Symbol puncturing	Interleaving block	Coded symbols (ksps)
16	20	0	6	8	1.5	$R = 1/6, k = 9$	8	1/5	1152	57.6
40	$20n_f$	0	6	8	$2.7/n_f$	$R = 1/6, k = 9$	4	1/9	1152	$57.6/n_f$
80	$20n_f$	0	8	8	$4.8/n_f$	$R = 1/6, k = 9$	2	–	1152	$57.6/n_f$
172	$20n_f$	0	12	8	$9.6/n_f$	$R = 1/6, k = 9$	1	–	1152	$57.6/n_f$
360	$20n_f$	0	16	8	$19.2/n_f$	$R = 1/6, k = 9$	1	–	2304	$115.2/n_f$
744	$20n_f$	0	16	8	$38.4/n_f$	$R = 1/6, k = 9$	1	–	4608	$230.4/n_f$
1512	$20n_f$	0	16	8	$76.8/n_f$	$R = 1/6, k = 9$	1	–	9216	$460.8/n_f$
3048	$20n_f$	0	16	8	$153.6/n_f$	$R = 1/6, k = 9$	1	–	18 432	$921.6/n_f$
6120	$20n_f$	0	16	8	$307.2/n_f$	$R = 1/6, k = 9$	1	–	36 864	$1843.2/n_f$

Table 4.56 FSCh bit rates for RC7 operation

Information bits	Frame (ms)	Reserved bit	CRC (F)	Reserved/tail	Total (kbps)	Convolutional/turbo encoder	Repetition	Symbol puncturing	Interleaving block	Coded symbols (ksps)
16	20	0	6	8	1.5	$R = 1/3, k = 9$	8	1/5	576	28.3
40	$20n_f$	0	6	8	$2.7/n_f$	$R = 1/3, k = 9$	4	1/9	576	$28.8/n_f$
80	$20n_f$	0	8	8	$4.8/n_f$	$R = 1/3, k = 9$	2	–	576	$28.8/n_f$
172	$20n_f$	0	12	8	$9.6/n_f$	$R = 1/3, k = 9$	1	–	576	$28.8/n_f$
360	$20n_f$	0	16	8	$19.2/n_f$	$R = 1/3, k = 9$	1	–	1152	$57.6/n_f$
744	$20n_f$	0	16	8	$38.4/n_f$	$R = 1/3, k = 9$	1	–	2304	$115.2/n_f$
1512	$20n_f$	0	16	8	$76.8/n_f$	$R = 1/3, k = 9$	1	–	4608	$230.4/n_f$
3048	$20n_f$	0	16	8	$153.6/n_f$	$R = 1/3, k = 9$	1	–	9216	$460.4/n_f$
6120	$20n_f$	0	16	8	$307.2/n_f$	$R = 1/3, k = 9$	1	–	18 432	$921.6/n_f$
12 264	$20n_f$	0	16	8	$614.4/n_f$	$R = 1/3, k = 9$	1	–	36 864	$1843.2/n_f$

Table 4.57 FSCh bit rates for RC8 operation

Information bits	Frame (ms)	Reserved bit	CRC (F)	Reserved/tail	Total (kbps)	Convolutional/turbo encoder	Repetition	Symbol puncturing	Interleaving block	Coded symbols (kbps)
21	20	1	6	8	1.8	$R = 1/4, k = 9$	8	–	1152	57.6
55	$20n_f$	1	8	8	$3.6/n_f$	$R = 1/4, k = 9$	4	–	1152	$57.6/n_f$
125	$20n_f$	1	10	8	$7.2/n_f$	$R = 1/4, k = 9$	2	–	1152	$57.6/n_f$
267	$20n_f$	1	12	8	$14.4/n_f$	$R = 1/4, k = 9$	1	–	1152	$57.6/n_f$
552	$20n_f$	0	16	8	$28.8/n_f$	$R = 1/4, k = 9$	1	–	2304	$115.2/n_f$
1128	$20n_f$	0	16	8	$57.6/n_f$	$R = 1/4, k = 9$	1	–	4608	$230.4/n_f$
2280	$20n_f$	0	16	8	$115.2/n_f$	$R = 1/4, k = 9$	1	–	9216	$460.8/n_f$
4584	$20n_f$	0	16	8	$230.4/n_f$	$R = 1/4, k = 9$	1	–	18 432	$921.6/n_f$
9192	$20n_f$	0	16	8	$460.8/n_f$	$R = 1/4, k = 9$	1	–	36 864	$1843.2/n_f$

Table 4.58 FSCh bit rates for RC9 operation

Information bits	Frame (ms)	Reserved bit	CRC (F)	Reserved/ tail	Total (kbps)	Convolutional/turbo encoder	Repetition	Symbol puncturing	Interleaving block	Coded symbols (kbps)
21	20	1	6	8	1.8	$R = 1/2, k = 9$	8	–	576	28.8
55	$20n_f$	1	8	8	$3.6/n_f$	$R = 1/2, k = 9$	4	–	576	$28.8/n_f$
125	$20n_f$	1	10	8	$7.2/n_f$	$R = 1/2, k = 9$	2	–	576	$28.8/n_f$
267	$20n_f$	1	12	8	$14.4/n_f$	$R = 1/2, k = 9$	1	–	576	$28.8/n_f$
552	$20n_f$	0	16	8	$28.8/n_f$	$R = 1/2, k = 9$	1	–	1152	$57.6/n_f$
1128	$20n_f$	0	16	8	$57.6/n_f$	$R = 1/2, k = 9$	1	–	2304	$115.2/n_f$
2280	$20n_f$	0	16	8	$115.2/n_f$	$R = 1/2, k = 9$	1	–	4608	$230.4/n_f$
4584	$20n_f$	0	16	8	$230.4/n_f$	$R = 1/2, k = 9$	1	–	9216	$460.8/n_f$
9192	$20n_f$	0	16	8	$460.8/n_f$	$R = 1/2, k = 9$	1	–	18 432	$921.6/n_f$
20 712	$20n_f$	0	16	8	$1036.8/n_f$	$R = 1/2, k = 9$	1	2/18	36 864	$1843.2/n_f$

- for frames with 360 bits or more, the turbo encoder can be used, otherwise, the convolutional coder should be used.

In Tables 4.57 and 4.58, related to configurations RC8 and RC9:

- for frames with up to 55 bits, n_f can be 1 or 2;

- for frames with more than 55 bits, n_f is 1, 2 or, 4;

- for frames with 552 bits or more, the turbo encoder can be used, otherwise the convolutional coder specified in the table should be used.

4.2.11.1 Turbo Encoder

The turbo encoder works with two convolutional coders in a recursive and systematic way. The two coders (called constituent encoders) are connected in parallel. A special interleaving block, called turbo interleaver, precedes the second convolutional coder, as in Figure 4.109.

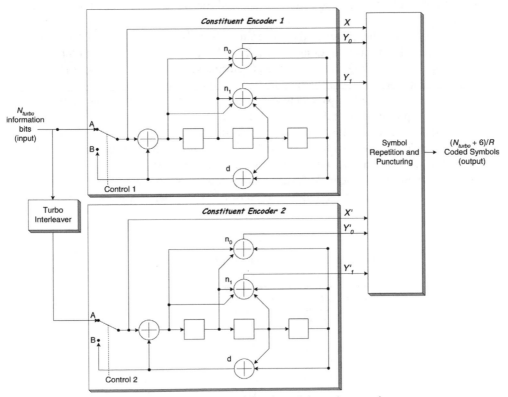

Figure 4.109 Basic configuration of the turbo encoder.

The turbo encoder codes all data bits, the frame quality indicator (CRC) and the two reserved bits inserted in the input data as padding bits during the organisation of each frame. Six additional padding bits are added at the turbo encoder output.

Considering that the sum of the information bits, CRC and reserved bits results in N_{turbo} bits, the turbo encoder output is $(N_{turbo} + 6)/R$ bits, where R represents the recursive convolutional coders coding rate (1/2, 1/3, 1/4 and 1/5) and 6 stands for the turbo generated padding bits.

Initially, all turbo encoder registers are set to zero to process any traffic channel frame. During encoding, the switches (control 1 and 2) are changed, re-configuring the coders according to the type of bits being processed (data bits or tail bits).

When controls 1 and 2 are in position A, as in Figure 4.109, there are N_{turbo} clock pulses processing the data bits. Then switch Control 1, on constituent encoder 1, changes to position B and clock pulses (that depend on the number of tail bits used) generate the first $3/R$ output bits of the turbo encoder. Upon completion, switch Control 2 also changes to position B and clock pulses trigger the passage of tail bits by the shift registers, generating the remaining $3/R$ output bits of the turbo encoder.

The bit reading sequence after processing always follows the order X, Y_0, Y_1, X', Y_0', Y_1', regardless of whether the output bits are data, padding or reserved bits. Even though all bits are read in the output, not all of them are transmitted.

Regarding the puncturing patterns in Table 4.59, each '0' represents a coded bit that must be excluded, whereas bits '1' represent a bit that must be kept. If the frame is bigger than the associated puncturing pattern, the pattern is repeated for the remaining bits until completing the frame.

Table 4.59 shows the patterns that determine bits to be kept and to be punctured. Patterns are read from top to bottom (X, Y_0, Y_1, X', Y_0', Y_1') and then from left to right.

For example, for a coding rate of 1/2, the puncturing sequence is '1 1 0 0 0 0 1 0 0 0 1 0', which means that the bits from the outputs X Y_0 X Y_0' are to be kept.

As for data bits, there are also puncturing patterns for padding bits (Table 4.60). Patterns are read from top to bottom and then from left to right.

Some outputs might have their bits repeated according to the coding rate; this situation is represented in Table 4.60 with numbers 2 and 3. Table 4.61 summarises the outputs considered to generate the $6/R$ bits related to the coded tail bits.

Table 4.61 shows that puncturing ensures that no coded bits are read from the constituent encoder 2 when determining the tail bits of constituent encoder 1, and vice versa.

Table 4.59 Puncturing patterns for coded data bits from the turbo encoder

Output	$R = 1/2$	$R = 1/3$	$R = 1/4$	$R = 1/5$
X	11	11	11	11
Y_0	10	11	11	11
Y_1	00	00	10	11
X'	00	00	00	00
Y_0'	01	11	01	11
Y_1'	00	00	11	11

Table 4.60 Puncturing patterns for tail bits from the turbo encoder

Output	$R = 1/2$	$R = 1/3$	$R = 1/4$	$R = 1/5$
X	111 000	222 000	222 000	333 000
Y_0	111 000	111 000	111 000	111 000
Y_1	000 000	000 000	111 000	111 000
X'	000 111	000 222	000 222	000 333
Y_0'	000 111	000 111	000 111	000 111
Y_1'	000 000	000 000	000 111	000 111

Table 4.61 Repetition and reading patterns at the constituent encoders output

Reading of coded tail bits	Rate 1/2	Rate 1/3	Rate 1/4	Rate 1/5
Constituent encoder 1	X Y_0	X X Y_0	X X Y_0 Y_1	X X X Y_0 Y_1
Constituent encoder 2	X' Y_0'	X' X' Y_0'	X' X' Y_0' Y_1'	X' X' X' Y_0' Y_1'

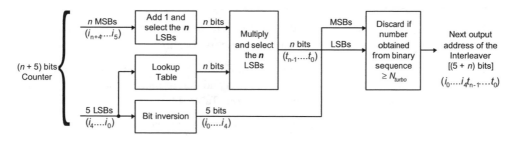

Figure 4.110 Turbo interleaver configuration.

Before passing through constituent encoder 2, bits are turbo interleaved. Figure 4.110 shows the interleaving of N_{turbo} bits (information bits, CRC and 2 reserved tail bits), numbered from 0 to $N_{turbo} - 1$.

The bit interleaving on the turbo encoder, also known as turbo interleaving, consists in writing the bits in a matrix with 2^5 lines (32) and 2^n columns, where n is obtained from Table 4.62. The parameter n corresponds to the lower integer that matches $N_{turbo} \leq 2^{n+5}$.

The steps to calculate the reading sequence of the bits in the turbo interleaver matrix are the following.

Table 4.62 Parameter n of the turbo interleaver

Block of N_{turbo} bits	Parameter n
378	4
402	4
570	5
762	5
786	5
1146	6
1530	6
1554	6
2298	7
2322	7
3066	7
3090	7
3858	7
4602	8
6138	8
9210	9
12 282	9
20 730	10

1. Reset the counter of $(n + 5)$ bits.

2. For the n most significant bits (n MSBs), add 1 to obtain a new value and discard the other bits, except the n less significant bits (n LSBs) of this new value, named N_1.

3. Use the five less significant bits (5 LSBs $\to i_4 \ldots i_0$) to calculate the index $c = (16 \times i_4 + 8 \times i_3 + 4 \times i_2 + 2 \times i_1 + 1 \times i_0)$ and determine, for the value n obtained before, what is the corresponding N_2 value from Table 4.63.

4. Multiply $N_{TI} = N_1 \times N_2$, and from this value keep only the n LSBs, generating the sequence $(t_{n-1} \cdots t_0)$,

5. Reverse the order of the counter's five LSBs.

6. Generate an output trial, consisting of the inversion of bits, creating a sequence $(i_0 \cdots i_4 t_{n-1} \cdots t_0)$. Accept this trial if the value obtained from these bits is less than N_{turbo}.

7. If the output trial is not approved, add the counter and repeat steps 2–6 until the turbo interleaver has processed all N_{turbo} bits.

Figure 4.111 presents the reading sequence for the turbo interleaver matrix.

Table 4.63 presents the lookup table values to be obtained from the index c, according to the parameter n used by the turbo interleaver.

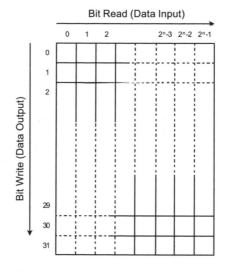

Figure 4.111 Reading sequence for the turbo interleaver matrix.

Table 4.63 Reference table for determining parameter N_2 of the turbo interleaver

Index (n)	4	5	6	7	8	9	10
0	5	27	3	15	3	13	1
1	15	3	27	127	1	335	349
2	5	1	15	89	5	87	303
3	15	15	13	1	83	15	721
4	1	13	29	31	19	15	973
5	9	17	5	15	179	1	703
6	9	23	1	61	19	333	761
7	15	13	31	47	99	11	327
8	13	9	3	127	23	13	453
9	15	3	9	17	1	1	95
10	7	15	15	119	3	121	241
11	11	3	31	15	13	155	187
12	15	13	17	57	13	1	497
13	3	1	5	123	3	175	909
14	15	13	39	95	17	421	769
15	5	29	1	5	1	5	349
16	13	21	19	85	63	509	71
17	15	19	27	17	131	215	557
18	9	1	15	55	17	47	197
19	3	3	13	57	131	425	499
20	1	29	45	15	211	295	409
21	3	17	5	41	173	229	259
22	15	25	33	93	231	427	335

Table 4.63 (*Continued*)

Index (n)	4	5	6	7	8	9	10
23	1	29	15	87	171	83	253
24	13	9	13	63	23	409	677
25	1	13	9	15	147	387	717
26	9	23	15	13	243	193	313
27	15	13	31	15	213	57	757
28	11	13	17	81	189	501	189
29	3	1	5	57	51	313	15
30	15	13	15	31	15	489	75
31	5	13	33	69	67	391	163

4.2.11.2   Puncturing

Coded bits from FSCh frames may be punctured, depending on the bit rate per frame and on the RC in use. There are two types of puncturing: convolutional code puncturing (described in Section 4.2.9.2) and turbo code puncturing. The coding process (convolutional or turbo) in use determines the type of puncturing. When using the convolutional encoder, Table 4.45 describes the puncturing patterns.

4.2.11.2.1   Turbo Code Symbol Puncturing

Turbo puncturing happens after the bits are turbo encoded. Puncturing execution depends on the encoder rate, on the radio configuration, and on the puncturing patterns specified in Table 4.64.

As in regular puncturing, a bit '0' means that associated symbol must be removed, whereas a bit '1' means that the symbol is kept. If the frame is bigger than the length of the associated pattern, the puncturing pattern is repeated until completing the frame length.

If using the turbo encoder, besides the puncturing executed during the encoding process, another bit exclusion happens after encoding. Therefore, two puncturing process are performed.

4.2.11.3   Interleaving

The interleaving process starts aligned with the beginning of FSCh frames, which can be 20, 40 or 80 ms long. After convolutional or turbo coding, and bits repetition/puncturing, the

Table 4.64 Patterns used for data bits turbo puncturing

Turbo rate	Puncturing rate	Puncturing patterns	Associated RC
1/2	2/18	'111110101111111111'	9
1/4	4/12	'110111011010'	5

coded bits are interleaved using forward-backwards BRO interleaving (Figure 4.58) for SR1 and complex cyclic-shift BRO interleaving (Figure 4.60) for SR3. Both types of interleaving are described in Section 4.2.4.1.

Table 4.17 presents parameters m and J that define the interleaving matrix size, whereas Tables 4.52–4.58 define the number of interleaved bits per frame according to the RC being used.

4.2.11.4 Scrambling

Scrambling of FSChs is equivalent to the process described for forward common control channels (Figure 4.63), where two long code chips are decimated for scrambling groups of $2M$ bits.

The public long code offset mask used to scramble FSCh coded bits is the same as that used for all traffic channels, and depends on the permuted *ESN* value of each MS, as in Figure 4.112.

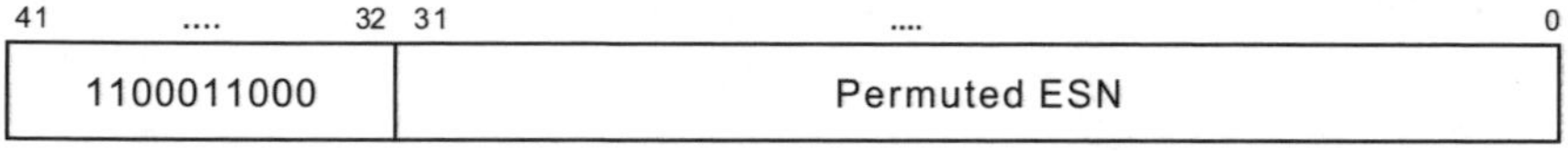

41	....	32	31	....	0
1100011000			Permuted ESN		

Figure 4.112 Public long code offset mask of forward supplemental channels.

FSChs, as any traffic channel, can also be scrambled by chips decimated from the PNLC after application of the *private* long code offset mask, described in the chapter on call processing.

4.2.11.5 De-multiplexing (Serial-to-Parallel Conversion)

FSChs (RC3–RC9) can operate in three modes: SR1 in non-TD mode (Figure 4.113), SR1 with OTD or STS (Figure 4.114) and SR3 MC (Figure 4.115). Each of these processes is described in Section 4.2.5.3.

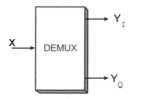

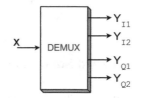

Figure 4.113 DEMUX operation for SR1 in non-TD mode.

Figure 4.114 DEMUX operation for SR1 in TD mode (OTD or STS).

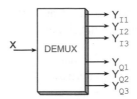

Figure 4.115 DEMUX operation for SR3 mode.

4.2.11.6 Channelisation and Spreading by Walsh Sequences and Complex Multiplier Spreading by PN Sequences

Table 4.65 shows the length of Walsh sequences used to spread and channelise FSChs, associated to the SR and transmission data rate.

BTSs assign the value for N, depending on the data rate transmitted by the channel. The higher the data rate, the smaller the Walsh code length assigned to FSChs. This ensures a smaller spreading factor, allowing more bits to be transmitted. The disadvantage, however, is that shorter Walsh code lengths provide lower protection during transmission.

Table 4.65 Length of Walsh sequences used by FSChs

SR	RC	Walsh code	Sequence index	Walsh code length (N)	Symbol rate after spreading (Msps)
SR1	3, 4 and 5	W_i^N	$1 \le i \le (N-1)$	$4, 8, 16, 32, 64, 128$	2.4576
SR3	6, 7, 8 and 9	W_i^N	$1 \le i \le (N-1)$	4, 8, 16, 32, 64, 128, 256	7.3728

This technique is used in cdma2000 to improve transmission data rates and to require smaller processing gain values (P_G). Therefore FSChs can only be used for higher data rates if users are located close to a BTS and have low mobility.

Employing shorter Walsh codes requires reducing the number of voice users active in the system. The chapter on radio resource management explains this implication in detail.

Because forward supplemental channels are designed for cdma2000 radio configurations (RC3–RC9), they can operate in any of the TD schemes. Figures 4.37, 4.69 and 4.70 illustrate Walsh code spreading and channelising, PN complex multiplier spreading and QPSK carrier modulation for FSChs in systems operating, respectively, with SR1 non-TD or SR3, SR1 OTD and SR1 STS.

4.2.12 Forward Supplemental Code Channel

The Forward Supplemental Code Channels (FSCCh) were developed for CDMA IS-95B systems to improve data rate transmissions during a call or while in mobile station control on the traffic channel state.

Table 4.66 Bit rate of FSCCHs frames

RC	Data rate (bps)	Reserved bit	Information bits	CRC (F)	Encoder tail bits	Bits/frame
1	9600	0	172	12	8	192
2	14400	1	267	12	8	288

Because this channel is CDMA IS-95B compatible, it is only applicable to RC1 and RC2 (SR1), operating with rates of 9.6 and 14.4 kbps, respectively. Each traffic channel can have up to seven forward supplemental code channels, allowing MSs to theoretically achieve transmission data rates of up to 115 kbps (7 FSCCHs and 1 FTCh).

FSCCHs are organised in 20 ms frames, as in Table 4.66. Figure 4.116 shows the frame structure of FSCCHs. In the figure, R represents a reserved bit (EIB); F, the Frame Quality Indicator (CRC) and T, the encoder tail bits ('0'). The encoder tail bits are used to decorrelate bits in successive frames during convolutional encoder processing, that is, the eight bits '0' 'clean' the convolutional encoder's shift registers.

R	Information Bits	F	T

Figure 4.116 Frame structure of forward supplemental code channels.

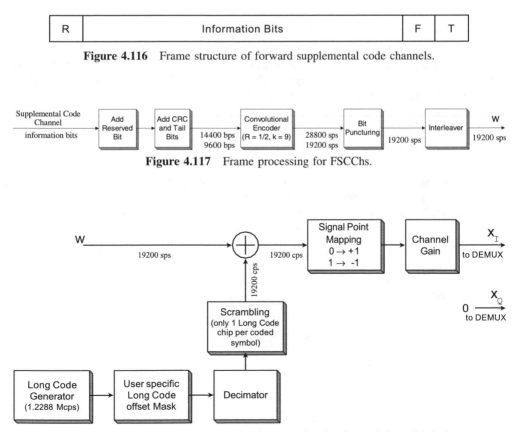

Figure 4.117 Frame processing for FSCCHs.

Figure 4.118 FSCCh frame processing between interleaving and de-multiplexing.

Figures 4.117 and 4.118 show a block diagram representing the FSCCh frame processing, which is quite similar to FFChs processing when transmitting with the maximum rate available for RC1 and RC2. FSCChs, however, do not have the insertion of power control sub-channels.

Tables 4.67 and 4.68 shows the bit rates during the processing of FSCChs for the two possible radio configurations (RC1 and RC2).

4.2.12.1 Convolutional Coding

FSCChs use the same convolutional encoder used for CDMA IS-95 channels, with coding rate $R = 1/2$ and $k = 9$, as in Figure 4.42.

4.2.12.2 Puncturing

Puncturing is only executed when operating with RC2. Because the data rate, in this situation, is 28800 bps, puncturing is used to equalise the number of coded bits prior to interleaving in 19.2 kbps (Figure 4.117).

The convolutional code symbol puncturing process is described in Section 4.2.9.2, and the puncturing pattern is given by Table 4.45.

4.2.12.3 Interleaving

After puncturing, the bits are also interleaved. Because the coded bit rate, at the interleaver input, is constant for both radio configurations (RC1 and RC2), the system uses the BRO interleaver for interleaving FSCChs (Figure 4.43). The interleaving process must start aligned to the beginning of FSCChs frames.

4.2.12.4 Scrambling

Figure 4.119 shows the public long code offset mask used to generate the offset prior to decimating chips for scrambling the FSCChs coded bits.

The scrambling process uses long code chips decimated with the same rate as the coded

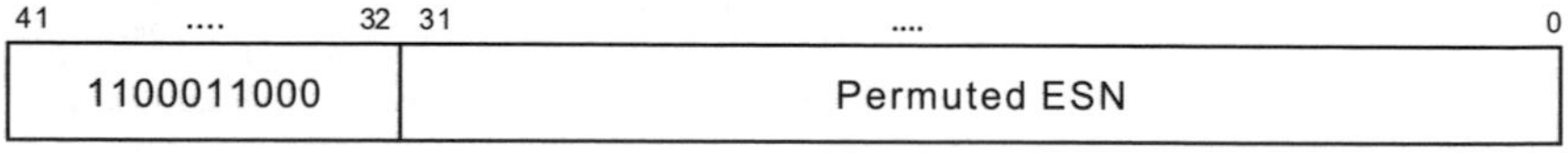

41	....	32	31	....	0
1100011000			Permuted ESN		

Figure 4.119 Public long code mask of forward supplemental code channels.

symbol rate after interleaving, i.e. 19 200 sps, that is, one chip at every 64. This process is described in Section 0.

4.2.12.5 De-multiplexing

Because this channel is compatible to CDMA IS-95B systems, using SR1 with RCs 1 and 2, only the non-TD mode is supported.

Table 4.67 Bit rates of FSCChs frames for RC1

Information bits	Frame (ms)	Reserved bit	CRC (F)	Tail bits	Total (kbps)	Convolutional coder	Bit puncturing	Interleaving block	Coded symbol rate after interleaving (ksps)
172	20	0	12	8	9.6	$R = 1/2, k = 9$	–	384	19.2

Table 4.68 Bit rates of FSCChs frames for RC2

Information bits	Frame (ms)	Reserved bit	CRC (F)	Tail bits	Total (kbps)	Convolutional coder	Bit puncturing	Interleaving block	Coded symbol rate after interleaving (ksps)
267	20	1	12	8	14.4	$R = 1/2, k = 9$	2/6	384	19.2

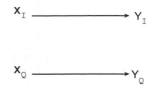

Figure 4.120   DEMUX operation of the FSCChs with RC1 and RC2 in non-TD mode.

The de-multiplexing process consists in relaying coded bits directly to the spreading and orthogonal modulation processes, as in Figure 4.120. For compatibility, no information is sent over the Q branch.

4.2.12.6   Channelisation and Spreading by Walsh Sequences and Complex Multiplier Spreading by PN Sequences

Walsh Codes of length 64 spread and channelise FSCChs coded bits, as detailed in Table 4.69.

Because FSCChs only operate with RCs 1 and 2, i.e. SR1, the transmission via air interface is performed as in Figure 4.121.

Table 4.69   Length of Walsh sequences used for FSCChs

SR	RC	Coded bit rate after interleaving (kbps)	Walsh code	Sequence index	Symbol rate after spreading (Msps)
SR1	1 and 2	19.2	W_i^{64}	$1 \leq i \leq 63$	1.2288

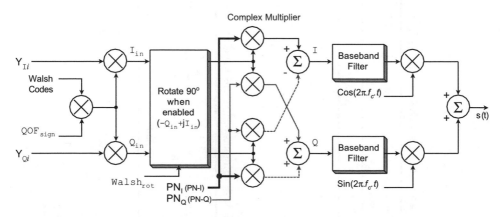

Figure 4.121   Spreading, channel multiplexing and complex PN code spreading process for forward supplemental code channels.

BIBLIOGRAPHY AND REFERENCES

1. C.S0002-C, Physical Layer Standard for cdma2000 Spread Spectrum Systems, Release C, 3GPP2, May 2002.
2. C.S0005-B, Upper Layer (Layer 3) Signaling Standard for cdma2000 Spread Spectrum Systems, Release B, 3GPP2, April 2002.
3. TIA/EIA-95-B, Mobile Station – Base Station Compatibility Standard for Dual-Mode Spread Spectrum Cellular System, November 1998.
4. CelTec/CelPlan, CDMA IS-95 and cdma2000 Systems – Training Course.
5. Lee, J. S. and Miller, L. E., *CDMA Systems Engineering Handbook*. Boston: Artech House, 1998.

5

Reverse Link Channels

BRUNO DE SOUZA ABREU XAVIER

This chapter is concerned with the reverse link channel structure of IS-95 and cdma2000 networks. The main logical and functional characteristics of each type of channel are surveyed. The behavior and processing of each channel are examined in detail.

5.1 CDMA IS-95 REVERSE LINK CHANNEL STRUCTURE

In CDMA IS-95 networks, MSs can transmit on one of two logical channels.

- Reverse Traffic Channel (RTCh),
- Reverse Access Channel (RACh).

Figure 5.1 presents the processing sequence of IS-95 reverse link channels.

Each channel has different functions. MSs use RAChs to access the system, for example, to respond to the system or to establish a call. MSs use RTChs only during a call, i.e. when users are already in communication. The following sections in this chapter describe each channel in detail.

5.1.1 Reverse Traffic Channel (RTCh)

The RTCh undergoes a complex process because of the need of voice quantisation and coding prior to channel processing. Besides, RTChs are not only used for voice traffic, but also for data, fax and signalling during a call. Figure 5.2 shows the reverse traffic channel processing block diagram for these networks.

5.1.1.1 Voice Sampling, Quantisation and Coding

Prior to transmission through the air interface, MSs must transform voice into digital symbols, i.e. digitise the voice signal, which must be then quantised and coded.

Designing CDMA 2000 Systems   L. Korowajczuk
© 2004 John Wiley & Sons, Ltd   ISBN: 0-470-85399-9

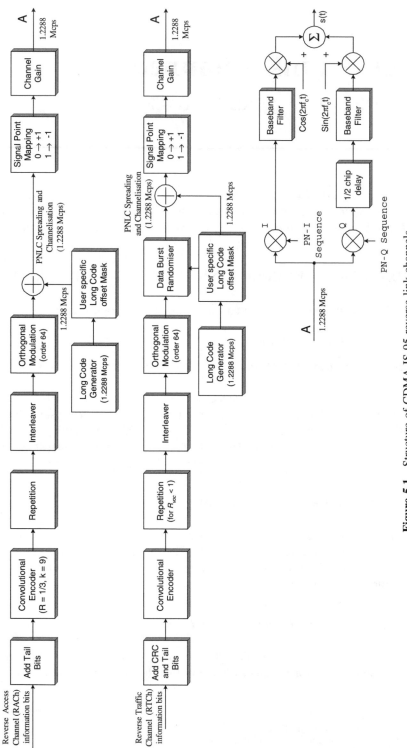

Figure 5.1 Structure of CDMA IS-95 reverse link channels.

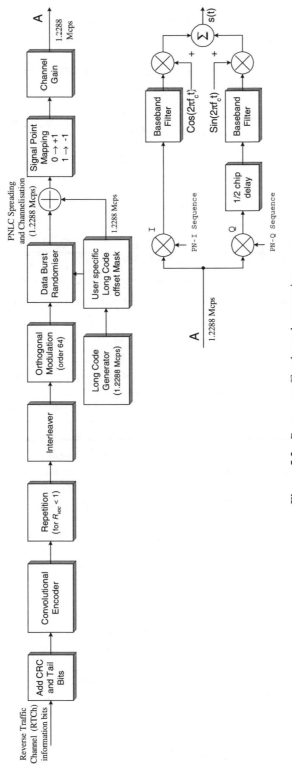

Figure 5.2 Reverse traffic channel processing.

To digitise the voice, a low pass filter first limits the voice signal in a frequency band ranging from 300 Hz to approximately 3400 Hz. This range comprises the most important speech characteristics.

Figure 5.3 shows the voice coding and decoding process for transmission and reception at mobile terminals, depicting the data rate at the vocoder.

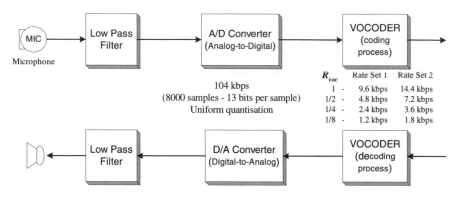

R_{voc}		Rate Set 1	Rate Set 2
1	-	9.6 kbps	14.4 kbps
1/2	-	4.8 kbps	7.2 kbps
1/4	-	2.4 kbps	3.6 kbps
1/8	-	1.2 kbps	1.8 kbps

104 kbps
(8000 samples - 13 bits per sample)
Uniform quantisation

Figure 5.3   Voice coding and decoding processes performed by the MS on forward and reverse traffic channels.

The A/D (Analog-to-Digital) converter at the mobile terminal samples the voice signal 8000 times per second, or 160 samples per 20 ms frame, and codes each sample using Pulse Code Modulation (PCM). Each sample is coded into a set of 13 bits that represent 8192 different levels divided into uniform steps, obtaining a total rate of 104 kbps.

This bit rate (104 kbps), however, is much higher than what is usually transmitted on cellular networks. To lower this rate, an additional coder, called a vocoder, processes data and generates a digitised voice signal. Vocoders employ DSP algorithms to extract essential speech and voice characteristics from the signal, compressing the number of bits. Vocoders have variable output rate, i.e. the number of bits at the output depends on the voice activity factor. Lesser voice activity implies less energy of the voice signal and fewer bits at the vocoder output.

Usually, there are four discrete voice activity levels, defined by three dynamically adjusted thresholds that depend on the background noise. This allows achieving higher bit rates only when there is more user voice activity, leading to good voice quality even when in noisy environments.

There are two basic vocoder Rate Sets (RS), RS1 and RS2, with maximal net rates of 8.6 kbps and 13.3 kbps, respectively. However, vocoders are sometimes identified by their gross bit rates, 9.6 kbps and 14.4 kbps. Vocoder rates (R_{voc}) can be 1 (full rate), 1/2 (half rate), 1/4 (quarter) and 1/8. The RTCh throughput depends on the combination of the vocoder rate (R_{voc}) and RS.

After the vocoder, Cyclic Redundancy Check (CRC) bits and encoder tail bits (8 bits) are added at the end of each frame. Rates per 20 ms frames and CRC circuits used on the reverse link traffic channels are the same as employed on forward link traffic channels.

Figure 5.4 presents the reverse traffic channel's frame structure. In the figure, R/E represents a reserved bit (Erasure Indicator Bit – EIB), F corresponds to frame quality indicators or CRC and T represents eight encoder tail bits '0'.

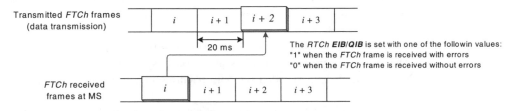

Figure 5.4   RFChs frame structure.

MSs only use the EIB when working with RS2 vocoders and when both RTCh and FTCh are continuously transmitting data to inform BTSs whether a frame has been detected with errors.

If an error detection occurs, resulting in a frame erasure, MSs must set the EIB to '1' when transmitting the second RTCh frame after the erroneous FTCh frame, as in Figure 5.5.

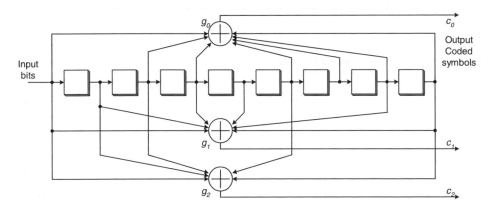

Figure 5.5   EIB determination and transmission procedure.

5.1.1.2   Convolutional Encoding and Coded Bits Repetition

To improve system protection, each 20 ms frame is convolutionally encoded using an encoder with rate 1/3 and constraint length $k = 9$, instead of using the half-rate encoder as in the forward link. This 1/3 rate encoder (Figure 5.6) provides extra protection to reverse link bits, which are usually 'weaker' (more vulnerable to interference) than forward link bits.

Figure 5.6   CDMA IS-95 reverse link convolutional encoder for RS1 vocoders.

Eight encoder tail bits ('0s'), inserted at the vocoder output, 'clean-up' (or re-start) the convolutional coder, allowing sequential 20 ms TCh frames to be independent of each other. In fact, the tail bits provide additional benefit to the Viterbi decoder algorithm, providing a known sequence of coder input bits at the end of the frame.

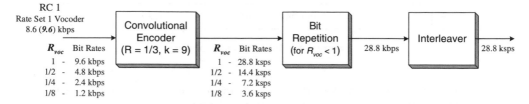

Figure 5.7 Convolutional encoding and coded bits repetition for 9.6 kbps (RS1) vocoders (radio configuration 1).

Figure 5.7 illustrates RTCh processing, showing convolutional coding, bit repetition and interleaving, and providing data rates obtained at each stage when using RS1 vocoders (9.6 kbps).

Table 5.1 presents the number of coded symbols (after convolutional coding), according to the R_{voc} in use. The table also provides bit rates after coded symbol repetition. Prior to interleaving, all 20 ms frames have 576 coded symbols, resulting in a maximal throughput of 28.8 kbps.

Table 5.2 shows a summary of data rates obtained at each stage during RTCh processing when using 9.6 kbps vocoders.

Table 5.1   Bit repetitions for 9.6 Kbps vocoders

Vocoder rates (R_{voc})	Data rate (bps)	Convolutional encoder output (sps)	Number of repetitions	Interleaving input rate (Ksps)
1	9600	28800		28.8
1/2	4800	14400	1	28.8
1/4	2400	7200	3	28.8
1/8	1200	3600	7	28.8

Table 5.2   Data rates for 9.6 kbps vocoders

Vocoder input		Vocoder rate (R_{voc})	Vocoder output		Convolutional encoder output		
PSTN PCM[a]	Linear coding		Net rate	Gross rate	Before repetition block	After repetition block	
1280	2080	1	172	192	576	576	Symbol/frame
64000	104000		8600	9600	28800	28800	sps
1280	2080	1/2	80	96	288	576	Symbol/frame
64000	104000		4000	4800	14400	28800	sps
1280	2080	1/4	40	48	144	576	Symbol/frame
64000	104000		2000	2400	7200	28800	sps
1280	2080	1/8	16	24	72	576	Symbol/frame
64000	104000		1000	1200	3600	28800	sps

[a]This column is provided for comparison purposes.

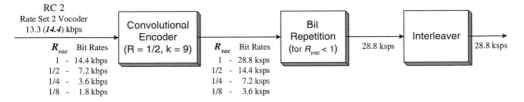

Figure 5.8 Convolutional encoding and coded bits repetition for 14.4 kbps (RS2) vocoders (radio configuration 2).

Figure 5.8 illustrates RTCh processing, showing convolutional coding, repetition and interleaving. The data rates in the figure are obtained at each stage when using RS2 vocoders (14.4 kbps).

Because the number of coded bits prior to interleaving must be 576, for both RS1 and RS2 vocoders, the number of coded symbols after the convolutional encoder has to be the same. Therefore, for 14.4 kbps vocoders, the convolutional encoder is the sameas that used in the forward link, i.e. a half-rate encoder ($R = 1/2$), as in Figure 5.9.

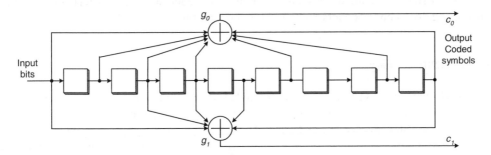

Figure 5.9 Convolutional coder ($R = 1/2$, $k = 9$) used by RTChs when employing RS2 vocoders.

Table 5.3 presents the number of coded symbols after convolutional coding and the rates after repetition.

For both vocoders, when $R_{voc} < 1$, all coded symbols pass through a repetition block. This block equalises the number of coded symbols, for all rates, for the interleaving process, i.e. 576 per 20 ms frame. When $R_{voc} = 1/2$, the number of repetitions is 1, that is, coded bits

Table 5.3 Bit repetitions for 14.4 Kbps vocoders

Vocoder rates (R_{voc})	Data rate (bps)	Convolutional encoder output (sps)	Number of repetitions	Interleaving input rate (Ksps)
1	14400	28800		28.8
1/2	7200	14400	1	28.8
1/4	3600	7200	3	28.8
1/8	1800	3600	7	28.8

Table 5.4 Main data rates for the 14.4 kbps vocoder

Vocoder input		Vocoder rate (R_{voc})	Vocoder output		Convolutional encoder output		
PSTN PCM	Linear coding		Net rate	Gross rate	Before repetition block	After repetition block	
1280	2080	1	267	288	576	576	Bits/frame
64000	104000		13350	14400	28800	28800	bps
1280	2080	1/2	125	144	288	576	Bits/frame
64000	104000		6250	7200	14400	28800	bps
1280	2080	1/4	55	72	144	576	Bits/frame
64000	104000		2750	3600	7200	28800	bps
1280	2080	1/8	21	36	72	576	Bits/frame
64000	104000		1050	1800	3600	28800	bps

are doubled prior to interleaving. For $R_{voc} = 1/4$ and 1/8, coded symbols are repeated 3 and 7 times, respectively.

Table 5.4 shows a summary of data rates obtained at each stage during RTCh processing when using 14.4 kbps vocoders and half-rate convolutional encoders.

5.1.1.3 Interleaving

The interleaving improves the reliability of coded transmitted signals. During interleaving, the bit transmission order is rearranged in a systematic way so that the process can be undone at the reception side. In case of burst errors during transmission, this procedure mitigates the damage due to consecutive errors, which otherwise would seriously degrade detection and decoding capabilities. Interleaving allows higher error detection capabilities and correction efficiency for the entire system.

Systems always use interleaving and convolutional encoding together, because the latter is not efficient for correcting burst errors. BTSs also use detected interleaving patterns to detect MS vocoder rate.

After convolutional encoding and bit repetition, all reverse traffic channels have 576 coded symbols per 20 ms frames to be compatible with the interleaving matrix, which consists of 18 columns and 32 rows, i.e. 576 cells.

Coded bits are written by columns, following a specific sequence, and are read by rows, also according to a pre-determined sequence. Figures 5.10 and 5.11 illustrate the interleaving process.

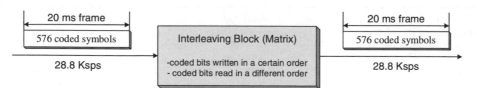

Figure 5.10 Interleaving process.

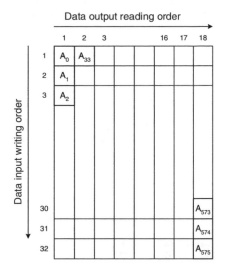

Figure 5.11 Interleaving matrix example.

Figure 5.11 shows a generic example of the interleaving matrix, depicting writing and reading directions. Table 5.8, in the next subsection, shows an example of the read/write process when using vocoders with rate $R_{voc} = 1/4$.

Coded bits are read from the interleaving matrix buffer by rows, according to the vocoder rate, as described in Table 5.5.

Table 5.5 Interleaving matrix reading sequence according to vocoder rate (R_{voc})

Vocoder rate (R_{voc})	Data rate (kbps)	Interleaver matrix rows reading sequence
1	9.6 or 14.4	1, 2, 3, 4, 5, 6, 7, 8, ..., 26, 27, 28, 29, 30, 31, 32
1/2	4.8 or 7.2	1, 3, 2, 4, 5, 7, 6, 8, 9, 11, 10, 12, 13, 15, 14, 16, 17, 19, 18, 20, 21, 23, 22, 24, 25, 27, 26, 28, 29, 31, 30, 32
1/4	2.4 or 3.6	1, 5, 2, 6, 3, 7, 4, 8, 9, 13, 10, 14, 11, 15, 12, 16, 17, 21, 18, 22, 19, 23, 20, 24, 25, 29, 26, 30, 27, 31, 28, 32
1/8	1.2 or 1.8	1, 9, 2, 10, 3, 11, 4, 12, 5, 13, 6, 4, 7, 15, 8, 16, 17, 25, 18, 26, 19, 27, 20, 28, 21, 29, 22, 30, 23, 31, 24, 32

5.1.1.4 Orthogonal Modulation using the Walsh Codes

Whereas the forward link uses Walsh codes to spread and channelise data, uniquely identifying each channel with a different code, the reverse link uses Walsh codes for 64-ary orthogonal modulation and for a type of spreading. Long code PN sequences (PNLCs) perform channelling and spreading functions in the reverse link.

Orthogonal modulation performs a type of Forward Error Correction (FEC) on reverse link channels. IS-95 networks use a Walsh matrix of order 64, i.e. 64 sets of 64-bit Walsh

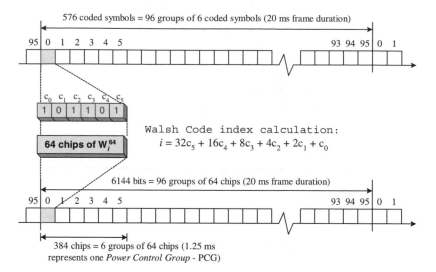

Figure 5.12 Orthogonal modulation: Walsh code definition for each set of 6 coded bits.

codes, and each of these codes can be mapped by a set of 6 bits. This sequence mapping process is part of the FEC and is known as a block code.

The orthogonal modulator groups the 576 coded symbols from the interleaving block into 96×6 sets for every 20 ms frame. Each of these 6-bit sets determines a Walsh Code index, as in eqn (5.1) and Figure 5.12.

Considering that c_0, c_1, c_2, c_3, c_4 and c_5 represents a set of six coded symbols, with c_5 being the MSB and c_0 the LSB, eqn (5.1) shows the index (i) calculation of a Walsh sequence (W_i^{64})

$$i = 32c_5 + 16c_4 + 8c_3 + 4c_2 + 2c_1 + c_0 \qquad (5.1)$$

Because each 6-bit set from the interleaver is associated to a 64-chip Walsh Code (W_0^{64} a W_{63}^{64}), each 20 ms frame has a total of 6144 chips, that is, a rate of 307.2 kcps.

On the reception side, the BTS synchronously correlates the signal received with the 64 Walsh codes available, selecting the code with the best correlation factor to retrieve the original six bits of data.

5.1.1.5 Data Burst Randomisation: Gating on/Gating off

Before transmission of the reverse traffic channel, to save energy (increasing MS battery lifetime), some bits are gated off in a process called *Data Burst Randomising* (DBR). This process occurs in a frame-by-frame basis and aims to avoid transmission of redundant information data bits, that is, MSs transmit data bits only once.

Frames are divided in 16 Power Control Groups (PCGs) of 1.25 ms each. The data burst randomising process gates on and off a certain number of PCGs per frame. The number of gated off bits depends on the vocoder rate. When $R_{voc} = 1$, all coded bits in a 20 ms frame are transmitted, providing a Transmission Duty Cycle (TDC) of 100%. For $R_{voc} = 1/2$, only half of the bits is transmitted, i.e. TDC of 50%. The same concept applies to other rates.

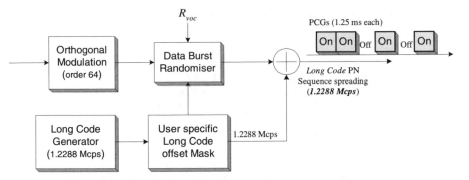

Figure 5.13   Power control groups gating off process.

PCGs have their transmission power reduced by 20–30 dB when gated off (Figure 5.13). Therefore, MSs save battery life and less interference is generated. By using this power control scheme, the system conserves capacity on the reverse link.

Figure 5.13 shows the Data Burst Randomiser (DBR), which pseudo-randomly defines the PCGs to be transmitted (gated on) or not (gated off). The DBR generates a mask to 'erase' redundant PCGs. This mask corresponds to the last 14 PNLC chips of the fifteenth PCG, in the latest 20 ms frame, as in Figure 5.14.

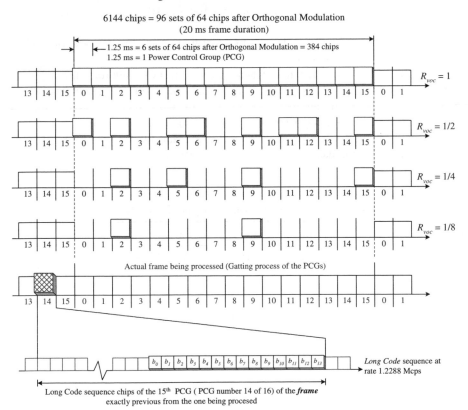

Figure 5.14   Redundant PCGs erasure (gate off) example.

Table 5.6 Power control groups transmission definition

Vocoder output rates (R_{voc})	Transmitted power control groups
1	0, 1, 2, 3, 4, 5, 6, 7, 8, 9, 10, 11, 12, 13, 14, 15
1/2	b_0, $(2 + b_1)$, $(4 + b_2)$, $(6 + b_3)$, $(8 + b_4)$, $(10 + b_5)$,$(12 + b_6)$, $(14 + b_7)$
1/4	$b_0.(1 - b_8)$, $(2 + b_1).b_8$, $(4 + b_2).(1 - b_9)$, $(6 + b_3).b_9$, $(8 + b_4).(1 - b_{10})$, $(10 + b_5).b_{10}$, $(12 + b_6).(1 - b_{11})$, $(14 + b_7).b_{11}$
1/8	$b_0.(1 - b_8).(1 - b_{12})$, $(2 + b_1).b_8.(1 - b_{12})$, $(4 + b_2).(1 - b_9).b_{12}$, $(6 + b_3).b_9.b_{12}$, $(8 + b_4).(1 - b_{10}).b_{13}$, $(10 + b_5).b_{10}.b_{13}$, $(12 + b_6).(1 - b_{11}).(1 - b_{13})$, $(14 + b_7).b_{11}.(1 - b_{13})$

These 14 chips are depicted as follows

$$b_0\, b_1\, b_2\, b_3\, b_4\, b_5\, b_6\, b_7\, b_8\, b_9\, b_{10}\, b_{11}\, b_{12}\, b_{13} \tag{5.2}$$

where

b_0 the 'oldest' bit (Most Significant Bit – MSB)
b_{13} the 'most recent' bit (Least Significant Bit – LSB)

Table 5.6 presents PCG gate off definitions according to possible vocoder rates (R_{voc}) for RC1 (9.6 kbps) and RC2 (14.4 kbps).

Table 5.6 and Figure 5.14 show that, when $R_{voc} = 1$, all PCGs are transmitted (0–15), i.e. no repetition is required during RTCh processing, therefore no redundant PCGs are generated. For $R_{voc} < 1$, the number of PCGs gated off is directly related to the number of coded symbol repetitions required at the convolutional encoder output, e.g. for $R_{voc} = 1/2$, eight PCGs are gated off; for $R_{voc} = 1/4$, twelve; and for $R_{voc} = 1/8$, fourteen.

Table 5.7 illustrates the data burst randomiser process.

Table 5.7 Algorithm for PCGs transmission randomising per 20 ms frame (example)

Transmitted power control groups (according to PN chip values)																Vocoder output
0	**1**	**2**	**3**	**4**	**5**	**6**	**7**	**8**	**9**	**10**	**11**	**12**	**13**	**14**	**15**	$R_{voc} = 1$
0	1	0	1	0	1	0	1	0	1	0	1	0	1	0	1	All PCGs are transmitted
b_0	b_1		b_2		b_3		b_4		b_5		b_6		b_7			$R_{voc} = 1/2$
0	1		0		1		0		1		0		1			Transmission of eight PCGs
	b_8			b_9				b_{10}				b_{11}				$R_{voc} = 1/4$
	0			1				0				1				Transmission of four PCGs
		b_{12}								b_{13}						$R_{voc} = 1/8$
																Transmission of two PCGs

Table 5.8 represents the interleaving matrix used for vocoders with $R_{voc} = 1/4$, showing that only redundant information is gated off during the DBR process. The table also indicates reading (R) and writing (W) sequence.

Each set of eight rows in Table 5.8 represents a 5 ms interval. The shaded rows represent the redundant information position in the interleaving matrix.

Table 5.8 Reverse link channels interleaving matrix when employing vocoder with rate $R_{voc} = 1/4$

PCG	W	R																			
0	1	1	1	9	17	25	33	41	49	57	65	73	81	89	97	105	113	121	129	137
	5	2	2	10	18	26	34	42	50	58	66	74	82	90	98	106	114	122	130	138
1	2	3	1	9	17	25	33	41	49	57	65	73	81	89	97	105	113	121	129	137
	6	4	2	10	18	26	34	42	50	58	66	74	82	90	98	106	114	122	130	138
2	3	5	1	9	17	25	33	41	49	57	65	73	81	89	97	105	113	121	129	137
	7	6	2	10	18	26	34	42	50	58	66	74	82	90	98	106	114	122	130	138
3	4	7	1	9	17	25	33	41	49	57	65	73	81	89	97	105	113	121	129	137
	8	8	2	10	18	26	34	42	50	58	66	74	82	90	98	106	114	122	130	138
4	9	9	3	11	19	27	35	43	51	59	67	75	83	91	99	107	115	123	131	139
	13	10	4	12	20	28	36	44	52	60	68	76	84	92	100	108	116	124	132	140
5	10	11	3	11	19	27	35	43	51	59	67	75	83	91	99	107	115	123	131	139
	14	12	4	12	20	28	36	44	52	60	68	76	84	92	100	108	116	124	132	140
6	11	13	3	11	19	27	35	43	51	59	67	75	83	91	99	107	115	123	131	139
	15	14	4	12	20	28	36	44	52	60	68	76	84	92	100	108	116	124	132	140
7	12	15	3	11	19	27	35	43	51	59	67	75	83	91	99	107	115	123	131	139
	16	16	4	12	20	28	36	44	52	60	68	76	84	92	100	108	116	124	132	140
8	17	17	5	13	21	29	37	45	53	61	69	77	85	93	101	109	117	125	133	141
	21	18	6	14	22	30	38	46	54	62	70	78	86	94	102	110	118	126	134	142
9	18	19	5	13	21	29	37	45	53	61	69	77	85	93	101	109	117	125	133	141
	22	20	6	14	22	30	38	46	54	62	70	78	86	94	102	110	118	126	134	142
10	19	21	5	13	21	29	37	45	53	61	69	77	85	93	101	109	117	125	133	141
	23	22	6	14	22	30	38	46	54	62	70	78	86	94	102	110	118	126	134	142
11	20	23	5	13	21	29	37	45	53	61	69	77	85	93	101	109	117	125	133	141
	24	24	6	14	22	30	38	46	54	62	70	78	86	94	102	110	118	126	134	142
12	25	25	7	15	23	31	39	47	55	63	71	79	87	95	103	111	119	127	135	143
	29	26	8	16	24	32	40	48	56	64	72	80	88	96	104	112	120	128	136	144
13	26	27	7	15	23	31	39	47	55	63	71	79	87	95	103	111	119	127	135	143
	30	28	8	16	24	32	40	48	56	64	72	80	88	96	104	112	120	128	136	144
14	27	29	7	15	23	31	39	47	55	63	71	79	87	95	103	111	119	127	135	143
	31	30	8	16	24	32	40	48	56	64	72	80	88	96	104	112	120	128	136	144
15	28	31	7	15	23	31	39	47	55	63	71	79	87	95	103	111	119	127	135	143
	32	32	8	16	24	32	40	48	56	64	72	80	88	96	104	112	120	128	136	144

In Table 5.8, for $R_{voc} = 1/4$, there are 144 coded symbols at the convolutional encoder output. Each one of them is repeated three times, prior to interleaving, to form a 576-coded symbol frame. Frames are written sequentially into the interleaving matrix by rows according to the following sequence (from W column)

> 1, 5, 2, 6, 3, 7, 4, 8, 9, 13, 10, 14, 11, 15, 12, 16, 17, 21, 18, 22, 19, 23, 20, 24, 25,
>   29, 26, 30, 27, 31, 28, 32

These numbers represent the sequence in which rows are filled with bits from the encoder and repetition blocks. That is, number 5, in the list (W column), represents the fifth row in the matrix for writing, but the second row for reading (R column).

Table 5.8 also presents the reading (R) sequence for the interleaving matrix, showing that each pair of rows, e.g. {1, 5}, {2, 6}, {3, 7}, {4, 8}, ..., {27, 31} and {28, 32}, on the W column, corresponds to one of the 16 PCGs in a 20 ms frame.

5.1.1.6   Long Code Spreading (PNLC Spreading) and Scrambling

After the data burst randomising process, the Long Code PN Sequence (PNLC) spreads RTChs. The spreading process consists of a modulo-2 sum (exclusive-OR) of the data bits (307.2 kbps) with PNLC chips (1.22898 Mcps), as in Figure 5.15. This sum results in an information sequence with 1.2288 Mcps prior to transmission.

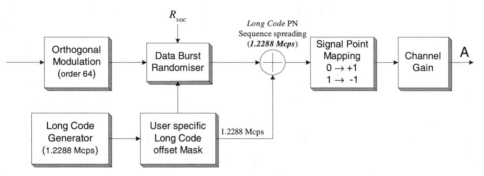

Figure 5.15   Reverse traffic channels long code spreading.

CDMA networks uniquely identify each reverse link channel by associating an offset mask to the spreading sequence. The masks consist of 42 bits that offset the $(2^{42}-1)$-chip PNLC sequence according to user characteristics. Two types of masks can be used: public and private.

The public long code offset mask is generated by a fixed set of bits combined with the MS permuted ESN, as in Figure 5.16.

Figure 5.16   Public long code offset mask used on RTChs.

There are two main reasons for using offset masks: multiple access (channelling) capabilities and encryption (privacy).

The *ESN* is permuted to avoid mis-interpretation of the system when mobile stations (MSs) with similar *ESNs* are present on the same area. The call processing and authentication chapter explains the *ESN* in detail.

Because offset masks are unique to each user, they also provide some encryption to the transmitted information.

The second type of mask is the private long code offset mask, which International Standards (ISs) do not describe. The description and formation of these masks is in a document called Annex A, whose distribution is controlled by US federal organisations, being restricted to general public. This mask aims to provide potential communication privacy to CDMA networks.

Information chips are then mapped into symbols (i.e. 0's (zeroes) are mapped into '+1' signals and 1's (ones) into '−1' signals), which are amplified according to the channel gain.

5.1.1.7 Phase and Quadrature Carrier Modulation

After long code spreading, PN sequences modulate RTChs in phase (PN-I) and in quadrature (PN-Q), as in Figure 5.17. Each information bit is copied to both branches, I and Q, with a chip rate of 1.2288 Mcps. Bits (from A) are modulo-2 summed to PN-I and PN-Q sequences. In the reverse link, both PN-I and PN-Q sequences always use zero phase offset. Chips in the Q branch are delayed by 1/2 chip to implement O-QPSK modulation (*Offset Quadrature Phase-Shift Keying*).

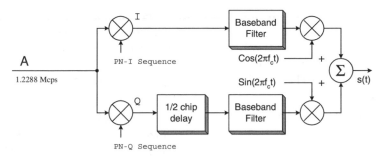

Figure 5.17 Phase and quadrature modulation and O-QPSK carrier modulation.

A baseband filter shapes the waveform prior to transmission to control the spectrum. The baseband coefficients are described in the IS-95 recommendations in Section 6.1.3.1.10.

The modulation of the filtered signal is performed using O-QPSK to generate carrier-modulated symbols for transmission.

The reverse link uses O-QPSK modulation, whereas the forward link uses QPSK. O-QPSK modulation symbols are obtained because of the 1/2 chip delay (406 901 ns) in the Q branch (Figure 5.18).

O-QPSK modulation is used in the reverse link to increase detection efficiency, avoiding 180° carrier phase transitions. This leads to a more linear signal, achieving better detection

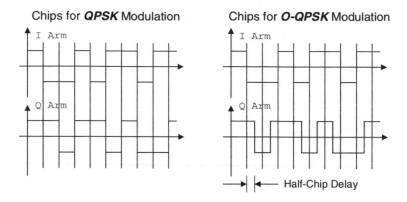

Figure 5.18 I and Q arms representation in QPSK and O-QPSK modulation schemes.

Table 5.9 Carrier phase mapping for O-QPSK modulation

I	Q	Phase
0	0	$\pi/4$
1	0	$3\pi/4$
1	1	$-3\pi/4$
0	1	$-\pi/4$

results and allowing the use of less expensive power amplifiers in MSs. More efficient transmission power amplifiers are extremely important to improve MS battery life.

Table 5.9 shows in phase (I) and quadrature (Q) carrier phase mapping for O-QPSK modulation used on reverse link channels.

Figure 5.19 compares phases and transitions diagram of QPSK and O-QPSK modulators.

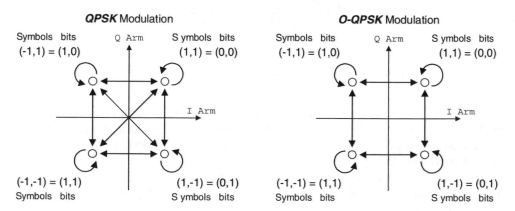

Figure 5.19 Phases and transition diagrams of QPSK (forward link) and O-QPSK (reverse link) modulators.

5.1.2 Reverse Access Channel

MSs use Reverse Access Channels (RACHs) when they are in the *idle state* and receive a paging message informing about an incoming call, or when they need to perform an autonomous registration or when users attempt to make a call. The RACh starts the communication with the BTS, transmitting *response* and/or *request* attempts, according to the procedure described in Figure 5.20.

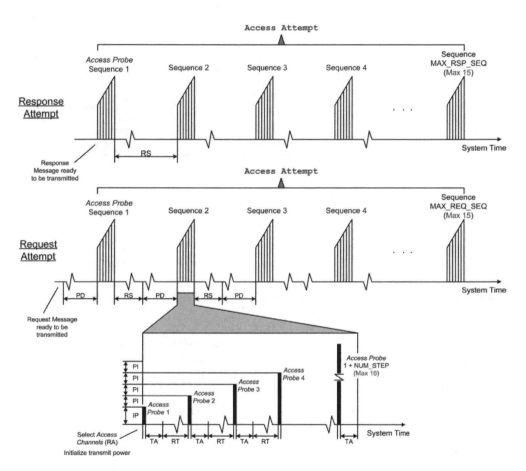

Figure 5.20 Reverse access channels transmission of *request/response* attempts.

The system may have up to 32 RACHs simultaneously assigned to each Paging Channel (PCh). MSs select a RACh using the ESN and a Hash function. This selection method allows several mobile terminals located within the coverage area of a BTS to randomly select RACHs during a request or response attempt.

Messages in the access probes, within access or response attempts, are transmitted at randomised intervals. This technique reduces the probability of collisions among messages transmitted from different MSs.

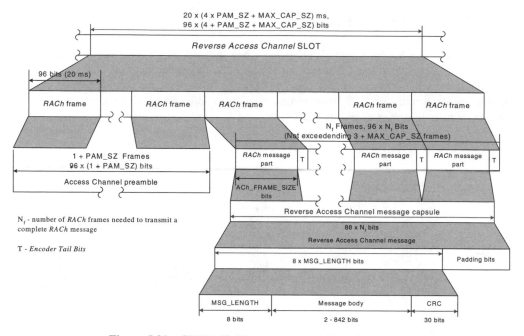

Figure 5.21 CDMA IS-95 reverse access channel structure.

RAChs are organised in slots, divided in 20 ms frames. The number of frames per RACh slot depends on the length of the message. Each RACh slot is divided into two main parts: preamble and message capsule, as in Figure 5.21.

The preamble allows the BTS to acquire RACh transmissions, distinguishing slots and enhancing synchronisation capabilities at the BTS receiver.

The message capsule consists of the message itself, which may be followed, or not, by padding bits. This capsule has three parts: the *MSG_LENGTH*, which stores the size of the message in bytes (coded into 8 bits); the message body; and 30 bits for Cyclic Redundancy Check (CRC) that functions as a message quality indicator.

Each input frame has 88 information bits, with a rate of 4.4 kbps. Eight encoder tail bits (T) are added to these frames, resulting in 96 bits per frame with a rate of 4800 kbps for each RACh, shown in Figure 5.22.

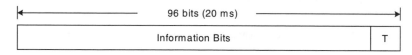

Figure 5.22 Reverse access channel frame structure.

Figure 5.23 shows the steps in RAChs processing.

The 96 bits in a frame pass through the convolutional encoder with rate 1/3, as in Figure 5.6, resulting in 14.4 ksps (288 coded symbols per frame), which are repeated, resulting in 576 coded symbols per 20 ms frame (28.8 ksps).

Table 5.10 presents the number of bits during RACh frame processing.

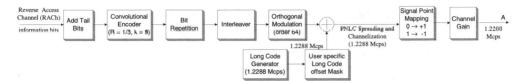

Figure 5.23   CDMA IS-95 reverse access channel bit processing.

Table 5.10   RACh bits during channel processing

Information bits	Tail bits	Bits per 20 ms frame	Convolutional encoder	Coded symbol repetition	Interleaving block
88	8	96	$R = 1/3, k = 9$	1	576

5.1.2.1   Interleaving Process

As for RTChs, the interleaver consists of a matrix with 18 columns and 32 rows, that is 576 cells. Coded symbols from the encoder are written by columns, as in Figure 5.24, and, at the output, they are read in rows according to a Bit Reversal Order (BRO) function.

The BRO function defines the sequence of rows to be read from the interleaving matrix, as follows:

- $x = 0, 1, 2, \ldots, N - 1$ – index of each row in the interleaving matrix;

- $BRO_m(x)$, or m bit-reversal order – returns the decimal value of the row index x after converting it to binary (with m bits) and reversing the bits order.

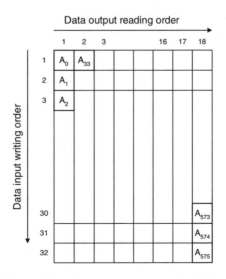

Figure 5.24   Interleaving process for RAChs.

Table 5.11 Example of the bit-reversal order function

X (row index)	Binary sequence ($m = 4$)	Reversed binary sequence	Decimal value for the reversed binary sequence ($\mathrm{BRO_m}(x)$)
0	0000	0000	0
1	0001	1000	8
2	0010	0100	4
3	0011	1100	12
4	0100	0010	2
5	0101	1010	10
6	0110	0110	6
7	0111	1110	14
8	1000	0001	1
9	1001	1001	9
10	1010	0101	5
11	1011	1101	13
12	1100	0011	3
13	1101	1011	11
14	1110	0111	7
15	1111	1111	15

As an example, Table 5.11 shows the result of a $\mathrm{BRO_4}(x)$ function for $m = 4$.

5.1.2.2   Orthogonal Modulation using Walsh Codes

RACh and RTCh coded bits are orthogonally modulated in the same way (Section 5.1.1.4), mapping every 6-coded symbol set into a 64-bit set corresponding to one of the 64 Walsh codes available. This modulation is a spreadinglike process because the bit rate at this stage is 307.2 Kbps (6144 bits per 20 ms frame).

5.1.2.3   Long Code Spreading (PNLC Spreading) and Scrambling

The long code PN sequence, or PNLC, performs the actual spreading for reverse link channels, as described in Section 5.1.1.6, and also scrambling (encryption).

Figure 5.25 shows the offset mask applied to the PNLC before spreading. In the figure, *ACN* corresponds to access channel number, PCN is the paging channel to which the RACh is associated, *BASE_ID* is the BTS identification number and *PILOT_PN* represents the pilot PN offset of the transmitting pilot.

41 33	32 28	27 25	24 9	8 0
110001111	ACN	PCN	Base_ID	Pilot_PN

Figure 5.25 RACh long code offset mask (42 bits).

5.1.2.4 *Phase and Quadrature Carrier Modulation*

After spreading, PN sequences modulate the bits in phase (PN-I) and in quadrature (PN-Q). Both PN sequences use zero offset and have a rate of 1.2288 Mcps prior to transmission, as in Section 5.1.1.7.

5.2 cdma2000 REVERSE LINK CHANNEL STRUCTURES

MSs can use the following channels in the reverse link of cdma2000 networks.

- Reverse Pilot Channel (RPiCh).

- Reverse Access Channel (RACh): same as CDMA IS-95 systems.

- Enhanced Access Channel (EACh).

- Reverse Common Control Channel (RCCCh).

- Reverse Fundamental Channel (RFCh): corresponds to CDMA IS-95 Reverse Traffic Channel (RTCh).

- Reverse Dedicated Control Channel (RDCCh).

- Reverse Supplemental Channel (RSCh).

- Reverse Supplemental Code Channel (RSCCh).

cdma2000 systems use a Pilot Channel (RPiCh) in the reverse link, which is an important improvement when compared to IS-95 systems. This logical channel, transmitted by all MSs using RC3–RC6, transmits power control bits for forward link channels, establishing a closed loop power control during the system access state.

Some of the channels used in IS-95 networks are improved in cdma2000 standards, for example, Reverse Traffic Channels (RTCh) are specialised and divided into three channels: Reverse Fundamental Channel (RFCh), Reverse Dedicated Control Channel (RDCCh) and Reverse Supplemental Channel (RSCh).

cdma2000 also implements additional logical channels to perform specific tasks, such as Reverse Common Control Channels (RCCCh) and Enhanced Access Channels (EACh).

Simultaneous transmission of multiple channels by means of Walsh Code multiplexing on the reverse link is also a unique characteristic of cdma2000.

5.2.1 Reverse Pilot Channel (RPiCh)

The Reverse Pilot Channel (RPiCh) is employed to help the BTS to detect MS transmissions (similar to a preamble). It can also be used to transmit power control bits or indicator bits, and to serve as power and timing reference (for coherent detection) for reverse link channels.

The pilot channel is multiplexed with several other reverse link channels using the Walsh code W_0^{32}, consisting of a certain number of 0's per 20 ms frame according to the Spreading Rate (SR) in use.

MSs only transmit RPiChs when operating with RC3–RC6, always before and while Enhanced Access Channels (EAChs), Reverse Common Control Channels (RCCChs) or any reverse traffic channels (fundamental, dedicated control or supplemental) are in use.

RPiChs can operate in two modes.

- As a preamble, when transmitted before EAChs, RCCChs or traffic channels (fundamental and reverse dedicated control channels only), without transmitting any power control bits.

- In a normal operation mode, when transmitted simultaneously with any other logical channel, with Forward Power Control (FPC) scheme enabled, transmitting power control bits for forward link channels and acting as timing reference for coherent detection.

Figure 5.26 shows the RPiCh basic structure. There are 16 Power Control Groups (PCG) of 1.25 ms each in every 20 ms frame of a RPiCh, resulting in $1536 \times N_{SR}$ chips, where $N_{SR} = 1$ for SR1 and $N_{SR} = 3$ for SR3.

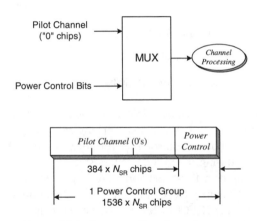

Figure 5.26   Reverse pilot channel structure.

Mobile terminals always transmit the first $1152 \times N_{SR}$ chips as '0s'. The remaining $384 \times N_{SR}$ chips depend on the operation mode. If the RPiCh is transmitted as a preamble, then all $384 \times N_{SR}$ chips are also '0s'.

In the normal mode, i.e. performing forward power control, the $384 \times N_{SR}$ chips are called Reverse Power Control Sub-Channel (RPCSCh) and can be different than '0'.

As an example, when the parameter which defines the forward power control operation mode, *FPC_MODE*, is set to '000', '001' or '010', each of the $384 \times N_{SR}$ chips repeats the power control bit, that is, if the power control bit is '1', all $384 \times N_{SR}$ chips '1' transmitted at the last quarter of a PCG are also '1'. When the *FPC_MODE* is '011', the $384 \times N_{SR}$ chips repeat the Erasure Indicator Bit (EIB).

All chips transmitted on the RPCSCh of a RPiCh PCG always have the same power transmission level. Chapter 7 discusses power control schemes in more detail.

5.2.1.1   RPiCh Transmission as a Preamble for the Reverse Traffic Channels

Two situations define when a RPiCh is transmitted as a preamble: if transmitted prior to EAChs and RCCChs, or prior to some reverse traffic channel. The transmission procedure as a preamble for EAChs and RCCChs is explained in detail in Sections 5.2.3.4 and 5.2.4.4.

As a preamble for reverse traffic channels, RPiChs must be transmitted prior to reverse fundamental or reverse dedicated common control channels, when operating with RC3–RC6, and not in gated mode.

In both cases, prior to EAChs or to RCCChs, MSs must transmit a certain number (*NUM_PREAMBLE*) of PCGs via RPiChs before the beginning of the other logical channel frames, as in Figure 5.27.

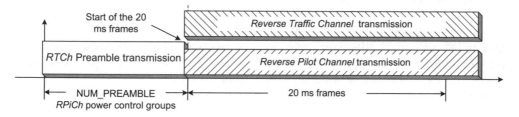

Figure 5.27   RPiCh transmission as a preamble prior to reverse traffic channels transmission.

5.2.1.2   Power Control on RPiChs–Normal Operation Mode

cdma2000 networks implement improved power control procedures when compared to CDMA IS-95. Improvements, however, are only valid for RC3–RC6 operation, i.e. cdma2000-1X (SR1) and cdma2000-3X (SR3).

During *system access state* (more details in the call processing chapter), a closed loop power control scheme improves the original open loop power control. The power control channel, called Forward Common Power Control Channel (FCPCCh), implements this new control scheme.

During *MS control on the traffic channels state* (conversation), the network performs power control over forward link channels in a much faster and accurate way.

All the improvements are achieved because of the RPiCh implementation, which transmits power control bits on reverse power control sub-channels (Section 5.2.1). All mobile stations must support two types of closed loop power control schemes: inner power control loop (or inner loop) and outer power control loop (or outer loop).

When operating in outer loop mode, every MS must estimate the 'target E_b/N_0' (or setpoint) required to achieve a specific Frame Error Rate (FER) for each logic channel. This setpoint can be understood as the reception signal level that MSs wants to receive from BTSs. This process is usually performed in a frame-by-frame basis.

In the inner loop mode, MSs compare E_b/N_0 of the received signals to the 'target E_b/N_0' set by the outer loop. This comparison defines the power control bit (0 or 1) that MSs send to BTSs through RPCSChs in a RPiCh. This process happens in a 1.25 ms basis, according to the current transmission characteristics of the channel associated to the MS.

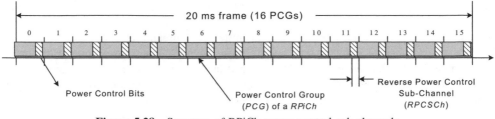

Figure 5.28   Structure of RPiCh power control sub-channels.

Zero as the power control bit indicates that the received signal level is below the target value, and the BTS must increase the associated channels transmission power on the forward link. If the bit is '1', the BTS must decrease the transmission power.

Besides power control bits, RPCSChs may also transmit Erasure Indicator Bits (EIBs) or Quality Indicator Bits (QIBs). These bits, however, are only transmitted when requested by the BTS. When transmitting EIBs or QIBs, all power control bits on the RPiCh frame must be set with the same value as defined for the EIB or the QIB.

Figure 5.28 presents the RPiChs structure showing PCGs and reverse power control sub-channels.

5.2.1.3   PCGs Gating on/offf

To improve power control performance and increase MS battery life, the BTS can instruct an MS to operate in gated mode, i.e. to gate off some PCGs of the RPiCh. The gating process only happens if the RPiCh is transmitted as a preamble to a logical channel or when no forward link channels, associated with the MS, are being transmitted and the MS is not transmitting any other reverse link channel.

An MS only performs the gating process if the parameter *PILOT_GATING_USE_RATING*, transmitted by the BTS, is set to 1. If this condition holds, the BTS must send another parameter to the MS to set the gating rate, *PILOT_GATING_RATE*.

Table 5.12 shows possible values for gating rates and PCGs to be transmitted according to each configuration in use.

Figure 5.29 depicts the distribution of PCGs transmitted in a 20 ms RPiCh frame according to gating rates in Table 5.12.

For any gating rate, gated PCGs are always arranged in such a way that PCGs are transmitted at 5 ms frame boundaries, i.e. at the end of each set of four PCGs. This

Table 5.12   Gating rates for RPiCh power control groups

PILOT_GATING_RATE	Gating rate	Number of transmitted PCGs	PCGs transmitted
00	1	16	0–15
01	1/2	8	1, 3, 5, 7, 9, 11, 13, 15
10	1/4	4	3, 7, 11, 15

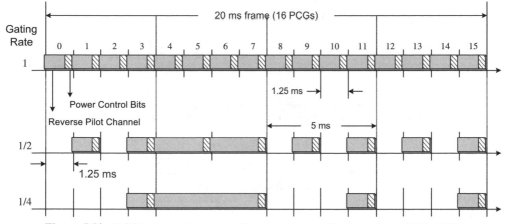

Figure 5.29 PCG transmission and gating process according to rates in Table 5.12.

characteristic assures that the RPiCh is always transmitted prior to the beginning of a new set of 5 ms frames (four PCGs), assisting the BTS in channel detection and acquisition. The standards, however, also established some specific configurations under which this rule is not necessarily true.

When the RPiCh is in gated mode operation and transmitting a reverse fundamental channel at rates 1500 bps for RC3 and RC5 or even at 1800 bps for RC4 and RC6, PCGs are repeated, as illustrated in Figure 5.30, occupying intervals of 2.5 ms. Because the RPiCh must always be transmitted simultaneously with any other logical channel, PCGs are only gated on during the transmission of the logical channel data.

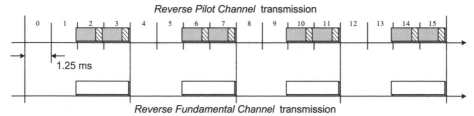

Figure 5.30 RPiCh PCG gating process during reverse fundamental channel transmission.

Another exception to this rule occurs when, in gated mode, PCGs must be transmitted/ repeated during the transmission of a Reverse Dedicated Control Channel (RDCCh), modifying gating patterns. Figures 5.31 and 5.32 present the RPiCh gating process during transmission of a RDCCh for 5 and 20 ms frames, respectively.

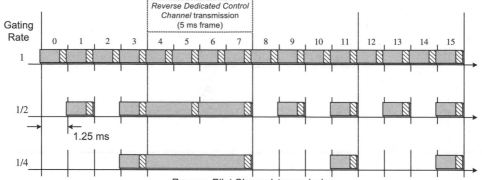

Figure 5.31 RPiCh PCG gating process during reverse dedicated control channel transmission (5 ms frames).

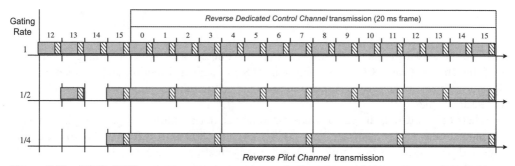

Figure 5.32 RPiCh PCG gating process during reverse dedicated control channel transmission (20 ms frames).

5.2.1.4 Multiplexing by Walsh codes, Complex Modulation by PN sequences and Carrier Modulation

Figure 5.33 shows pseudo-random noise sequences (PN-I, PN-Q, PNLC-I and PNLC-Q) spreading and modulating coded bits on the reverse link of cdma2000 systems. Logical channels used by mobile terminals are modulo-2 summed in baseband prior to phase and quadrature modulation.

The cdma2000 reverse link logical channels are multiplexed using Walsh codes similar to the forward link (Figure 5.33) prior to PN sequence modulation. Each channel may use Walsh codes with different lengths to improve system channel reliability.

cdma2000 networks separate logical channel processing into I and Q branches, i.e. some channels are processed in the I branch, such as RPiCh, RSCh 2 and RDCCh, whereas others are processed on the Q branch, such as RFCh, RSCh 1, EACh and RCCCh.

The reverse link of cdma2000 systems performs a special type of PN application process called *complex multiplier spreading*, as indicated in Figure 5.33.

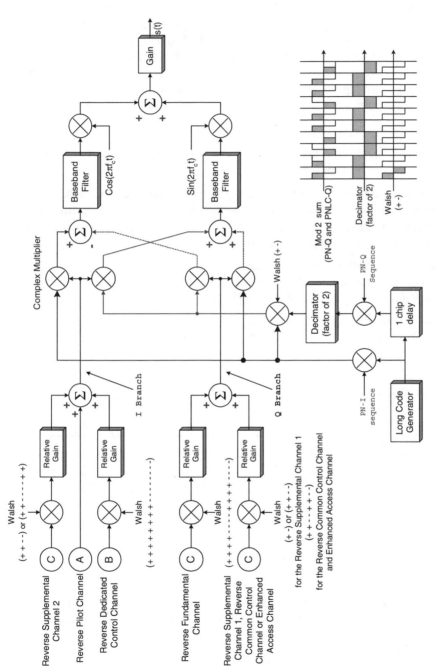

Figure 5.33 Reverse link channel modulation and transmission processes.

The *complex multiplier spreading* involves all PN sequences and all logical channels in use. A modulo-2 sum (implemented as a multiplier) combines PN-I and PNLC-I forming the in-phase spreading sequence. PNLC-Q, which is the PNLC-I sequence delayed in one chip, is modulo-2 summed with PN-Q. A factor-two decimator processes this result keeping only the first chip out of each set of two chips. Figure 5.33 shows the decimation process. Resulting chips have twice the length of PN-Q and PNLC-Q chips. After decimation, another modulo-2 sum combines these chips with Walsh code W_1^2 and with the in-phase spreading sequence, forming the quadrature-phase spreading sequence.

Some of the characteristics of PN spreading and modulation in cdma2000 are the following.

- Channels not in use (at a certain moment) or gated-off symbols are implemented by analogue zeroes, i.e. they do not interfere with modulation processes of other channels because they are not transmitted.

- Every time that an EACh or an RCCCh is in use, only one extra channel can be processed in parallel, the RPiCh. In this case, according to Figure 5.33, the I branch consists of only the RPiCh (bits '0' implemented as a +1 signal level), whereas the Q branch consists of the EACh or RCCCh during system access state.

- In case any reverse traffic channel is in use, while in MS control on the traffic channel state, the Q branch consists of the reverse fundamental channel whereas the I branch carries the RPiCh and optional RSChs or RDCCh.

- Long code generators must have a chip rate compatible with the spreading rate (SR) in use, i.e. 1.2288 Mcps for SR1 and 3.6864 Mcps for SR3.

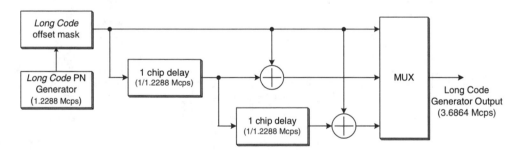

Figure 5.34 Long code PN generator (PNLC) for SR1 and SR3 and specific offset mask application.

cdma2000 systems require the application of a specific long code offset mask for each channel in use, as indicated in Figure 5.34.

- The 42-bit long code offset mask in use depends on the logical channel being processed and transmitted with the RPiCh, i.e. EACh, RCCCh or a reverse traffic channel.

5.2.2 Reverse Access Channel (RACh)

RAChs for cdma2000 systems have the same characteristics as in CDMA IS-95 systems (Section 5.1.2).

5.2.3   Enhanced Access Channels (EACh)

Mobile stations operating with RC3–RC6 (cdma2000 1X and 3X only) use EAChs to replace RAChs. EAChs perform tasks such as authentication, starting communication with a BTS and responding to BTS commands, i.e. response and request attempts. EAChs transmit short messages, such as signalling, MAC layer messages, paging responses and call origination messages, being also capable of transmitting data messages of moderate sizes.

EAChs can transmit three distinct types of binary information.

- *Enhanced access data*: Message data, such as channel negotiation, signalling, and MAC layer messages, transmitted to the BTS by the EACh while in basic access mode or by the Reverse Common Control Channel (RCCCh) while in reservation access mode.

- *Enhanced access header*: Origination message data transmitted while in reservation access mode, right after the enhanced access preamble.

- *Enhanced access preamble*: Non-information part of the EACh transmitted prior to any other data, when *EACH_PREAMBLE_ENABLE* = '1'. It consists of the transmission of the Reverse Pilot Channel (RPiCh) without power control information, to assist the BTS in the initial acquisition.

Table 5.13 presents enhanced access data and header frame structures.

Message bits (data and header) transmitted by the EACh are organised in frames with duration of 5, 10 or 20 ms. Figure 5.35 illustrates the frame structure, where F represents frame quality indicator bits or Cyclic Redundancy Check (CRC) and T represents encoder tail bits (eight '0' bits). The number of CRC bits depends on the number of information bits transmitted by the EACh message and on the frame duration, as in Table 5.13.

EAChs can operate in two distinct operation modes.

Table 5.13   EAChs frame structure configurations

Frame (ms)	Information type	Transmission rate (bps)	Information bits	CRC	Encoder tail bits	Total bits/frame
5	Header	9600	32	8	8	48
5	Data	38400	172	12	8	192
10	Data	19200	172	12	8	192
10	Data	38400	360	16	8	384
20	Data	9600	172	12	8	192
20	Data	19200	360	16	8	384
20	Data	38400	744	16	8	768

Information Bits	F	T

Figure 5.35   EACh frame structure.

- *Basic access mode (BAM)*: Transmission of the enhanced access preamble and the enhanced access data, quite similar to the operation of RAChs.

- *Reservation access mode (RAM)*: Transmission of the EACh preamble and header on EACh probes, 'reserving' the Reverse Common Control Channel (RCCCh) for transmission of EACh data. The FCPCCh performs a closed loop power control in this case.

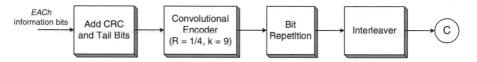

Figure 5.36 EACh data and header processing block diagram.

There may be up to 32 EAChs from different mobile users simultaneously assigned to each Forward Common Control Channel (FCCCh). Each MS selects an EACh using the *ESN* and the Hash function. This method allows several mobile terminals located within the same coverage area to randomly select EAChs to be used during enhanced access or response attempts. The maximal number of EACh slots for possible selection is 512 per channel in use, corresponding to the number of PN offsets available in cdma2000 systems.

Figure 5.36 presents a block diagram for EACh (data and header) processing.

Table 5.14 presents bit rates obtained during EACh processing (Figure 5.36).

Table 5.14 EACh data rates during channel processing

Frame (ms)	Data type	Total bits/frame	Transmission rates (bps)	Convolutional encoder rate	Symbol Repetition	Interleaving block size	Coded symbols after interleaving (Ksps)
5	Header	48	9600	$R = 1/4, k = 9$	3	768	153.6
5	Data	192	38400	$R = 1/4, k = 9$	0	768	153.6
10	Data	192	19200	$R = 1/4, k = 9$	2	1536	153.6
10	Data	384	38400	$R = 1/4, k = 9$	0	1536	153.6
20	Data	192	9600	$R = 1/4, k = 9$	3	3072	153.6
20	Data	384	19200	$R = 1/4, k = 9$	1	3072	153.6
20	Data	768	38400	$R = 1/4, k = 9$	0	3072	153.6

5.2.3.1 Convolutional Encoding and Repeating Process

Figure 5.37 shows the convolutional encoder for the EACh, with $R = 1/4$ and $k = 9$.

In some specific configurations, as detailed in Table 5.14, coded symbols at the encoder output must be repeated prior to interleaving to allow equalised rates among all channel data types.

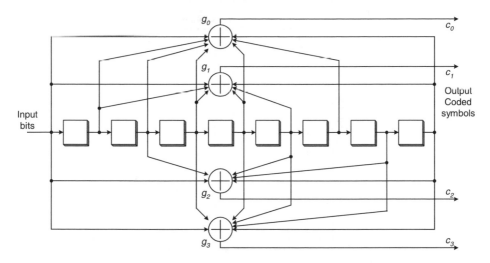

Figure 5.37   Convolutional encoder ($R = 1/4$, $k = 9$) employed for the EAChs.

5.2.3.2 BRO Interleaving

EACh coded symbols, corresponding to encoded EACh header or data, are interleaved using the Bit Reversal Order (BRO) interleaving process. Figure 5.24 provides a visualisation of BRO interleaving for SR1.

The interleaving process is performed in two basic steps:

- data writing (interleaver data input);

- data reading (data output).

According to parameters (m and J) from Table 5.15, to process sets of N cells, the interleaver builds a matrix with 2^m rows (0–$2^m - 1$) and with J columns (0–$J - 1$), as in Figure 5.38.

Bits are written in the matrix according to eqn (5.3).

$$A_i = 2^m(i \bmod J) + BRO_m(\lfloor i/J \rfloor) \tag{5.3}$$

Table 5.15   Parameters m and J used for building the interleaving matrix

N bits/frame block	m	J	N bits/frame block	m	J
384	6	6	576	5	18
768	6	12	2304	6	36
1536	6	24	4608	7	36
3072	6	48	9216	7	72
6144	7	48	18432	8	72
12288	7	96	36864	8	144

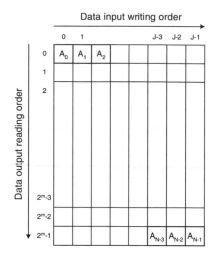

Figure 5.38   Bit-reversal order interleaving.

where

$i = 0, 1, 2, \ldots, N - 1$ index of each coded symbol in the interleaving input

$BRO_m(x)$ (or m bit-reversal   function that returns the decimal value of the

  order) m-length reversed binary sequence of x

$\lfloor x \rfloor$ truncates x to the lowest integer (e.g. returns 3 for

                  $x = 3.6$)

In eqn (5.3), $2^m(i \bmod J)$ represents the column where the input bit will be written, whereas $BRO_m(\lfloor i/J \rfloor)$ calculates the row. Table 5.11 shows an example of BRO interleaving for $m = 4$.

Coded symbols are read (output) by columns, starting from the first column on the left and reading consecutive columns to the right, as in Figures 5.38 and 5.39.

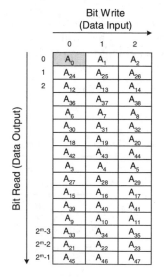

Figure 5.39   Example of bit positioning in the BRO Interleaver matrix for SR1.

Bit positioning in the interleaving matrix example of Figure 5.39 shows an interleaving matrix with $N = 48$, $m = 4$ and $J = 3$.

5.2.3.3 Multiplexing by Walsh Codes, Complex Multiplier by PN Sequences and Carrier Modulation

Lower insertion point C of Q branch in Figure 5.33 is used for the EACh transmission, processed using the Walsh code multiplexing and pseudo-random noise sequences (PN-I, PN-Q, PNLC-I and PNLC-Q) spreading, according to the *complex multiplier spreading* on the reverse link of cdma2000 systems.

Figure 5.40 presents the long code offset mask used for the EACh.

41	....	33	32	....	28	27	....	25	24	....	9	8	....	0
110001110			EACN			FCCCN			Base_ID			Slot_Offset		

Figure 5.40 Long code offset mask used for EAChs.

In the figure, EACN represents the EACh identity, FCCCN represents the FCCCh associated to the EACh in use, *Base_ID* corresponds to the BTS identification to which the message is being sent and *Slot_Offset* represents the slot offset information to assist the BTS to locate where the EACh message starts.

The offset mask in Figure 5.40 is used for encryption and channelisation purposes, as in Figure 5.33 of Section 5.2.1.4.

5.2.3.4 EACh Preamble Transmission Process

The preamble assists the BTS to detect and acquire transmitted channels, in this case, the EACh. It is always transmitted prior to EACh data or EACh header, as in Figure 5.41.

The Reverse Pilot Channel (RPiCh) is transmitted as the EACh preamble using a slightly higher power transmission level. No power control bits are transmitted (only 0's are transmitted) at this time.

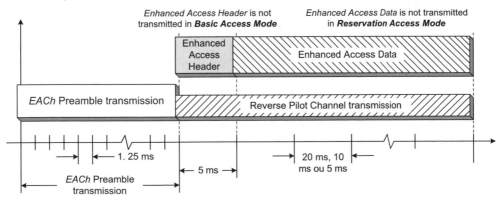

Figure 5.41 EACh preamble transmission and timing characteristics.

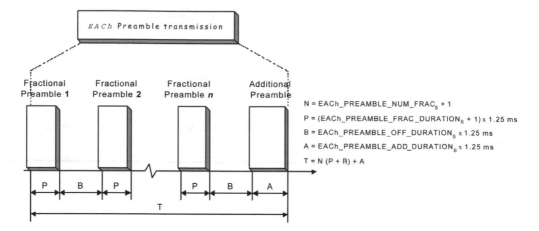

Figure 5.42   EACh preamble transmission, showing fractional and additional preambles.

The preamble is only transmitted when the parameter *EACH_PREAMBLE_ENABLE* is '1'. The preamble is a multiple of 1.25 ms, because it consists of several RPiCh 1.25 ms PCGs, with only zeroes instead of power control bits. EACh preamble transmission includes *n* fractional preambles and one additional preamble, where *n* is the parameter *EACH_PREAMBLE_NUM_FRAC* plus 1 (Figure 5.42). The access parameter message transmits the parameters presented in Figure 5.42 to MSs.

5.2.4   Reverse Common Control Channels (RCCCh)

RCCChs are generally used with EAChs, because they usually carry EACh data information bits. RCCChs transmit user and signalling information to BTSs, such as data required during call setup, service options and registration information. RCCChs are only used during *idle state* and *system access state*, i.e. while no reverse traffic channels are in use.

Reverse common control channels operate in *Reservation Access Mode* (RAM). This mode consists of the transmission of EACh preamble and header on EACh probes, and the RCCCh transmits EACh data using the closed loop power control associated to an FCPCCh on the forward link.

Each Forward Common Control Channel (FCCCh) and each Common Assignment Channel (CACh) can have up to 32 RCCChs (0–31), from different mobiles, associated with them simultaneously. FCCChs must always have at least one associated RCCCh. MSs select RCCChs using the ESN and a Hash function.

Information bits (EACh data) in RCCChs are organised in frames of 5, 10 or 20 ms, with a frame structure compatible to EAChs, as in Figure 5.43. In the figure, F represents frame

Information Bits	F	T

Figure 5.43   RCCCh frame structure.

Table 5.16 RCCCh possible frame structures

Frame (ms)	Transmission rate (bps)	Total bits/frame	Information bits	CRC	Encoder tail bits
5	38400	192	172	12	8
10	19200	192	172	12	8
10	38400	384	360	16	8
20	9600	192	172	12	8
20	19200	384	360	16	8
20	38400	768	744	16	8

quality indicator bits, or CRC, and T represents encoder tail bits (eight '0' bits). The number of CRC bits depends on the number of information bits transmitted, as in Table 5.16.

Table 5.16 presents RCCCh frame structures available. These structures are the same available for EAChs.

Figure 5.44 illustrates the steps in RCCChs channel processing, which is the same presented for EAChs.

Table 5.17 shows bit rates obtained during RCCChs processing in Figure 5.44.

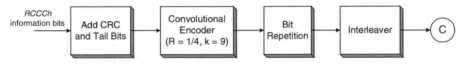

Figure 5.44 RCCCh channel processing block diagram.

5.2.4.1 Convolutional Encoding and Repeating Process

Figure 5.37 shows the convolutional encoder used for RCCChs, which is the same used for EAChs, with $R = 1/4$ and $k = 9$. Some radio configurations must go through a repeating process, as in Table 5.17, to equalise rates among all channel frame durations.

Table 5.17 Data rates during RCCCh processing

Frame (ms)	Total bits/frame	Transmission rates (bps)	Convolutional encoder	Symbol repetition	Interleaving block	Coded symbols after interleaving (Ksps)
5	192	38400	$R = 1/4, k = 9$		768	153.6
10	192	19200	$R = 1/4, k = 9$	1	1536	153.6
10	384	38400	$R = 1/4, k = 9$		1536	153.6
20	192	9600	$R = 1/4, k = 9$	3	3072	153.6
20	384	19200	$R = 1/4, k = 9$	1	3072	153.6
20	768	38400	$R = 1/4, k = 9$		3072	153.6

5.2.4.2   *Interleaving*

The interleaving method employed in RCCChs is the same applied to EAChs, i.e. Bit Reversal Order (BRO) interleaving, described in Section 5.2.3.2.

5.2.4.3   *Multiplexing by Walsh Codes, Complex Modulation by PN Sequences and Carrier Modulation*

Lower insertion point C of Q branches in Figure 5.33 is used for the RCCCh transmission, presenting the Walsh code multiplexing and pseudo-random noise sequences (PN-I, PN-Q, PNLC-I and PNLC-Q) spreading on the *complex multiplier* of the reverse link of cdma2000 systems.

PN spreading and modulation process for RCCChs are very similar to processes applied on EAChs (Figure 5.33 in Section 5.2.1.4). The main difference is the use of a specific RCCCh offset mask for long code sequences when in *reservation access mode* operation, as in Figure 5.45.

41 33	32 28	27 25	24 9	8 0
110001101	RCCCN	FCCCN	Base_ID	Pilot_PN

Figure 5.45   Long code offset mask used for RCCCh operating on reservation access mode.

In Figure 5.45, RCCCN represents the RCCCh identity, FCCCN represents the associated FCCCh number, *Base_ID* corresponds to the identification of the BTS to which the message is being sent and *Pilot_PN* indicates to the pilot PN offset of the transmitting BTS.

5.2.4.4   *RCCCh Preamble Transmission*

The preamble assists the BTS to detect and acquire the transmitted channel, in this case, RCCCh. The RCCCh preamble is always transmitted prior to EACh data, as in Figure 5.46.

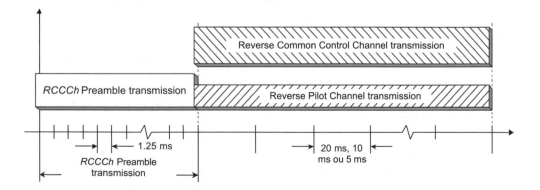

Figure 5.46   RCCCh preamble transmission and timing characteristics.

The Reverse Pilot Channel (RPiCh) transmits the RCCCh preamble using a slightly higher power transmission level. Power control bits are not transmitted (only 0's are transmitted instead).

The preamble is only transmitted when the parameter *EACH_PREAMBLE_ENABLE* is '1', with duration of multiple 1.25 ms intervals. RCCCh preamble transmission includes *n* fractional preambles and one additional preamble, where *n* is the parameter *EACH_ PREAMBLE_NUM_FRAC* plus 1 (Figure 5.47). The access parameter message transmits the parameters in Figure 5.47 to MSs.

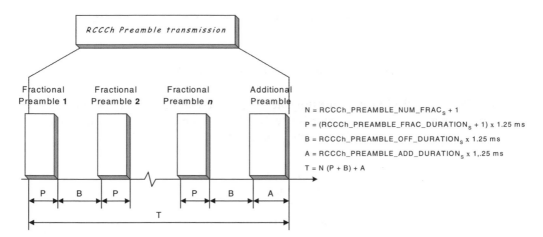

Figure 5.47 RCCCh preamble transmission, presenting the fractional and additional preamble configurations.

5.2.5 Reverse Fundamental Channel

Logical channel specialisation in cdma2000 networks divided IS-95 reverse traffic channel (RTCh) functions (such as transmission of voice, signalling and data) into three channels: Reverse Fundamental Channel (RFCh), Reverse Dedicated Control Channel (RDCCh) and the Reverse Supplemental Channel (RSCh).

RFChs have the same tasks as IS-95 RTChs to maintain compatibility (RC1 and RC2). For cdma2000 voice transmission is the main function, or preferential task, of this channel, although other types of data may also be transmitted.

RDCChs are responsible for signalling and user specific data, whereas RSChs transmit data at higher data rates. RDCChs and RSChs are explained in Sections 5.2.6 and 5.2.7, respectively.

For compatibility RFChs channel processing characteristics are different for RC1 and RC2 (CDMA IS-95 compliant) and for RC3–RC6 (cdma2000-1X and 3X compliant).

5.2.5.1 Channel Processing for RFChs Operating on RC 1-2

Reverse fundamental channel processing for RC1 and RC2 (Figure 5.48) is equivalent to IS-95 processing, described in Section 5.1.1.

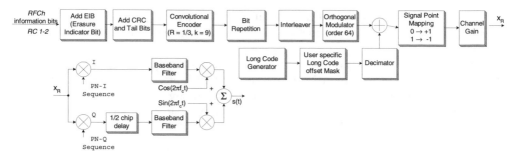

Figure 5.48 Reverse fundamental channel processing for RC1 and RC2 operation.

5.2.5.2 Channel Processing for RFChs Operating on RC3–RC6

Figure 5.49 shows the RFChs frame structure valid for all radio configurations. In the figure, R/E represents the Erasure Indicator Bit (EIB), F corresponds to frame quality indicators (CRC) and T represents encoder tail bits (eight '0' bits).

Figure 5.49 RFCh frame structure.

Table 5.18 presents RFCh bit rates obtained during channel processing when operating with RC3–RC6.

When on RC3–RC6, the EIB informs BTSs when FFChs or FDCChs frames are erased because of detection with errors or when they have not been transmitted.

When employing the power control scheme defined by the parameter $FPC\_MODE =$ '011', MSs must set all RPiCh 16 power control sub-channels with the same value as the EIB

Table 5.18 RFCh bit rates for RC3–RC6

RC	Bits/ frame	Frame (ms)	EIB	CRC	Tail bits	Total bits/frame	Information bit rates (Kbps)
	16	20	0	6	8	30	1.5
	40	20	0	6	8	54	2.7
3 and 5	80	20	0	8	8	96	4.8
	172	20	0	12	8	96	9.6
	24	5	0	16	8	48	9.6
	24	20	0	16	8	48	9.6
	21	20	1	6	8	36	1.8
4 and 6	55	20	1	8	8	72	3.6
	125	20	1	10	8	144	7.2
	267	20	1	12	8	288	14.4

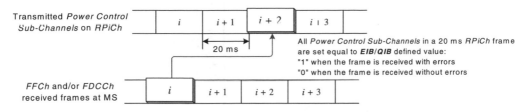

Figure 5.50 EIB transmission on RPCSChs when $FPC\_MODE =$ '011.

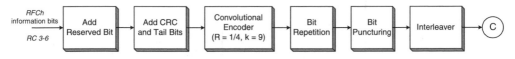

Figure 5.51 RFCh frame processing block diagram for RC3–RC6.

defined for the FFCh or FDCCh received immediately before. This process is illustrated in Figure 5.50.

Figure 5.51 shows the block diagram for reverse fundamental channel processing when operating on RC3–RC6.

FFChs use a convolutional encoder of rate $R = 1/4$ (Figure 5.37) for all RCs from 3–6. Tables 5.19 and 5.20 present data rates obtained during channel processing for RC3 and RC5 operation modes and for RC4 and RC6 operation, respectively.

An MS can discontinue transmission of the RFCh for up to three 5 ms intervals within a 20 ms frame, when operating in RC3–RC6. The decision of transmitting an RFCh or not is made on a frame-by-frame basis.

Because of high traffic demand and processing capabilities, the system may instruct, i.e. command, MSs to start RFCh frame transmissions with time offsets that are multiples of 1.25 ms. The parameter $FRAME\_OFFSET$ determines the number of 1.25 ms offsets in a zero-offset frame. Zero-offset frames start only when the variable system time is a multiple of 20 or 5 ms, depending on the frame size.

5.2.5.3   Puncturing for RFChs Operating on RC3–RC6 (Convolutional Code Puncturing)

Puncturing happens on RFCh frames right after bit repetition (see Tables 5.19 and 5.20). Because the information bits are convolutionally encoded, this process is also called *convolutional code puncturing*.

Table 5.21 shows the number of coded symbols to be punctured and the puncturing patterns, according to the RC in use and the convolutional encoder rate.

In the puncturing pattern, the bits '0' indicate that the coded symbol shall be deleted, while the bits '1' indicate that the coded bit should be kept. In order to know which coded symbols within a frame will be punctured, the puncturing pattern must be repeated as many times as needed so that all coded bits in a frame are processed.

Table 5.19 RFCh frame bit rates for RC3 and RC5

Information bits	Frame (ms)	Reservation bit	CRC (F)	Tail bits	Total (Kbps)	Convolutional encoder	Symbol repetition	Symbol puncturing	Interleaving block	Coded symbols after interleaving (Ksps)
24	5	0	16	8	9.6	$R = 1/4, k = 9$	1		384	76.8
16	20	0	6	8	1.5	$R = 1/4, k = 9$	15	1/5	1536	76.8
40	20	0	6	8	2.7	$R = 1/4, k = 9$	7	1/9	1536	76.8
80	20	0	8	8	4.8	$R = 1/4, k = 9$	3		1536	76.8
172	20	0	12	8	9.6	$R = 1/4, k = 9$	1		1536	76.8

Table 5.20 RFCh frame bit rates for RC4 and RC6

Information bits	Frame (ms)	Reservation bit	CRC (F)	Tail bits	Total (Kbps)	Convolutional encoder	Symbol repetition	Symbol puncturing	Interleaving block	Coded symbols after interleaving (Ksps)
24	5	0	16	8	9.6	$R = 1/4, k = 9$	1		384	76.8
21	20	1	6	8	1.8	$R = 1/4, k = 9$	15	8/24	1536	76.8
55	20	1	8	8	3.6	$R = 1/4, k = 9$	7	8/24	1536	76.8
125	20	1	10	8	7.2	$R = 1/4, k = 9$	3	8/24	1536	76.8
267	20	1	12	8	14.4	$R = 1/4, k = 9$	1	8/24	1536	76.8

Table 5.21 Puncturing patterns when using convolutional coding

Convolutional encoder rate	Puncturing rate	Puncturing pattern	Associated RC
1/4	8/24	'111010111011101011101010'	4 and 6
1/4	4/12	'110110011011'	4
1/4	1/5	'11110'	3 and 5
1/4	1/9	'111111110'	3 and 5
1/2	2/18	'111011111111111110'	6

5.2.5.4 Interleaving for RFChs Operating on RC3–RC6

As in Figure 5.51, after bit repetition and puncturing, coded bits are interleaved. The interleaving method used for RFChs operating on RC3–RC6 is the Bit Reversal Order (BRO) interleaving, described in Section 5.2.3.2.

5.2.5.5 Multiplexing by Walsh Codes, Complex Modulation by PN Sequences and Carrier Modulation

For RFChs, PN spreading and modulation processes are the same as applied to EAChs, although a distinct Walsh code is used to orthogonalise the spreading, as described in Section 5.2.1.4. Traffic channels can use two types of long code offset masks: private and public. In normal operation, the public mask is used. If additional protection is required, the private mask is used.

Figure 5.52 shows the structure of the public long code offset mask used in RFChs. All traffic channel long code masks use the permuted Electronic Serial Number (*ESN*). The *ESN* is a unique number associated with each mobile terminal. The chapter on call processing and authentication describes the *ESN* in detail.

41	40 39	....	37 36	....	32 31	....	0
11	000		11000		Permuted ESN		

Figure 5.52 Public long code mask for RFChs operating on RC3–RC6.

5.2.5.6 RFCh Preamble Transmission for RFChs Operating on RC3–RC6

As described in Section 5.2.1.1, the RPiCh is transmitted as a preamble for RFChs (Figure 5.53).

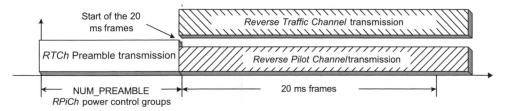

Figure 5.53 RPiCh transmission as a preamble for reverse fundamental channels.

5.2.6   Reverse Dedicated Control Channels (RDCCHs)

RDCCHs are reverse traffic channels used only for radio configurations 3–6, especially for transmission of system and user information data, such as signalling and control. RDCCHs are a system enhancement compared to IS-95 reverse traffic channels in terms of signalling and control during a call.

Because of its specific traffic characteristics (only control and/or signalling data), the MS can decide on discontinuous RDCCh transmission on a frame-by-frame basis, i.e. the transmission of the RDCCh may be enabled or disabled according to traffic needs, for example when transmitting user data on other traffic channel.

As for RFChs, the system may instruct the MS to start RDCCh transmission with frame offsets that are multiples of 1.25 ms, as defined by the *FRAME_OFFSET* parameter. Zero-offset frames start only when the variable system time is a multiple of 20 or 5 ms, depending on the frame size. Non-zero offset frame transmission must start at $1.25 \times FRAME\_OFFSET$ and $1.25 \times (FRAME\_OFFSET \bmod 4)$, respectively, for 20 and 1.25 ms frames. Frame interleaving must be aligned to each RDCCh frame.

5.2.6.1   Channel Processing for RDCCHs

Figure 5.54 shows the frame structure for RDCCHs. In the figure, R represents a reserved bit, F corresponds to frame quality indicators (CRC) and T represents encoder tail bits (eight bits '0').

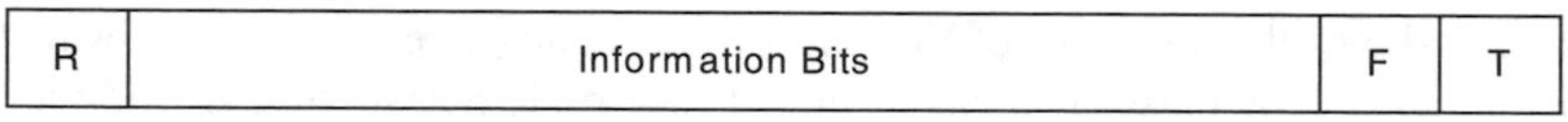

R	Information Bits	F	T

Figure 5.54   RDCCh frame structure.

Table 5.22   RDCCh possible frame structures according to the radio configuration (RC)

RC	Frame (ms)	Data rate (bps)	Total bits/frame	Reserved bit	Information bits	CRC (F)	Tail bits (T)
3–6	5	9600	48	0	24	16	8
3 and 5	20	9600	192	0	172	12	8
4 and 6	20	14400	288	1	267	12	8

The number of reserved bits, CRC bits and information bits varies according to the frame length and to the radio configuration. Table 5.22 presents the possible frame structures.

Figure 5.55 represents RDCCh channel processing using a block diagram and Table 5.23 provides the data rates obtained at each stage of processing.

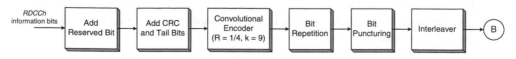

Figure 5.55   RDCCh frame processing block diagram.

Table 5.23   RDCCh bit rates during channel processing according to the radio configuration (RC)

RC	Frame (ms)	Total bits/ frame	Transmission rate (bps)	Convolutional encoder	Symbol repetition	Symbol puncturing	Interleaving block	Coded symbols after interleaving (Ksps)
3 to 6	5	48	9600	$R = 1/4$, $k = 9$	1		384	76.8
3 and 5	20	192	9600	$R = 1/4$, $k = 9$	1		1536	76.8
4 and 6	20	288	14400	$R = 1/4$, $k = 9$	1	8/24	1536	76.8

The convolutional encoder used has rate $R = 1/4$, as shown previously in Figure 5.37.

5.2.6.2   Puncturing

Puncturing is applied to coded symbols when using RC4 and RC6. Because RDCChs use a convolutional encoder for frame processing, the puncturing process corresponds to the convolutional code puncturing, described in Section 5.2.5.3. The puncturing pattern applied to RDCChs is always 8/24, as in Table 5.21, that is, eight coded symbols are removed from each set of 24.

5.2.6.3   Interleaving

As in Figure 5.55, after coded symbol repetition and puncturing, RDCChs frames are interleaved. The interleaving method used is the Bit Reversal Order (BRO), described in Section 5.2.3.2.

5.2.6.4   Multiplexing by Walsh Codes, Complex Modulation by PN Sequences and Carrier Modulation

PN quadrature spreading and modulation processes for RDCChs are equivalent to processes used for EAChs, depicted in Figure 5.33. Because RDCChs are traffic channels, they can use two types of long code offset masks: private and public. In normal operation, the public mask is used. If additional protection is required, the private mask is used. Figure 5.56 shows

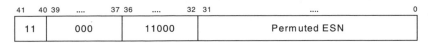

Figure 5.56   RDCCh long code offset mask.

the structure of the public long code offset mask used for both spreading rate configurations (SR1 and SR3).

5.2.7 Reverse Supplemental Channels (RSCh)

Like RDCChs, RSChs are reverse traffic channels, but used only for RC3–RC6. One or more of these channels transmitted by the same MS enable support of data at rates higher than the transmitted by RFChs.

These channels are enhancements of reverse traffic channels for data transmission. RSChs are the natural evolution of the multiple Reverse Supplemental Code Channels (RSCChs) provided for CDMA IS-95B systems for RC1 and RC2. When compared to these IS-95B channels, cdma2000 RSChs have much higher data transmission capacity. Another difference is that an MS can only use a maximum of two RSChs whereas, in IS-95, up to seven RSCChs can be transmitted by the MS.

RSChs are usually organised in 20 ms frames, but MSs can optionally operate with 40 and 80 ms frames. Figure 5.57 shows the frame structure for RSChs.

In the figure, R represents a reserved bit, F corresponds to frame quality indicators (CRC) and R/T represents a set of 8 bits that may be reserved bits or encoder tail bits, according to the encoder being used.

Figure 5.57 RSCh frame structure.

If using the convolutional encoder, the frame uses eight encoder tail bits (all with value '0'). Otherwise, if using a turbo encoder, the two first bits of the R/T block in the frame are '0s' and the remaining 6 bits are internally generated and inserted in the frame by the encoder. Turbo coders can only be used if a certain number of bits per frame, which depends on the RC, is being processed. In some radio configurations the initial reserved bit (R) may not be inserted.

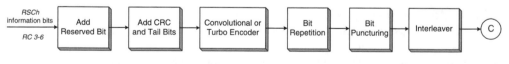

Figure 5.58 RSCh frame processing for RC3–RC6.

Figure 5.58 shows a block diagram representing RSCh frame processing. The use of turbo or convolutional encoder depends on the desired channel protection, prior to transmission, and on the number of bits processed per frame. PN modulation and transmission processes are equivalent to the processes in Figure 5.33.

Tables 5.24–5.27 present RSCh bit rates obtained during channel processing for RC3–RC6. In the tables, n_f represents the frame duration in multiples of 20 ms, n_f may be equal to 1, 2 or 4.

Table 5.24 RSCh bit rates during channel processing for RC3

Information bits	Frame (ms)	Reservation bit	CRC (F)	Reservation/ tail	Total (kbps)	Convolutional/ turbo encoder	Symbol repetition	Symbol puncturing	Interleaving block	Coded symbols (ksps)
16	20	0	6	8	1.5	$R = 1/4, k = 9$	15	1/5	1536	38.4
40	$20n_f$	0	6	8	$2.7/n_f$	$R = 1/4, k = 9$	7	1/9	1536	$38.4/n_f$
80	$20n_f$	0	8	8	$4.8/n_f$	$R = 1/4, k = 9$	3		1536	$38.4/n_f$
172	$20n_f$	0	12	8	$9.6/n_f$	$R = 1/4, k = 9$	1		1536	$38.4/n_f$
360	$20n_f$	0	16	8	$19.2/n_f$	$R = 1/4, k = 9$			1536	$76.8/n_f$
744	$20n_f$	0	16	8	$38.4/n_f$	$R = 1/4, k = 9$			3072	$153.6/n_f$
1512	$20n_f$	0	16	8	$76.8/n_f$	$R = 1/4, k = 9$			6144	$307.2/n_f$
3048	$20n_f$	0	16	8	$153.6/n_f$	$R = 1/4, k = 9$			12288	$614.4/n_f$
6120	$20n_f$	0	16	8	$307.2/n_f$	$R = 1/2, k = 9$			12288	$614.4/n_f$

Table 5.25 RSCh bit rates during channel processing for RC4

Information bits	Frame (ms)	Reservation bit	CRC (F)	Reservation/ tail	Total (kbps)	Convolutional/ turbo encoder	Symbol repetition	Symbol puncturing	Interleaving block	Coded symbols (ksps)
21	20	1	6	8	1.8	$R = 1/4, k = 9$	15	8/24	1536	76.8
55	$20n_f$	1	8	8	$3.6/n_f$	$R = 1/4, k = 9$	7	8/24	1536	$76.8/n_f$
125	$20n_f$	1	10	8	$7.2/n_f$	$R = 1/4, k = 9$	3	8/24	1536	$76.8/n_f$
267	$20n_f$	1	12	8	$14.4/n_f$	$R = 1/4, k = 9$	1	8/24	1536	$76.8/n_f$
552	$20n_f$	0	16	8	$28.8/n_f$	$R = 1/4, k = 9$		4/12	1536	$76.8/n_f$
1128	$20n_f$	0	16	8	$57.6/n_f$	$R = 1/4, k = 9$		4/12	3072	$153.6/n_f$
2280	$20n_f$	0	16	8	$115.2/n_f$	$R = 1/4, k = 9$		4/12	6144	$307.2/n_f$
4584	$20n_f$	0	16	8	$230.4/n_f$	$R = 1/4, k = 9$		4/12	12288	$614.4/n_f$

Table 5.26 RSCh bit rates during channel processing for RC5

Information bits	Frame (ms)	Reservation bit	CRC (F)	Reservation/tail	Total (kbps)	Convolutional/turbo encoder	Symbol repetition	Symbol puncturing	Interleaving block	Coded symbols (ksps)
16	20	0	6	8	1.5	$R = 1/4, k = 9$	15	1/5	1536	38.4
40	$20n_f$	0	6	8	$2.7/n_f$	$R = 1/4, k = 9$	7	1/9	1536	$38.4/n_f$
80	$20n_f$	0	8	8	$4.8/n_f$	$R = 1/4, k = 9$	3		1536	$38.4/n_f$
172	$20n_f$	0	12	8	$9.6/n_f$	$R = 1/4, k = 9$	1		1536	$38.4/n_f$
360	$20n_f$	0	16	8	$19.2/n_f$	$R = 1/4, k = 9$			1536	$76.8/n_f$
744	$20n_f$	0	16	8	$38.4/n_f$	$R = 1/4, k = 9$			3072	$153.6/n_f$
1512	$20n_f$	0	16	8	$76.8/n_f$	$R = 1/4, k = 9$			6144	$307.2/n_f$
3048	$20n_f$	0	16	8	$153.6/n_f$	$R = 1/4, k = 9$			12288	$614.4/n_f$
6120	$20n_f$	0	16	8	$307.2/n_f$	$R = 1/3, k = 9$			18432	$921.6/n_f$
12264	$20n_f$	0	16	8	$614.4/n_f$	$R = 1/3, k = 9$			36864	$1843.2/n_f$

Table 5.27 RSCh bit rates during channel processing for RC6

Information bits	Frame (ms)	Reservation bit	CRC (F)	Reservation/tail	Total (kbps)	Convolutional/turbo encoder	Symbol repetition	Symbol puncturing	Interleaving block	Coded symbols (ksps)
21	20	1	6	8	1.8	$R = 1/4, k = 9$	15	8/24	1536	76.8
55	$20n_f$	1	8	8	$3.6/n_f$	$R = 1/4, k = 9$	7	8/24	1536	$76.8/n_f$
125	$20n_f$	1	10	8	$7.2/n_f$	$R = 1/4, k = 9$	3	8/24	1536	$76.8/n_f$
267	$20n_f$	1	12	8	$14.4/n_f$	$R = 1/4, k = 9$	1	8/24	1536	$76.8/n_f$
552	$20n_f$	0	16	8	$28.8/n_f$	$R = 1/4, k = 9$			2304	$115.2/n_f$
1128	$20n_f$	0	16	8	$57.6/n_f$	$R = 1/4, k = 9$			4608	$230.4/n_f$
2280	$20n_f$	0	16	8	$115.2/n_f$	$R = 1/4, k = 9$			9216	$460.8/n_f$
4584	$20n_f$	0	16	8	$230.4/n_f$	$R = 1/4, k = 9$			18432	$921.6/n_f$
9192	$20n_f$	0	16	8	$460.8/n_f$	$R = 1/4, k = 9$			36864	$1843.2/n_f$
20712	$20n_f$	0	16	8	$1036.8/n_f$	$R = 1/2, k = 9$		2/18	36864	$1843.2/n_f$

For radio configurations 3 (Table 5.24) and 5 (Table 5.26), the following conditions apply.

- For frames with up to 40 bits n_f can be 1 or 2; for frames with more than 40 bits n_f can be 1, 2 or 4.

- Turbo encoders can only be used for frames with 360 input bits or more. Otherwise, a convolutional coder is used.

For radio configurations 4 (Table 5.25) and 6 (Table 5.27), the following conditions apply.

- For frames with up to 55 bits n_f can be 1 or 2; for frames with more than 55 bits n_f can be 1, 2 or 4.

- Turbo encoders can only be used for frames with 552 input bits or more. Otherwise, a convolutional coder is used.

5.2.7.1 Puncturing of Coded Symbols

Puncturing of reverse supplemental channels can be performed in two distinct ways, depending on the encoder. For convolutional encoders, the process (*convolutional code symbol puncturing*) is the same as the process used for RFChs, described in Section 5.2.5.3.

When using turbo encoders, a turbo code symbol puncturing process is applied. Table 5.28 shows the puncturing patterns used in this situation.

5.2.7.2 Interleaving

As in Figure 5.58, after coded symbol repetition and puncturing, RSChs are interleaved. The interleaving method used is the Bit Reversal Order (BRO), described in Section 5.2.3.2.

5.2.7.3 Multiplexing by Walsh Codes, Complex Modulation by PN Sequences and Carrier Modulation

PN quadrature spreading and modulation processes for RSChs are equivalent to the processes used for EAChs, depicted in Figure 5.33. Because RSChs are traffic channels, they can use two types of long code offset masks: private and public. In normal operation, the public mask is used. If additional protection is required, the private mask is used.

Table 5.28 Puncturing patterns when using turbo coding

Convolutional encoder rate	Puncturing rate	Puncturing pattern	Associated RC
1/2	2/18	'111110101111111111'	6
1/4	4/12	'110111011010'	4

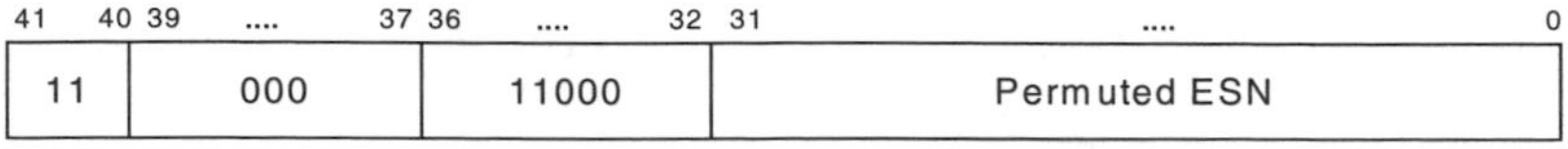

41	40 39	....	37 36	....	32 31	....	0
11	000		11000		Permuted ESN		

Figure 5.59   RSCh public long code offset mask.

Figure 5.59 shows the structure of the public long code offset mask used for both spreading rate configurations (SR1 and SR3).

Because RSChs are traffic channels, the system may request MSs to start transmission of RSCh frames with time offsets as multiple of 1.25 ms, as defined by the *FRAME_OFFSET* parameter. As in other traffic channels, zero-offset RSCh frames start only when system time is a multiple of 20, 40 or 80 ms. MSs can support delays of intervals multiple of 20 ms when working with 40 or 80 ms frames, as specified by *REV_SCH_FRAME_OFFSET*[i], where $i = 1$ represents the transmission offset associated to the primary RSCh and $i = 2$ for the secondary RSCh.

Non-zero-offset frames transmissions must start at $1.25 \times FRAME\_OFFSET$ for 20 ms frames and at $1.25 \times FRAME\_OFFSET + REV\_SCH\_FRAME\_OFFSET[i]$ for 40 and 80 ms. Frame interleaving processes must always be aligned to the beginning of each RSCh frame.

5.2.8   Reverse Supplemental Code Channels (RSCCh)

Reverse supplemental code channels were defined and created for CDMA IS-95 B systems to allow higher data traffic transmission. Therefore, in cdma2000, these channels are used only for RC1 and RC2 (SR1) for backward compatibility. An MS can use a maximum of seven channels simultaneously.

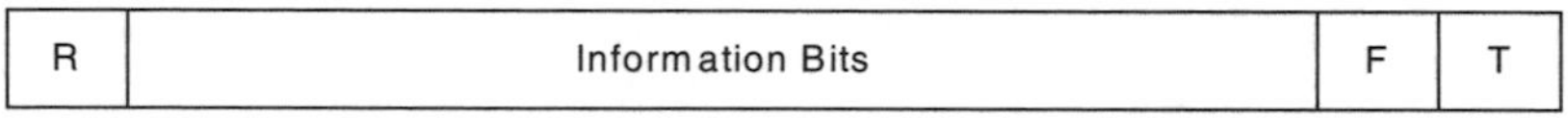

R	Information Bits	F	T

Figure 5.60   RSCCh frame structure.

Each RSCCh transmits maximal data rates of 9.6 or 14.4 kbps using 20 ms frames. The data rate depends on the radio configuration. Figure 5.60 shows the frame structure for RSCChs, where R represents a reserved bit, F corresponds to frame quality indicators (CRC) and T represents encoder tail bits (eight bits with value '0').

The number of reserved and information bits vary according to the radio configuration. Table 5.29 shows the possible frame structures for RSCChs.

Table 5.29   Possible RSCCh frame structures, according radio configuration

RC	Data rates (bps)	Reserved bit	Information bits	CRC (F)	Encoder tail bits	Bits/frame
1	9600	0	172	12	8	192
2	14400	1	267	12	8	288

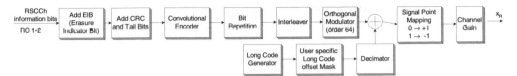

Figure 5.61 RSCh frame processing diagram block for RC1 and RC2.

Figure 5.61 shows a block diagram representing RSCCh channel processing. Processing of these channels is very similar to RTCh processing in IS-95 B systems, including the insertion of a reserved bit (EIB) for RC2 and the use of an orthogonal modulator of order 64 (using the Walsh codes).

Table 5.30 presents bit rates obtained during processing of RSCChs.

Table 5.30 Bit rates obtained during channel processing for RC1 and RC2

RC	Bits/frame	Total bits/ frame	Bit rates (Kbps)	Convolutional encoder	*Interleaving* block	Coded symbols after interleaving (Ksps)
1	172	192	9.6	$R = 1/3, k = 9$	576	28.8
2	267	288	14.4	$R = 1/2, k = 9$	576	28.8

5.2.8.1 Interleaving

As RSCChs are traffic channels compliant with CDMA IS-95 B systems, i.e. for terminals compatible only to SR1 (RC1 and RC2), the interleaving process is the same used for IS-95, described in Section 5.1.1.3. Interleaving always starts aligned to the beginning of RSCCh frames.

5.2.8.2 Orthogonal Modulation (of order 64)

After interleaving process, coded symbols are orthogonally modulated, as in Section 5.1.1.4.

5.2.8.3 PN Quadrature Spreading, Modulation and Transmission Through Air–Interface

Prior to the PN quadrature spreading and modulation, long code PN sequences, shifted with a user-specific offset mask, scramble coded bits. Figure 5.62 shows this long code mask

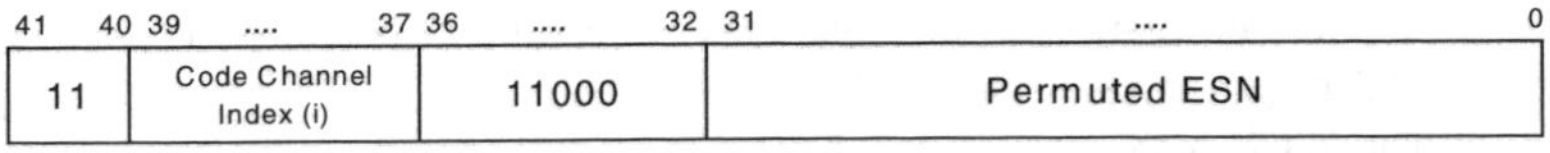

41	40 39	....	37 36	....	32 31	....	0
11	Code Channel Index (i)		11000		Permuted ESN		

Figure 5.62 Public long code offset mask used for RSCChs scrambling for RC1 and RC2.

Table 5.31 Specific carrier phase offset modulation during RSCChs transmission

RSCCh identity number	Additional transmission carrier phase offset
1 ('001')	$\pi/2$
2 ('010')	$\pi/4$
3 ('011')	$3\pi/4$
4 ('100')	0
5 ('101')	$\pi/2$
6 ('110')	$\pi/4$
7 ('111')	$3\pi/4$

(public). In the figure, code channel index (i) represents the RSCCh index number ($i = 1, \ldots, 7$) as a binary value (from '001' to '111'). The code channel index is also used to determine carrier phase modulation, as in Table 5.31.

Figure 5.63 shows a block diagram representing the modulation process for RSCChs.

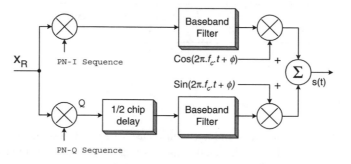

Figure 5.63 PN spreading, modulation and transmission through air interface for RSCChs.

RSCChs can also be transmitted using a frame offset delay, when MSs request the transmission to start with time offsets multiple of 1.25 ms, as defined by the *FRAME_OFFSET* parameter. Zero-offset RSCCh frames start when the system time is a multiple of 20 ms [1–11].

BIBLIOGRAPHY AND REFERENCES

1. C.S0002-C, 'Physical Layer Standard for cdma2000 Spread Spectrum Systems,' Release C, 3GPP2, May 2002.
2. C.S0005-B, 'Upper Layer (Layer 3) Signaling Standard for cdma2000 Spread Spectrum Systems,' Release B, 3GPP2, April 2002.
3. TIA/EIA-95-B, 'Mobile Station – Base Station Compatibility Standard for Dual-Mode Spread Spectrum Cellular System,' November 1998.
4. CelTec/CelPlan, 'CDMA IS-95 and cdma2000 Systems - Training Course.'
5. CDG, 'cdma2000 Overview'.

6. The Shosteck Group, '*Third Generation Wireless (3G): The Continuing Saga.*' Maryland: Wheaton, 2001, pp. 263–264
7. Recommendation ITU-R M.1079-1, 'Performance And Quality of Service Requirements for International Mobile Telecommunications-2000 (IMT-2000).'
8. CDG, 'Detailed Info: Standard Requirements for the Evolution of cdma2000.' May 30, 2000.
9. TIA/EIA-41-D (ANSI-41), 'Cellular Radio Telecommunications Intersystem Operations,' December, 1997.
10. TIA/EIA/95 (ANSI-95), 'Mobile Station – Base Station Compatibility Standard for Dual-Mode Wideband Spread Spectrum Cellular Systems,' Telecommunications Industry Association; May 1995.
11. TSB74, 'Support for 14.4 kbps Data Rates and PCS Interaction for Wideband Spread Spectrum Cellular Systems,' December, 1995.

6

Call Processing in CDMA Systems

BRUNO DE SOUZA ABREU XAVIER

This chapter describes the basic initialisation procedure for a Mobile Station (MS) to access the CDMA network. The chapter also describes call processing and the authentication process.

6.1 INTRODUCTION

The initialisation process consists of the steps that MSs execute to recognise, access and set up configuration parameters according to the CDMA network. These procedures include acquisition of the pilot channel, synchronisation, and reading of messages on control channels. In some situations, besides going through the whole initialisation process, MSs also need to perform registration and authentication processes to establish a call.

Call processing consists of the exchange of messages through which an MS and its server Base Transceiver Station (BTS) negotiate origination and termination of calls. This chapter describes the parameters exchanged in the messages and the channels used for call establishment.

Call processing procedures are very similar for cdma2000-1X and CDMA IS-95 systems because of backward compatibility requirements. This chapter presents, first, all procedures common for the two technologies. A second major focus of the chapter is to highlight certain cdma2000 exclusive procedures.

Even though the ideal evolution path of a CDMA IS-95 network would be IS-95 B and, then, cdma2000-1X, the majority of telecommunication companies have opted to upgrade from IS-95 directly to cdma2000-1X. Therefore, some of the processes highlighted as cdma2000-1X features are, in fact, IS-95B implementations.

6.2 CALL PROCESSING

To establish a call, an MS must go through several states. Each state consists of a set of commands and procedures that need to be executed according to configuration parameters stored internally in the MS or transmitted through messages. These include turning on the

Designing CDMA 2000 Systems   L. Korowajczuk et al.
© 2004 John Wiley & Sons, Ltd   ISBN: 0-470-85399-9

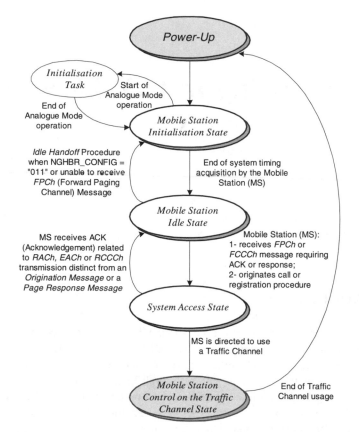

Figure 6.1   Processing states of a dual-mode AMPS/CDMA Mobile Station (MS).

terminal, for synchronisation to the network, 'reading' the overhead messages and up to setting up the configurations for the communication logical channel. The call processing states are the following (Figure 6.1).

- *Mobile station initialisation state*: the MS selects and executes system acquisition and synchronisation.

- *Mobile station idle state*: the MS monitors messages transmitted through signalling and control channels, such as Forward Paging Channels (FPChs) Forward Broadcast Control Channels (FBCCh), Forward Common Control Channels (FCCChs) and Quick Paging Channels (QPChs).

- *System access state*: the MS attempts to access the system, sending messages or responding to orders from the server BTS through a Reverse Access Channel (RACh) or, in cdma2000 systems, through an Enhanced Access Channel (EACh) and Reverse Common Control Channel (RCCCh).

- *Mobile station control on the traffic channel state*: on forward and reverse links, the communication link is established between an MS and a BTS during a call, using

Fundamental Channels (FChs) and Dedicated Control Channels (DCChs) to send and receive data messages and voice.

Figure 6.1 does not show all state transitions. Therefore, the MS may be re-directed to restart the *mobile station initialisation process* from any other state. The following sections explain the reasons for this and for distinct transitions through different states.

6.2.1 Mobile Station Initialisation State

As the first step in the MS *initialisation state*, the MS must select the operation band and system (technology) of operation, e.g. dual-mode terminals have to select AMPS or CDMA.

If there is a CDMA network capable of providing service when the MS selects CDMA as the technology of choice, the mobile station must start acquisition and synchronisation of the pilot channel for this system.

If the mobile station selects to operate on the analogue system (AMPS, for example), however, it must start the analogue operation with the *initialisation task* procedure, as in Figure 6.2. This procedure is described in detail in the CDMA standards and is not in the scope of this book.

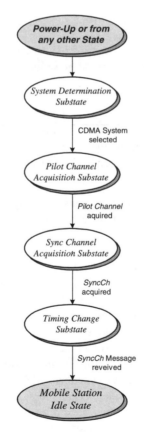

Figure 6.2 MS initialisation state.

The *mobile station initialisation state* is divided into the following sub-states (Figure 6.2).

- *System determination sub-state* (Section 6.2.1.1): Selection of the system and its main characteristics.

- *Pilot channel acquisition sub-state* (Section 6.2.1.2): When CDMA is selected, the MS acquires and synchronises itself to a BTS, synchronizing codes using the Forward Pilot Channel (FPiCh) transmitted by its server BTS as a timing reference.

- *Sync channel acquisition sub-state* (Section 6.2.1.3): The MS gathers more information about CDMA network timing and configuration parameters.

- *Timing change sub-state* (Section 6.2.1.4): The MS has a small period of time to adjust the PNLC code generator and internal clock to the system.

As a rule of thumb, every time the MS loses synchronisation, or does not receive expected orders or message confirmations (ACK), it has to restart the *mobile station initialisation state*. This can happen while in MS *idle state* (Section 6.2.2), or when performing *access attempts* while in *system access state* (Section 6.2.3), or even when a call is finished (Section 6.2.4).

6.2.1.1   System Determination Sub-state

In this sub-state, the MS must choose the system to be used. Each manufacturer defines the selection process according to its own criteria; however, the decision is always based on the following characteristics.

- Set of programmed preferences defined by the subscriber in the terminal (MS). Subscriber defined parameters for system selection may include the operation frequency (e.g. 1900 MHz PCS or 800 MHz cellular), operation band (*A* or *B*), digital and analogue capabilities and also operation preferences in terms of band (*A* or *B*) and type of system (analogue, digital).

- System programmed re-directing criteria transmitted by the system to the MS through the air interface. This only happens when the MS re-starts the MS *initialisation state*, coming from any other state.

If the selected system for operation is CDMA, in an 800 MHz frequency band, as an example, the MS must tune itself to the *setup carrier* (primary or secondary) of the selected operation band (*A* or *B*). It must also initiate synchronisation to the FPiCh transmitted by the *setup carrier*. Section 6.2.1.2 describes this process.

System determination sub-state tasks are executed when the MS is turned on (*power-up*) or when there is a re-direction from any other state.

Besides the initialisation processes described in this section, CDMA systems have another system determination process, performed whenever MSCs order BTSs to transmit specific messages through the FPCHs or FCCChs. These messages reconfigure the MS to operate under specific conditions, pre-programmed according to a system set of parameters. This process requires the MS to be already operating in *mobile station idle state*, described in Section 6.2.2.

Table 6.1 Setup carriers defined for bands A and B (800 MHz systems)

CDMA Carrier	A band number	B band number
Primary	283 (AMPS channel)	384 (AMPS channel)
Secondary	691 (AMPS channel)	777 (AMPS channel)

6.2.1.2 Pilot Channel Acquisition Sub-state

The first task for any MS in this sub-state is to set its de-modulator to the *setup carrier* of the selected operation band.

Setup carriers, called *primary* and *secondary*, are defined in both IS-95 and cdma2000 standards. They are the only two carriers whose central frequency is defined for the whole system. In the case of the 800 MHz systems, they are defined by 30 kHz channel numbers that represent the analogue system carrier centre frequencies. Table 6.1 presents carrier frequency assignments for setup carriers of both *A* and *B* operation bands.

The networks, however, can change the identification number of primary and secondary setup carriers. This modification must also be made in the MS memory. As an example, some telecom operators in Brazil have configured the primary setup carrier as AMPS channel number 160.

These carriers are called *setup carriers* because they are used as initial references to access the CDMA networks. As soon as an MS completes the initial system access, it can be re-directed to any other carrier available.

Each carrier of a CDMA system must transmit an FPiCh that is primarily used as initial timing reference. Pilot channels transmit a sequence of '0s' coded by Walsh code 0 (W_0^{64}) that are I/Q (quadrature) modulated by PN-I and PN-Q sequences. Because of this, the 'real' information transmitted by the pilot channels are the PN sequences themselves. Chapter 4 (*Forward Link*) explains FPiChs in more detail.

In *pilot acquisition sub-state*, the MS tunes to the *primary setup carrier* and synchronises its short PN sequence generators to the system PN sequence transmitted by the FPiCh. If the acquisition and synchronisation processes are finished before $T_{20m} = 15$ s, then the MS goes to the *sync channel acquisition sub-state* (Section 6.2.1.3). In case of failure during the *primary setup carrier* acquisition process, the mobile must restart the process, trying the *secondary setup carrier* this time. If the process fails again, the MS is automatically redirected to the *system determination sub-state* (Section 6.2.1.1).

Before internal and system PN sequences are synchronised, the MS has no idea of the phase offset of the short PN sequences, used to identify each BTS within the network, as will be further explained later. The synchronisation obtained for the short PN sequences is enough to allow the MS to understand (read) messages of the SyncCh during the *sync channel acquisition sub-state*, as described in Section 6.2.1.3.

Signals transmitted through a certain channel may only be de-spread and decoded at the receiver if both BTS and MS PN sequences are identical and perfectly synchronised. Figure 6.3 shows a simplified diagram representing synchronisation between transmitted and internally generated sequences. The synchronisation process is divided into two steps

- *Acquisition or sliding correlator process*: The receiver shifts, chip by chip, the phase of the internally generated sequence until it obtains a *partial alignment* of the sequences

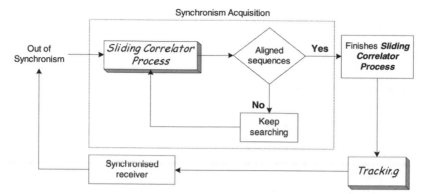

Figure 6.3　PN Sequence synchronisation acquisition flowchart.

with error smaller than one chip interval. Instead of comparing the transmitted and internally generated short PN sequences to verify and establish their synchronisation, the MSs perform partial sequence comparison, a process also known as *partial alignment*.

- *Tracking (fine tuning)*: After obtaining the partial alignment, the MS uses a control signal to achieve and maintain synchronisation. This signal controls the internal oscillator clock used to generate PN sequences at the MS.

Figure 6.4 presents a possible implementation of the acquisition circuit, in which the correlation between received and internally generated codes happens through successive phase shifts.

The correlation process consists of the multiplication of the signal received by the local code, followed by a bandpass filter. The filter characteristics depend on the signal bandwidth prior to spreading, that is, a narrowband signal. The signal is de-spread when the transmitted and locally generated codes are synchronised.

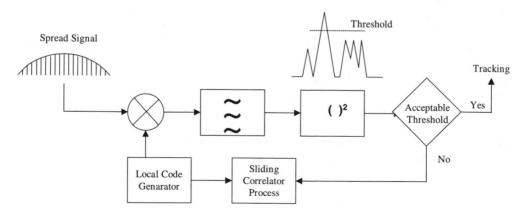

Figure 6.4　Synchronisation acquisition circuit.

After being de-spread, the signal is applied to a square-law detector producing a direct current (DC) component proportional to its power level, as in eqn (6.1)

$$(A \cos \omega_c t)^2 = (1/2)A^2 + (1/2)A^2 \cos(2\omega_c t) \tag{6.1}$$

In eqn (6.1), A corresponds to the amplitude of the de-spread signal. The second harmonic component is eliminated through appropriate filtering.

The DC component is used in the synchronisation detection. When both codes are synchronised, or almost, the DC level reaches a threshold and the local clock-delaying ceases and tracking begins. Sequence sliding usually occurs in 1/2- or 1-chip steps. Because of the selective nature of the autocorrelation function and of the sliding step size, when the threshold is achieved the error is under 1/2 or 1 chip interval.

A bandpass filter limits the clock-delaying rate after the square-law detector. As a rule of thumb, this rate is equal to the maximum data rate produced by the vocoder. This implies a limitation in the synchronisation speed, compromising performance for very long codes.

Figure 6.5 shows a tracking circuit diagram for a Delay Lock Loop (DLL). The signal received is multiplied by two versions of the PN sequence, shifted by one or two chips between each other.

Figure 6.6 shows the waveforms associated with the tracking circuit in Figure 6.5.

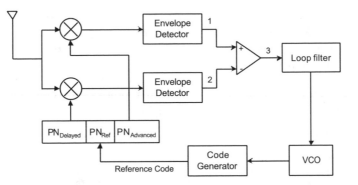

Figure 6.5 Synchronisation tracking circuit.

In the tracking circuit, shifted versions of the PN sequence are applied to two correlator circuits together with the received signal. Each correlator's output generates a 2-chip wide triangular function with peaks shifted with the same phase applied to the PN sequences. These signals are then summed, filtered and used to control the Voltage Controlled Oscillator (VCO), which generates the local code clock. Synchronisation happens at an average point between the correlation peaks.

6.2.1.3 Sync Channel Acquisition Sub-state

After acquiring the FPiCh, the MS tunes to the SyncCh using Walsh code 32 (W_{32}^{64}). If the MS is not able to read (decode) any message in less than 1 s (T_{21m}), it is automatically re-

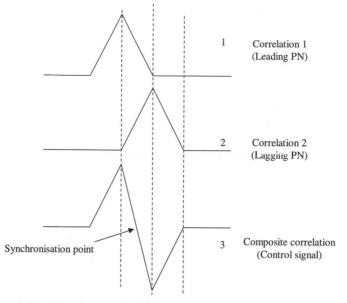

Figure 6.6   Waveforms associated to the synchronisation tracking circuit.

directed to the *system determination sub-state* (Section 6.2.1.1) to restart FPiCh acquisition, with a synchronisation acquisition failure indication.

The SyncCh message transmits network information parameters to MSs. Some of the parameters are the following:

- protocol revision level in use (*P_REV_IN_USE*), protocol revision level (*P_REV*) and minimum protocol revision level (*MIN_P_REV*);

- *SID* (System IDentification number);

- *NID* (Network IDentification number);

- FPiCh phase offset (*PILOT_PN*), to allow BTS identification;

- long code state (*LC_STATE*), 320 ms after the end of the 80 ms super-frame containing the SyncCh message, allowing MSs to synchronise their long code sequence generator;

- system time (*SYS_TIME*), 320 ms after the end of the super-frame containing the SyncCh message;

- FPCh data rate (P_{rat}).

Even if an MS reads the message before T_{21m}, specific system settings may force it to be re-directed to the *system determination sub-state* (Section 6.2.1.1) indicating incompatibilities, such as

- the MS supports a minimum protocol revision level smaller than the level defined by the BTS (*MIN_P_REV*);

- the MS does not support the FPCh P_{rat} data rate.

The MS is also informed if it has to tune another CDMA carrier within the network, the Radio Configurations (RC) in use and if the *Quick Paging Channels* (*QPChs*) are enabled.

6.2.1.4   Timing Change Sub-state

In this sub-state, the MS executes all the internal modifications required regarding its base timing passed through the SyncCh message, in order to be totally synchronised to its server BTS.

Procedures executed by the MS in this sub-state include adjustment of the long code sequence generator (*LC_STATE*), clocks for timing reference (*SYS_TIME*) and the FPiCh phase offset (*PILOT_PN*). Figure 6.7 illustrates some of these changes and settings.

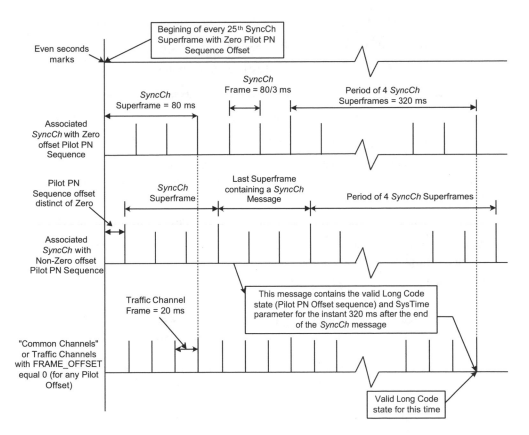

Figure 6.7   Long code sequence synchronisation and reference time.

This sub-state provides the MS enough time to adjust and synchronise all timing and code references to the system. After synchronisation, the MS starts the registration process, explained in detail in Section 6.4.

The MS tunes to the Primary Forward Paging Channel (called FPCh-P or FPCh-1) and enters the *mobile station idle state* (Section 6.2.2).

6.2.2   Mobile Station Idle State

In the *MS idle state*, the MS monitors the FPCh-P or QPChs using the data rate defined in the SyncCh message, 4.8 or 9.6 kbps.

FPCh-P overhead messages inform MSs of carriers transmitting FPChs and how many are transmitted in each of them (up to 7). When there is more than one FPCh, an MS selects a channel using the Hash algorithm, establishing the identity of the channel through which the system will send specific and directed messages and/or orders. The MS also selects a Reverse Access Channel (RACh) or Enhanced Access Channel (EACh) to communicate with the system while not in the *mobile station control* on the *traffic channel state* (Section 6.2.4).

While in MS *idle state*, the mobile executes the following procedures.

- Monitor the designated paging channel or (cdma2000) quick paging channels.

- Confirm reception of BTS transmitted messages.

- Initiate a call (MS originated call procedure), as in Section 6.3.1.

- Receive an incoming call (MS terminated call procedure), as in Section 6.3.2.

- Perform registration processes, as in Section 6.4.

- Search for other pilot channels and update the pilot list.

- Idle handoff processes, as in Section 6.2.2.4 (further details are also explained in Chapter 7, which treats power control, handoff processes and radio resource management).

- Update system data (parameters) transmitted by BTSs through overhead messages.

- Priority Access and Channel Assignment (PACA) cancellation procedures (in cdma2000 systems – originally defined in IS-95B).

- Power-off.

The Priority Access and Channel Assignment procedure, also known as PACA (Section 6.2.2.5), sets an MS originated call as a priority when there is a lack of resources, i.e. no traffic channels available.

Overhead messages are specialised messages in CDMA systems that carry system configuration parameters required by MSs during initialisation. Overhead messages include access parameters message, system parameters message, CDMA channel list message, extended system parameters message, global service redirection message and extended neighbour list message.

While in MS *idle state*, terminals may operate on *non-slotted* or on *slotted mode*. In the first case, i.e. *non-slotted mode*, MSs continuously monitor the FPCh because BTSs can transmit messages at any time. In *slotted mode*, MSs only read FPChs at specific time intervals, maintaining just a few active circuits most of the time to save battery.

Sections 6.2.2.1 to 6.2.2.5 describe the most common tasks performed in MS *idle state*.

6.2.2.1 *Forward Paging Channels or Forward Common Control Channels Monitoring*

FPCHs and FCCCHs are structured in 80 ms slots, with a total of 2048 slots. Channel monitoring depends on the operation mode of the MS: *slotted* or *non-slotted*.

An MS operating in slotted mode usually monitors FPCHs or FCCCHs during 1 or 2 slots at each cycle, according to the *SLOT_CYCLE_INDEX* parameter, which varies from 0 to 7. This parameter is defined using the Hash algorithm with the *International Mobile Station Identification Number* (*IMSI* – explained in detail in Section 6.6) and the maximum number of slots per FPCh or FCCCHs (2048) as inputs.

Equation (6.2) defines the duration of each reading cycle

$$T = 2^{SLOT\_CYCLE\_INDEX} \times 16 \text{ slots} = 2^{SLOT\_CYCLE\_INDEX} \times 1.28 \text{ seconds} \qquad (6.2)$$

The minimum cycle interval is 1.28 s, or 16 slots (in case the parameter *SLOT_CYCLE_INDEX* = 0), and the maximum is 163.84 s, or 2048 slots (for *SLOT_CYCLE_INDEX* = 7). Figure 6.8 illustrates this, showing one of the 128 sets of 16 slots (1.28 s), numbered from 0 to 15.

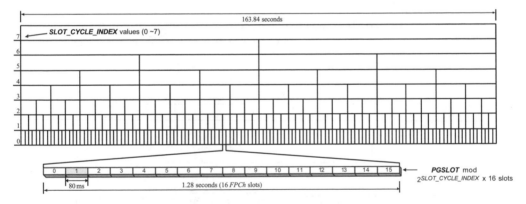

Figure 6.8   FPCh or FCCCh slot reading process.

The Hash algorithm also assigns a *PGSLOT* number to each MS, varying from 0 to 2047. This number defines the slot to be read by an MS. Each MS must read its assigned FPCh or FCCCh slot every $2^{SLOT\_CYCLE\_INDEX} \times 16$ slots. Thus, it becomes necessary to make a simple calculation to define the number, varying from 0 to 15, of the slot to be read, as in Figure 6.8.

As an example, consider an MS that has values *PGSLOT* = 561 and *SLOT_CYCLE_INDEX* = 1. In this case, performing the necessary calculation, the MS must read, every 2.56 seconds, the slot number 1, highlighted in Figure 6.8. Thus, it should read slot number 561, wait 2.56 seconds (32 slot intervals), read slot number 593 (that also consists of the same slot 1 in the set varying from 0 to 15) and so on.

The greater the *SLOT_CYCLE_INDEX*, the greater the interval between two slots assigned for the same MS and the less the battery power consumption. Smaller values lead to constant reading of FPCh or FCCCh slots and more battery power consumption. Thus, it is very common to find CDMA systems using *SLOT_CYCLE_INDEX* = 2.

Any MS operating in slotted mode can also monitor additional slots of their FPCh or FCCChs, to look for broadcast messages or broadcast pages, as described in the following section.

6.2.2.2   Quick Paging Channels Monitoring

QPChs are also structured in 80 ms slots, with a total of 2048 slots. Three types of signalling data are transmitted in quick paging channels: *paging indicators*, *configuration change indicators* and *broadcast indicators*, each consisting of a single bit.

Paging indicators inform MSs of messages in their assigned FPCh or FCCCh slot. Each MS that supports this feature is assigned two *paging indicators* in the associated QPCh.

Configuration change indicators inform all MSs on the coverage area of a BTS about any system configuration changes, causing MSs to tune to FPCh or FCCCh slots to receive the information.

Broadcast indicators inform MSs that a broadcast message will be transmitted on the associated FPCh, FCCCh or FBCCh. Section 6.2.2.3 describes the processes of receiving broadcast messages.

QPCh slots always start 20 ms before FPCh and FCCCh slots, as illustrated in Figure 6.9, thus the QPCh slot assigned to a MS starts 100 ms before the FPCh slot assigned for that

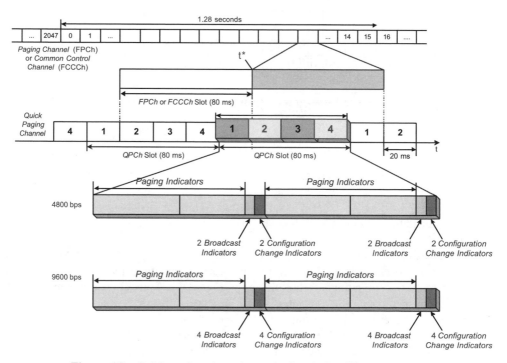

Figure 6.9   Quick paging channels monitoring during idle state.

same MS. This allows the MS to tune to the correct channel when a *paging*, a *broadcast* or a *configuration change indicator*, or both, indicate that there is a message for itself.

The QPCh slot is divided into four quarters of 20 ms each (1–4). Figure 6.9 shows the position paging indicators assigned to a specific MS within the QPCh structure. The Hash algorithm defines their position in the QPCh slot and the two paging indicators are sent according to one of the following rules.

- One indicator is sent in the first quarter (between $(t^* - 100)$ ms and $(t^* - 80)$ ms) and the other on the third quarter (between $(t^* - 60)$ ms and $(t^* - 40)$ ms).

- The first indicator is sent on the second quarter (between $(t^* - 60)$ ms and $(t^* - 40)$ ms) and another on the fourth quarter (between $(t^* - 40)$ ms and $(t^* - 20)$ ms).

The purpose of using QPChs is for MSs to monitor FPCHs or FCCCHs less frequently, increasing the *SLOT_CYCLE_INDEX* value, and reducing the period in which the receiver is on, therefore reducing battery power consumption.

6.2.2.3 Reception of Broadcast Messages

Broadcast messages are sent to all MSs simultaneously. The transmission protocol defined for FPCHs, FCCCHs and FBCCHs allows mobile stations to operate on slotted or non-slotted modes to receive broadcast messages.

Broadcast messages can be transmitted as a broadcast page or as an actual broadcast message. The latter consists of a data burst message containing the information to be transmitted. The first consists of an address, or slot number, in which the data burst message starts.

In both cases, if an MS is capable of monitoring the QPCh, it receives a *broadcast indicator* informing the existence of a broadcast message or page in one of the previously mentioned channels (FPCHs, FCCCHs or FBCCHs). If the MS is associated with an FPCh, the message is transmitted in the next FPCh slot after the end of the current QPCh slot.

Broadcast messages can be sent in two ways: *multi-slot* broadcast message transmission or *periodic* broadcast paging transmission.

When transmitted as a *multi-slot* broadcast transmission, the broadcast message is sent during a certain number of slots, long enough to allow all MSs operating on slotted mode to read it, as in Figure 6.10. Even if the message is only one slot long, it is transmitted during as many slots as defined by the *SLOT_CYCLE_INDEX* cycle. The message is transmitted in the slot that starts exactly after the current QPCh slot ends, depending on the logical channel the MS is assigned to, be it an FPCh or an FCCCh.

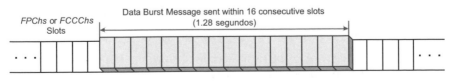
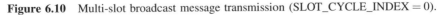

Figure 6.10 Multi-slot broadcast message transmission (SLOT_CYCLE_INDEX = 0).

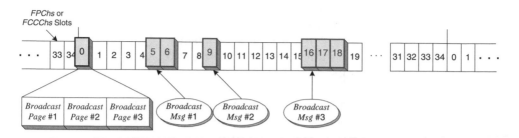

Figure 6.11 Periodic broadcast message transmission (BCAST_INDEX = 1).

When the broadcast message is transmitted as *periodic* broadcast paging transmission, an MS monitors specific slots according to the transmission of one or more broadcast pages, as in Figure 6.11. In this case, the broadcast messages cycle is $B + 3$ slots long, with B defined as

$$B = (2^{BCAST\_INDEX} \times 16) \text{ slots} \rightarrow 1 \leq BCAST\_INDEX \leq 7 \qquad (6.3)$$

In the periodic transmission mode, the broadcast indicator informs MSs that a broadcast page will be transmitted in the FPCh or FCCCh slot beginning after the current QPCh slot ends. If FPChs are assigned to mobile stations, the broadcast page informs the slots of the broadcast messages in the FPCh itself. If an FCCCh is assigned to the MS, the broadcast page informs the broadcast message slots transmitted in this same channel or on the FBCCh.

6.2.2.4 Idle Handoff Management and Procedures

Idle handoff occurs when an MS moves from the coverage area of one BTS to another while in idle state (Section 6.2.2). As soon an MS detects signals transmitted by the FPiCh of another server with power level sufficiently greater than the current server, it transmits a message to the system requesting idle handoff.

Handoff processes are described in detail in the following chapter, which also treats power control and radio resource management.

While in MS *idle state*, an MS searches for other FPiChs signals, evaluating their power levels. This search can be performed in distinct carriers or even in different operation band classes used by the system. The handoff process is explained in details in Chapter 7.

After executing idle handoff, an MS can only resume operation in non-slotted mode if the following conditions are satisfied:

- the MS supports QPCh monitoring;

- the MS knows that the server BTS supports the use of configuration change indicators;

- the MS verifies that configuration change indicators of the new server BTS are 'OFF';

- a period no longer than $T_{31m} = 600$ s has occurred since the MS received the last valid FPCh message from the new server BTS (T_{31m} corresponds to the maximal interval within which all parameters are considered valid for the CDMA system).

6.2.2.5 PACA Cancellation Procedure

This operation is performed when a user cancels the PACA or when the PACA timer expires because of lack of system resources, i.e. there are no traffic channels available at the moment. In both cases, the MS starts a PACA cancel operation and are re-directed to the update overhead information sub-state, within the *system access state*, described in Section 6.2.3.

6.2.3 System Access State

In the *system access state*, each mobile terminal can send messages to the server using a Reverse Access Channel (RACh) or Enhanced Access Channels (EACh). They can also receive messages or orders through a Forward Paging Channel (FPCh) or Forward Common Control Channel (FCCCh).

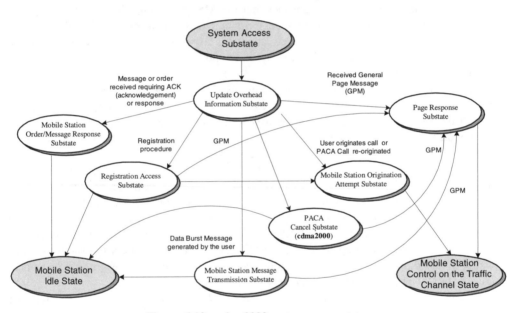

Figure 6.12   cdma2000 system access state.

The *system access state* is organised into sub-states, as shown in Figure 6.12. Each sub-state is described in detail in the following sections. The following topics summarise the main events in each sub-state.

- *Update overhead information sub-state* (Section 6.2.3.2): The MS monitors FPChs or FCCChs to update all system overhead messages.

- *Page response sub-state* (Section 6.2.3.3): The MS sends a page response message required by the server BTS.

- *Mobile station order/message response sub-state* (Section 6.2.3.4): The MS sends responses related to a specific message or order previously sent by the system.

- *Mobile station origination attempt sub-state* (Section 6.2.3.5): The MS transmits an origination message to the server BTS.

- *Registration access sub-state* (Section 6.2.3.6): The MS sends messages to start registration.

- *Mobile station message transmission sub-state* (Section 6.2.3.7): The MS transmits data burst messages to the system.

- *PACA cancel sub-state* (Section 6.2.3.8): The MS transmits PACA cancellation message to perform a PACA call cancellation (cdma2000 networks).

6.2.3.1 System Access Procedure

The *system access procedure* is very similar for both CDMA IS-95 and cdma2000 standards because of backward compatibility requirements.

The system access procedure establishes that the MS must transmit request or response messages using a randomly selected access channel (*RA*). This process is called *access attempt*. There are two types of access attempts: a *response attempt*, when MSs are transmitting a response message to the system, and a *request attempt*, when they are transmitting a request message.

The system sends some of the parameters involved in the selection process in the access parameters message. MSs send response messages to respond to a BTS message or order. Request messages, however, are autonomously sent to the system. The access channel used depends on the mobile terminal configuration; for CDMA IS-95 an RACh is used, whereas for cdma2000, either EAChs or RCChs can be used.

The transmission of request or response messages through the air interface does not assure successful access to the system. The process is only complete after the same message is sent a certain number of times or when the MS receives a reception acknowledgement (ACK).

Each access attempt may consist of one or more sub-attempts, each consisting of repeated transmissions of the same message. Each message transmission is called an *access probe*.

Each MS, when attempting to access the CDMA network, transmits the same message several times, that is, it transmits multiple *access probes*, or an access probe sequence. Figure 6.13 illustrates the *access attempts* of a mobile station.

The parameters *MAX_REQ_SEQ* and *MAX_RSP_SEQ* that vary from 1 to 15, determine the maximum number of *access attempts* (request or response). In each attempt, the MS can transmit up to 16 *access probes*, according to the expression $(1 + NUM\_STEP)$, as in Figure 6.13. That is, the MS can execute up to 15 *access attempts*, each with up to 16 *access probes*.

Here

PI   *Power Increment*, corresponds to the *PWR_STEP*
IP   *Initial open loop Power*, corresponds to the initial access probes signal power
PD   *Persistence Delay*, pseudo-randomly defined for the beginning of a new access attempt
RS   *sequence backoff*, interval between two consecutive attempts
RT   *probe backoff*, pseudo-randomly defined as the interval between two consecutive access probes, based on the *PROBE_BKOFF* parameter

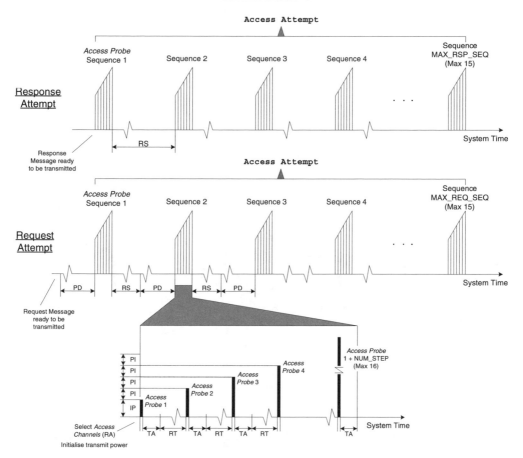

Figure 6.13 Transmission procedure of response and request attempts during access attempts.

TA   ACK *response timeout*, [*TA* = 80(2 + *ACC_TMO*) ms], maximal interval that MSs must wait for an acknowledgment confirmation for an access message transmitted previously

Each *access probe* consists of a preamble and an access message capsule, as in Figure 6.14.

Before the transmission of any access probe sequence, a reverse access channel must be selected, the transmission power of the first access probe must be determined and specific delays between access probes and between access attempts must be defined.

The MS must randomly select the access channel (*RA*), from 0 to *ACC_CHAN* channels available, depending on the terminal operation mode. In case of a CDMA IS-95 terminal, the RACh associated with the FPCh assigned to the terminal is used. In cdma2000 terminals, MSs are able to use EAChs if in *basic access mode*, or an EACh and an RCCCh if in *reservation access mode*. These operation modes are described with further details in Chapter 4 (Forward Link).

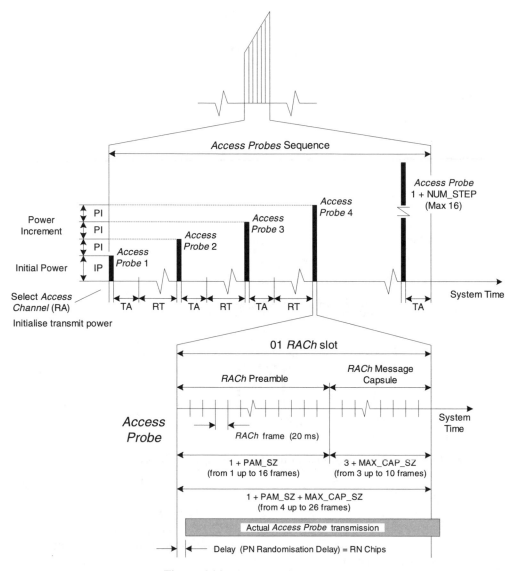

Figure 6.14   Access probes structure.

Both EACh and RCCCh are associated with the FCCCh assigned to the MS. In CDMA IS-95, up to 32 RACHs may be associated to a unique FPCh, whereas for cdma2000, up to 32 EAChs and 32 RCCChs are associated to a unique FCCCh.

Once the *RA* channel is selected, all access probes within the same sequence are transmitted on it. For any subsequent Access probe sequence, another *RA* channel may be selected.

Besides selecting the access channel, the MS must also configure the transmission power, as in Figures 6.13 and 6.14. The first access probe of each of the access attempts is transmitted with initial power (*IP*) defined as a function of the open loop power control (see Chapter 7). The subsequent access probes in the same sequence have the power increased according to a Power Increment (*PI*), given by *PWR_STEP* and other variables. In cdma2000, the transmission of an EACh access probe only occurs when a minimal E_c/I_0 transmission threshold level is achieved.

The MS must also define a set of specific delays: sequence backoff (*RS*), Persistence Delay (*PD*), ACK response timeout (*TA*), probe backoff (*RT*), and randomisation delay (*RN*).

Figure 6.13 shows sequence backoff (*RS*) and persistence delay (*PD*), which are estimated for the interval between consecutive access attempts. Both are randomly generated and the sum of *RS* and *PD* delays for request attempts can be quite similar to the *RS* delay specified during response attempts. ACK response timeout (*TA*) and probe backoff (*RT*) must be also defined between sequential access probes.

A randomisation delay (*RN*) is also configured before transmission of the access message capsule, as in Figure 6.14. This delay is also pseudo-randomly defined, using a process called PN randomisation, based on the parameter *PROBE_PN_RAN* provided to MSs in the access parameters message and on the MS ESN. *RN* delay range is within 0 to $2^{PROBE\_PN\_RAN} - 1$ chips, counted from the start of the reverse access slot interval.

Delays involved during the transmissions of access probes and access attempts are usually expressed as a function of 80 ms slot intervals, except for the *RN* parameter, which is expressed in terms of chip intervals.

The main reasons for generating pseudo-random delays and selecting the RA channel before each access probe sequence is to reduce the probability of message collisions at the reception side, i.e. to avoid the situation that messages sent by distinct MSs reach the BTS at the same time on the same carrier.

Figure 6.15 indicates the procedures executed by an MS to perform access attempts with IS-95 RAChs.

Figure 6.16 shows a diagram representing the procedures executed by an MS to perform access attempts on EAChs and RCCChs of cdma2000 systems. The overall process is similar to the process used to access IS-95 RAChs.

In both Figures 6.15 and 6.16, the number of probes and attempts is initially set to zero and, before the first access probe sequence, a sequence backoff delay (*RS*) interval (0 to *BKOFF*) is pseudo-randomly defined.

For request attempts, a persistence delay (*PD*) is generated, serving as a comparison between variables *RP* and *P*. The value of *P* depends on the following input parameters:

- type of message being transmitted, i.e. request for registration, for message transmission or any other request message;

- an MS specific persistence parameter (*PSIST*) and other parameters transmitted in the access parameters message, *REG_PSIST* and *MSG_PSIST*, stored in the MS memory.

RP is randomly generated in the MS, with value ranging from 0 to 1. To transmit an access probe, *P* must be smaller than *RP*. When *P* equals zero, MSs must end the access attempt, declare failure in the process, and update some of the internally stored values, such as *NID*, *SID*, *REG_ZONE* and *ZONE_TIMER*. MSs then return to the *system determination sub-state* (Section 6.2.1.1).

If *P* is greater than zero (from 0 to 1) and *RP* is less than *P*, then MSs can transmit the first access probe. Otherwise (*RP* is greater or equal to *P*), MSs must recalculate *RP* until the *RP* is less than *P*. This process only happens for the first access probe of a request attempt, as in Figures 6.13, 6.15 and 6.16.

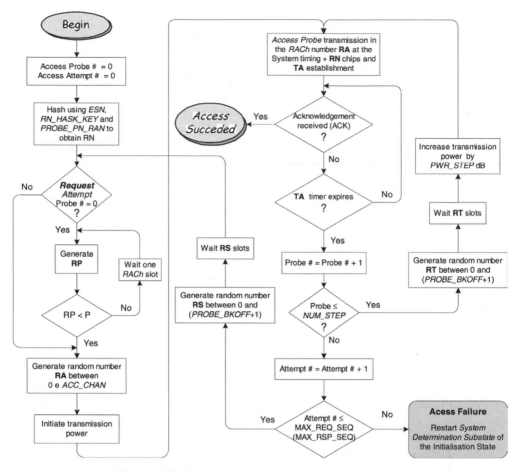

Figure 6.15 Reverse access channel procedure flowchart.

The next step is the choice of the reverse access channel (RACh, EACh or RCCCh), depending on the system operation mode. Channels vary from 0 to *ACC_CHAN* and are selected with the Hash algorithm. The parameter *ACC_CHAN* is passed to the MS in the access parameters message to define the number of access channels assigned to the FPCh or FCCCh.

For cdma2000 systems, before channel selection, the MS must evaluate if the E_c/I_0 transmission threshold level has been achieved, as in Figure 6.16.

After these first steps, the first access probe is then transmitted with Initial Power (*IP*), defined by the open loop power control. The MS must then define the ACK response timeout, $TA = (2 + ACC\_TMO) \times 80$ ms, which is the maximum interval valid for receiving an ACK response from the BTS. If ACK is not received before *TA*, the MS must establish and wait for probe backoff (*RT*) slots before transmitting the next access probe. Each subsequent access probe is then transmitted with increased power, defined by *PWR_STEP*.

PWR_STEP, *PROBE_BKOFF* and *EACH_PROBE_BKOFF* are transmitted to MSs in the access parameters message when required.

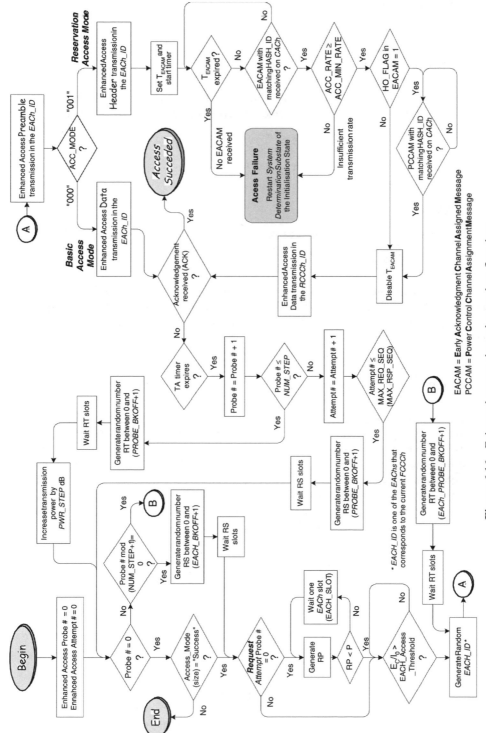

Figure 6.16 Enhanced access channel procedure flowchart.

If system access fails, the mobile terminal returns to the *system determination sub-state* (Section 6.2.1.1).

6.2.3.2   Update Overhead Information Sub-state

In this sub-state, the MS monitors FPChs or FCCChs waiting for messages containing network parameters. These types of messages are also known as *overhead messages*, which include

- access parameters message,

- system parameters message,

- CDMA channel list message,

- extended system parameters message,

- extended neighbour list message.

The system configuration update procedure succeeds when all *CONFIG_MSG_SEQ* parameters of all overhead messages received and stored at the MS are identical. Upon completion of the update process, the MS returns to the *mobile station idle state* (Section 6.2.2).

6.2.3.3   Page Response Sub-state

In the page response sub-state, the MS monitors FPChs or FCCChs to receive messages transmitted to it by the BTS. When the MS finds a General Page Message (GPM), it transmits a page response message. As soon as the BTS receives this page response message, it sends an authentication challenge order, requiring the MS to perform the authentication process (Section 6.2.3.6).

The MS must send each message in assured mode, i.e. requiring reception acknowledgement (ACK) from the BTS. When it receives this acknowledgement, the MS finishes any access attempt in progress, disables its transmitter and returns to the *mobile station idle state*.

The MS can also execute access handoff to use FPChs or FCCChs with better signal power levels. The MS verifies the access channels associated with the new pilot, FPCh or FCCCh, to transmit the new access probes in the following situations.

- The MS is waiting or just prior to the transmission of a response message to the BTS.

- After transmitting an access attempt and while waiting in *page response sub-state* (Section 6.2.3.3) or *mobile station origination attempt sub-state* (Section 6.2.3.4).

6.2.3.4   Mobile Station Order/Message Response Sub-state

In this sub-state, the mobile terminal must monitor FPChs or FCCChs searching for any oriented message or order, and send an appropriate response to this message, except in the case of overhead messages (described in Section 6.2.3.2) and response page message (described in Section 6.2.3.3).

The BTS may transmit an authentication challenge message, requiring the MS to perform an authentication process (as described in Section 6.2.3.6), still in this sub-state.

The MS must also send each message in assured mode, i.e. always requiring a confirmation of reception (ACK—Acknowledgement) by the BTS, as depicted in Section 6.2.3.3. Once the acknowledgement is received from the BTS, the MS finishes any access attempt in progress, disables its transmitter and returns to *mobile station idle state* (Section 6.2.2).

6.2.3.5   Mobile Station Origination Attempt Sub-state

Besides monitoring FPChs or FCCChs for messages, the MS can also transmit origination messages. When an origination indication initiates this sub-state, the MS transmits an origination message containing all digits pressed by the user without exceeding the message capsule size. When a PACA response indication initiates the sub-state, the MS must retransmit the origination message of the previous attempt, containing the digits pressed on the last message.

If the BTS responds with an authentication challenge message, the MS must execute authentication while in this sub-state.

The MS must also send messages in assured mode, i.e. requiring reception ACK from the BTS, as in Section 6.2.3.3. When the acknowledgement is received from the BTS, the MS finishes any access attempt in progress, disables its transmitter and returns to *mobile station idle state* (Section 6.2.2).

If the user powers-off the terminal in this state, the MS must abort any Access Attempt in progress and is redirected to the *system determination sub-state* (Section 6.2.1.1) with liberation indication.

6.2.3.6   Registration Access Sub-state

The MS sends a registration message to the BTS in this sub-state, using RAChs or EAChs, and starts monitoring FPChs or FCCChs. If the BTS responds with an authentication challenge message, the MS must execute authentication while in this sub-state.

The MS must also send messages in assured mode, i.e. requiring reception ACK from the BTS, as in Section 6.2.3.3. When the acknowledgement is received from the BTS, the MS finishes any access attempt in progress, disables their transmitter and returns to *mobile station idle state* (Section 6.2.2).

As in the *mobile station origination attempt sub-state*, if the user powers-off the terminal in this state, the MS must abort any access attempt in progress and is re-directed to the *system determination sub-state* (Section 6.2.1.1) with release indication.

6.2.3.7   Mobile Station Message Transmission Sub-state

In this sub-state, the MS transmits data burst messages to the BTS, while monitoring FPChs or FCCChs. The support to this sub-state is optional.

As in the previous sub-states, if the BTS responds with an authentication challenge message, the MS must execute authentication while in this sub-state.

The MS must send messages in assured mode, i.e. requiring reception ACK from the BTS, as in Section 6.2.3.3. When the acknowledgement is received from the BTS, the MS finishes any access attempt in progress, disables its transmitter and returns to *mobile station idle state* (Section 6.2.2).

If the MS is powered-off by the user, it must abort any access attempt in progress and is re-directed to *System Determination Sub-state* (Section 6.2.1.1) with release indication.

6.2.3.8   PACA Cancel Sub-state (Priority Access and Channel Assignment)

In the PACA cancel sub-state, the MS transmits a PACA cancel message to the BTS, and starts monitoring the FPCh or FCCCh. The BTS may respond to the message with an authentication request, in an authentication challenge message. In this case, the MS must execute authentication while in this sub-state.

As in the previous sub-states, the MS must send messages in assured mode, i.e. requesting reception ACK from the BTS, as in Section 6.2.3.3. When the acknowledgement is received from the BTS, the MS finishes any access attempt in progress, disables its transmitter and returns to *mobile station idle state* (Section 6.2.2).

If the MS is powered-off by the user, it must abort any access attempt in progress and is redirected to *system determination sub-state* (Section 6.2.1.1) with release indication.

6.2.4   Mobile Station Control in the Traffic Channel State

In this last call processing state, the MS establishes a communication path between itself and its serving BTS using traffic channels on both forward and reverse links.

This state is divided into sub-states, as illustrated in Figure 6.17. The sub-states can be summarised as follows.

- *Traffic channel initialisation sub-state* (Section 6.2.4.1): The MS evaluates and monitors the possibility of receiving messages (data) on a FTCh and starts transmitting on the corresponding reverse link channel.

- *Waiting for order sub-state* (Section 6.2.4.2): The mobile terminal waits for an alert with information message.

- *Waiting for mobile station answer sub-state* (Section 6.2.4.3): The MS waits for the user to answer the call or to perform another appropriate action.

- *Conversation sub-state* (Section 6.2.4.4): The MS transmits and receives traffic data frames to/from its serving BTS, according to current service configuration options. It may also perform gating operations of transmitted Reverse Pilot Channel (RPiCh).

- *Release sub-state* (Section 6.2.4.5): The MS releases all resources used during a call, disconnecting the current call and entering the *system determination sub-state* (Section 6.2.1.1).

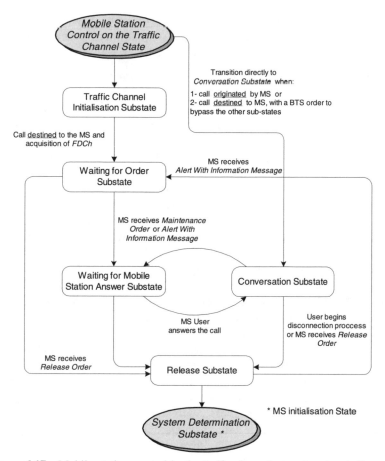

Figure 6.17   Mobile station control in the traffic channel state functional diagram.

6.2.4.1   Traffic Channel Initialisation Sub-state

The MS evaluates and monitors the possibility of receiving messages or orders on an FTCh and starts transmitting on the corresponding reverse link channel.

To perform this operation, some configurations must be set before transmission and reception of frames on forward and reverse links. Configurations are negotiated between BTS and MS, allowing them to assemble and read each other's traffic channel frames. A message exchange process allows them to establish and define configuration parameters, achieving a common service configuration.

The MS may request a default service configuration associated to a service option at call origination, or negotiate a new service configuration during traffic channel operation.

If negotiating a new service configuration, upon entering in the traffic channel initialisation sub-state, the MS performs the following tasks.

- Registration, described in Section 6.4.

- Power control initialisation on traffic channels, reporting Frame Error Rates (FER) statistics to the BTS. This is done periodically (within intervals defined by the BTS) or when pre-established FER thresholds are achieved.

- Initial service configurations definition, setting the correct traffic channel frame offsets and forward and reverse links long code public masks.

While in this sub-state, the MS constantly monitors the FTCh associated with the FPiCh selected in the active set. It also measures pilot channel signal levels, but does not transmit Pilot Strength Measurement Messages (PSMM). Other tasks in this sub-state include maintaining the autonomous registration timer (Section 6.4) and updating and reinitialising *TMSI* and *TMSI_CODE* timers (for registration procedures) if required.

When the MS requests a valid service configuration, i.e. acceptable for the BTS, both BTS and MS must operate according to it. If the service configuration requested is not allowed, the BTS rejects it and suggests another configuration. The MS is then in charge of accepting, rejecting or suggesting another configuration.

This process, known as *service option negotiation*, continues until both MS and BTS mutually accept or reject a service configuration. The configuration parameters negotiated between BTS and MS are known as the service configuration. These parameters include forward and reverse multiplex options, traffic channel configuration, traffic channel transmission rate and service option connections.

Forward and reverse multiplex options define the number of bits and distribution within frames transmitted in forward and reverse traffic channels, as described in chapters on forward link and reverse link. The main traffic data types are primary (voice only), secondary (data, fax) and signalling (control). Any traffic channel may have several multiplex options working together. These configurations determine transmission rates and frame structure in conjunction with Radio Configurations (RC).

Forward and reverse traffic channel configuration, required only for cdma2000, corresponds to RC and other attributes relevant for the correct setting of traffic channels, also dependent on Spreading Rates (SR). Distinct configurations between links are possible.

Forward and reverse traffic channel transmission rates depend on MSs' radio configurations and spreading rates. These parameters define transmission rates for forward and reverse links. The links can support distinct rates, which may or may not include all supported rates associated to the frame multiplex options defined.

Service option connections include service options used for FTChs and RTChs, allowing none, one, or more connections. When no service option connection is established, the MS sends null and signaling traffic data to the BTS through RTChs. The following parameters and procedures are associated with Service Option Connections:

- Service option negotiation, which defines how transmitted information bits must be processed either by BTS or MS;

- FTChs and RTChs traffic type definition, which defines traffic channel types in use to support the selected service option (unidirectional, bi-directional, only primary, primary and secondary, or primary and signalling). For cdma2000 networks, the connected service option can request Supplemental Channels (SChs) and Supplemental Code Channels (SCChs) for one link direction or for both directions, forward and reverse.

An appropriate multiplex option that supports supplemental channel operation is negotiated to attend this request.

- Associated service option connection reference, which establishes, in a unique way, the service option connection identification.

Besides these negotiable service configuration parameters, the BTS also establishes non-negotiable parameters that define the reverse pilot gating rate, forward and reverse power control parameters and logical to physical mapping, which consists of service reference identifier, logical resources, flags and priorities.

In case there is a failure during any of the tasks performed in this sub-state, the MS must disable its transmitter and enter in the *system determination sub-state* (Section 6.2.1.1), in the MS *initialisation state*. If the MS is powered-off, it is redirected to the *release sub-state* (Section 6.2.4.5). If all the tasks in this sub-state are successful, the MS enters the *waiting for order sub-state* (Section 6.2.4.2).

6.2.4.2   Waiting for Order Sub-state

In this sub-state, the MS waits for an alert with information message from the BTS before the sub-state timer ($T_{52m} = 5$ s) expires. If a message is received, the MS informs the subscriber of the message content and is re-directed to the *waiting for mobile station answer sub-state* (Section 6.2.4.3).

Other tasks that the MS performs in this sub-state include the following.

- Transmit Pilot Strength Measurement Mini-Messages (PSMMM) informing of modifications regarding signal power level. The MS sends the messages periodically or based on a threshold.

- Monitor Forward Fundamental Traffic Channels (FFCh).

- Analyse and search for new pilot channels and/or transmit messages informing the BTS of these new pilots, requesting, if needed, handoff procedures.

- Adjust transmission power and initialise the closed loop power control (Chapter 7).

- Execute a timer-based registration (Section 6.4).

The MS can also receive several other messages, and must act according to an appropriate procedure.

If any of these tasks fails or if the timer expires, the MS must disable its transmitter and enter the *system determination sub-state* (Section 6.2.1.1), in the *mobile station initialisation state*. If the MS is powered-off, it is re-directed to the *Release sub-state* (Section 6.2.4.5).

The process of sending mini-messages is a feature implemented in cdma2000 systems to reduce time of exchanging and processing messages prior to the start of some fundamental procedures, such as handoffs, registration, authentication and allocation of SChs, as an example.

6.2.4.3 Waiting for Mobile Station Answer Sub-state

In this sub-state, the mobile station waits until the subscriber performs one of the following:

- answer an MS terminated call (Section 6.3.2),

- request special action, such as, send this call to the voice mail or put it on hold.

CDMA systems have a timer designed specifically for when the subscriber does not answer an incoming call. The timer limit is $T_{53m} = 65$ s. If the timer expires before the subscriber answers the call or requests a specific action (e.g. voice mail, hold), the MS turns off its transmitter circuit and enters the *system determination sub-state* (Section 6.2.1.1), part of the *mobile station initialisation state*.

Other tasks that the MS performs in this sub-state include FTCh monitoring, transmission power adjustment according to power control procedures and handoff. The mobile may also receive other specific messages or orders, and shall respond appropriately to them.

If any of these tasks fails, the MS must disable its transmitter and enter the *system determination sub-state* (Section 6.2.1.1), in the MS *initialisation state*. If the MS is powered-off, it is redirected to the *release sub-state* (Section 6.2.4.5). If all the tasks in this sub-state are successful, the MS enters the *conversation sub-state* (Section 6.2.4.4).

6.2.4.4 Conversation Sub-state

During this state, the MS exchanges frames through FTChs and RTChs with one or more BTSs (soft or softer handoff) according to the service configuration.

The MS also performs RPiCh gating, depending on the operation mode in terms of power control processes, *active mode* or *control hold mode*. While in *active mode*, the MS is effectively transmitting, and the RPiCh is not gated. During *control hold operation mode*, even though the MS is not transmitting data, it may be waiting for a handoff process to be concluded or expecting a resource allocation for supplemental channel use, as depicted in Figure 6.18.

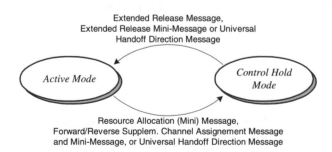

Figure 6.18 Conversation sub-state operation modes.

If the condition happened because of a mobile station originated call (Section 6.3.1) and not all the pressed digits were sent on the origination message, the MS must transmit an origination continuation message containing the remaining digits. This message must be transmitted in assured mode, i.e. it requires an acknowledgement message.

While in this sub-state, the MS also performs specific tasks, such as FTCh monitoring, transmission power adjustment according to power control procedures and handoff processes. The mobile may also receive other specific messages or orders, and shall respond appropriately to them.

In this sub-state the MS can also transmit a Data Burst Message to the BTS, through which the subscriber requests configuration changes such as

- new service configuration;

- transition of the long code mask from public to private;

- flash with information message in assured mode, requesting incoming calls to be held or redirected to voice mail;

- transmission of burst DTMF digits.

In case there is a failure during this sub-state, the MS must disable its transmitter and enter the *system determination sub-state* (Section 6.2.1.1), part of the *mobile station initialisation state*. If the MS is powered-off or the call disconnected, it is redirected to the *release sub-state* (Section 6.2.4.5).

6.2.4.5 Release Sub-state

In this sub-state, the MS releases all resources used during a call and confirms the disconnection.

If calls are terminated and the terminal is not powered-off, the MS is automatically directed to the system *determination sub-state* (Section 6.2.1.1).

As in previous sub-states, the MS also performs tasks such as FTCh monitoring, transmission power adjustment according to power control procedures and handoff processes. The mobile may also receive other specific messages or orders, and shall respond appropriately to them.

6.3 MESSAGES EXCHANGED DURING CALL ESTABLISHMENT

The following sections explain messages exchanged during a call in different scenarios.

6.3.1 MS Originated Call Scenario

Table 6.2 presents messages exchanged between MS and BTS when the MS tries to originate a call under service option 1 (only primary traffic, i.e. voice).

6.3.2 MS Terminated Call Scenario

Table 6.3 presents messages exchanged between MS and BTS when the MS terminates the call, i.e. the subscriber receives a call. This scenario also pertains to service option 1 (only primary traffic, i.e. voice).

Table 6.2   MS originated call procedure

Mobile Station (MS)	Channel	Base Transceiver Station (BTS)
Detection of user originated call and storage of digits pressed before 'Send'		
Transmission of the origination message	Access $\rightarrow$	Information processing and FTCh allocation
Establishment of the RTCh assigned by the system	Paging $\leftarrow$	Channel assignment message transmission
Reception of at least $N_{5m} = 2$ consecutive valid frames	Traffic $\leftarrow$	Null message transmission (null traffic channel data)
Preamble transmission	Traffic $\rightarrow$	RTCh acquisition
Processing of information received	Traffic $\leftarrow$	Base station acknowledge order transmission
Null message transmission (null traffic channel data)	Traffic $\rightarrow$	Processing of information received
Service option request message transmission	Traffic $\rightarrow$	Resource allocation for requested service option
Primary traffic processing according to service option 1	Traffic $\leftarrow$	Service option response order transmission for service option 1
Service connect message transmission	Traffic $\rightarrow$	Processing of information received
Optional: origination continuation message transmission	Traffic $\rightarrow$	Processing of information received
Optional: ring back tone in audio path	Traffic $\leftarrow$	Optional: alert with information message transmission (ring back tone)
Optional: removal of ring back tone in audio path	Traffic $\leftarrow$	Call is answered Optional: alert with information message transmission (turns-off ring back tone)
Users in conversation	Traffic $\leftarrow/\rightarrow$	Users in conversation

Table 6.3   MS terminated call procedure (MS receives a call)

Mobile Station (MS)	Channel	Base Transceiver Station (BTS)
Processing of information received	Paging $\leftarrow$	General Page Message (GPM) transmission
Page response message transmission	Access $\rightarrow$	Processing of information received and FTCh assignment
Establishment of the RTCh assigned by the system	Paging $\leftarrow$	Channel assignment message transmission
Reception of at least $N_{5m} = 2$ consecutive valid frames	Traffic $\leftarrow$	Null message transmission (null traffic channel data)
Preamble transmission	Traffic $\rightarrow$	Acquisition of RTCh
Processing of information received	Traffic $\leftarrow$	Base station acknowledge order transmission
Null message transmission	Traffic $\rightarrow$	Processing of information received

Table 6.3 (*Continued*)

Mobile Station (MS)	Channel	Base Transceiver Station (BTS)
Service option negotiation	Traffic ←	Message requesting service option 1 (service option negotiation)
Service option response message transmission	Traffic →	Processing of information received
Primary traffic processing according to service option 1	Traffic ←	Service option response message transmission
Service connect completion message transmission	Traffic →	Processing of information received
Ring is initiated to alert user	Traffic ←	Optional: alert with information message transmission
User answers call and ring is turned off		
Connect order transmission	Traffic →	Processing of information received
Users in conversation	Traffic ←/→	Users in conversation

Table 6.4 Message exchange during soft handoff

Mobile Station (MS)	Channel	Base Transceiver Station (BTS)
User in conversation with BTS A	Traffic ←/→	User on conversation with BTS A
BTS B FPiCh strength exceeds T_ADD		
Pilot Strength Measurement Message (PSMM) transmission	Traffic →	BTS A receives message BTS B starts transmission on FTCh and acquires RTCh
Extended handoff direction message processing	Traffic ←	BTSs A and B send Extended handoff direction message for soft handoff processing and usage of A and B
BTS B pilot acquisition and active set update $\{A, B\}$		
Handoff completion message transmission	Traffic →	BTSs A and B process information received
T_TDROP timer expires regarding RBS A FPiCh		
Pilot strength measurement message processing	Traffic →	BTSs A and B receive Pilot Strength Measurement Message (PSMM)
Extended handoff direction message processing	Traffic ←	BTSs A and B send extended handoff direction message to use BTS B FTCh
Active set $\{B\}$ update		
Handoff completion message transmission	Traffic →	BTSs A and B receive handoff completion message BTS A stops transmitting on FTCh and receiving of RTCh
Users in conversation with BTS B	Traffic ←/→	Users in conversation with BTS B

6.3.3   Call Processing Scenario During Soft Handoff

Table 6.4 presents messages exchanged during a call while in soft handoff. The Power Control, Handoff Processes and Radio Resource Management chapter explains some of the parameters presented in this sub-section, such as *T_ADD*, *T_DROP* and *T_TDROP*.

6.3.4   Priority Access and Channel Assignment Procedure

Priority call establishment occurs when the system does not have resources available to establish a call, i.e. there are no traffic channels available (Table 6.5). The MS informs the

Table 6.5   Message exchange in a PACA call procedure

Mobile Station (MS)	Channel	Base Transceiver Station (BTS)
User initiates priority call with an origination message transmission	Access →	System determines there are no TChs available and that this is a priority call (PACA call)
Processing of information received and non-slotted mode operation while waiting for channel assignment	Paging ←	PACA message transmission, informing user that the call has been queued as a PACA call
Processing of information received	Paging ←	PACA message periodic transmission to update user PACA. Call queue position
	Paging ←	PACA Message transmission to inform the user to reoriginate (PACA Call)
Origination message retransmission	Access →	Processing of information received and allocation of FTCh
Establishment of RTCh allocated by the system. Reception of at least $N_{5m} = 2$ consecutive valid frames	Paging ←	Channel assignment message transmission
Preamble transmission	Traffic →	Acquisition of RTCh
Processing of information received	Traffic ←	Base station acknowledgement order message transmission
Null message transmission	Traffic →	Processing of information received
Resource allocation according to service option. Processing of primary traffic according to service option	Traffic ←	Service connect message transmission for service option 1
Service completion message transmission, accepting the service option required by the system	Traffic →	Processing of information received
Distinct alert to users according to PACA call. Users answer call and ringing tone is removed.	Traffic ←	Optional: alert with information message transmission (distinct alert when PACA call)
Connect order message transmission	Traffic →	Processing of information received (remove PACA call)
Users in conversation	Traffic ←/→	Users in conversation

BTS whether the call is a priority call. If so, the BTS continues executing resource management until a traffic channel is available resuming the call establishment process.

6.4 REGISTRATION

Every mobile terminal communicates its location, slot cycle and identification to the network through the registration processes. Knowing the approximate MS location allows the system to easily page the mobile when a MS terminated call is requested. The system must be sure about MS's identification to avoid cloning and invalid authentication processes. The slot cycle timing (*SLOT_CYCLE_INDEX*) parameter allows the system to determine slots that will be monitored by the MS, indicating when to transmit specific and oriented messages.

Other parameters can also be sent through registration processes. CDMA IS-95 and cdma2000 standards foresee a total of nine and ten registration processes respectively.

- Power-up registration: occurs when the MS is turned on and enters the *mobile station idle state* (Section 6.2.2).

- Power-down registration: occurs when the MS is turned off. Before turning all circuits off, the MS communicates the power-off condition to the network.

- Timer-based registration: the MS must register according to a system pre-programmed timer.

- Distance-based registration: occurs when the MS reaches a pre-specified distance (or threshold) from a BTS, i.e. the distance between the current server and the BTS where the mobile last registered reaches the threshold.

- Zone-based registration: occurs based on internal system zone configurations (Section 6.4.1), when the MS enters a new zone, or while in roaming (Section 6.5).

- Parameter-change registration: occurs when internally stored parameters change or when the MS enters a distinct system.

- Ordered registration: occurs every time the system requests a registration process.

- Implicit registration: when the MS sends an origination message or responds to a general page message. The system may determine the MS approximate location.

- Traffic channel registration: when the MS registers while requesting a traffic channel allocation.

- User zone registration: this registration type occurs only in cdma2000 networks, when the MS enters or selects an active user zone (Section 6.4.2).

6.4.1 System Zones

Any service area, or system zone, may be divided into smaller zones, identified by specific parameters transmitted as part of the access parameters message, such as

- *SID* (System IDentification number);

- *NID* (Network IDentification number);

- *Reg_Zone*: identifies any specific system zone within the system;

- *Total_Zones*: defines the total of zones the MS must consider to perform an autonomous registration;

- *Zone_Timer*: timer assigned to each specific zone to instruct the MS to start registration within the current system zone.

A registration zone is composed of a group of BTSs that belong to a certain cellular system. When the MS moves from one zone to another in MS *idle state* (Section 6.2.2), it may initiate a registration process, indicating its new location and updating internally stored system access parameters.

Each cellular system provider defines zone areas. An adequate management and distribution of BTSs among the zones is necessary to achieve better system performance. Smaller zones lead to a great number of system zones, allowing easier and more efficient paging by transmitting the GPM. However, it may result in a greater number of registration processes because of zone changing. Larger system zones avoid this situation, but require GPM's transmission to a greater number of BTSs.

6.4.2   User Zones

The user zones feature allows any BTS or group of BTSs to offer specific and specialised services, or tiered services. These services are strictly related to the location of any MS to its serving BTS, and involve high transmission rates for cdma2000 systems.

A set of services with special characteristics is assigned to each user zone. These services are valid within a limited geographical area in which users registered to this zone have access to them. For example, Virtual Private Networks (VPN) or wireless LANs inside a company office.

User zones can be supported by public services (for common wireless subscribers) using the same frequencies of the CDMA system and also by private cellular systems operating at different frequencies.

There are two basic types of user zones.

- Broadcast user zones: These zones are identified to the MS in broadcast messages over FPChs within the coverage area of some BTSs.

- Mobile specific user zones: These zones are not broadcast by the BTS, instead they are identified by the MS through the comparison of some parameters transmitted on overhead messages to internally stored values, such as *SID*, *NID*, *BASE_LAT*, *BASE_LONG* and *BASE_ID*.

Broadcast user zones allow permanent and temporary user subscription. The temporary registration within these user zones allows the MS to use all services provided by a BTS while in its coverage area. Mobile specific user zones only allow permanently registered users to use the features and services provided by them.

Every MS that supports user zones may store a list of the systems' user zones, where a unique user zone ID (*UZID*) identifies each zone. The set of parameters required for zone identification is also included in this list.

While in *mobile station idle state*, described in Section 6.2.2, the MS must register every time it enters an active user zone, to update the *UZID* parameter. In situations where this process is not necessary, this user zone is defined as a passive user zone.

6.5   ROAMING

MSs are considered 'home' when they are located in their home systems, that is, where they are registered and allowed to operate. Therefore, MSs are 'roamers' when they are out of their home systems (Figure 6.19).

A cellular system is composed of BTSs, BSCs, MSCs and other entities. Thus it is sub-divided to allow a better control in terms of positioning, traffic, billing, configuration and administration. Subdivisions may also define target areas for specific services. Sub-sets of BTSs, BSCs and MSCs are grouped to form one or more networks within a system, as in Figure 6.20.

MSs have a list of one or more locations where they are in the 'home system', or on a 'non-roamer' status. An MS is roaming when one of the identification parameters (such as

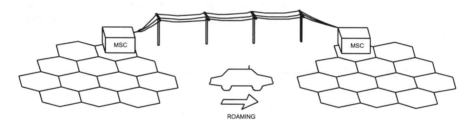

Figure 6.19   Roaming MS (visitor).

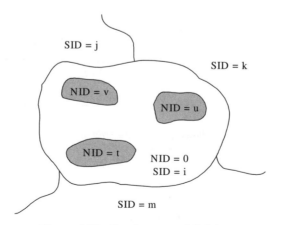

Figure 6.20   Service area subdivisions.

SID or *NID*) received in the system parameters message from a BTS differs from the home system parameters, SID_{HOME} and NID_{HOME}.

In some cases, the system may transmit the *NID* parameter as 65535 ($FFFF_{hex}$) instructing MSs to consider all *NID*s within a specific *SID* to be non-roamers. This allows MSs to interact with all the system BTSs without roaming.

NID and *SID* are also employed to allow call control actions, for example when MSs are set not to receive or originate calls while outside the home network or home system, or both cases.

6.6　THE AUTHENTICATION PROCESS

Authentication validates and provides a minimal level of security within a wireless system. Together with registration processes, it uniquely identifies each MS. Authentication and registration are used for privacy (cryptography) and to avoid cloned terminals.

The parameters used in the authentication process include the Mobile Identification Number (*MIN*) and the Electronic Serial Number (*ESN*). These parameters are transmitted in the air interface, through logic control channels, weakening the security of transmitted data because unauthorised agents are able to obtain these and other MS configuration data to clone terminals.

Authentication processes were created to avoid, or at least make it harder, to obtain these parameters, especially during transmission from BTS to MS. These processes use data known by mobile terminals and the system; therefore it does not have to be transmitted through the air interface. This data generates random variables through a process that is also not disclosed.

6.6.1　Air Interface Parameters

6.6.1.1　*Mobile Station Identification Number*

The International Mobile Station Identity (*IMSI*) number uniquely identifies all mobile stations within CDMA systems. Figure 6.21 shows the IMSI structure.

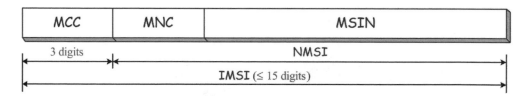

Figure 6.21　IMSI structure.

In the figure,

- Mobile Country Code (*MCC*) is country code where the MS is registered, composed of three numbers;

- Mobile Network Code (*MNC*) is network code where the MS is registered, or area code, usually composed of three numbers;

- Mobile Station Identification Number (*NSIN*) is MS identification number within its registration area, i.e. the phone number, usually composed of nine numbers;

- National Mobile Station Identity (*NMSI*) is a combination of *MNC* and *MSIN*, representing the telephone number and area code (prefix) within a country, composed of up to 12 numbers;

- International Mobile Station Identity (*IMSI*) is complete identification of a mobile station within CDMA systems, composed of up to 15 numbers.

Each MS has two distinct identifiers, called *IMSI_T* and *IMSI_M*, each one consisting of up to 15 digits. The *IMSI_M*, or MIN-based IMSI, is an IMSI that uses the Mobile Identification Number as the last 10 digits of the National Mobile Station Identity (*NMSI*). The *IMSI_T*, or True IMSI, uses an identification number given by the system in addition to the MIN. The type of *IMSI* used, M or T, depends on BTS capabilities.

In specific cases, the CDMA system may define a *Temporary Mobile Station Identity* (*TMSI*) number for a terminal. This procedure is performed for security purposes, because the TMSI is only used during an Authentication or Registration Process.

In case neither *IMSI* can be programmed, the MS sets the four Least Significant Bits (LSB) of the *IMSI* from the Electronic Serial Number (*ESN*), converting from binary to decimal through a modulo-10000 operation. The remaining digits are set to zero ('0').

6.6.1.2 Mobile Identification Number

The Mobile Identification Number (*MIN*) is a 34-bit long binary number, representing the ten least significant digits of the *IMSI*, also known as *IMSI_S*. If the *IMSI* has less than ten digits, the *MIN* LSBs are equal to the *IMSI*, whereas the Most Significant Bits (MSBs) are set as '0s' (zeros) to complete the ten digits.

This number represents the list number assigned to MSs. It is divided into two parts, MIN1 and MIN2, for processing purposes, as in Figure 6.22.

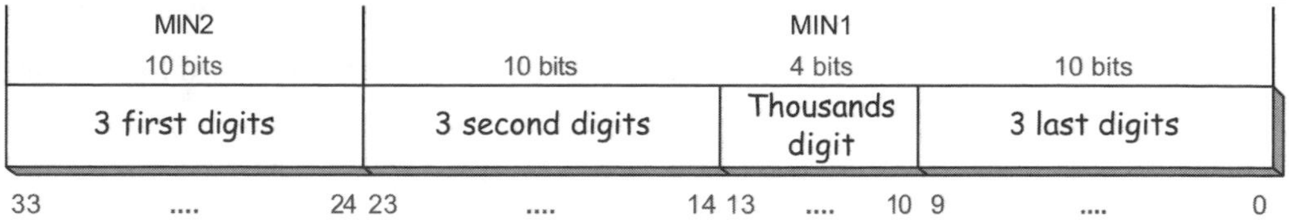

MIN2		MIN1	
10 bits	10 bits	4 bits	10 bits
3 first digits	3 second digits	Thousands digit	3 last digits
33 24	23 14	13 10	9 0

Figure 6.22 *MIN* structure when obtained from the *IMSI*.

The first three digits of the list number are mapped into ten bits, *MIN2*, using the following coding process:

a   Characterisation of the three digits as '$D_1D_2D_3$', '0s' (zeros) are set to 10 (ten);

b   Calculation of the following expression: $100 D_1 + 10 D_2 + D_3 - 111$;

Table 6.6 BCD conversion

Conversion decimal-to-binary (BCD)	
Decimal number	Binary sequence
1	0000000001
2	0000000010
3	0000000011
4	0000000100
998	1111100110
999	1111100111
Conversion of the thousands digit from decimal-to-binary (BCD)	
Thousands digit	Binary sequence
1	0001
2	0010
3	0011
4	0100
5	0101
6	0110
7	0111
8	1000
9	1001
0	1010

c   Binary conversion of the result using the Binary Coded Decimal (BCD) standard, as in Table 6.6.

The next three digits are mapped into the ten Most Significant Bits (MSB) of *MIN1*, following the same steps used for mapping *MIN2*.

The last four digits are mapped into the 14 least significant bits of MIN1 as follows:

a   The thousands digit is mapped into a four bits through BCD conversion, as in Table 6.6;

b   The last three digits are mapped into ten bits following the same steps used for mapping MIN2.

6.6.1.3   Electronic Serial Number

The *ESN* is a 32-bit long number that uniquely identifies each MS. This number is defined during manufacturing and can not be modified by subscribers. *ESN* modification requires special equipment usually not available to subscribers. Attempts to tamper with this number may lead to loss of terminal functionality, i.e. the MS becomes inoperative.

The 32 bits are organised as in Figure 6.23, in which E_{31} corresponds to the MSB.

Each MS terminal manufacturer has a fabrication code (*MFR*), used as the eight MSB of the *ESN* (bits 31 to 24). Bits 23 to 18 are reserved, initially set as zero. The 18 LSB (17 to 0)

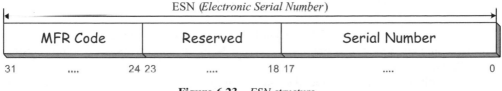

Figure 6.23 *ESN* structure.

must be unique to each manufactured terminal. When all combinations of these 18 bits are reached (approximately 262 000 terminals), manufacturers must inform the FCC and request another binary combination for the reserved bits.

The *ESN* is used in two ways: as the regular sequence, as in Figure 6.23, or as a permuted sequence, as follows:

$$E_0, E_{31}, E_{22}, E_{13}, E_4, E_{26}, E_{17}, E_8, E_{30}, E_{21}, E_{12}, E_3, E_{25}, E_{16}, E_7, E_{29}, E_{20}, E_{11}, E_2,$$
$$E_{24}, E_{15}, E_6, E_{28}, E_{19}, E_{10}, E_1, E_{23}, E_{14}, E_5, E_{27}, E_{18}, E_9$$

The permuted ESN is employed in scrambling for both forward and reverse links and also during channelling for the reverse link. The use of permuted *ESN*s avoids that terminals with similar *ESN* numbers present long code phase offsets close to each other, which may cause collision and problems regarding synchronisation, demodulation and decoding of received signals.

6.6.1.4 Station Class Mark

The Station Class Mark (SCM) is set in an 8-bit long sequence stored in the MS memory. The SCM provides some of the configuration parameters, such as power control, when operating in the idle state mode. Table 6.7 shows the digital representation of SCMs.

Table 6.7 Station class marks for current CDMA systems

Function	Bit(s)	Configuration	
Extended SCM indicator	7	Band classes 1, 4	1XXXXXXX
		Other bands	0XXXXXXX
Dual mode	6	CDMA only	X0XXXXXX
		Dual mode	X1XXXXXX
Slotted class	5	Non-slotted	XX0XXXXX
		Slotted	XX1XXXXX
IS-54 power class	4	Always 0	XXX0XXXX
25 MHz bandwidth	3	Always 1	XXXX1XXX
transmission	2	Continuous	XXXXX0XX
		Discontinuous	XXXXX1XX
Power class for band class 0	1-0	Class I	XXXXXX00
analogue operation		Class II	XXXXXX01
		Class III	XXXXXX10
		Reserved	XXXXXX11

6.6.1.5 *Home System and Network Identification*

This parameter, composed of 15 bits, identifies the MS home system, i.e. where the MS is registered and allowed to be operational. Figure 6.24 shows the structure of the home system (*HOME_SID*) number.

The two bits used to identify the MS home country, International Code (*INTL*), follow the criteria in Table 6.8:

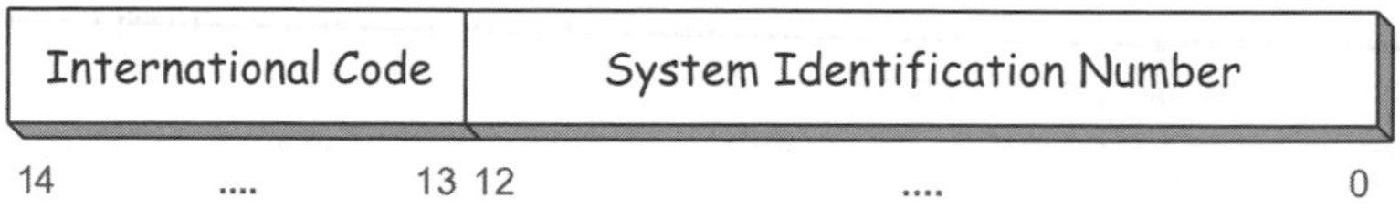

Figure 6.24 Home system (HOME_SID) structure.

The remaining bits, the system identification number (0 to 12), identify each system within a country. In the US, the FCC provides this identification (13 bits). The EIA/TIA document *Telecommunications Service Bulletin TSB29* (International Implementation of Cellular Radiotelephone Systems Compliant with ANSI/EIA/TIA-533) presents international requirements for System IDentification (*SID*) and Network IDentification (*NID*).

In CDMA systems, the system parameters message, on the Paging Channel (PCh) or on the Forward Common Control Channel (FCCh), provides these parameters to MSs. With these parameters MSs are able to identify the system where they are currently located and if they are in a roaming condition or not.

6.6.2 Secure Parameters

Authentication consists of information exchange procedures between an MS and the system for identification purposes. This process is only successful if the system validates that MS and BTS possess identical sets of Secret Shared Data (SSD). For authentication the MS must use either the *IMSI_M* or the *IMSI_T*. The BTS must comply with MS programming.

Figure 6.25 depicts, in a simplified diagram, the data exchange during authentication. The system stores some of the parameters, the MS stores others. Not all parameters are transmitted through the air interface to improve system security and to avoid cloning.

Table 6.8 Country codes for the home system identification (HOME_SID)

Bit 14	Bit 13	Country
0	0	United States
0	1	Remaining countries
1	0	Canada
1	1	Mexico

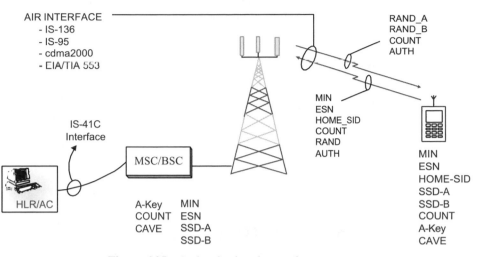

AIR INTERFACE
- IS-136
- IS-95
- cdma2000
- EIA/TIA 553

RAND_A
RAND_B
COUNT
AUTH

MIN
ESN
HOME_SID
COUNT
RAND
AUTH

IS-41C
Interface

MIN
ESN
HOME-SID
SSD-A
SSD-B
COUNT
A-Key
CAVE

MSC/BSC

HLR/AC

A-Key   MIN
COUNT ESN
CAVE   SSD-A
            SSD-B

Figure 6.25 Authentication data exchange.

Some parameters exchanged through the air interface have been described in the previous sections, although other very specialised parameters will be described in the following sections.

6.6.2.1 Secret Shared Data

The SSD consists of 128 bits divided into two sub-sets of 64 bits each, called *SSD_A* and *SSD_B*, stored in the semi-permanent memory of MSs. Subscribers do not have access to their values, however BTSs have a copy of these parameters stored at the Home Location Register/Authentication Center (HLR/AC). For a successful authentication process, the system must validate that both the MS and BTS have the same values for these parameters. Figure 6.26 depicts the SSD structure.

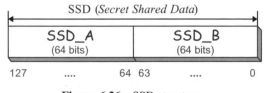

SSD (*Secret Shared Data*)

| SSD_A (64 bits) | SSD_B (64 bits) |

127   ....   64 63   0

Figure 6.26 SSD structure.

SSD_A and *SSD_B* are used for distinct purposes: *SSD_A* is mainly used for authentication, whereas *SSD_B* is employed in voice privacy and reliability for transmitting messages to other systems.

These parameters are never transmitted through the air interface. A special algorithm uses the SSD as an input to process information and transmits only the results through the air interface. Sometimes it may be necessary to generate/update SSD values, as described in Section 6.6.4.2, through the SSD update process.

6.6.2.2   Random Challenge Memory

The random challenge memory (RAND) consists of 32 bits stored in MS memory, used with the SSD and other parameters to authenticate MS call originations, terminations, and registrations.

The BTS usually transmits *RAND_A* and *RAND_B* to the MS through random challenge *A* and random challenge *B* action messages, on the overhead messages sent over the control channel.

6.6.2.3   Call History Parameter (COUNT)

COUNT is a modulo-64 counter, established only at the MS and at the AC. This counter is incremented every time the MS has a successful authentication process, receiving a parameter update order from the BTS.

6.6.2.4   A_Key

The main objective of the *A_Key* is to protect the SSD stored in the MS. The *A_Key* is not transmitted via the air interface. Instead it is stored at the permanent memory of the MS, not allowing subscribers to access or alter it. Any attempt to read or modify the *A_Key* leads to an inoperative condition for the MS.

The *A_Key* consists of a 64-bit sequence known only by the MS and the AC. During the SSD update process, the *A_Key* is used by both entities. If the current serving system is not the MS home system, it obtains a copy of the SSD, calculated and stored at the original AC, through the systems interoperation protocol, EIA/TIA IS-41 C.

As soon as the SSD update process (Section 6.6.4.2) is successfully concluded, the system can use the SSD for authentication purposes (Sections 6.6.4.3 and 6.6.4.5). If any failure occurs while executing the SSD update process, the system may initiate the unique challenge-response procedure, described in Section 6.6.4.1, before denying service to the MS.

6.6.3   Cellular Algorithms for Validation and Encryption

The document '*Common Cryptographic Algorithms*' describes the authentication algorithm known as Cellular Algorithms for Validation and Encryption (CAVE). Another document called '*Interface Specification for Common Cryptographic Algorithms*' describes the interface, input and output parameters for the algorithm.

The CAVE are employed for SSD update and for generation of authentication signatures (*AUTH*), using known parameters and random variables (*RAND*). These algorithms were initially employed by the U.S. Defense Department and later used, with government authorisation, by cellular equipment manufactures to reduce cloning.

U.S. International Traffic and Arms Regulation (ITAR) and export administration regulations control the description of authentication algorithms. The TIA acts as the organisation responsible for distributing the algorithm to equipment manufacturers. The *'Technology Transfer Control Plan'* document applied to *'Common Cryptographic Algorithms'* describes TIA procedures regarding algorithms distribution.

6.6.4 The Authentication Process

The following sections describe the several stages of the authentication process, starting from the SSD update process, including autonomous registrations, call originations and terminations.

6.6.4.1 Unique Challenge-Response Procedure

The MS initiates the Unique Challenge-Response Procedure (UCRP) over voice or control channels, analogue or digital.

The BTS generates the 24-bit random variable *RANDU*, and transmits it to the MS, as shown in Figure 6.27. The MS configures CAVE input parameters as shown in Figure 6.28. The 24 most significant bits of *RAND_CHALLENGE* are configured as *RANDU*, whereas the remaining eight LSB are the eight LSB of *MIN2*.

The CAVE are executed to obtain the *AUTHU_Signature*, which is stored as *RANDU* (18 bits). As soon as the MS receives *RANDU*, it calculates its *AUTHU* using the received *RANDU* and other internally stored parameters as input for CAVE, as indicated in Figure 6.28. The MS then transmits the calculated *AUTHU* to the BTS.

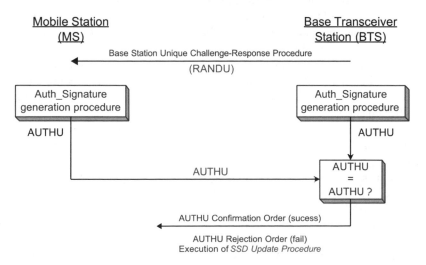

Figure 6.27 Unique challenge-response procedure.

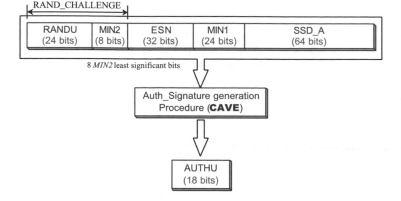

Figure 6.28 *AUTHU* for the unique-challenge-response procedure.

When the BTS receives the *AUTHU*, it compares this value to the value previously generated and internally stored. If the *AUTHU* is not validated, the BTS can deny service to the MS, denying further MS access attempts, 'dropping' the current call, or even initiating the SSD update process, as in Section 6.6.4.2.

6.6.4.2 SSD Update Process

The SSD Update Process (SSDUP) generates a new SSD using random variables, MS specific information and the *A_Key*. Because many transactions are involved, the process may initiate in one channel and end in another, as in Figure 6.29.

Figure 6.30 illustrates the steps in the SSD update process. This process is executed when the MS is turned on for the first time, right after being registered as an operational terminal, when the AC 'suspects' of the MS, and because of administrative procedures required by the service provider. The document '*User Interface for Authentication Key Entry*', TSB50, provides more details about this procedure.

The system always initiates the SSD update process when sending an SSD update message to any MS. This message contains the *RANDSSD* parameter configured with the same 56 bits as in the AC. The message can be sent either on the paging channel or on any forward traffic channel.

After receiving the message, the MS configures CAVE input parameters for the SSD generation process, as indicated in Figure 6.31. The document '*Interface Specification for Common Cryptographic Algorithms*' describes this configuration.

CAVE obtain new values for SSD, *SSD_A_NEW* and *SSD_B_NEW*. The results of the preceding process are stored as *RANDU* (18 bits). The MS then generates and selects a 32-bit random number *RANDBS* that is transmitted to the BTS in a base station challenge order on the access channel or on the reverse traffic channel.

The MS must use *RANDBS* as one of the inputs for the *Auth_Signature* generation procedure, together with *ESN*, *MIN1* and *SSD_A_NEW*, as in Figure 6.32. CAVE uses these parameters to obtain the *Auth_Signature* for the 18-bit random variable *AUTHBS*.

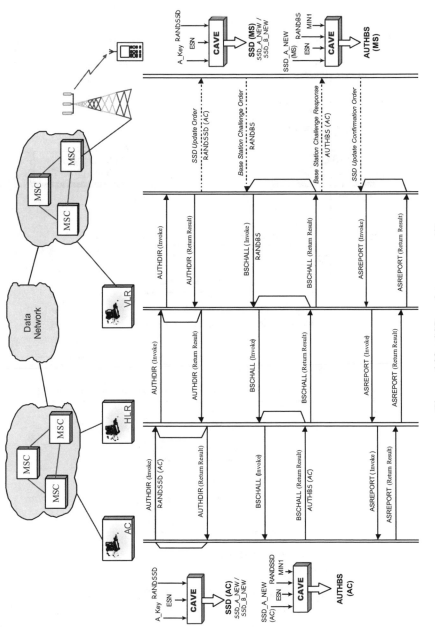

Figure 6.29 SSD update process (1).

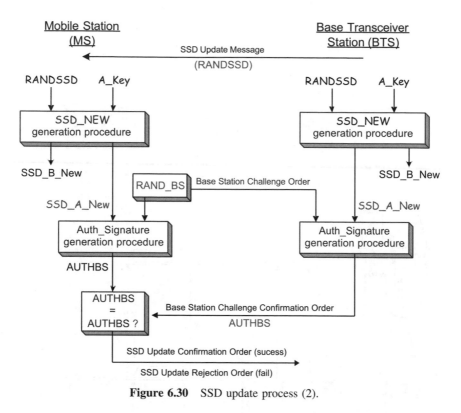

Figure 6.30 SSD update process (2).

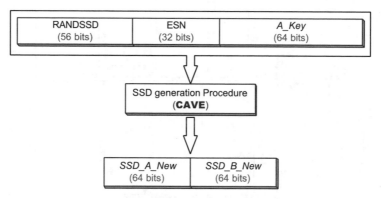

Figure 6.31 CAVE input parameters for the SSD generation process.

When the BTS receives the base station challenge order from the MS, it configures CAVE input parameters for the *AUTHBS* generation process, as portrayed in Figure 6.32. The document '*Interface Specification for Common Cryptographic Algorithms*' describes this configuration.

CAVE determine the *Auth_Signature* value for the 18-bit random variable *AUTHBS*. *AUTHBS* is then sent to the MS in a base station challenge confirmation order, for comparison purposes.

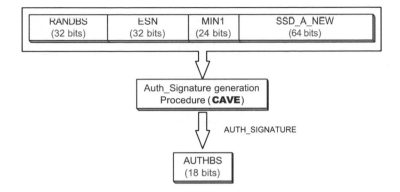

Figure 6.32 *AUTHBS* generation process.

Upon receipt of this order, the MS compares both *AUTHBS* values, the value received on the base station challenge confirmation order and the internally generated value. If both values coincide, the MS records the *SSD_A* and *SSD_B* as *SSD_A_NEW* and *SSD_B_NEW* in its semi-permanent memory and sends an SSD update order confirmation indicating success. If the SSD update process fails, the MS sends an SSD update rejection order message indicating failure and discards *SSD_A_NEW* and *SSD_B_NEW*.

As soon as the BTS receives the SSD confirmation order from the MS, it sets *SSD_A* and *SSD_B* received from the HLR/AC and sends a base station challenge confirmation order to the MS. If the MS does not receive this message within $T_{64m} = 10$ s, it discards *SSD_A_NEW* and *SSD_B_NEW* and terminates the SSD update process.

All the information exchange between MSCs, HLR and AC happens on the EIA/TIA IS-41 C interface.

6.6.4.3 Authentication During MS Registration

When an MS attempts to register by sending a registration message with *AUTH* set as '01', which means standard authentication procedure mode, the MS configures CAVE input parameters for *Auth_Signature* generation, as indicated in Figure 6.33. The document '*Interface Specification for Common Cryptographic Algorithms*' describes this configuration.

The CAVE calculates *AUTH_Signature* for the 18-bit random variable *AUTHR*, using *SSD_A*, *ESN* and *MIN1* as inputs. *AUTHR* is sent to the BTS in a registration message for comparison purposes. *RANDC* (eight most significant bits of *RAND*) and *COUNT*, stored in the MS memory, are also sent on the registration message.

Upon receiving the registration message, the BTS compares *RANDC* and *COUNT* with internally stored values, accounting for the *COUNT* associated with the received *MIN/ESN*. The BTS calculates the *AUTHR* in the same way as the MS, using the value they have stored for *SSD_A*, and then compares both *AUTHR* values.

If any of the comparisons result does not match, the BTSs denies service to the MS and marks the registration process as failed, requiring a unique challenge-response procedure or an SSD update procedure (Sections 6.6.4.1 and 6.6.4.2, respectively).

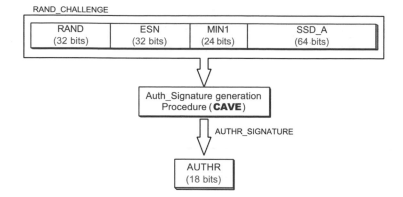

Figure 6.33 *AUTHR* generation for authentication during MS registration.

Otherwise, the process is considered successful and is terminated. Figure 6.34 presents the message exchange process between the system and MSs during authentication within an autonomous registration procedure.

The MS transmits the parameter *COUNT* to the BTS through the air interface. However, this step of the authentication process is optional. If executed, *COUNT* is passed to the authentication centre for comparison with the internally stored value.

COUNT is incremented in the MS memory each time an authentication process is performed. Thus, it represents the number of times the MS accesses the HLR/AC to perform authentication. The comparison of this value allows detecting fraud due to unauthorised *A_Key* acquisition, which allows cloning of terminals.

6.6.4.4 Authentication Process During an MS Originated Call Procedure

When the MS attempts to originate a call by sending an origination message with *AUTH* set as '01', which means standard authentication procedure mode, the MS configures CAVE input parameters for *Auth_Signature* generation, as in Figure 6.35. The document *'Interface Specification for Common Cryptographic Algorithms'* describes this configuration.

The *AUTH_DATA* consists, initially, of the *MIN1* 24 bits followed by the last digits (up to six) pressed by the subscriber. Each of these digits is converted into a set of 4 bits.

The CAVE then calculate *AUTH_Signature* for the 18-bit random variable *AUTHR*, using *SSD_A*, *ESN* and *RAND* as inputs. *AUTHR* is sent to the BTS in an origination message for comparison purposes. *RANDC* (eight most significant bits of *RAND*) and *COUNT*, stored in the MS memory, are also sent in the origination message.

Upon receiving the registration message, the BTS compares received *RANDC* and *COUNT* with internally stored values, accounting for the *COUNT* associated with the received *MIN/ESN*. The BTS calculates the *AUTHR* in the same way as the MS, using the value it has stored for *SSD_A*, and then compares both *AUTHR* values.

If any of the comparisons result does not match, the BTS denies service to the MS and may require a UCRP or an SSDUP (Sections 6.6.4.1 and 6.6.4.2, respectively).

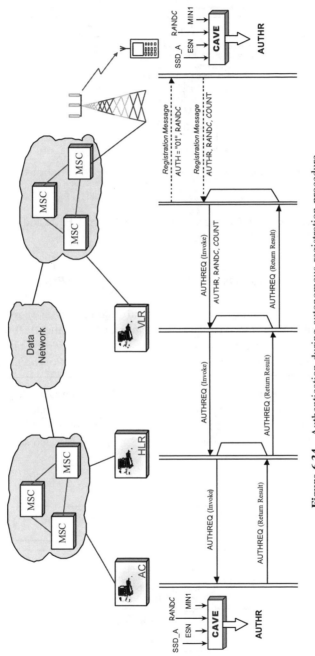

Figure 6.34 Authentication during autonomous registration procedure.

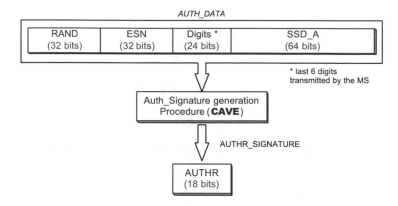
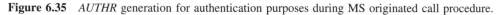

Figure 6.35 *AUTHR* generation for authentication purposes during MS originated call procedure.

Otherwise, the process is considered successful and BTS, MSC, HLR and AC proceed to traffic channel designation and related steps.

6.6.4.5 Authentication Process During MS Terminated Call Procedure

When the MS responds to a paging message (general paging message) by sending an origination message with *AUTH* set as '01', which means standard authentication procedure mode, the MS configures CAVE input parameters for *AUTH_Signature* generation, as in Figure 6.36. The document *'Interface Specification for Common Cryptographic Algorithms'* describes this configuration.

The *AUTH_DATA* consists of the *MIN1* 24 bits. The CAVE calculate *AUTH_Signature* for the 18-bit random variable *AUTHR*, using *SSD_A ESN*, and *RAND* as inputs. *AUTHR* is sent to the BTS in a page response message for comparison purposes. *RANDC* (eight most significant bits of *RAND*) and *COUNT*, stored in the MS memory, are also sent on the page response message.

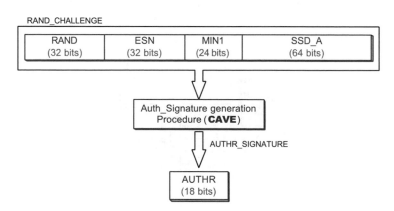

Figure 6.36 *AUTHR* generation for authentication purposes during MS terminated call procedure.

Upon receiving the registration message, the BTS compares *RANDC* and *COUNT* with internally stored values, accounting for the *COUNT* associated with the received *MIN/ESN*. The BTS calculates *AUTHR* in the same way as the MS, using the value it has stored for *SSD_A*, and then compares both *AUTHR* values.

If any of the comparison results does not match, the BTS denies service to the MS and may require a unique challenge-response procedure or an SSDUP (Sections 6.6.4.1 and 6.6.4.2, respectively).

Otherwise, the process is considered successful and BTS, MSC, HLR and AC proceed to traffic channel designation and related steps.

6.6.4.6 Authentication Process for PACA Cancellation

When the MS cancels a PACA call by sending a PACA call message with *AUTH* set as '01', which means standard authentication procedure mode, the MS configures CAVE input parameters for *Auth_Signature* generation, as indicated in Figure 6.37. The document '*Interface Specification for Common Cryptographic Algorithms*' describes this configuration.

The CAVE calculate *AUTH_Signature* for the 18-bit random variable *AUTHR*, using *RAND*, *ESN*, *MIN1* and *SSD_A* as inputs. *AUTHR* is sent to the BTS in a page response message for comparison purposes. *RANDC* (eight most significant bits of *RAND*) and *COUNT*, stored in the MS memory, are also sent in the Page Response Message.

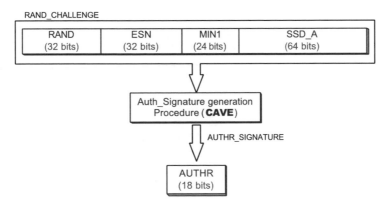

Figure 6.37 *AUTHR* generation for authentication purposes during PACA cancellation procedure.

Upon receiving the registration message, the BTS compares *RANDC* and *COUNT* with internally stored values, accounting for the *COUNT* associated with the received *MIN/ESN*. The BTS calculates *AUTHR* in the same way as the MS, using the value they have stored for *SSD_A*, and then compares both *AUTHR* values.

If any of the comparison results does not match, the BTS denies service to the MS and requires a unique challenge-response procedure or an SSD update procedure (Sections 6.6.4.1 and 6.6.4.2, respectively).

Otherwise, the process is considered successful and BTS, MSC, HLR and AC proceed to traffic channel designation and related steps [1–10].

BIBLIOGRAPHY AND REFERENCES

1. C.S0002-C, Physical Layer Standard for cdma2000 Spread Spectrum Systems, Release C, 3GPP2, May 2002.
2. C.S0005-B, Upper Layer (Layer 3) Signaling Standard for cdma2000 Spread Spectrum Systems, Release B, 3GPP2, April 2002.
3. TIA/EIA-95-B, Mobile Station – Base Station Compatibility Standard for Dual-Mode Spread Spectrum Cellular System, November 1998.
4. CelTec/CelPlan, CDMA IS-95 and cdma2000 Systems - Training Course.
5. Common Cryptography Algorithms, Revision C, 1997. An EAR-controlled document subjected to restrict distribution. Contact Telecommunications Industry Associations (TIA), Arlington, VA.
6. Interface Specifications for Common Cryptography Algorithms, Revision C, 1997. Contact Telecommunications Industry Associations (TIA), Arlington, VA.
7. Enhanced Cryptography Algorithms, TR45.AHAG 2001.
8. EIA/TIA/IS-54-B, Cellular System Dual-Mode Mobile Station – Base Station Compatibility Standard, April 1992.
9. EIA/TIA/IS-91, Mobile Station – Base Station Compatibility Standard for 800 MHZ Analog Cellular, October 1994.
10. TSB50, User Interface Authentication Key Entry, March 1993.

7

Power Control, Handoff and Radio Resource Management

BRUNO DE SOUZA ABREU XAVIER

This chapter describes three of the main processes in CDMA based networks: power control, handoff and radio resource management (RRM). Power control implementations are presented and analysed for IS-95 and cdma2000 networks. Handoff processes are also discussed for both types of network, with the description of the main configuration parameters and conditions involved. RRM is responsible for Walsh codes and radio configuration assignment. Parameters such as transmission rate, Walsh code length and encoder type all depend on RRM; therefore system capacity is directly influenced by RRM decisions.

7.1 INTRODUCTION

One of CDMA's main operational requirements is power control, which is extremely important for good performance and higher system capacity because an MS generates interference to all other users within the same BTS coverage area and to all users on neighbour cells. Therefore, the greater the number of active subscribers, the higher the interference level within the system. Thus, the lower the MS transmission power, the lower the total interference within the RF carrier bandwidth and the higher the system capacity.

CDMA systems implement power control techniques as a way to minimise interference, allowing the network to achieve higher capacity. Power Control processes for CDMA IS-95 and cdma2000 systems are quite similar. Sections 7.1.1 and 7.1.2 describe these processes in detail.

To execute power control, the system analyses network quality parameters. The most common parameter is the Frame Error Rate (FER). This parameter is also involved in the determination of other ratios used for system evaluation, such as E_c/I_0 (chip energy, E_c, to interference plus noise spectral density ratio) and E_b/I_0 (bit energy, E_b, to interference plus noise spectral density ratio).

Power control aims to quickly and efficiently estimate the minimum MS transmission power to achieve the required communication quality (based on FER statistics). Table 7.1

Designing CDMA 2000 Systems L. Korowajczuk et al.
© 2004 John Wiley & Sons, Ltd ISBN: 0-470-85399-9

Table 7.1   Maximum and minimum ERP according to MS classes

Band class	MS class	Minimum ERP	Maximum ERP
0 (800 MHz)	I	1 dB W (1250 mW)	8 dB W (6.3 W)
	II	−3 dB W (500 mW)	4 dB W (2.5 W)
	III	−7 dB W (200 mW)	0 dB W (1.0 W)
1 (PCS)	I	−2 dB W (630 mW)	3 dB W (2000 mW)
	II	−7 dB W (200 mW)	0 dBW (1000 mW)
	III	−12 dB W (63 mW)	−3 dB W (500 mW)
	IV	−17 dB W (20 mW)	−6 dB W (250 mW)
	V	−22 dB W (6.3 mW)	−9 dB W (125 mW)

presents power transmission configurations available for mobile terminals, showing maximum and minimum Effective Radiated Power (ERP) values for each of terminal class, according to the band being used (800 MHz and PCS at 1900 MHz). Terminal transmission power values are not fixed, but are limited within a range. This allows manufacturers to define their own terminal transmission power.

The establishment of the ideal minimum transmission power in each communication channel, for the forward link (BTS → MS) and also reverse link (MS → BTS), is a main objective of power control in CDMA networks. For example, achieving a certain *Quality of Service* (QoS), considering FER of about 1%, is a very hard task, especially because of user mobility and signal degradation due to propagation effects. The main benefits of power control in the CDMA system are the following:

- increase of system capacity;

- minimisation of near-far effect;

- increase of battery lifetime.

The random distribution of MSs within the coverage area of a BTS causes the 'near-far problem'. Because some subscribers are closer to the BTS than others, if all terminals transmitted with the same mean output power, the BTS would receive signals of closer users much stronger than signals of distant users. The main idea of power control schemes is to set MSs' transmission power to a minimum, so that BTSs receive signals from all MSs with a similar level.

The near-far problem can be illustrated in real life by a classroom where several students want to ask questions simultaneously. If they all use the same voice level, the teacher will probably understand only the ones closer to her.

Even though power control is used, due to propagation effects, such as multi-path, some transmission problems still occur. For example, on the forward link, all logical channels are processed and transmitted simultaneously and in phase; therefore an MS receives all channels with the same delays and propagation losses. However, these channels not only come from its server BTS, but also from neighbour cells, and because they may be using the same carrier, this may cause interference. The forward link is considered synchronous for a specific MS when receiving and de-multiplexing signals from only one BTS, and not for all BTSs within the system.

On the reverse link, however, the time reference for the signal of each MS is different, a condition known as asynchronous transmission. BTSs receive signals from MSs, with distinct phases, delays and fading characteristics because of the random distribution (location) of users within the BTS coverage area.

Almost all logical channels use power control techniques, which can be divided into two basic types: open loop and closed loop.

Open loop power control is usually implemented as the initial power control to give an estimate of minimum transmission power used for access channels during *system access state*, such as Reverse Access Channels (RAChs), Enhanced Access Channels (EAChs) and Reverse Common Control Channels (RCCChs). MSs measure the E_c/I_0 ratio per active/ candidate pilot set (see *Handoff Processes* for further details, Section 7.5) of the current carrier to estimate the reverse link transmission power.

Closed loop power control is usually implemented for traffic channels, which need a higher performance power control system. cdma2000 systems can also employ this type of power control even during the *system access state*. As a rule of a thumb, the MS initially uses the transmission power defined in the last message (from last access probe) transmitted on the *system access state*. The BTS determines a 'set point' power level, which is considered ideal for receiving signals within a certain QoS (based on FER statistics). This set point is periodically adjusted. The BTS performs power control by transmitting power control bits, which instruct MSs individually to increase or reduce transmission power according to the estimated set point value.

7.1.1   Overview of Power Control in CDMA IS-95 Systems

On CDMA IS-95 systems, power control schemes are applied to reverse access channels (RAChs) and forward and reverse traffic channels (FTChs and RTChs). As mentioned in Section 7.1, power control is implemented in two different ways (open and closed loops), depending on the function and logical channels in use.

IS-95 BTSs implement open loop power control for power adjustments for RAChs during *system access state*. Closed loop power control is used for power adjustments for TChs while in *mobile station control on the traffic channels*. The closed loop is divided into two processes: inner loop (faster and more accurate – up to 800 Hz power control rate) and outer loop (slower – 50 Hz power control update rate).

MSs also employ power control, transmitting from time to time a Power Measurement Report Message (PMRM) to the BTS, with a report about the received signal, indicating the QoS (FER). This is the forward link power control. The next sections in this chapter provide details of each type of power control implementation.

7.1.2   Overview of Power Control in cdma2000 Systems

cdma2000 introduced new features and processes to improve power control schemes, for voice and data transmissions, and consequently to increase system capacity.

The transmission of a Reverse Pilot Channel (RPiCh) allows coherent detection of the signals transmitted by an MS and serves as a timing and power reference on the reverse link. The RPiCh also makes forward link power control more accurate and faster because of the transmission of power control bits. The main applications of the RPiChs are presented in the following.

- During system access state procedures – Carries power control bits during the transmission of an EACh or RCCCh, serving as power, timing and phase reference for the BTS to detect signals transmitted on the reverse link.

- During a voice call – Carries power control bits for traffic channels on the forward link.

- During a call when data is transmitted – Carries Erasure Indicator Bits (EIB) and Quality Indicator Bits (QIBs) informing BTSs when the data transmitted on the forward traffic channels was not understood.

The introduction of Forward Common Power Control Channels (FCPCCHs) makes power control of reverse channels (EAChs and RCCCHs), during the system access state, easier and faster when compared to IS-95 systems.

As in IS-95, cdma2000 networks also have two implementations of power control, used in different situations. cdma2000 employs open loop power control for cdma2000-1X and cdma2000-3X networks, only for the estimation of reverse channels (EAChs and RCCCHs) first access probe transmission power during system access state. After that, the system uses closed loop power control to perform power adjustments for access channels (EAChs and RCCCHs), while in system access state, and TChs, while in mobile station control on the traffic channels state. The closed loop power control consists of two processes: inner and outer loops, which achieve up to 800 Hz of power control rates.

On the reverse link, only six Radio Configurations (RCs) are defined. RC1 and RC2 are employed to maintain backward compatibility to CDMA IS-95 systems, with the same characteristics (using only RACHs and RTCHs) and, consequently, employing the same power control schemes. RC3 – RC6 (cdma2000-1X and cdma2000-3X) make use of all the new features.

7.2 MAIN CHARACTERISTICS OF POWER CONTROL IN THE SYSTEM ACCESS STATE

Depending on the radio configuration (RC), an MS may use different channels in the system access state. For RC1 and RC2, an MS uses RACHs on the reverse link, whereas for RC3–RC6, it uses EAChs and RCCCHs on the reverse link and FCPCCHs on the forward link.

An MS uses these channels for *response attempts*, transmitting a response message to a BTS message (or order), or for *request attempts*, autonomously transmitting a request message to the system while registering or originating a call. The MS pseudo-randomly selects the access channel to be used. In CDMA IS-95 systems, up to 32 RACHs may be associated with a particular FPCh, whereas in cdma2000 up to 32 EAChs and 32 RCCCHs are associated with a unique FCCCh.

Regardless of the channel in use, power control processes are quite the same and consist of the transmission of access probes within a set of access attempts, as illustrated in Figure 7.1.

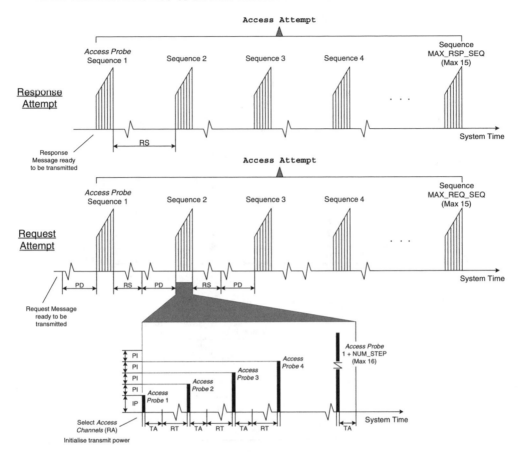

Figure 7.1 Power control on the system access state.

Here

PI   Power Increment (*PWR_STEP*)

IP   Initial Open Loop Power (initial access probes transmission power)

PD   Persistence Delay, pseudo-randomly defined for the beginning of a new access attempt

RS   Sequence Backoff (interval between two consecutive access attempts)

RT   Probe Backoff, pseudo-randomly defined as the interval between two consecutive access probes, based on the *PROBE_BKOFF* parameter

TA   ACK Response Timeout, maximum interval MSs must wait for an access message acknowledgement

An *access attempt* consists of re-transmitting the same message until a previously defined counter expires or until the MS receives an acknowledgement (ACK) for the message. Each access attempt is divided into one or more sub-attempts, which consist of transmissions of the same message. Each of these single message transmissions is called an *access probe*.

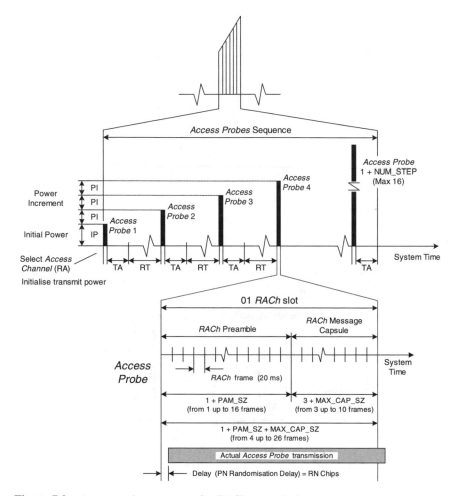

Figure 7.2   Access probes structure for RACh transmission on the system access state.

Each access probe consists of a preamble and an access message capsule, as depicted in Figure 7.2.

The maximum number of access attempts, for request or response, is given, respectively, by *MAX_REQ_SEQ* and *MAX_RSP_SEQ*, both parameters varying from 1 to 15. Within each attempt, MSs can transmit up to 16 access probes $(1 + NUM\_STEP)$, as in Figure 7.1.

The first probe is transmitted with the initial power (IP). Subsequent probes have their power incremented by PI. An *open loop power control* defines the IP value, i.e. the access probe initial mean output power, as in Table 7.2. In this process, the MS estimates the signal power received from the server BTS and evaluates the transmission power for the reverse link. The estimated value for IP depends on parameters transmitted in the system access message that describe the radio configuration (Figure 7.3).

Table 7.2 Offset power parameters, according to Spreading Rate (SR), Radio Configuration (RC) and band class

Band class	Spreading rate	Reverse channels	Offset power
0 (800 MHz)	1	Reverse Access Channel (RACh)	−73.0
		Reverse Traffic Channel (RTCh) for RC = 1 or 2	
		Enhanced Access Channel (EACh)	
		Reverse Common Control Channel (RCCCh)	−81.5
		Reverse Traffic Channel for RC = 3 or 4	
	3	Enhanced Access Channel	
		Reverse Common Control Channel	−76.5
		Reverse Traffic Channel for RC = 5 or 6	
1 (PCS)	1	Reverse Access Channel	−76.0
		Reverse Traffic Channel for RC = 1 or 2	
		Enhanced Access Channel	
		Reverse Common Control Channel	
		Reverse Traffic Channel for RC = 3 or 4	−84.5
	3	Enhanced Access Channel	
		Reverse Common Control Channel	
		Reverse Traffic Channel for RC = 5 or 6	−79.5

Table 7.2 shows one of these parameters, the power offset.

There is a high probability that the first access probes transmitted by a mobile terminal will not be received by the BTS. If the MS does not receive an ACK message within a predefined time limit (TA), the access probe must be retransmitted. The probe is also retransmitted if the MS receives a confirmation message indicating NOK, i.e. the BTS did not understand the data transmitted on the probe. In both situations, the MS only transmits a new access probe after the probe backoff timer (RT) expires.

The use of an *open loop power control* is indicated for cases in which there are high intensity signal level fluctuations and/or long-term fading, and during the establishment of the initial transmission power for an MS to access the system. An open loop scheme is also desirable in situations where there are shadow coverage areas and variations of the distance between MSs and their server BTSs.

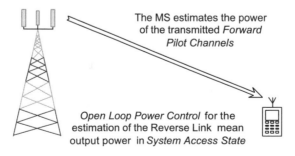

The MS estimates the power of the transmitted *Forward Pilot Channels*

Open Loop Power Control for the estimation of the Reverse Link mean output power in *System Access State*

Figure 7.3 Open loop power control.

In these situations, an MS monitors the power (and E_c/I_0 ratio) of the received Forward Pilot Channel (FPiCh) and increases, or reduces, the transmission power in steps of up to 0.75 dB, to achieve the desired QoS. The MS performs this process autonomously while in the system access state.

There are, however, disadvantages with the open loop power control. Short-term fading, such as Rayleigh fading, is not minimised because this type of control is relatively slow when compared to this type of fading, because it has to calculate the power average before determining the power adjustment value, therefore it is not sensitive to fading variations.

The estimated reverse link transmission power is based on propagation loss assumptions, equivalent for both forward and reverse links, and does not consider possible asymmetric propagation losses between the links, because of multi-path propagation effects, for instance.

The evaluation of forward link power may lead to errors, because it represents the sum of the power of all channels transmitted on the same carrier in use by the MS, of all BTSs transmitting close to where the MS is located (network service area). The MS may estimate inaccurate signal levels depending on the server BTS.

7.2.1   Power Control in the System Access State for CDMA IS-95 Systems

IS-95 RAChs, compatible with RC1 and RC2 of cdma2000 systems, use open loop power control to estimate the initial transmission power of the first access probe. Power control also helps MSs to reduce interference before the transmission of each access probe.

The access probe preamble of an RACh consists of 20 ms frames containing 96 bits '0' each. The frames are used to help the BTS to acquire the RACh during system access procedures.

The Mean Output Power (P_{RACh}) of the first access probe, i.e. the Initial Power (IP), is estimated by measuring the E_c/I_0 for the carrier with the strongest Forward Pilot Channel (FPiCh), which the MS is using, as illustrated in Figure 7.4. During an access probe sequence transmission, an MS may update the Mean Output Power (P_{RACh}) because of changes in the received mean input power (P_{in}).

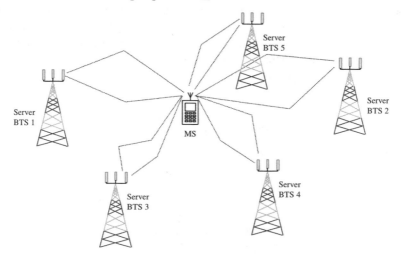

Figure 7.4   Possible FPiCh multi-path components.

Subsequent access probes within the same access probe sequence have the power incremented by a PI, according to the following expressions

$$P_{\text{RACh}} = \text{IP} + \text{PI}$$

$$P_{\text{RACh}} = -P_{\text{in}} + K + INIT\_PWR + (NOM\_PWR - 16NOM\_PWR\_EXT) + \text{PI}$$

Here

P_{RACh}	*Mean Output Power* (dBm) (mean RACh output power transmitted by the MS)
P_{in}	*Mean Input Power* (dBm) (mean power received from the strongest FPiCh)
K	*Offset Power* (values in Table 7.2), also known as turnaround constant; it is used as a compensation factor for access channel probes.
INIT_PWR	*Initial Power Offset* – transmitted on the access parameters message for the transmission of Reverse Access Channel (RACh), used for the first access probe.
NOM_PWR	*Nominal Transmit Power* – correction factor during open loop power estimation, transmitted in the access parameter message.
NOM_PWR_EXT	*Extended Nominal Transmit Power* – correction factor during open loop power estimation, transmitted in the access parameter message.
PI	*PWR_LVL* × *PWR_STEP*, represents the power increment.
PWR_STEP	*Power step* (0 – 7 dB) – transmitted on the access parameters message, it defines the transmission power increment between successive access probes within the same access probe sequence.
PWR_LVL	*Power level* (1 – *NUM_STEP*), transmitted on the access parameters message; it identifies the power level step (maximum value for *NUM_STEP* is 15).

7.2.2   Power Control in the System Access State for cdma2000 Systems

There are two main differences in system access procedures for MSs using RC3 – RC6 (cdma2000). The first is the transmission of a Reverse Pilot Channel (RPiCh) while EAChs or RCCChs are transmitted and also as a preamble. The RPiCh also transmits power control bits to perform power control over forward link channels.

The second difference is the use of a Forward Common Power Control Channel (FCPCCh) to allow faster and more efficient power control while in *system access state* (performing a *closed loop power control*). The use of FCPCChs minimises the 'near-far problem', consequently reducing the interference generated within a carrier. BTSs use this channel to transmit power control bits to reverse link channels.

Each BTS may have up to two FCPCChs. An FCPCCh can perform power control of 12, 24 or 48 users, with power control rates of 800, 400, or 200 Hz, respectively. The greater the number of simultaneous users in conversation controlled by a specific FCPCCh, the smaller the power control rates employed.

The overall power control process for cdma2000 is similar to IS-95 systems, consisting of the transmission of access probe sequences for a pre-defined number of access attempts. Parameters *MAX_REQ_SEQ* and *MAX_RSP_SEQ* define the maximum number of access attempts for request and response attempts, respectively. The number of probes is by *NUM_STEP*, also transmitted in the access parameter message.

7.2.2.1 Power Control using Reverse Pilot Channel (RPiCh)

Some of the functions of RPiChs include timing reference, power estimation and coherent phase detection of signals transmitted on the reverse link, and transmission of power control and Erasure Indicator Bits (EIB).

An MS transmits an RPiCh before and during the transmission of any reverse channel. When transmitted before a voice/data reverse channel, the RPiCh acts as the preamble for this channel. When transmitted with another channel, it acts as a reference and carries power control bits.

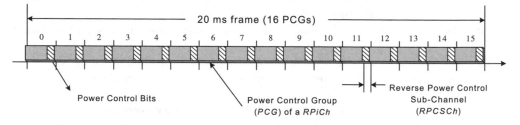

Figure 7.5 Structure of RPiCh power control sub-channels.

RPiChs are built with 20 ms frames, divided into 16 Power Control Groups (PCGs) of 1.25 ms each, as in Figure 7.5. This figure presents the structure of RPiCh frames, showing the 16 PCGs, each containing a Reverse Power Control Sub-Channel (RPCSCh).

Figure 7.6 shows that this channel transmits only zeroes and power control bits (when power control is enabled). The number of chips '0' within a frame is always $1536 \times N_{SR}$

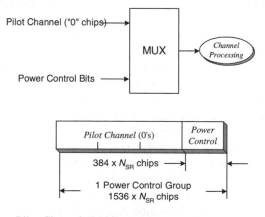

Figure 7.6 Reverse Pilot Channel (RPiCh) – structure of a Power Control Group (PCG).

Table 7.3 Gating rates of RPiCh power control groups

PILOT_GATING_RATE	Gating rate	Number of transmitted PCGs	PCGs transmitted
00	1	16	0 – 15
01	1/2	8	1, 3, 5, 7, 9, 11, 13, 15
10	1/4	4	3, 7, 11, 15

chips, where N_{SR} depends on the Spreading Rate (SR) in use (N_{SR} equals 1 for SR1 (cdma2000-1X) and 3 for SR3 (cdma2000-3X)).

The Reverse Power Control Sub-Channel (RPCSCh) corresponds to the last quarter of each power control group. Each RPCSCh consists of $384 \times N_{SR}$ chips with the same information of the power control bit, i.e. if the power control bit is '1' the RPCSCh has $384 \times N_{SR}$ chips '1'.

To improve power control performance and increase battery lifetime, a BTS may request an MS to perform gating of some power control groups. The gating process is only performed if no other logical channel is been simultaneously transmitted.

Two parameters define the gating process: *PILOT_GATING_USE_RATING* and *PILOT_GATING_RATE*. Both parameters are transmitted by the BTS; the first determines whether gating should be performed ('1' = yes, '0' = no); the second sets the gating rate with values presented in Table 7.3.

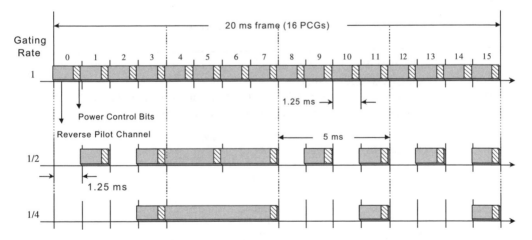

Figure 7.7 Power Control Groups (PCGs) transmission according to gating rates.

Figure 7.7 shows the distribution of PCGs in a 20 ms RPiCh frame according to gating rates presented in Table 7.3.

In Figure 7.7, regardless of the gating rate, PCGs 3, 7, 11 and 15 are always transmitted, i.e. PCGs positioned before 5 ms frame boundaries are always transmitted, aiding BTSs in acquiring reverse traffic channels. Section 7.3.3.4 describes 5 ms frames in more detail.

7.2.2.2   Access Probe of an Enhanced Access Channel

Enhanced Access Channels (EAChs) may operate in two different access modes: reservation and basic. In the reservation access mode, FCPCChs control the transmission power of the reverse common control channel associated to the MS. The access probes consist of an enhanced access preamble followed by an enhanced access header, and the enhanced access data, which is transmitted by the RCCCh using closed loop power control.

In the basic access mode, FCPCChs control EACh transmission power. The access probes consist of an enhanced access preamble followed directly by the enhanced access data, similar to the operation of Reverse Access Channels (RACh) in IS-95 systems.

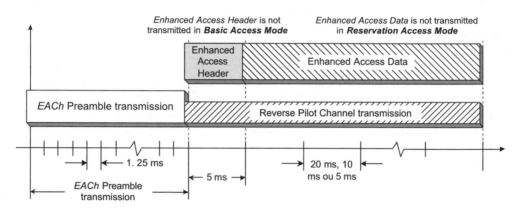

Figure 7.8   Access probe of an enhanced access channel (EACh).

The access probe of an EACh depends on the access mode operation. Figure 7.8 illustrates the structure of an EACh access probe.

The enhanced access preamble consists of the transmission of a set of RPiCh PCGs without power control bits, i.e. $1536 \times N_{SR}$ chips '0', as in Figure 7.9.

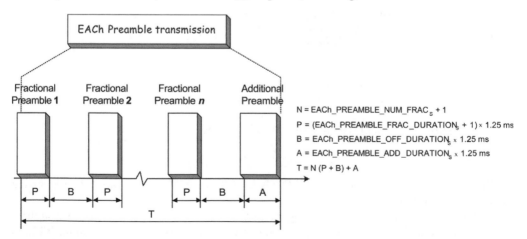

Figure 7.9   EACh preamble transmission.

The parameters in Figure 7.9 define the number of RPiCh PCGs (1.25 ms) transmitted during the EACh preamble. BTSs send these parameters to MSs through the access parameter message.

The definition of IP and PI to any subsequent transmission of EACh access probes is given by the following expression

$$P_{\text{EACh}} = \text{IP} + \text{PI}$$
$$P_{\text{EACh}} = -P_{\text{in}} + K + INIT\_PWR\_EACh + NOM\_PWR\_EACh + 6 + \text{PI}$$

where

P_{EACh}	*Mean output power* (dBm), mean EACh output power transmitted by the MS
P_{in}	*Mean input power* (dBm), mean power received from the strongest FPiCh
K	*Offset power* (values in Table 7.2), also known as turnaround constant; it is used as a compensation factor for preamble transmission of enhanced access channel probes
INIT_PWR_EACh	*Initial EACh Power Offset* – transmitted on the access parameters message for the transmission of EAChs, used for the first access probe
NOM_PWR_EACh	*Nominal EACh transmit power* – correction factor during open loop power estimations transmitted on the access parameters message
PI	*PWR_LVL × PWR_STEP_EACh*, represents the power increment
PWR_STEP_EACh	*Power step* (0 – 7 dB) – transmitted on the access parameters message; it defines the transmission power increment between successive access probes within the same access probe sequence
PWR_LVL	*Power level* (1 – *NUM_STEP*), transmitted on the access parameters message; it identifies the power level step (maximum value for *NUM_STEP* is 15).

7.2.2.3 Access Probe of Reverse Common Control Channel

Reverse Common Control Channels (RCCChs) operate in the reservation access mode, where FCPCChs control the transmission power of the RCCCh. Access probes consist of an enhanced access preamble followed by enhanced access data transmitted by RCCChs using closed loop power control. Figure 7.10 shows RCCChs access probe structure.

The reverse common control channel preamble consists of the transmission of a set of RPiCh PCGs without power control bits, i.e. $1536 \times N_{\text{SR}}$ chips '0', as in Figure 7.11.

The parameters in Figure 7.11 define the number of RPiCh PCGs (each 1.25 ms) transmitted during the reverse common control channel preamble. BTSs send these parameters to MSs through the access parameters message.

The definition of IP and PI to any subsequent transmission of RCCChs access probes is given by the following expression

$$P_{\text{RCCCh}} = \text{IP} + \text{PI}$$
$$P_{\text{RCCCh}} = -P_{\text{in}} + K + INIT\_PWR\_RCCCh + NOM\_PWR\_RCCCh + 6 + \text{PI}$$

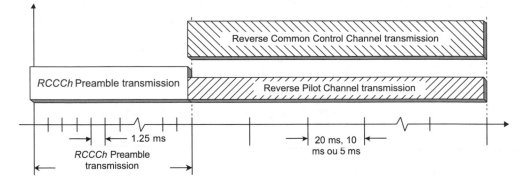

Figure 7.10   Access Probe of a Reverse Common Control Channel (RCCCh).

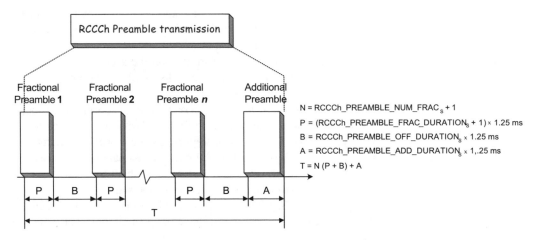

Figure 7.11   RCCCh preamble transmission.

where

P_{RCCCh}	*Mean output power* (dBm) (mean RCCCh output power transmitted by the MS)
P_{in}	*Mean input power* (dBm) (mean power received from the strongest FPiCh)
K	*Offset power* (values in Table 7.2), also known as turnaround constant; it is used as a compensation factor for preamble transmission of reverse common control channel probes
INIT_PWR	*Initial RCCCh Power Offset* – transmitted on the access parameters message for the transmission of RCCChs, used for the first access probe
NOM_PWR	*Nominal RCCCh Transmit Power* – correction factor during open loop power estimations transmitted in the access parameters message

PI   *PREV_CORRECTIONS*, power increment, corresponding to previous
                      power corrections due to closed loop power control plus
                      PWR_LVL × *PWR_STEP_EACh*, if there was a previous operation
                      of the enhanced access channel.

7.2.2.4 Power Control Performed by Forward Common Power Control Channels

The use of Forward Common Power Control Channels (FCPCChs) leads to the establishment of a closed loop power control. BTSs may use FCPCChs to transmit power control bits within power control sub-channels (one bit per sub-channel, one per MS), to control multiple Enhanced Access Channels (EAChs) and Reverse Common Control Channels (RCCChs) during system access state.

FCPCChs may also perform power control over other channels, such as the Reverse Acknowledgement Channel (RAckCh) or the Reverse Channel Quality Indicator Channel (RCQICh). The FCPCCh depends on the operation modes for each reverse channel in use, as described previously for the EAChs (Section 7.2.2.2) and for RCCChs (Section 7.2.2.3).

FCPCChs are structured in 20 ms frames divided into 16 PCGs of 1.25 ms each. Twelve power control bits are transmitted in each FCPCCh PCG, resulting in a total of 192 bits per 20 ms frame. Each FCPCCh is able to control up to 48 users simultaneously.

Because there are 12 power control bits in a PCG, each PCG can control up to twelve users. Two PCGs control from 13 to 24 users, and three PCGs control from 25 to 48 users. Therefore, the number of users controlled determines the power control rate: 800, 400 or 200 bps. These rates correspond to one power control bit per user every 1.25, 2.5 and 5 ms, respectively, as in Table 7.4.

Analysing the structure of PCGs, and considering that an FCPCCh is controlling 12 users (control rate of 800 bps), referred to as $u_1, u_2, \ldots, u_{12}$, the transmission sequence is always the same for every PCG. This condition allows MSs to be prepared to receive a power control bit within the current PCG, and then receive another bit every 1.25 ms, in that same position. This position index is known as the initial offset.

However misunderstandings can happen when reading power control bits, if, for example, there is a jamming interference being transmitted at every 1.25 ms. Because of this, depending on the interference signal duration, users whose power control bit would be transmitted at the same time as the interference will not be able to decode and understand power control transmitted by the BTS.

To avoid this type of problem and to assure that MSs are able to receive their respective power control sub-channels, cdma2000 implements a randomisation process for transmitting

Table 7.4 Number of FCPCCh power control sub-channels

Power control rate (bps)	Duration (ms)	Number (N) of sub-channels
800	1.25	12
400	2.50	24
200	5.00	48

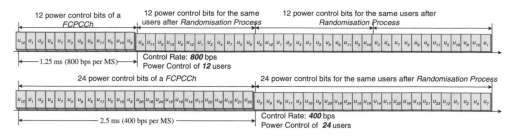

Figure 7.12   Randomisation process example: power control over 12 and 24 users.

power control bits. The process is performed every 1.25, 2.5 or 5 ms, depending on the power control rate. The idea is to change the power control bit transmission sequence for users within that interval. This process can be visualised in Figure 7.12 for power control rates of 800 bps (up to 12 users simultaneously) and 400 bps (up to 24 users simultaneously).

Randomisation is obtained by modulo-N adding a relative index to the previously assigned offset index, defining the effective transmission sequence (effective index) for the power control bits of an FCPCCh. The process is pseudo-random because MSs must be able to identify their corresponding power control sub-channel position within the PCG. Therefore, a well-known long code mask is used to define the set of chips that will be employed for the definition of relative and effective indices, according to the power control rate in use. Chapter 4 provides more details about the randomisation process.

7.3   POWER CONTROL IN MOBILE STATION CONTROL ON THE TRAFFIC CHANNEL STATE

In the mobile station control on the traffic channel state, MSs only use traffic channels to transmit voice/data traffic or signalling and control bits. The system uses closed loop power control schemes on the channels of both forward and reverse links. The initial transmission power for these channels has been previously established during the system access state.

The use of both closed loop power control processes, inner and outer loop, in combination may lead to fading compensations of 20 – 35 dB within a 20 ms interval, within a dynamic range of up to 80 dB.

This state also presents some differences in power control processes employed by CDMA IS-95 and cdma2000 systems, such as the existence of the reverse pilot channel and of specific power control modes for supplemental channels. Power control schemes may be applied to two logical channels simultaneously if necessary, for example when MSs are using two distinct traffic channels such as the Fundamental Channel (FCh), Dedicated Control Channel (DCCh) and/or Supplemental Channels (SChs).

7.3.1   Power Control in Mobile Station Control on the Traffic Channel State for CDMA IS-95 Systems

Because RC1 and RC2, defined in cdma2000 standards, are backward compatible to CDMA IS-95 systems, the characteristics and processes described in this section are also valid for these radio configurations in cdma2000.

7.3.1.1 Open Loop Power Control

The initial transmission power of reverse traffic channels is defined as the power used in the last Reverse Access Channel (RACh) probe, during the system access state, in which the MS has received a confirmation message (ACK). The transmission power is defined by using open loop power control, as in Figure 7.2. The following expression represents the transmission power calculation.

$$P_{\mathrm{RTCh}} = \mathrm{IP} + ACC\_CORRECTIONS + RL\_GAIN$$

where

P_{RTCh}	*mean output power* (dBm), mean RTCh output power transmitted by the MS
IP	$(-P_{\mathrm{in}} + K +$ interference corrections), representing the power of the last transmitted access probe
P_{in}	*mean input power* (dBm), mean power received from the strongest FPiCh
K	offset power (values in Table 7.2) used as a compensation factor for access channel probes
ACC_CORRECTIONS	$INIT\_PWR + (NOM\_PWR - 16.NOM\_PWR\_EXT) + PWR\_LVL \times PWR\_STEP$, represents the last power adjustments performed while in system access state.
INIT_PWR	*initial power offset* – transmitted on the access parameters message for the transmission of RAChs, used for the first access probe
NOM_PWR	*nominal transmit power* – correction factor during open loop power estimations
NOM_PWR_EXT	*Extended Nominal Transmit Power* – corresponds to another correction factor employed during open loop power control estimations
PI	$PWR\_LVL \times PWR\_STEP\_EACh$, represents the power increment
PWR_STEP	*power step* – defines the transmission power increment between successive access probes within the same access probe sequence
PWR_LVL	*power level* – identifies the power level step
RL_GAIN	additional power gain adjustment applied specifically to the reverse traffic channel, relative to the RACh transmitted power

After establishing the initial transmission power for the reverse traffic channel, both MS and BTS start the closed loop power control, described in the next section.

7.3.1.2 Closed Loop Power Control

The closed loop power control's main purpose is to minimise fast-fading effects (short-term fading, or Rayleigh fading) caused by signal multi-path propagation losses. This power control scheme better compensates forward and reverse link asymmetry.

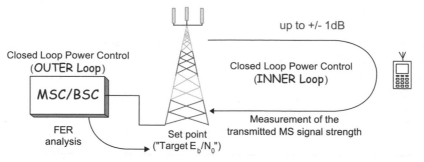

Figure 7.13   Closed loop power control.

The closed loop is divided into two processes: *outer* and *inner* power control loops. Figure 7.13 illustrates these processes.

Outer closed loop power control is performed between a BTS and its controller BSC or MSC every 20 ms, i.e. at the interval of each traffic channel frame. The BSC specifies a threshold value, called 'set point' or target E_b/I_0, by analyzing the Frame Error Rate (FER) required for the desired Quality of Service (QoS). This value is then transmitted to the BTS, which is then in charge of measuring, evaluating and comparing the signal transmitted by MSs to this threshold.

If there is no frame reception error, the BTS automatically reduces the target E_b/I_0 value, usually in steps of 0.2 or 0.3 dB. This reduction occurs frame-by-frame, that is, at every 20 ms. On the other hand, the BSC may raise the value by 3 up to 5 dB, informing the BTS about this increment amount. Figure 7.14 illustrates the target E_b/I_0 adjustment process.

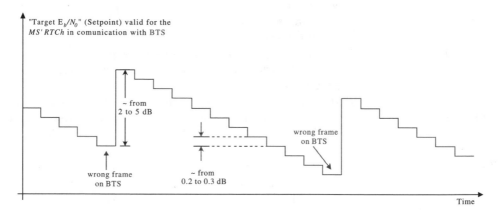

Figure 7.14   Obtaining the target E_b/N_0 during outer loop power control.

The inner loop happens between the BTS and an MS. The BTS compares the signal coming from an MS to the target E_b/I_0 determined in the outer loop. Using the FTCh power control sub-channels (one bit per sub-channel), the BTS instructs the MS to reduce or increase transmission power.

Power control sub-channels operate with a rate of 800 bps, or every 1.25 ms. This corresponds to 16 power control bits per frame, or 16 Power Control Groups (PCGs). Power steps vary from ± 0.25 to ± 1.0 dB. Figure 7.15 illustrates the power adjustment process.

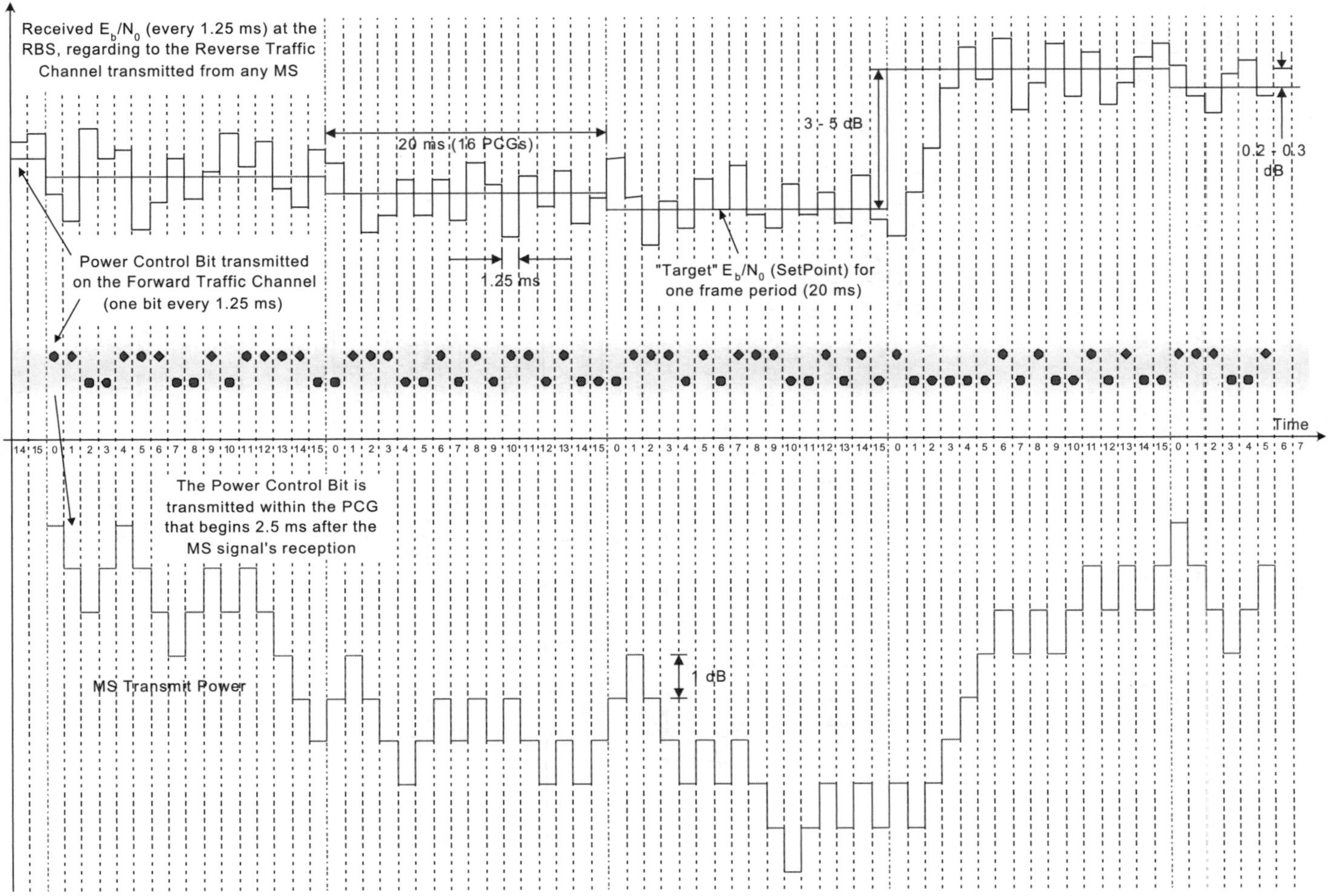

Figure 7.15 Closed loop power control in reverse traffic channels (RTChs – IS-95).

7.3.1.3 Gating Process of Reverse Traffic Channel PCGs (Data Burst Randomiser)

Figure 7.16 illustrates another technique used to reduce interference and to increase battery lifetime in CDMA systems, the data burst randomiser. This technique consists of gating off PGCs with redundant data from Reverse Traffic Channels (RTChs) operating with RC1 and RC2. Chapters 4 and 5 provide more information about traffic channels coding and processing.

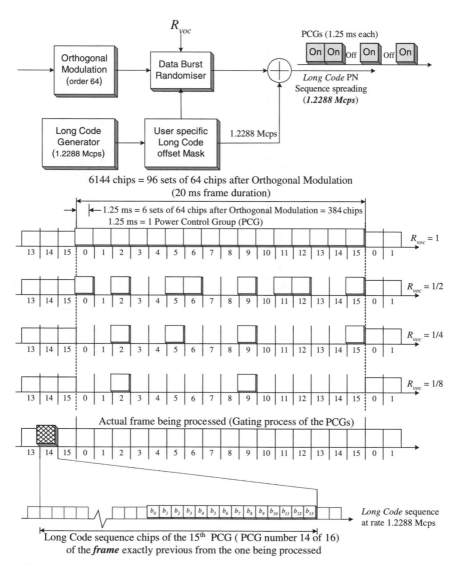

Figure 7.16 Data burst randomiser – example of gating process for RTCh PCGs.

The number of gated off PGCs depends on the voice activity factor, which influences the vocoder rate. The lower the vocoder rate (R_{voc}), the more the repetitions during channel processing. An MS uses chips of its locally generated long code sequence to determine the PCGs to be gated off. Therefore, the data burst randomiser defines, for each MS, the pseudo-random position of PCGs.

When $R_{voc} < 1$, the BTS is not able to listen to an MS during gated off intervals, causing it to transmit power control bits on FTCh power control sub-channels to request transmission power increase to the MS. This situation seems contradictory considering that the main purpose of the gating process is to reduce MSs transmission power. To avoid conflicts of this kind, the standards establish that an MS shall only accept and perform power control commands when they are received on the second PCG after the PCG transmitted on the RTCh, as illustrated in Figure 7.17.

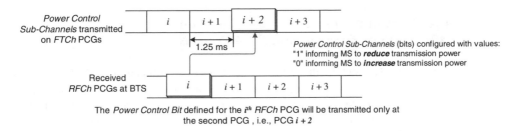

Figure 7.17   Transmission of power control sub-channel on FTCh PCGs.

7.3.1.4   Power Control for Forward Traffic Channels

Besides controlling the transmission power of mobile stations, BTSs also control their own transmission power. An MS evaluates the FTCh FER and, from time to time, reports this value on a Power Measurement Report Message (PMRM) to its server BTS. This information allows the BTS to adjust the power of the FTCh assigned to that MS, as in Figure 7.18.

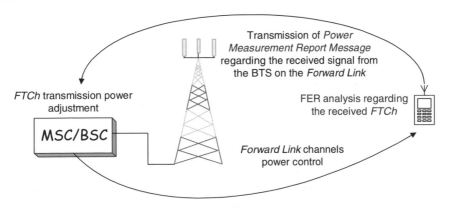

Figure 7.18   Power control of forward traffic channels (FTChs).

MSs report the FER to BTSs periodically, at a time interval defined by the system, or when a pre-established threshold is reached. The parameters used to establish the time interval and the threshold for FER report generation are transmitted in the system parameters message. MSs use the parameters *PWR_PERIOD_ENABLE* and *PWR_REP_FRAMES* to determine the generation of periodical reports, and *PWR_THRESH_ENABLE*, *PWR_REP_THRESH* and *PWR_REP_DELAY* for threshold triggered reports.

In the power measurement report message (PMRM), an MS transmits the number of frames measured (*PWR_MEAS_FRAMES*) and the total of frames with errors (*ERRORS_DETECTED*).

The great advantage of this power control scheme is that a BTS transmits only enough power to assure a certain QoS in its service area. This improves spectral re-use, allowing a greater number of BTSs to operate in the same frequency band used for the downlink.

7.3.2   Power Control in Mobile Station Control on the Traffic Channel State for cdma2000 Systems

For radio configurations 1 and 2, cdma2000 systems are fully compatible with IS-95 power control techniques. For RC3 – RC6, additional features are implemented to enhance power control. These features include the use of data specific channels, such as supplemental channels (SChs), and transmission of power control bits on the RPiCh (more efficient closed loop power control).

7.3.2.1   Open Loop Power Control for the Reverse Link

For RC3–RC6, the RPiCh is always transmitted before (preamble) and during the transmission of any reverse traffic channel, as in Figure 7.19.

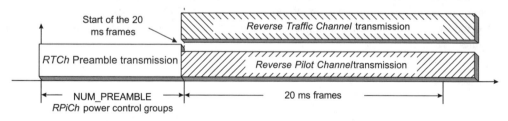
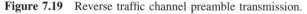

Figure 7.19   Reverse traffic channel preamble transmission.

The open loop power control defines the transmission power as in Figure 7.2:

$$P_{\text{RTCh}} = \text{IP} + ACC\_CORRECTIONS + RL\_GAIN$$

where

P_{RTCh} *mean output power* (dBm), mean output power transmitted for reverse traffic channels with RC3 – RC6

IP $(-P_{\text{in}} + K +$ interference corrections), representing the power of the last transmitted access probe

P_{in} *mean input power* (dBm), mean power received from the strongest FPiCh

K offset power (values in Table 7.2) used as a compensation factor for access channel probes

ACC_CORRECTIONS   *INIT_PWR + NOM_PWR + PWR_LVL × PWR_STEP*, represents the last power adjustments performed while in system access state

INIT_PWR_EACh or *INIT_PWR_RCCCh*   initial power offset – transmitted on the access parameters message for the transmission of EAChs or RCCChs, used for the first access probe

NOM_PWR_EACh or *NOM_PWR_RCCCh*   *nominal transmit power* – correction factor during open loop power estimations

PWR_STEP   *power step* – defines the transmission power increment between successive access probes within the same access probe sequence

PWR_LVL   *power level* – identifies the power level step

RL_GAIN   additional power gain adjustment applied specifically to the reverse traffic channel, relative to the EACh or RCCCh transmitted power

After the initial transmission power for a reverse traffic channel is determined through the *open loop*, both MS and BTS must be prepared to start the *closed loop power control*, described in the next section.

7.3.2.2 Closed Loop Power Control for the Reverse Link

The closed loop power control in this state is also divided into two sub-loops: inner and outer loop. Section 7.3.2.2 describes the complete closed loop process.

7.3.2.3 Closed Loop Power Control for the Forward Link

There are multiple modes of forward power control available in cdma2000 networks (*FPC_MODE* parameter). Table 7.5 presents these modes along with their control rate

Table 7.5 Characteristics of power control on forward traffic channels

Maximum power control rate (bps)	*FPC_MODE*	PCG (primary channels)	PCG (secondary channels)
800	000	0–15	
400/400	001	0–14 (even only)	1–15 (odd only)
200/600	010	1, 5, 9, 13	Remaining PCGs
50	011	0–15	
50	100	0–15	
50/50	101	0–14 (even only)	1–15 (odd only)
400/50	110	0–14 (even only)	1–15 (odd only)
All other values	Reserved	Reserved	Reserved

and channels affected by them. Both inner and outer closed loop power controls are used, but not for all configurations.

cdma2000 networks divide channels into two main groups: primary and secondary. Primary channels are used to transmit voice, data and/or signalling bits, such as Fundamental Channels (FChs) and Dedicated Control Channels (DCChs). Secondary channels transmit only data traffic, such as Supplemental Channels (SChs).

The RPiCh operation for power control bits transmission depends on the logic channel in use at the moment and also on the selected forward power control mode, with values presented in Table 7.5.

PCGs transmitting power control sub-channels for primary channels are also known as primary reverse power control sub-channels. When transmitting for secondary channels, the PCGs are known as secondary reverse power control sub-channels. There are seven different modes of forward power control (*FPC_MODE*), each with different configuration and characteristics.

- *FPC_MODE* = '*000*': All RPiCh PCGs transmit power control bits with a rate of 800 bps performing power control over forward link primary channels. In this mode, MSs must support outer loop power control for all forward traffic channels. Inner loop power control must be supported on primary reverse power control sub-channels only for forward fundamental and forward dedicated control channels.

- *FPC_MODE* = '*001*': The RPiCh transmits power control bits with a rate of 400 bps performing power control over primary channels on even numbered PCGs and over secondary channels on odd numbered PCGs. In this mode, MSs must support outer closed loop power control for all forward traffic channels. Inner loop power control must be supported on primary reverse power control sub-channels only for forward fundamental and forward dedicated control channels and on secondary sub-channels for the assigned forward supplemental channel.

- *FPC_MODE* = '*010*': The RPiCh transmits power control sub-channels with a rate of 200 bps performing power control over primary channels on PCGs 1, 5, 9 and 13, and with a rate of 600 bps for secondary channels on the remaining PCGs. In this mode, MSs must support outer closed loop power control for all forward traffic channels. Inner loop power control must be supported on primary reverse power control sub-channels only for forward fundamental and forward dedicated control channels and on secondary sub-channels for the assigned forward supplemental channel.

- *FPC_MODE* = '*011*': All 16 power control sub-channels in a 20 ms frame are set as a copy of the Erasure Indicator Bit (EIB), resulting in a power control rate of 50 bps. The EIB is transmitted in the second RPiCh frame after the end of the last Forward Fundamental Channel (FFCh) or Forward Dedicated Control Channel (FDCCh) received by MSs, as illustrated in Figure 7.20.

- *FPC_MODE* = '*100*': All 16 power control sub-channels in a 20 ms frame are set as a copy of the Quality Indicator Bit (QIB), resulting in a power control rate of 50 bps. The QIB is transmitted in the second RPiCh frame after the end of the current FDCCh, as in Figure 7.20. If the forward fundamental channel is transmitted while in this mode, the QIB is set as the EIB entering the mode *FPC_MODE* = '*011*'.

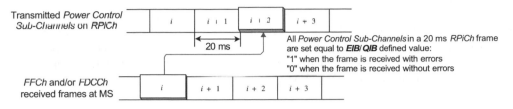

Figure 7.20   Transmission of the EIB/QIB on reverse power control sub-channels of RPiChs.

- *FPC_MODE* = '*101*': Only EIBs and/or QIBs are transmitted. Even RPiCh PCGs are used as QIBs with a rate of 50 bps, derived from FFChs or FDCChs. If the forward fundamental channel is transmitted, the QIB is set with the same value defined for the EIB. Odd PCGs shall transmit EIBs derived from the FSCh received by the MS. Because FSChs are structured in frames of 20, 40 and/or 80 ms, the effective power control rate for each type is respectively 50, 25 and 12.5 bps. Table 7.5 provides the maximum power control rates.

- *FPC_MODE* = '*110*': The RPiCh transmits power control bits with a rate of 400 bps performing power control over primary channels on even numbered PCGs and transmits EIBs on odd numbered PCGs derived from the Forward Supplemental Channel (FSCh). The transmission of EIBs shall start on the second 20 ms frame after the end of the corresponding FSCh. Because FSChs are structured in frames of 20, 40 and/or 80 ms, the effective power control rate for each type is respectively 50, 25 and 12.5 bps. In this mode, MSs must support outer closed loop power control only for forward fundamental and forward dedicated control channels. Inner loop power control must be supported on primary reverse power control sub-channels only for forward fundamental and forward dedicated control channels.

For radio configurations 3 – 6, EIBs indicate when MSs did not receive or will erase the last frame transmitted on the FFCh or FDCCh, whereas QIBs indicate signal quality of the FDCCh received by MSs. Whenever the FFCh is used, the QIB is set with the same value defined for the EIB.

7.3.2.4   *Gating Process of the Reverse Pilot Channel while in MS Control on the Traffic Channel State*

Section 7.2.2.1 shows that RPiChs may operate in gated mode, with gating rates as in Table 7.3 and depicted in Figure 7.7.

RPiChs may only be gated if no other reverse channel, such as Reverse Fundamental Channel (RFCh), Reverse Dedicated Control Channel (RDCCh) or Reverse Supplemental Channel (RSCh), is been simultaneously transmitted.

RPiChs PCGs sent before each 5 ms frame are always transmitted. These PCGs aid BTSs in acquiring transmitted reverse traffic channels (timing reference) and in the channel equalisation process, increasing correct decoding probabilities.

In cdma2000, for RC3 – RC6, there are two types of traffic frames: 5 and 10 ms. This new frame structure relies on the transmission of mini-messages, which are used for purposes that demand fast action or response, such as handoff requests, power measurement reports and channel allocation requests; 5 and 10 ms frames are commonly used by traffic channels and are described in detail in Chapters 3 and 4.

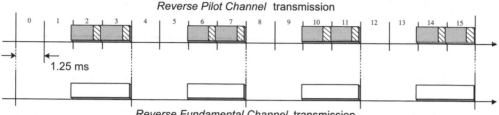

Figure 7.21   Gating of RPiChs' PCGs during transmission of reverse fundamental channels.

RPiCh PCGs are gated off in a different way during RFCh transmission when operating with a rate of 1500 bps for RC3 and RC5 or with rate of 1800 bps for RC4 and RC6 (Figure 7.21).

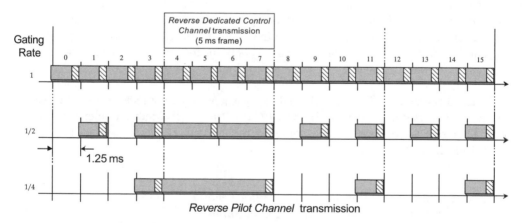

Figure 7.22   Gating of RPiChs PCGs during transmission of reverse dedicated control channels (5 ms frames).

During reverse dedicated control channel transmission for 5 and 20 ms frames, Figures 7.22 and 7.23 respectively, the RPiCh is present while the RDCCh is transmitted, but power control bits are transmitted in PCG intervals according to gating rate configurations, as specified in Table 7.3.

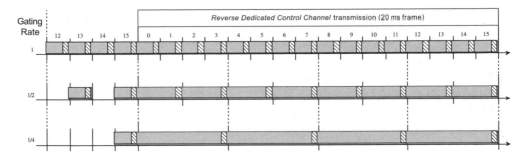

Figure 7.23 Gating of RPiChs PCGs during transmission of reverse dedicated control channels (20 ms frames).

7.4   INTRODUCTION TO HANDOFF PROCESSES

One of the most important tasks when planning wireless telecommunication systems is the definition of the target area. Regions within the target area that have enough signal quality to allow users to benefit from services offered by the network are known as the service area of the system.

Because of user mobility and the need of eliminating shadow areas within the service area, systems are designed with some overlap in the coverage areas between sites, i.e. regions where an MS may recognise more than one BTS as a possible server. Figure 7.24 illustrates this concept.

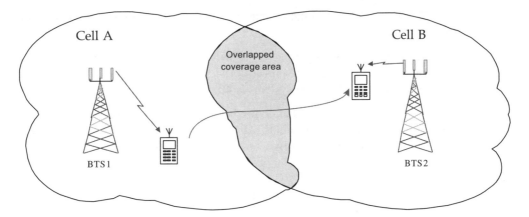

Figure 7.24 Handoff Process – the MS moves from one BTS coverage area to another.

Figure 7.24 illustrates a situation where the mobile terminal is moving from the coverage area of a BTS (cell A) to another (cell B). Considering a CDMA network, within cell A, the MS is receiving signals from BTS1, i.e. it has acquired and synchronised itself to the Forward Pilot Channel (FPiCh) transmitted by BTS1. When moving out of cell A (BTS1 coverage area) into cell B (BTS2 coverage area), regardless of being in a call or in idle state, the MS undergoes procedures that allow it to acquire the FPiCh transmitted by BTS2 while remaining connected to the system.

These procedures are called handoff processes. CDMA handoff processes have specific characteristics that distinguish them from the processes performed in GSM, TDMA and AMPS networks. The next sections of this chapter describe these characteristics in detail. The tasks, signal level measurements and system conditions that lead MSs to perform handoff are also explained.

7.5   MAINTENANCE OF PILOT SETS

To perform handoff processes, every MS measures and internally stores information about FPiChs present in its current location area. This information is always kept in four pilot sets, which are constantly updated by MSs. A pilot may be represented by the forward pilot channel sequence offset identification and associated frequency.

Each FPiCh in the system can only belong to one pilot set at a time. All pilots registered in any of the sets maintained by an MS use the same frequency, i.e. the pilots are transmitted on the same frequency assigned to the MS. The four pilot sets in CDMA systems are the following.

- *Active set*: pilots of CDMA carriers whose forward paging channels (FPChs) or forward common control channels (FCCChs) are being monitored while in idle state. During a call, these are the pilots associated with forward traffic channels assigned to the MS. The maximum number of pilots in the active set defined for cdma2000 systems is 6.

- *Candidate set*: pilots not currently in the active set but with enough signal power level to become an active pilot. The maximum number of pilots in the candidate set defined for cdma2000 systems is 10.

- *Neighbour set*: pilots that do not belong to the active or candidate sets but may also be considered candidates for a handoff process. The maximum number of pilots in the neighbour set defined for cdma2000 systems is 40.

- *Remaining set*: all remaining pilots that do not belong to any of the other sets but are in the area where the MS is located; or pilots that have already been in one of the previous sets but do not have enough signal power strength to be considered as a candidate for handoff.

Except for the first active FPiCh used by the MS to acquire the CDMA system, all the other pilots are communicated to the terminal by the BTS via the neighbour list message. Every MS is equipped with devices especially designed for digital signal processing (DSP) of forward link signals, as in Figure 7.25. These devices are usually grouped in sets of four, with three of them, called 'fingers', capable of synchronising and demodulating forward link multi-path signal components. The last device, called 'searcher', is used to search for pilots coming from other BTSs in the area where the MS is located. Some MSs are equipped with six fingers and two searchers.

In a rake receiver these fingers are used together to simultaneously demodulate the strongest multi-path signal components received. The multi-path signal components are

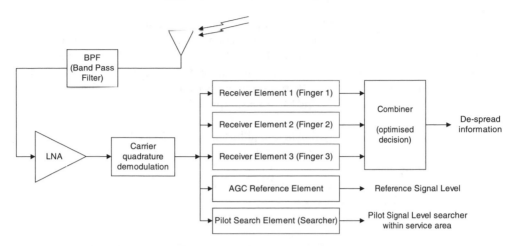

Figure 7.25   Fingers and searcher used for digital signal processing.

generated because of propagation effects and represent 'copies' of the same transmitted signal arriving at MS or BTS antennas with phase or amplitude differences, and a small time delay.

Both BTSs and MSs employ rake receivers, which are used to provide diversity reception gain. These receivers can compare multi-path signal components arriving from a single source or use the finger elements to receive signals from different sources, selecting the traffic channel with best quality, in a frame-by-frame basis, as in Figure 7.26.

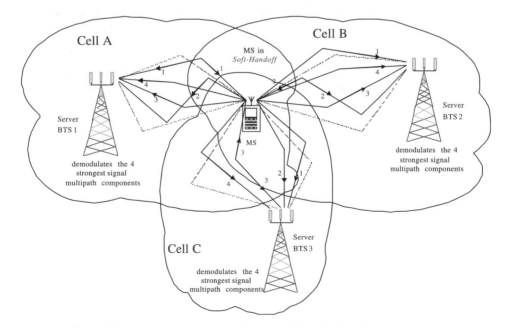

Figure 7.26   Rake receiver considering an MS with three finger elements.

Figure 7.26 shows an MS, during a call, simultaneously de-modulating signals from three BTSs, in a process known as soft handoff (see Section 7.6.2). The MS uses the searcher to estimate pilots' signal levels, evaluating the E_c/N_0 ratio. E_c/N_0 represents the ratio of the energy per chip (E_c) to the thermal noise plus interference spectral densities (N_0). There is no meaning in evaluating the E_b/N_0 ratio (energy per bit, E_b, to noise plus interference spectral power density ratio) for FPiChs, because this logical channel does not carry any data.

The searcher element can also evaluate the pilot sequence offset (*PN_Offset*) of the pilots detected. With this information, MSs can perform pilot set maintenance by moving them from one list to another.

MSs always intiate handoff processes in CDMA systems, in a procedure known as Mobile Assisted Handoff (MAHO). There are six main parameters involved in MAHO, each of them is described next.

- *T_ADD*: expressed in steps of 0.5 dB; it is used to control movement of pilots exceeding the E_c/N_0 target. Qualified pilots are moved from neighbour or remaining sets to active or candidate sets.

- *T_DROP* and *T_TDROP*: expressed, respectively, in dB and in seconds. These parameters define the E_c/N_0 detection threshold and the drop timer for pilots on active and candidate sets. If a pilot is in the active set, whose E_c/N_0 value is below *T_DROP* for a period greater than *T_TDROP*, it is moved to candidate or neighbour sets. Similarly, a pilot in the candidate set with E_c/N_0 value below *T_DROP* for a period greater than *T_TDROP* is moved to the neighbour or remaining sets. Table 7.6 shows *T_TDROP* values.

- *T_COMP*: controls the transference of pilots from the candidate to the active set, expressed in units of 0.5 dB. Every time the E_c/N_0 of a pilot in the candidate set exceeds an active set pilot by a multiple of *T_COMP*, this candidate pilot replaces the weakest pilot of the active set.

- *NEIGH_MAX_AGE*: when a pilot is moved to the neighbour set, an AGE timer is initiated for that pilot (one timer per pilot) and is incremented regularly up to *NEIGH_MAX_AGE*, defining the maximum retention time for that pilot in the set. If this timer expires, the pilot is moved to the remaining set.

Table 7.6   *T_TDROP* timer values

T_TDROP parameter	Timer expiration (s)	T_TDROP parameter	Timer expiration (s)
0	≤0,1	8	27
1	1	9	39
2	2	10	55
3	4	11	79
4	6	12	112
5	9	13	159
6	13	14	225
7	19	15	319

Table 7.7 Possible search window sizes

Search parameter	Search window size	Search parameter	Search window size
0	4	8	60
1	6	9	80
2	8	10	100
3	10	11	130
4	14	12	160
5	20	13	226
6	28	14	320
7	40	15	452

- *SRCH_WIN_A*, *SRCH_WIN_N* and *SRCH_WIN_R*: these parameters define pilots search window size, expressed in chip intervals according to Table 7.7 – for the active and candidate set (*SRCH_WIN_A*), neighbour set (*SRCH_WIN_N*) and remaining set (*SRCH_WIN_R*).

Figure 7.27 presents an example of pilot search window determining active, candidate, neighbour and remaining sets. In the example, search window configurations are defined as *SRCH_WIN_A* = 3 (10 chips), *SRCH_WIN_N* = 4 (14 chips) and *SRCH_WIN_R* = 6 (28 chips).

Table 7.8 provides recommended ranges and typical values used as handoff parameters.

Because of the handoff parameters configuration, sometimes pilots are constantly moved from active or candidate to neighbour sets and vice versa, this effect is known as 'ping pong'. Situations like this require a great number of messages to be transmitted to perform handoff. The greater the number of messages, the worse the system performance, because more time is needed to process all messages.

The need to improve handoff performance, reducing ping pong while increasing efficiency in channel assignments, caused the introduction of new procedures and parameters, establishing a dynamic handoff process. In this process, MSs evaluate pilots' E_c/N_0 and try to estimate its impact on the maintenance of pilot sets.

- *ADD_INTERCEPT* and *DROP_INTERCEPT*: respectively define intercept conditions to add or remove (drop) a pilot from the active set. These parameters are used to control pilots' transference between the active and candidate sets.

Table 7.8 Recommended range and typical values use as handoff parameters

Parameter	Recommended Range	Typical value
T_ADD	from −13 to −17 dB	−16 dB
T_DROP	from −13 to −20 dB	−20 dB
T_TDROP		5.0 s
T_COMP	≤3.0 dB (6 × 0.5 dB)	2.5 dB

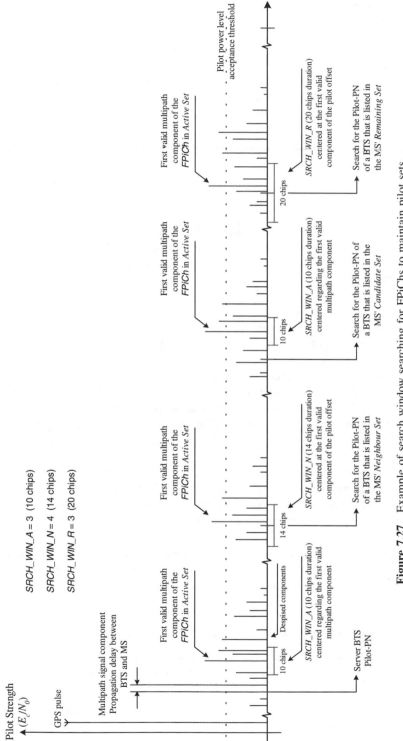

Figure 7.27 Example of search window searching for FPiChs to maintain pilot sets.

- *SOFT_SLOPE*: also used to determine pilots' E_c/N_0, defined as the slope in an inequality criterion to add or drop pilots from the active set.

An MS receives all handoff parameters in an FPCh system parameters message, or during a call in the in-traffic system parameters message, general handoff direction message, universal handoff direction message or extended handoff direction message.

Table 7.9 summarises the events and procedures that lead to pilot transference between sets stored in MSs.

If an MS receives a General Handoff Direction Message (GHDM), an Extended Handoff Direction Message (EHDM), or a Universal Handoff Direction Message (UHDM) through the traffic channel, it immediately updates the pilot sets.

MSs have an active participation in any handoff process. Every time they detect forward pilot channels (FPiCh) with E_b/N_0 value greater than T_ADD, they inform the MSC through a Pilot Strength Measurement Message (PSMM) or Extended Pilot Strength Measurement Message (EPSMM). Based upon this information, the MSC indentifies the BTSs associated to these pilots and directs MSs to perform handoff, assuming there is an available traffic channel.

A BTS may also transmit a Neighbour List Update Message (NLUM) to MSs within its coverage area, informing of natural or preferential neighbour pilots. Every time an MS

Table 7.9 Pilot sets maintenance, performed by every MS

Original set	Target set	Event or procedure
Active	Candidate	Reception of *General Handoff Direction Message* (GHDM), *Extended Handoff Direction Message* (EHDM) or *Universal Handoff Direction Message* (UHDM) without the inclusion of pilot information in a period inferior to *T_TDROP*
Active	Neighbour	Reception of GHDM, EHDM or UHDM, without the inclusion of pilot offset information, in a period greater than *T_TDROP*
Active	Remaining	Not performed
Candidate	Active	Reception of GHDM, EHDM or UHDM, including pilot offset information
Candidate	Neighbour	*T_TDROP* timer expiration or occurrence of *candidate set* overload
Candidate	Remaining	Not performed
Neighbour	Active	Reception of GHDM, EHDM or UHDM, including pilot offset information
Neighbour	Candidate	Pilot E_c/N_0 value exceeds *T_ADD*
Neighbour	Remaining	Pilot specific *AGE* timer exceeds *NEIGH_MAX_AGE* or occurrence of *neighbour set* overload
Remaining	Active	Reception of GHDM, EHDM or UHDM, including pilot offset information
Remaining	Candidate	Pilot E_c/N_0 value exceeds *T_ADD*
Remaining	Neighbour	Reception of *neighbour list update message*, including pilot offset information

receives an NLUM, it must update its neighbour set and all the AGE timers corresponding to the pilots in this set.

A timer is also associated and initiated for each active or candidate set pilot every time the E_c/N_0 value is smaller than T_DROP. If the E_c/N_0 value for this pilot is measured and continues with a value greater than T_DROP before expiration of T_TDROP, the timer is re-initialised and disabled. If the timer exceeds T_TDROP, the pilot is moved to the neighbour set.

7.5.1 Handoff Process example for CDMA IS-95 systems

Figure 7.28 shows an example of handoff and pilot set maintenance performed by CDMA IS-95 mobile terminals. The figure assigns a number to each stage of the handoff process.

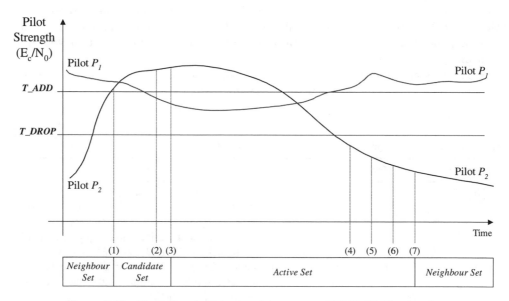

Figure 7.28 Handoff and pilot set maintenance on CDMA IS-95 networks.

At point (1), P_1 is the best server pilot. The mobile terminal measures P_2 during pilot set maintenance and verifies whether it exceeds T_ADD. The MS transmits a Pilot Strength Measurement Message (PSMM) to the current server BTS (P_1) informing the measured pilot signal level and corresponding PN_Offset and requesting handoff. The MS then transfers pilot P_2 to the candidate set.

At point (2), the current server (P_1), upon receiving the PSMM, informs the BSC or MSC about the handoff request and, if available, a traffic channel is assigned to the MS. The BSC, or MSC, then informs the BTS to send a General Handoff Direction Message (GHDM), an Extended Handoff Direction Message (EHDM) or a Universal Handoff Direction Message (UHDM) establishing the handoff.

If the new server belongs to the CDMA system and if there is a traffic channel available with the same frame offset characteristics available on the frequency in use by the MS, the BSC transmits a GHDM, EHDM or UHDM, establishing soft handoff (see Section 7.6.2).

If the new server only has traffic channels available on a different frequency, the BCS informs the involved BTSs to transmit a message establishing hard handoff (see Section 7.6.3).

If no CDMA traffic channel is available, the MSC verifies the availability of analogue voice channels. Analogue BTSs may have a pilot beacon (also known as a dummy pilot) that is used to inform the boundaries of CDMA networks. If an analogue traffic channel is available, the BCS informs the BTS to transmit a message establishing a CDMA-to-analogue hard handoff (see Section 7.6.4).

At point (3), for a soft handoff procedure, upon receiving the message from the BTS, the MS transfers P_2 to the active set and sends a Handoff Completion Message (HCM). The MS is then communicating with the two BTSs, using traffic channels from both.

At point (4), when performing pilot set maintenance, the MS verifies that P_2 drops below *T_DROP* and initialises a handoff drop timer to this pilot. At point (5), the handoff drop timer (*T_TDROP*) assigned to pilot P_2 expires and the MS transmits a PSMM to the BTS.

At point (6), upon receiving the PSMM, the BSC or MSC releases the allocated traffic channel, informing the BTS to transmit a GHDM, EHDM or UHDM to the MS. Then at point (7), the MS transfers P_2 to the neighbour set, transmitting a Handoff Completion Message (HCM) to the BTS and initialising the AGE timer for P_2.

Figure 7.29 presents a soft handoff example considering three pilots measured by the MS, showing the use of *T_ADD*, *T_DROP* and *T_TDROP*.

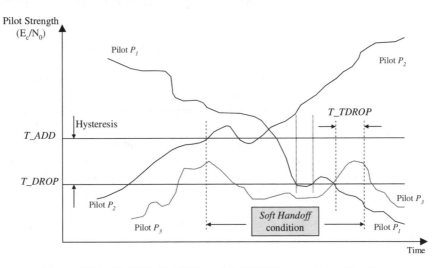

Figure 7.29 *T_ADD*, *T_DROP* and *T_TDROP* parameters utilisation.

In Figure 7.29, the MS is initially de-modulating traffic channels from the BTS using P_1, but it measures P_2 and verifies that its strength (E_c/N_0) exceeds *T_ADD*. The MS exchanges messages with the system requesting soft handoff.

The MS then verifies that P_1 drops below *T_DROP* and initialises a timer for this pilot (*T_TDROP*). If P_1 measurement exceeds *T_DROP* at any time, the MS resets and disables the handoff drop timer. This process avoids transferring the pilot from the active to the neighbour set.

When P_1 drops below *T_DROP* again, this time a period higher than *T_TDROP*, the MS transmits a PSMM and the pilot is transferred from active to neighbour. The MS is no longer

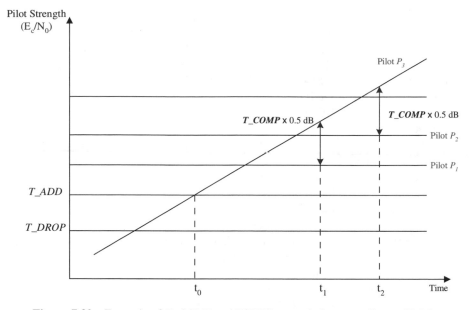

Figure 7.30   Example of *T_COMP* and PSMM transmission according to IS-95.

in soft handoff. In this example, because P_3 never exceeds *T_ADD*, the MS does not use any traffic channel from this BTS.

Figure 7.30 shows an example of *T_COMP* and Pilot Strength Measurement Message (PSMM) usage in CDMA IS-95 networks. In this case, the MS is in soft handoff, de-modulating traffic channels associated to two pilots, P_1 and P_2. The MS also indicates that the E_c/N_0 from a third pilot, P_3, is increasing.

At t_0, pilots P_1 and P_2 are in the active set and the MS transmits a PSMM moving P_3 to the candidate set, because its signal strength exceeds *T_ADD*. If P_3's strength exceeds pilot P_1's (E_c/N_0) by *T_COMP* $\times$ 0.5 dB, the MS transmits a PSMM requesting the addition of pilot P3 to the active set. The same happens if P_3's strength exceeds P_2's (E_c/N_0) by *T_COMP* $\times$ 0.5 dB.

These transmissions increase signalling and control traffic, reducing the system performance. cdma2000 overcomes this issue by adopting new procedures and protocols described in Section 7.5.3.

7.5.2   Handoff Process for cdma2000 Systems

One of the important improvements in cdma2000 systems is the increase in system performance during handoffs achieved by better handoff management, based on enhanced Radio Resource Management (RRM), avoiding the ping-pong effect, and by more efficient message exchanges (mini-messages) resulting in better pilot set management. The mini-messages allow faster channel assignment, improving RRM and fast system and MS responses to service negotiations. The additional handoff procedures implemented are based on the dynamic handoff process, which consists of a pilot trend analysis based on E_c/N_0 measurements. Figure 7.31 exemplifies some of these features.

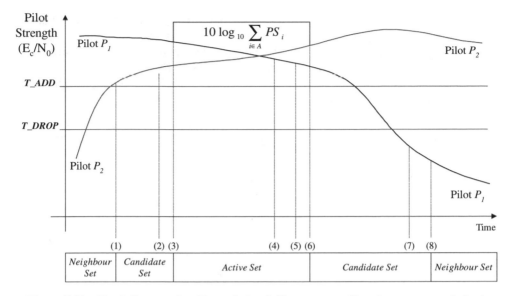

Figure 7.31   Handoff processing (dynamic handoff) – resource allocation process optimised.

The points highlighted in Figure 7.31 indicate message exchanges and procedures performed by the MS and BTS as described next.

1. The MS, using P_1, measures P_2 and verifies that it exceeds *T_ADD*. The MS then transfers P_2 to the candidate set.

2. Because P_2 exceeds $[(SOFT\_SLOPE/8) \times 10\log_{10}(PS_1) + ADD\_INTERCEPT/2]$, where PS_1 represents P_1's strength, the MS transmits a Pilot Strength Measurement Message (PSMM) to BTSs requesting handoff.

3. Upon receiving the PSMM and performing channel allocation procedures, the BSC or MSC controlling the BTSs sends a General Handoff Direction Message (GHDM) or an Extended Handoff Direction Message (EHDM) or a Universal Handoff Direction Message (UHDM) informing the MS to transfer P_2 to the active set and to transmit a Handoff Completion Message (HCM) establishing soft handoff.

4. The MS continues maintaining the pilot sets and verifies that P_1 tends to drop below $[(SOFT\_SLOPE/8) \times 10\log_{10}(PS_2) + DROP\_INTERCEPT/2]$. The MS then initiates a handoff drop timer to pilot P_1.

5. The timer assigned to P_1 expires and the MS transmits a PSMM to BTSs transferring P_1 to the candidate set.

6. The MS transfers P_1 to the candidate set when it receives a GHDM, UHDM or EHDM and transmits a handoff completion message to BTSs as a confirmation.

7. The MS verifies that P_1 drops below *T_DROP* and starts the handoff drop timer, *T_TDROP*.

8. The timer expires, i.e. exceeds *T_TDROP*, and the MS transfers P_1 to the neighbour set.

The main feature presented in Figure 7.31 is the pilot trend analysis based on continuous and frequent Pilot Strength (PS) measurements. MSs use searchers and enable a handoff drop timer for each pilot in the active and candidate sets. Active set pilots are arranged in ascending order in terms of signal strength, i.e. $PS_1 < PS_2 < PS_3 < \cdots < PS_{NA}$.

The example in Figure 7.31 can be generalised using the following expression, which determines the beginning of the handoff drop timer for a pilot P_i

$$10 \log_{10}(PS_i) < \max \left(\frac{SOFT\_SLOPE}{8} \log_{10} \left(\sum_{j>i} PS_j \right) + \frac{DROP\_INTERCEPT}{2}; -\frac{T\_DROP}{2} \right)$$

for $i = 1, 2, \ldots, PS_{NA-1}$ (pilots in the active set only).

Figure 7.32 shows an example of pilot set management related to the Pilot Strength Measurement Message (PSMM) transmission.

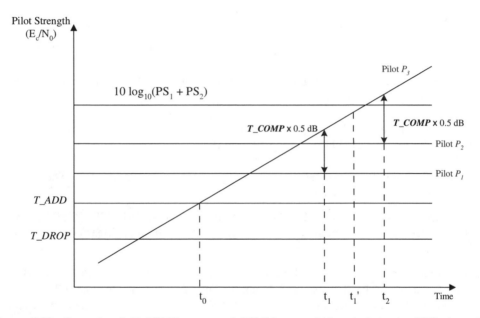

Figure 7.32 Example of *T_COMP* usage and PSMM transmission during cdma2000 dynamic handoff.

At t_0, P_3 reaches *T_ADD*, but the PSMM is not transmitted because eqn (7.1) is not satisfied

$$\{10 \log_{10}(PS_3) > [(SOFT\_SLOPE/8) \times 10 \log_{10}(PS_1 + PS_2) + ADD\_INTERCEPT/2]\} \tag{7.1}$$

At t_1, P_3 is above *T_ADD* and exceeds P_1 by *T_COMP* $+ 0.5$ dB. The PSMM, however, is still not transmitted because eqn (7.1) has not been satisfied.

The PSMM is transmitted for both t'_1 and t'_2. At t'_1, eqn (7.1) is satisfied, representing that P_3 tends to increase and to remain greater than *T_ADD*. At t_2, besides satisfying the expression, P3 exceeds P_2 by *T_COMP* $+ 0.5$ dB, becoming the strongest pilot in active set.

This example illustrates that there is a reduction in the transmission of messages reporting pilot strength, such as the PSMM. It also shows pilot strength trend evaluation, intended to avoid the ping-pong effect during channel assignment and pilot set maintenance.

7.6 HANDOFF TYPES

While in conversation (during a call), all signalling and control information are passed to an MS through the traffic channel. To allow mobility between cells, several handoff types are defined in the standards, such as the soft handoff, hard handoff and CDMA-to-analogue handoff.

The CDMA-to-CDMA soft handoff is specifically defined for CDMA systems, and applies to an MS that starts communication with a BTS (sector) while still connected to another. In this situation, the MS is using channels from both BTSs, i.e. communicating with two or more BTSs simultaneously. MSs are capable of communicating with as many BTSs as the number of finger elements in the terminal receiver, unless limited by the system. For example, an MS equipped with three fingers is able to perform soft handoff with up to three BTSs. The soft handoff process provides diversity for traffic channels on forward and reverse links, but may also reduce system capacity. Section 7.6.2 provides a detailed description of the soft handoff process.

In the CDMA-to-CDMA hard handoff, an MS disconnects from one BTS and switches to another. This happens when the MS transitions between disjoint BTS sets, such as BTSs using different frequency assignments, traffic channel frame offsets or band classes. Some CDMA networks are capable of bypassing some of these situations due to evolutions and improvements achieved by some manufacturers.

In the CDMA-to-analogue handoff, an MS must stop using the CDMA traffic channel and start using an analogue voice channel. To perform this handoff, the MS must go through the procedures defined in the initialisation task of the MS initialisation state, described in Chapter 6.

Besides these three main types of handoff, an MS can also perform handoff processes such as the softer handoff, idle handoff and access handoff. The last two types are not performed while in conversation.

The softer handoff is a specialisation of the soft handoff in which MSs execute handoff between sectors of the same BTS.

The idle handoff is performed while in idle state, when the MS is moving from one BTS to another and starts de-modulating the forward paging channel (FPCh) of the new BTS.

The access handoff happens in the system access state, when the MS is performing access attempts. The MS transmits access probes assigned to access channels of different BTSs and receives channel acknowledgements on the respective FPChs or forward common control channels (FCCChs).

7.6.1 Soft Handoff

The soft handoff process depends on specific conditions to occur. First, the candidate BTS must use the same CDMA frequency, frame offset and band as the current server. Both BTSs, candidate and current server, must be controlled by a single MSC.

While in soft handoff, the MS simultaneously demodulates traffic channels associated with pilots from different BTSs. A pilot is only removed from the active set if its signal strength drops below *T_DROP* for a period greater than *T_TDROP*, if the strength tends to continuously lower, or if there is a traffic overload in the BTS. In the latter case, the system may opt for releasing some of the traffic channels used for soft handoff.

BTSs involved in the soft handoff send traffic channel frames received to their associated BSC or MSC, which are responsible for choosing the best quality frames, on frame-by-frame basis.

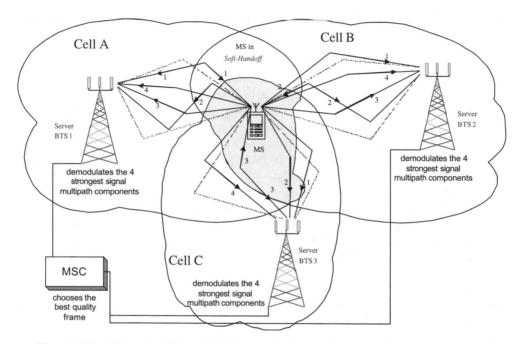

Figure 7.33   Either the MSC or BSC chooses the best quality frame from different BTSs.

Figure 7.33 illustrates the soft handoff process considering a BTS with four finger elements, each de-modulating multi-path signal component.

Soft handoffs usually occur at cell boundary regions, where the forward and reverse links are more fragile, i.e. received signal levels at MSs and BTSs have poorer quality because of propagation effects. The use of rake receivers may increase reception signal quality, providing forward and reverse link diversity gain. Soft handoffs allow a seamless transition between cells. Users do not experience the momentary interruption ('click') that results from 'hard' handoffs. An important characteristic of this type of handoff is that MSs can perform transmission power control.

7.6.1.1   Power Control While in Soft Handoff

One of the main objectives of power control is to reduce the interference caused by MSs on the reverse link. It also preserves battery life, because MSs transmit at the minimum power level necessary to maintain the link.

Since power control schemes adjust the transmission power of MSs throughout the coverage area, so that base stations receive their signals with similar levels. MSs located farther from the BTS must transmit with higher power levels, as illustrated in Figure 7.34.

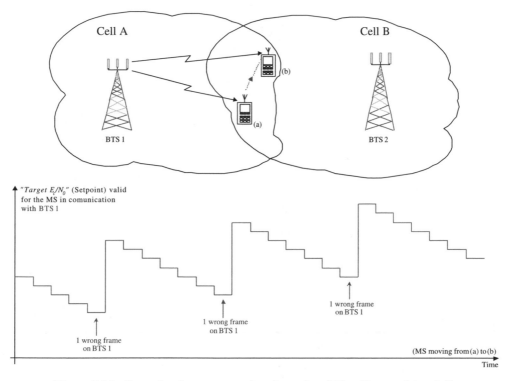

Figure 7.34   Example of power control performed on MSs without soft handoff.

Figure 7.34 shows an example of MS 'target E_c/N_0' definition during a call. The graph starts showing the value defined when the MS is located at position 'a'. The target E_c/N_0 tends to increase when the MS moves towards cell boundaries (position 'b'). To simplify the example, in Figure 7.34, the terrain is assumed to be flat and path propagation losses increase according to the distance between MS and BTS.

During soft handoff, BTSs perform an optimised power control scheme. All BTSs in communication with an MS perform independent power control on the traffic channel, transmitting power control bits and defining specific target E_c/N_0 values according to the corresponding path loss from the MS to the BTS.

The MS only increases its transmission power if all BTSs send power control sub-channels requesting the increase. Otherwise, the MS reduces the transmit power for the next Power Control Group (PCG), as depicted in Figure 7.35. In other words, the MS sets its transmission power to satisfy the BTS requesting the lower target E_c/N_0, as in Figure 7.36 (thick lines).

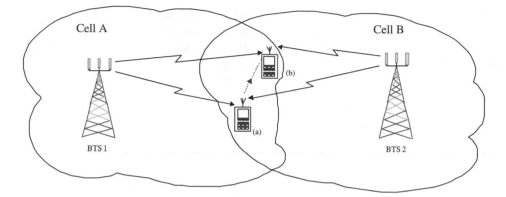

Figure 7.35   MS moving from point (a) to point (b) in a region covered by two BTSs.

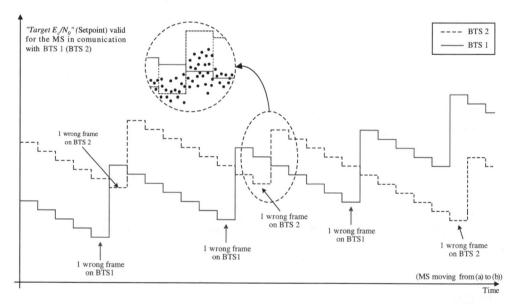

Figure 7.36   Power control during handoff when the MS is moving from point (a) to point (b).

7.6.1.2   Soft Handoff Effects on System Capacity

While in soft handoff, even though power control schemes cause positive effects, there is a negative impact on system capacity. Two or more traffic channels are simultaneously assigned for every MS performing soft handoff, one per BTS involved. This means that radio resources from two or more BTSs are used in one call at the same time, reducing system capacity.

Figure 7.37 presents as example with three BTSs with overlaps in their coverage areas, allowing soft handoff. The example considers uniform traffic distribution and a capacity of

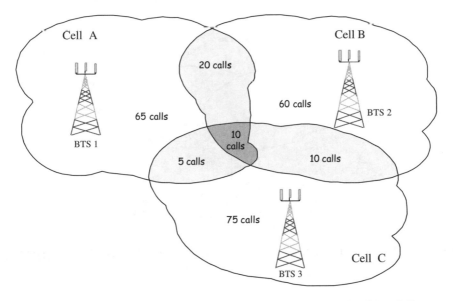

Figure 7.37 Example of system capacity reduction because of soft handoff.

100 simultaneous calls per BTS (100 traffic channels), leading to a maximum of 300 calls simultaneously.

Assuming that all calls are taking place simultaneously, in Figure 7.37 there are

- 20 mobile terminals in soft handoff between cells A and B;
- 5 mobile terminals in soft handoff between cells A and C;
- 10 mobile terminals in soft handoff between cells B and C;
- 10 mobile terminals in soft handoff between all cells, i.e. A, B and C;
- 200 mobile terminals performing a call in only one cell, i.e. A, B or C.

The total number of calls is 245 out of a total of 300 possible calls; therefore the system capacity efficiency factor in the example is defined as 245/300 = 81.67%. The reduction in system capacity due to soft handoff may be even greater in situations involving a higher number of overlapping cells and wider soft handoff areas.

Because system capacity is highly influenced by interference in CDMA networks, a better MS power control may reduce the effects of soft handoff on the system capacity efficiency factor.

7.6.2 Hard Handoff

During hard handoff, an MS may experience a brief interruption ('click') in the connection caused by tasks such as changing traffic channel frame offset timing or CDMA frequency assignment. The MS must execute hard handoff when one of the following conditions occur.

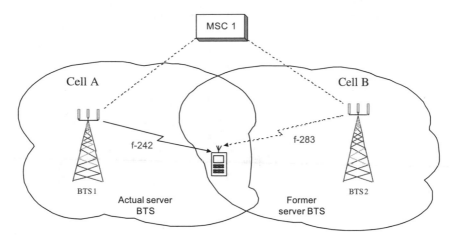

Figure 7.38   CDMA-to-CDMA *hard handoff due to unavailability of traffic channels using the same* *frequency.*

- There are no traffic channels available on the candidate BTS using the same CDMA frequency as the MS before requesting handoff (Figure 7.38). This process is also known as CDMA inter-frequency handoff.

- The candidate BTS and the current server belong to different BSCs or MSCs as in Figure 7.39.

- The candidate BTS and the current server transmit using different frame offsets (Figure 7.40). The *FRAME_OFFSET* parameter indicates the beginning of traffic channel frames in relation to the system timing reference. As an example, this parameter is given in units of 1.25 ms for forward fundamental traffic channels. Thus, for FTChs, the minimal and maximal delays caused by *FRAME_OFFSET* for 20 ms frames are 0 and 18.75 ms.

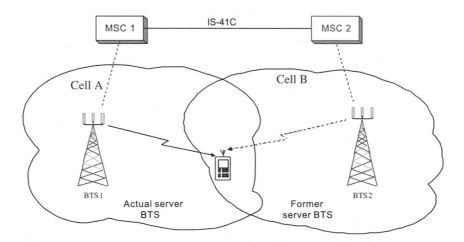

Figure 7.39   CDMA-to-CDMA *hard handoff because BTSs belong to distinct BSCs or MSCs.*

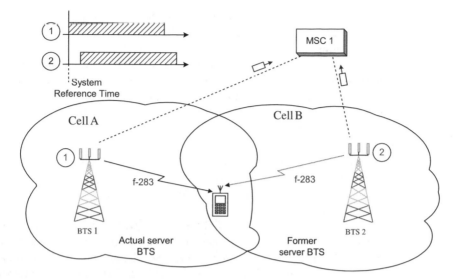

Figure 7.40 CDMA-to-CDMA hard handoff because of traffic channels using distinct frame offset parameters.

7.6.3 CDMA-to-Analogue Handoff

On the CDMA-to-analogue hard handoff, a dual-mode mobile terminal is directed to use an analogue voice channel instead of a digital CDMA traffic channel. In this situation, current traffic channels are disconnected when the MS tunes to analogue system frequencies.

This type of handoff occurs mostly on CDMA system boundaries, where BTSs usually use both analogue and CDMA digital systems, or in situations where the MS moves towards a single-mode, analogue BTS, as portrayed in Figure 7.41.

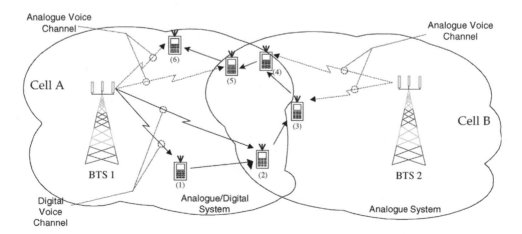

Figure 7.41 CDMA-to-AMPS hard handoff and AMPS-to-AMPS hard handoff executed by a dual-mode MS.

Once a CDMA-to-analogue hard handoff happens, the MS can not perform an in-call analogue-to-CDMA handoff, as this is not defined for CDMA systems (see Section 7.6.5). The dual-mode mobile station must continue and conclude the current call on the analogue system, even if it returns to a dual system coverage area, as in Figure 7.41. Upon completion of the call, an MS located in a CDMA-served area can re-acquire the CDMA system, as described in Chapter 6. While in communication on an analogue system, CDMA terminals lose all benefits of a CDMA wireless system, such as diversity due to use of two or more traffic channels and power control.

To perform a CDMA-to-analogue hard handoff, analogue BTSs must be equipped with hardware used to transmit a signalling CDMA pilot, known as pilot beacon. This equipment consists of a short PN sequence generator with a rate of 1.2288 Mcps, synchronised and time aligned to the CDMA system via a GPS (Global Positioning System) receiver.

The pilot beacon is used to inform MSs of CDMA system boundaries. Beacons are transmitted as a normal Forward Pilot Channel (FPiCh), but with specific characteristics, discussed in the forward link chapter. The '0s' transmitted by the pilot beacon are spread by Walsh code 0 (W_0^{64}) and PN-I and PN-Q sequences (1.2288 Mcps), and QPSK modulated prior to transmission through the air interface, as in Figure 7.42. The transmission happens on a CDMA carrier with the same frequency employed by MSs on the CDMA or CDMA/analogue BTS. This equipment does not transmit any other CDMA logical channel.

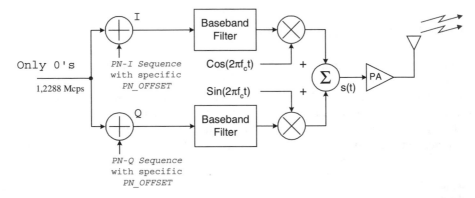

Figure 7.42   Signalling pilot, or pilot beacon, used to inform CDMA system boundaries to MSs.

Analogue BTSs transmitting the pilot beacon are considered as regular CDMA BTSs for pilot PN offset planning purposes (*PN_OFFSET*).

Because all handoff processes are MS initiated (MAHO – Mobile Assisted Handoff), the pilot beacon function is to deceive dual-mode terminals, operating in digital mode, to request handoff to single-mode analogue BTSs, as in Figure 7.43.

When MSs detect a pilot beacon with strength greater than *T_ADD*, they may verify its identity on the candidate set. MSs then request handoff by transmitting a Pilot Strength Measurement Message (PSMM), informing *PN_OFFSET* and measured E_c/N_0 strength, to the server BTS. Upon receiving the PSMM, the BTS informs the BSC or MSC about the handoff request.

The MSC checks if the BTS belongs to an analogue system and if there are analogue voice channels available. If there are available channels, the system assigns an analogue channel to the dual-mode MS allowing the current call to be continued.

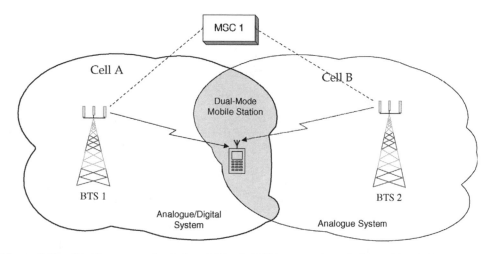

Figure 7.43   Pilot beacon used to inform MSs of CDMA system boundaries and handoff to analogue system.

Finally, the MSC transmits a General (GHDM), Universal (UHDM) or Extended Handoff Direction Message (EHDM) informing MSs to start operating in analogue mode.

7.6.4   Analogue-to-CDMA Hard Handoff Process

Because of equipment constraints, different carrier characteristics and CDMA synchronisation complexities, mobile stations are not capable of re-acquiring the CDMA system while communicating in analogue mode. Therefore, this type of handoff is not defined either on CDMA IS-95 or on cdma2000 standards.

7.6.5   Idle Handoff

Idle handoff is a handoff process performed while in MS idle state. In this state, MSs are constantly searching for pilots to maintain the pilot sets. The only pilots in the active set are those whose associated forward paging channels (FPChs) or forward common control channels (FCCChs) are being monitored.

If the mobile terminal moves from one BTS to another, the searcher element detects and measures the new pilot's strength, as in Figure 7.44.

When a pilot E_c/N_0 is greater than that of the current pilot, the MS tunes to the new BTS setup carrier (primary or secondary) to demodulate the information transmitted on the SyncCh, FPCh and FCCCh.

7.6.6   Access Handoff and Access Probe Handoff

Access handoff only occurs in the system access state, when MSs are waiting for a response from the BTS (*MS origination attempt sub-state*) or before the transmission of an origination message (*MS origination attempt sub-state*).

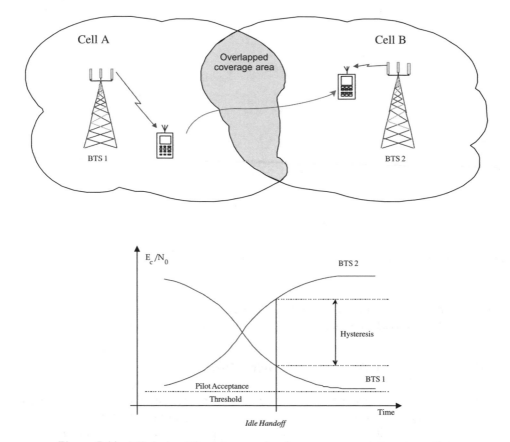

Figure 7.44   MS during idle state – moving from one coverage area to another.

As in the MS idle state, MSs are constantly searching for pilots to maintain the pilot sets. Only one pilot is in the active set in this situation, the pilot whose associated FPCh or FCCCh is being monitored.

An MS de-modulates the FPCh or FCCCh of the strongest pilot (higher E_c/N_0) while in mobile station idle state or when attempting to access the system (system access state).

For this type of handoff to happen, prior to the transmission of any access probe of an access attempt, the MS tunes to the strongest pilot and updates all system and access parameters transmitted on the respective FPCh or FCCCh overhead messages. The MS then continues transmitting probes (for responses or requests) on the reverse access channel (RACh, EACh or RCCCh) associated with the new monitored FPCh or FCCCh.

7.6.7   Softer Handoff

Softer handoff is not either on CDMA IS-95 or on cdma2000 standards. It is considered to be an AT&T proprietary concept, representing a specialisation of the soft handoff. MSs are considered to be in softer handoff when located in a region covered by two or more sectors of the same base station, each using a distinct pilot *PN_OFFSET*, as in Figure 7.45.

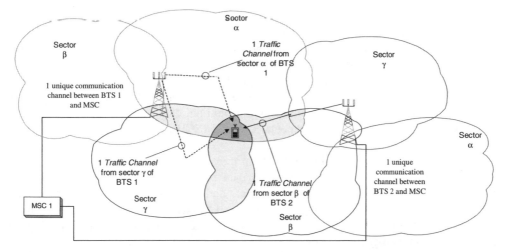

Figure 7.45 Mobile station in soft and softer handoff simultaneously.

For an MS, softer handoff 'feels like' a soft handoff, because it is de-modulating forward CDMA channels in the same frequency, each with an FPiCh using a specific *PN_OFFSET* identity, as if transmitted from different BTSs.

In Figure 7.45, BTS1 internally selects the best quality frame received from each of the sectors, transmitting only this frame to the MSC. The MSC selects the best frame received from BTS1 and BTS2. For the softer handoff, the BTS is responsible for the first frame selection prior to transmission to the MSC. The case illustrated in the figure is known as a soft-softer handoff, because it is the combination of soft and softer handoffs. The MSC treats this situation as a soft handoff involving three BTSs, i.e. assigns one traffic channel per BTS sector.

7.7 INTRODUCTION TO RADIO RESOURCE MANAGEMENT

Another concept directly related to system capacity and operational capabilities is Radio Resource Management (RRM). Efficient RRM is critical for the success of 3G wireless data in a multi-user environment, especially for cdma2000 standards, which define logical channels with transmission rates of up to 153.6 kbps for 1X and with even higher rates for EVDO and EVDV. The cdma2000 physical layer includes a number of major improvements intended to provide higher data rates and better spectral efficiency when compared to second-generation CDMA systems.

The new generations of wireless systems must consider two types of users (voice and data) and their corresponding traffic when analysing system capacity. But capacity by itself is not enough to provide a good system; Quality of Service (QoS) must also be considered and optimised.

cdma2000-1X voice capacity analysis is quite similar to the analysis of CDMA IS-95 systems, although it presents different key parameter values, such as different and not so hard E_b/N_0 requirements for both links. These new parameters give cdma2000 an Erlang/Hz capacity of up to twice the value of CDMA IS-95 systems, for example, up to 26 Erlangs per 1.23 MHz carrier for an RS1 vocoder (9.6 kbps). Improved E_b/N_0 requirements are obtained from air interface enhancements such as advanced convolutional coding, faster power control and reverse pilot channel transmission.

For data users, a burst-mode capability is defined to allow better interference management and capacity utilisation. For heavy data users, a Fundamental Channel (FCH) is always allocated, providing signalling capabilities and a minimal data rate of 9.6 kbps. An active packet data call, handled by a Supplemental Channel (SCH), can last for the duration of a data burst.

Data service can be asymmetric, i.e. different transmission rates may be assigned on the forward and reverse links. Hence, a dynamic data burst allocation infrastructure is used, aiming to maximise system resources management and the bandwidth use efficiency in both links, sharing system capacity among all users. The data rate and duration of each data burst is dynamically determined by the RRM, depending on the system load, level of interference and resource availability.

Data-burst allocation for the forward link is triggered when data gets back-logged on the network side of the system. On the reverse link, data is stored in mobile memory until it reaches a level when the MS requests an RSCh to the system, triggering the burst allocation procedure.

Transmission latency, resource availability and service coverage are the main aspects that contribute to a satisfactory end-user data experience. Very complex processing is expected within the RRM due to the competition between multiple user demands and to the self-regulating delay-sensitive nature of upper layer data protocols, such as TCP.

7.7.1 Logical Channels General Configuration Characteristics

To properly manage a cdma2000 network and to achieve good QoS and maximum capacity (according to available system resources), the following characteristics must be considered.

- *Spreading Rate (SR)* in use by the mobile terminal (main operation mode).

- *Logical channel* in use, which defines the type of traffic being transmitted by the mobile terminal (voice or data).

- *Radio Configuration (RC)* of the mobile terminal.

- *Vocoder Rate Set (RS)* 9.6 kbps for RS1 or 14.4 kbps for RS2.

- *Walsh code* selection, which determines the spreading factor to be used by the selected logical channel (Walsh length and index), also influencing data traffic rate and system reliability.

- *Encoder type* and rate (convolutional or turbo), which, in conjunction to the selection of a Walsh Code length, influences system reliability and security (coding) and implies in a specific spreading process.

Each of these characteristics affects system operation in a different way. Table 7.10 shows some of these aspects on the forward link of cdma2000 systems.

Table 7.10 cdma2000 forward link radio configurations (RCs) and spreading rates (SRs)

RC	Associated SR	Data rate (Kbps)	Forward error correction (FEC) (R)	Modulation
1	1	1.2; 2.4; 4.8; **9.6**	1/2	BPSK pre-spreading
2	1	1.8; 3.6; 7.2; **14.4**	1/2	BPSK pre-spreading
3	1	1.2; 1.35; 1.5; 2.4; 2.7; 4.8; **9.6**; 19.2; 38.4; 76.8; 153.6	1/4	QPSK pre-spreading
4	1	1.2; 1.35; 1.5; 2.4; 2.7; 4.8; **9.6**; 19.2; 38.4; 76.8; 153.6; 307.2	1/2	QPSK pre-spreading
5	1	1.8; 3.6; 7.2; **14.4**; 28.8; 57.6; 115.2; 230.4	1/4	QPSK pre-spreading
6	3	1.2; 1.35; 1.5; 2.4; 2.7; 4.8; **9.6**; 19.2; 38.4; 76.8; 153.6; 307.2	1/6	QPSK pre-spreading
7	3	1.2; 1.35; 1.5; 2.4; 2.7; 4.8; **9.6**; 19.2; 38.4; 76.8; 153.6; 307.2	1/2	QPSK pre-spreading
8	3	1.8; 3.6; 7.2; **14.4**; 28.8; 57.6; 115.2; 230.4; 460.8	1/4 (20 ms) 1/3 (5 ms)	QPSK pre-spreading
9	3	1.8; 3.6; 7.2; **14.4**; 28.8; 57.6; 115.2; 230.4; 259.2; 460.8; 518.4; 1036.8	1/2 (20 ms) 1/3 (5 ms)	QPSK pre-spreading
10	1	81.6; 158.4; 163.2; 312.0; 316.8; 326.4; 465.6; 619.2; 624.0; 633.6; 772.8; 931.2; 1238.4; 1248.0; 1545.6; 1862.4; 2476.8; 3091.2	1/5	QPSK, 8-PSK, 16-QAM

The Spreading Rate (SR) and Radio Configuration (RC) are strictly related to the terminal operational mode. The SR defines whether the MS is operating in CDMA IS-95, cdma2000-1X or 3X, whereas the RC defines a set of procedures, such as spreading and coding properties, according to the base Rate Set (RS) of the mobile station. The data rates shown in bold in Table 7.10 represent the basic RSs used on CDMA systems, 9.6 kbps for radio configurations 1, 3, 4, 6 and 7, and 14.4 kbps for RC2, RC5, RC8 and RC9.

The symbol transmission rate is always 1.2288 Msps on each carrier on the forward link through the air interface, regardless of the system type (IS-95, cdma2000) and of the modulation scheme (QPSK, for example).

Figure 7.46 shows a generic block diagram representing logical channel processing in cdma2000 systems, valid for both 1X and 3X operation modes. Both I and Q branches have the same rate of 1.2288 Msps prior to QPSK RF carrier modulation and transmission through the air interface, regardless of the information bit rate (R_{info}) of the logical channel in use. In the example, the basic information bit rates considered are the maximal vocoder rates, i.e. 8.6 kbps for RS1 and 13.3 kbps for RS2.

The top part of Figure 7.46 shows the basic processing blocks of any logical channel. Information bits (R_{info}) are structured in frames in a process that may include CRC (Cyclic Redundancy Check) and encoder tail bits, resulting in a total of R_b bits.

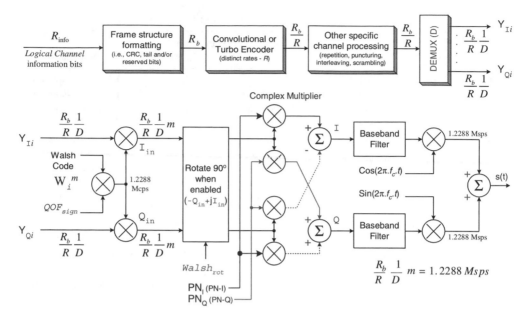

Figure 7.46   Basic channel-processing block diagram.

The bits then go through a coding process in the convolutional or turbo encoder (the latter only used on supplemental channels). The encoders may have coding rates (R) of 1/2, 1/3, 1/4, 1/5 or 1/6, which means 2, 3, 4, 5 or 6 coded bits at the output for each bit at the input.

The block showing 'other specific channel processing' represents processes such as bit repetition and puncturing, which are used to normalise the number of bits prior to spreading, obtaining the maximum allowed for each type of logical channel. Repetition is used when transmitting with rates lower than the maximum supported vocoder rates (R_{voc}), such as 1/2 (4.8 or 7.2 bps), 1/4 (2.4 or 3.6 kbps) and 1/8 (1.2 or 1.8 kbps). Interleaving and scrambling are also part of this block but do not modify the number of bits within frames.

The DEMUX (serial-to-parallel converter) divides the data stream into subsets (D) that are processed in I and Q according to the operational mode in use, i.e. two subsets for SR1 non-TD operation, four when using SR1 with TD (OTD or STS) and six for SR3 mode.

The previous steps explain the selection processes for Walsh code length (m) and encoder rate (R_{Enc}) and type. In the example in Figure 7.46, the information rate through a Forward Dedicated Control Channel (FDCCh) using SR1 in non-TD operation mode is $R_b = 9.6$ kbps.

The system uses a convolutional encoder with $R = 1/4$ for RC3 operation. After coding, there is a rate of $R_b/R = 9600/(1/4) = 38400$ bps. The DEMUX then splits this data stream into $D = 2$ sets, each with 19.2 kbps, for I and Q. To achieve 1.2288 Msps prior to QPSK carrier modulation, only a Walsh code with length $m = 64$ or less may be used.

If RC4 operation is selected, the system uses a convolutional encoder with rate $R = 1/2$, resulting in $R_b/R = 9600/(1/2) = 19200$ bps, which is half of the total when in RC3 mode. The same DEMUX process can achieve the rate of 1.2288 Msps per branch when using a Walsh Code of length $m = 128$.

Table 7.11 Rates during channel processing for distinct radio configuration in SR1 non-TD operation.

RC	Data rate R_b (bps)	Encoder Rate R	Rate after coding (bps) R_b/R	DEMUX bits per branch $D = 2$	Walsh Code length $(m \rightarrow W_i^m)$	Rate after spreading (bps) $[(R_b/R)/D]m$ (Msps)
3	9600	1/4	38400	19200	64	1.2288
4	9600	1/2	19200	9600	128	1.2288

Table 7.11 summarises these examples.

The encoder rate selection affects channel protection as a forward error correction (FEC) scheme (indirectly as a 'spreading-like' process). The greater the encoder rate (R), the stronger the channel protection, i.e. a $R = 1/4$ encoder that generates four coded bits per input bit has stronger encoding and decoding capabilities when compared to a $R = 1/2$ coder.

The Walsh Code length selection also directly influences system capacity. Figure 7.47 presents the Walsh code tree and the Walsh code generation procedure.

Even though Walsh codes are strictly used for multiplexing (channel identification) purposes on the forward link, they also contribute in a sense to the spreading factor. The greater the Walsh code length, the greater the number of Walsh sequences available for other logical channels in the system. Increasing the length of the Walsh codes (m) requires the use of encoders with smaller output rates. The use of Walsh codes with smaller lengths, however, enables more protection through the use of a more robust (higher rate) encoder.

It is a task of the RRM to make this decision and instruct the terminal of which type of encoder and Walsh code length to use. For example, in a communication link with high interference level, the RRM should instruct the terminal to use a more robust encoder and allocate a Walsh code with smaller length for the terminal, improving system reliability. An example of this situation is when MSs are located far from the server BTS or when there are too many active subscribers in the system.

If a good quality link is established, however, the RRM function may instruct the terminal to reduce coding protection and assign a Walsh code with greater length. This scenario usually happens to terminals located close to a BTS or when just a few users are active in the system.

In summary, the RRM function must be able to select the best combination of Walsh code length and encoder rate for each MS-BTS link in the system, according to link specifications and considering system capacity and performance needs.

7.7.2 Possible Walsh Code Allocation Conflicts

Another task of RRM is the allocation of particular Walsh codes. As described in Chapter 3, Figure 7.48 shows the Walsh codes formation pattern, where N represents the order of the Walsh matrix.

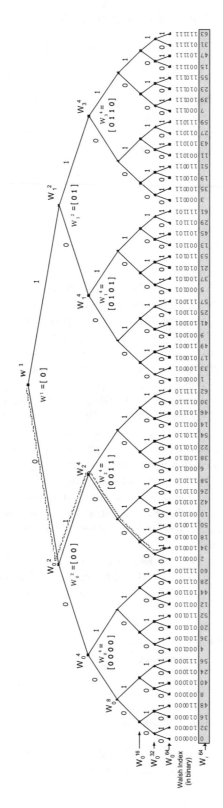

Figure 7.47 Walsh code tree (construction of the Walsh codes and inter-relations).

$$M_{2N} = \begin{bmatrix} M_N & M_N \\ \hline M_N & M_N \end{bmatrix}$$

Figure 7.48 Walsh matrix formation rule.

This process can be also be visualised in the Walsh tree (Figure 7.47), in which an initial Walsh code of length 1 (formed by a bit '0') is used to generate all other codes. Because of this generation pattern, Walsh sequences of order N are also said to be parents of the Walsh codes of length $2N$.

Some Walsh codes are uniquely assigned to specific CDMA logical channels, such as W_0^{64} for the Forward Pilot Channel (FPiCh), W_{32}^{64} for the Sync Channel (SyncCh) and W_1^{64} for the Forward Paging Channel (FPCh).

Table 7.12 shows an example of a possible Walsh code conflict. The example considers a Forward Supplemental Channel (FSCh) for RC3 in SR1 non-TD operation mode with an information bit rate of $R_b = 153.6$ kbps, employing an encoder with rate $R = 1/4$. After encoding, the DEMUX processes a total of 614.4 kbps, dividing the data stream into two sets of 307.2 kbps, for I and Q branches. According to these settings, the maximum Walsh code length for spreading and multiplexing is $m = 4$.

Table 7.12 Rates during FSCh channel processing for RC3 in SR1 non-TD operation

RC	Data rate R_b(bps)	Encoder Rate R	Rate after coding (bps) R_b/R	DEMUX bits per branch $D = 2$	Walsh Code length $(m \rightarrow W_i^m)$	Rate after spreading (bps) $[(R_b/R)/D]m$ (Msps)
3	153600	1/4	614400	307200	4	1.2288

Even though the Walsh code length has been determined, not all codes (W_i^4) may be assigned. Figure 7.49 shows the fixed Walsh codes assigned to the FPiCh (W_0^{64}), SyncCh (W_{32}^{64}) and FPCh (W_1^{64}). The RRM function must respect this assignment. The figure also shows that W_0^4 is used to generate the Walsh codes for the FPiCh, and SyncCh, whereas W_1^4 is used to generate the code for FPCh. Therefore, the RRM function can assign Walsh codes W_2^4 or W_3^4 for this FSCh, as long as these codes are not being used to generate the code used by any other logical channel that might be in use.

If the RRM function can not allocate any of these codes, it tries to assign a distinct Walsh code with length $m = 8$, requiring that all logical channel settings be re-configured. For example, the RRM may diminish the information data rate maintaining the same encoder rate or modify the encoder rate to transmit the maximum data rate by the associated FSCh.

7.7.3 RF Engineering for Data Users

Data transmission (data calls) over high-speed data packet channels on cdma2000 systems may be characterised as either always-on or on-demand traffic. Each of these modes of traffic has sets of configuration parameters that have been chosen to improve data transmission rates and system capacity and performance, optimising resource usage.

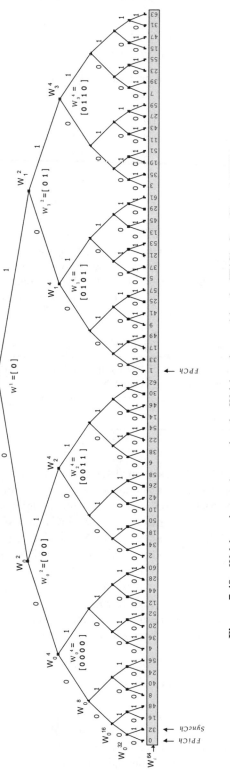

Figure 7.49 Walsh code tree (presenting the Walsh codes used by the FPiCh, SyncCh and FPCh).

The operation in the always-on mode relies on the fact that users do not have to issue any command or configuration to transmit or receive data. For example, a user who wants to browse the Internet only needs to communicate the URL of the web page he wants to view. All connection procedures are transparently performed.

The operation in the on-demand mode is based on the fact that a high-speed data logical channel must be assigned to each subscriber when the user requests a high-speed packet data channel.

The RRM performed for these processes must consider the amount of user data stored in memory, the transmission history of the data session for the user, the data rate available for transmission, the service class provided for the user, the current interference background level in the system and the mobile's RF conditions.

Data rates and assignment intervals (scheduling) of the high-speed packet data channel must be defined on a frame-by-frame basis, for each data burst, between the MS and its server BTS.

7.7.3.1   Always On

To accomplish the always-on mode in a cdma2000 system, MSs must keep a constant low-rate data connection with the BTS (using a fundamental channel) to maintain their data calls. The associated fundamental channel for each MS is used to provide infrequent signalling frames, and occasionally aids in data transmission.

A fundamental channel (FCh) is assigned to each user intending to perform a data call, on both the forward and reverse links, before the high-rate connection starts. User data transmissions are queued up for service until one of the high-speed data packet channels is available. Because traffic has a bursty nature, the time sharing of resources is not readily apparent to the end user.

If no high-speed data packet channel is available at a certain time, for example, due to lack of resources, the data transmission can also be performed using the associated FCh. Therefore, when available transmission rates are too low, FChs are configured to reduce their rates according to the data source activity, achieving rates similar to $R_{voc} = 1/8$. This procedure helps to avoid interference to other users in the system, as illustrated in Figure 7.50.

If there is no data traffic for transmission for a certain period, the user is considered to be in a 'dormant' state, however the PPP connection settings are maintained throughout the state. The user goes to the active state as soon as another data has to be transmitted or is received.

7.7.3.2   On-Demand Transmission

The on-demand mode implemented by the RRM function consists of the assignment of data logical channels to specific users during a certain time. Transmission rates are set according to users' needs aiming to maximise system data throughput, as in Figure 7.50.

The main idea is to time share system resources among active users throughout the air interface, employing packet-switched instead of circuit-switched connections.

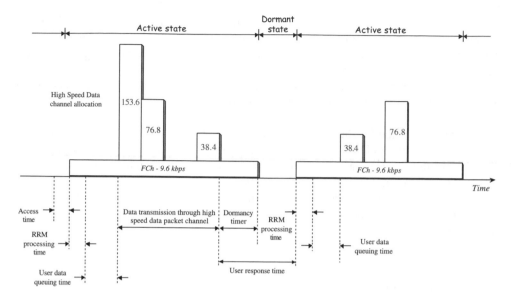

Figure 7.50 RRM allocation periods for high-speed data packet channels.

The performance criteria used by the RRM to optimise the system includes data throughput, average wait time, probability of transmission delay and average length of the user data queue. User traffic statistics also directly affect system performance. These statistics are a function of the data applications employed (for example service class type such as E-mail, web browsing, etc.) and of user behaviour (for example time used to read and/or to act between transmissions).

The main factors considered by the RRM for the assignment of the high-speed packet data channels on the forward link include the following.

- The fraction of the total BTS amplifier power required to transmit the data channel. This fraction is determined by user environment conditions (fast and slow fading effects), propagation losses between MS and server BTS, interference level experienced by the MS, expected Frame Error Rate (FER) for the possible rates and turbo coding support for the assigned high-speed packet data channel.

- The fraction of the total amplifier power required to transmit other voice and data channels with the minimal required QoS.

- The number of resources allocated for users in handoff, for voice and data calls.

- The amount of data stored in memory to be transmitted for associated MSs.

RRM scheduling procedures must prevent one or more users from monopolising system resources for a large period of time. The RRM may decide to maintain a given channel rate and assignment duration if there are enough system resources available and if there is still data in the transmit memory. However, if there is lack of resources, the channel assignment is terminated.

Rate and channel assignment are made on a frame-by-frame basis. When choosing whether to maintain a given configuration, during the channel continuation analysis the RRM may also decide to increase (if the maximum has not yet been achieved) or reduce the transmission rate.

Most of these considerations must also be analysed for the reverse link. The main factors considered by the RRM for high-speed packet data channel assignments by MSs include the following.

- Maximum power available for high-speed data packet channel transmission, due to terminal battery constraints.

- System load produced by the transmission of assigned channels. The transmission includes consideration of user environment conditions (fast and slow fading effects), propagation losses between MS and BTS, expected Frame Error Rate (FER) and turbo coding support for the assigned high-speed packet data channel.

- The fraction of the total MS amplifier power required to transmit other voice and data channels with the minimal QoS required.

- Terminal resources (channel elements and processing capabilities) allocated to all handoff procedures during a call.

- The amount of data to be transmitted stored in the MS memory.

The RRM tasks can be summarised as a set of call admission procedures, load balancing analysis and channel and transmission rate assignments intended to optimise system's resource utilisation, capacity and performance [1–10].

BIBLIOGRAPHY AND REFERENCES

1. C.S0002-C, 'Physical Layer Standard for cdma2000 Spread Spectrum Systems,' Release C, 3GPP2, May, 2002.
2. C.S0005-B, 'Upper Layer (Layer 3) Signaling Standard for cdma2000 Spread Spectrum Systems,' Release B, 3GPP2, April, 2002.
3. TIA/EIA-95-B, 'Mobile Station – Base Station Compatibility Standard for Dual-Mode Spread Spectrum Cellular System,' November, 1998.
4. CelTec/CelPlan, 'CDMA IS-95 and cdma2000 Systems – Training Course'.
5. 'Common Cryptography Algorithms,' Revision C, 1997, An EAR-controlled document subjected to restrict distribution. Contact Telecommunications Industry Associations (TIA), Arlington, VA.
6. 'Interface Specifications for Common Cryptography Algorithms,' Revision C, 1997. Contact Telecommunications Industry Associations (TIA), Arlington, VA.
7. 'Enhanced Cryptography Algorithms,' TR45.AHAG 2001.
8. EIA/TIA/IS-54-B, 'Cellular System Dual-Mode Mobile Station – Base Station Compatibility Standard,' April, 1992.
9. EIA/TIA/IS-91, 'Mobile Station – Base Station Compatibility Standard for 800 MHZ Analogue Cellular,' October, 1994.
10. TSB50, 'User Interface Authentication Key Entry,' March, 1993.

8

EVDO and EVDV

ARLINDO MOREIRA FARTES FILHO and
CRISTINE KOROWAJCZUK

This chapter describes the technologies considered to be the most realistic evolution path for cdma2000-1X systems, EVDO and EVDV. EVDO is specified by IS-856 standards (High Rate Packet Data – HRPD) and EVDV by R.S.0026. This chapter presents and analyses the main requirements and characteristics from these standards.

8.1   EVDO REQUIREMENTS (IS-856)

Although the evolution path of 3G cdma2000-1X systems is projected to be cdma2000-3X systems, the most realistic path for non-real-time services is the data-optimised system known as EVDO as mentioned in Chapter 2.

EVDO's main objective is to implement a reliable and spectrally efficient packet wireless system compatible to current Internet usage, that is, a system that provides high data transfer capacity (up to 2.4 Mbps) to current services, such as browsers, E-mail, and downloads-uploads (HTTP or FTP). Besides high throughput, EVDO also offers improved security resources, low energy consumption and low cost. The technology not only improves current services but also anticipates future systems by converging to an ALL IP configuration, therefore enabling mobility among packet data IP systems (e.g. Wi-FI, cdma2000 and UMTS).

EVDO is optimised for data transfer, offering higher transmission capacity in the forward link through a single channel contention access, which is a pipeline that transports packets to all active connected users in a time slotted basis, identifying each user by a CDMA cover (the user information is spread with the active user identification). The reverse link is totally based on CDMA to perform optimised asymmetric transactions (rates are lower in the reverse link) instead of a single pipeline, therefore allowing power control and interference minimisation.

IS-856 High Rate Packet Data (HRPD) standards specify the main characteristics of EVDO systems, establishing the requirements of the technology. Other standards, such as IS-864 and IS-866, specify performance requirements for the network and mobile terminals.

Designing CDMA 2000 Systems   L. Korowajczuk et al.
© 2004 John Wiley & Sons, Ltd   ISBN: 0-470-85399-9

The requirements presented in IS-856 include general, data rate, radio environment, interoperability support and authentication support requirements, which are all provided next.

8.1.1 General Requirements

The general requirements establish that EVDO systems must enable reliable and spectrally efficient packet data services because of the user demand for economical, high-speed, wireless Internet services. The systems must also support a large practical number of users simultaneously running HRPD applications.

Because it is mainly a data-oriented technology, it must enable data-optimised operation for non-real-time high-speed packet data services, including asymmetric data rate services. The standards also establish that the technology should allow use of multiple and concurrent packet sessions on the same network.

EVDO must support deployment in all frequency bands where cellular and PCS systems are deployed, providing an always-on capability (e.g. no need to dial-in and no user action required, for PDSN log-on process).

Use of traffic balancing mechanisms for all EVDO frequency channels. Traffic balancing means an equal and uniform traffic distribution not only within the several carriers transmitted by a sector, but also among the sectors, because the system instructs terminals to use servers with the best quality/congestion ratios.

8.1.2 Data Rate Requirements

As per data requirements, EVDO must enable a range of information data rates suitable for HRPD applications as specified in the following tables.

According to 3GPP2 S.R0023 standards (high speed data enhancements for cdma2000-1X data only requirements), Table 8.1 presents the initial data rate requirements for EVDO. These requirements evolved into 3GPP2 C.S0024 (TIA IS-856) cdma2000 high rate packet data air-interface specification standards, which presented a set of maximum data rates provided by the technology (Table 8.2).

8.1.3 Radio Environment Requirements

In terms of radio environment requirements, EVDO should provide a graceful evolution from IS-95A/B or cdma2000-1X to minimise the impact on terminals and on infrastructure. The idea is to achieve an economical evolution. Operators should use existing cell/sector

Table 8.1 EVDO data rate minimum requirements

	Forward link (to user)	Reverse link (from user) (Kbps)
Vehicular peak data rate	1.25 Mbps	144
Vehicular average data rate	600 Kbps	144
Pedestrian/fixed peak data rate	2 Mbps	144

Table 8.2 EVDO (IIRPD) data rates

Forward link	Reverse link (Kbps)
38.4 Kbps	9.6
76.8 Kbps	19.2
153.6 Kbps	38.4
307.2 Kbps	76.8
614.4 Kbps	153.6
1.2288 Mbps	
1.8432 Mbps	
2.4572 Mbps	

configurations, including the use of existing antenna types, without performing cell splitting. The purpose is to deploy EVDO on existing BTS equipment, which can operate several RF channels, including a mix of IS-95, cdma2000-1X and EVDO.

Even though EVDO should operate on a separate data-optimised channel, the coverage area for the system should closely align to the area covered by IS-95 and cdma2000-1X networks. EVDO out-of-band emissions must conform to cdma2000-1X requirements.

EVDO networks must support both mobile and fixed users, providing bandwidth efficiency by dynamically allocating resources such as power, code space and time slots, and by adapting to the time-varying nature of shadowing and fast fading processes.

The network should make full use of the capability of maximising spectral efficiency through Adaptive Modulation and Coding (AMC) techniques, by selecting the Modulation and Coding Scheme (MCS) that best matches the channel environment for the required data rate. These techniques include a collection of methods referred to as link adaptation, which consist of fast feedback channel state information, adaptive modulation, incremental redundancy, repetition coding, time diversity adaptation, hybrid ARQ, selection diversity and multi-user diversity.

Link adaptation consists of determining which is the most efficient transmission condition for a link. This includes

(1) fast detection of the channel transmission conditions (in terms of server signal level, interference and traffic capacity);

(2) adaptive modulation;

(3) scheduler strategy, where the type of demand, number of users and time of usage determine which users are served first.

The adaptive modulation chooses the best modulation type (QPSK, 8PSK or 16QAM), according to RF channel conditions, to achieve the best bandwidth efficiency. This technique is also responsible for determining the size of the data block to be transmitted according to the associated robustness in terms of codification type, amount of data repetition and data spreading over time.

The reception side of a link may detect errors in the received data and request re-transmission Automatic Repeat and reQuest (ARQ). The hybrid ARQ process consists of adding codes to the retransmitted information, so that the data is redundant to what was previously sent but it is not an exact copy. This strategy reduces the number of re-transmissions necessary.

Incremental redundancy, also known as type II hybrid ARQ, is another technique used to correct transmission errors. In this case, the information is transmitted with parity bits. If errors occur, new parity bits are transmitted and computed with the previously received information, producing a stronger code at each re-transmission. The family of codes is generated by puncturing the mother code.

The code repetition consists of the repetition of information, redundancy, to improve recovery capacity during short periods of interference.

Time diversity is a transmission method in which signals representing the same information are sent over the same channel at different times. Systems subject to burst error conditions commonly apply this method at intervals adjusted to be longer than an error burst.

Site Selection Diversity Transmission (SSDT) is a macro-diversity method used in soft handoff mode. Activated by the network, the main objective of the method is to use the best serving cell for downlink transmission, thus reducing interference caused by multiple transmissions in soft handoff mode.

Multi-user diversity is a technique where the overall system throughput is improved by scheduling a user with strong and small interference and noise power.

8.1.4   Interoperability Support Requirements

Interoperability support requirements represent the characteristics that EVDO networks must have to be able to work with existing networks, such as cdma2000-1X and IS-95. These requirements include the need of having an air interface operating with a radio access network (RAN) designed according to the interoperability specification [A.S0001].

EVDO networks must also have the capability of inter-operating (including handoff) with cdma2000-1X channels for packet services, and with GSM-MAP systems, which allow the interaction with GSM systems at data transport network level. This capability includes having an embedded and transparent way to redirect user invoked voice/real-time services to the cdma2000-1X network.

8.1.5   Authentication Support Requirements

The access to network (or service) facilities and resources are restricted by authentication keys. EVDO RANs shall enable authenticated access terminal (AT) identifiers for hybrid EVDO/IS-2000 devices. This identifier may be the same as the mobile station identifier authenticated by IS-2000 for the same hybrid device (e.g. IMSI). This allows IS-2000 and EVDO RANs to coordinate the operation of hybrid devices.

Until the access terminal identity has been authenticated, the network may prevent any end-user IP data traffic from being exchanged and may deny access to any dedicated RAN resources (i.e. resources supporting user data transfer to or from the PDSN, which do not include power and rate control).

The network should provide means to minimise the time and total transmission power required to authenticate hybrid devices to consequently reduce the interference and resource consumption that invalid devices can impose on the RAN during authentication.

8.2   EVDV REQUIREMENTS (R.S0026)

Even though EVDO is a plausible 3G path for data services, the technology does not provide integrated voice capacity. Therefore another technology, EVDV, was introduced. EVDV can reach even higher data transmission rates, supporting both voice and data.

EVDV networks are specified by the R.S0026 (3GPP2), which, besides establishing their general requirements, also defines their main requirements in terms of data rate, radio environment, compatibility and interoperability support.

8.2.1   General Requirements

EVDV systems shall be an extension and incorporate all aspects of existing cdma2000-1X features, functions, applications and services, that is, EVDV standards maintain all voice and packet data capabilities of cdma2000-1X and EVDO specifications. The voice quality must meet or exceed that of cdma2000 systems. EVDV systems must supersede EVDO data services by providing real-time capability, delivering traffic in three distinct modes: real-time, non-real-time and mixed, that is real- and non-real-time on the same carrier. EVDV networks must support voice over IP (VoIP).

The technology must support deployment in all frequency bands where cellular and PCS systems are deployed, providing an always-on capability. EVDV networks must be able to deal with asymmetric and symmetric data rate services and use traffic balancing mechanisms for all frequency channels. The system must support existing cdma2000 vocoders as well as GSM full-rate and AMR vocoders.

The EVDV systems' goal is to double voice service spectral efficiency of cdma2000-1X systems, while also supporting multiple and concurrent packet sessions. Whereas IS95 uses fixed handoff thresholds (Tadd and Tdrop), IS2000 and EVDO networks use adaptive handoff thresholds with varying Tadd/Tdrop according to the current and candidate server signal strengths. EVDO networks must also allow handoff of voice and data services between an EVDV radio channel and another radio channel operating in accordance with IS-2000 standards. This requirement applies every time the two radio channels support an equivalent data or voice service, even if there are differences in QoS or data rate. A mobile station that supports EVDV does not necessarily have to support EVDO; however, if this holds, then the mobile station shall support handoff of commonly supported data services between EVDO and EVDV radio channels.

This technology must perform quality of service (QoS) control to achieve higher voice quality standards when in the highest quality mode. QoS parameters include traffic priority, minimum acceptable data rate, maximum permitted data loss rate (FER/BER) and maximum allowable latency or delay. The network shall be able to adapt data transmission speeds and other operating parameters to maximise system capacity, while satisfying QoS constraints, for a given mix of RF conditions and system workload.

Table 8.3   EVDV data rate minimum requirements

	Forward link (to user)	Reverse link (from user)
Vehicular peak data rate	2.4 Mbps	1.25 Mbps
Vehicular average data rate	600 Kbps	600 Kbps
Pedestrian/fixed peak data rate	2.4 Mbps	2.4 Mbps

The standards establish that, when compared to IS2000 systems, EVDO must reduce power consumption to one-half, extending active transmission time and increasing standby time by 100%.

The system must support different traffic types with varying QoS constraints including voice, video and data combined on a single radio channel or segregated onto separate radio channels.

8.2.2   Data Rate Requirements

Regardless of the system load, systems supporting EVDV, operating in a non-constrained QoS packet service only environment, must provide peak and system-wide average data rates at least compatible to EVDO specifications. Table 8.3 shows the minimum data rate requirements for EVDV networks whereas Tables 8.4 and 8.5 present, respectively, the data

Table 8.4   EVDV data rates proposed in 1Xtreme

Forward link (Kbps)	Reverse link (Kbps)
$76.8 \times 1 \times N$	9.6×2^P
$115.4 \times 1 \times N$	460.8
$76.8 \times 2 \times N$	$P = 0, \ldots, 6$
$115.4 \times 2 \times N$	
$76.8 \times 4 \times N$	
$115.4 \times 4 \times N$	
$N = 1, \ldots, 14$	

Table 8.5   EVDV data rates in the IS2000 standard

Forward link (Kbps)	Reverse link (Kbps)
81.6×2^P	9.6
158.4×2^P	19.2
312.0×2^P	38.4
465.6×2^P	76.8
619.2×2^P	153.6
772.8×2^P	307.2
$P = 0, 1, 2$	614.2

rates associated with 1Xtreme (Motorola) and RC10 (IS2000) implementations of this technology.

8.2.3 Radio Environment Requirements

EVDV must provide a graceful evolution from IS-95A/B, cdma2000-1X or EVDO minimising the impact on terminals and infrastructure. Because EVDV systems are an evolution of EVDO, they have to comply with the same radio environment requirements proposed for the latter (see Section 8.1.3).

As in EVDO networks, EVDV must also enable Adaptive Modulation and Coding (AMC) techniques. This change implies the migration from a unique modulation technique (QPSK), as in cdma2000-1X, to multi-modulation (QPSK, 8PSK, 16QAM).

8.2.4 Compatibility Requirements

A radio channel that supports EVDV can provide service to EVDO mobile stations or other MSs that conform to cdma2000 standards (C.S0001 through C.S0006), to its precursors (e.g. TIA/EIA-95-B) and to ancillary standards. This is possible because EVDV standards define a backward compatible class of mobile devices, which supports all these standards and technologies. When the specifications of the mobile and the base station differ, services are provided in the quality level associated with the limiting element. EVDV must be compatible with the IS-2000 chip rate and band plan, also supporting existing frame lengths. The latter requirement, however, does not preclude the addition of new frame lengths.

8.2.5 Interoperability Support Requirements

An open RAN, designed according to interoperability specification [A.S0001], supports EVDV systems. The network must provide interoperability (including handoff) with cdma2000-1X channels, for packet services, and with GSM-MAP systems.

8.3 EVDO REFERENCE MODEL

8.3.1 Architecture Reference Model

The EVDO architecture reference model represents the system as a connection of the access terminal to the access network through the air interface. The Access Terminal (AT) is a physical entity that provides data connectivity to users through an access network. It can be associated with another device, for example a computer interface card, or be self-contained, such as personal digital assistants (PDAs). ATs are equivalent to Mobile Stations (MSs) in cdma2000 standards.

The Access Network (AN) is a physical entity that provides data connectivity between a Packet-Switched Data Network (PDSN) and the AT. The AN corresponds to base stations

(BSs) in cdma2000 standards and are composed of sectors. A sector is the part of the network that provides one CDMA channel.

Figure 8.1 represents the relationship among EVDO reference model elements.

The last part of the reference model is the air interface, a logical entity where the information between AT and AN flows. This flow of information is known as a link. Information from ANs to ATs is sent on the forward link; conversely, from ATs to ANs, it is sent on the reverse link. Each link has multiple channels, configured in terms of structure, frequency, modulation, encoding and power output level.

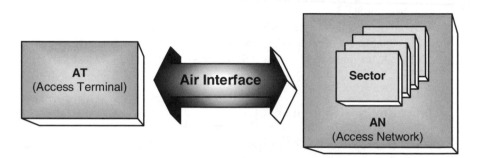

Figure 8.1   EVDO architecture reference model.

The air interface is organised in seven layers each of them with specific tasks. This allows modifications to be made only in a single layer without affecting the whole air interface structure. The seven layers are the following.

- Application layer – provides the default signalling and default packet applications that transport air interface protocol messages and user data, respectively.

- Stream layer – multiplexes distinct application streams.

- Session layer – provides address management, maintenance services and protocol negotiation and configuration.

- Connection layer – establishes and maintains air link connection.

- Security layer – provides authentication and encryption services.

- Medium access control (MAC) layer – defines procedures used to transmit and receive over the physical layer.

- Physical layer – specifies channel structure, frequency, power output, modulation and encoding for forward and reverse channels.

8.3.2   Protocols

The different layers can be transmitted over the air interface and perform all services required to allocate, manage and control system resources (hard and soft). The functions of a layer may be executed using one or multiple protocols. However, even though a single layer may contain multiple protocols, each of them can be individually negotiated to better accommodate network requirements and availability.

Figure 8.2 shows a block diagram illustrating the air interface organisation layers and the protocols associated with each of them in EVDO networks.

EVDO protocols shown in Figure 8.2 are briefly described next. The next sections of this chapter describe each of the layers in detail.

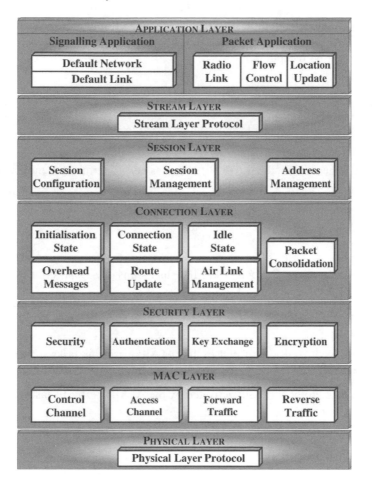

Figure 8.2 EVDO layered protocol structure.

Application Layer

- Signalling application: allows protocols from two distinct entities to exchange messages through the air interface. It consists of two protocols: Signalling Network Protocol (SNP) and Signalling Link Protocol (SLP). The SNP provides signalling message transmission services and the SLP provides mechanisms to improve message delivery.

- Default packet application: allows the transmission of packets from the access terminal to the access network. It consists of three protocols: Radio Link Protocol (RLP), Location Update Protocol (LUP) and Flow Control Protocol (FCP). The RLP retransmits streams and detects duplicates, the LUP manages mobility and the FCP manages data flow.

Stream Layer

- Stream layer protocol: multiplexes up to four parallel application streams transmitted over the air interface by adding a stream header (number of the stream) for transmission and removing it on reception.

Session Layer

- Session Configuration Protocol (SCP): allows negotiation of protocols and configuration parameters used in a session.
- Address Management Protocol (AMP): manages access terminal identifiers (ATIs).
- Session Management Protocol (SMP): controls the activation and parameter configuration of other protocols in the session layer.

Connection Layer

- Air Link Management Protocol (ALMP): maintains the connection state in access terminals and networks.
- Initialisation State Protocol (ISP): activated by the ALMP, helps with access network acquisition.
- Idle State Protocol (IDP): activated by the ALMP, performs tasks associated with idle access terminals.
- Connected State Protocol (CSP): activated by the ALMP, performs tasks associated with active access terminals.
- Route Update Protocol (RUP): keeps track of access terminals location.
- Packet Consolidation Protocol (PCP): broadcasts essential parameters over the control channel.
- Overhead Messages Protocol (OMP): consolidates and prioritises packets for transmission.

Security Layer

- Security Protocol (SP): divided into default and generic security protocols. The default protocol transfers packets between the authentication protocol and the MAC layer, whereas the generic protocol provides and computes a cryptosync.
- Key Exchange Protocol (KEP): exchanges security keys for encrypting and authentication applied to AT/AN.

- Authentication Protocol (AP): authenticates AT/AN traffic.

- Encryption Protocol (EP): encrypts AT/AN traffic.

MAC Layer

- Control Channel MAC Protocol (CCMP): builds control channel MAC layer packets out of one or more security layer packets, adding the AT address to transmitted packets. It contains rules for AN transmission and packet scheduling on the control channel, AT acquisition of the control channel, and AT control channel MAC layer packet reception.

- Access Channel MAC Protocol (ACMP): contains the rules for AT transmission timing for the access channel and AT transmission power characteristics for the access channel.

- Forward Traffic Channel MAC Protocol (FTCMP): contains the rules for operation of the forward traffic channel, determining whether the channel is using variable or fixed rate.

- Reverse Traffic Channel MAC Protocol (RTCMP): contains the rules for operation of the reverse traffic channel.

Physical Layer

- The physical layer does not use protocols as it refers to the physical transmission of data. This layer handles forward and reverse links and determines the channel structure, frequency relations, modulation and encoding types and output power control and maintenance.

Each protocol or application (application layer protocol) has a type and a sub-type identifier for an inside layer specific function. Table 8.6 lists type and subtype values for the different protocols in EVDO networks.

8.3.2.1 Instances

Protocols can have multiple independent instantiations, each instance being classified as InConfiguration or InUse. The InConfiguration instance is associated with the binary value '1'. This instance is related to configuration procedures. It starts at the beginning of operations and also during operations when reconfigurations are needed. The InUse instance is associated with the binary value '0'. When not in configuration, the operational state is always associated with this instance.

8.3.2.2 Interfaces

There is communication between different protocols in the same physical entity (AT or AN) or between physical entities through the same protocol. This communication is performed through one of the interfaces defined in the standard: headers and messages, commands, indications or public data. Figure 8.3 illustrates the different interface elements.

Table 8.6 Type and sub-type identifiers

Type	Sub-type	Sub-type Prefix	Element name	Element type	Type constant	Layer
0x00	0	Default	Physical layer	Protocol	N_{PHYtype}	Physical
0x01	0	Default	Control channel MAC	Protocol	N_{CCMPtype}	MAC
0x02	0	Default	Access channel MAC	Protocol	N_{ACMPtype}	MAC
0x03	0	Default	Forward traffic channel MAC	Protocol	$N_{\text{FTCMPtype}}$	MAC
0x04	0	Default	Reverse traffic channel MAC	Protocol	$N_{\text{RTCMPtype}}$	MAC
0x05	0	Default	Key exchange	Protocol	N_{KEPtype}	Security
0x05	1	DH	Key exchange	Protocol	N_{KEPtype}	Security
0x06	0	Default	Authentication	Protocol	N_{APtype}	Security
0x06	1	SHA-1	Authentication	Protocol	N_{APtype}	Security
0x07	0	Default	Encryption	Protocol	N_{EPtype}	Security
0x08	0	Default	Security	Protocol	N_{SPtype}	Security
0x08	1	Generic	Security	Protocol	N_{SPtype}	Security
0x09	0	Default	Packet consolidation	Protocol	N_{PCPtype}	Connection
0x0a	0	Default	Air link management	Protocol	N_{ALMPtype}	Connection
0x0b	0	Default	Initialisation state	Protocol	N_{ISPtype}	Connection
0x0c	0	Default	Idle state	Protocol	N_{IDPtype}	Connection
0x0d	0	Default	Connected state	Protocol	N_{CSPtype}	Connection
0x0e	0	Default	Route update	Protocol	N_{RUPtype}	Connection
0x0f	0	Default	Overhead messages	Protocol	N_{OMPtype}	Connection
0x10	0	Default	Session management	Protocol	N_{SMPtype}	Session
0x11	0	Default	Address management	Protocol	N_{ADMPtype}	Session
0x12	0	Default	Session configuration	Protocol	N_{SCPtype}	Session
0x13	0	Default	Stream	Protocol	N_{STREAM}	Stream
0x14	1	Default	Signalling (Stream 0)	Application	N_{APP0type}	Application
0x15	1	AN default	Packet (Stream 1)	Application	N_{APP1type}	Application
0x15	2	Service default	Packet (Stream 1)	Application	N_{APP1type}	Application
0x16	1	AN default	Packet (Stream 2)	Application	N_{APP2type}	Application
0x16	2	Service default	Packet (Stream 2)	Application	N_{APP2type}	Application
0x17	1	AN default	Packet (Stream 3)	Application	N_{APP3type}	Application
0x17	2	Service default	Packet (Stream 3)	Application	N_{APP3type}	Application

Headers and Messages

Headers and messages are used when protocols pass information to their peers on the other side of the link, that is, when protocols are executed in two different physical entities. These elements are required on all implementations, and are used for the inter-operation of entities distributed throughout the system. Messages can have the same name even when used by different protocols; for example the ConfigurationRequest message is used for all protocols but follows distinct procedures in each of them.

Each protocol/application has messages and procedures related to the InConfiguration and InUse instances. InConfiguration instances usually process configuration messages whereas InUse instances process non-configuration messages and procedures.

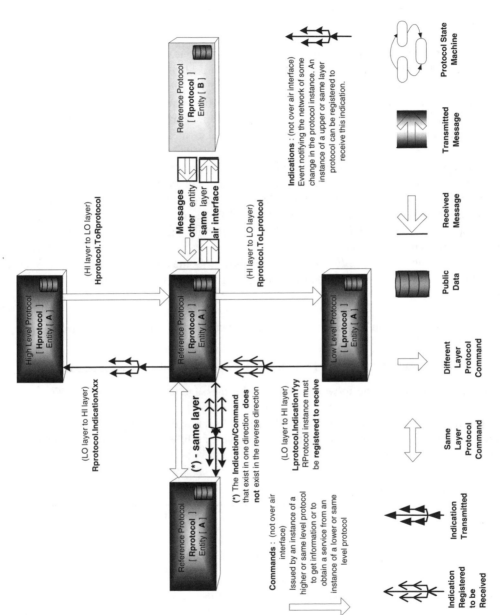

Figure 8.3 Interface elements.

Table 8.7   Message description representation sample

Protocol	Message name	ID	Physical channel	Delivery	Address	Priority
			Description			

A message name and ID identify each message. Because the same message can be processed by different protocols, to identify a message it is also necessary to associate it with a protocol. The type of physical channel transmitting the message, the type of delivery required (best effort or reliable), the addressing mode (broadcast, multicast or unicast), and the priority are the main parameters that define a message. Table 8.7 shows a sample of the representation used in this chapter to describe the main messages exchanged in EVDO systems.

In Table 8.7, the protocol column indicates with which protocol the message is associated. Message name and ID, respectively, show the name and numeric ID of the message within the protocol. The physical channel column lists the channels that transmit the message.

Delivery represents Signalling Link Protocol (SLP) requirements. Two delivery modes can be used: best effort, where the message is sent only once, and reliable, where the message is re-transmitted if necessary.

The addressing mode (Address) determines how the message is going to reach an individual physical entity or a group of physical entities. There are three different addressing modes.

(1) Unicast – used on the forward or reverse link when sending a message to a specific AN or AT.

(2) Multicast – used on the forward link to send a message to a selected group of ATs.

(3) Broadcast – used on the forward link to send a message to a wide group of ATs.

A priority is also assigned to messages. Priority values range from 0 to 255; the lower the value, the higher the priority. This chapter also provides a brief description of the function performed by each message.

Table 8.8   Simple attribute format

Field	Bit size	Description
Length	8	Length of the attribute excluding the length field
AttributeID	Protocol dependent	Unique identity of the attribute in the protocol
	One or more of the following parameters	
Attribute value	Attribute dependent	Value of the attribute with an implicit length indication, such as a fixed length integer or a null terminate string
		$\cdots$
Reserved	Variable	The length of this field is the smallest value that will make the attribute record octet aligned. Value is '0'.

Table 8.9 Complex attribute format

Field	Bit size	Description
Length	8	Length of the attribute excluding the length field
AttributeID	Protocol dependent	Unique identity of the attribute in the protocol
One or more of the following parameters		
ValueID	Protocol dependent	Identifies the attribute values following this field. To be unique in this structure ValueID must be incremented for each new set of values
Each ValueID is followed by this same record structure		
Attribute value	Attribute dependent	Value of the attribute with an implicit length indication, such as a fixed length integer or a null terminate string
		. . .
Reserved	Variable	The length of this field is the smallest value that will make the attribute record octet aligned. Value is '0'

Messages are many times used for sending attributes to configure system parameters. These attributes are kept in attribute records that follow a specific format so that if the recipient of the message does not recognise the attribute it can discard the current record and read the next one.

There are three types of attributes: the simple attribute, which contains a single value; the attribute list, which contains different suggested values for the same attribute identifier; and the complex attribute, which contains multiple values required by a particular attribute identifier. The attribute identifier determines the type of the attribute.

Tables 8.8 and 8.9 describe the format of an attribute list and a complex attribute, respectively. The simple attribute can be considered as a special case of the attribute list, that is, a list with a single value.

InConfiguration Messages

The messages processed in the InConfiguration instance are the same for all EVDO protocols, except for the session configuration protocol. To negotiate a mutually acceptable configuration the protocols use ConfigurationRequest and ConfigurationResponse messages. Table 8.10 provides a brief description of each of them.

In Table 8.10, SMP is session management protocol, FTC is forward traffic channel and RTC is reverse traffic channel.

A generic configuration protocol provides means to negotiate protocol parameters. The negotiation consists of the initiator sending an attribute with one or more allowed values and the responder selecting one of the values offered. The complete process consists of the following steps.

1. The initiator sends one or more ConfigurationRequest messages with a list of acceptable values for each attribute organised in descending order of preference.

Table 8.10 InConfiguration messages summary

Protocol	Message name	ID	Physical channel	Delivery	Address	Priority
All protocols, except for SMP	ConfigurationRequest	80	FTC RTC	Reliable	Unicast	40

This message transmits a set of attributes with a list of possible values to the addressed protocol/application instance

All protocols, except for SMP	ConfigurationResponse	81	FTC RTC	Reliable	Unicast	40

This message has the same attributes as the ConfigurationRequest message but with the values configured by the instance addressed

2. The responder responds to the message received within 2 s (TTurnaround) unless specified otherwise. It selects an acceptable value for each of the attributes in the message. If there is no acceptable value or if the attribute is not recognised, the responder skips the attribute. The selected values are sent in a single ConfigurationResponse message.

3. The initiator compares both messages (request and response). If there is an attribute in the request message that is not defined in the response, the initiator assumes that the missing attribute is using the fall-back value.

4. In case the initiator requires no further negotiation or configuration of negotiated protocols and if the value of any of the parameters sent in the ConfigurationRequest message is NULL, than a failure is declared.

Both the ConfigurationRequest and ConfigurationResponse messages follow a specific format. If the receiver does not recognise a configuration message, it discards the message. If it does not recognised fields within the message, it discards any field other than that described in Table 8.11. In both situations, the receiver may log the messages for diagnostic purposes.

The key exchange protocol, in the security layer, also processes additional InConfiguration messages through its sub-type, the DH (Diffie–Hellman) key exchange protocol. Table 8.12 summarises these messages, providing a brief description of each of them.

InUse Messages

The protocols listed next do not have InUse instance messages.

- Stream layer protocol – stream layer.
- Packet consolidation protocol – connection layer.

Table 8.11 ConfigurationRequest message description

Field	Bit size	Description
MessageID	8	Unique numerical identity of the message in the protocol
TransactionID	8	Unique numerical identity of the transaction. Each Configuration-Request message sent has this value incremented. Configuration-Response messages use the ID of the concerning request message.
		Zero or more of the following record
AttributeRecord	Attribute dependent	Defined in Tables 8.8 and 8.9.

Table 8.12 Additional InConfiguration messages summary – DH key exchange protocol

Protocol	Message name	ID	Physical channel	Delivery	Address	Priority
DH key exchange	KeyRequest	0	FTC	Reliable	Unicast	40
The AN sends this message to start the key exchange session						
DH key exchange	KeyResponse	1	RTC	Reliable	Unicast	40
The AT sends this message responding the KeyRequest message						
DH key exchange	ANKeyComplete	2	FTC	Reliable	Unicast	40
The AN sends this message responding the ANKeyResponse message						
DH key exchange	ATKeyComplete	3	RTC	Reliable	Unicast	40
The AT sends this message responding the ANKeyComplete message						

- Default/generic security protocols – security layer.
- Default encryption/default key exchange protocols – security layer.
- Default/SHA-1 authentication protocols – security layer.

The InUse instance messages of the other EVDO protocols vary according to the layer. Tables 8.13–8.18 show a summary of InUse instance messages respectively for the application, session, connection, MAC and physical layers.

The following tables indicate protocols and channels using acronyms. The following is a list of these acronyms and their explanation.

AC	Access Channel
ACMP	Access Channel MAC Protocol
ALMP	Air Link Management Protocol
AMP	Address Management Protocol
CC	Control Channel (which includes CCsyn and CcsynSS in the connection layer)

Table 8.13 InUse messages summary for the application layer

Protocol	Message Name	ID	Physical Channel	Delivery	Address	Priority
SLP	Reset	0	FTC	Best	Unicast	40
The AN sends this message to reset the SLP at AT						
SLP	ResetAcked	1	RTC	Best	Unicast	40
The AT sends this message to complete the reset of the SLP						
RLP	Reset	0	FTC RTC	Reliable	Unicast	50
The AT/AN sends this message to reset the radio link protocol						
RLP	ResetAck	1	FTC RTC	Reliable	Unicast	50
The AT/AN sends this message to reset the radio link protocol						
RLP	Nak	2	FTC RTC	Best	Unicast	50
The AT/AN sends this message to request the re-transmission of one or more octects						
LU	LocationRequest	3	CC FTC	Best	Unicast	40
The AN sends this message to get information about AT location						
LU	LocationNotification	4	AC RTC[Reliable]	Best	Unicast	40
The AN sends this message to answer the LocationRequest message, or due to a handoff						
LU	LocationAssignment	5	CC FTC	Best	Unicast	40
The AN sends this message to update AT location information						
LU	LocationComplete	6	AC RTC[Reliable]	Best	Unicast	40
The AT sends this message to acknowledge the LocationAssignment message						
FC	XonRequest	7	AC RTC	Best	Unicast	40
The AT sends this message to request transition to the open state						
FC	XonResponse	8	CC FTC	Best	Unicast	40
The AN sends this message to respond to the XonRequest message						
FC	XoffRequest	9	AC RTC	Best	Unicast	40
The AT sends this message to request the transition to the close state						
FC	XoffResponse	10	CC FTC	Best	Unicast	40
The AN sends this message to respond to the XoffResponse message						
FC	DataReady	11	CC FTC	Best	Unicast	40
The AT sends this message to indicate the existence of data waiting for transmission						
FC	DataReadyAck	12	AC RTC	Best	Unicast	40
The AN sends this message to acknowledge the DataReady message						

Table 8.14 InUse messages summary for the session layer

Protocol	Message name	ID	Physical channel	Delivery	Address	Priority
SMP	SessionClose	1	FTC RTC CC AC	Best	Unicast	10
The AT/AN send this message to terminate a session						
SMP	KeepAliveRequest	2	FTC RTC CC AC	Best	Unicast	10
The AT/AN send this message to verify if a peer is alive						
SMP	KeepAliveResponse	3	FTC RTC CC AC	Best	Unicast	10
The AN sends this message to respond to the KeepAliveRequest message						
AMP	UATIRequest	0	AC	Best	Unicast	40
The AT sends this message to request a UATI assignment or re-assignment by the AN						
AMP	UATIAssignment	1	CC FTC	Best	Unicast	40
The AN sends this message to assign or re-assign a UATI to an AT						
AMP	UATIComplete	2	AC RTC[Reliable]	Best	Unicast	40
The AT sends this message to acknowledge the AN of the UATIAssigment message reception						
AMP	HardwareIDRequest	3	CC FTC	Best	Unicast	40
The AN sends this message to get the HardwareID information from the AT						
AMP	HardwareIDResponse	4	AC RTC[Reliable]	Best	Unicast	40
The AT sends this message to respond to the HardwareIDRequest message						
SCP	ConfigurationStart	1	FTC	Best	Unicast	40
The AN sends this message to start a session configuration process						
SCP	ConfigurationComplete	0	FTC RTC	Best	Unicast	40
The AT/AN send this message to indicate the conclusion of negotiating procedures						
SCP	ConfigurationRequest	80	FTC RTC	Reliable	Unicast	40
This message sends a set of attributes, with a list of possible values, to the addressed protocol/application instance. (Only instance where this message is not associated with the InConfiguration mode.)						
SCP	ConfigurationResponse	81	FTC RTC	Reliable	Unicast	40
This message has the same attribute list as the ConfigurationRequest message with values chosen by the called instance. (Only instance where this message is not associated with the InConfiguration mode.)						

Table 8.15 InUse messages summary for the Connection Layer

Protocol	Message name	ID	Physical channel	Delivery	Address	Priority
ALMP	Re-direct	0	FTC CC	Best	Unicast	40

The AN sends this message to re-direct the AT away from the current network with the possibility of indicating a set of other networks

| ISP | Sync | 0 | CCsyn | Reliable | Broadcast | 40 |

The AN sends this message as basic network and timing information

| IDP | Page | 0 | CCsynSS | Best | Unicast | 40 |

The AN sends this message to instruct the AT to request a connection

| IDP | ConnectionRequest | 1 | AC | Best | Unicast | 40 |

The AT sends this message to request a connection

| IDP | ConnectionDeny | 2 | CC | Best | Unicast | 40 |

The AN sends this message to deny a connection

| CSP | ConnectionClose | 0 | FTC RTC | Best | Unicast | 40 |

The AT/AN send this message to close the connection

| RUP | RouteUpdate | 0 | AC RTC[Reliable] | Best | Unicast | 40 |

The AN sends this message to redirect the AT away from the current network with the possibility of indicating a set of other networks

| RUP | TrafficChannel Assignment | 1 | FTC CC | Reliable | Unicast | 40 |

The AN sends this message to manage the AT's active set

| RUP | TrafficChannelComplete | 2 | RTC | Reliable | Unicast | 40 |

The AT sends this message to answer to the TrafficChannelAssignment message

| RUP | ResetReport | 3 | FTC | Reliable | Unicast | 40 |

The AN sends this message to reset RouteUpdate transmission rules at the AT

| RUP | NeighborList | 4 | FTC | Reliable | Unicast | 40 |

The AN sends this message to the AT in Connected State to inform neighbour sectors

| RUP | AttributeOverride | 5 | FTC | Best | Unicast | 40 |

The AN sends this message to override values configured

| RUP | AttributeOverideResponse | 6 | RTC | Best | Unicast | 40 |

The AT sends this message to override the configured attributes with the values in this message

| OMP | QuickConfig | 0 | CCsynSS | Best | Broadcast | 40 |

The AN sends this message to indicate change in the contents of the overhead messages and also to provide frequently changing information

| OMP | SectorParameters | 1 | CCsynSS | Best | Broadcast | 40 |

The AN sends this message to send sector specific information to the AT

Table 8.16 InUse messages summary for the MAC layer

Protocol	Message name	ID	Physical channel	Delivery	Address	Priority
ACMP	ACAck	0	CC	Best	Unicast	10

The AN sends this message to acknowledge an Access Channel MAC layer capsule

ACMP	AccessParameters	1	CC	Best	Broadcast	30

The AN issues this message to send access channel information to ATs

FTCMP	FixedModeEnabled	0	RTC	Best	Unicast	40

The AT sends this message to indicate a transition to the fixed rate state

FTCMP	FixedModeOff	1	RTC	Best	Unicast	40

The AT sends this message to inform the AN, that it is not able to receive packets from the sector specified in the last FixedModeEnabled message

RTCMP	RTCAck	0	FTC	Reliable	Unicast	10

The AN sends this message to inform the AT that the reverse link traffic channel was acquired

RTCMP	BroadcastReverse RateLimit	1	CC	Best	Broadcast	40

The AN sends this message to control the transmission rate on the reverse link

RTCMP	UnicastReverse RateLimit	2	FTC	Reliable	Unicast	40

The AN sends this message to control the transmission rate on the reverse link for a particular AT

CCSyn	Control Channel Synchronous, synchronised to the control channel cycle (includes CCsynSS)
CCSynSS	Control Channel Synchronous Sleep State, CCSyn for the Sleep State
CSP	Connected State Protocol
FC	Flow Control Protocol
FTC	Forward Traffic Channel
FTCMP	Forward Traffic Channel MAC Protocol
HardwareID	Information about the Hardware ID
IDP	Idle State Protocol
ISP	Initialisation State Protocol
LU	Location Update Protocol
OMP	Overhead Messages Protocol
RLP	Radio Link Protocol
RTC	Reverse Traffic Channel
1RTC[Reliable]	the transmission mode is reliable for the RTC
RTCMP	Reverse Traffic Channel MAC Protocol
RUP	Route Update Protocol
SCP	Session Configuration Protocol
SLP	Signalling Link Protocol
SMP	Session Management Protocol
UATI	Unicast Access Terminal Identifier

Table 8.17 InUse messages used on physical layer assigned channels

Forward traffic channel	Reverse traffic channel
SLP.Reset	SLP.ResetAcked
RadioLinkProtocol.Reset	RadioLinkProtocol.Reset
RadioLinkProtocol.ResetAck	RadioLinkProtocol.ResetAck
RadioLinkProtocol.Nak	RadioLinkProtocol.Nak
LocationUpdate.LocationRequest	LocationUpdate.LocationNotification
LocationUpdate.LoacationAssignment	LocationUpdate.LocationComplete
FlowControl.XonResponse	FlowControl.XonRequest
FlowControl.XoffResponse	FlowControl.XoffRequest
FlowControl.DataReady	FlowControl.DataReadyAck
SessionManagement.SessionClose	SessionManagement.SessionClose
SessionManagement.KeepAliveRequest	SessionManagement.KeepAliveRequest
SessionManagement.KeepAliveResponse	SessionManagement.KeepAliveResponse
AddressManagement.UATIAssignment	AddressManagement.UATIComplete
AddressManagement.HardwareIDRequest	AddressManagement.HardwareIDResponse
SessionConfiguration.ConfigurationComplete	SessionConfiguration.ConfigurationComplete
SessionConfiguration.ConfigurationStart	SessionConfiguration.ConfigurationRequest
SessionConfiguration.ConfigurationRequest	SessionConfiguration.ConfigurationResponse
SessionConfiguration.ConfigurationResponse	ConnectedState.ConnectionClose
AirLinkManagement.Redirect	RouteUpdate.RouteUpdate
ConnectedState.ConnectionClose	RouteUpdate.TrafficChannelComplete
RouteUpdate.TrafficChannelAssignment	RouteUpdate.AttributeOverrideResponse
RouteUpdate.ResetReport	DHKeyExchange.KeyResponse
RouteUpdate.NeighborList	DHKeyExchange.ATKeyComplete
RouteUpdate.AttributeOverride	ForwardTrafficChannelMAC.FixedModeEnable
DHKeyExchange.KeyRequest	ForwardTrafficChannelMAC.FixedModeOff
DHKeyExchange.ANKeyComplete	
ReverseTrafficChannelMAC.RTCAck	
ReverseTrafficChannelMAC.UnicastReverseRateLimit	

8.3.2.2.1 Commands

 Protocols use commands to communicate with other protocols within the same entity. Commands can be sent from a higher to a lower level protocol, or between protocols in the same layer; the communication, however, only happens in one direction. One protocol instance sends a command to another protocol in the form of *Protocol.Command*. Commands are usually associated with services; for example ReverseTrafficChannelMAC.Activate is a command for changing the AN/AT from the inactive to the setup state.

Commands activate specific procedures in the addressed protocol instance. Most protocols support the following basic commands.

- Activate – instructs protocols to transition from inactive to another state.

- Deactivate – instructs protocols to go to the inactive state. Some protocols may not transition immediately because of cleanup requirements.

- Open/close – instructs protocols to open/close sessions or connections.

Table 8.18 InUse messages used on physical layer public channels

Control channel	Acess channel
LocationUpdate.LocationRequest	LocationUpdate.LocationNotification
LocationUpdate.LocationAssignment	LocationUpdate.LocationComplete
FlowControl.XonResponse	FlowControl.XonRequest
FlowControl.XoffResponse	FlowControl.XoffRequest
FlowControl.DataReady	FlowControl.DataReadyAck
SessionManagement.SessionClose	SessionManagement.SessionClose
SessionManagement.KeepAliveRequest	SessionManagement.KeepAliveRequest
SessionManagement.KeepAliveResponse	SessionManagement.KeepAliveResponse
AddressManagement.UATIAssignment	AddressManagement.UATIRequest
AddressManagement.HardwareIDRequest	AddressManagement.UATIComplete
AirLinkManagement.Redirect	AddressManagement.HardwareIDResponse
InitializationState.Sync	IdleState.ConnectionRequest
IdleState.Page	RouteUpdate.RouteUpdate
IdleState.ConnectionDeny	
RouteUpdate.TrafficChannelAssignment	
OverheadMessages.QuickConfig	
OverheadMessages.SectorParameters	
AccessChannelMAC.ACAck	
AccessChannelMAC.AccessParameters	
ReverseTrafficChannelMAC.BroadcastReverseRateLimit	

Table 8.19 Summary of commands

	Protocol name	Layer	Commands 1	2	3	4
Default	Control channel	MAC	Activated	Deactivated		
Default	Access channel	MAC	Activated	Deactivated		
Default	Forward traffic channel	MAC	Activated	Deactivated		
Default	Reverse traffic channel	MAC	Activated	Deactivated		
Default	Air link management	Connection	Open connection	Close connection		
Default	Initialisation state	Connection	Activated	Deactivated		
Default	Idle state	Connection	Activated	Deactivated	Open connection	Close
Default	Connected state	Connection	Activated	Deactivated	Close connection	
Default	Route update	Connection	Activated	Deactivated	Open	Close
Default	Overhead messages	Connection	Activated	Deactivated		
Default	Session management	Session	Activated	Deactivated		
Default	Address management	Session	Activated	Deactivated	Update UATI[a]	
Default	Session configuration	Session	Activated	Deactivated		

[a] The AddressManagement.UpdateUATI command is defined for the address management protocol according to the standard IS-856 version 4.0, but it is not actually used.

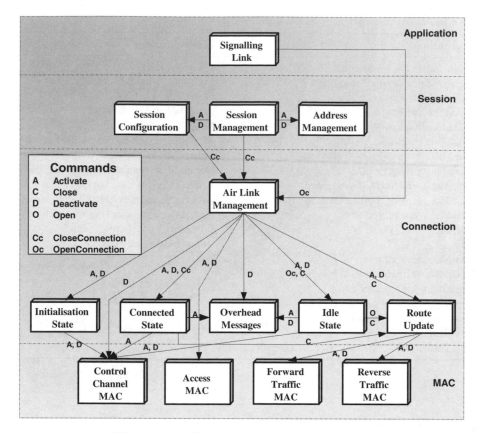

Figure 8.4 InUse instance commands interaction.

Table 8.19 presents a summary of commands used in EVDO systems.

Figure 8.4 represents commands interaction when in the InUse instance.

8.3.2.2.2   Indications

Lower level protocols use *indications* to convey information within the same entity. A protocol instance must register to receive indications from another protocol. The indication is written in the form of *Protocol.Command*. Indications are usually associated with information such as state or critical data change; for example ReverseTrafficChannelMAC.LinkAcquired is an indication generated after the AT receives an RTCAck message from the AN indicating change from the setup to the open state. Table 8.20 presents a summary of the indications within an EVDO system.

8.3.2.2.3   Public Data

The *public data* is a common resource used to share information in the same physical entity between protocols/applications, regardless of their being in the same layer or not. The public data has a unique name that does not depend on the protocol identification. For example, during the ReverseAccessMAC channel procedure, probes are sent in power steps, which conform to the public data PowerSetup.

Table 8.20 Summary of indications

Protocol name[a]		Layer	Indications						
			1	2	3	4	5	6	7
Dfl	Control channel	MAC	Sup. failed						
Default	Access channel	MAC	Sup. failed	Trans. successful	Trans. aborted	Trans. failed	TxStart	TxEnded	
Default	Forward traffic channel	MAC	Sup. failed						
Default	Reverse traffic channel	MAC	Sup. failed	Link acquired					
SHA-1	Authentication	Security	Failed						
Default	Initialisation state	Conn.	Network acquired						
Default	Idle state	Conn.	Conn. opened	Conn. failed					
Default	Connected state	Conn.	Conn. closed						
Default	Route update	Conn.	Conn. lost	Network lost	IdleHO	Active setup updated	Assignment rejected	Conn. initiated	Conn. opened
Default	Overhead messages	Conn.	AN re-directed	Sup. failed	Updated				
Default	Session management	Session	Session opened	Session closed					
Default	Address management	Session	Opened	UATI release	UATI assigned	Failed	Subnet change		
Default	Session configuration	Session	Re-conf.	Failed					
Default	Radio link protocol	App.	Conn. opened						
Default	Signalling link	App.	Reset	Reset acknowledged					

[a] In the table, Transmission (Trans.), Connection (Conn.), Supervision (Sup.) and Application (App.).

8.3.2.3 States

Protocols can also be in different states. Each state represents a different behaviour of the protocol caused by a change in the environment. A protocol can be designed to characterise this behaviour, or actions, as a set of transitions between one state and another state. When a protocol is not being used, it is set in the inactive state. If it has only one more state besides inactive, this state is referred to as active state. Otherwise, each of the active states receives an individual name.

 The multiple states of protocols can be represented using a state machine. Each instance has to be in one of the states of the machine associated with the protocol at any given time. For example, if an AN can communicate with

Table 8.21   Access Terminal (AT) states summary

	Protocol name	Layer	State1	State2	State3	State4
				AT		
Default	Control channel	MAC	Inactive	Active		
Default	Access channel	MAC	Inactive	Active		
Default	Forward traffic channel	MAC	Inactive	Variable rate	Fixed rate	
Default	Reverse traffic channel	MAC	Inactive	Setup	Open	
Default	Air link management	Connection	Initialization	Idle	Connected	
Default	Initialisation state	Connection	Inactive	Network determination	Pilot acquisition	Synchronisation
Default	Idle state	Connection	Inactive	Monitor	Sleep	Connection setup
Default	Connected state	Connection	Inactive	Open		
Default	Route update	Connection	Inactive	Idle	Connected	
Default	Overhead messages	Connection	Inactive	Active		
Default	Session management	Session	Inactive	AMP setup	Open	
Default	Address management	Session	Inactive	Setup	Open	
Default	Session configuration	Session	Inactive	AT initiated	AN initiated	Open
Default	Flow control	Application	Open	Close		

multiple ATs, there is an independent state associated with each AT instance. The transition from one state to another, and consequent indication, can be triggered by a command, an indication or by the expiration of a timer set in the previous state transition. Table 8.21 shows a summary of states in EVDO systems (also see Table 8.22).

Table 8.22   Access Network (AN) states summary

	Protocol name	Layer	State1	State2	State3	State4
				AN		
Default	Control channel	MAC	Inactive	Active		
Default	Access channel	MAC	Inactive	Active		
Default	Forward traffic channel	MAC	Inactive	Variable rate	FixedRate	
Default	Reverse traffic channel	MAC	Inactive	Setup	Open	
Default	Air link management	Connection	Initialisation	Idle	Connected	
Default	Initialisation state	Connection	Inactive	Network determination	Pilot acquisition	Synchronisation
Default	Idle state	Connection	Inactive	Monitor	Sleep	Connection setup
Default	Connected state	Connection	Inactive	Open	Close	
Default	Route update	Connection	Inactive	Idle	Connected	
Default	Overhead messages	Connection	Inactive	Active		
Default	Session management	Session	Close	AMP setup	Open	
Default	Address management	Session	Inactive	Setup	Open	
Default	Session configuration	Session	Inactive	AT initiated	AN initiated	Open
Default	Flow control	Application	Open	Close		

8.3.2.4 Message Encapsulation and Transmission Overhead

The transmission of a message incurs an overhead caused by the addition of headers and padding bits, which are required for security reasons and to guide the message to the correct recipient. Each layer adds an overhead to the original message. The next sections describe this overhead and in which situations it is required.

8.3.2.4.1 Application Layer

All access terminals and networks compliant with EVDO standards must support two types of default application: default signalling application and default packet application.

The default signalling application allows a protocol to communicate with its peer element in another entity. This application consists of a message and a link layer protocol, called signalling network protocol and signalling link protocol, respectively.

The default packet application allows octet re-transmission and duplicate detection using the radio link protocol. This application is also responsible for providing mobility between service networks through a location update protocol. Each of these applications is described in detail in the next sections.

Default Signalling Application

The default signalling application primarily uses two protocols, the Signalling Network Protocol (SNP) and the Signalling Link Protocol (SLP). The first provides signalling message transmission services, allowing message exchange between protocols, routing the message to the protocol type and instance defined in the SNP header. The latter is a link layer protocol with signalling message mechanisms for message fragmentation and reliable/best-effort delivery. The SLP is sub-divided into two sub-layers.

- SLP-D: Delivery sub-layer (reliable or best effort).

- SLP-F: Fragmentation sub-layer.

Figure 8.5 illustrates the relationship among these protocols.

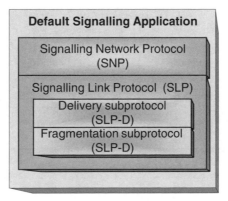

Figure 8.5 Default signalling application protocols SNP, SLP, SLP-D and SLP-F relationship.

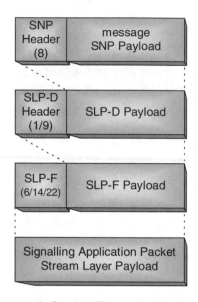

Figure 8.6   Packet capsule for signalling application layer (not fragmented).

Figures 8.6 and 8.7 illustrate two message encapsulation scenarios: non-fragmented and fragmented SNP packets, respectively.

As depicted in Figures 8.6 and 8.7, a header is associated with each payload block during message encapsulation. The information in these headers varies according to the protocol. Tables 8.23–8.25 show the header description for SNP, SLP-D and SLP-F, respectively.

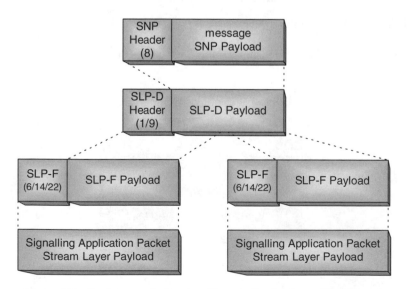

Figure 8.7   Packet capsule for signalling application layer (fragmented).

Table 8.23 Signalling network protocol header

SNP header fields	Bit size	Description
InConfigurationProtocol	1	'1'indicates the message is destined to the InConfiguration instance of the protocol identified by the type field
Type	7	Indicates the addressed protocol, as in Table 8.6

Table 8.24 Signalling link protocol (delivery sub-protocol) header

SLP-D header fields	Bit size	Description
FullHeaderIncluded	1	'1' indicates the presence of the following SLP-D header fields
AckSequenceValid	0 or 1	'1' indicates that the AckSequenceNumber is valid
AckSequenceNumber	0 or 3	Depicts the first reliable-delivered SLP-D payload not acknowledged yet
SequenceValid	0 or 1	'1' indicates that the SequenceNumber is valid
SequenceNumber	0 or 1	Depicts the number of the current reliable-delivered SLP-D payload

Table 8.25 Signalling link protocol (fragmentation sub-protocol) header

SLP-F header fields	Bit size	Description
Reserved	4	All four bits of this field are '0'
Fragmented	1	'1' indicates fragmented operation and the following fields are defined
Begin	0 or 1	'1' indicates that this fragment contains the begin of the SLP-D packet
End	0 or 1	'1' indicates that this fragment contains the last part of the SLP-D packet
SequenceNumber	0 or 6	Number of the fragment within the transmitted sequence
OctectAlignmentPad	1	Additional bit with value '0' used when transmitting SLP-D fragment with header, to keep both SLP-D and SLP-F headers with a size of 14 or 22 bits (size modulo $8 = 6$)

In Tables 8.24 and 8.25, the existence of fields that may assume more than one value (0 or 'n') depends on the previous header field. For example, AckSequenceValid is only in the header if the FullHeaderIncluded field is configured as '1'.

Default Packet Application

The default packet application comprises three independent protocols, the radio link protocol (RLP), the Location Update Protocol (LUP) and the Flow Control Protocol (FCP). The first provides octet aligned data stream retransmission and duplicate detection.

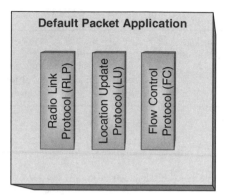

Figure 8.8 Default packet application protocols RLP, LU and FC relationship.

The second is a mobility management protocol that provides location update procedures and messages. The latter performs packet flow control, establishing procedures to enable and disable data traffic flow.

Figure 8.8 illustrates the relationship among these protocols.

Figure 8.9 illustrates the default packet application encapsulation.

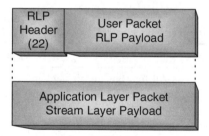

Figure 8.9 Packet capsule for packet application layer.

Table 8.26 describes the field in the RLP header associated with the payload.

8.3.2.4.2 Stream Layer

The stream layer has a single protocol, the Stream Layer Protocol (SLP), which multiplexes the application streams coming from different ATs and provides configuration messages to map applications to streams.

There are four streams within the stream layer. In the multiplexing process, Stream 0 is always assigned to the signalling application (SNP and SLP). Streams 1 to 3 can be assigned to applications with different QoS requirements (packet applications).

Table 8.26 Radio link protocol header

RLP header field	Bit size	Description
Sec	22	Sequence number of the first transmitted octet

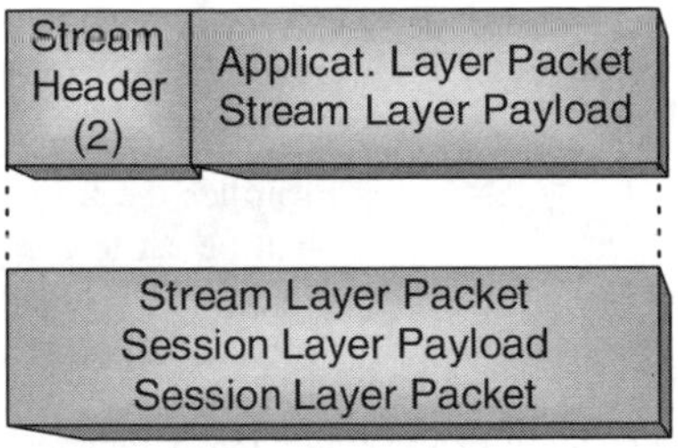

Figure 8.10   Stream layer capsule.

The mapping of the application to a stream is performed using the stream header associated with the packet coming from the application layer, as in Figure 8.1 (see Figure 8.10).

Table 8.27 describes the header used by the stream layer to associate the application layer packet to a given stream.

Table 8.27   Stream header

Stream header field	Bit size	Description
Stream	2	Stream number (0 for signalling and 1, 2 or 3 for packet applications)

8.3.2.4.3   Session Layer

The session layer comprises three inter-related protocols: the Session Management Protocol (SMP), the Session Configuration Protocol (SCP) and the Address Management Protocol (AMP). The session layer is responsible for establishing, maintaining and controlling a common state that is shared by the AN and AT to allow communication. Sessions are opened when actions begin and close when actions are concluded.

The session management protocol controls other session layer protocols (SCP and AMP) using activate and de-activate commands. It also assures that a non-closed session is still valid through a keep alive mechanism (KeepAliveRequest and KeepAliveResponse). This protocol is also responsible for closing a session (SessionClose).

The main function of the session configuration protocol is to negotiate and provision protocols used during the session, also negotiating the session parameters for these protocols. Session parameters are the attributes and the internal parameters that define the state of each protocol. The commands used in this negotiation process are Configuration-Start, ConfiturationComplete, ConfigurationRequest and ConfigurationResponse.

A session state information record is used for transferring the session parameters from the source to the target access network (3GPP2 A.S0007 1xEV-DO Inter-Operability Specification (IOS) for cdma2000 access network interfaces). Tables 8.28 and 8.29 describe this record. If an attribute is not in this record, the target AN assumes the default values for the missing attributes. Default attribute values vary for each protocol.

Table 8.28 Session state information record

Field	Bit size	Description
FormatID	8	Identifies the format of the fields within the record
Reserved	1	Shall be set to zero
ProtocolType	7	Protocol type identifier according to Table 8.6
ProtocolSubtype	16	Protocol sub-type identifier according to Table 8.6
One or more of the following parameter records		
ParameterType	8	Configured according to Table 8.29, determines the content of the ParameterType specific record
ParameterType specific record	Variable	If set to 0x00, the record is set to the attribute associated with the protocol identified by (ProtocolType, ProtocolSubtype). Otherwise, this record follows the structure specified by (ProtocolType, ProtocolSubtype)

Table 8.29 Encoding of the ParameterType field

Field value	Meaning
0x00	Consists of a complex or a simple attribute. The ValueID field of the complex attribute shall be set to zero
All other values	ParameterType specific record are protocol dependent

The address management protocol provides the AT address (identifier) management service, keeping the Unicast Access Terminal Identity (UATI) updated through the following messages: UATIRequest, UATIAssignment and UATIComplete.

The AMT also maintains the AT hardware identity database. This identity is typically an Electronic Serial Number (ESN) but may also be a hardware identifier such as the MAC address, depending on the HardwareIDType parameter. The following messages are used for this task: HardwareIDRequest and HardwareIDResponse (which provide the HardwareID and HardwareIDType).

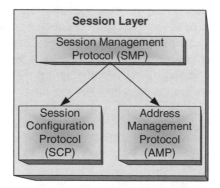

Figure 8.11 Session layer protocols relationship.

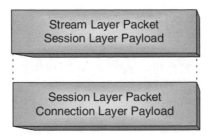

Figure 8.12 Session layer capsule.

Figure 8.11 illustrates the relationship among the protocols within the session layer.

The session layer does not cause any overhead to the packet coming from the stream layer as depicted in Figure 8.12.

8.3.2.4.4 Connection Layer

The connection layer comprises seven protocols: the Air Link Management Protocol (ALMP), the Initialisation State Protocol (ISP), the Idle State Protocol (IDP), the Connected State Protocol (CSP), the Route Update Protocol (RUP), the Overhead Message Protocol (OMP) and the Packet Consolidation Protocol (PCP).

This layer is responsible for controlling the state of the air link being used by the AN and ATs. A connection is associated with a previously opened session and can be opened and closed multiple times during this session. Some of the main functions of this layer include management of network acquisition, of connections, of the neighbour list and of the radio link. This layer also updates ATs with information about the AN and maintains a database with the approximate location of ATs.

This layer uses four protocols to deal with the state of the connection. The air link management protocol is the core of the connection layer and performs the overall state machine management applied to an AT/AN during a connection. This protocol can be in three different states and activates another of the state management protocols depending on the state it is in. The main commands used to execute these tasks are Activate, Deactivate, OpenConnection and CloseConnection.

The initialisation state protocol is used for network acquisition also supporting synchronisation during this process. The main message used by this protocol is the Sync message.

The idle state protocol executes the functions required by ATs that have acquired the network but do not have an open connection. These functions include keeping track of the ATs location for efficient paging and opening a connection. The messages used to perform these tasks are Connection Request, ConnectionDeny and Page.

The last state management protocol is the connected state protocol, which is used when the AT has an open connection. The main tasks of this protocol are to manage the radio link and to close the connection (ConnectionClose).

Besides the state management protocols this layer also has three other protocols. The route update protocol maintains the route between ATs and AN by keeping track of the ATs location and maintaining the radio link. This protocol also performs supervision on the pilots. Some of the messages used to perform these functions are RouteUpdate, TrafficChannelAssignnent, TrafficChannelComplete, ResetReport, NeighborList, AttributeOverride and AttributeOverrideResponse.

The overhead messages protocol broadcasts essential network parameters over the control channel. These parameters are shared by protocols in all layers. The messages used by this protocol include QuickConfiguration and SectorParameters.

The last protocol in this layer is the packet consolidation protocol, responsible for packet encapsulation and prioritisation as a function of assigned priority and target transmission channel.

Figure 8.13 illustrates the relationship among all the protocols described in this section.

The connection layer prioritises and encapsulates transmitted data received from the session layer and forwards it to the security layer. The connection layer can encapsulate packets in two different ways, known as Formats A and B.

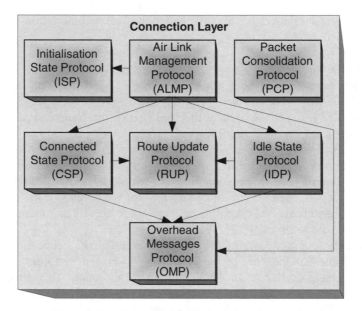

Figure 8.13   Connection layer protocols relationship.

The packet format type is passed with the packet to the lower layers. The maximum size of the packet that can be encapsulated in this layer depends on the security protocols negotiated and on the physical layer channel that will transmit the packet.

Format A packets are maximum length packets that contain only one session layer packet and do not have connection layer headers or padding. Format A provides an extra byte of payload per packet.

Format B packets contain one or more session layer packets and have one or multiple connection layer headers, which are placed in front of each session layer packet. This format may also add padding to create maximum length packets.

Figures 8.14 and 8.15 show the possible packet capsules generated in the connection layer.

The connection header used in Format B packets contains one single field described in Table 8.30.

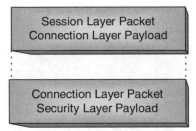

Figure 8.14 Connection layer capsule (Format A).

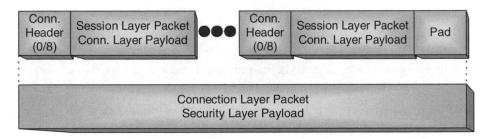

Figure 8.15 Connection layer capsule (Format B).

The type of the connection layer packet format is declared in the MAC header for the control channel and access channel MAC protocols and in the MAC trailer for the forward and reverse traffic channel MAC protocol.

8.3.2.4.5   Security Layer

The security layer authenticates and encrypts the information transmitted over the air interface through control, access, forward traffic and reverse traffic channels. There are four types of protocol in this layer, with their relationship illustrated in Figure 8.16.

One of the protocols in this layer is the Security Protocol (SP), responsible for generating the common data (cryptosync) needed in the authentication and encryption protocols. This protocol has two sub-types: the default security protocol and the generic security protocol. The first provides means to communicate with the MAC layer and does not add any overhead to the authentication packet. The latter adds the cryptosync value as a header to the authentication packet when the communication with the MAC layer involves authenticated or encrypted information.

Cryptosync (cryptosynchronisation) is a method for allowing the state of a cipher to change with each frame by synchronising states at both ends. In wireless systems, there are two common methods of encryption which use cryptosync.

Table 8.30 Connection header

Connection header field	Bit size	Description
Length	0 or 8	Exists only in Format B capsules

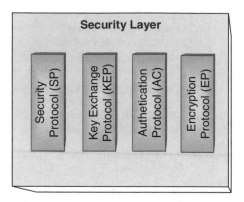

Figure 8.16   Security layer protocols relationship.

- Stream encryption: uses cryptosync to generate a random stream (keystream) to be processed with the plaintext generating the ciphered text.

- Block encryption: uses cryptosync and the entropy of the plaintext to provide the ciphered text with unique encryption.

Another protocol used in the security layer is the Key Exchange Protocol (KEP), which establishes the procedures used to exchange security keys between ATs and AN for the encryption and authentication processes. This protocol also does not add any overhead (header or trailer) to the packet. There are two sub-types of this protocol: the default key exchange protocol and the Diffie–Hellman (DH) key exchange protocol.

The default key exchange protocol does not provide any services and is selected when the default authentication protocol and the default encryption protocol are selected. The DH key exchange protocol provides procedures based on the Diffie-Hellman key exchange algorithm (internet key exchange), which generates 160-bit long encryption keys. The keys generated include the following.

- Authentication.

  FACAuthKey for the forward link assigned channel (traffic).

  RACAuthKey for the reverse link assigned channel (traffic).

  FPCAuthKey for the forward link public channel (control).

  RPCAuthKey for the reverse link public channel (access).

- Encryption.

  FACEncKey for the forward link assigned channel (traffic).

  RACEncKey for the reverse link assigned channel (traffic).

  FPCEncKey for the forward link public channel (control).

  RPCEncKey for the reverse link public channel (access).

The basic key exchange process starts with the AN sending a KeyRequest message to the AT using an AN public key (ANPubKey) and computation timeout value ($T_{\text{KEPKeyCompAN}}$).

The AN starts a timer with $T_{\text{KEPANResponse}}$, waiting for a KeyResponse message. If the timer expires, the AN declares a failure and aborts the process.

Otherwise, the AT sends a KeyResponse message with an AT public Key (ATPubKey) and an AT computation timeout value ($T_{\text{KFPKeyCompAT}}$). Here also the AT starts a timer with $T_{\text{KEPKeyCompAN}}$ waiting for a ANKeyComplete message. If the timer expires, the AT declares a failure and aborts the process.

If the timer does not expire, the AN starts a timer with $T_{\text{KEPKeyCompAT}}$, sends a ANKeyComplete message to the AT with an arbitrary seed (Nonce), a time stamp value (TimeStampShort) and a 160-bit KeySignature. Again it starts a timer with $T_{\text{KEPSigCompAN}}$ waiting for the ATKeyComplete message (after AT signature computation). If both timers expire or if the AT sends an ATKeyComplete with a value '0' in the result field, failure is declared and the process is aborted.

If the timers do not expire, the AT validates the KeySignature sent in the ANKeyComplete message and sends an ATKeyComplete message with the field result indicating success ('1') or failure ('0').

The last two protocols used in this layer are the encryption protocol and the authentication protocol. The first is composed only of the default encryption protocols that provide rules followed by ATs and AN to encrypt the traffic information to and from the MAC layer according to negotiation procedures. This protocol adds no header or trailer to the connection layer packet.

The second, the authentication protocol, provides rules to authenticate the traffic information to and from the MAC layer according to negotiation procedures. This protocol

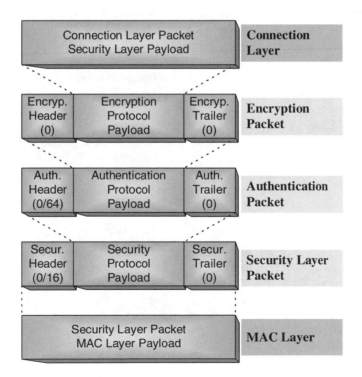

Figure 8.17 Security layer capsules.

Table 8.31   SAH-1 authentication header

SAH-1 authentication header field	Bit size	Description
ACPAC	64	Access channel packet authentication code

Table 8.32   Generic security header

Generic security header field	Bit size	Description
CryptosyncShort	16	Crypt[15:0], where $Crypt = (SysTime - (SysTime[15 : 0] - pCryptS)$ $\times \bmod \ 2^{16}) \bmod 2^{64}$ [15:0] represents the 16 least significant bits Crypt is the Cryptosync 64 bit basic encryption pCryptS is the previous CryptoSyncShort value SysTime is the current CDMA system time in 80 ms units

can use header/trailer to provide a digital signature of the encryption packet. This protocol has two protocols sub-type: the default authentication protocol, which only transfers packet between encryption protocol and security protocol; and the SHA-1 authentication protocol, which provides means for authentication of access channel MAC layer packets by applying the SHA-1 hash function to message bits (ACPAC – Access Channel MAC Layer Packet Authentication Code). The access terminal shall construct the message bits for computing the ACPAC based on the ACAuthKey (Access Channel MAC Authentication Key), the security layer payload, the cryptosync and the sector ID.

Figure 8.17 illustrates packet encapsulation in the security layer.

The headers that may be added by the authentication and security protocols are described, respectively, in Tables 8.31 and 8.32.

8.3.2.4.6   MAC Layer
There are four protocols defined for this layer, one for each channel: forward traffic channel, reverse traffic channel, control channel and access channel. Figure 8.18 illustrates the

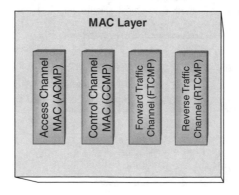

Figure 8.18   MAC layer protocols relationship.

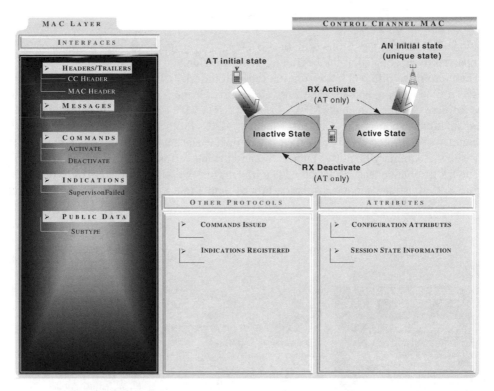

Figure 8.19 Control channel MAC protocol summary.

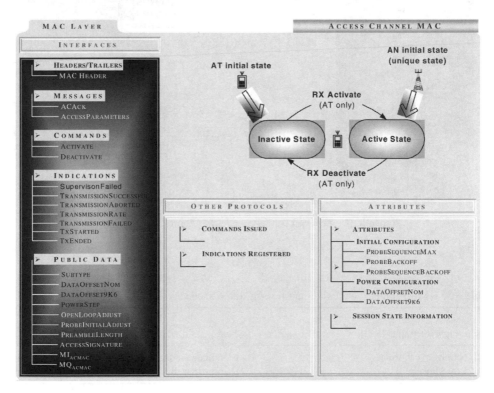

Figure 8.20 Access channel MAC protocol summary.

relationship among the protocols. In the MAC layer each of the protocols can be independently negotiated at the beginning of the session.

There are three different encapsulation types in the MAC layer. The selected type varies on the protocol being applied. The Forward Traffic Channel MAC Protocol (FTCMP) and the Reverse Traffic Channel MAC Protocol (RTCMP) both use the traffic channel encapsulation, whereas the Control Channel MAC Protocol (CCMP) and the Access Channel MAC Protocol (ACMP) have their own encapsulation methods. Figures 8.19–8.22 show a summary of states, interfaces and attributes of each of these protocols.

The traffic channel encapsulation adds a trailer with two bits to the security layer packet. Figure 8.23 illustrates the encapsulation process and Table 8.33 describes each trailer bit.

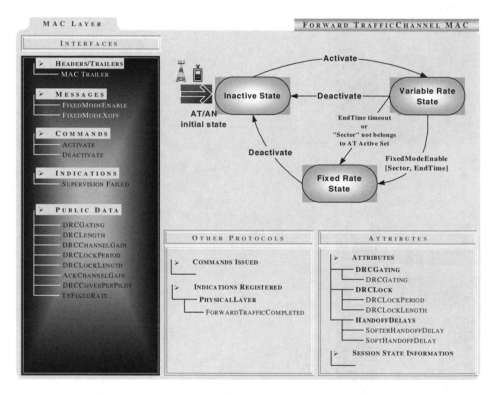

Figure 8.21 Forward traffic channel MAC protocol summary.

Table 8.33 MAC trailer for forward and reverse traffic channels

Forward and reverse Traffic channel trailer fields	Bit size	Description
ConnectionLayerFormat	1	'1' indicates connection layer packet Format B
MACLayerFormat	1	'1' indicates that the MAC layer payload is valid

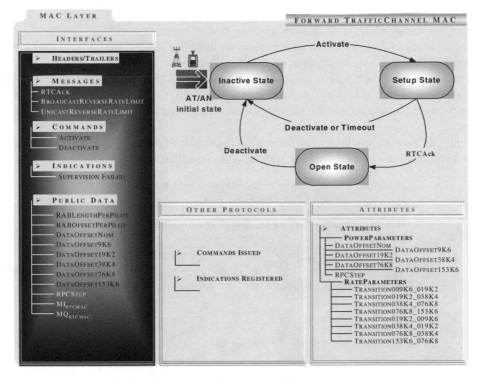

Figure 8.22   Reverse traffic channel MAC protocol summary.

The access channel capsule defines the following additional elements.

- FCS (32 bits): frame check sequence calculated based on the standard CRC-CCITT generator polynomial: $g(x) = x^{32} + x^{26} + x^{23} + x^{22} + x^{16} + x^{12} + x^{11} + x^{10} + x^8 + x^7 + x^5 + x^4 + x^2 + x^1 + 1$.

- PAD: bits with value '0' to complete the minimum multiple of 232 bits containing the message.

- Reserved (2 bits): bits with value '0'.

The encapsulation performed in the MAC layer is different according to the link direction. In the transmit direction, security layer packets are received as the MAC layer payload. The

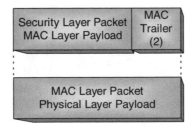

Figure 8.23   MAC layer capsule (forward and reverse traffic channel).

MAC layer header is added to the control and access channel packet capsules. A control channel header is added to the control channel packet capsules, whereas a frame checking sequence is added to the access channel packet capsules. The MAC layer trailer is added to the forward and reverse traffic channel packet capsules. Padding bits are also included in the control and access channel packet capsules. The resulting packet is forwarded to the physical layer.

In the receive direction, MAC packets are received from the physical layer, and the MAC layer payload that constitutes the security layer packet is extracted and sent to the security layer.

Figure 8.24 illustrates the access channel encapsulation. Tables 8.34 and 8.35 describe the access channel encapsulation header and the ATI record fields, respectively.

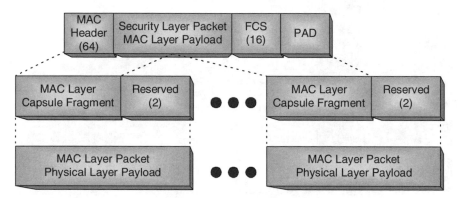

Figure 8.24   MAC layer capsule for access channel.

Table 8.34   MAC header for access channel

Access channel MAC header	Bit size	Description
Length	8	Length of all MAC capsule, including the header
SessionConfigurationToken	16	Indicates the negotiated parameters and selected protocols
SecurityLayerFormat	1	'1' indicates that the security layer applies security algorithm
ConnectionLayerFormat	1	'1' indicates use of Format B on the connection messages
Reserved	4	All bits with value '0'
ATIRecord	2 or 34	Access terminal identifier record is defined as in

Table 8.35   ATIRecord

ATIRecord fields	Bit size	Description
ATIType	2	'00' for BATI: Broadcast ATI '01' for MATI: Multicast ATI '10' for UATI: Unicast ATI '11' for RATI: Random ATI
ATI	0 or 32	32: if MATI, UATI or RATI 00: if BATI

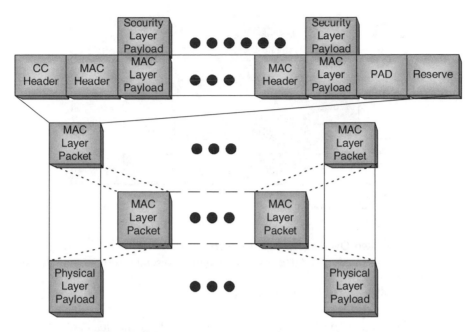

Figure 8.25 MAC layer capsule for control channel.

The control channel capsule defines the following additional elements.

- PAD: bits with value '0' to complete the minimum multiple of 1002 bits containing the message (including the reserved bits).

- Reserved (2 bits): all bits with value '0'.

Figure 8.25 illustrates the control channel encapsulation and Table 8.36 describes the encapsulation header.

Table 8.36 MAC header for control channel

Control channel header	Bit size	Description
SynchronousCapsule	1	'1' indicates that this capsule is synchronous (synchronised with the control channel cycle – every 256 slots)
FirstPack	1	'1' indicates that this is the first packet of the capsule
LastPack	1	'1' indicates that this is the last packet of the capsule
Offset	2	Indicates the number of slots from the beginning of the control channel cycle
SleepStateCapsuleDone	1	'0' if the MAC layer packet belongs to a CCsyncSS (control channel synchronous capsule for sleep state) and is not the last packet
Reserved	2	All bits with '0' value

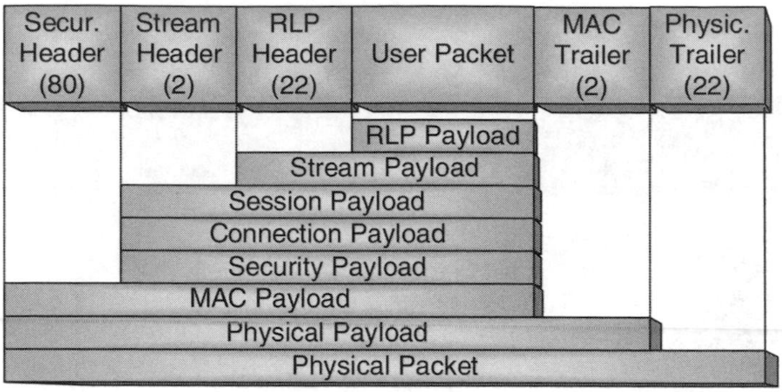

Figure 8.26 Forward traffic channel overhead for secure transmission.

8.3.2.4.7 Accumulated Overhead (For All Layers)

Figure 8.26 illustrates the accumulated overhead required to send information in the forward traffic channel with security.

The physical trailer consists of the following.

- FCS (16 bits): frame check sequence calculated based on the standard CRC-CCITT generator polynomial: $g(x) = x^{16} + x^{12} + x^5 + 1$.

- Tail (6 bits): all bits with a '0' value.

According to Figure 8.26, for a packet with 1024 bits, there is an overhead of 128 bits (12.5%). To transmit the same information without any specific security protocol this overhead is less than 5%. Considering that there are at least eight control channel slots in a 256 slots frame, a maximum ratio of 248/256 is dedicated to traffic slots, the expected overhead becomes 18% for transmission with security and 7.7% without any special security.

8.3.2.5 *Timers and Malfunction Detection*

Timers are used to keep actions under control, avoiding locks or over-rides. Some protocols use timers with the numeric constants specified in Table 8.37. The first letters in the timer name identify the protocol to which it relates.

ATs have an additional timer, which is independent of other functions, that runs continuously whenever power is applied to the transmitter. When this watchdog timer expires, it indicates a malfunction and inhibits the AT from transmitting. The maximum time allowed for expiration of this timer is two seconds.

8.3.3 CDMA System Time

CDMA systems adopt a reference timing, which is the same throughout the whole network. This timing uses the Global Positioning System (GPS) time and is synchronous with the

Table 8.37 Numeric constants for timers

Constant name	Description	Value
$T_{SLPWaitAck}$	Retransmission timer for a reliable-delivery SLP-D packet	400 ms
T_{SLPAck}	Time for the receiver to acknowledge an arriving reliable-delivery SLP-D packet	200 ms
$N_{SLPAttempt}$	Maximum number of attempts for sending a reliable-delivery SLP-D packet	3
$T_{RLPAbort}$	Time to wait for a re-transmission of an octet requested in a Nak message	500 ms
$T_{RLPFlush}$	Time to wait before re-transmitting the last transmitted octet	300 ms
$T_{FCResponse}$	Time period within which the AT and AN are to respond to flow control messages	1 s
$T_{SMPMinClose}$	Minimum recommended timer setting for Close	300 s
$N_{SMPKeepAlive}$	Maximum number of keep alive transactions within $T_{SMPClose}$	3
$T_{ADMPATResponse}$	Time to receive UATIAssignment after sending UATIRequest	120 s
$T_{ADMPAddress}$	The duration of time that the AT declares an address match if it receives a message that is addressed using either the old or the new UATI	180 s
$T_{ISPSync}$	Sync message transmission period	1.28 s
$T_{ISPPilotAcq}$	Time to acquire the pilot in the AT	60 s
$T_{ISPSyncAcq}$	Time to acquire the Sync message in AT	5 s
$T_{IDPATSetup}$	Maximum AT time in the connection setup state	2.5 s
$T_{IDPANSetup}$	Maximum access network time in the connection setup state	1 s
$N_{IDPSleep}$	Sleep period expressed in $T_{ACMPCycleLen}$ units	12
$T_{CSPClose}$	AN timer waiting for a responding ConnectionClose message	1.5 s
$N_{RUPActive}$	Maximum size of the active set	6
$N_{RUPCandidate}$	Maximum size of the candidate set	6
$N_{RUPNeighbour}$	Minimum size of the neighbour set	20
$T_{OMPQCSupervision}$	QuickConfig supervision timer expressed in $T_{ACMPCycleLen}$ units	12
$T_{OMPSPSupervision}$	SectorParameters supervision timer expressed in $T_{ACMPCycleLen}$ units	12
$N_{OMPSectorParameters}$	The recommended maximum number of control channel cycles between two consecutive SectorParameters message transmissions	4
$T_{KEPSigCompAN}$	Time to receive ATKeyComplete after sending ANKeyComplete	3.5 s
$T_{KEPSigCompAT}$	Time to send ATKeyComplete after receiving ANKeyComplete	3 s
$T_{KEPANResponse}$	Time to receive KeyResponse after sending KeyRequest	3.5 s
$T_{KEPATResponse}$	Time to send KeyResponse after receiving KeyRequest	3 s
$T_{CCMPSupervision}$	Control channel supervision timer value expressed in $T_{ACMPCycleLen}$ units	12
$T_{ACMPCycleLen}$	Length of control channel cycle	256 slots
$T_{ACMPAPSupervision}$	AccessParameters supervision timer	$12 \times T_{ACMPCycleLen}$
$T_{ACMPATProbeTimeout}$	Time to receive an acknowledgement at the AT for a probe before sending another probe	128 slots
$T_{ACMPANProbeTimeout}$	Maximum time to send an acknowledgement for a probe at the AN	96 slots

Table 8.37 (*Continued*)

Constant name	Description	Value
$T_{ACMPTransaction}$	Time for ATl to wait after a successful transmission before returning a TxEnded indication	1 s
$N_{ACMPAccessParameters}$	Recommended maximum number of slots between transmission of two consecutive AccessParameters messages	$3 \times T_{ACMPCycleLen}$
$T_{FTCMDRCSupervision}$	DRC supervision timer	240 ms
$T_{FTCMPRestartTx}$	Reverse channel re-start timer expressed in $T_{ACMPCycleLen}$ units	12
$N_{FTCMPRestartTx}$	Number of consecutive slots of non-null rate DRCs to re-enable the reverse traffic channel transmitter once it is disabled due to DRC supervision failure	16
$T_{RTCMPATSetup}$	Maximum time for the AT to transmit the reverse traffic channel in the setup state	1.5 s
$T_{RTCMPANSetup}$	Maximum time for the AN to acquire the reverse traffic channel and send a notification to the AT	1 s

Universal Coordinated Time (UTC). The GPS time and the UTC, however, differ by a number of seconds. Even though CDMA systems do not adjust clocks to match this difference, they keep track of it to maintain the synchronisation.

Figure 8.27 illustrates the CDMA system time process. The figure shows the AN offset pilot PN sequences and the AT common short-code PN sequences for I and Q channels. The CDMA system time is the absolute time referenced at the AN antenna, offset by the time delay of the transmission.

8.3.4 Synchronisation and Timing

When the AT acquires the pilot channel and receives the Sync message transmitted on the control channel, it establishes a time reference for transmission of chips, symbols, slots, frames and system timing. The AT time reference is used for the transmission of reverse traffic and access channels.

- This timing reference is set to be within ± 1 μs of the time of occurrence of the earliest arriving multi-path component being used for demodulation, regardless of which pilot this component belongs to.

- The rate of change for timing corrections shall not be faster than 203 ns (1/4 PN chip) per 200 ms, not slower than 305 ns (3/8 PN chip) per second.

The AN (sector) also uses the system time as a reference for all its time-critical transmission components, including pilot PN sequences, slots and Walsh functions. The

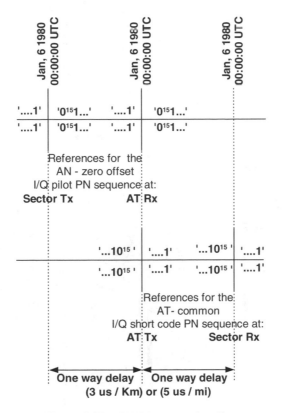

Figure 8.27 CDMA system time line.

synchronisation to the system time is obtained through reliable external means at each sector.

- The sector must keep transmitting within ±10 μs of system time for a period of, at least, 8 h, in case of losing external time source.

- It radiates the pilot PN sequence within ±3 μs of system time and the pilot PN sequence within ±10 μs of system time.

- The rate of change for timing corrections shall not be faster than 102 ns (1/8 PN chip) per 200 ms.

8.4 CHANNEL STRUCTURE

Figure 8.28 illustrates the channel structure of EVDO systems. In the figure, values in [⋯] indicate the number of chips in the Time Division Multiplex (TDM) phase. Channels

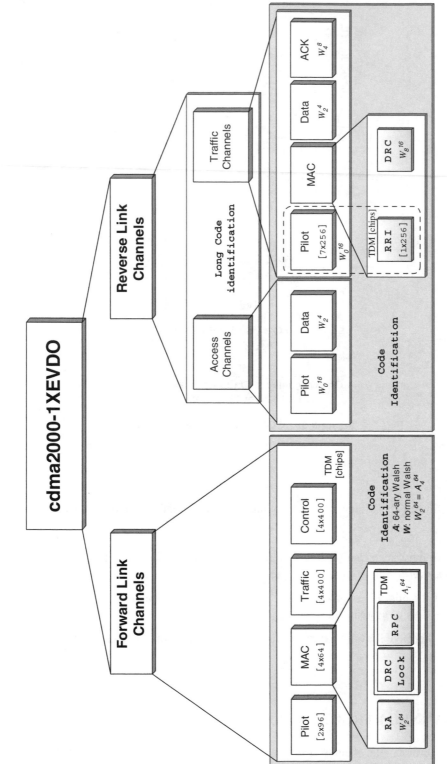

Figure 8.28 EVDO air interface channel structure.

identified by Walsh codes have the code indicated. Some channels, such as control and forward traffic, have no sub-channels multiplexed by Walsh codes. The forward channel is a pipe channel, time-division multiplexed to accommodate four sub-channels: pilot, MAC, traffic and control.

In this channel structure, the pilot channel is a channel that irradiates constant power but without any information (unmodulated signals with binary code '0'). The functions of this channel include evaluation of the power level of the current server and of neighbouring cells, evaluation of BER, help in coherent detection, maintenance of the connected route by triggering routing messages and monitoring of cell with the control channel. There is only one pilot channel in the forward link but there are two in the reverse link, one for access channels and other for traffic channels.

The Medium Access Control (MAC) channel is dedicated to media access control activities. This channel is sub-divided into three sub-channels in the forward link and two sub-channels in the reverse link.

In the forward link, the three MAC sub-channels are the Reverse Activity (RA) channel, the Reverse Power Control (RPC) channel, and the Data Rate Control (DRC) lock channel. The RA is a broadcast channel that indicates the reverse channel activity level. The RPC controls the power of the reverse channel for an AT. The DRC lock determines if the AN can receive the DRC sent by the AT.

In the reverse link, the first MAC sub-channel indicates the rate of the reverse traffic data channel, thus it is known as the Reverse Rate Indicator (RRI) channel. The second subchannel is the DRC channel which is addressed to all users as a broadcast, and indicates the rate and the sector from which the AT wants to receive the forward traffic channel, indicating to the reverse link MAC algorithm whether the AT data rate has to be reduced (due to unacceptably high load levels) or increased (due to low load levels).

The forward traffic channel carries information to a specific AT. This channel is used as a non-dedicated resource prior to AT authentication. After successful authentication, the channel can be used as a dedicated resource. In the reverse link, the traffic channel is subdivided into pilot, MAC, data and Acknowledge (ACK) channels. The pilot and MAC channels are described in the previous paragraphs. The AT uses the ACK channel to indicate success or failure in the reception of the forward traffic channel. The reverse link data sub-channels contain data from MAC layer packets.

The control channel is a broadcast channel of the forward link used by the AN to transmit broadcast messages or signalling information.

The reverse link access channel is used by ATs to initiate communication with the AN or to respond to an AT directed message. These channels are subdivided into a pilot and a data sub-channel.

Each of the channels in the forward and reverse links is described in detail in the following sections.

8.4.1 Reverse Link Channels

The reverse link is divided into access and traffic channels, which are, in turn, sub-divided into more specific logical channels. Figure 8.29 illustrates the reverse link channel structure.

Each time slot is defined by 2048 PN consecutive chips, which, for a 1.2288 Mcps EVDO system, represent a 600 Hz slot ratio (1.67 ms for each time slot). Reverse link channel

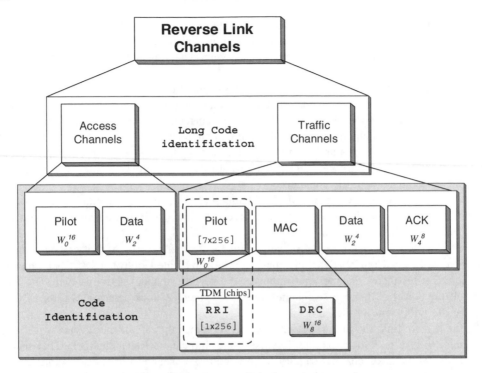

Figure 8.29   Reverse link channel structure.

frames are divided into 16 consecutive slots aligned to shifts of the zero-offset PN sequence and to the system time on even-second ticks (32768 chips = 26.67 ms).

Reverse access and traffic channels use the long-code identification system. The in-phase and quadrature-phase long codes are generated from a sequence, called the long-code generating sequence, and modulo-2 multiplied by two different masks (M_I and M_Q). These masks vary according to the channel used for transmission by the access terminal. The following sections describe the masks used for each of the reverse link channels. Chapter 3 provides a detailed description of the long-code generation process.

8.4.1.1 Access Channel

The access channel is part of the reverse link in EVDO systems and is sub-divided into the pilot and the data channels. Figure 8.30 shows the access channel structure at the physical layer level. The following sections describe the pilot and data channels in detail.

Access Channel Identification

The access channel is identified at the access network by the access long-code PN sequence (Figure 8.31), which is produced using the 42-bit long MI_{ACMAC} and MQ_{ACMAC} masks, which are given as public data of the access channel MAC protocol.

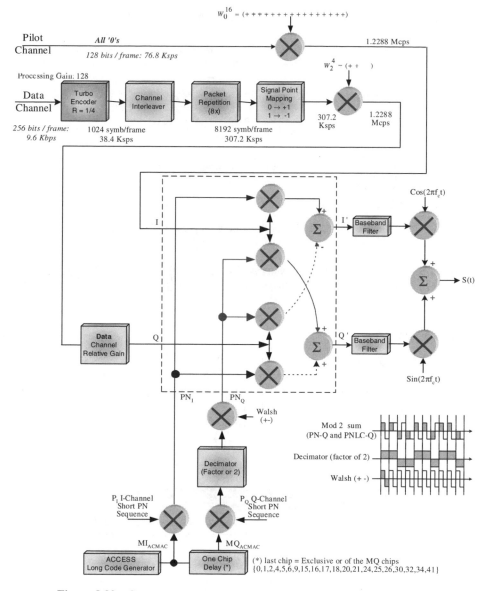

Figure 8.30 General structure of the access channel at the physical layer.

The 2 most significant bits of the MI_{ACMAC} mask are set to '1', the following 8 bits represent the access cycle number and the remaining 32 bits are a permutation of the colour code and the last 24 bits of the sector ID. Both the sector ID and the colour code are given as public data of overhead messages protocol.

The MQ_{ACMAC} mask is derived from the MI_{ACMAC}. The 41 most significant bits of the first mask are the 41 least significant bits of the other, i.e. MQ_{ACMAC} is MI_{ACMAC} shifted to

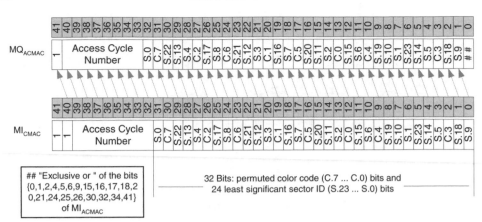

Figure 8.31   Masks for the access long-code PN sequence.

the left, as in IS-95 and cdma2000 systems. The least significant bit of MQ_{ACMAC} is the exclusive OR of bits {0, 1, 2, 4, 5, 6, 9, 15, 16, 17, 18, 20, 21, 24, 25, 26, 30, 32, 34, 41} of MI_{ACMAC}.

Access Channel Packet

Access channel physical layer packets are 256-bit long. Each packet carries one access channel MAC layer packet with a Frame Check Sequence (FCS) and encoder tail bits (Figure 8.32). The data channel contained in the access channel is used, at the physical layer, to transport these packets. Access channel physical layer packets are transmitted at a fixed data rate of 9.6 kbps (256 bits per frame of 16 slots).

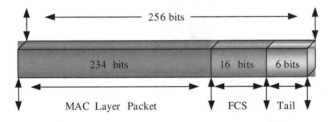

Figure 8.32   Configuration of access channel packet at physical layer.

Access Channel Function

Access terminals use the access channel to initiate communication with the network or to respond to an AT directed message. An access probe consists of a preamble and a capsule. During the preamble only the access pilot channel is transmitted. The capsule, or packet transmission, transmits both the access pilot and data channels. The transmission power of the pilot channel varies from the preamble to the actual packet transmission so that the overall power is kept constant, as illustrated in Figure 8.33.

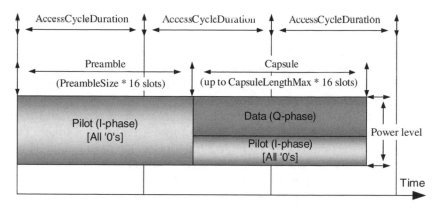

Figure 8.33 Access probe timing.

Access channel cycles are the time instants at which the access terminal may start an access probe. Probes may only begin at slots when T mod (AccessCycleDuration) = 0. Access channel MAC protocol messages (Table 8.39) sent in the control channel defines operational parameters such as AccessCycleDuration, PreambleLength and CapsuleLength-Max.

Before sending the first probe of the sequence, the access terminal performs two tests in this order: the silence period test and the persistence test. In the silence period test, the terminal waits until the beginning of the first access channel cycle to verify that the transmission of the access probe does not overlap with the reverse link silence interval.[1]

The persistence test is executed only after the silence period test is completed. In this test, the terminal compares an element of the APersistence vector, which varies according to the AT class, to a pseudo-random number. If the value from the vector is greater then the random number the test succeeds. If the test succeeds or if the number of consecutive unsuccessful tests exceeds (4/(APersistence vector element)) then the terminal can begin transmitting in this access channel cycle, otherwise the tests must be repeated, starting from the silence period test.

The contents of the data channel can not change during the probe sequence. The power level during an access probe is constant and is defined by open-loop estimation. The power level increases in PowerStep steps between consecutive probes.

Before transmitting a probe, the terminal generates a pseudo-random number, which is a uniformly distributed integer random number between 0 and ProbeBackoff. This number is added to a constant y (initial value '0') producing y_{total}. The time interval for sending the next probe is τ_P slots after the end of the current probe, where $\tau_P = T_{ACMPATProbeTimeoout} + (y_{total} \times \text{AccessCycleDuration})$. If any portion of this next probe will overlap with the reverse link silence interval, then another random number is generated and y_{total} is re-calculated to determine a new τ_P.

If the access terminal receives an ACAck message or if a given number of probes (ProbeNumStep) has already been transmitted within a sequence, the terminal does not send

---

[1] The reverse silence link interval represents the duration of the ReverseLinkSilenceDuration frames that starts at times T where the following equation is satisfied: T mod $(2048 \times 2^{ReverseLinkSilencePeriod} - 1) = 0$. ReverseLink-SilenceDuration and ReverseLinkSilencePeriod are given as public data by the overhead messages protocol.

another probe within the same sequence. The interval between probe sequences (inter-sequence backoff) is also calculated using a random number. The terminal generates an integer random number (k) between 0 and ProbeSequenceBackoff. To determine the number of inter-sequence slots, the terminal multiplies this number by the AccessCycle Duration and adds this total to $T_{\text{ACMPATProbeTimeoout}}$, i.e. $\tau_P = (k\text{AccessCycleDuration}) + T_{\text{ACMPATProbeTimeoout}}$.

Each probe within a sequence is transmitted with an increased power level. The initial power is $X_0 = (\text{mean Rx pilot power}) + \text{OpenLoopAdjust} + \text{ProbeInitialAdjust}$. The power at step n is $X_0 + (n - 1)$ PowerStep. PowerStep has a resolution of in resolution of 0.5 dB and is transmitted as public data of the default access channel MAC protocol.

During an access probe sequence transmission, the AT may update the mean output power due to changes in the received mean input power. It is possible to estimate the mean output power of the first access probe, i.e. its Initial Power (IP) by measuring the power level for the carrier in which the AT is operating and taking the forward pilot channel with the best power level in the area where the AT is located, as presented in Figure 8.34.

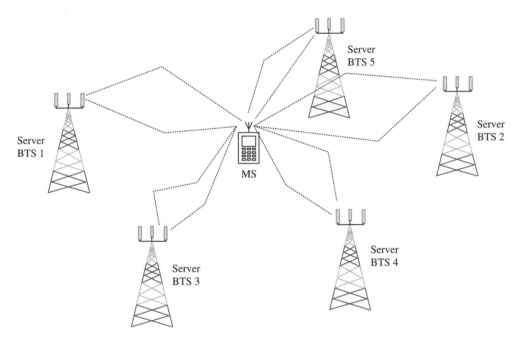

Figure 8.34   Possible reverse link components originating at the AT.

The access network determines ProbeNumStep, which may range from 1 to 15. The AN also establishes a maximum number of probe sequences that can be transmitted for a single access attempt (ProbeSequenceMax), this value also ranges from 1 to 15. The transmission of probes stops when the terminal receives an ACAck message, the protocol receives a de-activate command, or the maximum number of probes within a sequence (ProbeNumStep) has been transmitted. Figure 8.35 illustrates an access attempt, also showing the structure of access sequences.

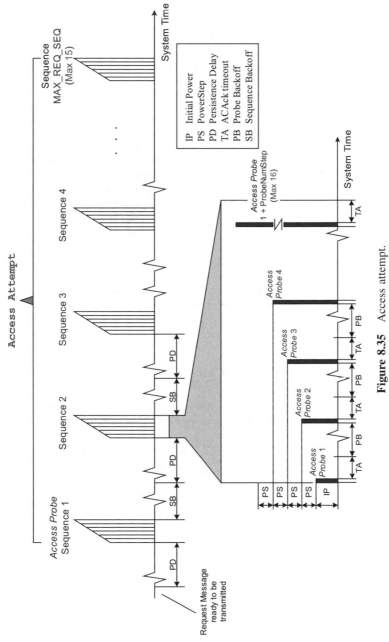

Figure 8.35 Access attempt.

Table 8.38 Attribute related to the access channel behavior

Protocol	Attribute name		ID
ACMP	InitialConfiguration		0

Field	Bit size	Default	Description
Length	8	N/A	Length of the attribute excluding the length field
AttributeID	8	N/A	Identification of the attribute within the protocol
		One or more of the following record:	
ValueID	8	N/A	Identification of the value within this attribute list
ProbeSequenceMax	4	3	Maximum number of probe sequences for a single access attempt, set by AN in a range of 1 to 15. It is equivalent to ProbeNumSteps in the Access-Parameters message
ProbeBackoff	4	4	Upper limit of the inter-probe backoff set by AN in units of AccessCycleDuration, representing the time interval between access probes
ProbeSequenceBackoff	4	8	Upper limit of the inter-probe sequence backoff set by AN in units of AccessCycleDuration representing the time interval between sequences of access probes
Reserved	4	N/A	The AN sets this field to 0

The Access Channel MAC Protocol (ACMP) attribute related to access probes is the InitialConfiguration, described in Table 8.38.

The two messages that determine the access channel behaviour are ACAck and AccessParameters, both sent by the access channel MAC protocol. Tables 8.39 and 8.40 describe these messages.

The access channel not only has its behaviour guided by messages, but it also transmits messages generated by several layers. Table 8.41 provides a list of the messages transmitted over this channel.

Table 8.39 ACAck message identification and description

Protocol	Message name	ID	Physical channel	Delivery	Address	Priority
ACMP	ACAck	0	CC	Best	Unicast	10

The AN sends this message acknowledging a valid access channel MAC layer capsule aborting the access probes

Field	Bit size	Description
MessageID	8	Identity of this message in this protocol, set by the AN

Table 8.40 AccessParameters message identification and description

Protocol	Message name	ID	Physical channel	Delivery	Address	Priority
ACMP	AccessParameters	1	CC	Best	Broadcast	30

The AN issues this message to send Access Channel information to ATs

Field	Bit size	Description
MessageID	8	Identity of this message in this protocol, set by the AN
AccessCycleDuration	8	Number of slots defining the access channel cycle, set by the AN
AccessSignature	16	Value changed each time the AccessParameters message changes
OpenLoopAdjust	8	Negative value of the AT nominal power in the open-loop power estimate, it is set by the AN and expressed in units of 1 dB
ProbeInitialAdjust	5	Correction factor set by the AN to be used by the AT in the open-loop power estimate for initial transmission on the access channel. It is expressed as a complement of two value in units of 1 dB
ProbeNumStep	4	Maximum number of access probes designated by the AN that can be transmitted by the AT in a single probe sequence. The value range from 1 to 15
PowerStep	4	Probe power increase step set by the AN in a resolution of 0.5 dB
PreambleLength	3	Length, in frames, of the access probe preamble, set by the AN
CapsuleLengthMax	4	Maximum number of frames in an Access Channel Capsule, set by the AN in a range from 2 to 15
APersistence	N*6	Vector with N elements (AT class) indicating the persistence probability number used in the persistence test
Reserved	variable	Bits set to '0' to align the message in an octet basis

Table 8.41 Messages transmitted over the access channel

Access channel	Layer
LocationUpdate.LocationNotification	Application
LocationUpdate.LocationComplete	Application
FlowControl.XonRequest	Application
FlowControl.XoffRequest	Application
FlowControl.DataReadyAck	Application
SessionManagement.SessionClose	Session
SessionManagement.KeepAliveRequest	Session
SessionManagement.KeepAliveResponse	Session
AddressManagement.UATIRequest	Session
AddressManagement.UATIComplete	Session
AddressManagement.HardwareIDResponse	Session
IdleState.ConnectionRequest	Connection
RouteUpdate.RouteUpdate	Connection

8.4.1.1.1   Pilot Channel

Pilot Channel Identification

After the access channel is identified, the pilot channel is extracted using Walsh function 0 with 16 chips: $W_0^{16} = (+ + + + + + + + + + + + + + + +)$ (Figure 8.36).

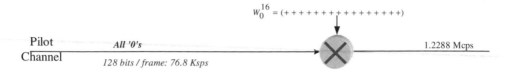

Figure 8.36   Pilot channel at access channel.

Pilot Channel Function

The AT transmits only un-modulated symbols with a binary value of '0' on the pilot Channel. This channel is continuously transmitted on the I channel using the 16-chip Walsh function 0, while the access channel is transmitting.

8.4.1.1.2   Data Channel

Data Channel Identification

After the access channel is identified, the data channel is extracted using Walsh function 2 with 4 chips: $W_2^4 = (+ + - -)$ (Figure 8.37).

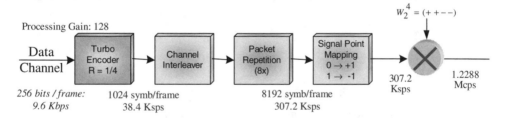

Figure 8.37   Data channel at access channel.

Data Channel Function

Multiple access channel physical layer packets can be transmitted on the data channel during every access probe. The packets are transmitted at a fixed data rate of 9.6 Kbps on the Q channel using the 4-chip Walsh function 2, and are preceded by a preamble where only the pilot channel is transmitted.

8.4.1.2   Reverse Traffic Channel

The reverse traffic channel is part of the reverse link in EVDO systems and is subdivided into pilot, MAC, data and ACK channels. Figures 8.38 and 8.39 show the reverse traffic

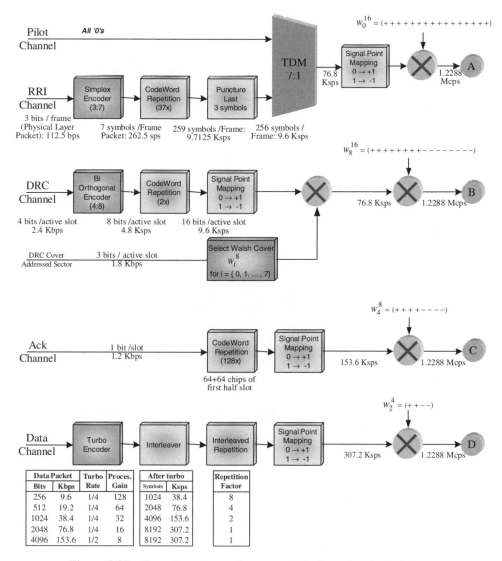

Figure 8.38 General structure of reverse traffic channel at physical layer.

channel structure at the physical layer level. The following sections describe each of the sub-channels in detail.

Reverse Traffic Channel Identification

The reverse traffic channel is identified at the access network by the reverse traffic long-code PN sequence, which is produced using the 42-bit long MI_{RTCMAC} and MQ_{RTCMAC} masks (Figure 8.40), which are given as public data of the access channel MAC protocol.

The 10 most significant bits of the MI_{RTCMAC} mask are set to '1', the remaining 32 bits are permuted Access Terminal Identifier (ATI) bits.

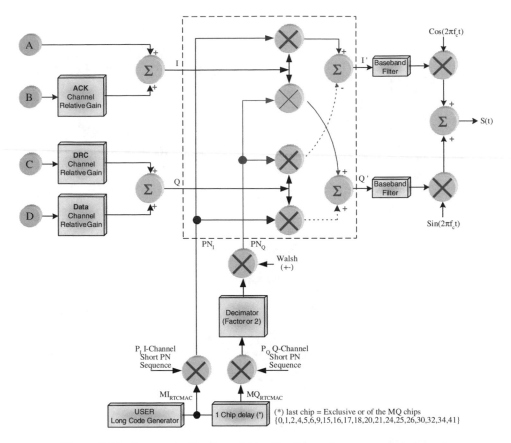

Figure 8.39 Reverse traffic channel identification and power-control elements.

The MQ_{RTCMAC} mask is derived from the MI_{RTCMAC}. The 41 most significant bits of the first mask are the 41 least significant bits of the other, i.e. MQ_{RTCMAC} is MI_{RTCMAC} shifted to the left, as in IS-95 and cdma2000 systems. The least significant bit of MQ_{RTCMAC} is the exclusive-OR of bits {0, 1, 2, 4, 5, 6, 9, 15, 16, 17, 18, 20, 21, 24, 25, 26, 30, 32, 34, 41} of MI_{RTCMAC}.

Reverse Traffic Channel Packet

Reverse traffic channel physical layer packet length is a power of two (2^n), with n varying from 8 to 12. Each packet carries one access channel MAC layer packet with variable length, a 16-bit Frame Check Sequence (FCS), and six encoder tail bits (Figure 8.41).

Reverse Traffic Channel Function

The terminal uses the reverse traffic channel to transmit user-specific traffic or signalling information to the network. The reverse traffic channel consists of a pilot channel, a MAC channel (subdivided into an RRI and a DRC channel), an ACK channel and a data channel.

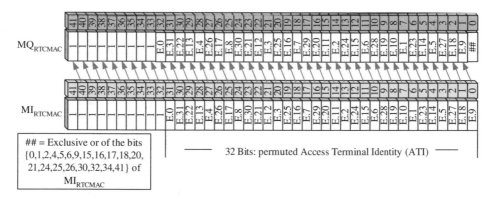

Figure 8.40 presents the masks. The box reads:

= Exclusive or of the bits {0,1,2,4,5,6,9,15,16,17,18,20, 21,24,25,26,30,32,34,41} of MI$_{RTCMAC}$

—— 32 Bits: permuted Access Terminal Identity (ATI) ——

Figure 8.40 Masks for the reverse traffic long-code PN sequence.

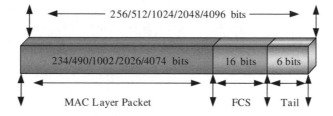

256/512/1024/2048/4096 bits

234/490/1002/2026/4074 bits | 16 bits | 6 bits

MAC Layer Packet   FCS   Tail

Figure 8.41 Configuration of reverse traffic channel packet at physical layer.

The transmission rate supported on the data channel is specified by the reverse traffic channel MAC Protocol and can be 9.6, 19.2, 38.4, 76.8 or 153.6 kbps. The data channel gain relative to the reverse traffic pilot channel depends on the data rate.

Two attributes of the reverse traffic channel MAC protocol are used to configure the reverse traffic channel: the PowerParameterAttribute and the RateParameterAttribute. Tables 8.42 and 8.43 describe these parameters.

The messages sent by the reverse traffic channel MAC protocol that determine the reverse traffic channel behaviour are RTCAck, BroadcastReverseRateLimit and UnicastReverse-RateLimit, respectively, described in Tables 8.44–8.46.

The rate control is performed based on the following attributes and messages.

- CurrentLimit: from BroadcastReverseRateLimit and UnicastReverseRateLimit messages.

- CurrentRate: current transmission rate.

- CombinedBusyBit: set to '1' if any of the sectors included in the ActiveSet indicates a value '1' for the Reverse Activity Bit (RAB).

- MaxRate: calculated according Table 8.48.

Table 8.42 PowerParameterAttribute description

Protocol	Attribute name		ID
RTCMP	PowerParameterAttribute		0

Field	Bit size	Default	Description
Length	8	N/A	Length of the attribute excluding the length field
AttributeID	8	N/A	Identification of the attribute within the protocol
One or more of the following record			
ValueID	8	N/A	Identification of the record within this attribute list
DataOffsetNom	4	0	Power level for the nominal offset, expressed as complement of 2, in steps of 0.5 dB
DataOffset9K6	4	0	Power level for 9.6 Kbps rate, expressed as complement of 2, in steps of 0.25 dB
DataOffset19k2	4	0	Power level for 19.2 Kbps rate, expressed as complement of 2, in steps of 0.25 dB
DataOffset38k4	4	0	Power level for 38.4 Kbps rate, expressed as complement of 2, in steps of 0.25 dB
DataOffset76K8	4	0	Power level for 76.8 Kbps rate, expressed as complement of 2, in steps of 0.25 dB
DataOffset153K6	4	0	Power level for 153.6 Kbps rate, expressed as complement of 2, in steps of 0.25 dB
RPCStep	2	1	Reverse power-control step: '00' for 0.5 dB, '01' for 1 dB. Any other value is invalid.
Reserved	6	N/A	The AN sets this field to 0

Table 8.43 RateParameterAttribute description

Protocol	Attribute name		ID
RTCMP	RateParameterAttribute		0

Field	Bit size	Default	Description
Length	8	N/A	Length of the attribute excluding the length field
AttributeID	8	N/A	Identification of the attribute within the protocol
One or more of the following record			
ValueID	8	N/A	Identification of the record within this attribute list
Transition009k6_019k2	8	0x30	Probability of increasing the terminal transmission rate from 9.6 kbps to 19.2 kbps[a]
Transition019k2_038k4	8	0x10	Probability of increasing the terminal transmission rate from 19.2 kbps to 38.4 kbps[a]
Transition038k4_076k8	8	0x08	Probability of increasing the terminal transmission rate from 38.4 kbps to 76.8 kbps[a]
Transition076k8_153k6	8	0x08	Probability of increasing the terminal transmission rate from 76.8 kbps to 153.6 kbps[a]

Table 8.43 (*Continued*)

Protocol	Attribute name			ID
RTCMP	RateParameterAttribute			0

Field	Bit size	Default	Description
Transition019k2_009k6	8	0x10	Probability of decreasing the terminal transmission rate from 19.2 kbps to 9.6 kbps[a]
Transition038k4_019k2	8	0x10	Probability of decreasing the terminal transmission rate from 38.4 kbps to 19.2 kbps[a]
Transition076k8_038k4	8	0x20	Probability of decreasing the terminal transmission rate from 76.8 kbps to 38.4 kbps[a]
Transition153k6_076k8	8	0xFF	Probability of decreasing the terminal transmission rate from 153.6 kbps to 76.8 kbps[a]

[a] Probability expressed in units of 1/255. The terminal must support all the valid values specified by these fields.

Table 8.44 RTCAck message identification and description

Protocol	Message name	ID	Physical channel	Delivery	Address	Priority
RTCMP	RTCAck	0	FTC	Reliable	Unicast	10

The AN sends this message to notify that it has acquired the reverse traffic channel

Fields	Bit size	Description
MessageID	8	Identification of this message within this protocol, set by the AN

Table 8.45 BroadcastReverseRateLimit message identification and description

Protocol	Message name	ID	Physical channel	Delivery	Address	Priority
RTCMP	BroadcastReverseRateLimit	1	CC	Best	Broadcast	40

The AN issues this message to control the data rate in the reverse link

Fields	Bit size	Description
MessageID	8	Identification of this message within this protocol, set by the AN
RPCCount	6	Maximal number of Reverse Power Control (RPC) supported by the sector
		RPCCount occurrences of the following field
RateLimit	4	The nth occurrence is related to the AT related to MACIndex $(64-n)$ according to Table 8.47
		. . .
Reserved	Variable	Bits set to '0' to align the message in an octet basis

Table 8.46   UnicastReverseRateLimit message identification and description

Protocol	Message name	ID	Physical channel	Delivery	Address	Priority
RTCMP	UnicastReverseRateLimit	2	FTC	Reliable	Unicast	40

The AN issues this message in order to control de data rate in the reverse link for a particular AT

Fields	Bit size	Description
MessageID	8	Identification of this message within this protocol, set by the AN
RateLimit	4	The nth occurrence is related to the AT related to MACIndex $(64-n)$ according to Table 8.47
Reserved	4	Bits set to '0' to align the message in an octet basis

Table 8.47   Rate limit indicator

Data rate (Kbps)	RateLimit
0	0
9.6	1
19.2	2
38.4	3
76.8	4
153.6	5
Invalid	Other values

Table 8.48   MaxRate calculation

Current rate (kbps)	Combined busy bit	Condition	MaxRate if true (kbps)	MaxRate if false (kbps)
0	'0'	True	9.6	N/A
9.6	'0'	$x <$ Transition009k6_019k2	19.2	9.6
19.2	'0'	$x <$ Transition019k2_038k4	38.4	19.2
38.4	'0'	$x <$ Transition038k4_076k8	76.8	38.4
76.8	'0'	$x <$ Transition076k8_153k6	153.6	76.8
153.6	'0'	False	N/A	153.6
0	'1'	False	N/A	9.6
9.6	'1'	False	N/A	9.6
19.2	'1'	$x <$ Transition019k2_009k6	9.6	19.2
38.4	'1'	$x <$ Transition038k4_019k2	19.2	38.4
76.8	'1'	$x <$ Transition076k8_038k4	38.4	76.8
153.6	'1'	$x <$ Transition153k6_076k8	76.8	153.6

The reverse traffic channel is also responsible for transporting messages from several layers. Table 8.49 provides a list of these messages.

8.4.1.2.1   Composite Pilot Channel (Pilot and RRI channels)
The pilot channel has the same functions as that in cdma2000-1X standards. The main difference is that in EVDO, the pilot channel is time-division multiplexed with the MAC

Table 8.49 Messages transmitted on the reverse traffic channel

Reverse traffic channel	Layer
SLP.ResetAcked	Application
RadioLinkProtocol.Reset	Application
RadioLinkProtocol.ResetAck	Application
RadioLinkProtocol.Nak	Application
LocationUpdate.LocationNotification	Application
LocationUpdate.LocationComplete	Application
FlowControl.XonRequest	Application
FlowControl.XoffRequest	Application
FlowControl.DataReadyAck	Application
SessionManagement.SessionClose	Session
SessionManagement.KeepAliveRequest	Session
SessionManagement.KeepAliveResponse	Session
AddressManagement.UATIComplete	Session
AddressManagement.HardwareIDResponse	Session
SessionConfiguration.ConfigurationComplete	Session
SessionConfiguration.ConfigurationRequest	Session
SessionConfiguration.ConfigurationResponse	Session
ConnectedState.ConnectionClose	Connection
RouteUpdate.RouteUpdate	Connection
RouteUpdate.TrafficChannelComplete	Connection
RouteUpdate.AttributeOverrideResponse	Connection
DHKeyExchange.KeyResponse	Security
DHKeyExchange.ATKeyComplete	Security
ForwardTrafficChannelMAC.FixedModeEnable	MAC
ForwardTrafficChannelMAC.FixedModeOff	MAC

Reverse Rate Indicator (RRI) channel (Figure 8.42). This multiplexed channel, however, is still known as the pilot channel. The RRI channel is part of the MAC channel.

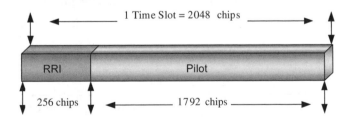

Figure 8.42 Reverse pilot channel TDM structure (RRI and Pilot channels).

Composite Pilot Channel Identification

After the reverse traffic channel is identified, the composite pilot channel is extracted using Walsh function number 0 with 16 chips: $W_0^{16} = (+ + + + + + + + + + + + + + + +)$. The first 256 chips define the RRI channel and the last 1792 chips, the pilot channel.

Composite Pilot Channel Function

The pilot channel transmits un-modulated symbols with a binary value of '0'. This channel is time multiplexed with the RRI channel on the in-phase quadrature (I channel). Both pilot and RRI channels are transmitted at the same power level.

The RRI channel indicates the data rate at which the AT can receive the reverse traffic data channel. This means that, in EVDO, the terminal determines the reverse link data rate and not the network as usual in IS95 and cdma2000.

Each frame has the data rate represented by a 3-bit RRI symbol (16-slot physical layer packet). A simplex encoder encodes each RRI symbol into a 7-bit codeword as in Figure 8.43. Each codeword is repeated 37 times and the last three symbols are punctured. The resulting

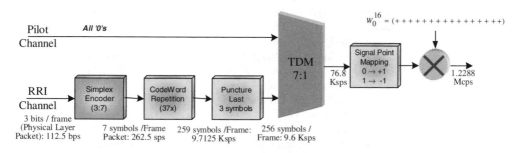

Figure 8.43   Pilot channel at access channel.

256 binary symbols per physical layer packet are time-division multiplexed with pilot channel symbols. The multiplexed sequence producess 256 RRI chips per slot after being spread with Walsh function 0. RRI chips are time-division multiplexed into the first 256 chips of every slot as in Figure 8.42. When no physical layer packet is transmitted on the reverse traffic channel, the access terminal transmits the zero data rate RRI codeword on the RRI Channel, as indicated in Table 8.50.

The RRI codeword is based on Walsh code length 8, removing the left most symbol ('0') of each code as depicted in Table 8.50.

Table 8.50   Reverse channel data rate indicator.

Data rate (Kbps)	RRI index	RRI symbol	RRI codeword
0	0	000	0000000
9.6	1	001	1010101
19.2	2	010	0110011
38.4	3	011	1100110
76.8	4	100	0001111
153.6	5	101	1011010
Reserved	6	110	0111100
Reserved	7	111	1101001

8.4.1.2.2 Data Rate Control Channel (DRC)

Data Rate Control Channel Identification

After the reverse traffic channel is identified, the DRC channel is extracted using Walsh function number 8 with 16 chips: $W_8^{16} = (+ + + + + + + + - - - - - - - -)$.

Data Rate Control Channel Function

The terminal uses the DRC channel to indicate the selected serving sector and the requested forward channel data rate to the network. The DRC channel shall be transmitted on the cross-quadrature (Q channel) as illustrated in Figure 8.44.

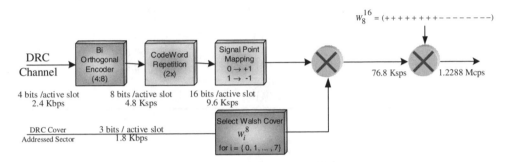

Figure 8.44 DRC channel (reverse traffic channel) structure.

Even though the DRC channel is part of the reverse link, the three parameters used to determine the DRC channel behaviour are defined as public data of the forward traffic channel MAC protocol: DRCCover, DRCLength and DRCGating.

DRCCover defines the cover mapping used to map the requested data rate into a 4-bit DRC value. The DRC channel transmission is spread using an 8-ary Walsh function corresponding to the selected serving sector.

The DRC channel transmits at a data rate of 600/DRCLength values per second; when DRCLength is greater than one, the DRC value and DRCCover inputs in Table 8.51 are repeated for consecutive DRCLength slots. The access terminal sets the DRC to the maximum value that channel conditions permit for the sector the DRC refers to. If nor even the lowest rate is permitted, the terminal uses the null rate.

DRC values are block encoded to generate bi-orthogonal codewords with 8 bits, as in Table 8.52. Each codeword is transmitted twice per slot. Each bit of a repeated codeword is spread by an 8-ary Walsh function (i) defined in Table 8.53, where (i) equals DRCCover, and represents the sector from the active list required to send data. Each Walsh chip of the 8-ary Walsh function is further spread by the 16-chip Walsh function 8. The DRC channel transmits each DRC value over DRCLength slots.

The parameter DRCGating determines whether the access terminal supports gated transmission. This parameter may assume two values, where '1' indicates that the terminal is capable of gated DRC transmissions. When DRC transmissions are gated, each DRC symbol is transmitted over the last slot of every DRCLength slots. Slots where the DRC channel is not gated off are called active slots.

Table 8.51   DRC and forward traffic data rate mapping

DRC index	DRC value	Codeword (bi-orthogonal)	Data rate (Kbps)	Packet slots
0	0000	00000000	Null	
1	0001	11111111	38.4	16
2	0010	01010101	76.8	8
3	0011	10101010	153.6	4
4	0100	00110011	307.2	2
5	0101	11001100	307.2	4
6	0110	01100110	614	1
7	0111	10011001	614	2
8	1000	00001111	921.6	2
9	1001	11110000	1228.8	1
10	1010	01011010	1228.8	2
11	1011	10100101	1843.2	1
12	1100	00111100	2457.6	1
13	1101	11000011	Invalid	
14	1110	01101001	Invalid	
15	1111	10010110	Invalid	

Table 8.52   DRC cover codeword

Walsh function	Codeword
W_0^8	00000000
W_1^8	01010101
W_2^8	00110011
W_3^8	01100110
W_4^8	00001111
W_5^8	01011010
W_6^8	00111100
W_7^8	01101001

The forward traffic channel MAC protocol establishes the forward traffic channel transmission time corresponding to a DRC symbol, which starts at the mid-slot point. Figures 8.45 and 8.46 illustrate this timing for non-gated and gated transmission, respectively.

The DRC channel operates in a 600 Hz time-slotted basis to match the timing structure of the forward link and operate at 1.67 ms intervals. The estimation of the DRC value needed to achieve this depends not only on the C/I calculated from the received pilots but also on the attempt to predict a future value, counter-balancing the effect of possible increases on the mobile speed. This is required, because an increase on the speed causes the channel coherence time to decrease and increases the difficulty to track changes in the forward link radio environment. Because of the TDM structure in the EVDO forward link, the AT does

Table 8.53 Reverse traffic data rates

Parameter	Reverse traffic channel data rate (Kbps)				
	9.6	19.2	38.4	76.8	153.6
Reverse Rate Indicator (RRI) index	1	2	3	4	5
Bits per physical layer packet	256	512	1024	2048	4096
Physical layer packet duration–frame (slots)	16	16	16	16	16
Frame time (ms)	26.667	26.667	26.667	26.667	26.667
Turbo encoder rate	1/4	1/4	1/4	1/4	1/2
Symbols after turbo encoder	1024	2048	4096	8192	8192
Symbol rate after turbo encoder (Ksps)	38.4	76.8	153.6	307.2	307.2
Interleaved packet repetition factor	8	4	2	1	1
Modulation symbol rate (after repetition) (Ksps)	307.2	307.2	307.2	307.2	307.2
Modulation type	BPSK	BPSK	BPSK	BPSK	BPSK
Effective coding rate	32	16	8	4	2
Walsh spread factor	4	4	4	4	4
PN chips/packet bits or processing gain	128	64	32	16	8

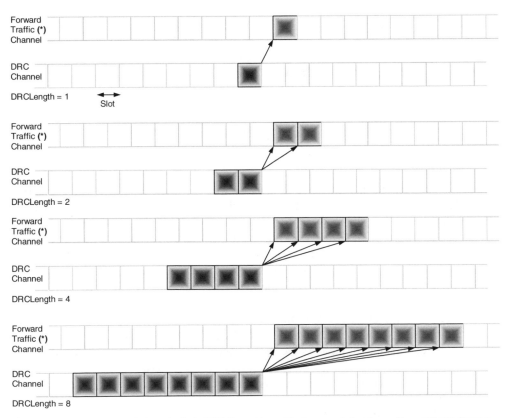

(*) The end of a continuous set of information in the DRC Channel is used for the new set of continuous Forward Traffic Channel

Figure 8.45 DRC timing for non-gated transmission.

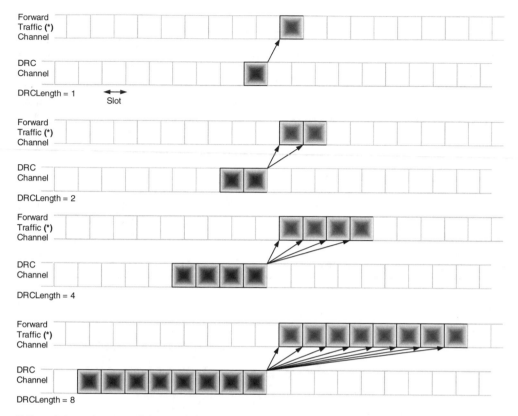

Figure 8.46   DRC timing for gated transmission.

not need to subtract channels transmitted together with the pilot, as in IS95 and cdma2000. The AT has a precise time to extract the pilot, which results in a high SNR allowing a fast C/I estimation.

The access terminal must follow a specific set of rules and procedures to send DRCs. The terminal uses DRCLength slots to transmit a single DRC. The DRC value and/or cover may change in slots T, when $(T + 1 - \text{FrameOffset}) \bmod \text{DRCLength} = 0$. If the access terminal supports gated transmission, it transmits the DRC over one slot, starting at slots T where $(T + 2 - \text{FrameOffset}) \bmod \text{DRCLength} = 0$.

There are two types of DRC cover: null and sector cover. When the the AT uses the DRC cover index 0 the DRC cover is known as the null cover; when it uses a cover based on the index associated with one of the possible serving sectors of the terminal, it is known as the sector cover.

If current DRC cover is a sector cover, then the next cover can only be the same sector cover or a null cover. The AT can only use a sector cover corresponding to a different sector when it has been determined that packets received from the new sector will not overlap in time with packets received from the first sector. The AT may inhibit transmission of data from the AN by covering the DRC with the null cover.

After the AT finishes sending the DRC to a sector at a given slot n specifying a requested rate (r), the sector transmits a preamble at rate r between slots $n + 1$ through $n + \text{DRCLength}$. If the terminal detects the preamble, it continues to receive the entire packet using the requested rate. If the AT terminal receives a DRCLock bit '0' from a sector, it stops pointing its DRC to that sector.

Not only the transmission of DRCs has rules but also the DRC processing by the network. To begin transmitting a packet to a sector at slot T, the network must use the rate specified by the DRC whose reception was completed in slot $T - 1 - ((T - \text{FrameOffset}) \bmod \text{DRCLength})$. After the transmission is initiated, the network must continue transmitting until it receives the indication of *PhysicalLayer.ForwardTrafficCompleted*.

8.4.1.2.3 ACK Channel

ACK Channel Identification

After the reverse traffic channel is identified, the ACK Channel is extracted using Walsh function 4 with 8 chips: $W_4^8 = (+ + + + - - - -)$.

ACK Channel Function

The terminal uses the ACK channel to acknowledge the successful reception of a physical layer packet transmitted on the forward traffic channel to the network. This acknowledgment works as a handshake, which is required for implementing hybrid ARQ. Figure 8.47 illustrates the structure of the ACK channel.

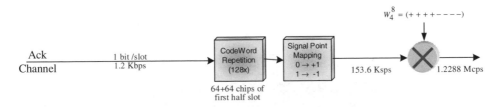

Figure 8.47 ACK channel (reverse data channel) structure.

The AT transmits an ACK channel bit in response to every forward traffic channel slot associated with a preamble directing the message to the terminal. However, a maximum of one redundant positive ACK is transmitted in response to a slot that is a continuation of the physical layer packet that has been successfully received. The remaining time, the ACK channel is gated off.

The ACK Channel is BPSK modulated. A bit '0', acknowledges a successful reception of the physical layer packet, whereas a '1' indicates failure on reception (NAK). A forward traffic channel physical layer packet is considered successfully received if the Frame Check Sequence (FCS) checks.

For a physical layer packet transmitted in slot n on the forward channel, the corresponding ACK channel bit is transmitted in slot $(n + 3)$ on the reverse channel as in Figures 8.48 and 8.49. The figures also show that the ACK channel transmission happens in the first half of the slot and lasts for 1024 PN chips.

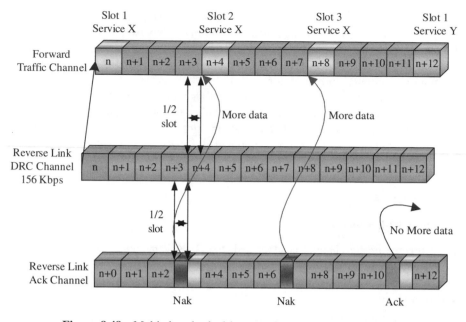

Figure 8.48 Multi-slot physical layer packet with early termination.

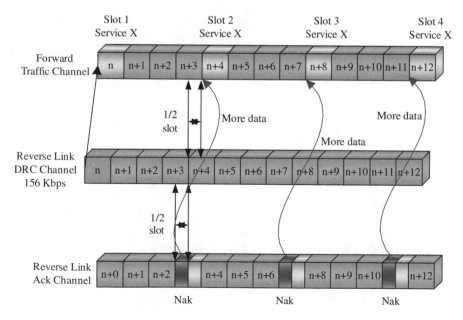

Figure 8.49 Multi-slot physical layer packet with normal termination.

In Figure 8.48, the early termination of the required 153.6 Kbps transmission implies that the AT successfully de-coded the data with only 3 out of the 4 required slots, providing a 204.8 Kbps data rate and allowing higher bandwidth efficiency.

When the AN finishes the transmission (transmitted all slots or received a positive ACK response), the physical layer returns a ForwardTrafficCompleted indication.

8.4.1.2.4   Data Channel

Data Channel Identification

After the reverse traffic channel is identified, the data channel is extracted using Walsh function 2 with 4 chips: $W_2^4 = (+ + --)$.

Data Channel Function

The access terminal uses the data channel to transmit traffic contents to the network according to Table 8.53.

Data transmissions begin at the FrameOffset slot within a frame. The reverse traffic channel MAC protocol defines this parameter, FrameOffset, as public data. All data transmitted on this channel is encoded, block interleaved, sequence repeated and orthogonally spread by the 4-chip Walsh function 2, as illustrated in Figure 8.50.

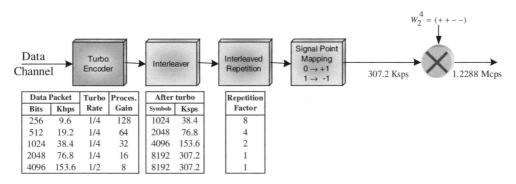

Figure 8.50   Data channel (reverse traffic channel) structure.

8.4.2   Forward Link Channels

The forward channel time division multiplexes four logical channels: pilot, MAC, traffic and control channels. The MAC channel is subdivided into three sub-channels: Reverse Activity (RA) channel, Reverse link Power Control (RPC) channel and Data Rate Control (DRC) lock channel.

The forward link channel employs the same time-slot structure defined for the reverse link (Section 8.4.1). Three channels transmitted with the same power compose each slot: pilot, MAC and traffic or control channels (Figure 8.51). None of the forward channels uses a power-control mechanism. Control and traffic channels are mutually exclusive, i.e. they

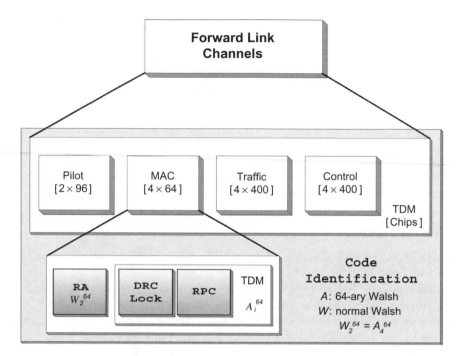

Figure 8.51 Forward channel structure.

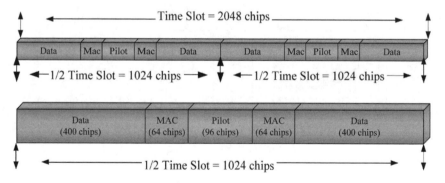

Figure 8.52 TDM channel distribution in an ACTIVE time slot.

always belong to different time slots. Time slots are time-division multiplexed according to the structures presented in Figures 8.52 and 8.53.

Physical Layer Packets

The traffic or the control channel uses the data segment to transport packets from the MAC layer as payload. Figures 8.54 and 8.55 illustrate the configuration of a packet transmitted,

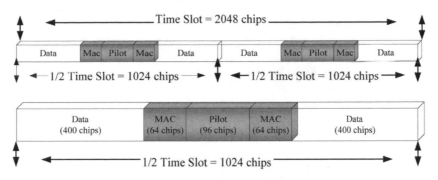

Figure 8.53 TDM channel distribution in an INACTIVE time slot.

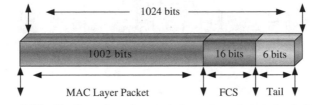

Figure 8.54 Configuration of control channels packet at physical layer.

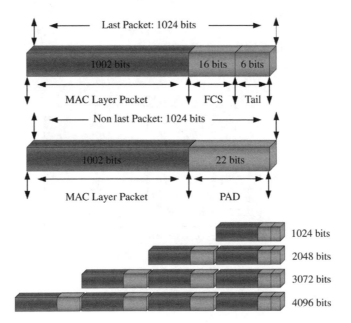

Figure 8.55 Configurations of forward traffic channels packet at physical layer.

respectively, by the control and traffic channels. In these figures, PAD represents the
auxiliary stuffing bits used to keep a homogeneous structure; Tail represents encoder tail
bits, i.e. bits generated at the end of coding process to flush internal registers and FCS stands
for frame check sum. The ratio between received frames with invalid FCS and all transmitted
frames is known as Frame Error Rate (FER).

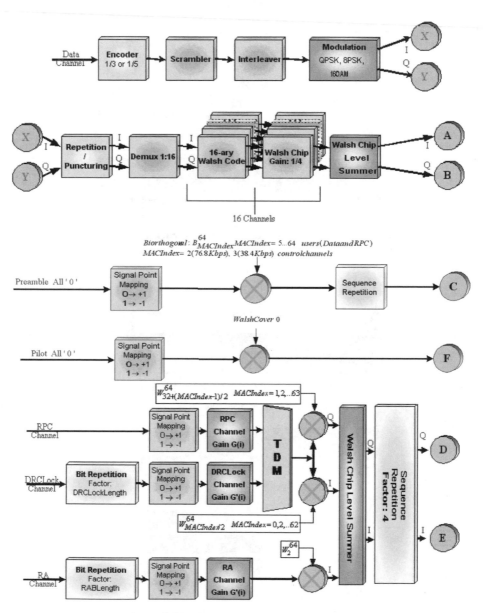

Figure 8.56 Forward link channel elements.

Channel Multiplexing

Figures 8.52 and 8.53, shown earlier, illustrate the composition of an EVDO time slot in the forward link. As depicted in the figure the pilot channel uses 192 chips of a slot, whereas the MAC channel uses 256. Traffic and control channels use the remaining chips, which are subdivided into preamble and data. The preamble may require from 64 to 1024 chips, depending on the data rate and modulation scheme. These factors also affect the data block that can have 1024, 2048, 3072 or 4096 bits.

Table 8.59 associates these parameters with the data block and preamble configuration.

Figures 8.56 and 8.57 illustrate, respectively, the structure and multiplexing of forward link channels.

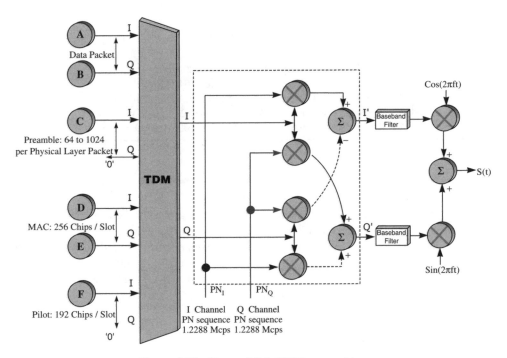

Figure 8.57 Forward link TDM composition.

8.4.2.1 Medium Access Control (MAC) Channel

The MAC channel provides support for scheduling different users onto the forward link by informing users about reverse link receiver capabilities of the system through three sub-channels: the Reverse Link Power-Control Channel (RPC), the Reverse Activity Channel (RA) and the DRCLock Channel.

The RPC and DRCLock are addressed to an individual active access terminal, whereas the RA is broadcast to all users. The RA indicates the reverse link load to the reverse link MAC algorithm to reduce or increase the terminal data rate. The DRCLock provides information

Table 8.54   RA, DRCLock and RPC modulation parameters

Parameter	MAC channels		
	DRCLock	RPC	RA
Period (slots)	DRCLockPeriod	1 (except DRCLock stolen slots)	1
Rate (bps)	SlotRate/(DRCLockPeriod × DRCLockLength)	SlotRate × (1 − 1/DRCLockPeriod)	SlotRate/RABLength
Bit repetition	DRCLockLength	1	RABLength
Modulation rate (sps)	MACRate × DRCLockPeriod	MACRate × (1 − 1/DRCLockPeriod)	MACRate
PN chips/bit	MACChips × DRCLockLength	MACChips	MACChips × RABLength

$$\text{ChipRate} = 1.2288 \text{ Mcps}$$
$$\text{SlotSize} = 2048 \text{ chips}$$
$$\text{SlotRate} = \text{ChipRate/SlotSize} = 600 \text{ slots/s}$$
$$\text{MACCoverLength} = \text{MACSlice} = 64 \text{ chips}$$
$$\text{MACRepetition} = 4 \text{ times (each side of pilot: before and after)}$$
$$\text{MACChips} = \text{MACSlice} \times \text{MACRepetition} = 256 \text{ chips per slot}$$
$$\text{MACRate} = \text{MACRepetition} \times \text{SlotRate} = 2400 \text{ MAC channel/s}$$

about network acknowledgement of the data rate requested at the DRC channel in the reverse link. Table 8.54 describes the modulation of each of these channels.

The channels within the MAC channel are Walsh channels, orthogonally covered (multiplexed) and BPSK modulated on a particular phase (in-phase or quadrature) of the carrier. Each Walsh channel is identified by a MACIndex value (0–63). The network uses the MACIndex to identify the target access terminal. Table 8.55 shows the association between MACIndexes and the addressed elements.

8.4.2.1.1   Reverse Activity (RA) Channel

RA Channel MACIndex

The RA cover uses W_2^{64} as the bi-orthogonal modulation with MACIndex = 4 (Table 8.55).

Table 8.55   MACIndex addressing

MACIndex	Addressed element
0	Used by pilot cover
1	Reserved
2	Broadcast: control channel at 76.8 Kbps
3	Broadcast: control channel at 38.4 Kbps
4	Broadcast: RA cover
5–63	Unicast: for up to 59 simultaneous active users

RA Channel Function

The RA transmits the Reverse Activity Bit (RAB) over RABLength successive slots. The RAB is related to the load of the addressed sector. RABLength is the number of continuous slots with the same RAB (8, 16, 24 or 32 slots) (Figure 8.58).

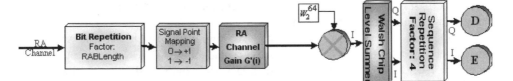

Figure 8.58 Reverse activity channel.

The transmission of an RAB starts in the slot T where (T mod RABLength = RABOffset). RABOffset is the initial slot transmitting the RAB. RABLength and RABOffset are established by the TrafficChannelAssignment message and belong to the route update protocol attributes. There is a RABLength/RABOffset value for each sector in the active set of an AT. The RAB is transmitted four times in a slot (after and before each pilot).

8.4.2.1.2 Data Rate Control (DRC) Lock Channel

DRCLock Channel MACIndex

The DRCLock cover varies according to the MACIndex selected (Table 8.55). There is one cover for even numbered and another for odd numbered MACIndexes:

$$W^{64}_{MACIndex/2} \quad MACIndex = 0, 2, \ldots, 62$$
$$W^{64}_{32+(MACIndex-1)/2} \quad MACIndex = 1, 3, \ldots, 63$$

DRCLock Channel Function

The RPC and DRCLock bit streams are addressed to one of the sectors within the terminal's active set. Both channels are TDM multiplexed and transmitted over the forward MAC channel as illustrated in Figure 8.59.

The route update protocol knows whether more than one sector is transmitting the same DRCLock bit by the SofterHandoff public data value. The AT must combine identical

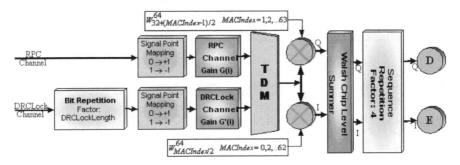

Figure 8.59 RPC and DRCLock channels.

DRCLock channels and must select just one DRCLock bit from each set of identical DRCLock channels.

After the connection is established, the AN continuously transmits '0' or '1' in the DRCLock bits to the addressed AT on the DRCLock channel. The AT processes the DRCLock bits received on the DRCLock channel as specified by the forward traffic channel MAC protocol. If the DRCLock bit is '0' then the AT must redirect the next DRC channel to another sector changing the DRC cover.

The DRCLock bit is transmitted by the AN in every system slot T where $(T - \text{FrameOffset}) \mod (\text{DRCLockPeriod}) = 0$. The value of the DRCLock bit may change only in slot T when $(T - \text{FrameOffset}) \mod (\text{DRCLockLength} \times \text{DRCLockPeriod}) = 0$.

DRCLockPeriod and DRCLockLength are defined in the forward traffic channel MAC protocol as a DRC attribute, described in Table 8.56.

Figure 8.60 shows an example of DRC allocation.

Table 8.56 Attributes related to DRCLock

Protocol	Attribute name		ID
FTCMP	DRCLock		1
Field	Bit Size	Default	Description
Length	8	N/A	Length of the attribute excluding the length field
AttributeID	8	N/A	Numerical identity of the attribute in the protocol (1)
One or more of the following record (one for each sector in the active set of the AT)			
ValueID	8	N/A	Identity of the record in this attribute list
DRCLockPeriod	1	1	The AT sets this field with a value '0' for 8 slots '1' for 16 slots to specify the time interval between transmission of two consecutive DRCLock bit transmissions on the Forward MAC Channel.
DRCLockLengh	2	'01'	The AN shall set this field with the number of times the DRCLock bit must be repeated '00' for 4 times '01' for 8 times '10' for 16 times '11' for 32 times
Reserved	5	N/A	The AN sets this field to 0.

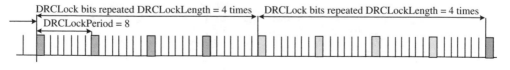

DRCLock bits repeated DRCLockLength = 4 times   DRCLock bits repeated DRCLockLength = 4 times
DRCLockPeriod = 8

Figure 8.60 DRCLock example.

8.4.2.1.3 Reverse Power-Control (RPC) Channel

RPC Channel MACIndex

The RPC cover used by the RPC channel varies according to the MACIndex selected (Table 8.55). There is one cover for even numbered and another for odd numbered MACIndexes:

$$W^{64}_{MACIndex/2} \quad MACIndex = 0, 2, \ldots, 62$$
$$W^{64}_{32} + (MACIndex - 1)/2 \quad MACIndex = 1/3, \ldots, 63$$

RPC Channel Function

After the connection is established, the AN continuously transmits the RPC bit indicating that the AT's power level must be increased ('0') or decreased ('1') according to the estimated quality of service.

RPC channels are transmitted in DRCLockPeriod -1 slots out of every DRCLockPeriod slots, where DRCLock is part of the DRC Attribute of the forward traffic channel MAC protocol, so RPC is transmitted in every system slot T with

$$(T - \text{FrameOffset}) \bmod (\text{DRCLockPeriod}) \neq 0$$

The RPC bit is transmitted four times (before and after the pilot channel) in bursts of 64 chips.

8.4.2.2 Pilot Channel

Pilot Channel MACIndex

The pilot cover uses W^{64}_0 as the bi-orthogonal modulation with MACIndex $= 0$ (Table 8.55).

Pilot Channel Function

The pilot channel (Figure 8.61) has the same functions as in IS-95/cdma2000-1X, being used by the access terminal for initial acquisition, phase and time recovery, and maximal-ratio combining. In EVDO, however, the pilot is time multiplexed with duration of 96 chips centred at the mid-point of the half slot. This channel also provides means for the estimation of the received C/I required to estimate the DRC (see Section 8.4.1.2.2).

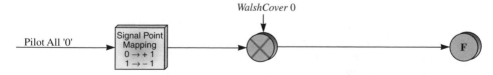

Figure 8.61 Forward pilot channel.

8.4.2.3   Forward Traffic and Control Channels

Forward Traffic and Control Channel MACIndex

The cover for the preamble of these channels is 32 bits long and varies according to the MACIndex (Table 8.55) associated with it:

$$W^{32}_{MACIndex/2} \quad MACIndex = 0.2, \ldots, 62, \text{which encompasses } W^{64}_i \ i = 0, 1, \ldots, 31$$

$$\overline{W}^{32}_{(MACIndex-1)/2} \quad MACIndex = 1, 3, \ldots, 63$$

Forward Traffic and Control Channel Functions

The forward traffic and control channels have two specific protocols at the MAC layer: forward traffic channel MAC protocol and control channel MAC protocol. These channels are divided into two parts at the air interface: preamble and data (Figure 8.62).

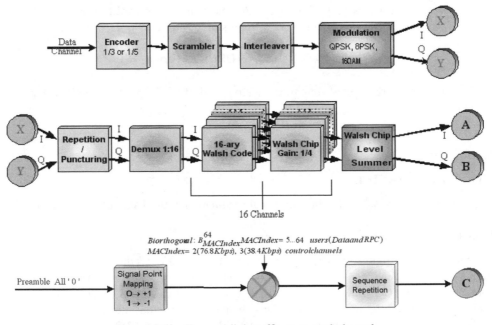

Figure 8.62   Forward link traffic or control channel.

The preamble is similar to a header where all bits equal '0' and with a cover identifying the user or the control channel data rate. The preamble size depends on the number of 1024-bit blocks (packet fragments), on the repetition factor (which is at least equal to 2) and on the 32-symbol bi-orthogonal Walsh cover by each preamble repetition as in Table 8.57. The preamble sequence assists the AT to synchronise the variable rate transmission; therefore it is transmitted at the beginning of physical layer forward traffic and control channels.

Table 8.57 Preamble repetition factor and chip size

Packet size bits	Number of slots	Data rate (bps)	Preamble walsh size	Packet fragments	Preamble repetition	Preamble size
1024	16	38 400	32	1	32	1024
1024	8	76 800	32	1	16	512
1024	4	153 600	32	1	8	256
1024	2	307 200	32	1	4	128
1024	1	614 400	32	1	2	64
2048	4	307 200	32	2	4	128
2048	2	614 400	32	2	2	64
2048	1	1 228 800	32	2	2	64
3072	2	921 600	32	3	2	64
3072	1	1 843 200	32	3	2	64
4096	2	1 228 800	32	4	2	64
4096	1	2 457 600	32	4	2	64

$$\text{ChipRate} = 1.2288 \text{ Mcps}$$
$$\text{SlotSize} = 2048 \text{ chips}$$
$$\text{SlotRate} = \text{ChipRate} / \text{SlotRate} = 600 \text{ slots/s}$$
$$\text{DataRate} = \text{PacketSize} \times \text{SlotRate}/\text{NumSlots}$$
$$\text{PacketFragments} = \text{PacketSize}/1024$$
$$\text{PreambleRepetition} = 2 \times \max(1, \text{NumSlots}/\text{PacketFragments})$$
$$\text{PreambleSize} = \text{PreambleWalshSize} \times \text{PreambleRepetition}$$

The data part of the channel contains the transmitted information modulated, coded and distributed in slots. Table 8.58 shows the relation between data rate and number of data chips.

Table 8.59 summarizes the symbols transformation during encoding and modulation.

According to Table 8.58, the forward traffic and control channels' physical layer packets can be transmitted in 1–16 slots. The modulation characteristics of the control channel correspond to the characteristics of the traffic channel operating at the same data rate. The transmission of multiple slots is based on a four-way inter-laced process equivalent to the method described in the reverse link access channel (Section 8.4.1.1).

After symbol repetition and puncturing, the resulting stream is demultiplexed into 16 in-phase and in-quadrature parallel streams, each with a distinct 16-ary Walsh function (76.8 Kcps) and a relative Walsh channel gain of 1/4 ($\sqrt{16}$). The Walsh coded streams are summed together and result in a single in-phase and in-quadrature stream with a 1.2288 Mcps chip rate.

The forward traffic and control channels are mutually exclusive and the control channel transmission is regulated on a 256 slot cycle basis (ControlChannelCycle public data at the MAC layer) with a duration (MACChannelLength public data at the MAC layer) of 16 slots (data rate of 38.4 Kbps) or 8 slots (data rate of 76.8 Kbps).

The control channel is transmitted by synchronous or asynchronous capsules, where the synchronisation of the capsules is according to the ControlChannelCycle (numeric

Table 8.58 Data rate and channels chip size

Packet size bits	Number of slots	Data rate (bps)	Processing gain	Total chips	MAC chips	Pilot chips	Preamble size	Data chips
1024	16	38 400	32	32 768	4096	3072	1024	24 576
1024	8	76 800	16	16 384	2048	1536	512	12 288
1024	4	153 600	8	8192	1024	768	256	6144
1024	2	307 200	4	4096	512	384	128	3072
1024	1	614 400	2	2048	256	192	64	1536
2048	4	307 200	4	8192	1024	768	128	6272
2048	2	614 400	2	4096	512	384	64	3136
2048	1	1 228 800	1	2048	256	192	64	1536
3072	2	921 600	4/3	4096	512	384	64	3136
3072	1	1 843 200	1/2	2048	256	192	64	1536
4096	2	1 228 800	1	4096	512	384	64	3136
4096	1	2 457 600	1/2	2048	256	192	64	1536

$$\text{ChipRate} = 1.2288 \text{ Mcps}$$
$$\text{SlotSize} = 2048 \text{ chips}$$
$$\text{SlotRate} = \text{ChipRate}/\text{SlotSize} = 600 \text{ slots/s}$$
$$\text{PilotSlice} = 96 \text{ chips}$$
$$\text{PilotRepetition} = 2 \text{ times (1 per half slot)}$$
$$\text{MACSLice} = 64 \text{ chips}$$
$$\text{MACRepetition} = 4 \text{ times (each side of pilot: before and after)}$$
$$\text{PreambleSize as in Table 8.57}$$
$$\text{DataRate} = \text{PacketSize} \times \text{SlotRate}/\text{NumSlots}$$
$$\text{ProcessingGain} = \text{ChipRate}/\text{DataRate}$$
$$\text{TotalChips} = \text{SlotSize} \times \text{NumSlots}$$
$$\text{MACChips} = \text{MACSlice} \times \text{MACRepetition} \times \text{NumSlots}$$
$$\text{PilotChips} = \text{PilotSlice} \times \text{PilotRepetition} \times \text{NumSlots}$$
$$\text{DataChips} = \text{TotalChips} - \text{MACChips} - \text{PilotChips} - \text{PreambleChips}$$

constant $= 256$) slots. There is at least one synchronous capsule active, and once the possible control channel data rate is 38.4 Kbps (16 slots) or 76.8 Kbps (8 slots), this capsule will be transmitted in a four-way inter-laced basis and will steal at least 8 slots out of 256 frame slots. This means that the maximum effective data rate will be 31/32 of the nominal data rate, as illustrated in Figures 8.63 and 8.64.

The two messages that determine the forward traffic channel behaviour are FixedModeEnable and FixedModeOff, both sent by the forward traffic channel MAC protocol. Tables 8.60 and 8.61 describe these messages.

The forward traffic channel MAC protocol determines the forward traffic channel addressing and data rate. There is one instance of this protocol in the network for each access terminal. The protocol may operate in three different states: inactive, variable rate, or fixed rate.

The procotol is only in the inactive state when it is not assigned to any AT. In the variable rate state, the AN transmits at the rate dictated by the Data Rate Control (DRC) channel,

Table 8.59 Symbols in the encoding/modulation processes

Packet size bits	Number of slots	Data rate (bps)	Modulation	Code rate	Encoded symbols	Modulation symbols	Data chips	Repetition factor	Punctured symbols	Partial symbols	Effective code rate
1024	16	38400	QPSK	1/5	5120	2560	24576	10	1024	1536	1/48
1024	8	76800	QPSK	1/5	5120	2560	12288	5	512	2048	1/24
1024	4	153600	QPSK	1/5	5120	2560	6144	3	1536	1024	1/12
1024	2	307200	QPSK	1/5	5120	2560	3072	2	2048	512	1/6
1024	1	614400	QPSK	1/3	3072	1536	1536	1	0	0	1/3
2048	4	307200	QPSK	1/3	6144	3072	6272	3	2944	128	8/49
2048	2	614400	QPSK	1/3	6144	3072	3136	2	3008	64	16/49
2048	1	1228800	QPSK	1/3	6144	3072	1536	1	1536	1536	2/3
3072	2	921600	8PSK	1/3	9216	3072	3136	2	3008	64	16/49
3072	1	1843200	8PSK	1/3	9216	3072	1536	1	1536	1536	2/3
4096	2	1228800	16-QAM	1/3	12288	3072	3136	2	3008	64	16/49
4096	1	2457600	16-QAM	1/3	12288	3072	1536	1	1536	1536	2/3

$ChipRate = 1.2288$ Mcps

$SlotSize = 2048$ chips

$SlotRate = ChipRate / SlotSize = 600$ slots/s

$ModulationSymbols = [QPSK = 2; 8PSK = 3; 16\text{-}QAM = 4]$

$DataChips$ as in Table 8.58

$DataRate = PacketSize \times SlotRate/NumSlots$

$EncodedSymbols = PacketSize/CodeRate$

$ModulatedSymbols = EncodedSymbols/ModulationSymbols$

$RepetitionFactor = Ceilling(DataChips/ModulatedSymbols)$

$PuncturedSymbols = ModulatedSymbols \times RepetitionFactor - DataChips$

$PartialSymbols$ (transmit. in the last incomplete repetition) $= ModulatedSymbols - PuncturedSymbols$

$EffectiveCodeRate = PacketSize/(DataChips \times ModulationSymbols)$

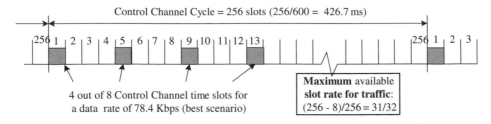

Figure 8.63 Control channel cycle and the effective maximum data rate.

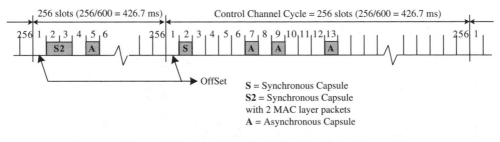

Figure 8.64 Location of control channel capsules.

Table 8.60 FixedModeEnable message identification and description

Protocol	Message name	ID	Physical channel	Delivery	Address	Priority
FTCMP	FixedModeEnable	0	RTC	Best	Unicast	40

The AT sends this message indicating a transition to the fixed rate state

Fields	Bit size	Description
MessageID	8	Identity of this message in this protocol set by the AN
TCAMessageSequence	8	The AT sets this field to the MessageSequence field of the TrafficChannelAssignment message specifying the association between the DRCCover field in this message and the sector in the active set.
DRCCover	3	The DRC cover set by AT is associated with the sector in its active set from which it wants to receive packets on the forward traffic channel.
DRCValue	4	The AT sets this field to one of the valid DRC values in Table 8.51 to indicate the rate at which it wants to receive packets.
EndTime	16	The AT sets this field to the least significant 16 bits of the system time in units of slots until which (inclusive) it requests to remain in the Fixed Rate State.
Reserved	1	Bits set to '0' to align the message in an octet basis

Table 8.61 FixedModeOff message identification and description

Protocol	Message name	ID	Physical channel	Delivery	Address	Priority
FTCMP	FixedModeOff	1	FTC	Best	Broadcast	40

The AT sends this message informing the AN that is not able to receive packets from the sector specified in the last FixedModeEnabled message.

Fields	Bit size	Description
MessageID	8	Identity of this message in this protocol set by the AN

transmitted by the AT on the reverse link. In the fixed rate state, the rate used for the traffic channel transmission to an AT coming from one particular sector is constant.

Not only the behavior of these channels is guided by messages, but these channels are also used to transmit messages coming from several layers. Tables 8.62 and 8.63 show, respectively, the messages carried by the control and forward traffic channels.

Table 8.62 Messages related to the control channel

Control channel	Layer
LocationUpdate.LocationRequest	Application
LocationUpdate.LoacationAssignment	Application
FlowControl.XonResponse	Application
FlowControl.XoffResponse	Application
FlowControl.DataReady	Application
SessionManagement.SessionClose	Session
SessionManagement.KeepAliveRequest	Session
SessionManagement.KeepAliveResponse	Session
AddressManagement.UATIAssignment	Session
AddressManagement.HardwareIDRequest	Session
AirLinkManagement.Redirect	Connection
InitializationState.Sync	Connection
IdleState.Page	Connection
IdleState.ConnectionDeny	Connection
RouteUpdate.TrafficChannelAssignment	Connection
OverheadMessages.QuickConfig	Connection
OverheadMessages.SectorParameters	Connection
AccessChannelMAC.ACAck	MAC
AccessChannelMAC.AccessParameters	MAC
ReverseTrafficChannelMAC.BroadcastReverseRateLimit	MAC

Table 8.63 Messages related to traffic channels

Forward traffic channel	Layer
SLP.Reset	Application
RadioLinkProtocol.Reset	Application
RadioLinkProtocol.ResetAck	Application
RadioLinkProtocol.Nak	Application
LocationUpdate.LocationRequest	Application
LocationUpdate.LoacationAssignment	Application
FlowControl.XonResponse	Application
FlowControl.XoffResponse	Application
FlowControl.DataReady	Application
SessionManagement.SessionClose	Session
SessionManagement.KeepAliveRequest	Session
SessionManagement.KeepAliveResponse	Session
AddressManagement.UATIAssignment	Session
AddressManagement.HardwareIDRequest	Session
SessionConfiguration.ConfigurationComplete	Session
SessionConfiguration.ConfigurationStart	Session
SessionConfiguration.ConfigurationRequest	Session
SessionConfiguration.ConfigurationResponse	Session
AirLinkManagement.Redirect	Connection
ConnectedState.ConnectionClose	Connection
RouteUpdate.TrafficChannelAssignment	Connection
RouteUpdate.ResetReport	Connection
RouteUpdate.NeighborList	Connection
RouteUpdate.AttributeOverride	Connection
DHKeyExchange.KeyRequest	Security
DHKeyExchange.ANKeyComplete	Security
ReverseTrafficChannelMAC.RTCAck	MAC
ReverseTrafficChannelMAC.UnicastReverseRateLimit	MAC

8.5 AIR INTERFACE ENCODING

8.5.1 Frame Check Sum

The Frame Check Sum (FCS) field is used in the MAC layer capsule and in the physical layer packet (control and traffic channels on the forward link and access and traffic channels on the reverse link). In the MAC layer capsule the FCS has 32 bits and is generated using the standard CRC-CCITT generator polynomial indicated in eqn (8.1). In the physical layer packet, the FCS has 16 bits and is generated using the standard CRC-CCITT polynomial in eqn (8.2)

$$g(x) = x^{32} + x^{26} + x^{23} + x^{22} + x^{16} + x^{12} + x^{11} + x^{10}$$
$$+ x^8 + x^7 + x^5 + x^4 + x^2 + x^1 + 1 \tag{8.1}$$
$$g(x) = x^{16} + x^{12} + x^5 + 1 \tag{8.2}$$

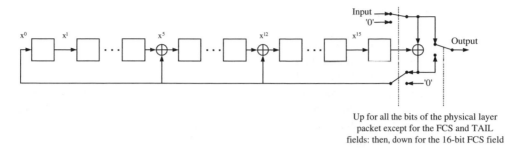

Figure 8.65 Frame Check Sum (FCS) generator for the access channel MAC layer capsule.

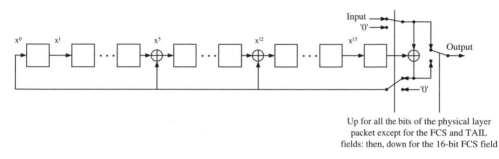

Figure 8.66 Frame Check Sum (FCS) generator for the physical layer packet.

Figure 8.65 illustrates the FCS generator for the MAC layer capsule and Figure 8.66 for the physical layer packet. The following procedure is followed to generate the FCS.

1. All shift-register elements are initialized as '0's.

2. All switches are set in the up position.

3. For the MAC layer capsule, the register is clocked once for each bit of the capsule except for the FCS and padding bits. The capsule is read from the Most Significant Bit (MSB) to the Least Significant Bit (LSB), starting at the MAC layer header. For the physical layer packet, the register is clocked once for each bit of the packet except for the FCS and tail bits. The physical layer packet is read from the MSB to the LSB.

4. Switches are set in the down position causing the output to be a modulo-2 addition with a '0' and the successive shift-register inputs to be '0's.

5. The register is clocked additional 32 times for the MAC layer capsule or 16 times for the physical layer packet (corresponding to the number of FCS bits).

6. The output bits constitute all fields of the capsule/packet except for the padding/tail bits.

8.5.2 Turbo Encoding

Assuming that there are N_{turbo} bits, the turbo encoder encodes them into N_{turbo}/R bits plus an output tail sequence with $R/6$ bits, where R is the encoder code rate: 1/2 or 1/4 for the reverse link and 1/3 or 1/5 for the forward link.

Figure 8.67 illustrates the turbo encoder, which consists of two recursive convolutional encoders connected in parallel, known as the constituent encoders. A turbo interleaver precedes the second constituent encoder. A common constituent code, whose transfer function is described in eqn (8.3), is used for both the turbo codes

$$G(D) = \left[1 \; \frac{n_0(D)}{d(D)} \; \frac{n_1(D)}{d(D)}\right] \tag{8.3}$$

where

$$d(D) = 1 + D^2 + D^3$$
$$n_0(D) = 1 + D + D^3$$
$$n_1(D) = 1 + D + D^2 + D^3$$

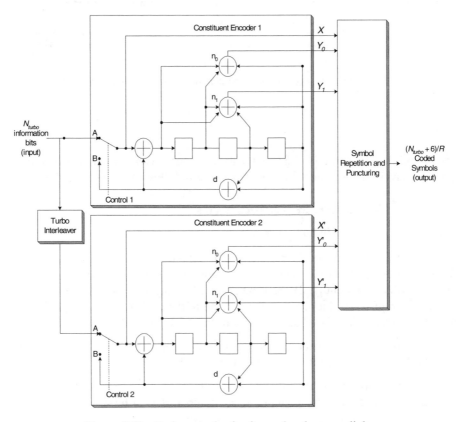

Figure 8.67 Turbo encoder for forward and reverse links.

Table 8.64 Puncturing patterns for data bits on forward and reverse links

Output	Forward link		Reverse link	
	1/3	1/5	1/2	1/4
X	1	1	11	11
Y_0	1	1	10	11
Y_1	0	1	00	10
X'	0	0	00	00
Y'_0	1	1	01	01
Y'_1	0	1	00	11

At the beginning of the encoding process, the states of the constituent encoder registers are set to zero. The constituent encoders are then clocked with the switches in the positions shown in Figure 8.67. The constituent encoders are clocked N_{turbo} times with the switches in the up positions. The outputs of the encoders are punctured, according to Table 8.64, to achieve the desired number of output symbols: $(N_{turbo} + 6)/R$. A bit '0' in the puncturing pattern means that the corresponding symbol must be deleted, whereas a '1' indicates that the symbol remains. The constituent encoder data output for each bit period is the sequence X, Y_0, Y_1, X', Y'_0, Y'_1.

The turbo encoder also generates $6/R$ tail output symbols after the encoded data output sequence is completed. After the encoders were clocked N_{turbo} times with the switches in the up positions (to complete the data sequence), the first constituent encoder is clocked three times with its switch in the down position, while the second encoder is not clocked, and puncturing and repetition (Table 8.65) are applied to generate the first $3/R$ tail output symbols.

Table 8.65 Puncturing patterns for tail bits at forward and reverse links

Output	Forward link		Reverse link	
	1/3	1/5	1/2	1/4
X	111000	111000	111000	111000
Y_0	111000	111000	111000	111000
Y_1	000000	111000	000000	111000
X'	000111	000111	000111	000111
Y'_0	000111	000111	000111	000111
Y'_1	000000	000111	000000	000111

To generate the last $3/R$ tail output symbols, the second constituent encoder is clocked three times with its switch in the down position while the first encoder is not clocked. Puncturing and repeating are applied to the resulting constituent encoder output symbols to complete the tail output symbols. Table 8.66 indicates which outputs are used to compose the tail output sequences depending on the rate being used.

Table 8.66　Tail output symbols

	Forward link		Reverse link	
Output symbols	1/3	1/5	1/2	1/4
First three tail bit periods	XXY_0	$XXY_0Y_1Y_1$	XY_0	XXY_0Y_1
Last three tail bit periods	$X'Y'Y'_0$	$X'Y'Y'_0Y'_1Y'_1$	$X'Y'_0$	$X'Y'Y'_0Y'_1$

The turbo interleaver block interleaves the input data fed to the second constituent encoder. This interleaving can be understood as a function where the input bits are written sequentially into an array and read out using a sequence of addresses defined by the procedure in Figure 8.68.

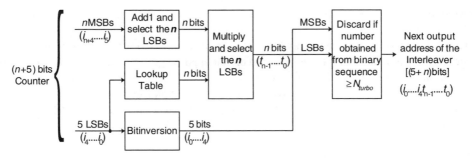

Figure 8.68　Turbo interleaver output address.

Given the sequence of input addresses from 0 to $N_{turbo} - 1$, the turbo interleaver parameter n is determined as being the smallest integer that allows $N_{turbo} \leq 2^{n+5}$. Table 8.67 shows the interleaving parameter for the different physical layer packet sizes.

After determining n, an $(n + 5)$-bit counter is initialised as 0. The n MSBs are extracted from the counter and added to '1'. From this new value, only the n LSBs are kept (result1).

Table 8.68 is used to obtain the n-bit output using the value of n and a read address (table index) equal to the five LSBs of the counter (result2).

Result1 and result2 are then multiplied, and all bits except for the n LSBs are discarded (result3). The five LSBs of the counter are bit reversed (result4).

A tentative output address is formed using result4 as the MSBs and result3 as the LSBs. This address is accepted if it is less than N_{turbo}; otherwise, it is discarded. If the address is discarded, the counter is incremented and the process re-starts at the point where the n MSBs are extracted from the counter. The process is repeated until all N_{turbo} interleaver output addresses are obtained.

Table 8.67 Turbo interleaver parameter n

Physical layer packet size	Interleaver block size N_{turbo}	Interleaver parameter (n)	Link
256	250	3	Reverse
512	506	4	Reverse
1024	1018	5	Both
2048	2042	6	Both
3072	3066	7	Forward
4096	4090	7	Both

Table 8.68 Turbo interleaver look up table

Table index	Entries				
	$N = 3$	$N = 4$	$N = 5$	$N = 6$	$N = 7$
0	1	5	27	3	15
1	1	15	3	27	127
2	3	5	1	15	89
3	5	15	15	13	1
4	1	1	13	29	31
5	5	9	17	5	15
6	1	9	23	1	61
7	5	15	13	31	47
8	3	13	9	3	127
9	5	15	3	9	17
10	3	7	15	15	119
11	5	11	3	31	15
12	3	15	13	17	57
13	5	3	1	5	123
14	5	15	13	39	95
15	1	5	29	1	5
16	3	13	21	19	85
17	5	15	19	27	17
18	3	9	1	15	55
19	5	3	3	13	57
20	3	1	29	45	15
21	5	3	17	5	41
22	5	15	25	33	93
23	5	1	29	15	87
24	1	13	9	13	63
25	5	1	13	9	15
26	1	9	23	15	13
27	5	15	13	31	15
28	3	11	13	17	81
29	5	3	1	5	57
30	5	15	13	15	31
31	3	5	13	33	69

This procedure is equivalent to writing the counter values into a 2^5-row by 2^n-column array by rows, shuffling the rows according to a bit-reversal rule, permuting the elements within each row according to a row-specific linear congruential sequence, and reading out tentative output addresses by column. In this case, the linear congruential sequence rule is $x(i+1) = (x(i) + c) \bmod 2^n$, where $x(0) = c$ and c is a row-specific value from a table lookup.

8.5.3 Channel Interleaving

Channel interleaving follows different procedures in the forward and reverse links. In the forward link, channel interleaving consists of symbol re-ordering followed by symbol permuting. In the reverse link, the interleaving happens through a bit-reversal channel interleaver.

The forward link channel interleaving process starts with symbol re-ordering. After the encoder scrambles the turbo encoder data and the tail output symbols, the symbols are re-ordered following two steps.

First the scrambled data and tail turbo encoder output symbols are de-multiplexed into three ($UV_0V'_0$) or five ($UV_0V_1V'_0V'_1$) sequences, according to the encoder rate 1/3 or 1/5, respectively (Table 8.69). These symbols are sequentially distributed among these sequences as a loop, starting at the first sequence (U) going to the last (V'_0 or V'_1) and restarting at the first one again. Then, the sequences are read in the order they appear, that is, starting by the first sequence (U).

After symbol re-ordering, the symbols are permuted in three bit-reversal interleaver blocks with rate-1/5 coding and in two separate blocks with rate-1/3 coding.

The sequence of interleaver output symbols follows the following procedure.

1. The input sequence is written into an array of K rows and M columns, where K and M vary according to the block (U or V) being interleaved and with the packet size, as in Table 8.70. Symbols are written by rows, starting on the top row and going from left to right.

2. The columns are labeled with an index j, from 0 to $M-1$, starting on the left-most column. Then, end-around shift the symbols of each column downward by $j \bmod K$ for the U block and by $j/4 \bmod K$ for the V_0/V_0 and V_1/V_1 blocks.

Table 8.69 Scrambled turbo encoder output and symbol re-ordering demultiplexer symbol sequences

Type of sequence	Symbol sequence	
	$R = 1/5$	$R = 1/3$
Turbo encoder data output sequence	$XY_0Y_1Y'_0Y'_1$	$XY_0Y'_0$
Turbo encoder constituent encoder 1 tail output sequence	$XXY_0Y_1Y_1$	XXY_0
Turbo encoder constituent encoder 2 tail output sequence	$X'X'Y'_0Y'_1Y'_1$	$X'X'Y'_0$
De-multiplexer output sequence	$UV_0V'_0V_1V'_1$	$UV_0V'_0$

Table 8.70 Channel interleaver parameters

Physical layer packet size	U block interleaver parameters		V block inteleaver parameters	
	K	M	K	M
1024	2	512	2	1024
2048	2	1024	2	2048
3072	3	1024	3	2048
4096	4	1024	4	2048

3. The columns are then re-ordered in a way that column j is moved to column indicated by the bit-reversed value of j (BRO(j)). For example, for $M = 512$, BRO(6) $= 192$.

4. The output symbols are read out by columns strating from the left-most column, reading columns from top to bottom.

As another type of channel interleaving, the bit-reversal channel interleaver, used in the reverse link, can be described as a process where the sequence of symbols to be interleaved is written into a linear sequential array (0 to $2^L - 1$) and read out from a sequence of addresses generated by the bit-reversal address generator, illustrated in Figure 8.69.

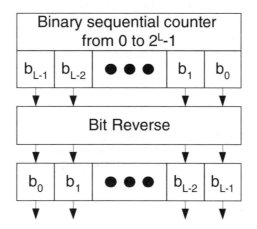

Figure 8.69 Channel interleaver address generator.

The ith interleaved symbol is read out from the array element at address A_i that satisfies

$$A_i = \text{Bit\_Reversal}(i, L) \tag{8.4}$$

where $i = 0$ to $2^L - 1$ and Bit_Reversal(i, L) indicates the bit-reversed L-bit value of i. That is, if i is expressed in the binary form of $i = b_{L-1}b_{L-2}\ldots b_1b_0$, where $b_k = 0$ or 1, b_0 is the LSB and b_{L-1} is the MSB, $A_i = b_0b_1\ldots b_{L-2}b_{L-1}$.

8.5.4   Access Terminal Common Short-Code PN Sequences

The common short-code PN sequences used by access terminals is the zero offset of I and Q PN sequences with a period of 2^{15} chips (32768). The chip rate for the short-code PN sequence is 1.2288 Mcps, with a period of 26.666 ms (32768/1 228 800), that is, exactly 75 PN sequence repetitions every 2 seconds. The generator polynomials for the in-phase (I) and quadrature-phase (Q) sequences are the following

$$P_I(x) = x^{15} + x^{13} + x^9 + x^8 + x^7 + x^5 + 1 \tag{8.5}$$

$$P_Q(x) = x^{15} + x^{12} + x^{11} + x^{10} + x^6 + x^5 + x^4 + x^3 + 1 \tag{8.6}$$

The maximum length linear feedback shift-register sequences {I(n)} and {Q(n)} based on these polynomials have length of $2^{15} - 1$ and can be generated by the following linear recursions

$$I(n) = I(n-15) \oplus I(n-10) \oplus I(n-8) \oplus I(n-7) \oplus I(n-6) \oplus I(n-2) \tag{8.7}$$

$$Q(n) = Q(n-15) \oplus Q(n-12) \oplus Q(n-11) \oplus Q(n-10) \oplus Q(n-9)$$
$$\oplus Q(n-5) \oplus Q(n-4) \oplus Q(n-3) \tag{8.8}$$

In eqns (8.7) and (8.8), I(n) and Q(n) are binary and the additions are modulo 2. To obtain the I and Q common short-code PN sequences (of period 2^{15}), an additional '0' is inserted in the {I(n)} and {Q(n)} sequences after 14 consecutive '0' outputs (this occurs only once in each period), causing the short-code PN sequences to have one run of 15 consecutive '0' outputs instead of 14. The initial state of the access terminal common short-code PN sequences is the first '1' following the 15 consecutive '0' outputs.

8.5.5   Long Codes

The in-phase and quadrature-phase long codes, U_I and U_Q, are generated from a long-code generating sequence using two different 42-bit masks M_I and M_Q respectively, which depend on the AT transmitting channel (Figure 8.70). The generating sequence is based on the following polynomial

$$p(x) = x^{42} + x^{35} + x^{33} + x^{31} + x^{27} + x^{26} + x^{25} + x^{22} + x^{21} + x^{19} + x^{18}$$
$$+ x^{17} + x^{16} + x^{10} + x^7 + x^6 + x^5 + x^3 + x^2 + x + 1 \tag{8.9}$$

Masks M_I and M_Q are public data of the MAC layer and vary according to the transmitted channel.

- Access channel: MI_{ACMAC} and MQ_{ACMAC} (for access long codes).

- Reverse traffic channel: MI_{RTCMAC} and MQ_{RTCMAC} (for user long codes).

The long-code generator is re-loaded with the hexadecimal value 0x24B91BFD3A8 at the beginning of every period of short codes, that is, a period of 2^{15} PN chips.

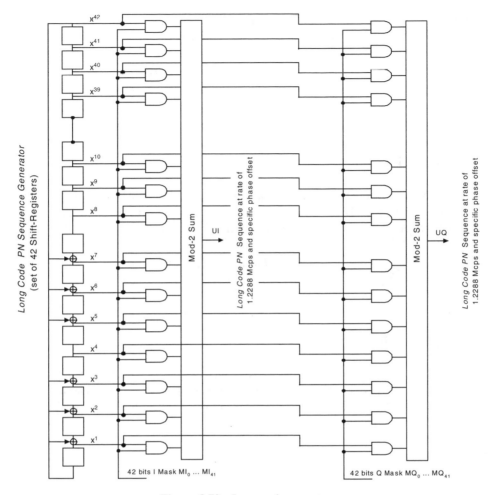

Figure 8.70   Long-code generator.

8.6   MODULATION/ENCODING

8.6.1   Orthogonal Encoding

EVDO systems only use orthogonal encoding in the reverse link through Walsh codes, which are described in detail in Chapter 3.

8.6.2   Bi-orthogonal Encoding

Bi-orthogonal cover encoding is applied by EVDO systems in the preamble bits of the forward traffic and control channels. The set of the 64 codewords, used to identify up to 59 active ATs (MACIndex $= 5, \ldots, 63$) or the data rate of the control channel

(MACIndex = 2, 3), is constructed from an orthogonal set of 32-bit long Walsh codes concatenated with the bit-wise negation of these 32 codes as follows:

$$W^{32}_{MACIndex/2} \quad MACIndex = 0, 2, \ldots, 62$$

$$\overline{W}^{32}_{(MACIndex-1)/2} \quad MACIndex = 1, 3, \ldots, 63$$

8.6.3 64-ary Encoding

A 64-ary code encoding is used to address up to 59 active AT MAC channels (MACIndex = 5,...,63). The set of the 64 codewords is constructed from an inter-laced orthogonal set of 64-bit long Walsh codes as follows:

$$W^{64}_{MACIndex/2} \quad MACIndex = 0, 2, \ldots, 62$$

$$W^{64}_{32+(MACIndex-1)/2} \quad MACIndex = 1, 3, \ldots, 63$$

8.6.4 Symbol Modulation

Symbol modulation is only used in the forward link of EVDO systems. The output of the channel interleaver is passes through a modulator that outputs an in-phase and a quadrature stream of modulated values. Depending on the data rate, the modulator generates symbols in one of the following modulations:

- QPSK,

- 8-PSK,

- 16-QAM.

8.6.4.1 Quadrature Phase Shift Keying (QPSK) Modulation

According to EVDO standards, QPSK modulation is just applied to forward link physical layer packets with sizes of 1024 or 2048 bits. Two successive channel interleaver output symbols are grouped to form QPSK modulation symbols. Each group of two adjacent block interleaver output symbols, $x(2i)$ and $x(2i+1)$ $(i = 0, \ldots, M-1)$, is mapped into a complex modulation symbol $(mI(i), mQ(i))$ as specified in Table 8.71. Figure 8.71 shows the signal constellation of the QPSK modulator, where $S_0 = x(2k)$ and $S_1 = x(2k+1)$.

8.6.4.2 8-PSK Modulation

According to EVDO standards, QPSK modulation is just applied to forward link physical layer packets with sizes of 3072 bits, three successive channel interleaver output symbols are

Table 8.71 QPSK modulation table

Interleaved symbols		Modulation symbols	
$S_1 = x(2k+1)$	$S_0 = x(2k)$	$M_I(k)$	$M_Q(k)$
0	0	D	D
0	1	−D	D
1	0	D	−D
1	1	−D	−D
	$D = 1\sqrt{2}$		

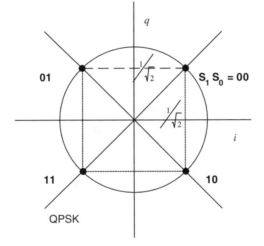

Figure 8.71 QPSK constellation.

grouped to form 8-PSK modulation symbols. Each group of three adjacent block interleaver output symbols, $x(3i)$, $x(3i+1)$ and $x(3i+2)$ $(i = 0,\ldots,M-1)$, is mapped into a complex modulation symbol $(mI(i), mQ(i))$ as specified in Table 8.72. Figure 8.72 shows the signal constellation of the 8-PSK modulator, where $S_0 = x(3k)$, $S_1 = x(3k+1)$ and $S_2 = x(3k+2)$.

8.6.4.3 16-QAM Modulation

According to EVDO standards, QPSK modulation is just applied to forward link physical layer packets with sizes of 4096 bits, four successive channel interleaver output symbols are grouped to form 16-QAM modulation symbols. Each group of four adjacent block interleaver output symbols, $x(4i)$, $x(4i+1)$, $x(4i+2)$ and $x(4i+3)$ $(i = 0,\ldots,M-1)$, is mapped into a complex modulation symbol $(mI(i), mQ(i))$ as specified in Table 8.73. Figure 8.73 shows the signal constellation of the 16-QAM modulator, where $S_0 = x(4k)$, $S_1 = x(4k+1)$, $S_2 = x(4k+2)$ and $S_3 = x(4k=3)$.

Table 8.72 8-PSK modulation table

Interleaved symbols			Modulation symbols	
$S_2 = x(3k+2)$	$S_1 = x(3k+1)$	$S_0 = x(3k)$	$M_I(k)$	$M_Q(k)$
0	0	0	C	S
0	0	1	S	C
0	1	0	−S	C
0	1	1	−C	S
1	0	0	−C	−S
1	0	1	−S	−C
1	1	0	S	−C
1	1	1	C	−S
$C = \cos(\pi/8) = 0.8239$; $S = \sin(\pi/8) = 0.3827$				

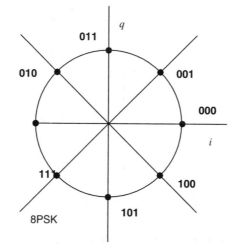

Figure 8.72 8PSK constellation.

8.7 POWER CONTROL

EVDO systems employ a shared TDM pipeline technique in the forward link as a data-optimisation strategy. This approach links the same channel to many different users; therefore it does allow power-control techniques to be employed. Thus, EVDO forward link channels, including the forward traffic and MAC channels, do not have any type of power control.

In contrast, the reverse link channel is totally based on CDMA techniques, and, as in cdma2000 systems, the power control can be applied in the reverse link. Power control is a fundamental feature to achieve good performance and allow higher density of users.

Table 8.73 16-QAM modulation table

$S_2 = x(4k + 3)$	$S_2 = x(4k + 2)$	$S_1 = x(4k + 1)$	$S_0 = x(4k)$	$M_I(k)$	$M_Q(k)$
0	0	0	0	3A	3A
0	0	0	1	3A	A
0	0	1	0	3A	−A
0	0	1	1	3A	−3A
0	1	0	0	A	3A
0	1	0	1	A	A
0	1	1	0	A	−A
0	1	1	1	A	−3A
1	0	0	0	−A	3A
1	0	0	1	−A	A
1	0	1	0	−A	−A
1	0	1	1	−A	−3A
1	1	0	0	−3A	3A
1	1	0	1	−3A	A
1	1	1	0	−3A	−A
1	1	1	1	−3A	−3A

$$D = 1\sqrt{10} = 0.3162$$

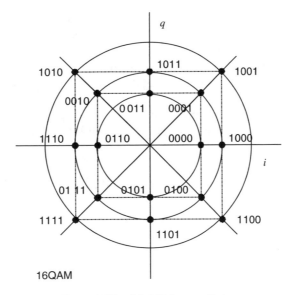

Figure 8.73 16-QAM constellation.

Because each user represents an interference source to the others, more active the users, more the interference. The power-control feature controls the interference level in the reverse link by estimating the best transmission power level and reacting to power-control directions sent by the network as the Reverse Power-Control (RPC) information in the forward MAC channel. Congestion control is managed through the activity level provided as the Reverse Activity (RA) information transmitted by the network in the forward MAC channel.

Power control increases the system's capacity by increasing the number of possible active users (simultaneous) within the same frequency band. It also minimises the effect and occurrence of the 'near-far problem' and increases the lifetime of batteries.

The 'near-far problem' occurs because of the random distribution of the ATs within the coverage area of the network. Some of the terminals are closer to the server than others. If all terminals transmitted with the same mean output power, the power level of the ones closer to the server would be higher than that of those far from it.

In the reverse link of a single carrier EVDO, all terminals with a session open with the same server, use the same long-code phase offset, identifying the server, and are, therefore, interferers to each other. Without power control it would be impossible for the network to properly de-code signals from terminals that are located far from this it.

EVDO is a packet data oriented system; therefore there is, usually, a great number of terminals with opened sessions with the server network. These terminals become active from time to time, transferring information throughout the air interface. As they use the air interface only during the data exchange, the air channels can be shared between several users in a very optimised approach, which usually accommodates more users than a circuit-switched (voice) system, because, in this type of systems, even during periods of silence the resources are still associated with the subscriber.

Power control must consider QoS and mobility requirements. In IS-95 and cdma2000 systems, the BS performs power control, whereas, in EVDO, the user terminal executes power control. The main idea of power-control schemes is to determine the minimum terminal transmission power level with which it is able to reach its server with a power level similar to other terminals, allowing the network to properly de-code each of the users.

There are two types of power-control techniques applied in EVDO systems to minimise interference in the reverse link logical channels: open-loop estimation and closed-loop correction.

- Open-loop estimation: This method is usually implemented as an initial power-control scheme, which estimates the minimum transmission power, used by the reverse access channel during the access probe attempts phase. The AT sends access probe sequences with increasing power level steps until acknowledged by the network. The initial power level value transmitted at the reverse power channel is based on the level received from the forward pilot channel.

- Closed-loop correction: This method is usually implemented by traffic channels, in which a higher performance power control is needed. The AT initial power level is a function of the level received from the forward pilot channel, from the reverse pilot channel at the last access probe and from the required data rate. The network performs periodic adjustments of the reverse pilot channel power level by changing power in steps (increasing or decreasing). Thus, the AN transmits power-control bits that inform the terminal to increase or reduce transmission power, according to the estimated power level.

8.7.1   Open-Loop Estimation

The open-loop estimation process consists of the transmission of access probes within a set of access attempts, as in Figure 8.75 (Section 8.4.1.1 describes the complete process of access probes transmission). During the preamble, the reverse pilot channel power is increased to keep the overall channel power level constant as illustrated in Figure 8.74 (Section 8.4.1.1 describes the preamble power adjustment in detail).

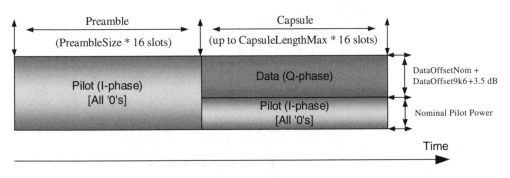

Figure 8.74   Access probe timing.

The increasing factor applied to the reverse pilot channel to match the preamble and capsule total powers is

$$DataOffsetNom + DataOffset9k6 + 3.6\,Db \tag{8.10}$$

In eqn (8.10), DataOffsetNom is a factor representing the ratio of the nominal data channel power to the nominal reverse pilot power, expressed as a complement of two in steps of 0.5 dB. DataOffset9K6 represents the ratio of the data channel power at 9.6 Kbps (reverse access channel data rate) to the nominal data channel power, expressed as a complement of two in steps of 0.25 dB.

In Figure 8.75,

- IP is the initial power value evaluated as (mean Rx pilot power) + OpenLoopAdjust + ProbeInitialAdjust, where

  OpenLoopAdjust is a negative value of the terminal's nominal power set by the network and expressed in units of 1 dB, which is defined in the AccessParameters message;

  ProbeInitialAdjust represents a correction factor set by the network as the initial transmission power adjust level and expressed as a complement of 2 in steps of 1 dB, which is defined in the AccessParameters message at the access channel MAC protocol.

- PS is the power step defined by the PowerStep field in the AccessParameters message.

- PD is the time spent in persistent tests and a pseudo-randomly generated delay between attempts.

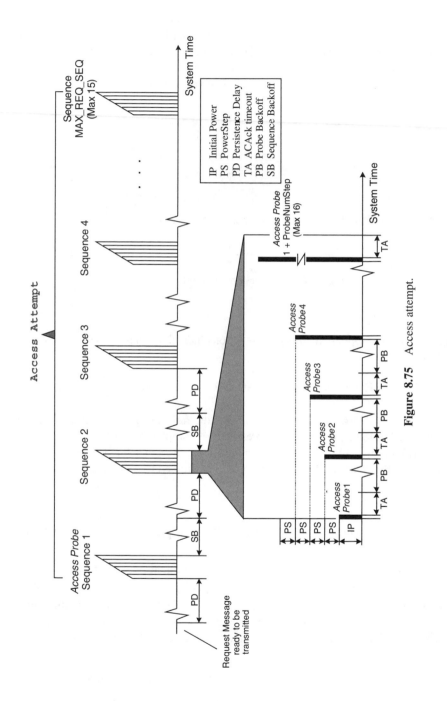

Figure 8.75 Access attempt.

- TA is the time acknowledgement timeout, which is the time waited to receive an acknowledgement for an access probe at the terminal before starting a new access probe cycle.

- PB is the probe backoff set by the network. This value is a pseudo-randomly generated delay between 0 and ProbeBackoff value. ProbeBackoff is part of the access channel MAC protocol attribute and is expressed in units of AccessCycleDuration.

- AccessCycleDuration is defined in the AccessParameters message at the access channel MAC protocol.

- SB is the probe sequence backoff set by the network. This parameter is a pseudo-randomly generated delay after the last access probe in a sequence and before the persistence tests.

- ProbeNumStep is the maximum number of access probes assigned by the network that can be transmitted by the terminal in a single probe sequence. Possible values range from 1 to 15. This parameter is defined in the AccessParameters message at the access channel MAC protocol.

- PreambleLength is the length, in frames, of the access probe preamble. It is set by the network and defined in the AccessParameters message at the access channel MAC protocol.

- CapsuleLengthMax is the maximum number of frames in an access channel capsule. It is set by the network in a range going from 2 to 15. This parameter is defined in the AccessParameters message at the access channel MAC protocol.

The access probe initial mean output power (IP) is defined by open-loop power control, as illustrated in Figure 8.76. In this process, the AT estimates the signal power received from the server and evaluates the transmission power for the reverse link.

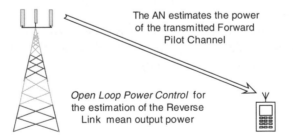

The AN estimates the power
of the transmitted Forward
Pilot Channel

Open Loop Power Control for
the estimation of the Reverse
Link mean output power

Figure 8.76 Open-loop power control.

Open-loop power control should be used in the cases where there is high intensity signal level fluctuations and/or long-term fading, and during the establishment of the initial transmission power for system acquisition. This type of power control also brings advantages to shadow coverage areas and variations of the distance between the terminal and server. This process, however, is not perfect. Short-term fading effects are not minimised, because this type of control is relatively slow in relation to the signal level variation during this kind of fading.

In this power-control technique, the terminals continuously monitor the power level of the received forward pilot channel and increase or reduce their transmission power to match the expected value. The main in-convenience of this technique is that the estimated reverse link transmission power is based on equivalent propagation loss assumptions for both the forward and reverse links, not considering asymmetric propagation effects between the links.

8.7.2 Closed-Loop Correction

An additional open-loop power-control process, known as closed-loop correction, is performed by ATs during the transmission of the reverse traffic channel and the determination of the output power needed to support the data channel, the DRC channel and the ACK channel.

The output power of RRI, data, DRC and ACK channels, part of the reverse traffic channel, is relative to the reverse pilot power level and vary according to their relative gains. Because the RRI channel is time multiplexed with the reverse pilot channel, both channels use the same power level. The data channel, however, has relative gain factors defined according to the selected data rate. Table 8.74 indicates these gains, Section 8.4.1.2 provides more details about the configuration messages that determine this factor.

The DRC channel gain is determined by the DRCChannelGain parameter, whereas the ACK channel gain is determined by the ACKChannelGain parameter. Both parameters are defined in the MAC layer.

The reverse traffic channel only employs the closed-loop power-control scheme. The initial transmission power has already been established during the access probe sequence. The initial power is equivalent to the mean output power of the pilot channel at the end of the last access channel probe. This power is adjusted with the difference in the forward link mean received signal power from the end of the last access channel probe to the start of the reverse traffic channel transmission. The accuracy of this adjustment, as determined by the access and reverse traffic channel MAC protocols is ± 0.5 dB or 20% of the change (in dB), whichever is greater.

Prior to the application of access probe corrections and closed-loop power-control corrections, the access terminal's open-loop mean output power of the pilot channel, X_0, should be within ± 6 dB of the value given by

$$X_0 = -\text{mean received power (dBm)} + \text{OpenLoopAdjust} + \text{ProbeInitialAdjust} \quad (8.11)$$

Table 8.74 Relative power levels vs data rate

Data rate (Kbps)	Data channel gain relative to pilot (dB)
0	(Data channel is not transmitted)
9.6	DataOffsetNom + DataOffset9k6 + 3.75
19.2	DataOffsetNom + DataOffset19k2 + 6.75
38.4	DataOffsetNom + DataOffset38k4 + 9.75
76.8	DataOffsetNom + DataOffset76k8 + 13.25
153.6	DataOffsetNom + DataOffset153k6 + 18.50

In eqn (8.11), OpenLoopAdjust is defined in the AccessParameters message at the access channel MAC protocol. This parameter represents a negative value of the AT nominal power set by the AN and expressed in units of 1 dB. The ProbeInitialAdjust parameter is defined in the AccessParameters message at the access channel MAC protocol and represents a correction factor set by the AN as initial transmission power adjust level. It is expressed as a complement of 2 in units of 1 dB.

OpenLoopAdjust + ProbeInitialAdjust varies from

- −81 to −66 dB for ATs running in frequency under a 1.0 GHz band;

- −100 to −69 dB for ATs running in frequency over 1.7 GHz band.

If the terminal can not transmit at the output power level requested for a specific transmission rate, the AT reduces the power of the DRC and ACK channels accordingly. The maximum power reduction corresponds to gating off the channels. If the ACK channel remains active, its power reduction only occurs after the DRC channel has been gated off. The terminal performs power reduction within one slot of determining that it is not capable of transmitting at the requested level.

The terminal adjusts the mean pilot output power level in response to each power-control bit received on the Reverse power-control (RPC) channel. The change in power depends on the RPCStep parameter defined as a reverse traffic channel MAC protocol attribute (Section 8.4.1.2).

8.8 SCHEDULING

Because EVDO systems are specialised in data transfers, some specific characteristics of the service have to be considered when optimising this type of network. First of all the network must be rate control oriented, that is, it should be optimised for packet data transfers. Packet data rates can be different for each type of service, presenting a different peak value depending on the channel's SNR condition. The peak data rate demands the maximum number of resources (time slots) available, and the data rate for each time slot varies according to the channel conditions. The time slot data rate for a given service and the AN-sector server can be changed in a 600 Hz basis, according to the SNR and the signal level strength sensed by the AT.

The concept of data system optimisation is directly related to the spectral efficiency of the system, therefore, throughput and system overload control are usually the main focus of optimisation algorithms. Also this optimisation must take into account the asymmetric nature of the service, usually requiring higher data transfer rates (throughput) in the downlink than in the uplink.

Throughput optimisation is achieved by activating users according to their channel SNR condition, their service demanded data rate, and to the bandwidth available. To make the best use of the available bandwidth, EVDO systems can transmit in the direct and reverse links with variable and different data rates, matching the asymmetric service data rate requirement.

System overload control is performed by temporarily removing some active users from the system. Their critical session information and resources, such as session state and IP address, are maintained but other users are allowed to become active to avoid system blocking and to promote a fair distribution of resources. This technique is possible because, unlike circuit-switched systems, packet-switched services are only blocked if the required resources are unavailable for a long time (depending on the required QoS), otherwise they can wait on a queue to be scheduled, allowing the use of resources shared among other active users. At EVDO the AN sector is fed back to the AT the cell loading allowing this AT to control the user access to the system in order to minimise the overload.

The decision of which service to remove during overload can be extremely complex. Most algorithms base the decisions on system usage, temporarily removing services activated for longer periods or with high traffic to allow other services access. Many algorithms can be implemented to decide on this, and, like most engineering decisions, the solution is a matter of tradeoffs.

After an AT is connected to the network, it can be on the active or dormant state. The entity that determines the AT's state is the scheduler. Some of the criteria followed by the EVDO scheduler to determine the AT's state, are the following.

- Preference for active ATs with improving signal quality, which need fewer resources, in spite of ATs experiencing signal quality degradation.

- Temporary dormant state for some high rate ATs to activate more ATs consequently activating a higher number of services, i.e. minimising the time a service stays in a dormant state.

- Temporary (milliseconds) dormant state for ATs with low data rate due to channel's condition and activation of ATs in better channel's condition to increase spectrum performance.

The main task of the scheduler is to find the optimal time to transmit data to an AT, aiming to achieve a higher average data rate. The priority in the scheduler is based on a combination of the AT's channel condition based on SNR and the elapsed time since an AT was served last. The AT with worst SNR has its priority increased, as much time it stays in-active. This is necessary to guarantee that this AT will be scheduled some time.

Many techniques can be used to determine the scheduler behaviour. One of the most common is the 'proportional fairness scheduling algorithm', which consists of a C/I weighted approach aiming to attend all users and to keep the better user's average throughput.

The TDM nature of the forward channel allows fast C/I characterisation from the pilot channel, because there is no need to subtract the effect of other channels as in regular IS-95 and cdma2000 systems. The main idea behind this scheduling algorithm is to keep the average data rate for a given RF condition higher than the requested data rate. This is possible by determining the appropriate moment to deliver data to the AT, that is, data is only delivering when the instantaneous conditions of the environment suggest a data rate higher than the expected.

8.8.1 Throughput Optimisation

Voice oriented systems such as IS-95 are designed to give all users equal access and performance, employing direct link RF power control to assure reliable communication in

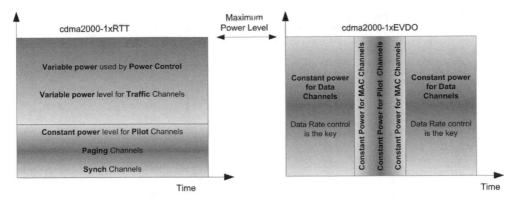

Figure 8.77   1X RTT power control and IS-856 rate control.

varying RF environments. Data oriented systems with a common shared resource in the direct link, as IS-856, provide full power to each AT for a 1.67 ms time slot.

Figure 8.77 shows the transmission burst to a single mobile consists of traffic and overhead information that includes synchronisation and control messages. During the time slot each mobile receives the highest data rate it can receive according to its SNR.

The scheduler is responsible to fairly distribute time slots among ATs. Depending on the implementation, the scheduler can intelligently allocate time slots to optimise throughput. Access terminals present a varied range of SNRs because of the different RF environment characteristics. The scheduler may take advantage of this variation by analysing SNR peaks and valleys and transmitting during expected peak SNR periods in an approach known as user diversity.

8.8.2   Spectral Efficiency

The more noisy the channel environment, the lower the data rate capacity for a given limit of error rate. Transmission, however, can be assured in greater error rate environment by employing information redundancy methods such as FEC and ARQ. The control of SNR and redundancy directly affects the channel data rate capacity.

All modulation techniques have an error rejection characteristic. Each coding scheme introduces redundancy, requiring additional bandwidth. Automatic re-transmission of un-recovered blocks also steals bandwidth. Therefore, to have a spectral efficient system the main goal is to have control on the channel SNR to avoid wasting bandwidth. Data services with segmented communication (packet switched) manage the SNR using data rate control.

Because CDMA has an additional mechanism affecting the transmission capacity, the spreading factor, CDMA systems usually express SNR together with the influence of the spreading processing gain, the E_b/N_o. In cdma2000, the link adaptation chooses among a set of Spreading factors and Modulation and Coding Schemes (SMCS). In IS-95 and EVDO, however, the spreading factor can not be modified.

Table 8.75 EVDO forward link modulation and coding schemes

Data rate (Kbps)	Slots	Code rate	Modulation type	Packet size	Effective code rate
38.4	16	1/5	QPSK	1024	1/48
76.8	8	1/5	QPSK	1024	1/24
153.6	4	1/5	QPSK	1024	1/12
307.2	2	1/5	QPSK	1024	1/6
614.4	1	1/3	QPSK	1024	1/3
307.2	4	1/3	QPSK	2048	8/49
614.4	2	1/3	QPSK	2048	16/49
1228.8	1	1/3	QPSK	2048	2/3
921.6	2	1/3	8-PSK	3072	16/49
1843.2	1	1/3	8-PSK	3072	2/3
1228.8	2	1/3	16-QAM	4096	16/49
2457.6	1	1/3	16-QAM	4096	2/3

EVDO provides non-real-time CDMA packet data services and, therefore, can change SMCS between packet bursts. Because this technology uses a common RF channel to all users, there is rate control but no power control at the downlink. The BS transmits at full power to ensure that the MS receives the highest SNR possible. Errors are lowered reducing the transmitted data rate according to the MS SNR information. The uplink employs both rate and power control.

Table 8.75 shows the available combinations of modulation and coding schemes for the downlink. The Effective Code Rate considers the repetition and puncturing effects.

The data rate and number of slots used to transmit is determined by the channel condition. When conditions are unfavourable, a lower data rate with more slots is used. The modulation technique, in turn, varies with the data rate to provide a more reliable communication channel.

The data rate in EVDO networks continuously changes because RF channel conditions change. The transmission rate in the downlink is determined by the RF signal strength received by the AT. The access terminal is constantly sending data rate requests to the base station based on the SNR. The mobile tries to predict future channel conditions based on past and current SNR values.

On reverse link, however, the system supports only one modulation, BPSK, but can still vary coding rates (Table 8.76) to accommodate different channel conditions. The BPSK modulation was selected for the reverse link because the mobile station has very limited RF

Table 8.76 EVDO reverse link coding schemes

Data rate (Kbps)	Packet size	Code rate	Modulation type
9.6	256	1/4	BPSK
19.2	512	1/4	BPSK
38.4	1024	1/4	BPSK
76.8	2048	1/4	BPSK
153.6	4096	1/2	BPSK

power and antenna gain, therefore requiring a modulation technique that can deal with harsh RF channel conditions.

8.8.3 Retransmission Algorithm (HARQ)

Multi-slot packet transmissions from the AN to ATs follow a four-slot interlacing structure, where the slots of a physical layer packet are transmitted with three other time slots between them, as in Figure 8.78. The four-way inter-lace technique adds efficiency to the 'proportional fairness algorithm'.

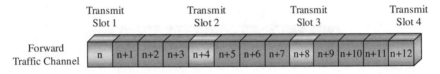

Figure 8.78   Four-slot inter-lacing for a 153.6 Kbps transmission from AN to AT.

The four-slot inter-lacing method allows the receiving AT to decode the partially received packet and to return an acknowledgement message to the AN via the ACK channel before receiving the next slot. If a positive acknowledgement is received, the remaining slots are not transmitted and may be allocated for other data packets, possibly destined to different users. Table 8.77 shows the number of packet slots depending on the DRC index and on the data rate.

Table 8.77   DRC index, slots and data rate

DRC index	Data rate (Kbps)	Packet slots
0	Null	
1	38.4	16
2	76.8	8
3	153.6	4
4	307.2	2
5	307.2	4
6	614	1
7	614	2
8	921.6	2
9	1228.8	1
10	1228.8	2
11	1843.2	1
12	2457.6	1
13	Invalid	
14	Invalid	
15	Invalid	

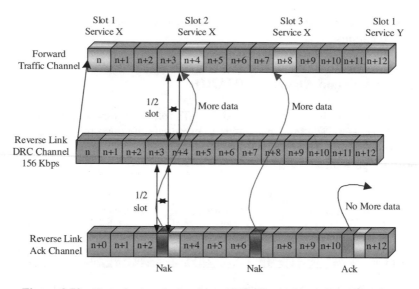

Figure 8.79   Four-slot interlacing for a 153.6 Kbps with early termination.

The early termination of multi-slot packets, or a type II hybrid-ARQ procedure, increases the effective throughput of the system. In Figure 8.79, the packet transmission of four-slot packets of a service with 153.6 Kbps ends with the ACK after the transmission of the third slot, i.e., because of the inherent redundancy the AT has correctly decoded the packet using only three slots, which means that a packet with 1024 bits was transmitted using only 3×16.7 ms of the system time, consequently increasing the transmission rate efficiency by 33% (204.8 kbps).

Note that the four-slot inter-lacing structure creates four independent ARQ streams that can be used to transmit packets to a single or multiple terminals at the same time.

Figure 8.80 shows the Probability Distribution Function (PDF) of the DRC index, the resulting data rate distribution (curve and histogram) at an environment for a requested service with 614.4 Kbps data rate at 2% FER.

Figure 8.81 shows the resulting simulated DRC requests and averages, the 614.4 Kbps required service data rate corresponds to DRC values 6 and 7 (Table 8.77). All averages are calculated based on the last 100 slots dedicated to the user; the moving average represents the behaviour of a round-robin scheduler (lower rate), the intermediate rate curve results from the use of the fairness algorithm only and the higher rate results from the fairness algorithm together with the four-way inter-laced process.

Figure 8.82 shows the filtered configuration for the fairness algorithm, with only the DRC requests resulting in a data rate greater than the required one (614.4 Kbps).

In Figure 8.83, the dotted lines represent the weighted average according to the serviced DRC.

Figure 8.84 shows the effect of the four-way inter-laced feature. The outlined shapes represent the resulted data rate; the dots represented the requested DRC data rate whereas the arrows show the resulting data increase in the data rate due to the four-way interlaced process. The difference in these two values is due to the communication channel diversity.

DRC distribution for 614.4 Kbps for FER 2%

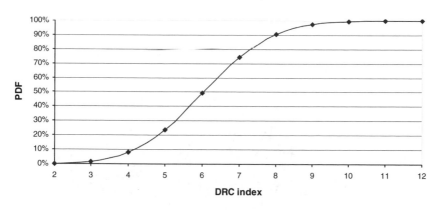

Requested Rate at environment for 614.4 Kbps for FER 2%

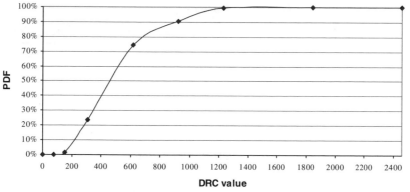

DRC rate for 614.4 Kbps for FER 2%

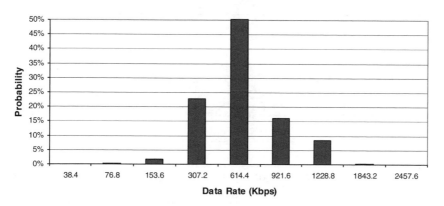

Figure 8.80 DRC distribution for a 614.4 Kbps environment.

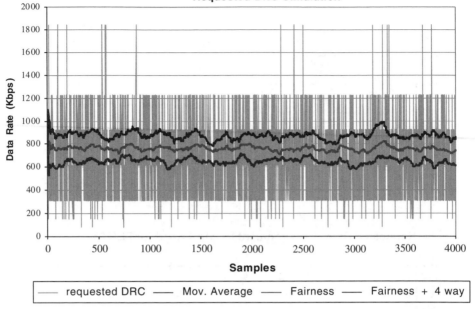

Figure 8.81 Averages from the simulation.

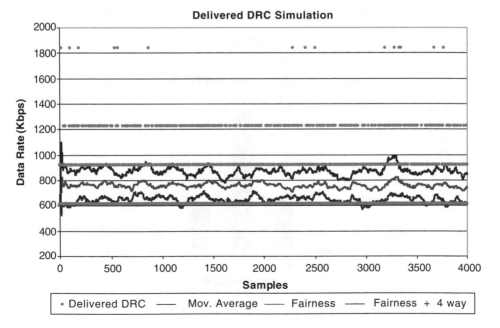

Figure 8.82 Filtered DRC requests.

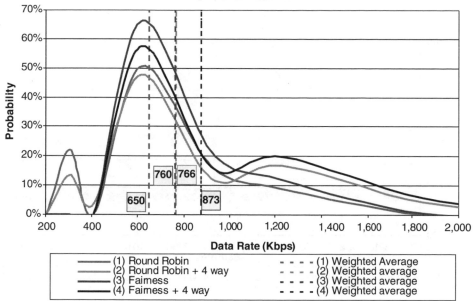

Data Rate Distribution

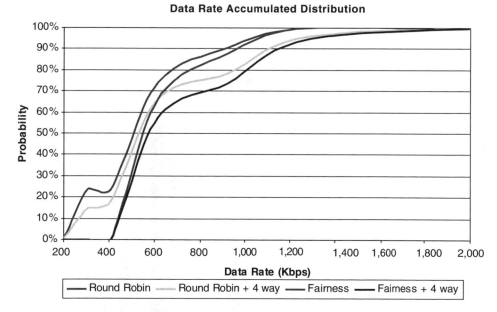

Figure 8.83 Resulting data rate simulation distribution.

Figure 8.85 shows a histogram representing the instantaneous peak data rate distribution for a round-robin strategy with and without the four-way interlaced process and the fairness strategy with and without the four-way interlaced process. The lower data rates show more activity for the round-robin algorithm, whereas the proportional fairness algorithm impacts

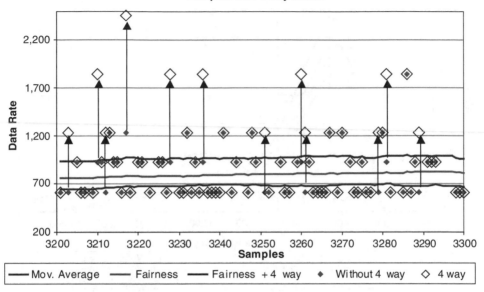

Figure 8.84 Four-way interlace impact over the average.

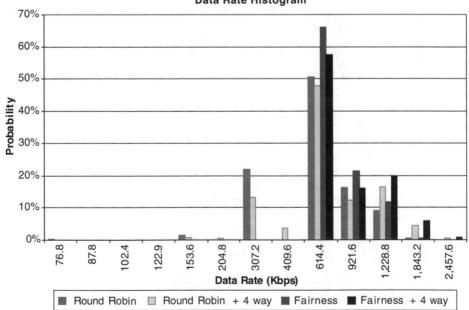

Figure 8.85 Data rate histogram.

on high data rates, and the fairness with four-way interlaced results in an even higher data rate.

In summary, the proportional fairness Scheduler with the four-way interlaced process takes advantage of the short-term channel variation to increase throughput and maintain the grade of service fairness over longer periods of time.

BIBLIOGRAPHY AND REFERENCES

1. IS-856 cdma2000 High Rate Packet Data Air Interface Specification 3GPP2 C.S0024.
2. High-Speed Data Enhancements for cdma2000 1x – Integrated Data and Voice 3GPP2 S.R0026.
3. High-Speed Data Enhancements for cdma2000 1x – Data Only 3GPP2 S.R0023.
4. IS-866 High Rate Packet Data Access Terminal Minimum Performance Standards 3GPP2 C.S0033.
5. IS-864 High Rate Packet Data Access Network Minimum Performance Standards 3GPP2 C.S0032.
6. IS-835 Wireless IP Network Standard 3GPP2 P.S0001-B.
7. Signalling Conformance Specification for High Speed Packet Data Networks TIA-919.
8. Interoperability Specification (IOS) for High Rate Packet Data (HRPD) Access Network Interfaces 3GPP2 A.S0007-A v2.0.
9. 1xEV: 1x EVolution IS-856 TIA/EIA Standard – Airlink Overview Qualcomm, Inc.
10. Wu, Q. and Esteves, E., The cdma2000 High Rate Packet Data System, Qualcomm, Inc.
11. Technical Overview of 1xEV-DV, Technical White paper Motorola.
12. Parry, Richard, cdma2000 1xEV-DO: A 3G Wireless Internet Access System, Submitted to IEEE Potentials, July 2002.
13. Kumaran, K. and Qian, L., Uplink Scheduling in CDMA Packet-Data Systems, IEEE INFOCOM 2003.
14. Jalali, A., Padovani, R. and Pankaj, R., Data Throughput of CDMA-HDR a High Efficiency-High Data Rate Personal, Communication Wireless System, Qualcomm, Inc.
15. Bender, P., Black, P., Grob, M., Padovani, R., Sindhushayana, N. and Viterbi, A., CDMA/HDR: A Bandwidth Efficient High Speed Wireless Data Service for Nomadic Users, Qualcomm, Inc.
16. Hahm, S., Lee, H. and Lee, J., A minimum-Bandwidth Guaranteed Scheduling Algorithm for Data Services in CDMA/HDR System, Seoul National University/Handong Global University.
17. Borst, S. and Whiting, P., Dynamic Rate Control Algorithms for HDR Throughput Optimization, Bell Labs, Lucent.
18. All IP 1xEV-DO Wireless Data Network, Technical White Paper Airvana.
19. Mandyam G. and Fry, G., 1XTREME: A Step Beyond 3G, Nokia Research Center.
20. Derryberry, R., Ma, l. and Rong, Z., Voice and Data Performance of the cdma2000-1xEV-DV systems, Nokia Research Center.
21. cdma2000 Packet Data Migration to High Speed Data, Bell Labs, Lucent.
22. Yang, J., Tin, N. and Khandani, A., Adaptive Modulation and Coding in 3G Wireless Systems, Coding & Signal Transmission Laboratory, University of Waterloo, Canada; Bell Mobility, Canada.
23. Esteves, E., Black, P. and Gurelli, M., Link Adaptation Techniques for High-Speed Packet Data in Third Generation Cellular Systems, Qualcomm, Inc.
24. Chang, K. and Han, Y., QoS-Based Adaptive Scheduling For a Mixed Service in HDR System, Information and Communications University, Korea.
25. Choi, Y. and Han, Y., A Channel-based Scheduling Algorithm for cdma2000 1xEV-DO System, Information and Communications University, Korea.
26. Black, P. and Wu, Q., LINK BUDGET OF cdma2000 1xEV-DO WIRELESS INTERNET ACCESS SYSTEM, Qualcomm, Inc.

9

Radio Network Engineering Fundamentals

LEONHARD KOROWAJCZUK

This chapter examines the use of wireless communication channels and the techniques used to design wireless networks, in particular CDMA. It adds to the information presented in the previous chapters where the CDMA standards were reviewed. Here we will discuss how these standards are implemented in a wireless communication channel and analyse issues related to the design of these systems. We will start analysing the theoretical limits of information transfer and compare them to the existing technology implementations. Next, detailed hardware block diagrams will be presented to materialise the standards recommendations and to serve as a base for the analysis of each block of the wireless communication channel. These blocks will be analysed from the transmit stage to the receive stage, block by block as the signal passes though them. Both communication directions will be analysed at the same time for each block. Specific topics will be analysed at each block with the depth required by it and with more detailed information in a later section, as required. The overall goal is to provide insight for the network designer to allow him to take design decisions.

9.1 DESIGN PRINCIPLES

The goal of a communication system is to transfer information from a source to a destination through a communication channel that connects them. All communication channels have limited capacity and costs associated with them. This motivated the development of techniques to maximise capacity and minimise cost.

The diagram in Figure 9.1 shows the basic blocks that constitute a wireless communication system. This chapter describes their basic functionality, implementation and design aspects.

Figure 9.1 represents a simplex (uni-directional) communication link, that is, the information flows only in one direction. The link is divided in a transmit stage, a channel, a receive stage and a link management stage. The transmit stage prepares the information for transmission over the channel to maximise the correct retrieval probability by the target

Designing CDMA 2000 Systems   L. Korowajczuk et al.
© 2004 John Wiley & Sons, Ltd   ISBN: 0-470-85399-9

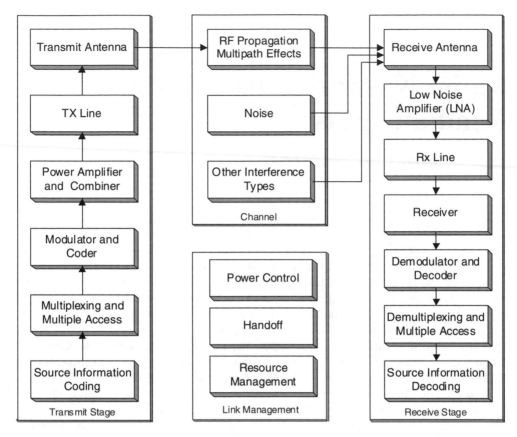

Figure 9.1   Wireless communication channel.

receive stage. The link management stage assures the best conditions for the channel operation and maximises the channel throughput for the services being transported.

Mobile communications, however, are typically duplex, that is, information is simultaneously exchanged in both directions. In actual systems, wireless communications do not happen directly between two users, as part of the link still depends on landline circuits. The wireless terminal at the landline side is called base, Base Station (BS), radio base station, or Base Transceiver Station (BTS). The wireless terminal at the user side is a mobile terminal and is known as mobile or Mobile Station (MS).

Figure 9.2 shows a wireless duplex (bi-directional) channel as it is usually labelled in a CDMA system. The path or link that goes from the BTS to the MS is called Forward Link (FL). Other technologies may refer to this link as downlink (for voice) or downstream (for voice and data). In the other direction, the path from the MS to the BTS is called the Reverse Link (RL) or uplink and upstream in other technologies.

The block diagram of Figure 9.1 applies equally for both the link directions but the information characteristics and the channel parameters may vary significantly from one direction to the other. In this chapter, the analysis moves from the transmit stage to the receive stage, analysing both links in each topic.

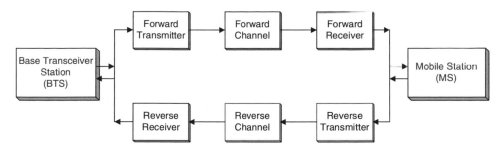

Figure 9.2   Wireless duplex channel.

Many users share the communication channel and the source information of each user has to be retrieved at its associated destination. The channel is limited to the spectrum bandwidth allocated to it. The designer's goal is to provide solutions that maximise channel capacity within a desired communication quality level.

A wireless network designer has to be able to predict the performance of different configurations of the network. Because it is impossible to emulate every detail of the solutions used, it is important to create simplified models that capture the essence of each solution in terms of the network impact.

The goal in this chapter is to analyse each of the network design issues and propose simplified models that lead to a good prediction of network performance. Some of the material presented in previous chapters is summarised or presented in a different way for the sake of completeness and to stress some important aspects of the network design.

9.1.1   Communication Channel Capacity

The communication channel information transmission capacity is limited. Claude E. Shannon [22, 23] calculated its maximum theoretical value in 1948. Shannon's calculations considered the channel being affected by noise with a uniform power distribution (W/Hz) over the entire channel bandwidth, defined as Additive White Gaussian Noise (AWGN). He developed a formula that gives the channel maximum information transfer capacity

$$C = B \log_2 \left(1 + \frac{S}{N} \right) = B \log_2(1 + SNR) \qquad (9.1)$$

where

C   channel capacity (bit/s)
B   bandwidth (Hz)
S   received signal power (W)
N   average noise power (W)
SNR   Signal-to-Noise Ratio (S/N)

Shannon's calculations are quite an achievement considering that his simple equation applies to any communication channel and any information type (analogue or digital). Figure 9.3

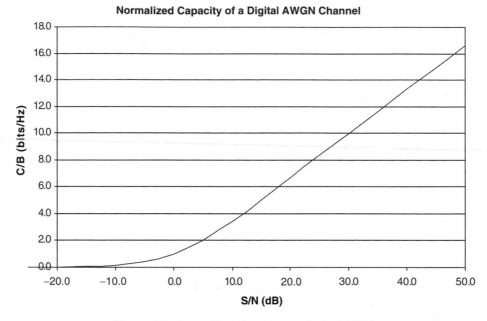

Figure 9.3   Normalised channel capacity in AWGN.

shows the channel capacity normalised in bits/Hz. The chart indicates that the maximum achievable channel capacity is 1 bit/Hz when the S/N ratio is 0 dB. Other noise distributions usually reduce the channel capacity even further.

It is theoretically possible to transmit information free of errors at a rate R_b (bit rate) smaller than the channel capacity when the channel is affected only by Gaussian noise. The ultimate goal of many researchers is to reach this theoretical limit. Several solutions have been proposed, coming each time closer to this limit. Many of these solutions have been

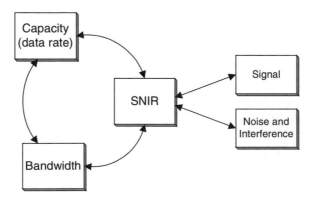

Figure 9.4   Information transfer trade-off.

adopted by CDMA standards and were described in the previous chapters. This chapter covers mainly practical aspects sometimes not emphasised enough in theoretical analysis.

Equation (9.1) expresses that a communication channel can be modelled by a trade-off among three parameters as illustrated in Figure 9.4. Each of the elements in this trade-off is described in the following sections.

The services offered to network users define the characteristics of the information to be transferred. The required channel capacity is dependent on the ability of the technology to eliminate redundant information and consequently reduce the required data rate. The transmission information data rate determines the channel capacity requirement, which should be considered at the source of the information.

The channel has to accommodate the required capacity in the available bandwidth by providing an adequate Signal-to-Noise plus Interference Ratio (SNIR). The SNIR is the combination of signal, noise, and interference. Increasing the signal above a certain value has a direct impact on network costs because it requires more hardware or a higher number of BTS. On the other hand, noise and interference reduction also impact on cost. It is up to the technology to optimise this relationship. This chapter analyses how different technologies approach this problem.

9.1.1.1   Capacity (Data Rate)

The term *Data Rate* is basically employed by digital systems, in which the information to be transmitted from the source to the destination is mapped into discrete forms, such as bits. The quantity of bits transmitted per second represents the data rate.

Digital communication systems employ several techniques to minimise the data rate required to transmit the source information as coding, quantising and modulation schemes, such as Pulse Code Modulation (PCM). Voice digitisation came later and brought several advantages to digital systems, including information reconstruction instead of the simple amplification of disturbed signals.

To achieve an acceptable quality standard, voice should be sampled at least at 8000 times per second. Each sample should be quantised, compressed, and mapped into, at least, 8 bits. This sampling results in a data rate of 64 000 bits per second (bps), i.e. 64 kbps.

Improvement in network capacity requires lower rates, which can be achieved by removing redundant information through source encoding. This is the case of a voice signal that may have its data rate reduced from 64 to 9.6 kbps with a minimum loss of quality.

Some applications, such as voice and video, require a constant data rate, whereas other applications may only require a minimum rate (web browsing, E-mail, etc.). Some types of information require a total integrity of the data whereas others accept a certain amount of errors.

Certain wireless technologies provide fixed data rates for user channels, which is not very efficient. Speech has a channel use ratio of less than 50% whereas user data generally has a very low usage rate. Time Division Multiple Access (TDMA) signals have a constant rate and do not vary with the amount of information transferred whereas CDMA systems change the source rate according to the amount of information to be transferred, increasing efficiency. This characteristic of CDMA is the basic reason why these networks offer two to three times more capacity than TDMA networks, a difference that is even more noticeable with the increase of user data.

9.1.1.2   Bandwidth

Every wireless system has a certain amount of spectrum assigned to it. This amount is referred to as bandwidth. Regulatory agencies define the total system bandwidth according to specific applications of the bandwidth. A network can use all or part of the total bandwidth assigned to the technology depending on the license obtained by the network operator.

Link Duplexing

The network bandwidth has to be used to provide communications for the forward and reverse link and the way this is done is called link duplexing. It can be done using Frequency Division Duplex (FDD) or Time Division Duplex (TDD), in order to provide a full-duplex service.

In FDD, the network bandwidth is divided in two blocks: one for the forward and one for the reverse link. To allow the separation of transmit and receive signals, there must be a significant frequency separation between these blocks (45 MHz typically). Therefore, this technique can only be used in systems that have a large block of spectrum available. Figure 9.5 shows FDD assignments for cellular 800 MHz and PCS 1900 MHz in the USA.

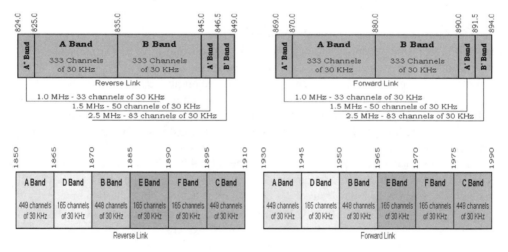

Figure 9.5   Example of available bandwidths for FDD cellular (800 MHz) and PCS (1900 MHz) wireless systems.

In TDD, the same network bandwidth is used for forward and reverse links but at different moments in time. There must be some sort of synchronisation and a guard-time is required to avoid collisions. This technique is preferred when the system has a limited spectrum.

Channel Multiplexing

The available bandwidth has to accommodate a large number of users. Each user capacity requirement is a fraction of the total capacity, constituting an information channel.

Information channels are carried together on the available bandwidth through the use of multiplexing techniques.

Design reasons oblige the bandwidth to be split into smaller blocks that can be more easily handled by the hardware circuitry as hardware linearity issue increases with the bandwidth size. This need implies the creation of frequency channels or carriers that are associated with information channels. In general, all carriers have the same bandwidth and are assigned to radios in the BTS and MS. Recently proposed standards, such as the cdma2000 3X, allow[1] different carrier bandwidths for forward and reverse links.

Each carrier can carry one or more information channels, which are multiplexed over the assigned bandwidth employing different multiple access technologies, with channels identified by frequencies, timings or codes. When a specific frequency uniquely identifies an information channel, the access technique is called Frequency Division Multiple Access (FDMA). When a different time interval is assigned to each channel, the technique is called Time Division Multiple Access (TDMA). When channel identification is based on codes, the technique is called Code Division Multiple Access (CDMA). These three multiplexing techniques are illustrated in Figure 9.6.

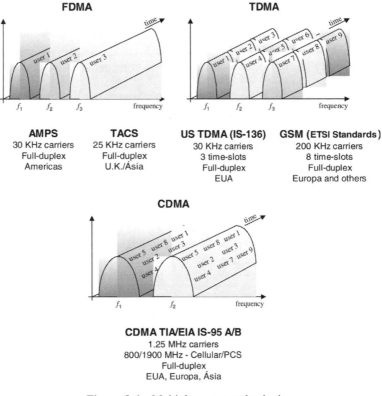

Figure 9.6 Multiple access technologies.

[1] AMPS – Advanced Mobile Phone System; NADC – North Digital Cellular; FDD – Frequency Division Duplex; FDMA – Frequency Division Multiple Access; TDMA – Time Division Multiple Access; GSM – Global System for Mobile communication; CDMA – Code Division Multiple Access; WCDMA – Wideband CDMA

A set of rules, called a standard, defines the combination of duplexing and multiplexing methods. Each standard defines what is known as a communication technology. The main wireless technologies today are AMPS (FDD-FDMA), NADC IS-136 (FDD-FDMA/TDMA), GSM (FDD-FDMA/TDMA), CDMA IS-95 or IS-2000 (FDD-FDMA-CDMA) and WCDMA (FDD-FDMA-CDMA or TDD-FDMA-CDMA). Section 9.3.2 describes these technologies and the multiplexing methods in more detail.

AMPS and TDMA carriers have a nominal bandwidth of 30 kHz of which the information channel uses only 10 kHz. The rest of the bandwidth is required to separate the information channels through filtering. GSM uses a nominal bandwidth of 200 kHz of which the information channel uses 160 kHz. CDMA uses a bandwidth of 1,250 kHz and WCDMA a bandwidth of 5,000 kHz.

9.1.1.3 SNIR

To achieve a good transmission performance, the signal must be received within a certain level when compared to the ambient noise, i.e. the signal must have a good Signal-to-Noise and Interference Ratio (SNIR, S/N or SNR).

The S/N parameter, also referred to as Signal-to-Noise Ratio (SNR), is the ratio between signal strength (S) and noise (N), which may or may not include an interference component. Therefore, the term SNIR (Signal-to-Noise–plus–Interference Ratio) is sometimes used to remove this ambiguity, indicating that interference is considered. This is the case in this chapter.

The minimum theoretically required SNIR is calculated using Shannon's equation after defining the data rate and bandwidth. Practical systems, however, usually require an SNIR higher than the theoretical. Each technology usually requires a different SNIR, which depends on the desired quality, defined partially by the technology and partially by marketing requirements.

Signal-to-Noise and Interference Ratio (SNIR)
The SNIR expresses the ratio between signal strength and noise and interference. The signal strength represents the absolute analogue signal power level measured at any desired point, more specifically at the reception point of interest. Although the measured value corresponds to a value expressed in Watts (W), it may also be represented in dBm, which corresponds to the comparison of the original value in Watts to a 1 mW reference signal as in equation (9.2)

$$P_{dBm} = 10 \log_{10}\left(\frac{P[W]}{1[mW]}\right) = 10 \log_{10}(P[W]) - 10 \log_{10}(1 \times 10^{-3}[W])$$
$$= 30 + 10 \log_{10}(P[W]) \tag{9.2}$$

where

P_{dBm}	signal power level, in dBm
$P[W]$	signal power level, in Watts

The received signal strength can be improved in different ways: increasing the transmit power, mitigating fading or repeating the information.

The interference and noise part of the SNIR comprises everything that can mask the original signal, including ambient noise, inter-symbol interference, multi-path interference and interference from other transmitters.

Interference can be greatly reduced by restricting the transmitted power (power control) used by each communication in the network to the minimum value required to establish the communication. Other strategies may also be employed with the purpose of reducing interference, such as the use of transmission and reception filters, better coding and decoding schemes and reduction of the overlap between cells (cell shaping). Circuit noise can be reduced by proper design of the input stages and filtering.

E_b/N_0 Ratio

In digital transmission systems, the SNIR is usually replaced by the normalised ratio E_b/N_0 (bit energy per noise spectral power density). The relation between these two criteria is calculated in eqns (9.3)–(9.5).

$$E_b = ST_b = \frac{S}{R_b} \tag{9.3}$$

where

E_b Received bit energy (W·s)
T_b Bit time duration (s)
R_b Bit rate (bps)

$$N_0 = \frac{N}{B} \tag{9.4}$$

where

N Noise power uniformly distributed over the entire spectrum (W).

The S/N ratio is related to E_b/N_0 by the ratio of R_b/B as in eqn (9.5). This equation shows that S/N and $E_b N_0$ are interchangeable parameters and are used as reference for evaluating network quality

$$\frac{S}{N} = \frac{E_b}{N_0} \frac{R_b}{B} \tag{9.5}$$

The capacity for a digital system can be expressed as in eqn (9.6).

$$C = B \log_2 \left(1 + \frac{E_b}{N_0} \frac{R_b}{B}\right) \tag{9.6}$$

Maximum channel utilisation occurs when $R_b = C$. Figure 9.7 shows the normalised capacity (C/B) of a digital communication channel indicating that capacity reaches zero for an E_b/N_0 of approximately -1.5896 dB.

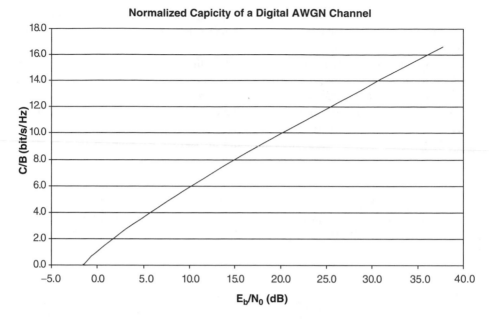

Figure 9.7   Normalised capacity of an AWGN digital channel.

9.1.2   Technology Performance

Practical implementations of different technologies require specific SNIR values. Table 9.1 compares the effective capacity of the most common wireless technologies for voice application. The SNIR presented in the table are values typically used in network design: 17 dB SNIR for TDMA and 12 dB SNIR for GSM. Even though an SNIR value is given for CDMA, this technology is most commonly analysed in terms of E_b/N_0. The typical value for this variable in real CDMA (IS-95) networks is 7 dB for voice. The number of channels per carrier for CDMA is also derived from typical applications.

The SNIR in Table 9.1 includes all aspects related to hardware and software therefore reflecting the real performance of each sector cell in the field, including the signalling

Table 9.1   Technology capacity comparison

Technology	Carrier bandwidth (Hz)	Practical SNIR (dB)	E_b/N_0 (dB)	Shannon capacity per carrier (bps)	Source rate (bps)	Channels per carrier	Carrier rate (bps)	Efficiency (%)
TDMA	30 000	17.00	17.18	170 273	9 600	3	28 800	16.91
GSM	200 000	12.00	14.84	814 917	13 000	8	104 000	12.76
CDMA IS-95	1 250 000	−1.61	7.00	946 535	8 600	20	172 000	18.17
cdma2000 (1XRTT)	1 250 000	0.15	7.00	1 281 412	8 600	30	258 000	20.13

overhead (synchronisation, access, paging and control). This value does not provide comparison of the number of BTS required to obtain such coverage.

The efficiency value expresses the percentage of the theoretical capacity that each technology can handle. GSM has the lowest efficiency due to its higher bit rate vocoder. cdma2000 (1XRTT) presents the best efficiency. Shannon's limit is calculated considering AWGN noise, whereas the performance of real systems is calculated using the actual noise distribution, which imposes a much larger burden (4–5 dB) than AWGN.

Table 9.1 shows that CDMA IS-95 has a slightly larger capacity than TDMA. This increase in efficiency is explained by two CDMA characteristics: full bandwidth usage and variable data rate. TDMA requires spacing between channels, for the filters to be able to separate them, consequently only 50% of the bandwidth is actually used to transmit signals, whereas in CDMA the whole bandwidth is used. In TDMA, the user channel has a fixed data rate, whereas in CDMA, the data rate varies with speech pauses that represent approximately 50% of the transmission time (Section 9.1.1.1).

This analysis of CDMA efficiency, however, may be deceiving because it considers a constant source rate of 8600 bps to allow the comparison with other technologies that use lesser-evolved vocoders. CDMA efficiency, in fact, is smaller than the one indicated, because the actual rate is about half of the one considered due to pauses in conversation that are encoded at a lower rate by variable rate vocoders. When compared to the Shannon limit, CDMA efficiency is approximately 20%, mainly because in practical systems the noise is not AWGN. cdma2000 has a higher efficiency and future technologies are expected to have even higher ones.

9.2   CDMA EQUIPMENT BLOCK DIAGRAM

The first part of this book provides a detailed description of the CDMA standardisation but little was said about how this is implemented in real networks. This chapter assumes that the reader is already familiar with the channel structure of the CDMA technology. Figure 9.8 presents a block diagram of the constituents of a typical CDMA network. This section describes a typical implementation of IS-95 using an Enhanced Variable Rate Coder (EVRC) vocoder.

The objective of a wireless network is to connect mobile stations (MS) to other mobile stations or to users in a Public Switched Telephone Network (PSTN) or in a Packet Data Serving Node (PDSN). To achieve this objective, mobile stations are connected to the BTS (Base Transceiver Station) through the standardised air-interface. One or more BTS may be located in the same place, known as a site, sharing the same infrastructure (e.g. tower, energy supply and shelter).

A BTS may have one (omni-directional or omni BTS) or multiple sectors (sectorised BTS). The typical number of sectors used in existing CDMA networks is three. The advantage of sectorisation is that it concentrates several cells (sectors) in the same site, optimising hardware usage and minimising the number of physical sites and infrastructure.

BTS sectors are also known as sector cells or sectors (ST). A group of sectors forms the BTS or cell. Each sector may be constituted of multiple transceivers that have the same RF settings (but different frequencies) and, consequently, have the same coverage. The multiple sectors within a BTS may be pointing their antennas toward different directions (α, β and γ) to provide service in different areas. Figure 9.9 illustrates the possible structure of a BTS.

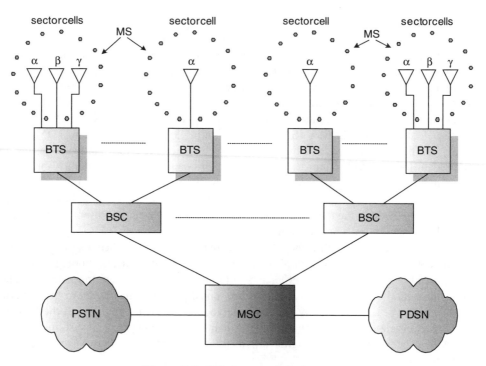

Figure 9.8 Wireless network diagram.

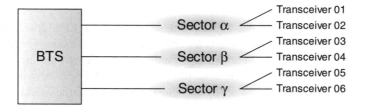

Figure 9.9 BTS sectors and transceivers structure.

This chapter uses the term *sector* to refer to the RF coverage of the set of transceivers that belong to a BTS sector, and the term *cell* to refer to the coverage of all sectors that belong to a BTS as illustrated in Figure 9.10.

The BTS is the bridge between the air-interface and the landline. The connection between BTS and landline is made through a Base Station Controller (BSC). The BSC holds common processing equipment for several BTSs, because it wouldn't be economical to distribute this equipment. Voice and data calls coming from the BSC/BTS are connected, respectively, to the PSTN and PDSN through the Mobile Switching Center (MSC). The MSC is the brain of the mobile network, storing information and controlling all connections.

Figure 9.11 illustrates the geographical distribution of BTS. The figure shows the theoretical cell grid and the real coverage provided by each cell. Cell sizes vary with traffic and can range from 100 m to tens of kilometers.

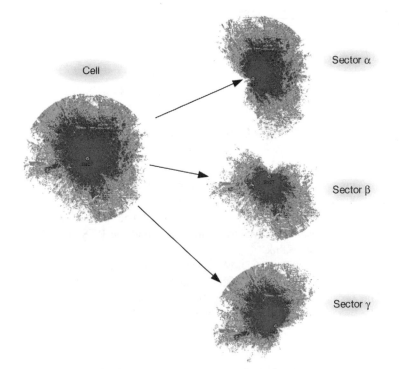

Figure 9.10 Cell and sector definition.

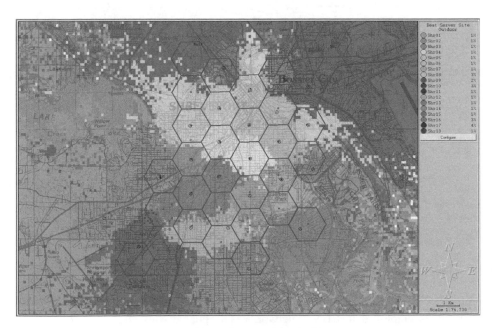

Figure 9.11 Wireless network diagram.

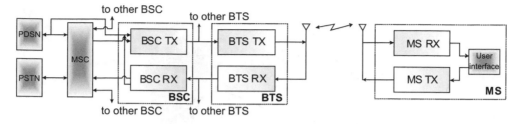

Figure 9.12 Wireless network block diagram.

Figure 9.12 shows a simplified block diagram of a wireless network. The top part (BSC TX → BTS TX → MS RX) represents the BSC and BTS transmit circuits and the MS receive circuits that are part of the forward link. The bottom part (MS TX → BTS RX → BSC RX) represents the MS transmit circuit and the BTS and BSC receive circuits that are part of the reverse link.

As explained in Chapter 3, CDMA systems employ a number of codes during the communication process. Figure 9.13 shows the sequence of CDMA codes applied in the network. The next sections describe the blocks in the diagram of Figure 9.12 in detail indicating the usage of the codes shown in Figure 9.13.

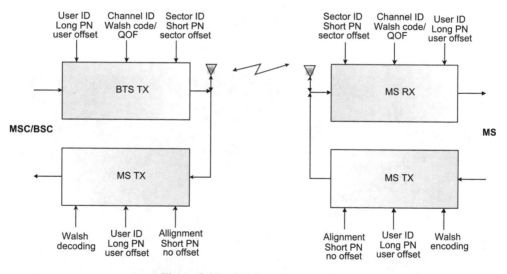

Figure 9.13 CDMA code sequence.

9.2.1 Network Block Diagram Description

The previous chapters of this book presented the description and the function of all elements and resources employed in CDMA networks. The goal of this chapter is to tie all this together to allow readers to understand the flow of a CDMA system.

The user information is coded in CDMA in traffic channels, a different code being assigned to each channel, allowing for the multiplexing and multiple access to the air-interface. Each channel information is transmitted at different power levels depending on the information rate being transmitted, the path loss between transmitter and receiver and the interference level at the receiver. During full speech, a larger power is transmitted then during speech pauses (1/8 of the power at full speech) and this diminishes the overall interference. Besides the user information, additional data has to be sent to establish and control the communication. Further channels have to be added like the pilot, paging and access channels described in previous chapters. These additional channels represent an overhead that diminishes network capacity, so they are carefully designed during the formulation of the technology and experience gained with the IS-95 networks allowed the more efficient design of voice and data carrying cdma2000 technology.

As explained before, a CDMA wireless network is composed of a forward and a reverse link that are described below.

Forward Link (FL)
The pilot channel in the forward link establishes the maximum sector footprint because only MS that receive the pilot signal above a certain level can be connected to it. The use of this pilot provides synchronization for the receiver by indicating the start and end of each code, and improves detection capability (coherent detection). In coherent detection the alignment of the code is known and this makes the detection much more robust. The overhead due to the pilot is relatively low as the BTS hosts many user channels and its impact is diluted.

Forward link traffic channels are power adjusted to meet the requirements of the target MS. The long PN code, offset by the channel user key, scrambles the information data allowing only the intended recipient to recover the information. Each channel is then identified and orthogonally spread by a Walsh code. The short PN code, offset by the sector assigned PN offset, is applied on top of the Walsh code, to identify the signals from each sector and to provide alignment recovery capabilities for multipath situations.

Reverse Link (RL)
In traditional CDMA systems, the RL does not use a pilot signal because it could represent up to 50% overhead considering that an MS has only one traffic channel. The lack of pilot requires a noncoherent detection, which is a much less efficient than the coherent detection used in the forward link. Therefore, the RL requires additional protection to the signal. Error encoding is more robust in the RL, even though it represents an additional overhead. In cdma2000, however, MS can have more than one traffic channel; therefore, the technology implements an embedded pilot to improve RL without causing such a large overhead.

The RL was expected to be the bottleneck of networks because of the noncoherent detection, but, in the majority of cases, the forward link proved to be the bottleneck. This is mainly due to the extra protections added to overcome the lack of pilot as described below.

The lack of pilot channel, in IS-95, means that the start and end of each code is not known in principle and a noncoherent detection has to be made in which every single position has to be tested, although there are algorithms that optimise this search.

In principle, Walsh codes usage is not required in the RL for the purpose of multiple access because there is only one traffic channel. However, these codes are used as a form of 64-ary orthogonal modulation that provides additional detection capabilities at the receiver. The Walsh code spreads the information data and encodes it into a multi-dimensional space (orthogonal modulation or error correction, like any M-ary modulation system using binary symbols) providing robustness to the decoding process. The long PN code, offset by the user key, spreads the data, provides security against eavesdropping, and allows the identification of individual users (MS). This is possible due to the large number of combinations available in this extensive code. Short PN codes with no offset are applied to the Q and I channels to provide quadrature diversity and alignment, adding another layer of detection capability at the receiver.

Call Flow

The MSC receives an incoming call and analyses the signalling (generally SS7 common channel). The MSC searches the network to determine the location of the destination MS. The call is then routed to the appropriate PCM link while the MS is instructed to connect to a specific RF channel. For a voice call, a vocoder is assigned to convert the wave-form coding of the PCM DS0^2 to the appropriate speech-coding scheme. For a data call, the information is stored and is ready to be routed to the BTS.

A BTS channel element (CE) is then assigned and the information is routed to/from it by a packet connection. The MS is directed to look for a specific offset of the short PN code (sector ID) and a channel defined by a Walsh code. In both the cases, the BTS and MS use the MS identification (IMSI) to scramble/un-scramble the signal.

An outgoing call requests a channel to the MSC. This channel is identified, in the forward link, by the short PN code offset and a Walsh code. In the RL, the channel is identified by the long PN code offset by the user key.

9.2.1.1   Network Elements Description

The Network Elements as seen in *Figure 9.12* can be divided in two clusters: Mobile Switching Center (MSC), Base Station Controller (BSC), Base Transceiver Station (BTS) and Mobile Station (MS).

These two clusters allow for the establishment of calls as described above briefly under "Call Flow".

An incoming call request is received by the MSC and the signalling is analysed (generally SS7 common channel signalling is used). The destination MS is searched and located in the network. The call is then routed to the appropriate PCM link while the MS is instructed to connect to a specific RF channel. For a voice call, a vocoder is assigned in the path to convert the waveform coding of the PCM DS0 to the appropriate speech-coding scheme. For a data call, the information is stored ready to be routed to the BTS. A BTS channel element (CE) is then assigned and the information is routed to/from it by a packet connection. The MS is directed to look for a specific offset of the short PN code (sector ID) and a channel defined by a Walsh code. In both the cases, the BTS and MS will use the MS identification (IMSI) to scramble/unscramble the signal.

An outgoing call will request for a channel to the MSC that will be assigned in the forward link and identified by the short PN code offset and a Walsh code and the RL channel will be identified by the long PN code offset by the user key.

---

^{2} Digital signal X (DSX) is a term for the series of standard digital transmission rates or levels. It is based on the bandwith normally used for one telephone voice channel, DS0 (a transmission rate of 64 kbps).

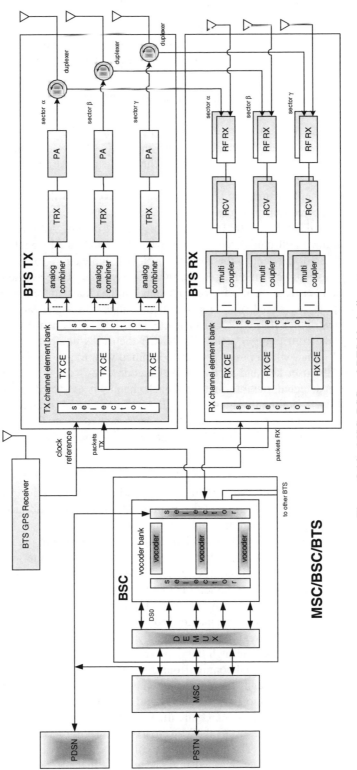

Figure 9.14 MSC, BSC and BTS block diagram.

Once this Call Flow procedure is performed, the transmit and receive paths are established. The MSC will establish a connection to the PSTN or to another MSC/BSC or to the PDSN and instruct the mobile to connect to a specific RF channel.

Mobile Switching Center (MSC)/Base Station Controller (BSC)/Base Transceiver Station (BTS)

This section describes the left part of Figure 9.12 which is shown in more detail in Figure 9.14. It shows the interconnections between the MSC, BSC and BTS.

The MSC connects to the PSTN through PCM links and to the PDSN through packet links.

For transmission, the Transmission Channel Element (TXCE), in the BTS, encodes the signal adding error correction, provides encryption, allows multiple access through Walsh codes, adjusts gain and identifies the BTS sectors using PN offsets. All channels within a sector are then summed and modulated by the carrier. The modulated signal is sent to the power amplifier, filtered and sent to the antenna through a duplexer.

For reception, a diversity scheme is used and two received signals are available, one from the duplexer and another from the receive-only antenna. Both signals are sent to RF receivers (RF RX) that amplify the signal. Each receiver demodulates the signal and sends it to a multicoupler that distributes it to all RX channel elements (RXCE). Each RXCE maximally combines the signals from the two paths. The RXCE then uses non-coherent detection to detect the encoded signal, which is successively combined with the short PN code, the long PN code offset by the user key and all the 64 Walsh codes.

After the phase is detected, the information of each channel is extracted through a similar process, tracking each multipath signal at a bit level. A Viterbi decoder extracts the information bits and sends them to the BSC vocoder assigned to it. Recovered voice is sent to the PSTN or to another MSC/BSC and data is sent to the PDSN.

Mobile Station (MS)

This section describes the right part of Figure 9.12, which is shown in more detail in Figure 9.15. It shows a detailed diagram of the mobile station. The MS interfaces directly to the user through its display, keypad, microphone and speaker.

In the transmit direction, the signal from the microphone is digitised in the vocoder and a CRC is added. The signal is encoded to add error correction and block interleaved to allow the error spreading at the destination. The Walsh code mapping is used for orthogonal modulation in the channel encoder. The long PN code, offset by the user key, is combined with this signal to scramble it and identify the user. The short PN code is applied to increase the detection capability. The signal is then amplified and power controlled through an open-loop and a closed-loop scheme. Next, it is modulated by the carrier and power amplified. The signal is then sent to the antenna through the duplexer.

In the receive direction, the signal coming from the antenna is separated by the duplexer and sent to the receive RF front end. There it is filtered by a band pass filter and amplified by an LNA. Next, it is demodulated to an IF frequency and sent to the receiver channel element where it has its gain adjusted by an AGC circuit. Each RX channel element chooses the strongest signal and will be instructed by the MSC to extract specific sector information (PN offset) and some of its multipath. The channel information is extracted next using the Walsh code assigned to the channel. The detected digital signal is then decoded recovering the

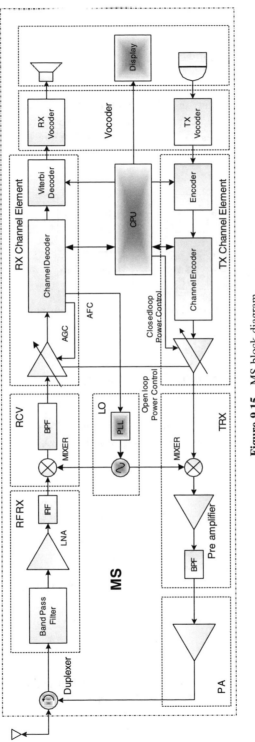

Figure 9.15 MS block diagram.

information sent by the BTS. The vocoder generates the analog speech signals that are sent to the speaker.

9.2.1.2   Forward Link

This section describes the main network elements used in the forward link. These elements are presented in Figures 9.14 and 9.15 as part of the MSC/BSC/BTS group and of the MS respectively. Figure 9.16 illustrates the forward link flow. Each of the blocks in this figure is described in detail next.

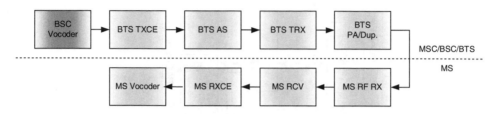

Figure 9.16   Forward link network elements.

Forward Link Transmission from the BSC to the MS

BSC Vocoder
Vocoders are centralised in the BSC to minimise interconnection requirements between BTS and the BSC and being used more efficiently as a pool. Figure 9.17 shows the BSC transmit vocoder block diagram.

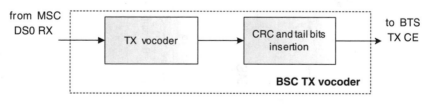

Figure 9.17   BSC TX vocoder.

The block depicted as the BSC Transmission (TX) vocoder comprises the actual vocoder and an additional circuit that adds a 12 Cyclic Redundancy Check (CRC) bits and 8 tail bits to each 20 ms voice frame. The CRC bits are used to calculate the Frame Erasure Rate (FER) that serves as a quality indicator of the information transfer. A FER of 10^{-3} is considered acceptable for voice transmission.

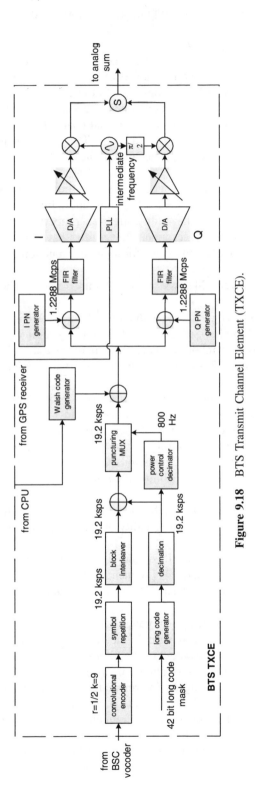

Figure 9.18 BTS Transmit Channel Element (TXCE).

BTS Transmit Channel Element (BTS TXCE)

Figure 9.18 shows the BTS Transmit Channel Element (TXCE). The TXCE receives the signals from the vocoder in one of two different sets of rates RC1 and RC2. RC2 is used by older MSs and is kept for compatibility reasons. The traffic channel has 20 ms frames with rates of 1.2, 2.4, 4.8 or 9.6 kbps when in RC1 and rates of 1.8, 3.6, 7.2 and 14.4 kbps when in RC2. Additional channel elements generate the overhead channels, pilot, sync and paging. The pilot channel is composed only of '0's (no data). The synch channel has a rate of 1.2 kbps with 26.66 ms frames. The paging channel has rates of 4.8 and 9.6 kbps, with 20 or 80 ms super-frames.

A half rate $(r = 1/2)$ convolutional encoder encodes each frame with constraint length (k) 9 code to form a stream of 19.2 ksps. In the RC2 configuration, bits are punctured (deleted) to achieve the 19.2 kbps relying on the error correction properties of the convolutional code to restore the deleted bits in the reception. Slower rates have their frames repeated to match the output rate and this repetition allows for a power reduction as the redundancy is increased.

The frame is block interleaved (24 rows by 16 columns = 384 symbols) as part of the mechanism to spread burst errors.

The TXCE also generates a long PN code with a rate of 1.2288 Mcps, offset according to the MS identification (IMSI number), which is used as an encryption key. This code is then decimated (sampled every 64th symbol) to match the rate of 19.2 kbps. This decimated PN sequence is combined (modulo 2 summed) with the interleaved signal to provide the required encryption.

The PN sequence is also further decimated (sampled every 24th symbol) to reach 800 Hz. This new decimated sequence is used to control the writing of power control bits, by puncturing (overwriting) the output sequence of each traffic channel.

The output sequence has introduced errors by the power control puncturing described above, 1 in 24 symbols for RC1 and 9 in 24 bits for RC2. The convolutional decoder, in the receiver, eliminates these errors, but its capability of eliminating other errors is compromised mainly for RC2.

To identify which channel is transmitting this data, the information within the channel is combined with a Walsh sequence. The orthogonality of Walsh codes allows the information to be recovered in the receiver.

After the channel is Walsh code identified, the signal is split into two branches and combined with one of two short (15 stages) PN sequences (I and Q) at a rate of 1.2288 Mcps. Both sequences are offset with the assigned BTS sector identification to specify where the signal is coming from.

After the signal is split and identified, each branch is wave-shaped using a digital lowpass FIR (Finite Impulse Response) filter, leaving only the base band components. A digital-to-analogue converter (D/A) converts the output of the filter to an analogue signal. The gain of this signal is adjusted according to network requirements. Finally, the signals are modulated in quadrature (phase 0 and $\pi/2$) using an intermediate frequency (4–12 MHz typically) generated by a Phase Locked Loop (PLL) that is synchronised by the BTS GPS (Global Positioning System) circuitry. This intermediate frequency is used as intermediate stage before the modulation by the carrier frequency, to eliminate noise and intermodulation products using a filter in a frequency range where the filtering circuits can be implemented effectively. The values of the intermediate frequency are not defined in the standard and vary among manufacturers.

The output of both branches is summed and the signal from each channel element is send to the BTS Analogue Combiner.

BTS Analogue Combiner (BTS AC)
This stage adds the analogue signals of all channels at the Intermediate Frequency (IF) level and sends them to the transmitter. The IF is a designer's choice; usual values range from 4 to 12 MHz. Figure 9.19 illustrates the analogue sum stage.

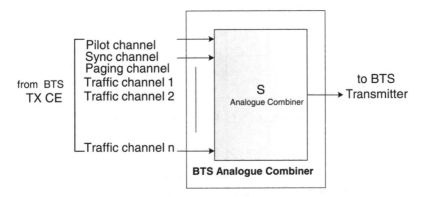

Figure 9.19 BTS TX Analogue Combiner (AC).

BTS Transmitter (TRX)
Figure 9.20 shows the transmitter part of the BTS transceiver. The transmitter modulates the IF signal coming from the analogue sum using the CDMA carrier (800 or 1900 MHz). The signal is then filtered by a band pass filter and pre-amplified.

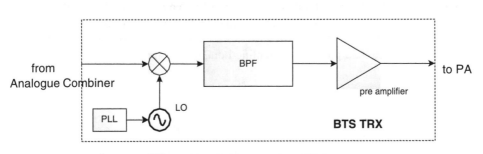

Figure 9.20 BTS transmitter (TRX).

BTS Power Amplifier and Duplexer
The Power Amplifier (Figure 9.21) must have a high gain (about 40 dB), be highly linear, output a high power signal (10–40 W), and be able to handle high power peaks (50–200 W).

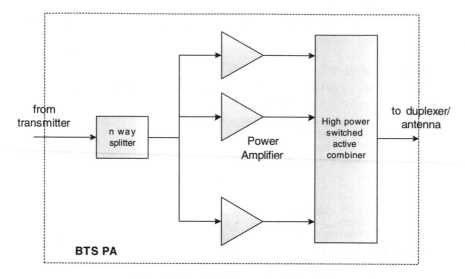

Figure 9.21   BTS Power Amplifier (PA).

These characteristics may require the combination of several amplifiers in parallel. The signal from the transmitter is split among all amplifiers and a high power switched combiner combines the output in an efficient manner.

The amplified signal is transmitted through the antenna. If the same antenna is used for transmission and reception, a duplexer is required. This element combines transmit and receive signals using band pass filters and a circulator. This isolates both signals allowing a single antenna to be used.

Forward Link Reception at the MS

MS Receive RF Front End (RF RX)
The signal received by the MS in the forward link is filtered by a Band Pass Filter (BPF) and amplified by a low noise amplifier (LNA). An image rejection filter (IRF) assures that image signals and noise at the image signal frequencies are removed to avoid interference with the demodulation. Figure 9.22 presents a simplified diagram of the RF receiver of the MS.

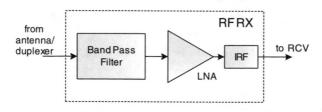

Figure 9.22   MS RF receiver (RF RX).

MS Receiver (RCV)

The next stage on the signal reception is the down-conversion of the signal to the IF frequency by a signal generated by a Local Oscillator (LO). After the down-conversion, the signal passes through another BPF, as illustrated in Figure 9.23.

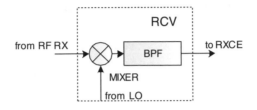

Figure 9.23   MS receiver (RCV).

MS Receive Channel Element

Figure 9.24 shows the MS Receive Channel Element (RXCE). The signal coming from the receiver is extracted by the 'I' (In phase) and 'Q' (Quadrature) signals generated by the IF local oscillator. Each branch is filtered and digitised by an A/D converter. An Automatic Gain Control (AGC) circuit adjusts the gains so that proper levels are handled by the circuitry.

The digitised signal of both the branches is connected to a PN searcher and three traffic correlators. The PN searcher scans and measures all PN sequences to detect the strongest one. The correlators are rake fingers that are used to detect signal multipaths and correlate to the Walsh codes of the original data. The output of these fingers is combined in a Maximum Ratio Combiner (MRC) and modulo 2 added to the long PN sequence. The offset of this sequence is determined based on the IMSI (International Mobile Subscriber Identity) number. The PN sequence is used to extract the information sent to that specific user. The extracted data is de-interleaved and decoded by a Viterbi Maximum Likelihood Decoder (MLD).

Figure 9.25 shows a more detailed diagram of the pilot searcher. The received signal is sent to both I and Q branches were the signal is compared (module 2 summed) with the 'I' and 'Q' PN codes. The pilot signal is composed of all zeroes and is spread by Walsh code 0 (all zeros) so the expected value for all chips is zero (that corresponds to +1). This means that the received signal is a sequence of PN codes. The PN codes are shifted chip by chip and the result of the comparison is analysed. The shift that results in a peak value corresponds to the alignment of the two sequences and establishes the information bit synchronism. An early–late detector performs fine chip synchronisation by continuously comparing the received sequence with the PN code, advanced (early) and delayed (late) by one chip. The output of 'I' and 'Q' channels is summed, the peak is detected and the AFC signal is sent to other PN generators in the channel element.

Figure 9.26 shows a more detailed diagram of the traffic correlator. The set of correlators is similar to the pilot searcher because each of them tunes to a different PN code or multipath. The pilot signal detection happens as in the PN searcher but the channel estimator also calculates the phase and strength related to the traffic signal. After being combined with

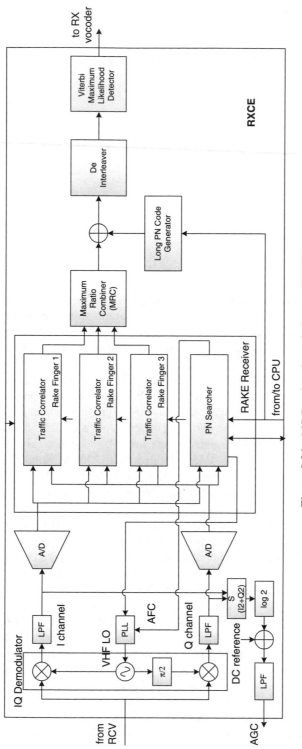

Figure 9.24 MS Receive channel element (RXCE).

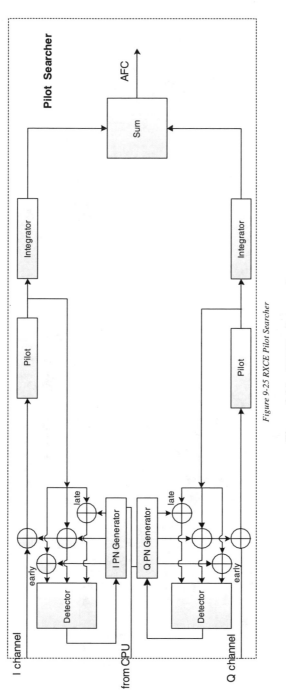

Figure 9-25 RXCE Pilot Searcher

Figure 9.25 RXCE pilot searcher.

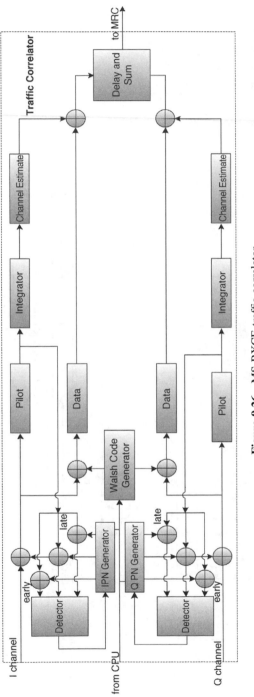

Figure 9.26 MS RXCE traffic correlator.

the PN code, the received signal is combined with the Walsh code assigned to that correlator (finger). The output of this combination is delayed to allow all fingers to be summed in phase.

The information of each finger is sent to the MRC where it is delayed and weight summed. The long PN, with an offset based on the user key (IMSI based), is combined with the MRCs output to remove scrambling. The signal is then de-interleaved. A Viterbi Maximum Likelihood Decoder (MLD) analyses the de-interleaved data, and the information bits are extracted and sent to the MS vocoder.

MS Receive (RX) Vocoder
The last stage in the forward link is the receive vocoder. This element is not aware of the data rate used for coding; therefore, it processes the signal using all four possible data rates and extracts the CRC. The best CRC determines the data rate of the speech. Frames with errors that exceed a preset limit are erased and the FER counter is updated. Figure 9.27 shows a diagram illustrating the MS receive vocoder.

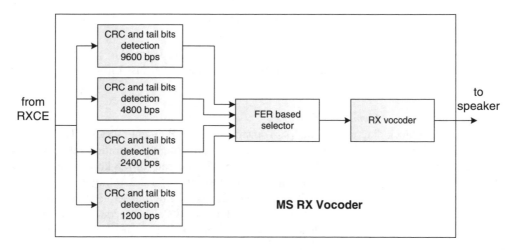

Figure 9.27   MS receive vocoder.

9.2.1.3   Reverse Link

This section describes the main network elements used in the reverse link. These elements are presented in Figurs 9.14 and 9.15 as part of the MSC/BSC/BTS group and of the MS respectively. Figure 9.28 illustrates the reverse link flow from the MS microphone to the DS0 sent to the PSTN. Each of the blocks in this figure is described in detail next.

Reverse Link Transmission from the MS to the BSC

MS Transmit (TX) Vocoder
Figure 9.29 shows the MS transmit vocoder block diagram. This element is composed of the vocoder itself and an additional circuit that adds the 12 Cyclic Redundancy Check (CRC)

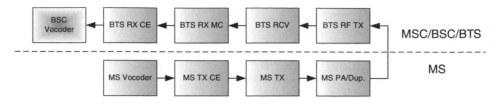

Figure 9.28 Reverse link network elements.

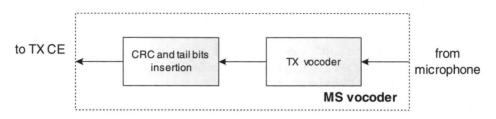

Figure 9.29 MS TX vocoder.

bits and the 8 tail bits to each 20 ms voice frame. The CRC bits are used to calculate the FER that serves as a quality indicator of the transmission. A FER of 10^{-3} is considered acceptable for voice transmission.

MS Transmit Channel Element (TXCE)
Figure 9.30 is a block diagram for the MS Transmit Channel Element (TXCE). The vocoder sends information to the TXCE using one of the two possible configuration rates, RC1 or RC2, depending on the channel. The access channel uses a rate of 4.8 kbps with a 20 ms frame. The traffic channel in the RC1 configuration has rates of 1.2/2.4/4.8/9.6 kbps and, in RC2, rates of 1.8/3.6/7.2/14.4 kbps with a frame of 20 ms. The output sequence is punctured (bits are deleted) and has missing symbols, 8 in 24 bits for RC2. The decoder in the receiver will eliminate these errors by replacing the deleted bits, but its capability of eliminating other errors is compromised.

Each frame is encoded by a convolutional encoder using a rate (r) of 1/3 with constrain length (k) 9 code to form a stream of 28.8 ksps. In RC2 configuration, the bits are punctured to achieve 28.8 kbps. Slower rates, RC1, have their frames repeated to match this output rate.

The frame is block interleaved (32 rows by 18 columns = 576 symbols) as part of the mechanism to distribute burst errors. A Walsh encoder maps every 6 bits of the output sequence into one of 64 orthogonal Walsh codewords. This provides a type of spreading and improves detection in the receiver. A data burst randomiser is used to delete some of the repetitions of the lower rates in a quasi-random manner.

The TXCE also generates a long PN code with a rate of 1.2288 Mcps, offset according to the MS electronic serial number (ESN), which is used as an encryption key and to uniquely identify the user. The signal is then split into two branches, where it is combined with two short (15 stages) PN sequences (I and Q) at a rate of 1.2288 Mcps.

Each branch is then filtered by a digital FIR (Finite Impulse Response) filter, leaving only the base band components. A digital-to-analogue converter (D/A) converts the output of this filter to an analogue signal. The signal gain is adjusted according to network requirements.

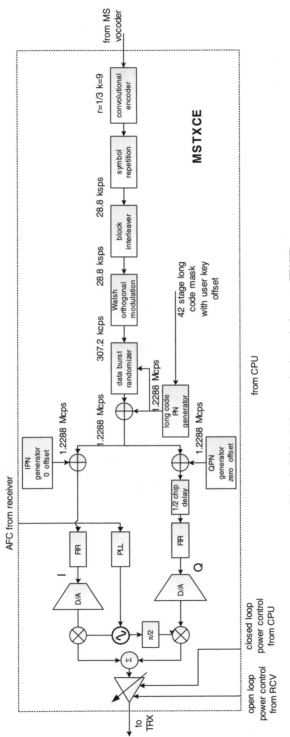

Figure 9.30 MS transmit channel element (TXCE).

Finally, the signal is modulated in quadrature using an intermediate frequency and the output is summed. The Phase Locked Loop (PLL) that generates the intermediate frequency is controlled by and Automatic Frequency Control (AFC) signal that comes from the receiver and synchronises both frequencies.

The output power is controlled by an open-loop power control based on the received signal strength and a closed loop power control based on messages transmitted by the forward link to the CPU.

MS Transmitter

Figure 9.31 shows the MS transmitter element. The transmitter modules the IF signal with the CDMA carrier (800 or 1900 MHz range). The modulated signal is pre-amplified and filtered using a BPF (Band Pass Filter).

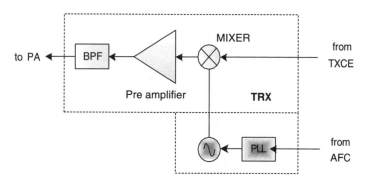

Figure 9.31   MS transmitter (TRX).

MS Power Amplifier and Duplexer

The power amplifier (Figure 9.32) has to be very linear, with a distortion of −60 dB in relation to carrier power with power capacity of approximately 30 dBm. A duplexer is required to isolate transmit and receive signals in the MS antenna.

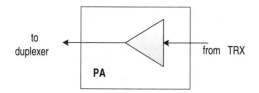

Figure 9.32   MS power amplifier (PA).

Reverse Link Reception at the BTS

BTS RF Receiver (RX)

Figure 9.33 shows the BTS RF receiver. The signal received by the BS is first filtered using a BPF. A low noise amplifier (LNA) then amplifies the filter output. An image rejection filter

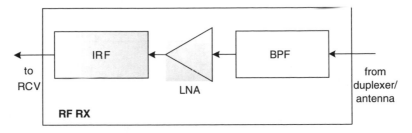

Figure 9.33 BTS RF Receiver (RF RX).

(IRF) assures that image signals and noise at the image signal frequencies are removed avoiding interference in the de-modulation process.

BTS Receiver (RCV)

The BTS receiver downconverts the output of the IRF to the IF frequency. A BPF filters the downconverted signal as illustrated in Figure 9.34 and sends it to the Receive Channel Element (RXCE) through the BTS Multicoupler (MC).

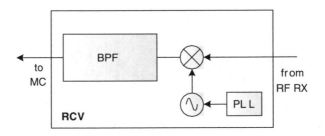

Figure 9.34 BTS receiver.

BTS Multi-Coupler

The multi-coupler (MC) distributes the signal from the receiver among all the receive channel elements as illustrated in Figure 9.35. There are two multi-couplers, one for each receive antenna, so diversity is provided in the receive path minimising multi-path fading.

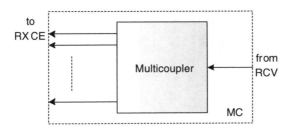

Figure 9.35 BTS RX multicoupler (MC).

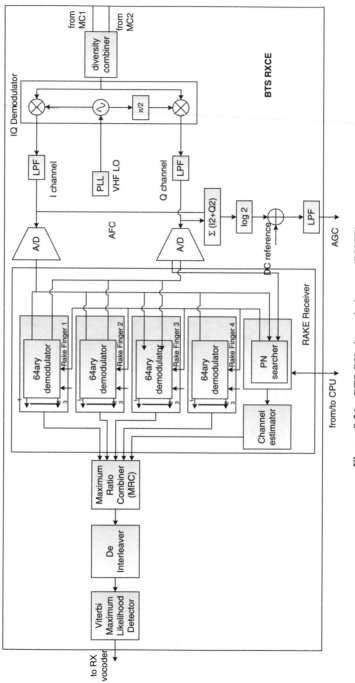

Figure 9.36 BTS RX channel element (RXCE).

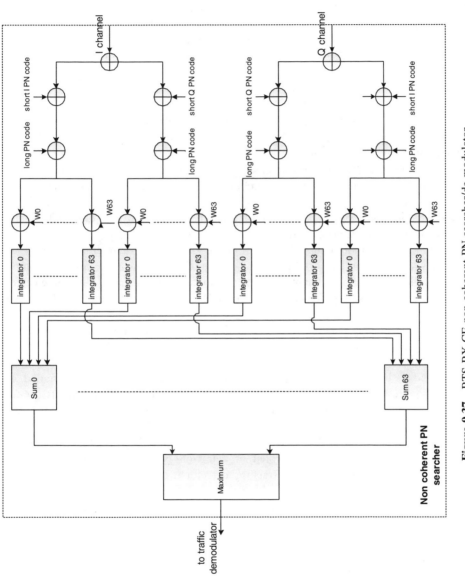

Figure 9.37 BTS RX CE non-coherent PN searcher/de-modulator.

BTS Receiver Channel Element (RXCE)

Figure 9.36 illustrates the BTS Receive Channel Element (RXCE). The signals from the two multicouplers are maximally combined and demodulated using the 'I' (in phase) and 'Q' (quadrature) signals from the local oscillator to extract the base-band signal.

The output of the demodulator is filtered with a Low Pass Filter (LPF) and sent to an A/D to become a digital signal. The digital output of both the converters is then connected to the PN searcher that extracts the phase information and to four 64-ary demodulators.

The CPU programs the searcher and the demodulators so that the desired channel information is looked for. The searcher and the demodulators use the same short PN, long PN with user offset and Walsh code simultaneously to find the signal phase during a preamble in which the receiver knows the Walsh-encoded data *a priori.*

An approximate phase value can be derived from the forward link as signals, sent by the MS, are synchronised to it. The unknown is the round trip delay that may correspond to up to 245 chips (four information bits) for a 30 km link. This information is sent by the CPU to the searcher minimising the search window.

Each demodulator is a rake receiver finger. The outputs of all fingers are combined in a Maximum Ratio Combiner (MRC). The combined data is then de-interleaved and a Viterbi decoder is used to extract the original information.

In this example, the designer elected to use a parallel correlator bank to detect which of the 64 Walsh codes has been sent after synchronisation, preambles, etc.

Figure 9.37 shows a more detailed diagram of the PN searcher/demodulator. The 'I' and 'Q' channels lose some of their orthogonality because of reflections in the transmission path and some of the energy of one channel is converted to the other channel. Combining each 'I' and 'Q' signals with both short PN sequences allows for the recovery of all the cross quadrature energy.

The new signals demodulated with the short PN sequences are then combined with the long PN code offset by the user ID. After this combination, each of the four signals is further combined with each of the 64 Walsh codes available. The result of this combination is integrated and the integrations of the four paths using the same Walsh code are summed. The strongest signal resulting from this sum is considered as representing the originally encoded value of the strongest sector and is used as a reference to synchronise the signals in the traffic demodulator.

Figure 9.38 shows a more detailed diagram of the 64-ary demodulators. Because the reverse link does not have a pilot, coherent detection is not available. The signal is synchronised to the PN searcher and a tracking detector (early–late comparator) is used to adjust for multipaths and pursue fast signal variations to a fraction of a chip. After this initial step, the procedure followed by the 64-ary demodulators is equivalent to what was described for the PN searcher (Figure 9.37).

BSC Receive (RX) Vocoder

The last stage in the reverse link is the BSC vocoder. This element is not aware of the data rate used for coding; therefore, it processes the signal using all four possible data rates and extracts the CRC. The best CRC determines the data rate of the speech. Frames with errors that exceed a preset limit are erased and the FER counter is updated. Figure 9.39 shows a diagram illustrating the BSC receive vocoder.

The output of the BSC RX vocoder is sent to the MSC through the PCM transmit (PCM TX).

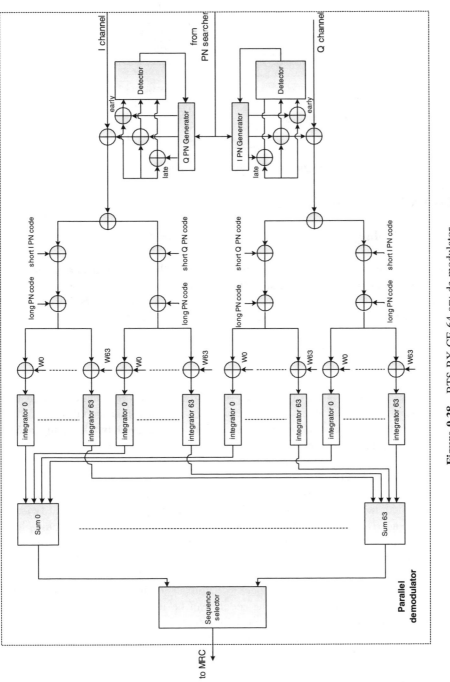

Figure 9.38 BTS RX CE 64-ary de-modulator.

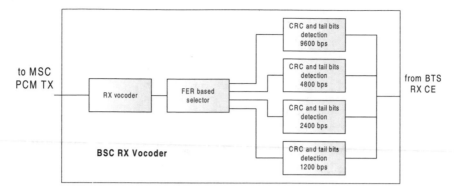

Figure 9.39   BSC RX vocoder.

9.3   TRANSMIT STAGE

9.3.1   Source Coding

In the communication process, the information at the source has to be transferred through a communication channel to the receiving side. In wireless systems, this information is source-coded to minimise redundancies and, consequently, the data rate. The source information is divided in two basic categories with different properties: speech and data.

9.3.1.1   Speech

Traditionally speech (voice) is the main source of information in a communication channel and it was soon determined that it had many redundancies that could be removed consequently diminishing data rate requirements. This is achieved by digitising speech using waveform coders or vocoders.

9.3.1.1.1   Waveform Coder

Voice bandwidth can be limited from 300 to 3400 Hz with a minimum loss of quality. Harry Nyquist established, in 1928, that no information is lost if a signal is sampled at twice its maximum frequency. Therefore, voice can be sampled at 6800 Hz. Practical circuits, however, require a slightly higher sampling frequency (8000 Hz) because of filter characteristics.

Each sample amplitude value is then digitally coded (mapped to digital values). Experimentation determined that, when linear mapping is used, 4096 levels are required to represent these samples, which is equivalent to 12 bits. This fine resolution is required to reproduce small speech signals. Non-linear coders can be used with increasing steps for larger signals requiring only a total of 256 levels (8 bits) to adequately reproduce amplitude variations. Different compression curves are used in Europe (A law) and in the USA (μ law). This process results in a 64 kbps data rate for digitised voice.

9.3.1.1.2   Vocoder

Vocoders (voice coders) use speech properties to further reduce rates required for data transmission. These are based on speech physiology and try to replicate the speaker's voice

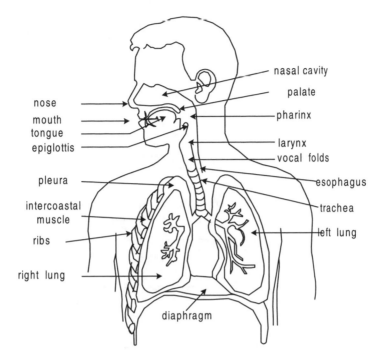

Figure 9.40 Lungs and vocal tract.

by modelling the speech mechanism formed by the lungs and vocal tract. Figure 9.40 shows the lungs and the vocal tract elements involved in speech. Figure 9.41 shows a schematic diagram of the speech apparatus.

Speech is generated by the passage of air from the lungs through the vocal tract. The diaphragm pressures (energy) the lungs and the air under pressure is sent through the trachea and larynx, where the vocal folds, or vocal cords, are located. The air energy is transformed into sound at this point. The vocal folds play an important role in the resulting sound as they generate the frequency components that will be modulated by rest of the vocal tract. Relaxed vocal folds produce a noise-like, un-voiced sound, which has many frequency components with similar amplitude. If the folds are stiff, that is, tensed by the muscles of the epiglottis, they vibrate and the sound is voiced with predominant frequencies that determine the pitch. The volume of air flowing through the pipes determines the volume of speech.

Three cavities play a key role in transforming this sound in speech, by changing their size and openings: the nasal cavity, the mouth and the pharynx cavity. The nasal cavity has approximately a constant resonance, whereas the mouth can change its resonance significantly using tongue and lips movements. These movements create formants that provide the speech characteristic to the sound coming from the larynx.

Human speech is divided in voiced segments (vowels) and un-voiced segments (consonants). Voiced segments are produced by the vibration of the vocal cords and shaped by the vocal tract (throat, mouth and nose) with a fine frequency modulated by the formant structure (cavities of the vocal tract). Un-voiced segments are noise-like and shaped by the air passage through the lips and teeth. Consonants such as 'm', 'n' and 'l' use the vocal tract to shape the sound, whereas others, such as the sibilants 's', 'sh', the fricatives 'f', 'ph' and

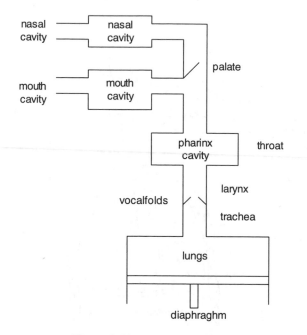

Figure 9.41   Speech apparatus.

the explosives 'b', 'p', 'k', 'd' use only flow control. Shaping the vocal formants produces vowels. The vowel 'e' (as in eel) has formants at 300 and 23000 Hz. The 'a' (as in car) has formants at 500 and 1900 Hz. The 'o' has formants at 300 and 700 Hz.

Vocoders reproduce the vocal tract, as in Figure 9.42, by replacing the lungs by an energy generator. Voiced sounds are created by a tone generator and unvoiced sounds by a noise

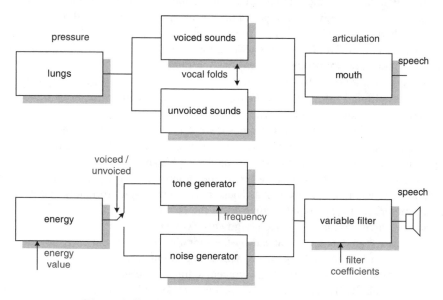

Figure 9.42   Speech apparatus modelled by a vocoder.

generator. A variable filter formats the energy at the output reproducing the speech. Vocoders do not transmit a single voice sample; instead they send the parameters required to dynamically model the user vocal tract. They represent voice segments by analysing speech characteristics and presenting the findings in digital form.

A typical voice signal segment has duration between 2.5 and 20 ms and is defined by the excitation source characteristics (binary voice decision, fundamental pitch frequency and gain) and by the formant filter characteristics (detected by several band filters within the voice band).

Many different models were created to implement vocoders. Figure 9.43 shows a diagram representing a typical vocoder implementation.

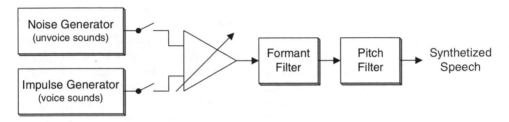

Figure 9.43   Typical vocoder diagram.

Several vocoders were proposed and adopted by the different technologies. Table 9.2 presents a summary of the most popular vocoders.

The most common vocoders employ Linear Predictive Coding (LPC) as a basis. This algorithm considers that speech is produced by energy sent through a variable diameter tube. The air that comes through the opening of the glottis vocal folds produces the signal characterised by its intensity (loudness) and frequency (pitch). The tube that represents the vocal tract (throat and mouth) is characterised by its resonance, which is referred to as

Table 9.2   Most common vocoders[a]

Development Year	Technology	Vocoder	Algorithm	Data rate (kbps)
1972	PCM	PCM	Pulse code modulation	64.0
1984	PHS	ADPCM (G.721)	Adaptive PCM	32.0
1991	CDMA	QCELP-13 (IS-733)	Qualcomm code excited linear prediction	1.1–13.3
1991	GSM	RPE-LTP	Regular pulse excitation–long term prediction	13.0
1992	TDMA/NADC	VSELP	Vector summed excitation linear prediction	8.0
1993	CDMA	QCELP-8 (IS-96A)	Qualcomm code excited linear prediction	0.8–8.6
1999	CDMA	EVRC	Enhanced variable rate coder (RCELP–relaxation CELP)	0.8–8.6

[a]Refer to the glossary for a list of the acronyms used in this table.

formant. This process is performed on short speech intervals of 20–30 ms. CDMA systems use frames of 20 ms.

In the analysis process, the LPC estimates the formants (through inverse filtering) and removes its effect from the speech signal, generating a signal known as residue. The parameters for the formants and residue are recorded and transmitted. In the speech synthesis, the residue is generated and filtered by the formant parameters, resulting in a speech signal.

A set of differential equations (linear predictors) that expresses each speech sample as a linear combination of previous samples reproduces the formants. The coefficients of the differential equations are the formant parameters and are obtained by minimising the mean square error between the predicted and the actual signal.

The characterisation of the residue is important because it defines how the speech sounds. This may require as much information as a waveform coder does because of how complex the residue can be for sounds that are a mix of frequencies and noise. The residues can be classified in relatively few categories and the idea of using a codebook was suggested to excite the formant filter. This scheme is called Code Excited Linear Prediction (CELP).

The characterisation of speech pauses and end of phonemes does not require as many information as for characterising full speech. Variable Rate Coders (VRC) explore this fact by varying the amount of information produced during different speech phases.

The vocoders' relative performance is subjective because these are developed based on perceptual criteria. The most common subjective method used is the Mean Opinion Score (MOS) whose criteria are presented in Table 9.3. This score is based scores assigned to the recovered signal based on the user's listening effort.

Table 9.4 shows typical scores obtained using the MOS method when analysing the typical vocoders. These scores were obtained for an FER of less than 1% from a compilation of different sources.

Table 9.3   Mean Opinion Score (MOS) criteria

Scale	Quality	Listening effort
5	Excellent	No effort
4	Good	No appreciable effort
3	Fair	Moderate effort
2	Poor	Considerable effort
1	Bad	Non-understood

Table 9.4   Vocoder data rate[a]

Technology	Output rate (kbps)	Vocoder	MOS
PCM	64	PCM	4.3
PHS	32	ADPCM	4.1
NADC	8	VSELP	3.7
GSM	13	RPE-LTP	3.5
CDMA	0.8–13	QCELP	3.9
CDMA	0.8–8.6	EVRC	3.6

[a]Refer to the glossary for a list of the acronyms used in this table.

Table 9.5 Vocoders data rate

CDMA vocoder	Rate	RS1 – EVRC bits/frame	bps	RS2 – QCELP bits/frame	bps
Voiced segments	FR (full rate)	171	8600	267	13.350
Voiced segments	HR (half rate)	80	4000	125	6250
Voiced/un-voiced segments	QR (quarter rate)	0	0	55	2750
Un-voiced/pause segments	ER (eighth rate)	10	800	21	1050
Un-voiced/pause segments	Blank	10	500	0	0

In CDMA, the vocoder data rate varies with the content of each voice segment and the rate set (RS1 or RS2), as indicated in Table 9.5.

FR (full rate) and ER (one eighth rate) occur more frequently than HR (half rate) and QR (quarter rate), which occur mainly at speech transitions (between speech and pause).

To complement the vocoder information, an error detection code is added so an FER also called frame erasure rate can be calculated at the receiver. This allows the muting of voice and even call termination when errors get large. The error detection code is a Cyclic Redundancy Code (CRC) and is added to the FR (12 bits) and the HR (8 bits). Additionally eight tail zeros are added at the end of each frame to indicate its end. Table 9.6 shows the final rates for encoded voice in CDMA.

Table 9.6 Data rate for encoded voice in CDMA

Vocoder rate (Rvoc)	RS 1 – EVRC bits/frame	CRC	Tail bits	bps	RS 2 – QCELP bits/frame	Reserved	CRC	Tail bits	bps
FR (1)	172	12	8	9600	267	1	12	8	14400
HR (1/2)	80	8	8	4800	125	1	10	8	7200
QR (1/4)	40	0	8	2400	55	1	8	8	3600
ER (1/8)	16	0	8	1200	21	1	6	8	1800
Blank	10	0	8	0	0	0	0	0	0

Another factor that characterises a vocoder is the Voice Activity Factor (VAF), shown in Table 9.7. This factor is calculated by defining the vocoder's average data rate and determining which fraction of the maximum rate it represents. In a normal conversation, the speaker talks about 3/8 (0.375) of the time, whereas pauses represent 5/8 (0.625) of the time. Table 9.7 gives the average rate and the VAF for each vocoder. A conservative factor for VAF is 0.5.

Table 9.7 Data rate for encoded voice in CDMA

	RS 1 – EVRC	RS 2 – QCELP
Average data rate (bps)	4725	6525
Voice activity factor (VAF)	0.49	0.45

Table 9.8 Data rates in 3G (cdma2000)

| | Forward link | | Reverse link | |
	Voice (bps)	data (bps)	Voice (bps)	data (bps)
RC1	9600		RC1 9600	
RC2	14 400		RC2 14 400	
RC3	9600	153 600	RC3 9600	153 600
RC4	9600	307 200		
RC5	14 400	230 400	RC4 14 400	230 400
RC6	9600	307 200	RC5 9600	614 400
RC7	9600	614 400		
RC8	14 400	460 800	RC6 14 400	1 036 600
RC9	14 400	1 038 600		

9.3.1.2   Data

Data can be classified into two types: system data and user data.

System data is vital for the control of the network and call processing. Therefore, it is important to assure that all messages are correctly received. Error detection codes are added to assure integrity of messages and compelled protocols are used to interchange these messages. Such protocols expect an acknowledgement for every message sent, if this confirmation is not received within a certain time, the message is repeated. The system data is usually treated as an overhead and considered as a percentage of the traffic messages.

User data was very limited in second-generation systems but plays a major role in the third generation. Table 9.8 shows data rates offered for each type of radio configuration.

When transmitting data, the network employs error detection codes and monitors the bit error rate (BER) to manage the quality of the information transmitted. If this quality is not within pre-established thresholds the connection is dropped.

Frames received with errors are erased and not reproduced by the vocoder. A counter of erased frames is kept and represents the Frame Erasure Rate (FER). Acceptable FER values for speech range from 1% to 3% because MOS deteriorates fast for higher FER. Data transfer has to be mainly error free and the user application is expected to take care of the data integrity, through error correction, compelled protocols, or Automatic Repeat Request (ARQ) methods.

As in system data transmission, compelled protocols control applications that require an acknowledgement for each message. Applications that only request the repetition when a block of data was received with error usually use ARQ. Due to the application level protocols data applications can work with higher FER. The ideal FER value for the data application has to be determined comparing the overhead required to lower the FER with the overhead of re-transmitting the messages in error.

All these methods used to preserve the integrity of the data add an overhead to the transmitted data and should be considered in the design. The overhead depends on the FER and should be dimensioned for the FER limits. Each application may have a different overhead. Table 9.9 shows typical values that may be considered in the design.

Table 9.9 Protocol overhead

Error correction method	Typical overhead (%)
ARQ	10
Compelled	20
Error correction codes	30

9.3.2 Multiplexing and Multiple Access in DS-CDMA

In a communication network, several information sources are transmitting, at the same time, to different destinations. This requires mixing and separating all these signals in a process called multiplexing. Some of the issues in multiplexing are that information cannot be lost during the process and there has to be privacy among users, that is, unintended users should not understand the information. The information to be multiplexed comes from some of multiple possible sources and this multiple access is performed in DS-CDMA by assigning different codes to each active source.

In a point to multi-point network, this process is achieved by multiplexing outgoing signals at the origin and broadcasting them to all receiving points. In the other direction, signals coming from different origins must be identified and extracted.

In a multiple access network, the receiver must be able to determine the transmission origin to properly extract the information. On a typical network, BTS and channels have to be identified by the MS on the forward link, whereas in the reverse link, the MS has to be uniquely identified by the BTS.

An addressing scheme similar to the IP (Internet Protocol) could be used, but this would be highly inefficient for large data transfers. Instead, a resource assignment scheme is used in the existing technologies. Multiple access schemes must address features such as authentication of MS to avoid fraud and information scrambling to avoid eavesdropping. Each technology addresses multiple access in a different way.

- Frequency Division Duplex/Frequency Division Multiple Access (FDD/FDMA) systems dedicate a pair of frequencies (carrier and channel) for each origin–destination pair. This is easily implemented but ties resources (bandwidth) permanently during a connection, even if there is no information being interchanged. Part of the bandwidth is wasted as guard bands, which are required to allow filters to properly separate the channels. This, by itself, wastes approximately 75% of the system's capacity. No authentication or scrambling is inherent in this access methodology. This functionality has to be provided by the analogue or digital channel being transmitted.

- Frequency Division Duplex/Time Division Multiple Access (FDD/TDMA) systems assign a time-slot (channel) to each origin–destination pair. The FDD pair of frequencies (carrier) can handle (multiplex) several of these time-slots (channels). Because less filtering guard bands are required per channel, this technique represents an improvement over FDMA, but the resources (bandwidth) are still permanently tied during a connection. Approximately 60% of the capacity is lost in this process. The use of digital technology in the TDMA part facilitates authentication and scrambling techniques but these are still not embedded in this access methodology.

Table 9.10   Applying codes to two channels

	Bits	Decimal code
Channel 1	0	1
	1	2
Channel 2	0	3
	1	6

- Frequency Division Duplex/Code Division Multiple Access (FDD/CDMA) systems assign a code (channel) to each origin–destination pair. The FDD pair of frequencies (carrier) can handle (multiplex) several codes (channels). This process also offers an improvement over FDMA because less filtering guard bands are required per channel and it also improves over TDMA because there is not a fixed assignment of bandwidth for a channel. In CDMA, an increase in power (code usage) characterises the usage of each channel. In this technology, transmit and receive power limits become capacity bottlenecks.

The CDMA multiplexing method is complex, and understanding of its basic concepts is important because the technology is not completely intuitive. The following example illustrates this by showing a very basic network using decimal codes.

Suppose that the network is transmitting the information (0s and 1s) of two channels simultaneously. First, each channel is associated with a code: channel 1 with decimal codes 1 and 2, and channel 2 with decimal codes 3 and 6. To send the information, the codes are added and this sum is sent to the receiver. The receiver must be able to retrieve the transmitted information regardless of which channels were transmitted, channel 1, 2 or both. The use of codes allows determining what combination has been sent by inspecting the received sum value, because each combination results in a unique value. Tables 9.10 and 9.11 show the possible transmission combinations.

The process shown in this example is called Direct Sequence Code Division Multiple Access (DS-CDMA). In this process, the multiplexing or spreading code spreads one or more information bits. CDMA systems use binary codes to spread the information.

Table 9.11   Multiplexing two channels

Channel 1	Channel 2	Decimal code
0	x	1
1	x	2
x	0	3
x	1	6
0	0	4
0	1	7
1	0	5
1	1	8

The bits of the spreading code are called *chips* to differentiate them from the information bits and information symbols. Several chips represent one information bit (depending on the spreading rate).

The use of codes to multiplex channels is based on a very simple but powerful principle that relies on the existence of orthogonal codes and/or quasi-orthogonal functions. The mathematics behind these codes is quite complex, but can be summarised as a simple orthogonality principle.

- The multiplexing of n channels requires a set of n codes to form a set of orthogonal vectors in n-space. Those vectors are orthogonal between themselves if any point in the n-space may be expressed as only one linear combination of these vectors.

This vector space (v) is a set of elements (binary codes) that satisfy the following operations.

- The commutative rule: $v + u = u + v$.

- The distributive rule: $v(u + w) = vu + vw$.

- The associative rule: $(vu)w = v(uw)$.

Code orthogonality can be expressed by the correlation between codes. Correlation is the measure of how similar the codes are and the higher the similarity, the higher the correlation. To check the correlation, the code sequences are compared bit by bit. A '+1' is assigned to every matching bit (agreement), otherwise a '−1' is assigned (disagreement). The comparison results are summed and normalised (divided) by the number of bits in the code

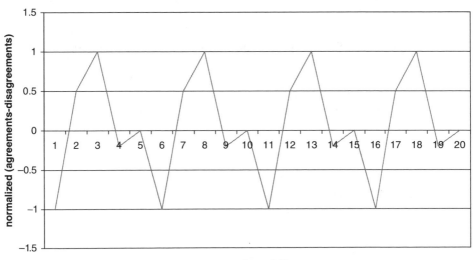

Figure 9.44 Normalised correlation factor.

sequence. Codes can be correlated to shifted versions of themselves (auto-correlation) or to other codes and their shifted versions (cross-correlation). Therefore, codes orthogonal to shifted versions of themselves are called auto-orthogonal codes, whereas a set of codes in which codes are orthogonal to each other is classified as a set of cross-orthogonal codes.

Figure 9.44 shows a normalised correlation plot graphically displaying how sequences or codes correlate to each other. This graph shows only the correlation of one sequence against others, therefore, it is not enough to give a complete picture for all the possible code combinations. Fortunately, for many codes, the correlation is similar between all pairs and it suffices to represent only one relation. A correlation of '0' indicates that the codes have a high correlation factor; whereas '+1' or '−1' show that they are not correlated.

The auto-correlation compares how different shifts of the same code correlate between them. This property can be used to synchronise two sequences by comparing an incoming sequence to different shifts of a sequence locally generated.

The following is a summary of the correlation properties of orthogonal codes. Section 3.2.2 provides a more detailed explanation of the correlation characteristics of spread spectrum systems such as CDMA.

- Two identical codes only have agreements, that is, a total match (zero shift auto-correlation = 1): $v\,v = 1$.

- Two orthogonal codes have the same number of agreements and disagreements: $v\,u = 0$.

- Two quasi-orthogonal codes have a similar number of agreements and disagreements.

- Two non-orthogonal codes have a different number of agreements and disagreements.

These simple properties allow the detection of a specific code sequence out of a sum of several sequences as indicated in Table 9.12. The sum of vectors is done assigning a value +1 to binary 0 and −1 to binary 1. The multiplication of vectors is done by multiplying by +1 the sum value for a binary 0 in the code chip and multiplying it by −1 for binary 1 in the code chip.

In the example (Table 9.12), orthogonal vectors v, u and w are summed. The task in hand is to extract from this sum the vector v and the vector z (not present). Each vector represents a specific code.

The sum of vectors is multiplied by vector v in the first column and by vector z in the second column. The distributive property is applied followed by the orthogonality property. For the column where vector v was used as a multiplier, a correlation is found and the sum is equal to 1, while for the column that was multiplied by vector z, no correlation is found and the sum is 0, meaning that the code is absent.

Table 9.12   Detecting a code vector in a sum of vectors

	Extracting vector v	Extracting vector z
Sum of vectors	$v + u + w$	$v + u + w$
Multiplying by a vector	$v(v + u + w)$	$z(v + u + w)$
Using the distributive rule	$vv + vu + vw$	$zv + zu + zw$
Sum	$1 + 0 + 1 = 1$	$0 + 0 + 0 = 0$

A direct consequence of using codes for multiplexing and multiple access is that the data rate is increased by the application of the code, resulting in a frequency spreading of the signal, increasing its bandwidth. The detection of the information in the CDMA received signal involves a substantial difference to the traditional way of detecting signals due to this spreading. We know that a certain code was transmitted and we look for its pattern in the received signal. The received chips are combined with the known pattern and the results are summed for the entire bit of information.

The traditional way of analysing $S/(N+I)$ (power ratio) does not apply here, as we will be comparing the signal information over the detection period against the noise of each chip period. We will use energy then instead of power. The total signal energy over the period is expressed as E_b (bit energy). This bit energy represents the sum of all the desired code chips energy. The total noise and interference is considered over the period of a chip and is expressed as I_0.

This method requires the transmission of higher rates and, consequently, it is bandwidth inefficient. This loss, however, is dramatically recovered with the spreading property. The original information has its bandwidth spread but the amount of the information is still the same. As the information is kept constant and the bandwidth is increased, Shannon's formula indicates that the SNIR requirement decreases as shown in Figure 9.45. This behaviour allows a closer reuse of frequencies, and in CDMA the same frequencies are reused at every cell. The chart shows that the spreading gain (relative increase in SNR) drops for larger spreadings. A good compromise in the relation spreading factor x gain is the use of a 64 spreading factor.

Two families of codes are the cornerstone of existing DS-CDMA systems, the Walsh codes and the PN codes, both are described next.

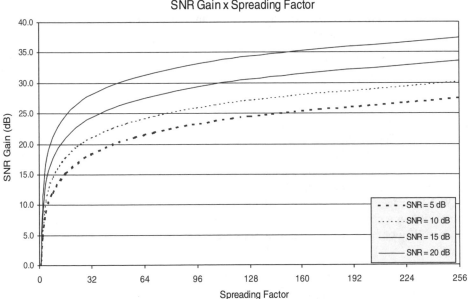

Figure 9.45 Spreading factor SNIR gain.

9.3.2.1 Walsh Codes

Walsh codes are a set of orthogonal codes used for multiple access in CDMA systems. An n-bit word has only n combinations that are orthogonal between them and they can be generated using Walsh functions. Words are generated by repeating the matrix in Figure 9.46, which implies words are always 2^n bits long. Section 3.3 (Chapter 3) describes the generation of Walsh codes in detail.

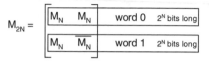

Figure 9.46 Walsh matrix formation.

The complement of each of the Walsh codes is also orthogonal to the set and this allows us to use one code to code bit 0 and its complement to code bit 1.

Each bit of the information data stream for a specific user is represented by one (Table 9.13) orthogonal code of the set. Zeroes are represented by the code itself (multiply by $+1$), while ones are represented by the inverted code (multiply by -1). Further multiplications are done to adjust to the power requirements.

Users may require one or more sessions and each session will use one Walsh code. The codes resulting from different sessions are just added up. In the receive side, the composite signal is combined with the desired orthogonal code and the information is retrieved.

The use of Walsh codes presents two limitations.

1. They are very limited in number and the spreading becomes very large with the increase of multiple access required.

2. They are not orthogonal to shifted versions of themselves and lose their properties in the presence of multi-path signals.

IS-95 uses 64 bit long Walsh codes to provide multiple access to different channels in a sector. 1XRTT uses 128 bit long Walsh codes to provide multiple access to different user sessions and different channels in a sector. 3XRTT uses Walsh codes with lengths of up to 256 bits. An entire Walsh code is used to spread each bit of information. Thus the length of the Walsh code used depends on the data rate. In a general way, this length is equal to the ratio given by chip rate/data rate.

A Walsh code can be considered as a family in which a root code produces several children. When a code is used its children or immediate parents cannot be used. This limits the availability of codes when high data rates are used. In 1XRTT, this problem was solved

Table 9.13 Coding information bits

Bit 0	0000000000000000
Bit 1	1111111111111111

by adding a set of auxiliary codes, known as Quasi Orthogonal Functions (QOFs) to complement the set of Walsh codes. These codes are orthogonal between themselves but only quasi orthogonal to Walsh codes. QOFs are described in detail in Section 3.4 (Chapter 3). The use of quasi-orthogonal codes diminishes the effectiveness of spreading because the peak energy created by the code matching is partially neutralised by the energy of partial correlations.

Orthogonal codes can be classified as follows.

- *Cross-orthogonal codes*: sets of different codes that are orthogonal between themselves (zero shift cross correlation $= 0$).

- *Auto-orthogonal codes*: orthogonal to shifted version of themselves (auto-correlation $= 0$).

Walsh codes are cross-orthogonal but not auto-orthogonal, that is, they are orthogonal to each other but not between shifted versions of themselves, therefore, they lose their properties in the presence of multipath signals. Table 9.14 shows auto- and cross-correlation of 32-bit long Walsh codes. Only the first 8 codes of the 32 are shown in the table.

Walsh codes are used to identify channels in the forward direction but they cannot be applied directly because of their sensitivity to time-shifts. Auto-orthogonality is required to extract the codes from delayed versions of themselves, which is common in multipath propagation environments. CDMA systems provide auto-orthogonality by applying another level of coding on top of the Walsh codes, the PN (pseudo-noise) codes. These additional codes also provide additional multiaccess recognition because Walsh codes can identify channels but not BTSs.

9.3.2.2 PN Codes

PN (pseudo-random or pseudo-noise) codes satisfy the auto-correlation property, which makes them good candidates for MS identification and capable of solving scrambling and alignment issues. PN codes are based on PN sequences and have the following properties for a sequence of length $P = 2^n$.

- Balance property: the number (n) of 1s and 0s in the sequence is the same.

- Run property: there are $2^n/2$ runs of consecutive 1s or 0s of length 1, $2^n/2^2$ runs of length 2, $2^n/2^3$ runs of length 3 and so on.

- Correlation property: if a sequence is compared, chip by chip, to a shifted version of itself, the sum of the number of agreements and disagreements is close to 0.

PN codes are generated from linear feedback shift register (LFSR) sequences. These sequences are generated by polynomials of order n. When using irreducible polynomial, the sequences will present good auto-orthogonality, and the number of agreements between a sequence and its shifted versions will always be equal to '-1'. These LFSR sequences are called PN sequences and their length is given by $(2^n - 1)$. Sections 3.2.3 and 3.2.4 (Chapter 3) describe the generation of PN sequences in detail. This chapter summarises the

Table 9.14 Orthogonality of Walsh codes

index integer	index sequence		1	2	3	4	5	6	7	8	9	10	11	12	13	14	15	16	17	18	19	20	21	22	23	24	25	26	27	28	29	30	31	32	Normal. sum
																Walsh sequence																		
0	000	W0	0	0	0	0	0	0	0	0	0	0	0	0	0	0	0	0	0	0	0	0	0	0	0	0	0	0	0	0	0	0	0	0	
1	001	W1	0	1	0	1	0	1	0	1	0	1	0	1	0	1	0	1	0	1	0	1	0	1	0	1	0	1	0	1	0	1	0	1	
2	010	W2	0	0	1	1	0	0	1	1	0	0	1	1	0	0	1	1	0	0	1	1	0	0	1	1	0	0	1	1	0	0	1	1	
3	011	W3	0	1	1	0	0	1	1	0	0	1	1	0	0	1	1	0	0	1	1	0	0	1	1	0	0	1	1	0	0	1	1	0	
4	100	W4	0	0	0	0	1	1	1	1	0	0	0	0	1	1	1	1	0	0	0	0	1	1	1	1	0	0	0	0	1	1	1	1	
5	101	W5	0	1	0	1	1	0	1	0	0	1	0	1	1	0	1	0	0	1	0	1	1	0	1	0	0	1	0	1	1	0	1	0	
6	110	W6	0	0	1	1	1	1	0	0	0	0	1	1	1	1	0	0	0	0	1	1	1	1	0	0	0	0	1	1	1	1	0	0	
7	111	W7	0	1	1	0	1	0	0	1	0	1	1	0	1	0	0	1	0	1	1	0	1	0	0	1	0	1	1	0	1	0	0	1	
auto correlation	W0 - W0		1	1	1	1	1	1	1	1	1	1	1	1	1	1	1	1	1	1	1	1	1	1	1	1	1	1	1	1	1	1	1	1	1
cross correlation	W0 - W1		1	-1	1	-1	1	-1	1	-1	1	-1	1	-1	1	-1	1	-1	1	-1	1	-1	1	-1	1	-1	1	-1	1	-1	1	-1	1	-1	0
cross correlation	W0 - W2		1	1	-1	-1	1	1	-1	-1	1	1	-1	-1	1	1	-1	-1	1	1	-1	-1	1	1	-1	-1	1	1	-1	-1	1	1	-1	-1	0
cross correlation	W0 - W3		1	-1	-1	1	1	-1	-1	1	1	-1	-1	1	1	-1	-1	1	1	-1	-1	1	1	-1	-1	1	1	-1	-1	1	1	-1	-1	1	0
cross correlation	W0 - W4		1	1	1	1	-1	-1	-1	-1	1	1	1	1	-1	-1	-1	-1	1	1	1	1	-1	-1	-1	-1	1	1	1	1	-1	-1	-1	-1	0
cross correlation	W0 - W5		1	-1	1	-1	-1	1	-1	1	1	-1	1	-1	-1	1	-1	1	1	-1	1	-1	-1	1	-1	1	1	-1	1	-1	-1	1	-1	1	0
cross correlation	W0 - W6		1	1	-1	-1	-1	-1	1	1	1	1	-1	-1	-1	-1	1	1	1	1	-1	-1	-1	-1	1	1	1	1	-1	-1	-1	-1	1	1	0
cross correlation	W0 - W7		1	-1	-1	1	-1	1	1	-1	1	-1	-1	1	-1	1	1	-1	1	-1	-1	1	-1	1	1	-1	1	-1	-1	1	-1	1	1	-1	0
auto correlation	W1- W1(1)		-1	-1	-1	-1	-1	-1	-1	-1	-1	-1	-1	-1	-1	-1	-1	-1	-1	-1	-1	-1	-1	-1	-1	-1	-1	-1	-1	-1	-1	-1	-1	-1	-1
auto correlation	W1- W1(2)		1	1	1	1	1	1	1	1	1	1	1	1	1	1	1	1	1	1	1	1	1	1	1	1	1	1	1	1	1	1	1	1	1
auto correlation	W1- W1(3)		-1	-1	-1	-1	-1	-1	-1	-1	-1	-1	-1	-1	-1	-1	-1	-1	-1	-1	-1	-1	-1	-1	-1	-1	-1	-1	-1	-1	-1	-1	-1	-1	-1
auto correlation	W1- W1(4)		1	1	1	1	1	1	1	1	1	1	1	1	1	1	1	1	1	1	1	1	1	1	1	1	1	1	1	1	1	1	1	1	1
auto correlation	W1- W1(5)		-1	-1	-1	-1	-1	-1	-1	-1	-1	-1	-1	-1	-1	-1	-1	-1	-1	-1	-1	-1	-1	-1	-1	-1	-1	-1	-1	-1	-1	-1	-1	-1	-1
auto correlation	W1- W1(6)		1	1	1	1	1	1	1	1	1	1	1	1	1	1	1	1	1	1	1	1	1	1	1	1	1	1	1	1	1	1	1	1	1
auto correlation	W1- W1(7)		-1	-1	-1	-1	-1	-1	-1	-1	-1	-1	-1	-1	-1	-1	-1	-1	-1	-1	-1	-1	-1	-1	-1	-1	-1	-1	-1	-1	-1	-1	-1	-1	-1
auto correlation	W1- W1(8)		1	1	1	1	1	1	1	1	1	1	1	1	1	1	1	1	1	1	1	1	1	1	1	1	1	1	1	1	1	1	1	1	1

generation process but focuses mainly on their properties and impact on the network. For this type of analysis, shorter sequences are sufficient; therefore, the examples in this chapter use 32-chip long PN sequences.

Two types of circuit configurations can be used to generate PN sequences: a Simple Shift Register Generator (SSRG) and a Modular Shift Register Generator (MSRG), illustrated in Figures 9.47 and 9.48. Even though both circuits generate PN sequences, the sequences generated by each of them are not the same, even when the same generator polynomial is used. The two sequences have the same sequence period but different offsets.

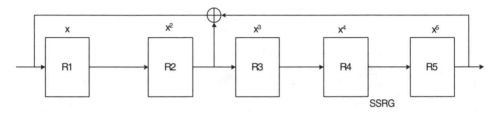

Figure 9.47 PN code generation circuit $(1 + x^2 + x^5)$.

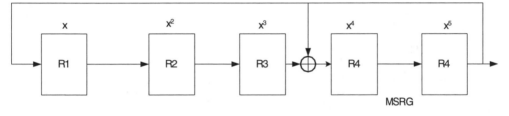

Figure 9.48 PN code generation circuit $(1 + x^3 + x^5)$.

Figure 9.47 shows an SSRG circuit implementing the irreducible polynomial $f(x) = 1 + x^2 + x^5$. The sequence obtained with this configuration has 31 bits $(2^5 - 1)$: 0000 1010 1110 1100 0111 1100 1101 001.

Figure 9.48 shows an MSRG circuit implementing the irreducible polynomial $f(x) = 1 + x^3 + x$. The sequence obtained with this configuration also has 31 bits: 1111 0011 0100 1000 0101 0111 0110 001.

Table 9.15 shows a 31 bit long PN sequence generated by the SSRG and all its possible shifts. Table 9.16 shows the correlation between sequence PN1 and its shifts. The last column of the table shows the normalised sum of all agreements and disagreements, that is, the sum of '+1s' and '−1s' divided by the number of chips (31). Sequence PN1 has a sum of 31 (all agreements), which corresponds to a normalised sum of 1, whereas all other correlations have a sum of −1, that is, one disagreement more than the total number of agreements, corresponding to a normalised sum of −0.032 (−1/31). This characterises this sequence as being quasi-orthogonal to all shifts of itself. Figure 9.49 shows the correlation graphically, depicting the energy level obtained at different relative shifts.

Table 9.15   31 bit long PN sequences

PN1	0	0	0	0	1	0	1	0	1	1	1	0	1	1	0	0	0	1	1	1	1	1	0	0	1	1	0	1	0	0	1
PN2	1	0	0	0	0	1	0	1	0	1	1	1	0	1	1	0	0	0	1	1	1	1	1	0	0	1	1	0	1	0	0
PN3	0	1	0	0	0	0	1	0	1	0	1	1	1	0	1	1	0	0	0	1	1	1	1	1	0	0	1	1	0	1	0
PN4	0	0	1	0	0	0	0	1	0	1	0	1	1	1	0	1	1	0	0	0	1	1	1	1	1	0	0	1	1	0	1
PN5	1	0	0	1	0	0	0	0	1	0	1	0	1	1	1	0	1	1	0	0	0	1	1	1	1	1	0	0	1	1	0
PN6	0	1	0	0	1	0	0	0	0	1	0	1	0	1	1	1	0	1	1	0	0	0	1	1	1	1	1	0	0	1	1
PN7	1	0	1	0	0	1	0	0	0	0	1	0	1	0	1	1	1	0	1	1	0	0	0	1	1	1	1	1	0	0	1
PN8	1	1	0	1	0	0	1	0	0	0	0	1	0	1	0	1	1	1	0	1	1	0	0	0	1	1	1	1	1	0	0
PN9	0	1	1	0	1	0	0	1	0	0	0	0	1	0	1	0	1	1	1	0	1	1	0	0	0	1	1	1	1	1	0
PN10	0	0	1	1	0	1	0	0	1	0	0	0	0	1	0	1	0	1	1	1	0	1	1	0	0	0	1	1	1	1	1
PN11	1	0	0	1	1	0	1	0	0	1	0	0	0	0	1	0	1	0	1	1	1	0	1	1	0	0	0	1	1	1	1
PN12	1	1	0	0	1	1	0	1	0	0	1	0	0	0	0	1	0	1	0	1	1	1	0	1	1	0	0	0	1	1	1
PN13	1	1	1	0	0	1	1	0	1	0	0	1	0	0	0	0	1	0	1	0	1	1	1	0	1	1	0	0	0	1	1
PN14	1	1	1	1	0	0	1	1	0	1	0	0	1	0	0	0	0	1	0	1	0	1	1	1	0	1	1	0	0	0	1
PN15	1	1	1	1	1	0	0	1	1	0	1	0	0	1	0	0	0	0	1	0	1	0	1	1	1	0	1	1	0	0	0
PN16	0	1	1	1	1	1	0	0	1	1	0	1	0	0	1	0	0	0	0	1	0	1	0	1	1	1	0	1	1	0	0
PN17	0	0	1	1	1	1	1	0	0	1	1	0	1	0	0	1	0	0	0	0	1	0	1	0	1	1	1	0	1	1	0
PN18	0	0	0	1	1	1	1	1	0	0	1	1	0	1	0	0	1	0	0	0	0	1	0	1	0	1	1	1	0	1	1
PN19	1	0	0	0	1	1	1	1	1	0	0	1	1	0	1	0	0	1	0	0	0	0	1	0	1	0	1	1	1	0	1
PN20	1	1	0	0	0	1	1	1	1	1	0	0	1	1	0	1	0	0	1	0	0	0	0	1	0	1	0	1	1	1	0
PN21	0	1	1	0	0	0	1	1	1	1	1	0	0	1	1	0	1	0	0	1	0	0	0	0	1	0	1	0	1	1	1
PN22	1	0	1	1	0	0	0	1	1	1	1	1	0	0	1	1	0	1	0	0	1	0	0	0	0	1	0	1	0	1	1
PN23	1	1	0	1	1	0	0	0	1	1	1	1	1	0	0	1	1	0	1	0	0	1	0	0	0	0	1	0	1	0	1
PN24	1	1	1	0	1	1	0	0	0	1	1	1	1	1	0	0	1	1	0	1	0	0	1	0	0	0	0	1	0	1	0
PN25	0	1	1	1	0	1	1	0	0	0	1	1	1	1	1	0	0	1	1	0	1	0	0	1	0	0	0	0	1	0	1
PN26	1	0	1	1	1	0	1	1	0	0	0	1	1	1	1	1	0	0	1	1	0	1	0	0	1	0	0	0	0	1	0
PN27	0	1	0	1	1	1	0	1	1	0	0	0	1	1	1	1	1	0	0	1	1	0	1	0	0	1	0	0	0	0	1
PN28	1	0	1	0	1	1	1	0	1	1	0	0	0	1	1	1	1	1	0	0	1	1	0	1	0	0	1	0	0	0	0
PN29	0	1	0	1	0	1	1	1	0	1	1	0	0	0	1	1	1	1	1	0	0	1	1	0	1	0	0	1	0	0	0
PN30	0	0	1	0	1	0	1	1	1	0	1	1	0	0	0	1	1	1	1	1	0	0	1	1	0	1	0	0	1	0	0
PN31	0	0	0	1	0	1	0	1	1	1	0	1	1	0	0	0	1	1	1	1	1	0	0	1	1	0	1	0	0	1	0

These PN sequences can be orthogonalised by adding a zero (0) at the end each sequence (shift). This extra chip is required anyway to make the sequence length a multiple of 2, causing the new correlation values to be 0. This is convenient in terms of orthogonality, but is costly to implement as each sequence becomes a different code, that is, the sequences are not only a shifted version of the original sequence, these are actually a different sequence. This makes the detection process much more complicated.

IS-95 implemented this orthogonality by adding the extra zero to the end of the longest sequence of zeroes, creating a code that is then shifted as shown in Table 9.17. This implementation reduces the circuitry required for the detection of the different shifts, but makes the orthogonality worst as seen in Table 9.18 and Figure 9.50. The auto-correlation peaks can reach 40% of the coincidence peak, although the average of all shifts is still small.

Table 9.16 Auto correlation against PN1 sequence

PN1	1	1	1	1	1	1	1	1	1	1	1	1	1	1	1	1	1	1	1	1	1	1	1	1	1	1	1	1	1	1	1	1.000
PN2	-1	1	1	1	-1	-1	-1	-1	-1	1	1	-1	-1	1	-1	1	1	-1	1	1	1	1	-1	1	-1	1	-1	-1	-1	1	-1	-0.032
PN3	1	-1	1	1	-1	1	1	1	1	-1	1	-1	1	-1	-1	-1	1	-1	-1	1	1	1	-1	-1	-1	-1	-1	1	1	-1	-1	-0.032
PN4	1	1	-1	1	-1	1	-1	-1	-1	1	-1	-1	1	1	1	-1	-1	-1	-1	-1	1	1	-1	-1	1	-1	1	1	-1	1	1	-0.032
PN5	-1	1	1	-1	-1	1	-1	1	1	-1	1	1	1	-1	1	1	-1	1	-1	-1	-1	1	-1	1	1	1	-1	-1	-1	-1	-1	-0.032
PN6	1	-1	1	1	1	1	-1	1	-1	1	-1	-1	-1	1	-1	-1	1	1	1	-1	-1	-1	-1	-1	1	1	-1	-1	1	-1	1	-0.032
PN7	-1	1	-1	1	-1	-1	-1	1	-1	-1	1	1	1	-1	-1	-1	-1	-1	1	1	-1	-1	1	-1	1	1	-1	1	1	1	-1	-0.032
PN8	-1	-1	1	-1	-1	1	1	1	-1	-1	-1	-1	-1	1	1	1	-1	-1	1	-1	1	1	-1	1	1	1	-1	1	-1	1	-1	-0.032
PN9	1	-1	-1	1	1	1	-1	-1	-1	-1	1	1	-1	-1	1	-1	1	1	-1	1	1	1	-1	1	-1	1	-1	-1	-1	-1	-1	-0.032
PN10	1	1	-1	-1	-1	-1	-1	1	1	-1	-1	1	-1	1	1	-1	1	1	1	1	-1	1	-1	1	-1	-1	-1	1	-1	-1	1	-0.032
PN11	-1	1	1	-1	1	1	1	1	-1	1	-1	1	-1	-1	-1	1	-1	-1	1	1	1	-1	-1	-1	-1	1	1	-1	-1	1	1	-0.032
PN12	-1	-1	1	1	1	-1	-1	-1	-1	1	1	1	-1	-1	1	-1	1	1	-1	1	1	1	1	-1	1	-1	1	-1	-1	-1	1	-0.032
PN13	-1	-1	-1	1	-1	-1	1	1	1	1	1	-1	-1	-1	-1	-1	1	1	1	-1	1	-1	1	1	1	-1	-1	-1	1	1	1	-0.032
PN14	-1	-1	-1	-1	-1	1	1	-1	-1	1	-1	1	1	-1	1	1	1	1	1	-1	1	-1	1	-1	-1	-1	1	-1	1	1	1	-0.032
PN15	-1	-1	-1	-1	1	1	-1	-1	1	-1	1	1	-1	1	1	1	1	-1	1	-1	1	-1	-1	-1	1	-1	-1	1	1	1	-1	-0.032
PN16	1	-1	-1	-1	1	-1	-1	1	1	1	-1	-1	-1	-1	-1	1	1	-1	-1	1	-1	1	1	1	1	1	-1	1	1	-1	-1	-0.032
PN17	1	1	-1	-1	1	-1	1	1	1	-1	1	1	1	1	-1	1	-1	1	-1	-1	-1	1	-1	1	1	-1	-1	-1	-1	-1	-1	-0.032
PN18	1	1	1	-1	1	-1	1	-1	-1	-1	1	-1	-1	1	1	1	-1	-1	-1	-1	-1	1	1	-1	-1	1	-1	1	1	-1	1	-0.032
PN19	-1	1	1	1	1	-1	1	-1	1	-1	-1	-1	1	-1	-1	1	1	1	-1	-1	-1	-1	-1	1	1	-1	-1	1	-1	1	1	-0.032
PN20	-1	-1	1	1	-1	-1	1	-1	1	1	-1	1	1	1	-1	1	-1	1	-1	-1	-1	1	-1	1	1	1	-1	-1	-1	-1	-1	-0.032
PN21	1	-1	-1	1	-1	1	1	-1	1	1	1	1	-1	1	-1	1	-1	-1	1	-1	-1	1	1	-1	-1	-1	-1	1	1	-1	1	-0.032
PN22	-1	1	-1	-1	-1	1	-1	-1	1	1	1	-1	-1	-1	-1	-1	1	1	-1	1	-1	1	1	1	1	1	1	1	1	-1	1	-0.032
PN23	-1	-1	1	-1	1	1	-1	1	1	1	1	-1	1	-1	-1	-1	-1	1	1	-1	1	-1	1	1	1	-1	-1	-1	-1	1	1	-0.032
PN24	-1	-1	-1	1	1	-1	-1	1	-1	1	1	1	-1	1	1	1	1	-1	1	-1	1	-1	-1	-1	1	-1	-1	1	1	-1	-1	-0.032
PN25	1	-1	-1	-1	-1	1	1	-1	-1	1	-1	1	1	1	1	1	1	-1	1	-1	-1	-1	1	-1	-1	-1	1	-1	-1	1	1	-0.032
PN26	-1	1	-1	-1	1	1	1	-1	-1	-1	-1	1	-1	-1	1	-1	1	1	1	1	1	1	1	-1	1	1	-1	-1	-1	-1	-1	-0.032
PN27	1	-1	1	-1	1	-1	-1	-1	1	-1	-1	1	1	-1	1	-1	-1	-1	1	-1	-1	1	1	1	1	1	1	1	-1	1	1	-0.032
PN28	-1	1	-1	1	1	1	1	1	1	1	-1	1	-1	1	-1	-1	-1	1	-1	-1	1	1	1	-1	-1	-1	-1	1	1	-1	1	-0.032
PN29	1	-1	1	-1	-1	-1	1	-1	-1	1	1	1	1	-1	-1	-1	-1	1	1	-1	1	1	-1	1	1	-1	1	1	1	1	-1	-0.032
PN30	1	1	-1	1	1	1	1	-1	1	-1	1	-1	-1	-1	1	-1	-1	1	1	1	-1	-1	-1	-1	-1	1	1	-1	-1	1	-1	-0.032
PN31	1	1	1	-1	-1	-1	-1	-1	1	1	-1	-1	1	-1	1	1	-1	1	1	1	1	-1	1	-1	1	-1	-1	-1	1	-1	-1	-0.032

The auto-correlation of PN codes makes them ideal for use in multi-path environments to detect shifted copies of the same signal and, consequently, provide space diversity. Another advantage is that large sequences can be generated, with many shifted versions available to generate many codes. Because of the large number of codes, these sequences are suited for authentication and scrambling. PN codes are used for identifying different sectors and MSs, detecting multi-path signals, authentication and scrambling. Sequences generated by different polynomials usually do not present good cross-orthogonality, but it is still possible to find PN codes that are orthogonal to each other.

The received signal is a result of a superposition of many waveforms coming from different locations. Summing the signal module two to a specific sequence is equivalent to do a chip by chip comparison against all the sequences that form the signal. Agreements are mapped to 1 and disagreements to -1. The resulting values are integrated for the

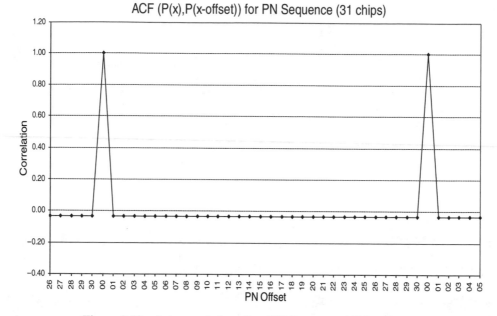

Figure 9.49 Auto correlation of an SSRG generated PN sequence.

complete sequence and compared to a reference. If the same sequence is found in phase with the searched sequence then a high positive value is obtained. This value is masked by correlation values from other sequences.

The use of quasi-orthogonal codes diminishes the effectiveness of the spreading as the peak energy created by the code matching, will be partially neutralised by the energies of partial correlations.

Decimation

Abridged versions of PN codes can be obtained by sampling them at a regular rate, a process called decimation. These decimated codes produce shifted versions of the same sequence, but at lower chip rate and can be used in applications that require pseudo-randomness at lower rates as in the 19.2 kbps data scrambling of traffic channels.

9.3.2.2.1   Partial Correlation

The identification of MSs requires many codes and, consequently, the PN code assigned to do this should be very long. In CDMA, this code is 4.398×10^{12} bits long. Because the sequence is so long, it is impossible to wait for the whole sequence to finish to be able to find its beginning and synchronise to it. However, it is possible to estimate the correlation by analysing only part of the sequence. This process is called partial correlation.

Let's assume a sequence of P chips of which we will correlate on *M* chips only. Using the balance property of the PN sequence, it is possible to compute statistically the partial correlation function. It will give over M bits a normalised correlation of 1 when the sequences are lined up and $-1/P$ when they are not synchronised. The number of chips used in the partial correlation is a trade-off between the time to acquire the synchronism and the noise margin desired.

Table 9.17 PN code with 32 bit period

PN1	0	0	0	0	1	0	1	0	1	1	1	0	1	1	0	0	0	1	1	1	1	1	0	0	1	1	0	1	0	0	1	0
PN2	0	0	0	0	0	1	0	1	0	1	1	1	0	1	1	0	0	0	1	1	1	1	1	0	0	1	1	0	1	0	0	1
PN3	1	0	0	0	0	0	1	0	1	0	1	1	1	0	1	1	0	0	0	1	1	1	1	1	0	0	1	1	0	1	0	0
PN4	0	1	0	0	0	0	0	1	0	1	0	1	1	1	0	1	1	0	0	0	1	1	1	1	1	0	0	1	1	0	1	0
PN5	0	0	1	0	0	0	0	0	1	0	1	0	1	1	1	0	1	1	0	0	0	1	1	1	1	1	0	0	1	1	0	1
PN6	1	0	0	1	0	0	0	0	0	1	0	1	0	1	1	1	0	1	1	0	0	0	1	1	1	1	1	0	0	1	1	0
PN7	0	1	0	0	1	0	0	0	0	0	1	0	1	0	1	1	1	0	1	1	0	0	0	1	1	1	1	1	0	0	1	1
PN8	1	0	1	0	0	1	0	0	0	0	0	1	0	1	0	1	1	1	0	1	1	0	0	0	1	1	1	1	1	0	0	1
PN9	1	1	0	1	0	0	1	0	0	0	0	0	1	0	1	0	1	1	1	0	1	1	0	0	0	1	1	1	1	1	0	0
PN10	0	1	1	0	1	0	0	1	0	0	0	0	0	1	0	1	0	1	1	1	0	1	1	0	0	0	1	1	1	1	1	0
PN11	0	0	1	1	0	1	0	0	1	0	0	0	0	0	1	0	1	0	1	1	1	0	1	1	0	0	0	1	1	1	1	1
PN12	1	0	0	1	1	0	1	0	0	1	0	0	0	0	0	1	0	1	0	1	1	1	0	1	1	0	0	0	1	1	1	1
PN13	1	1	0	0	1	1	0	1	0	0	1	0	0	0	0	0	1	0	1	0	1	1	1	0	1	1	0	0	0	1	1	1
PN14	1	1	1	0	0	1	1	0	1	0	0	1	0	0	0	0	0	1	0	1	0	1	1	1	0	1	1	0	0	0	1	1
PN15	1	1	1	1	0	0	1	1	0	1	0	0	1	0	0	0	0	0	1	0	1	0	1	1	1	0	1	1	0	0	0	1
PN16	1	1	1	1	1	0	0	1	1	0	1	0	0	1	0	0	0	0	0	1	0	1	0	1	1	1	0	1	1	0	0	0
PN17	0	1	1	1	1	1	0	0	1	1	0	1	0	0	1	0	0	0	0	0	1	0	1	0	1	1	1	0	1	1	0	0
PN18	0	0	1	1	1	1	1	0	0	1	1	0	1	0	0	1	0	0	0	0	0	1	0	1	0	1	1	1	0	1	1	0
PN19	0	0	0	1	1	1	1	1	0	0	1	1	0	1	0	0	1	0	0	0	0	0	1	0	1	0	1	1	1	0	1	1
PN20	1	0	0	0	1	1	1	1	1	0	0	1	1	0	1	0	0	1	0	0	0	0	0	1	0	1	0	1	1	1	0	1
PN21	1	1	0	0	0	1	1	1	1	1	0	0	1	1	0	1	0	0	1	0	0	0	0	0	1	0	1	0	1	1	1	0
PN22	0	1	1	0	0	0	1	1	1	1	1	0	0	1	1	0	1	0	0	1	0	0	0	0	0	1	0	1	0	1	1	1
PN23	1	0	1	1	0	0	0	1	1	1	1	1	0	0	1	1	0	1	0	0	1	0	0	0	0	0	1	0	1	0	1	1
PN24	1	1	0	1	1	0	0	0	1	1	1	1	1	0	0	1	1	0	1	0	0	1	0	0	0	0	0	1	0	1	0	1
PN25	1	1	1	0	1	1	0	0	0	1	1	1	1	1	0	0	1	1	0	1	0	0	1	0	0	0	0	0	1	0	1	0
PN26	0	1	1	1	0	1	1	0	0	0	1	1	1	1	1	0	0	1	1	0	1	0	0	1	0	0	0	0	0	1	0	1
PN27	1	0	1	1	1	0	1	1	0	0	0	1	1	1	1	1	0	0	1	1	0	1	0	0	1	0	0	0	0	0	1	0
PN28	0	1	0	1	1	1	0	1	1	0	0	0	1	1	1	1	1	0	0	1	1	0	1	0	0	1	0	0	0	0	0	1
PN29	1	0	1	0	1	1	1	0	1	1	0	0	0	1	1	1	1	1	0	0	1	1	0	1	0	0	1	0	0	0	0	0
PN30	0	1	0	1	0	1	1	1	0	1	1	0	0	0	1	1	1	1	1	0	0	1	1	0	1	0	0	1	0	0	0	0
PN31	0	0	1	0	1	0	1	1	1	0	1	1	0	0	0	1	1	1	1	1	0	0	1	1	0	1	0	0	1	0	0	0
PN32	0	0	0	1	0	1	0	1	1	1	0	1	1	0	0	0	1	1	1	1	1	0	0	1	1	0	1	0	0	1	0	0

The number of chips used in the partial correlation is a trade-off between the time to acquire the synchronism and the desired noise margin. Figures 9.51 and 9.52 show the partial auto correlation for four different segments of 8 bits in a 32 bit code respectively for a Maximal Length Sequence (MLS)[3] and a non-MLS sequence. The correlation peak at position 00 is barely distinguishable, showing that, in this example, the idea was to reduce the time to acquire synchronism.

Figures 9.53 and 9.54 show the same comparison but for two segments of 16 bits in a 32-bit code. In this example, the peak is more clearly identifiable, showing that the idea was to reduce the noise margin for synchronisation.

---

[3] By definition, MLSs are the longest sequences that can be generated by a certain arrangement of shift-registers or similar delay elements. Section 3.2 (Chapter 3) provides more details about MLSs.

Table 9.18   Auto-correlation of PN1 code

PN1	1	1	1	1	1	1	1	1	1	1	1	1	1	1	1	1	1	1	1	1	1	1	1	1	1	1	1	1	1	1	1	1	1.00
PN2	1	1	1	1	-1	-1	-1	-1	-1	1	1	-1	-1	1	-1	1	1	-1	1	1	1	1	-1	1	-1	1	-1	-1	-1	1	-1	-1	0.00
PN3	-1	1	1	1	-1	1	1	1	1	1	-1	1	-1	1	-1	-1	-1	1	-1	-1	1	1	1	-1	-1	-1	-1	1	1	-1	-1	1	0.00
PN4	1	-1	1	1	-1	1	-1	-1	-1	1	-1	1	1	1	1	-1	-1	-1	-1	-1	1	1	-1	-1	1	1	-1	1	1	-1	1	1	0.00
PN5	1	1	-1	1	-1	1	-1	1	1	1	-1	1	1	1	1	-1	1	-1	1	-1	-1	-1	1	-1	-1	1	1	-1	-1	-1	-1	-1	0.00
PN6	-1	1	1	-1	-1	1	-1	1	-1	1	-1	-1	-1	1	-1	1	1	1	1	-1	-1	-1	-1	-1	1	1	-1	-1	1	-1	1	1	-0.13
PN7	1	-1	1	1	1	1	-1	1	-1	-1	-1	1	1	1	-1	-1	-1	-1	-1	1	1	1	-1	-1	1	1	1	-1	1	1	1	-1	0.13
PN8	-1	1	-1	1	-1	-1	-1	1	-1	-1	-1	-1	-1	1	1	-1	-1	1	-1	1	1	-1	1	1	1	1	-1	1	-1	1	-1	-1	-0.13
PN9	-1	-1	1	-1	-1	1	1	1	1	-1	-1	-1	1	1	-1	1	-1	1	1	1	1	1	1	-1	1	-1	1	-1	-1	-1	-1	-1	0.00
PN10	1	-1	-1	1	1	1	-1	-1	-1	-1	1	1	-1	1	1	1	-1	1	1	1	1	-1	1	-1	1	1	-1	-1	-1	-1	1	1	0.00
PN11	1	1	-1	-1	-1	-1	-1	1	1	1	-1	-1	1	-1	-1	-1	1	-1	-1	1	1	1	-1	-1	-1	-1	1	1	-1	1	1	-1	-0.25
PN12	-1	1	1	1	-1	1	1	1	1	1	-1	1	-1	1	-1	1	1	1	-1	1	1	1	-1	1	-1	1	-1	-1	-1	-1	1	-1	0.13
PN13	-1	-1	1	1	1	-1	-1	-1	-1	-1	1	1	-1	-1	1	1	1	-1	-1	1	1	1	-1	1	1	1	-1	1	-1	-1	1	-1	0.00
PN14	-1	-1	-1	1	-1	-1	1	1	1	-1	-1	-1	-1	-1	1	1	1	1	-1	1	-1	1	1	-1	-1	1	-1	-1	1	1	1	-1	-0.13
PN15	-1	-1	-1	-1	-1	1	1	1	-1	1	1	1	1	-1	1	1	1	1	-1	1	1	-1	-1	-1	1	-1	-1	1	1	1	-1	-1	-0.13
PN16	-1	-1	-1	-1	1	1	-1	-1	1	1	1	1	-1	1	1	1	-1	-1	1	-1	1	1	1	1	1	-1	1	1	-1	1	1	1	0.13
PN17	1	-1	-1	1	1	-1	-1	1	1	1	1	-1	-1	-1	-1	-1	1	1	-1	1	-1	1	1	1	1	-1	-1	-1	-1	-1	-1	1	-0.25
PN18	1	1	-1	-1	1	-1	1	1	-1	1	1	1	1	-1	1	-1	1	-1	-1	-1	-1	1	1	-1	-1	1	-1	1	1	-1	1	1	0.13
PN19	1	1	1	-1	1	-1	1	-1	-1	-1	1	-1	-1	1	1	1	-1	-1	-1	-1	-1	-1	1	1	-1	1	-1	1	1	1	1	-1	-0.13
PN20	-1	1	1	1	1	-1	1	-1	1	-1	-1	-1	1	-1	-1	1	1	1	-1	-1	-1	1	-1	-1	1	1	1	-1	-1	-1	-1	-1	-0.13
PN21	-1	-1	1	1	-1	-1	1	-1	1	1	1	1	1	1	1	1	-1	1	1	1	-1	1	1	1	1	1	-1	-1	-1	-1	1	1	0.00
PN22	1	-1	-1	1	-1	1	1	1	1	1	1	-1	1	-1	1	1	1	-1	1	-1	-1	-1	1	1	1	1	1	1	1	-1	1	-1	0.13
PN23	-1	1	-1	-1	-1	1	-1	-1	1	1	1	-1	-1	-1	-1	-1	1	1	1	-1	1	1	1	-1	-1	-1	1	1	-1	-1	1	1	-0.25
PN24	-1	-1	1	-1	1	1	-1	1	1	1	1	1	-1	1	-1	1	-1	1	1	-1	1	1	1	1	-1	1	1	1	-1	-1	-1	-1	0.00
PN25	-1	-1	-1	1	1	-1	-1	1	-1	1	1	1	-1	1	1	1	1	1	1	-1	-1	-1	1	-1	-1	1	-1	-1	1	1	1	1	0.00
PN26	1	-1	-1	-1	-1	-1	1	1	-1	-1	1	-1	1	1	-1	1	1	1	1	-1	1	-1	-1	-1	-1	-1	-1	-1	-1	-1	-1	1	-0.13
PN27	-1	1	-1	-1	1	1	1	1	-1	-1	-1	-1	-1	1	1	-1	-1	1	1	-1	1	1	-1	1	1	1	1	1	-1	1	1	1	0.13
PN28	1	1	1	-1	1	-1	-1	-1	1	-1	-1	1	1	1	-1	-1	-1	-1	1	1	-1	-1	1	1	1	1	1	1	1	1	-1	-1	-0.13
PN29	-1	1	-1	1	1	1	-1	1	1	1	1	1	-1	1	1	-1	1	-1	-1	-1	-1	-1	1	1	1	-1	-1	1	-1	1	1	1	0.00
PN30	1	-1	1	-1	-1	-1	1	-1	-1	1	1	1	1	-1	-1	1	-1	-1	-1	1	1	1	-1	-1	1	1	1	1	1	1	-1	1	0.00
PN31	1	1	-1	1	1	1	1	1	-1	1	-1	1	-1	-1	-1	1	-1	1	1	1	-1	-1	-1	-1	-1	1	-1	-1	1	1	1	1	0.00
PN32	1	1	1	-1	-1	-1	-1	-1	1	1	-1	-1	1	-1	1	1	-1	1	1	1	1	-1	1	-1	1	-1	-1	-1	1	-1	-1	1	0.00

Orthogonality Factor

In the network, when the MS is trying to detect a channel, it not only receives the codes (shifts) being used by the serving sector BTS, but also receives components from other sectors. The net result is a residual noise that increases the interference. This residual noise can be expressed by an *orthogonality factor* (α), which expresses the amount of the transmitted power that leaks through non-orthogonality. An orthogonality of 1 means that the signals are orthogonal, a smaller α indicates the degree of non-orthogonality in terms of interfering power.

Orthogonality factors vary from 0.2 to 0.6 in real life systems and have a large impact on the network performance. It is very difficult to measure this factor but it can be estimated by comparing predicted and field results. A factor of 0.5 is a good compromise for green-field networks for which no practical data exists. The analysis done here holds in real life where

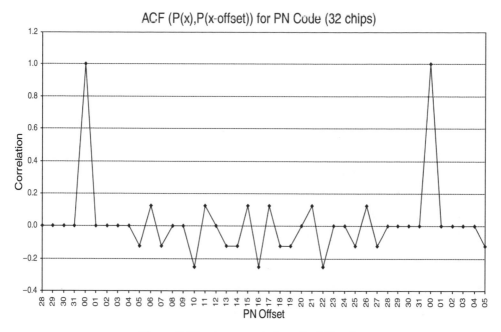

Figure 9.50   Auto correlation of a PN code.

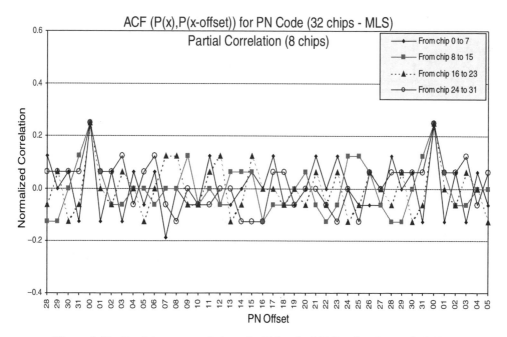

Figure 9.51   Partial auto-correlation of a PN code (MLS) using 8 out of 32 bits.

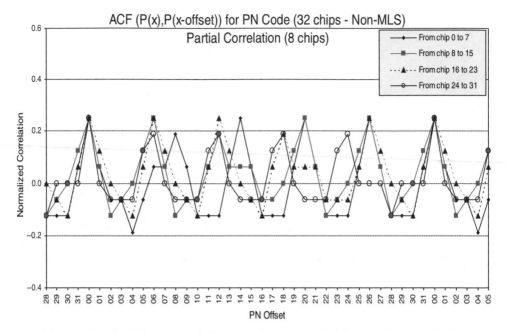

Figure 9.52   Partial auto correlation of a PN code (non-MLS) using 8 out of 32 bits.

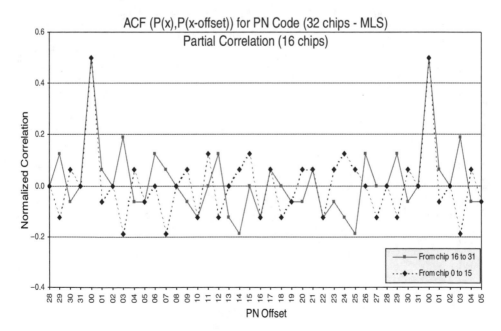

Figure 9.53   Partial auto-correlation of a PN code (MLS) using 16 out of 32 bits.

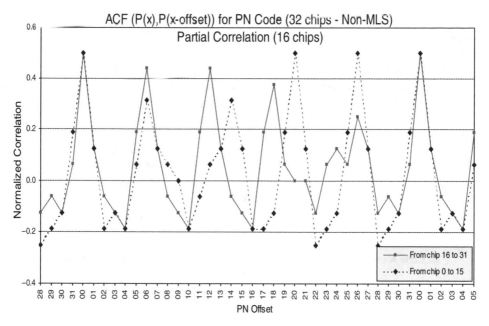

Figure 9.54 Partial auto-correlation of a PN code (non-MLS) using 16 out of 32 bits.

the output has its pulse shaped (waveforms are rounded) to minimise intersymbol interference.

The calculation of the SNIR at the output of the Rake finger, has to consider this non-orthogonal signals coming from the BTS and the combining can be expressed as

$$SNIR = \frac{R_c}{R_b} \frac{P_{total}}{P_{sector}(1 - \alpha) + P_{neighbour} + N} \tag{9.7}$$

where

SNIR	signal-to-noise-plus-interference ratio
R_c	chip rate (also compared to Spreading Rate – SR)
R_b	bit rate
P_{total}	total received power
P_{sector}	total received power from sector, supposedly orthogonal
α	orthogonality factor
$P_{neighbor}$	total received power from neighbors
N	received noise power.

A complete and detailed analysis of the PN sequence theory and Walsh codes is available in Ref. [1].

9.3.2.2.2   PN Offset

Section 4.2.2 in Chapter 4 describes the pilot channel used in IS-95 systems. Section 4.2.2.5 briefly explains the PN offset ands its usage. This section analyses the issues caused by delay when detecting the pilot and what factors should be considered in PN planning.

BTS are synchronised through Global Positioning System (GPS) synchronised clocks. The relationship between shifted versions of a PN sequence is kept within a fraction of a chip. The signals received from several base stations are delayed by the propagation time and lose their relative shift. A path difference of 244.14 m (0.8138 μs) leads to a difference of one chip, which may cause confusion when determining the actual shift of a sequence. For example, in Figure 9.55, the base stations use different short PN codes but the mobile perceives them with the same code because of the delay. This issue can be avoided by spacing the offsets that can be used in codes. The PN code is 32768 chips long and 512 offsets of 64 chips were established to be used as PN offsets (PN 0 to PN 511).

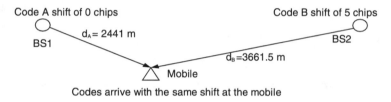

Figure 9.55   PN offset.

Each BTS has a unique PN offset associated with it (from the point of view of the mobile) allowing their signals to be distinguished from one another. As all BTSs are synchronised by GPS, they all use the same reference and know exactly when the PN sequences should start.

Mobiles do not have this reference and have to create their own reference using the strongest pilot received. This reference is distorted by the travelling time from the BTS to the mobile (propagation delay). All of this brings uncertainties to the detection process and makes it a statistical process.

Figures 9.56 and 9.57 show an example of PN offset determination.

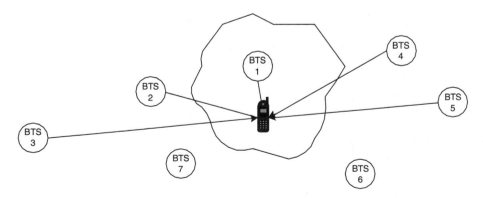

Figure 9.56   PN offset determination.

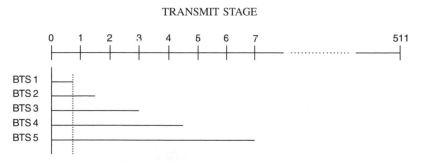

Figure 9.57   Relative PN offset.

The mobile receives its pilot from BTS1 with an offset of 0.75 chips caused because of propagation delay. A pilot coming from BTS2 is received with a relative shift of 0.75 instead of 1.5 and so on. These differential delays, illustrated in Figure 9.57, must be considered when assigning PN offsets.

The proper assignment of PN offsets requires building a tri-dimensional matrix. In this matrix, '0' offset is assumed for all BTSs. The network is divided by a geographical grid of x by y meter (usually $x = y = 30$ m to 90 m) and each grid is called a pixel.

For each network pixel, all candidate pilots $(E_c/I_o > T_{add})$ and a percentage of possible candidate pilots $(E_c/I_o > T_{drop})$ are analysed considering the relative delay to the main pilot.

The matrix has all the serving sectors (interfered) in one axis and all the candidate sectors (interfering) in the other. The relative offset for each pair represents the third dimension and it can vary from 0 to 511 offsets. The value stored for each offset is the amount of traffic of each pixel that has the relationship between the two sectors.

For each pair, main to candidate pilot, the relative offset is calculated and the pixel traffic is stored in a distribution diagram for the resulting offset. This 3D matrix is illustrated in Figure 9.58.

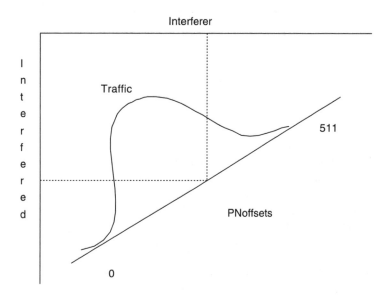

Figure 9.58   PN offset matrix.

This matrix can be used as an input to planning tools that will assign the best possible offsets to each sector.

PN planning is facilitated by the abundance of codes but some network relationships are hard to visualise and an automatic tool is required to properly plan them. *CelOptima* is one of the tools available in the market that implements the method described in Figure 9.58. Figure 9.59 shows the configuration screen for PN planning from this tool.

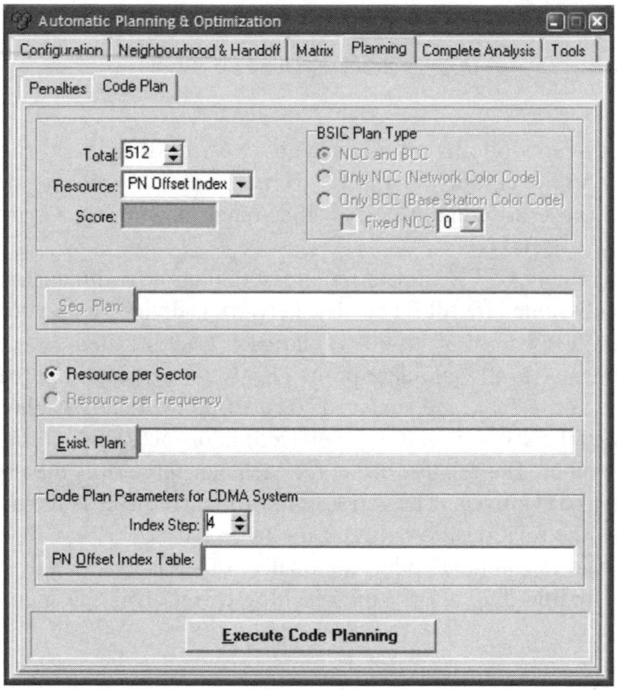

Figure 9.59   PN offset dialogue screen.

9.3.2.3   Simulation of a DS-CDMA System

We will examine now the complete process, the way it is implemented in forward link of IS-95, but using shorter codes. The scrambling done by the long PN code is not shown or can be considered already done for the input bits.

The reverse link is similar although the Walsh code is not used as a channel ID but as an orthogonal modulation and there are two levels of PN coding, the long PN followed by Walsh modulation and by the short PN.

An example of the complete DS-CDMA process is shown in Tables 9.19–9.27.

The information bits to be transmitted are encoded by the convolutional encoder, interleaved, scrambled by the long code and punctured by the power control bits resulting in stream of symbols at 19.2 ksps. The user data is then ready to be combined with the Walsh code.

Table 9.19 Information symbols for different channels

	Input bit	Input bit
Channel 1	0	1
Channel 2	0	1
Channel 3	1	−1
Channel 4	1	−1
Channel 5	0	1
Channel 6	0	1
Channel 7	1	−1
Channel 8	1	−1

Table 9.19 shows one bit of each of the eight channels that are going to be transmitted. These bits are mapped to '+1' and '−1' to facilitate the implementation of logical operations.

Table 9.20 shows eight different 32 bit long Walsh codes (W0 to W7) that can be assigned to the different channels. Table 9.21 shows the 32 bits of each Walsh code modulo 2 added to the corresponding information symbol.

Table 9.22 shows eight 32 bit long PN codes. These codes are modulo 2 summed with the output of the Walsh code combination resulting in Table 9.23. Each BTS has a different PN code assigned. If the channels have the same PN code they are assigned to the same BTS, otherwise, they belong to different BTSs.

Even though the channels are separately coded, they are all transmitted together. The chips of all channels presented in Table 9.23 are summed (Tx signal) to form the transmit signal as shown in Table 9.24. This sum is made per chip, that is, chip 1 of all channels are summed and transmitted, and then chip 2 of all channels are summed and transmitted. Each channel has its analogue level adjusted through power control so the required level reaches the receiver. In this example, the adjustment was done after the sum to simplify the spreadsheet.

The Tx signal in Table 9.24 was power adjusted to simulate the power adjustment of each channel so the required E_b/N_0 is reached at the receive side. In the example shown E_b/N_0 has a preset value of 13 dB, which is equivalent to a SNIR of −2 dB. Multi-path is simulated by delaying the transmitted signal respectively by 1, 2 and 3 chips (three multi-path signals) and applying a power loss. Noise is also added with a random seed, allowing different combinations to be tested. Finally, the transmitted sequence is summed to the sequences that simulate these effects (power control, multipath and noise).

The last row in Table 9.24 is the sum of all other rows in the table and is equivalent to the received signal in terms of energy proportionality. No propagation loss was considered as it does not change the relative strengths of the signals.

The detection of information in CDMA systems is substantially different than the traditional way of detecting signals. In CDMA, the system knows that a certain code was transmitted and looks for its pattern in the received signal. The received chips are combined with the known pattern and the results are summed for the entire bit of information. Tables 9.25 and 9.26 show, respectively, the received signal being modulo 2 added with the PN and Walsh codes associated with the channel to be extracted.

Table 9.20   Walsh coding

k=5	index	1	2	3	4	5	6	7	8	9	10	11	12	13	14	15	16	17	18	19	20	21	22	23	24	25	26	27	28	29	30	31	32
W0	0	0	0	0	0	0	0	0	0	0	0	0	0	0	0	0	0	0	0	0	0	0	0	0	0	0	0	0	0	0	0	0	0
W1	1	0	1	0	1	0	1	0	1	0	1	0	1	0	1	0	1	0	1	0	1	0	1	0	1	0	1	0	1	0	1	0	1
W2	2	0	0	1	1	0	0	1	1	0	0	1	1	0	0	1	1	0	0	1	1	0	0	1	1	0	0	1	1	0	0	1	1
W3	3	0	1	1	0	0	1	1	0	0	1	1	0	0	1	1	0	0	1	1	0	0	1	1	0	0	1	1	0	0	1	1	0
W4	4	0	0	0	0	1	1	1	1	0	0	0	0	1	1	1	1	0	0	0	0	1	1	1	1	0	0	0	0	1	1	1	1
W5	5	0	1	0	1	1	0	1	0	0	1	0	1	1	0	1	0	0	1	0	1	1	0	1	0	0	1	0	1	1	0	1	0
W6	6	0	0	1	1	1	1	0	0	0	0	1	1	1	1	0	0	0	0	1	1	1	1	0	0	0	0	1	1	1	1	0	0
W7	7	0	1	1	0	1	0	0	1	0	1	1	0	1	0	0	1	0	1	1	0	1	0	0	1	0	1	1	0	1	0	0	1

Table 9.21   Walsh coded channels

		1	2	3	4	5	6	7	8	9	10	11	12	13	14	15	16	17	18	19	20	21	22	23	24	25	26	27	28	29	30	31	32
CH1	W0	1	1	1	1	1	1	1	1	1	1	1	1	1	1	1	1	1	1	1	1	1	1	1	1	1	1	1	1	1	1	1	1
CH2	W1	1	-1	1	-1	1	-1	1	-1	1	-1	1	-1	1	-1	1	-1	1	-1	1	-1	1	-1	1	-1	1	-1	1	-1	1	-1	1	-1
CH3	W2	1	1	-1	-1	1	1	-1	-1	1	1	-1	-1	1	1	-1	-1	1	1	-1	-1	1	1	-1	-1	1	1	-1	-1	1	1	-1	-1
CH4	W3	1	-1	-1	1	1	-1	-1	1	1	-1	-1	1	1	-1	-1	1	1	-1	-1	1	1	-1	-1	1	1	-1	-1	1	1	-1	-1	1
CH5	W4	1	1	1	1	-1	-1	-1	-1	1	1	1	1	-1	-1	-1	-1	1	1	1	1	-1	-1	-1	-1	1	1	1	1	-1	-1	-1	-1
CH6	W5	1	-1	1	-1	-1	1	-1	1	1	-1	1	-1	-1	1	-1	1	1	-1	1	-1	-1	1	-1	1	1	-1	1	-1	-1	1	-1	1
CH7	W6	1	1	-1	-1	-1	-1	1	1	1	1	-1	-1	-1	-1	1	1	1	1	-1	-1	-1	-1	1	1	1	1	-1	-1	-1	-1	1	1
CH8	W7	1	-1	-1	1	-1	1	1	-1	1	-1	-1	1	-1	1	1	-1	1	-1	-1	1	-1	1	1	-1	1	-1	-1	1	-1	1	1	-1

Table 9.22 PN codes

	1	2	3	4	5	6	7	8	9	10	11	12	13	14	15	16	17	18	19	20	21	22	23	24	25	26	27	28	29	30	31	32
PN1	0	0	0	0	1	0	1	0	1	1	1	1	1	0	0	0	1	1	1	1	1	0	0	0	1	1	1	1	0	0	1	0
PN2	0	0	0	0	0	0	1	1	0	1	1	1	0	1	0	0	0	0	1	1	1	0	1	0	0	1	1	1	0	0	1	1
PN3	1	0	0	0	0	1	0	0	1	0	1	1	0	0	0	1	0	0	0	1	1	1	1	1	0	0	0	0	1	0	0	0
PN4	0	1	0	0	0	0	0	1	0	0	0	0	1	1	1	1	0	0	0	0	1	1	0	1	1	1	0	0	1	0	1	0
PN5	0	0	1	0	0	0	1	0	1	0	1	0	1	1	0	1	1	1	0	0	0	1	1	1	1	0	1	0	0	1	0	1
PN6	1	0	0	1	0	0	0	0	0	0	0	1	0	0	1	1	0	1	0	0	0	0	1	0	1	1	0	0	0	1	0	1
PN7	0	1	1	0	0	0	0	0	0	1	1	0	1	0	1	1	1	0	1	1	1	0	0	0	1	1	1	1	0	0	1	0
PN8	1	0	1	0	0	1	0	0	0	0	1	1	0	1	0	1	1	1	0	1	1	0	0	0	1	1	1	1	1	0	0	1

Table 9.23 PN coded BTS

		1	2	3	4	5	6	7	8	9	10	11	12	13	14	15	16	17	18	19	20	21	22	23	24	25	26	27	28	29	30	31	32
BS1	PN1	1	1	1	1	-1	1	-1	1	-1	-1	-1	1	1	1	1	1	1	-1	-1	1	-1	-1	1	1	-1	-1	1	-1	-1	1	-1	1
BS2	PN2	1	-1	1	1	-1	1	1	1	-1	1	-1	1	1	1	-1	1	1	-1	-1	1	-1	1	-1	-1	1	-1	1	-1	-1	1	-1	1
BS3	PN3	1	-1	1	-1	-1	1	1	1	1	1	-1	1	1	-1	-1	-1	-1	-1	1	-1	1	1	-1	1	1	-1	1	-1	1	1	-1	1
BS4	PN4	-1	-1	1	-1	1	-1	1	1	-1	-1	1	-1	1	-1	-1	-1	-1	-1	1	1	1	1	1	1	1	1	1	1	1	1	1	-1
BS5	PN5	1	1	-1	1	-1	-1	-1	1	-1	-1	1	1	1	1	1	1	-1	-1	-1	1	1	1	1	1	-1	-1	1	-1	1	1	1	1
BS6	PN6	-1	-1	1	1	-1	1	1	1	1	1	1	-1	1	1	-1	1	1	1	1	1	1	-1	1	1	-1	1	-1	1	-1	-1	-1	1
BS7	PN7	-1	1	1	1	-1	1	-1	1	-1	1	-1	1	1	1	1	-1	1	-1	1	-1	-1	1	-1	-1	1	-1	-1	1	-1	1	-1	1
BS8	PN8	1	1	-1	-1	1	1	-1	1	-1	1	-1	1	1	-1	1	-1	1	-1	-1	1	-1	1	-1	1	1	-1	-1	1	-1	1	-1	-1

Table 9.24 Transmitted and received signals

TX signal	2	0	4	2	0	-2	-2	0	2	2	-2	-2	0	2	4	-4	-4	2	4	2	0	6	2	0	2	-2	4	-4	0	-2	2	2	2	-2	0	-2	0	2	0	4	15.0
power control	2.9	0.0	5.8	2.9	0.0	-2.9	-2.9	0.0	2.9	2.9	-2.9	-2.9	0.0	2.9	5.8	-5.8	-5.8	2.9	5.8	2.9	0.0	8.6	2.9	0.0	2.9	-2.9	5.8	-5.8	0.0	-2.9	2.9	2.9	2.9	-2.9	0.0	-2.9	0.0	2.9	0.0	5.8	21.5
multipath	0.0	1.0	0.5	0.3	1.0	-0.5	-0.5	1.0	-0.5	-1.0	0.5	0.5	-0.5	1.0	0.5	-1.0	1.0	-0.5	0.5	0.0	-0.5	0.5	0.0	0.5	-0.5	1.5	-1.0	0.0	1.0	0.5	-0.5	0.0	0.5	0.0	-0.5	1.0	-0.5	0.5	1.0	0.0	3.7
multipath	0.5	0.3	-0.5	0.0	-0.3	0.3	0.3	0.0	-0.3	0.3	0.8	0.0	-0.3	0.3	0.0	-0.3	0.0	0.3	0.0	-0.3	0.3	0.0	0.3	0.5	0.5	0.8	0.0	0.3	0.3	-0.3	0.0	0.0	0.3	-0.3	-0.3	0.0	0.3	0.5	0.0	0.5	1.9
multipath	0.3	0.1	-0.3	0.3	0.3	-0.1	0.0	0.4	-0.1	-0.1	0.0	0.1	0.3	0.1	0.1	0.1	-0.1	0.4	-0.3	-0.1	0.1	0.1	0.3	0.0	0.0	0.1	0.0	0.3	0.0	0.1	0.1	0.1	0.3	0.0	-0.1	0.0	0.1	0.1	0.3	0.3	0.9
noise	-0.2	0.6	-0.7	-0.1	0.5	0.3	-1.0	-0.4	0.6	0.3	-0.4	0.2	0.1	-0.6	0.3	-0.1	0.3	0.1	0.1	0.1	-0.4	0.5	0.2	0.5	0.1	0.0	0.6	0.3	1.0	0.6	-0.3	-0.2	0.0	0.0	0.9	0.3	0.0	0.2	0.0	0.0	2.7
total interference	0.5	2.0	-0.9	0.2	2.0	-0.3	-1.6	1.0	0.0	-0.1	-0.3	0.4	0.8	-0.8	1.1	-0.2	0.5	0.0	0.1	0.0	1.0	0.8	0.4	0.3	-0.2	0.4	-1.0	0.0	0.6	0.3	-0.3	0.5	0.2	0.4	-0.3	1.0	-0.2	0.7	1.5	0.3	4.3
RX signal	3.4	2.0	4.8	-5.6	5.4	-4.6	5.6	-2.6	-1.9	9.1	3.7	-0.3	3.3	-2.5	4.2	-5.8	-2.9	3.2	3.1	0.6	-0.3	2.5	0.7	4.4	1.3	6.0															21.1

Table 9.25 Received PN codes

BS1	PN1	3.4	2.0	4.8	1.9	5.6	5.4	4.6	5.6	2.6	0.4	1.9	9.1	-3.7	0.3	3.3	-2.5	4.2	5.8	0.8	2.9	2.8	-3.2	-2.4	3.1	-0.6	0.3	-3.1	2.5	0.7	4.4	-1.3	6.0
BS2	PN2	3.4	2.0	4.8	1.9	5.6	5.4	4.6	5.6	-2.6	0.4	1.9	-9.1	3.7	0.3	-3.3	-2.5	4.2	-5.8	0.8	2.9	2.8	-3.2	2.4	3.1	-0.6	0.6	3.1	-2.5	-0.7	4.4	1.3	-6.0
BS3	PN3	-3.4	2.0	4.8	1.9	5.6	5.4	4.6	5.6	2.6	-0.4	1.9	-9.1	-3.7	-0.3	-3.3	2.5	4.2	-5.8	-0.8	2.9	2.8	-3.2	2.4	-3.1	0.6	0.6	3.1	2.5	0.7	-4.4	1.3	6.0
BS4	PN4	3.4	-2.0	4.8	1.9	-5.6	5.4	4.6	-5.6	-2.6	0.4	-1.9	-9.1	-3.7	0.3	3.3	2.5	-4.2	-5.8	-0.8	-2.9	2.8	-3.2	2.4	-3.1	-0.6	-0.3	-3.1	2.5	-0.7	4.4	-1.3	6.0
BS5	PN5	3.4	2.0	-4.8	1.9	-5.6	5.4	4.6	5.6	2.6	-0.4	1.9	9.1	-3.7	0.3	-3.3	-2.5	-4.2	5.8	-0.8	-2.9	-2.8	-3.2	2.4	-3.1	-0.6	0.3	-3.1	-2.5	-0.7	4.4	1.3	-6.0
BS6	PN6	-3.4	2.0	4.8	-1.9	5.6	5.4	4.6	5.6	-2.6	0.4	-1.9	-9.1	3.7	0.3	-3.3	2.5	4.2	5.8	0.8	-2.9	-2.8	3.2	2.4	-3.1	0.6	-0.6	3.1	-2.5	0.7	-4.4	-1.3	6.0
BS7	PN7	3.4	-2.0	4.8	1.9	5.6	-5.4	4.6	5.6	-2.6	-0.4	1.9	9.1	-3.7	-0.3	-3.3	2.5	-4.2	-5.8	0.8	2.9	-2.8	3.2	-2.4	-3.1	0.6	0.3	-3.1	2.5	0.7	4.4	-1.3	-6.0
BS8	PN8	-3.4	2.0	-4.8	1.9	-5.6	-5.4	-4.6	-5.6	-2.6	-0.4	-1.9	-9.1	3.7	0.3	3.3	2.5	-4.2	5.8	-0.8	2.9	2.8	-3.2	-2.4	3.1	0.3	-0.3	3.1	-2.5	-0.7	4.4	1.3	-6.0

Table 9.26 Received Walsh codes

																																	sum
channel 1	W0	3.4	2.0	4.8	1.9	5.6	5.4	4.6	5.6	2.6	0.4	1.9	9.1	-3.7	0.3	3.3	-2.5	4.2	5.8	0.8	2.9	2.8	-3.2	-2.4	3.1	-0.6	0.3	-3.1	2.5	0.7	4.4	-1.3	6.0	68
channel 2	W1	3.4	-2.0	4.8	-1.9	5.6	-5.4	4.6	-5.6	2.6	-0.4	1.9	-9.1	-3.7	-0.3	3.3	2.5	4.2	-5.8	0.8	-2.9	2.8	3.2	-2.4	-3.1	-0.6	-0.3	-3.1	-2.5	0.7	-4.4	-1.3	-6.0	37
channel 3	W2	-3.4	2.0	4.8	-1.9	-5.6	5.4	4.6	-5.6	-2.6	0.4	1.9	-9.1	3.7	-0.3	3.3	-2.5	4.2	5.8	-0.8	2.9	-2.8	-3.2	-2.4	3.1	0.6	-0.3	-3.1	2.5	-0.7	-4.4	-1.3	-6.0	-32
channel 4	W3	-3.4	-2.0	4.8	1.9	-5.6	-5.4	4.6	5.6	-2.6	-0.4	1.9	9.1	3.7	0.3	3.3	2.5	4.2	-5.8	-0.8	-2.9	-2.8	3.2	-2.4	-3.1	0.6	0.3	-3.1	-2.5	-0.7	4.4	-1.3	-6.0	-17
channel 5	W4	3.4	2.0	-4.8	1.9	5.6	5.4	-4.6	5.6	2.6	0.4	-1.9	9.1	-3.7	0.3	-3.3	-2.5	-4.2	5.8	0.8	2.9	-2.8	-3.2	2.4	3.1	-0.6	0.3	3.1	2.5	0.7	4.4	1.3	6.0	33
channel 6	W5	3.4	-2.0	-4.8	1.9	5.6	-5.4	-4.6	5.6	2.6	-0.4	-1.9	-9.1	-3.7	-0.3	-3.3	2.5	-4.2	-5.8	0.8	-2.9	-2.8	3.2	2.4	-3.1	-0.6	-0.3	3.1	-2.5	0.7	-4.4	1.3	-6.0	39
channel 7	W6	-3.4	2.0	-4.8	1.9	-5.6	5.4	-4.6	5.6	-2.6	0.4	-1.9	-9.1	3.7	-0.3	-3.3	-2.5	-4.2	5.8	-0.8	2.9	-2.8	-3.2	2.4	3.1	0.6	-0.3	3.1	2.5	-0.7	-4.4	1.3	-6.0	-64
channel 8	W7	-3.4	-2.0	-4.8	1.9	-5.6	-5.4	-4.6	5.6	-2.6	-0.4	-1.9	9.1	3.7	0.3	-3.3	2.5	-4.2	-5.8	-0.8	-2.9	-2.8	3.2	2.4	-3.1	0.6	0.3	3.1	-2.5	-0.7	4.4	1.3	6.0	-21

Table 9.27 Received information

Channel 1	1	0
Channel 2	1	0
Channel 3	−1	1
Channel 4	−1	1
Channel 5	1	0
Channel 6	1	0
Channel 7	−1	1
Channel 8	−1	1

All chips of a bit are summed to integrate the energy over the whole code and a threshold is established based on the correlation statistics. A value above the threshold represents a $+1$ (zero) and a negative a -1 (one) information symbol and the symbols are then mapped to binary in Table 9.27.

Digital systems replace the traditional SNR measure for rating performance with the normalised bit energy per noise spectral power density ratio (E_b/N_0). Section 9.1.1.3 shows that considering not only noise but also interference can enhance the SNR analysis. The same concept applies to the E_b/N_0 ratio, which can be enhanced to consider interference. Equation (9.5) from Section 9.1.1.3 can be adapted to express E_b/N_0 in terms of the signal-to-noise ratio

$$\frac{E_b}{N_0} = \left(\frac{S}{N}\right)\left(\frac{B}{R_b}\right) \tag{9.8}$$

Using an analogy, noise can be replaced by noise plus interference power spectral density (I_0) in the E_b/N_0 ratio, becoming an E_b/I_0 ratio, which is given by eqn (9.9).

$$\frac{E_b}{I_0} = \left(\frac{S}{N+I}\right)\left(\frac{B}{R_b}\right) \tag{9.9}$$

where

I_0 noise-plus-Interference power spectral density (W/Hz)

I interference power uniformly distributed over the entire spectrum (W)

Using 13 dB E_b/I_0 the output result (received information) is consistent with the input (transmitted information) even when the bits are changed, channels are assigned to different Walsh codes and PN code assignments are changed. It is possible in the spreadsheet to model channels of the same base station or channels of different base stations, by manipulating the codes.

Using 7 dB E_b/N_0 the output result (received information) shows sporadic errors due to the non-complete orthogonality of the PN codes, the multi-path interference and the noise spikes. These errors most probably can be corrected by the Viterbi decoder using the redundancy added by the convolutional encoder. This is a typical value used for voice detection in real systems.

It can be seen in the example above that the apparent $S/(N+I)$ is -8 dB while E_b/I_0 is 7 dB. At this level we will find occasional errors in the received signal due to the imperfect orthogonality of the PN codes, the multi-path interference and the noise spikes.

The errors stop for a $S/(N+I)$ of -2 dB equivalent to an E_b/I_0 of 13 dB. When using this E_b/I_0 value, the output result is consistent with the input even when bits are changed, channels are assigned to different Walsh codes and PN code assignments are changed.

The detection threshold is very complex to calculate and is generally obtained through statistical simulations in different scenarios. For voice transmission at 9600 bps, the error rate is between 10^{-1} and 10^{-3} for an E_b/I_0 of 7 dB, and between 10^{-6} and 10^{-8} for an E_b/I_0 of 13 dB.

We know that a certain code was transmitted and we look for its pattern in the received signal. The received chips are combined with the known pattern and the results are summed for the entire bit of information.

The traditional way of analysing $S/(N+I)$ (power ratio) does not apply here as we will be comparing the signal information over the detection period against the noise of each chip period. We will use energy then instead of power. The total signal energy over the period is expressed by E_b (bit energy) and is the sum of the energies of all the desired code chips. The total noise and interference is considered over the period of a chip and is expressed as I_0.

Because the energy of the signal is accumulated over the detection period, the code can be understood as providing a detection gain. The detection period is 1 symbol or 64 chips. It is initially assumed that there is no overhead for error correction. In this case, the detection or processing gain is shown in Table 9.28. These processing gain only applies if there is interference present. There is no advantage if just noise is present.

This means that each rate requires a different E_b/I_0 value. Because of the possibility of trading off quality with bandwidth, when applying error correction codes, the required value of E_b/I_0 can be reduced by trading it off with the bandwidth.

Sometimes the information has to be repeated twice because of framing requirements doubling the apparent data rate. The integration over the code chips, however, can be done

Table 9.28 Processing gain

Processing gain	Bits	Data rate (bps)
30.1	1024	1200
27.1	512	2400
24.1	256	4800
21.1	128	9600
18.1	64	19200
15.1	32	38400
12.0	16	76800
9.0	8	153600
6.0	4	307200
3.0	2	614400

ovei the two periods maintaining the overall processing gain. The higher the processing gain, the lower the energy that can be used to transmit the information.

The simple and rudimentary example illustrates well the trade-offs that had to be taken when designing the technology. A 6 dB gain in E_b/N_0 may compensate a 3 dB loss in capacity for error correction. This becomes more apparent in cdma2000 systems, which can use different rates. In these systems the trade-offs vary resulting in different settings for each rate.

The network designer has to understand the trade-offs and know how to mitigate them in the design. Some network gains can only be applied to the limit of the existing impairment. As an example, a diversity scheme may improve fading, but this improvement is limited by the amount of fading remaining after the action of the rake fingers, so a 3 dB fading gain may be restricted to 1 dB.

9.3.3   Modulator and Coder

9.3.3.1   Phase-Shift Keying (PSK) Modulation

As illustrated in Figure 9.1 after the source information is coded and multiplexed, it has to be modulated on the carrier to be transmitted. Several modulation schemes are available and each technology has chosen one or multiple schemes for transmitting the information.

Sets of bits are combined into symbols and assigned to carrier states (phase × energy) forming a constellation. Constellations can be represented in polar form, showing the phase and magnitude in the same diagram. The distance between the constellation points represents how much noise the modulation can accommodate.

A rectangular representation of the polar diagram (Figure 9.60) is generally used with axis 'I' lying on the 0° phase and the 'Q' axis on the 90° phase. The I/Q diagrams are very useful as they mirror how the signals are created in the real world. In Figure 9.61 the carrier

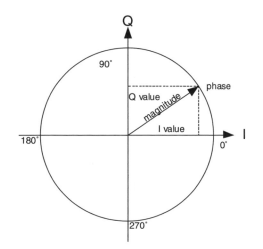

Figure 9.60   Polar and rectangular constellation diagram.

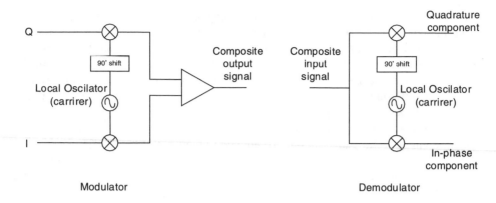

Figure 9.61 I and Q modulator/demodulator.

modulates directly the 'I' branch and is shifted by 90° before modulating the 'Q' branch. The output of both branches is then summed. The reverse is applied in the de-modulator.

A phase-shift keying (PSK) modulator assigns different phases to symbols. Figure 9.62 shows constellations for MPSK schemes in which M represents the number of phases allowed and is here represented for 2, 4 or 8 phases. For $M = 2$, symbol 1 is associated with phase 0° and symbol 0 with $-180°$. For $M = 4$ the phases are 0°, 90°, 180° and 270°, but other constructions are possible, like the one shown in the same figure that has phases 45°, 135°, 225° and 315° and is considered having an offset of 45°. For $M = 4$ two bits are assigned to each symbol. Table 9.29 maps the number of bits per symbol for the main PSK schemes.

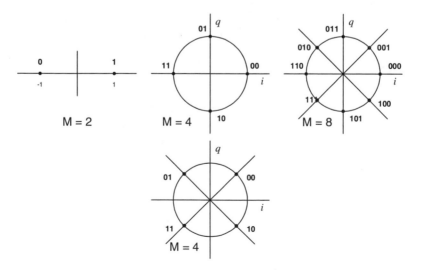

Figure 9.62 PSK modulation.

Table 9.29 MPSK

		PSK				
States	M	2	4	8	16	32
Bits/Symbol	bit	1	2	3	4	5

CDMA systems employ QPSK, OQPSK, 8QAM and 16 QAM, which are illustrated in Figure 9.63. QPSK means Quadrature (90° separation) PSK and is equivalent to the 4PSK schemes described above. OQPSK means Offset Quadrature phase-shift keying and in this modulation the I and Q bit streams are offset in their relative alignment by 1 bit (half a symbol) and this causes the carrier amplitude not to cross zero. 8QAM means 8-state quadrature amplitude modulation and corresponds to 3 I and 3 Q values with 3 bits per symbol. 16QAM means 16-state Quadrature Amplitude Modulation and corresponds to 4 I and 4 Q values with 4 bits per symbol. More detailed descriptions about modulation can be found in the bibliography.

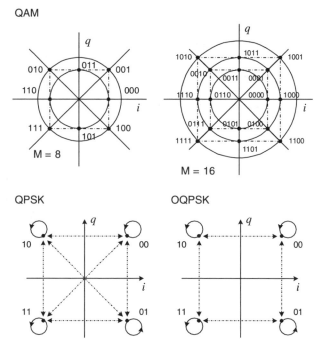

Figure 9.63 CDMA modulations.

The performance of each modulation type is briefly analysed next so the trade-offs between them can be understood. Figures 9.64 and 9.65 show, respectively, the Bit Error Rate (BER) and E_b/N_0 in the presence of white noise against S/N for the several MPSK modulation schemes.

BER x S/N for MPSK

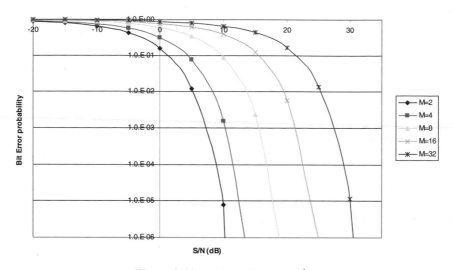

Figure 9.64   BER $\times$ S/N for MPSK.

BER x E_b/N_0 for MPSK

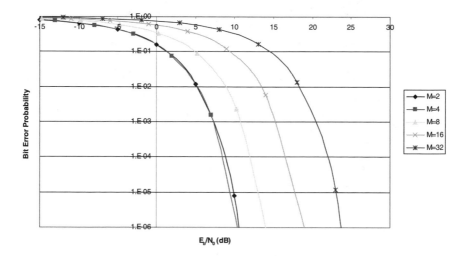

Figure 9.65   BER $\times$ E_b/N_0 for MPSK.

The error detection on voice is done by the Frame Error Rate (FER) that is detected at the vocoder over a whole frame. A frame with error may have one or multiple erroneous bits. Figure 9.66 shows the relation between FER and E_b/N_0 for each of the modulation schemes presented before. The chart indicates that there is an exponential relationship between SNIR and BER. A small variation of SNIR over a specific range drastically changes the FER.

FER x Eb/N₀ for MPSK

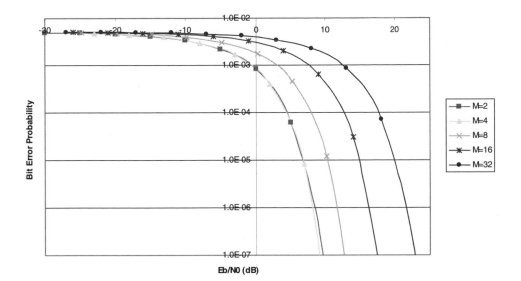

Figure 9.66 FER $\times E_b / N_0$ for MPSK.

The capacity of a channel varies depending on the modulation scheme selected. Figure 9.67 shows the capacity that can be achieved with Multiple Phase-Shift Keying (MPSK) for a BER of 10^{-5} in an AWGN channel. The chart also shows the curve for Shannon's channel capacity limit. QPSK ($M = 4$) is the modulation that comes closer to the

Shannon limit x MPSK Modulation in AWGN

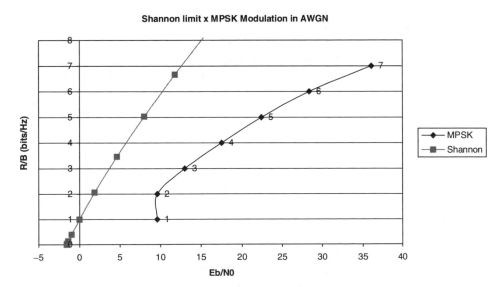

Figure 9.67 MPSK modulation and Shannon capacity limit in AWGN environment.

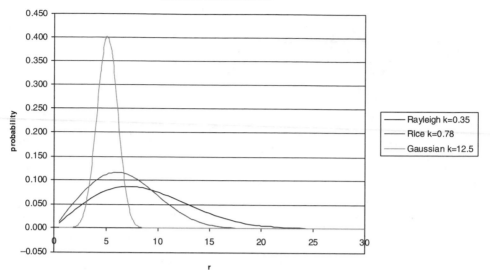

Figure 9.68   Signal, noise and interference distributions.

Shannon limit curve when we consider, just the modem. For other types of noise distribution, like the ones resulted from multi-path the modem performance becomes less efficient.

In real life systems the SNIR varies due to signal variations and non-uniform noise when interference is present. Signal and noise variations can be represented by different k parameters of the Ricean distribution as shown in Figure 9.68. The k parameter will be covered in Section 9.4.2.6.1. The three curves have the same area and the Gaussian distribution presents the smallest variation from the average. The larger the deviation the larger is the probability of an error. Figure 9.69 shows the BER for a Rayleigh distribution compared with the Gaussian distribution where it can be seen that a significant better E_b/N_0 is required for the same BER performance and the curve is nearly linear (in dB).

The Gaussian distribution typically occurs when there is Non-Line-Of Sight (NLOS) between transmitter and receiver while Rayleigh distribution occurs when there is a predominant line-of-sight (LOS) component between them. The Ricean distribution occurs in between the extreme situations.

According to the chart in Figure 9.69, whereas a 1 dB SNIR improvement is required to reduce BER by a factor of 10 in a Gaussian noise environment, an improvement of 10 dB SNIR is required to reduce the BER by the same amount in a Rayleigh environment.

The elimination of errors requires large SNIR values that would be impractical to be achieved in the network design. It is possible though to reduce the error rate to small values with a moderate SNIR even for a Rayleigh environment. The remaining errors can then be removed using error-correction codes. Error-correction codes allow the operation at a lower SNIR but require an increase in the data rate. The trade-off between the increased data rate and reduced SNIR is approached in different ways depending on the technology. The next section discusses error-correction codes used in CDMA.

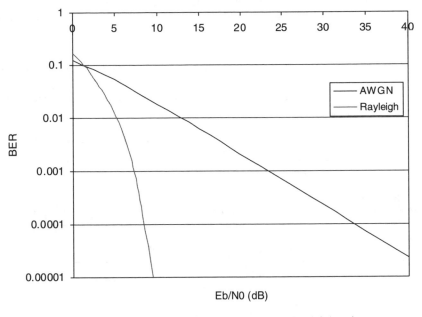

Figure 9.69 BER for QPSK with AWGN and Rayleigh noise.

9.3.3.2 Coding

The large dips of SNIR in a signal that follows a Rayleigh distribution curve cause errors that can be corrected using error-correction codes. Many codes have been proposed with this purpose. CDMA uses codes that belong to a class of forward error-correction codes (FECs) known as convolutional.

9.3.3.2.1 Convolutional Coding and Viterbi Decoding
Convolutional codes are a group of error-correction codes that continuously encode the information data and generate n symbols for every k input bits according to a certain pattern. The code rate (R) is defined by eqn (9.10) and corresponds to a (k, n) code. The encoding process is also related to a constraint length K factor that expresses how many (m) sets of k input bits are evaluated to generate the output symbols

$$R = k/n \tag{9.10}$$

$$K = k(m - 1) \tag{9.11}$$

where

R code rate
k number of bits input per encoder (clock) cycle into the convolutional encoder
n number of output symbols generated for each set of k input bits

K constraint length

m number of encoder cycles an input bit is retained and used for encoding after it first appears at the input (= number of shift register stages +2), it is called also encoder memory length

Figure 9.70 shows the convolutional encoders employed on the CDMA IS-95 forward and reverse links, with rates $R = 1/2$ (top) and $R = 1/3$ (bottom). cdma2000 systems use these convolutional encoders; however they also have additional types of encoders, which are described in Chapters 4 and 5.

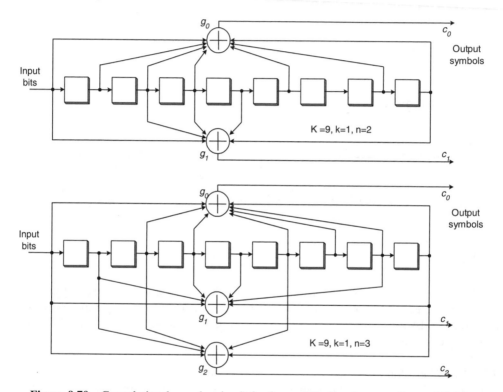

Figure 9.70   Convolutional encoder circuit for forward (top) and reverse (bottom) links.

In IS-95, each stage of the encoder stores only one bit ($k = 1$) and uses eight stages ($K = 8$), resulting in a code rate $R = k/K$. For each input bit two (forward link) or three (reverse link) output symbols are generated. The output bits are labelled $c_0, c_1, \ldots, c_n$.

The convolutional encoder is a finite state machine that can be represented by a state diagram. Received bits can be mapped to a Trellis diagram or a tree diagram that depicts the state changes for each input bit. These diagrams can be extended to all possible combinations of the received bits for a certain frame duration (word).

The word that best describes the received word (closest match) has the maximum likelihood of being the original sequence. The number of combinations increases very quickly with the number of bits in the word. Viterbi developed an algorithm that minimises the number of combinations to be analysed. The distance between the received word and each path in the Trellis diagram can be done using Euclidean distance (soft decoding) or Hamming distance (hard decoding). Euclidean distance is the energy difference between two codes while the Hamming distance represents the minimum number of bits that has to be changed to convert one-bit string into the other. Hamming distance is faster to calculate but the Euclidean distance provides a more accurate result.

Convolutionally Correcting Errors
The following example illustrates the operation of a convolutional coder and the Viterbi based decoder.

A two stage shift register in Figure 9.71 implements a convolution encoder with a code rate of $R = 1/2$ and a constraint length of $K = 3$. We will encode a word of length 11: '1 0 0 1 1 1 0 1 1 0 0'.

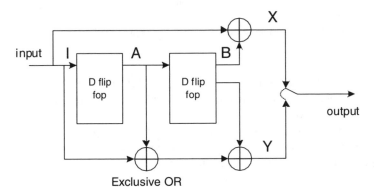

Figure 9.71 Convolutional encoder $R = 1/2, K = 3$.

The shift register is initialised always with all zeroes and the word is applied to the input. It is shifted at every clock cycle by the D flip flops. The outputs 'X' and 'Y' are a combination of the input 'I' and the outputs 'A' and 'B' of the flip flops. The outputs 'X' and 'Y' are then multiplexed to generate the output symbols.

Table 9.30 shows the input, the previous state, the outputs and the next state of the circuit. This is expressed by a Trellis diagram in Figure 9.72. The shift-register states are represented

Table 9.30 Convolutional encoder state table

Input	Previous state		Output		Next state	
I	A	B	Y	X	A	B
0	0	0	0	0	0	0
0	0	1	1	1	0	0
0	1	0	1	0	0	1
0	1	1	0	1	0	1
1	0	0	1	1	1	0
1	0	1	0	0	1	0
1	1	0	0	1	1	1
1	1	1	1	0	1	1

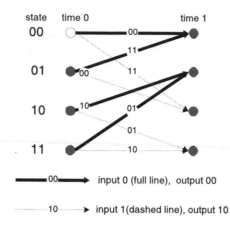

Figure 9.72　Trellis diagram for all states of a convolutional decoder.

by the dots and input 1 is represented by a dotted line while a full line represents input 0. The value on the line represents the output value. Figure 9.72 represents all possible transitions from one state to the next.

Figure 9.73 displays the transmission of the desired word applying a Trellis diagram. The output symbols are listed in it. The strength of the convolutional encoder resides on the fact that not all paths are possible and that the encoding of one-bit depends on the previous ones.

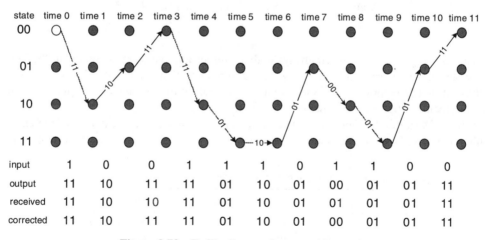

	time 0	time 1	time 2	time 3	time 4	time 5	time 6	time 7	time 8	time 9	time 10	time 11
input	1	0	0	1	1	1	0	1	1	0	0	
output	11	10	11	11	01	10	01	00	01	01	11	
received	11	10	10	11	01	10	01	01	01	01	11	
corrected	11	10	11	11	01	10	01	00	01	01	11	

Figure 9.73　Trellis diagram for a specific word.

The received symbols can be received with errors but the receiver does not know the number of errors and their position. The only certainty at the receiver is that the initial state was all zeroes and that the last $K - 1$ bits are all zeroes and are purposely introduced by adding tail bits to the encoded sequence. Based on this it is possible to draw all possible combinations in the Trellis diagram and choose the one that has a minimum difference to the received symbols, using either the Hamming distance (hard decision) or the Euclidian distance (soft decision). For 11 bits the number of combinations is $4 \times 2^{11} = 8192$. This number grows quadratically with the number of bits. Viterbi proposed an algorithm that reduces the number of paths to be analysed by eliminating paths with large errors and a way to distinguish between paths with the same error by using a trace-back method. More details about the Viterbi algorithm can be found in the bibliography.

9.3.3.2.2 Turbo Coding

The error correction performance of convolutional coders can be improved by using turbo codes. They use two convolutional encoders so the paths can now be resolved from two different prospectives. Turbo codes come closer to the Shannon limit than simple convolutional codes. The price to pay, however, is delay and processing effort. Practical systems employ a compromise to reduce the delay and effort but still providing gain over a convolutional coder.

Turbo coders are advanced forward error correcting algorithms that mix the un-coded data stream with two independently encoded data streams as in Figure 9.74. The two encoded streams are weakly correlated because one of them passes through an interleaver. The larger the interleaver, the more independent the data streams, but larger the delay.

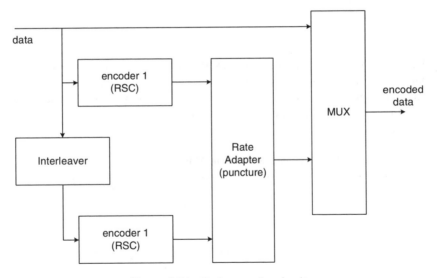

Figure 9.74 Turbo encoder circuit.

In the decoding process the two encoded streams are separately decoded using soft decoding and sharing the information obtained from each one over a number of iterations as illustrated in Figure 9.75.

Chapters 4 and 5 present a detailed description of the turbo encoder used in CDMA.

9.3.3.2.3 Coding Gain

Convolutional symbols are generated in CDMA for every 20 ms frame (word). Tail bits are added to flush the encoder for the next frame and provide the encoder with a known final

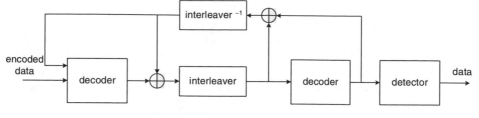

Figure 9.75 Turbo decoder circuit.

Table 9.31 Coding gain

| | | | | Coding gain (dB) BER | | | | | |
k	n	K	Decoding	1×10^{-3}	1×10^{-4}	1×10^{-5}	1×10^{-6}	1×10^{-7}	1×10^{-8}
1	2	9	Soft	4.7	5.5	6	6.5	6.8	6.9
1	3	9	Soft	5.7	6.2	6.7	7.2	7.5	7.6

state. The coding process reduces the SNIR requirements for a specific error rate performance and this may be considered as an equivalent signal gain or coding gain. Table 9.31 and Figure 9.76 show the coding gain for the convolutional coders used in the forward and reverse link in a white noise (AWGN) environment.

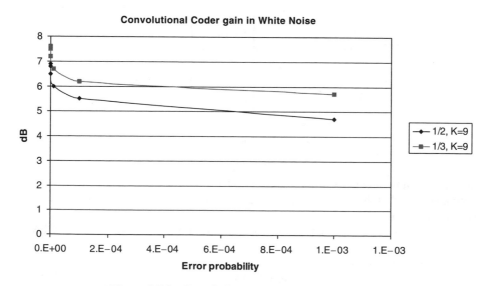

Figure 9.76 Convolutional coder gain in white noise.

The overall performance of the modulation and coding schemes over a certain type of communication channel is very complex to be fully modeled mathematically and the best way to obtain statistical results is through simulation.

In reality, the modulator, coder, interleaver and its counterparts on the receive side work together to retrieve the information. The best way to characterise them is to use a network simulator and generate the E_b/N_0 requirements for the different error rates, being it the BER (Bit Error Rate), BLER (Block Error Rate) or FER (Frame Error Rate). Typical E_b/N_0 values are calculated for different services and data rates.

9.3.3.3 Interleaving

Encoders can correct only a certain number of errors in each set but, if the errors appear in bursts, the encoder may even increase them. Therefore it is important to avoid blocks of contiguous errors. Interleavers are used to spread errors more uniformly, separating them and providing time for the encoder to correct them one at a time.

The interleaving mechanism consists in writing a block of bits (20 ms frame) in a sequence and reading them in a different sequence. This technique implies memory requirements and introduces a delay, as indicated in Table 9.32.

An interleaver is characterised by how much it can separate consecutive errors. The longer the error burst the less separation is provided. Tables 9.33 and 9.34 show the separation produced by forward and reverse link interleavers, respectively.

Table 9.32 Interleaving memory requirements

	Rows	Columns	Memory requirement (bit)	Interleaver delay (ms)
Sync channel	16	8	256	53.4
All direct traffic channels	24	16	768	40.0
Access channel	32	18	1152	40.0
Reverse traffic 9600	32	18	1152	40.0
Reverse traffic 4800	16	18	1152	40.0
Reverse traffic 2400	8	18	1152	40.0
Reverse traffic 1200	4	18	1152	40.0

Table 9.33 Error separation in the forward link

	Minimum error separation (forward link)							
	$B \geq 97$	$49 \leq B \leq 96$	$8 \leq B \leq 48$	$B = 7$	$B = 6$	$B = 5$	$3 \leq B \leq 4$	$B = 2$
Sync channel	1	1	1	1	1	1	2	4
Forward traffic channel	1	2	3	16	32	64	64	64

Table 9.34   Error separation in the reverse link

	Minimum error separation reverse link					
	$B \geq 145$	$73 \leq B \leq 144$	$20 \leq B \leq 72$	$B = 19$	$B = 18$	$B \leq 17$
Access	1	2	3	8	16	32
Reverse traffic 9600	1	1	1	1	31	31
Reverse traffic 4800	1	1	1	1	15	15
Reverse traffic 2400	1	1	1	1	7	7
Reverse traffic 1200	1	1	1	1	3	3

9.3.3.4   Adaptive Modulation and Coding Schemes

cdma2000 systems are able to accommodate different services because they offer several data rates. Some services require a minimum data rate whereas others are more flexible and can be transmitted whenever there is bandwidth availability. This feature requires a resource management (RM) entity to allocate the different services according to their requirements and to the network availability.

Service requirements usually define a minimum throughput data rate and a maximum delay. The resource management goal is to transmit all the information by maximising the channel use at every instant while respecting services' requirements.

Table 9.35   Forward link radio configurations

SR	RC	Forward link radio configurations Data rates (bps)				Convolutional	Turbo
SR1	RC1[a]	1200	2400	4800	9600	1/2	n
SR1	RC2	1800	3600	7200	14 400	1/2	n
SR1	RC3[a]	1500	2700	4800	9600	1/4	y
		19 200	38 400	76 800	153 600	1/4	y
SR1	RC4	1500	2700	4800	9600	1/2	y
		19 200	38 400	76 800	153 600	1/2	y
					307 200	1/2	y
SR1	RC5	1800	3600	7200	14 400	1/4	y
		28 800	57 600	115 200	230 400	1/4	y
SR2	RC6	1500	2700	4800	9600	1/6	y
		19 200	38 400	76 800	153 600	1/6	y
					307 200	1/6	y
SR2	RC7[a]	1500	2700	4800	9600	1/3	y
		19 200	38 400	76 800	153 600	1/3	y
				307 200	614 400	1/3	y
SR2	RC8	1800	3600	7200	14 400	1/4,1/3 (5 ms frames)	y
		28 800	57 600	115 200	230 400	1/4,1/3 (5 ms frames)	y
					460 800	1/4,1/3 (5 ms frames)	y
SR2	RC9	1800	3600	7200	14 400	1/2,1/3 (5ms frames)	y
		28 800	57 600	115 200	230 400	1/2,1/3 (5ms frames)	y
				460 800	1 036 800	1/2,1/3 (5ms frames)	y

[a]Mandatory.

Table 9.36 Reverse link radio configurations

SR	RC	Reverse link radio configurations Data rates (bps)							Convolutional	Turbo
SR1	RC1[a]				1200	2400	4800	9600	1/3	n
SR1	RC2				1800	3600	7200	14 400	1/2	n
SR1	RC3[a]	1200	1350	1500	2400	2700	4800	9600	1/4	y
					19 200	38 400	76 800	153 600	1/4	y
								307 200	1/2	y
SR1	RC4				1800	3600	7200	14 400	1/4	y
					28 800	57 600	115 200	230 400	1/4	y
SR2	RC5[a]	1200	1350	1500	2400	2700	4800	9600	1/4	y
					19 200	38 400	76 800	153 600	1/4	y
							307 200	614 400	1/3	y
SR2	RC6				1800	3600	7200	14 400	1/4	y
				28 800	57 600	115 200	230 400	460 800	1/4	y
								1 036 800	1/2	y

[a]Mandatory.

The different data rates offered in cdma2000 are referred to as radio configurations (RC). Tables 9.35 and 9.36 show, respectively, the radio configurations available for the forward and reverse links.

The resource manager analyses service requirements and network availability, choosing the best radio configuration to be used to transmit the data. To perform this analysis the manager must know the E_b/N_0 requirements of each RC.

These requirements are complex because they are a function of the overall information transfer process, and are influenced by spreading, code orthogonality loss, multi-path fading, interference, decoding efficiency, error correction efficacy and acceptable performance on the received side.

Because of the complexity of CDMA systems, these requirements can be better obtained from simulations. Even so there are hundreds of possible combinations requiring the derivation of statistically representative numbers. Minimum requirements of E_b/N_0 are given for some configurations in item 9.6.4.

9.3.4 Power Amplifier (PA) and Combiner

Transmit Signals (TRX) have to be combined and amplified before being transmitted. This task is done by the power amplifier and combiner stage. The circuit impedances can be represented by its Thevenin equivalent. The output impedance of each TRX (R_{source}) has to be matched to the transmission line and to the antenna impedance (R_{load}) to maximise the power transfer as shown in Figure 9.77.

9.3.4.1 PAs Configuration and Characteristics

The combination of signals coming from different TRX requires impedances to be matched. Several combining techniques can be used to achieve this and the most common techniques are described next.

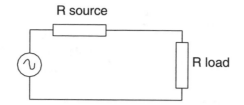

Figure 9.77   Power amplifier impedance matching.

One possible technique is the hybrid combiner, shown in Figure 9.78. This combiner is based on circulators[4] and a bridge circuit (hybrid) that provides good isolation between the inputs, and with a theoretical loss of 3 dB (close to 3.5 dB in practical circuits) for every two inputs that are combined. The overall loss is given by

$$L = l_h \times \text{roundup} \ (\log i / \log 2) \tag{9.12}$$

where

L overall loss in hybrid combining systems
l_h hybrid loss
i number of inputs

The hybrid circuit is detailed in Figure 9.79, signals at port A and B are sent to port C, but are isolated from each other.

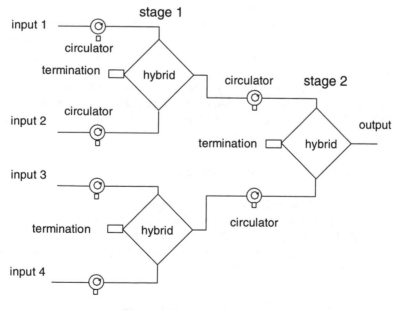

Figure 9.78   Hybrid combiner.

---

[4] An RF circulator is a three-port, passive RF device made of magnets and ferrite material, which is used to control the direction of signal flow in an RF circuit.

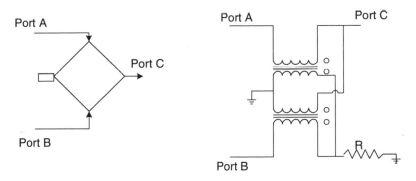

Figure 9.79 Hybrid circuit.

Another possible technique is the filter combiner, shown in Figure 9.80. This combiner uses the impedance variations between the pass band and the rejection band of filters to isolate between stages. A minimum separation is required between carriers for the filters to isolate between themselves. This separation limits the frequency plan. The filter and impedance mismatch causes a loss that typically varies between 0.5 and 1 dB.

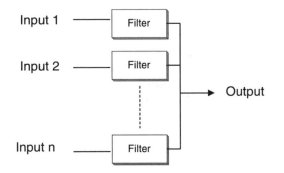

Figure 9.80 Filter combiner.

After signals are combined, they have to be amplified. Two output configurations are possible for power amplifiers: the Single Carrier Power Amplifier (SCPA) and the Multi-Carrier Power Amplifier (MCPA), respectively, shown in Figures 9.81 and 9.82.

In the SCPA, each carrier is connected directly to the Power Amplifier (PA). The PA output is then combined in a high-power combiner. The combiner is generally an air or a ceramic filter that presents low impedance for the power amplified frequencies of each carrier and larger impedance for out-of-band frequencies. When small separations are required a hybrid combiner is used and in this case due to the large loss of this circuit an antenna is used in general for every two TRX.

In the MCPA, all carriers are combined in a low-power combiner and then send to the PA. This configuration requires a highly linear amplifier because the different signals can inter-modulate between themselves.

SCPAs are simpler to implement because inter-modulation caused by non-linearity is less critical than in MCPAs.

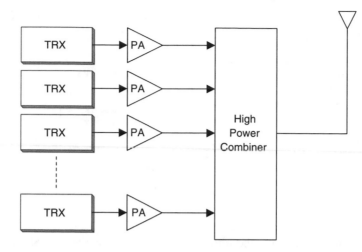

Figure 9.81   Single carrier power amplifier.

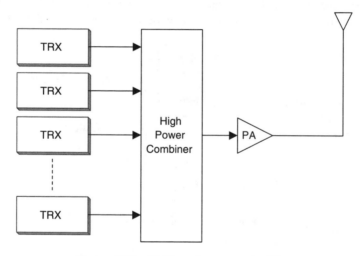

Figure 9.82   Multi-carrier power amplifier.

Linearity is a very important characteristic of an amplifier. The lack of linearity results in inter-modulation distortion at the output. Every pair of frequencies (f_1, f_2) present at the input results in a set of frequencies at the output following the combinations below

$$F_s = \pm M f_1 \pm N f_2, \qquad (9.13)$$

where, F_s are the inter-modulation products at the output and M and N are integers larger than 1 and the order of the product is the sum of $M + N$.

The inter-modulation product power depends on the amount of non-linearity present and the third-order component $(M + N = 3)$ is generally the most significant.

Table 9.37 Peak-to-average power ratio for different modulations

Technology	Modulation	Peak-to-average ratio (dB)
Analog	FM	0
GSM	GMSK	1.5
TDMA	$\pi/4$ DQPSK	3.5
CDMA	QPSK/DSSS	13
W-CDMA	QPSK/DSSS	9

The amplifier linearity is measured by the inter-modulation distortion present at the output when compared to the carrier power. This distortion is expressed in terms of dBc (dB in relation to carrier power).

SCPAs have less stringent requirements as their bandwidth is smaller. For these type of PAs, a distortion of −65 dBc is acceptable. MCPAs are required to have an inter-modulation distortion of approximately −70 dBc.

Advanced amplifiers use adaptive linear technology to minimise inter-modulation. This technique pre-distorts the input and compensates the output based on the deviation from the input.

PAs not only have different linearity characteristics but also differ in terms of power. Different modulation techniques have different power average to power peak ratios. As depicted in Table 9.37, CDMA has the highest ratio when compared to other, traditional technologies. This is because the CDMA forward signal is in reality a multi-carrier signal, each Walsh code being functionally equivalent to a carrier. The peak handling capabilities of a CDMA power amplifier have to be 20 times (13 dB) its average power to accommodate the peak-to-average ratio.

The power-handling capability of a power amplifier is defined by its 1 dB compression point (P1dB) and its third-order inter-modulation distortion defined by the Output Intercept

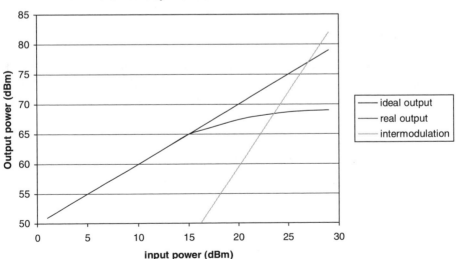

Figure 9.83 Power amplifier characteristics.

Table 9.38 Power amplifier characteristics

	SCPA	MCPA	Unit
Output power	50	200	W
P1dB	50	50	dBm
RF gain	50	50	dB
IP3	59	60	dBm
Efficiency	40	35	%
Output impedance	50	50	Ω
Return loss	12	12	dB
Inter-modulation distortion	−65	−70	dBc[a]
Out-of-band spurious	−60	−65	dBc

[a] dB in relation to the carrier power.

Point (IP3) or the Input Intercept Point (IIP3) as shown in Figure 9.83. The 1 dB compression point is where the output power deviates 1 dB from its ideal curve. The IP3 point is a theoretical point in which the inter-modulation power is equal to half of the output power. This point is calculated by extrapolating the PA curves for lower powers. For a class A amplifier a rule of the thumb is to consider the IP3 as 10 dB above the 1 dB compression point.

Figure 9.83 shows that the amplifier output is not linear. This non-linearity increases with the power level. To operate in a linear region designers 'back-off' the operating point from the 1 dB compression point by the peak-to-average ratio and by a margin in dB (3–6 dB), making the amplifiers bulkier and more expensive.

Table 9.38 summarises the typical characteristics of power amplifiers, which use Microwave Monolithic Integrated Circuit (MMIC) technology and are usually based on GaAs semiconductor devices.

CDMA mobile power amplifiers are limited to a peak transmitted power of 28 dBm (650 mW) with and average power 3 dB lower than the peak value (325 mW).

9.3.4.2 PA Mounts

Power amplifiers can be optionally mounted on top of a mast, in the Mast Head Unit (MHU), together with the receive side Low Noise Amplifier (LNA) constituting the Radio Frequency Front End (RFFE), shown in Figure 9.84. This solution minimises the transmission and

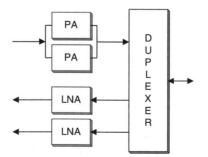

Figure 9.84 Radio frequency front end.

reception losses and requires smaller cables but it can only be used for small units due to size and weight issues.

Coverage issues may require a different kind of assembly that provides a simulcast solution in which several antennas transmit the same signal. This is used, as an example, in low-traffic areas along a road. The same signal is distributed to several antennas through optical cables at a digital IF level and fed to remote RFFEs. This arrangement constitutes a Remote Radio Unit (RRU), shown in Figure 9.85.

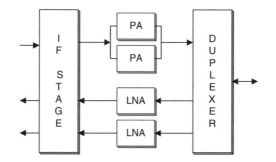

Figure 9.85 Remote radio unit.

9.3.5 Transmission Line

The connection between the Power Amplifier (PA) and the antenna is made using a transmission line (or RF feeder). Even though this feeder is made of passive components it should be analysed very carefully. This transmission line should match the impedances of the source (PA) and the load (antenna). Reflections can occur because of impedance mismatches and can add destructively depending on the length of the transmission line.

The transmission line should have a length that is an entire multiple of the wavelength to avoid reflections. The propagation speed in this line may vary significantly from the speed of light and the wavelength should be adjusted accordingly (Figure 9.86).

The transmission line efficiency is very important and depends mainly on impedance matching. A Voltage Standing Wave Ratio (VSWR) characterises this matching. This ratio is defined as the ratio of the highest impedance to the lowest impedance, i.e. it is a measure of the impedance mismatch between the transmission line and its load. The higher the VSWR value is, the greater the mismatch is. The best way to adjust the mismatch is to minimise the VSWR through adjustments and measurements.

The formulas below are commonly used to express the mismatch in a transmission system

$$\text{VSWR} = Z_h/Z_i \tag{9.14}$$

The reflected power in percentage is expressed by

$$P_R = (\text{VSWR} - 1)^2/(\text{VSWR} + 1)^2 \tag{9.15}$$

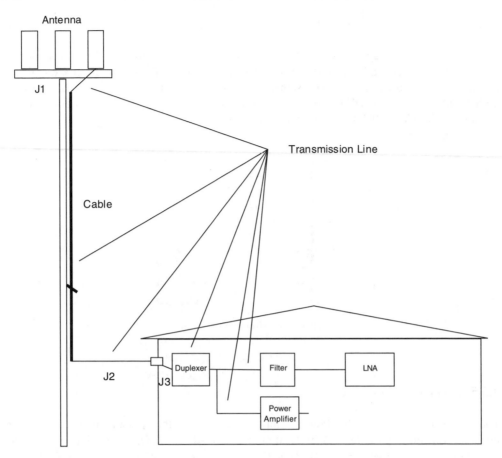

Figure 9.86 Transmission line elements.

It is generally expressed as return loss and represents the part of the energy that is reflected back in dB,

$$R_L = 20 \, \log_{10}((\text{VSWR} - 1)/(\text{VSWR} + 1)) \qquad (9.16)$$

The transmission loss is calculated in dB by

$$T_L = -10 \, \log_{10}(1 - P_R) \qquad (9.17)$$

Z_h highest impedance
Z_i lowest impedance
P_R reflected power
R_L return loss
T_L transmission loss.

Voltage Standing Wave Ratio (VSWR) can be understood as a measure of the impedance mismatch between the transmission line and its load. The higher is the VSWR value, the greater the mismatch. The minimum VSWR will always be unity, i.e. which corresponds to a perfect impedance match.

The combination of the original wave travelling the transmission line or cable (towards/from the antenna) and the reflecting wave is called a standing wave. The ratio of these two waves is known as the standing wave ratio.

To completely understand the previous definition we need to define what impedance is. The impedance corresponds to the ratio of voltage to current in the measured equipment. Each equipment, be it the antenna or the transmitter (source) has an intrinsic impedance (Z). The usual value for impedance, in cables and antennas, for example, is 50 Ω, however other values, such as 75 and 120 Ω, can also be found in the industry. Almost all radio equipments are built for an impedance of 50 Ω.

If any of the parts in the circuit presents impedance different from the nominal value, due to, for example, bad connections or incorrect antenna length, the transmission power radiated from the antenna is not the maximum power. Instead, part of the wave is reflected back down the transmission line, possibly causing damage to the transmitter equipment. The amount of the wave reflected back depends on how bad the mismatch is. Therefore the impedance for the entire transmission elements chain must be approximately the same. An impedance mismatch not only results in loss of power but can also have catastrophic implications, such as destroying the circuitry.

For example, if an antenna has absorbed 50% of the signal and the remaining 50% is reflected back through the transmission line, there is a return loss of -3.0 dB. A return loss of -10.0 dB (90% absorbed and 10% reflected) is considered a good quality standard for a transmission line and connected antenna.

The transmission line can be modelled by the diagrams in Figure 9.87. The diagram on the left shows a balanced line (no connection to the ground) and the one on the right shows an unbalanced line (one side connects to ground).

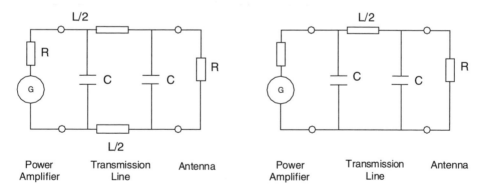

Figure 9.87 Balanced and unbalanced inter-connection from PA to antenna.

The simplified diagrams in Figure 9.87 are divided in three parts: the source (PA), the cable and the load (antenna). The impedance mismatch is defined by the impedance ratio of each component. The nominal impedances are of approximately 50 Ω and the cable capacitances vary from 20 to 80 pF/m.

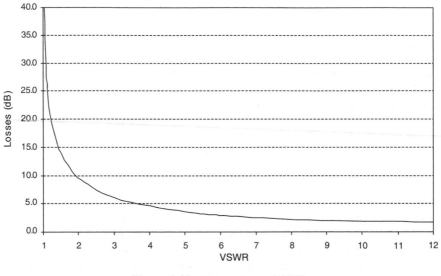

Figure 9.88   Return loss × VSWR.

The diagram in Figure 9.88 shows the return loss in dB for different VSWRs.

The diagram in Figure 9.89 shows the percentage of reflected power for different VSWRs. If the mismatch is large, the reflected power is very high and can cause damage to the components.

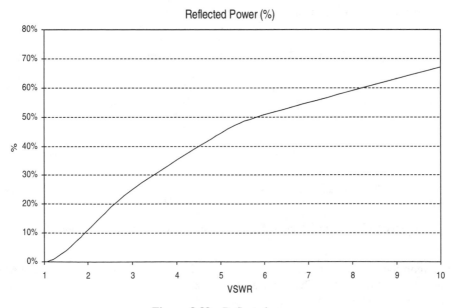

Figure 9.89   Reflected power.

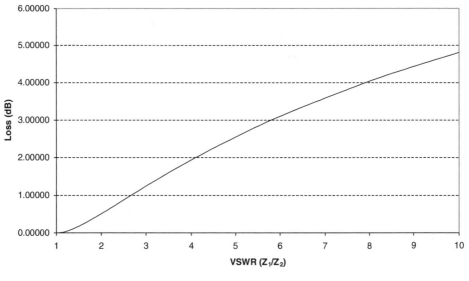

Figure 9.90 Transmission loss.

The diagram in Figure 9.90 shows the transmission loss in dB. VSWRs in the range of 1–1.5 are acceptable.

The transmission line elements are cables, accessories and duplexers/ diplexers and they are described next.

9.3.5.1 Duplexers and Diplexers

Antennas and cables may be shared between transmit and receive circuits or between transmit signals at different frequencies. This is possible through the use of duplexers and diplexers, respectively. The connection of the PA to the antenna is illustrated in Figure 9.91. Filters are applied to the PA output to assure that no out-of-band signals are transmitted.

A duplexer is required if the antenna is being shared between transmitter and receiver. Because transmit and receive frequencies are different band pass filters can be used to avoid

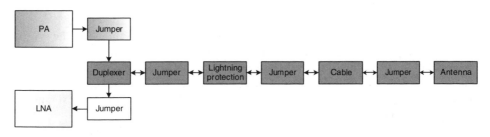

Figure 9.91 Power amplifier to antenna connection.

interference between the bands and a circulator is used as an RF power isolator between them.

A circulator is a three-port device that conducts power with minimum attenuation in one direction and produces high attenuation in the other direction. The third port is used as a load. The following are typical values for loss and impedance of circulators.

- Forward direction insertion loss: 0.25 dB.

- Reverse direction insertion loss: 35 dB.

- Impedance: 50 Ω.

The addition of the band pass filters provides the following typical characteristics to a duplexer.

- Insertion loss: 1.5 dB.

- Isolation: 50 dB.

- Impedance: 50 Ω.

- Return loss: 12 dB.

Advanced duplexers are used in mobile units because of space requirements. Microelctromechanical Systems (MEMS) technology is being used to design switches and variable capacitors and inductors. A volume reduction of 10 times is achieved with this technology and its use should spread to other areas of the network being an option to software radios.

A diplexer is required when two transmitters are connected to the same antenna (e.g. transmitting at 800 and 1900 MHz). Diplexers do not need a circulator because the transmission frequencies are far apart and both signals are transmitted in the same direction. Typical characteristics of duplexers are the following.

- Insertion loss: 2 dB

- Isolation: 40 dB

- Impedance: 50 Ω

- Return loss: 10 dB

9.3.5.2   Cables and Accessories

Cables and accessories also cause transmission losses. The connection from the PA to the antenna is usually long and provides considerable losses.

Jumpers are flexible cables of small diameter that connect the PA to the duplexer, then to the main cable and finally to the antenna. The loss caused by jumpers varies according to their diameter. Typical jumper diameters are 1/4, 3/8, 1/2, 5/8 and 7/8 inches. Figure 9.92 shows typical jumper losses.

The main cable (or feeder cable) is a thicker cable that is more rigid and presents less loss. As in jumpers, the cable loss is also affected by its diameter. Typical diameters for these

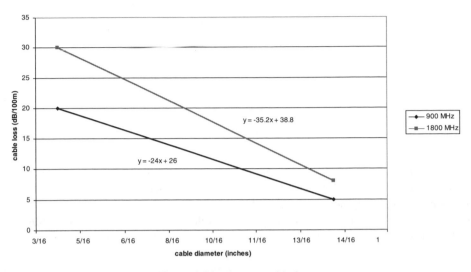

Figure 9.92 Jumper cable loss.

cables are 7/8, 1, 1 1/8, 1 1/4, 1 5/8, 2, and 2 1/4 inches. Figure 9.93 shows typical cable losses. The cable diameter is generally selected to maintain cable loss between 3 and 4 dB.

Jumpers and the main cable are coaxial cables. Figure 9.94 illustrates the coaxial cable structure. These cables have a centre conductor of diameter d, an isolation material of relative permeability ε and an external metallic shield of diameter D.

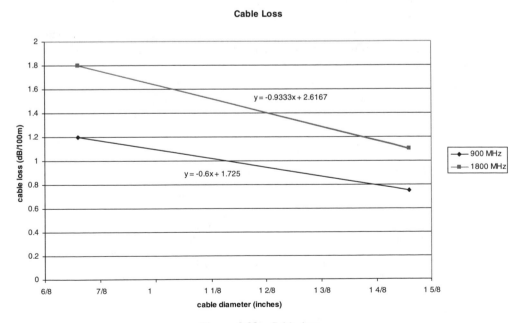

Figure 9.93 Cable loss.

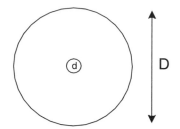

Figure 9.94 Coaxial cable structure.

Because the cable length has to be adjusted to an entire multiple of the wavelength to minimise reflections, it is necessary to determine the wavelength [eqns (9.18) and (9.19)]. Cable impedances are typically of $50 \pm 1\,\Omega$.

Equation (9.18) determines the propagation velocity, which in turn determines the wavelength depending on the transmission frequency, as in eqn (9.19).

$$v = c/\sqrt{\varepsilon} \qquad (9.18)$$
$$\lambda = v/f \qquad (9.19)$$

v propagation velocity (m/s)
c speed of light (m/s)
ε medium permittivity (F/m)
f frequency (Hz)

Table 9.39 show the typical materials used in coaxial cables with the associated propagation speed and resulting wavelength.

Connectors are used to interconnect equipment, jumpers and cables. Special connectors are used to provide lightning protection. Figure 9.95 shows typical connector losses for cables and jumper connections.

For higher frequencies waveguides have to be used to limit the transmission loss. Waveguides are hollow metallic tubes with a circular or rectangular cross section and a diameter proportional to the wavelength and present smaller losses than cables but are much more expensive.

Table 9.39 Coaxial cable characteristics

| Dielectric | ε | $k = 1/\mathrm{sqrt}(\varepsilon)$ | Frequency (MHz) | |
| | | | 900 | 1800 |
			Wavelength (m)	
Air	1	1.00	0.33	0.17
Teflon (TFE)	2.03	0.70	0.23	0.12
Polypropylene	2.25	0.67	0.22	0.11
Polyethylene	2.3	0.66	0.22	0.11
Nylon	4.2	0.49	0.16	0.08

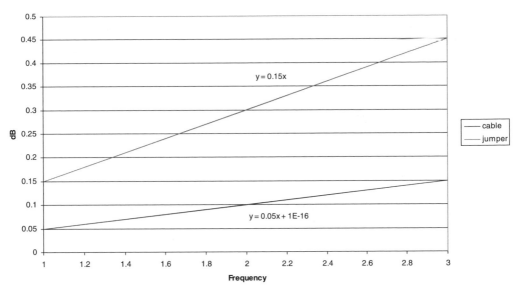

Figure 9.95   Connector loss.

9.3.6   Transmit Antenna

The RF carrier power has to be coupled to the space and this requires a coupling device. This device is the antenna and its function is to couple in the most efficient way the power from the transmitter to the radiating waves that will travel in space. Antennas are transducers that convert RF fields into alternating currents and vice versa.

Maxwell showed in his equations that only charges in non-uniform motion produce radiation. In the antenna this non-uniform motion is provided by the undulating waveform of the carrier.

9.3.6.1   Antenna Characteristics

Near and Far Field
An electromagnetic flux is the presence of force fields in a medium and represents the flow of energy through a surface. The energy in the proximity of the antenna appears to be stored on a reactive component and then it is released to a radiating field. This initial region encompasses the antenna itself and is called near field or Fresnel region. The near field nominal radius is calculated as in eqn (9.20).

$$R = 2L^2/\lambda \qquad (9.20)$$

where

R fresnel region radius (m)
L antenna size (m)
λ carrier wavelength (m).

For the 900 MHz band, the radius of the near field region is approximately 10.7 m and for the 1800 MHz band, it is 5.3 m.

After the near field region, there is a radiating region, which is called far field or Fraunhofer region. Figure 9.96 shows the flux in the near and far fields.

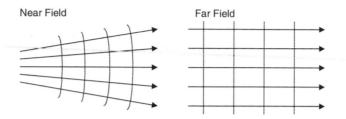

Figure 9.96   Near and far field signal propagation characteristic.

Impedance

The impedance of the antenna varies according to the antenna type but it always has to match the power amplifier, the transmission line and the space impedance. Maxwell's equations allow the calculation of the wave impedance for free space.

$$Z = \frac{|E|}{|H|} = \sqrt{\frac{\mu}{\varepsilon}} = 120\pi\,\Omega \cong 377\,\Omega \tag{9.21}$$

where

μ_0 free-space permeability $= 1.25 \times 10^{-6}$ H/m
ε_0 free-space permissivity $= 8.84 \times 10^{-12}$ F/m.

Polarisation

The polarisation of an antenna expresses the orientation of electric flux lines in an electromagnetic field. It can be constant or it can rotate with each wave cycle. The electrical field is parallel to the antenna orientation therefore vertical antennas transmit and receive vertically polarised waves and horizontal antennas, horizontally polarised waves. Elliptically or circular polarised antennas rotate the polarisation 360 degrees with each wave cycle. Section 9.4.1.2.1 provides a more detailed explanation of wave polarisation.

Loss

Antennas are dissipative elements because the energy that is not transferred to space is reflected back or dissipated on the antenna itself. The VSWR of the radiating set is key to determine how much energy is not radiated. Even with a good VSWR there is a limit of power an antenna can handle. This limit is specified by the manufacturer and is typically around 500 W.

9.3.6.2 Antenna Types

There are several different antenna types available in the market, but the two basic types are the isotropic and the dipole antennas.

9.3.6.2.1 Isotropic antennas

An isotropic antenna transmits the energy equally (outward from the antenna) to all directions. This antenna cannot be practically implemented; it is only used as a reference for practical implementations.

The power received by an antenna is proportional to the power density on the wave sphere. The antenna has a receiving aperture that corresponds to the amount of energy that it can capture from the incoming wave. This aperture is equivalent to the area of the sphere where this energy was distributed. Equation (9.22) shows how to determine the received power based on the antenna aperture and the power density

$$P_R = A_e S \tag{9.22}$$

where

P_R Received power (W)
A_e Antenna effective aperture (m^2)
S Power density (W/m^2)

$$S = P_R/4\,\pi d^2\,(\mathrm{W/m^2}) \tag{9.23}$$

The gain of an antenna is expressed by the ratio of the effective aperture and the square of the wavelength

$$G = 4\,\pi A_e/\lambda^2 \tag{9.24}$$

An isotropic antenna gain is 1 by definition, and its receiving effective aperture is

$$A_i = \lambda^2/4\,\pi \tag{9.25}$$

The effective aperture corresponds to 88.42 cm^2 for a 900 MHz frequency.

The antenna radiation pattern is represented by a radiation diagram, which is measured in test fields for several azimuth and elevation angles. In the test field, or anechoic chamber, the antenna is placed in front of a known receiver and it is rotated horizontally (azimuth) and vertically (elevation) to characterise its radiation patterns. Antenna vendors publish these data as the radiation pattern of the antennas. In an isotropic antenna the wave front radiates as a spherical wave. Figures 9.97 and 9.98 show the radiation pattern of an ideal isotropic antenna.

The antenna pattern can also be represented as a text file providing the data collected for the several angles of azimuth and elevation. Table 9.40 presents a possible file format.

The abbreviations used in this example are the following:

Mod   Antenna model, usually determined by the manufacturer
Man   Manufacturer
Dig   Responsible for digitising the antenna data
Dsc   Antenna type description

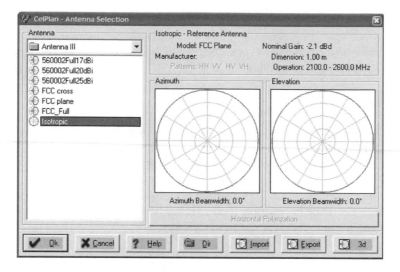

Figure 9.97   Isotropic antenna radiation pattern.

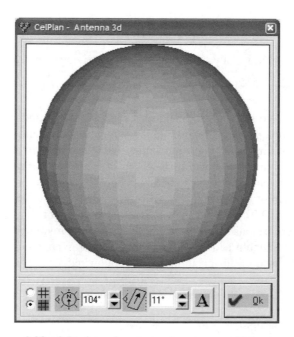

Figure 9.98   3 D view of an isotropic antenna radiation pattern.

Table 9.40   Antenna diagram in text format

[CelPlan DT ANT]	
Mod	Isotropic
Man	Not available
Dig	Y
Dsc	Reference Antenna
Ngn	−2.14 dBd
Abw	0.00
Ebw	0.00
Mnf	2100.00 MHz
Mxf	2600.00 MHz
Siz	1.00 m
Inc	1.0°
Azm	Gain
0	−2.14
1	−2.14
2	−2.14
. . .	. . .
359	−2.14
Elv	Gain
0	−2.14
1	−2.14
2	−2.14
. . .	. . .
359	−2.14

Ngn	Nominal gain in dBd
Abw	Azimuth beamwidth
Ebw	Elevation beamwidth
Mnf	Minimum operation frequency
Mxf	Maximum operation frequency
Siz	Antenna size
Inc	Angle increment for azimuth and elevation points
Azm	Antenna gain for each azimuth
Elv	Antenna gain for each elevation

9.3.6.2.2   Dipole Antennas

The most commonly implemented antennas are half-length dipoles. Dipoles are formed by two straight electrical conductors measuring, from end to end, a multiple of a wavelength (1/2 wavelength for example) and connected at the centre to a feed line as shown in Figure 9.99. The RF current maximum is at the centre of the antenna and the minimum is at the ends, whereas the voltage maximum is at the ends and the minimum at the centre.

A dipole antenna is inherently balanced, and requires a balanced line to be connected to it. A coaxial cable can be connected through a balun (contraction of balanced to unbalanced) transformer. The balun can perform also impedance matching functions.

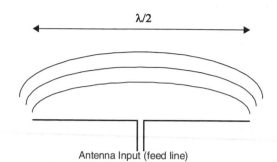

Figure 9.99   Half-wave dipole.

A dipole antenna should be placed at least $1/2$ wavelength above the ground and several wavelengths away from electrical conducting elements, such as towers, wires and other antennas. Dipoles can be oriented vertically, horizontally or at a slant. The polarisation of the electromagnetic field radiated by a dipole corresponds to the orientation of the elements.

Theoretically one polarisation should not receive signals from another type, but, in practice, signals change polarisation because of reflections and diffractions, which may cause some of the originally different polarisation signal to be received by the antenna. A polarisation loss should therefore be added to account for this effect. Typical values for this loss range from 0.25 to 0.3 dB.

The half-wavelength dipole gain is 1.64. This corresponds to an effective aperture (A_e) of 1431.17 cm^2 for a frequency of 900 MHz. The impedance of this dipole can be calculated by eqn (9.26)

$$Z = 73 + j\,42.5\,\Omega \tag{9.26}$$

It is possible to concentrate the energy to be transmitted in a smaller area, providing a gain to the antenna. The result is the effective radiated power (ERP), which is expressed in relation to the isotropic antenna (dBi) or to the half-wave dipole (dBd). The relation between the isotropic and the dipole antenna gains is given by eqn (9.28). A gain expressed in relation to isotropic antennas (dBi) has to be decreased by 2.148 dB to be expressed in relation to a dipole antenna (dBd).

$$G_i = 1.64\,G_{hwd} \tag{9.27}$$

$$0\,\mathrm{dB}_d = 2.148\,\mathrm{dBi} \tag{9.28}$$

G_i gain of an isotropic antenna
G_{hwd} gain of a half-wave dipole
dBd   dB relative to a half-wave dipole antenna
dBi   dB relative to an isotropic antenna

Figures 9.100 and 9.101 show typical diagrams representing dipole antennas. The figure presents what is referred to as an omni antenna. The main characteristic of this type of antenna is that it has a uniform gain in all azimuth directions.

Frequently it is convenient to direct the RF energy to a specific direction, making the dipole antenna directional. This is done by adding metallic reflective elements parallel to the main element or by combining antennas in close proximity.

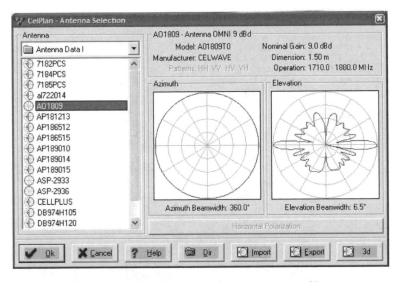

Figure 9.100   CelPlan omni antenna patterns 2D.

The directivity of an antenna is expressed by the ratio of the radiation intensity in a given direction to the radiation averaged over all directions. A directive antenna pattern is formed by a main lobe and side lobes as illustrated in Figures 9.102 and 9.103. Directive antennas are also called sector antennas in wireless communications because they form a sector in a cell.

Figure 9.101   CelPlan omni antenna pattern 3D.

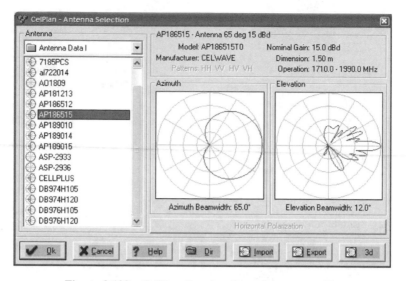

Figure 9.102   CelPlan directional antenna patterns 2D.

The main lobe of the antenna has a maximum gain at a specific angle, which becomes the reference for the antenna. The antenna beamwidth is characterised by the angles in which the maximum gain drops by 3 dB (half-power angles). The side lobes are an effect of the construction of the antenna and are generally undesired.

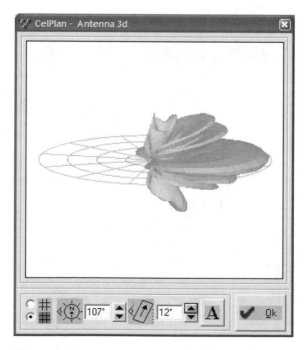

Figure 9.103   CelPlan directional antenna patterns 3D.

Λ directional antenna can also be characterised by a front to back ratio, which gives the ratio between the main lobe and side lobes energies. Some vendors use the ratio between the main lobe gain and the strongest side lobe gain, whereas others limit this to back lobes. Some vendors use the gain precisely at the back of the antenna, what can be very misleading.

The antenna bandwidth is defined by its maximum and minimum operating frequencies divided by its central frequency as in eqn (9.29)

$$A_{bw} = (Mxf - Mnf)/f_c \qquad (9.29)$$

9.3.6.2.3 Cross-polarized Antennas

At higher frequencies, where directional antennas are used, the diffracted and reflected components are limited so that polarization can be considered to transmit different information on each polarity of the wavefront. Figures 9.104 and 9.105 show the azimuth and elevation patterns of an omni-antenna when transmitting, respectively, in vertical or horizontal polarisations. The two patterns depicted in each box indicate the reception polarisation, e.g., VH means transmitted vertically, received horizontally.

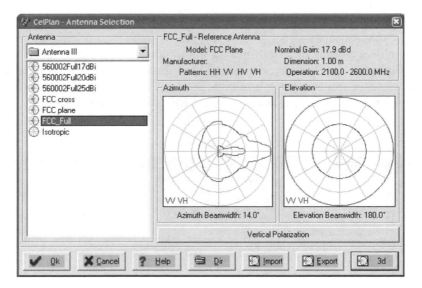

Figure 9.104 Azimuth and elevation diagrams for the vertical polarisation of an omni antenna.

Figures 9.106 and 9.107 show the azimuth and elevation patterns of a directional antenna when transmitting, respectively, in vertical or horizontal polarizations. Figure 9.108 shows the 3D visualisation of a directive antenna cross-polarized pattern.

Table 9.41 shows the cross-polarised diagram in text format. The file has four columns to describe each of the polarisation cases. Values are expressed in dB in relation to a dipole (dBd).

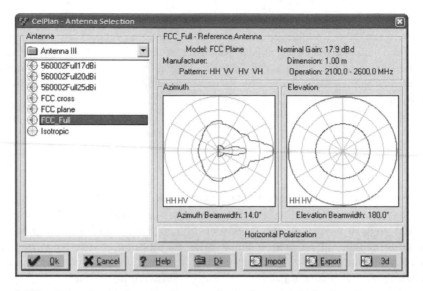

Figure 9.105 Azimuth and elevation diagrams for horizontal polarisation of an omni antenna.

9.3.6.2.4   Other Antennas

Another family of antenna commonly used in wireless communications is the horn antennas shown in Figure 9.109. These antennas have a horn shape that concentrates the energy of a large into a smaller area and then to an electric current. Figure 9.110 shows that parabolic or dish antennas use a reflector to increase its area of RF energy concentration, at the same time

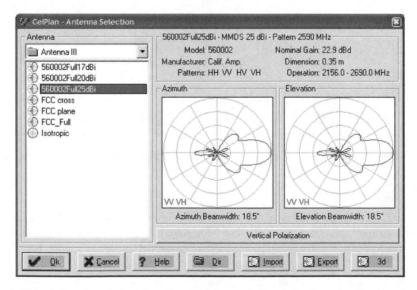

Figure 9.106 Azimuth and elevation diagrams for vertical polarisation of a directional antenna.

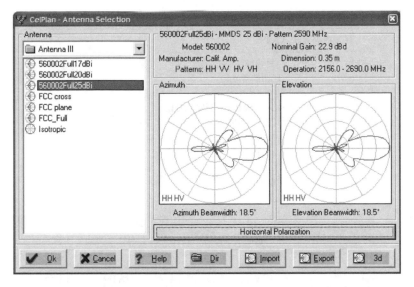

Figure 9.107 Azimuth and elevation diagrams for the horizontal polarisation of a directional antenna.

providing high directionality and gain. This type of antenna is generally used for micro-wave links.

9.3.6.2.5 Antenna Tilt

Antenna vertical tilting is a powerful way of changing the cell footprint and, consequently, minimising interference. Antenna tilt is measured using the horizontal plane as reference

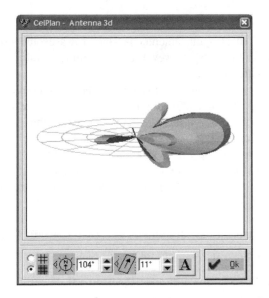

Figure 9.108 CelPlan cross-polarised antenna patterns in 3D.

Table 9.41 Antenna cross-polarized diagram in text format

[CelPlan DT ANT]			
Mod:	560002		
Man:	Calif. Amp.		
Dig:	X		
Dsc:	MMDS 25 dBi – Pattern 2590 MHz		
Ngn:	22.86 dBd		
Abw:	18.50 ☐		
Ebw:	18.50 ☐		
Mnf:	2156.00 MHz		
Mxf:	2690.00 MHz		
Siz:	0.35 m		
Inc:	1.0		
Dgr:	1	1	1	1
Azm	HH	VV	HV	VH
0	22.85	22.83	−17.94	−17.94
1	22.77	22.74	−18.09	−17.55
2	22.62	22.59	−18.3	−17.17
359	22.86	22.86	−17.78	−18.29
Elv	HH	VV	HV	VH
0	22.85	22.83	−17.94	−17.94
1	22.77	22.74	−18.09	−17.55
2	22.62	22.59	−18.3	−17.17
359	22.86	22.86	−17.78	−18.29

(0 degrees). Downtilt indicates the antenna is slant towards the ground, whereas uptilt indicates the opposite. The industry, however, usually employs only the term downtilt, using negative values to indicate uptilts.

Tilting can be performed mechanically by slanting the antenna mount by some degrees or electrically by changing the antenna pattern. These two tilt techniques, mechanical or electrical, have fundamental differences. This section analyses the example of an antenna set at 30 m above terrain, over flat earth, not considering the area morphology (land usage). The same antenna type is used in the four scenarios analysed but a different combination of mechanical and electrical tilts is applied in each case. Figures 9.111 to 9.114 present the footprint obtained in each scenario. The propagation predictions are displayed on the same area over a 2 km × 2 km grid.

- Scenario 1 –0° mechanical downtilt and 0° electrical downtilt (Figure 9.111).

- Scenario 2 –10° mechanical downtilt and 0° electrical downtilt (Figure 9.112).

- Scenario 3 –0° mechanical downtilt and 10° electrical downtilt (Figure 9.113).

- Scenario 4 –10° mechanical downtilt and 10° electrical downtilt (Figure 9.114).

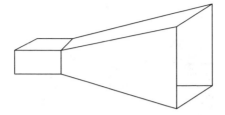

Figure 9.109 Horn antenna element.

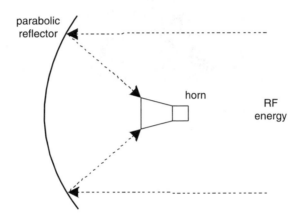

Figure 9.110 Parabolic antenna.

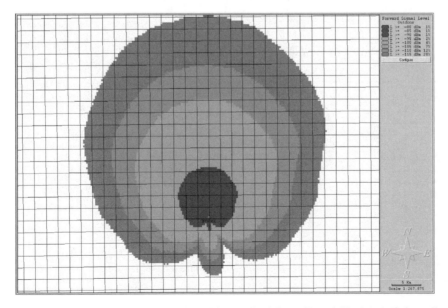

Figure 9.111 Footprint obtained with 0° mechanical downtilt and 0° electrical downtilt.

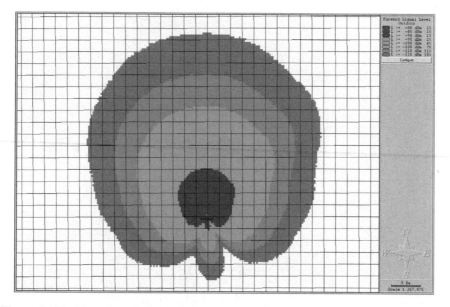

Figure 9.112 Footprint obtained with 10° mechanical downtilt and 0° electrical downtilt.

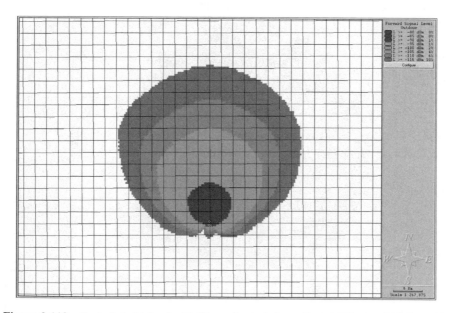

Figure 9.113 Footprint obtained with 0° mechanical downtilt and 10° electrical downtilt.

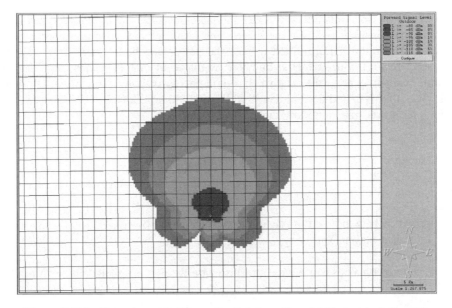

Figure 9.114 Footprint with 10° mechanical downtilt and 10° electrical downtilt.

The analysis of these scenarios indicates that the mechanical downtilt reduces the front lobe coverage while slightly increasing the back lobe coverage. The electrical downtilt produces better results by reducing the side lobes also. The combination of mechanical and electrical downtilts presents a differentiated shape that may be useful in some cases.

9.4 CHANNEL

9.4.1 Radio Frequency (RF) Propagation

After the information is prepared for transmission in the transmit stage (Figure 9.1), the next step is the radio frequency propagation that will take this information to the reception side.

9.4.1.1 RF Propagation Mechanisms

RF propagation is extremely complex in a real life environment. The best way to understand it is starting by the analysis of basic propagation mechanisms, such as free space propagation, reflection, refraction and diffraction.

9.4.1.1.1 Free Space Propagation
Power is radiated from an isotropic antenna uniformly in all directions and is spread over a sphere surface without any losses due to the medium. This consideration allows the calculation of the energy density on the sphere surface for any distance.

The area of the sphere is given by $4\pi d^2$ and increases very rapidly with the distance (d), which corresponds to the radius of the sphere. This implies in a very large variation in power magnitude. The ratio of transmitted to received power can be as large as 10^{12} and it is cumbersome to work with such large numbers. To make calculations easier, Bell[5] proposed a logarithmic unit to express the relative change in energy or power magnitude between two signals. This unit is called decibel and is widely used today. Equation (9.30) shows how to calculate the power transmitted by the antenna in decibels

$$P_{dB} = 10 \log_{10}\left(\frac{P\ [\text{W}]}{P_{\text{Ref}}\ [\text{W}]}\right) \tag{9.30}$$

where

P power to be expressed in relation to the reference power (W)
P_{Ref} reference power (W)

Decibels can use 1 mW as a reference, in which case the unit is called dBm as in eqn (9.31).

$$P_{dBm} = 10 \log_{10}\left(\frac{P\ [\text{W}]}{0.001\ [\text{W}]}\right) = 10 \log_{10}\left(\frac{P\ [\text{W}]}{1\ [\text{mW}]}\right) \tag{9.31}$$

Equation (9.22), in Section 9.3.6.2 indicates that it is possible to express the received power in terms of spectral density and receive effective aperture of an isotropic antenna. These variables can be replaced by eqns (9.23) and (9.25) to express received power in terms of transmitted power, wavelength and distance.

$$P_R = S.A_i = \frac{P_T}{4\pi d^2}\frac{\lambda^2}{4\pi} = P_T\left(\frac{\lambda}{4\pi d}\right)^2 \tag{9.32}$$

The path loss ratio or propagation loss is the ratio of the transmitted to the received power. By applying eqn (9.32) in this analysis, it is possible to express path loss in terms of wavelength and distance.

$$L = \frac{P_T}{P_R} = 1 \Big/ \left(\frac{\lambda}{4\pi d}\right)^2 = \left(\frac{4\pi d}{\lambda}\right)^2 \tag{9.33}$$

$$L_{dB} = 10 \log_{10}\left(\frac{4\pi d}{\lambda}\right)^2 = 20 \log_{10}\left(\frac{4\pi d}{\lambda}\right) \tag{9.34}$$

Equation (9.34) can also be expressed in dB units, considering that the distance is in km and the frequency is in MHz. The use of dB units, allow the received power to be calculated by subtracting the loss from the transmitted power, as in eqn (9.36).

$$L_{dB} = 32.44 + 20 \log_{10} f_{MHz} + 20 \log_{10} d_{Km} \tag{9.35}$$

$$P_R(dBm) = P_T(dBm) - L_{db} \tag{9.36}$$

---

[5] Alexander Graham Bell (1847–1922) invented the telephone in 1876 and proposed an unit to measure an compare sounds levels, the Bel. A decibel equates to 1/10 of a Bel.

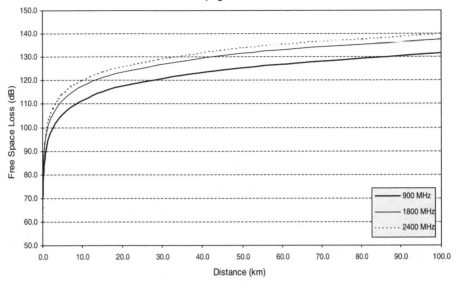

Figure 9.115   Propagation loss in a linear scale with distance (up to 100 km).

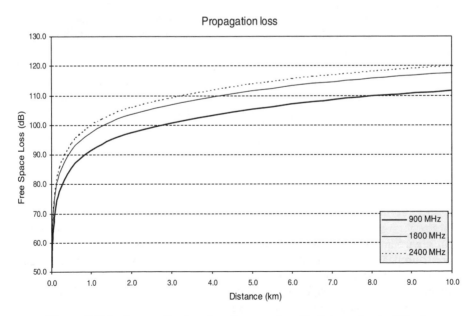

Figure 9.116   Propagation loss in a linear scale with distance (up to 10 km).

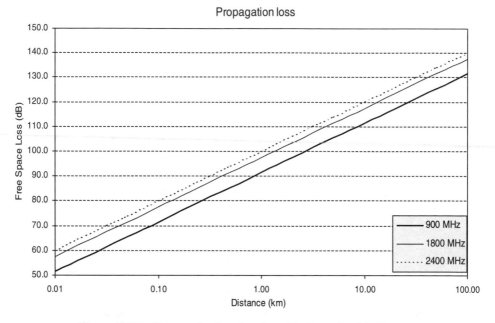

Figure 9.117 Propagation loss in a logarithmic scale with distance.

Figures 9.115 and 9.116 show the free space loss over distance. Figure 9.117 also shows the loss over distance but using a logarithmic scale. The path loss varies at a constant slope of 6 dB per octave (double distance) or 20 dB per decade (ten times distance), increasing with the frequency because of the isotropic antenna receive effective aperture, which is proportional to the wavelength. The free space propagation has a constant slope of 20 dB/decade.

In practical systems, antennas have a gain in the intended direction of the propagation, a fact that does not apply to isotropic antennas. These gains, for transmit and receive antennas, must be added to the formula when calculating the received power as in eqn (9.37). This equation is the Friis formula, which is expressed in dB units.

$$P_R[dB_m] = P_T[dB_m] - (32.44 + 20 \log_{10} f_{\text{MHz}} + 20 \log_{10} d_{\text{Km}}) + G_T[dB_i] + G_R[dB_i]$$

$$(9.37)$$

9.4.1.2 Reflection

In a real environment, transmitted RF waves are reflected and refracted by the different media in its propagation path. To analyse the effect of these parameters in the propagation it is necessary to understand the electromagnetic properties of the propagation medium.

Maxwell formulated, in 1865, the basic laws of electro-magnetism. These laws specify the relationship between magnetic and electric fields and can be summarised as follows.

- An electric field is produced by a time varying magnetic field.

- A magnetic field is produced by a time varying electrical field or a moving electrical charge.

- Magnetic and electrical field lines are continuous. Electrical fields may start and end on charges.

Maxwell's laws allow the analysis of the electromagnetic properties of the propagation medium. A propagation medium is basically defined by its permeability, permittivity and conductivity. Additional parameters can be derived from these to characterise the medium properties.

Table 9.42 lists the main parameters used to characterise propagation media. The parameters are defined in the first three columns and the formulas are given in the next columns for an isolator and a conductor media. The next column calculates the values for an isolator with $(\sigma/\omega\varepsilon)^2 = 0.0044$. The last column gives the units used for the different parameters. Table 9.43 shows the additional basic assumptions used to calculate these parameters.

Table 9.42 Propagation medium parameters

Parameter	Value or formula	Relationship Isolator $(\sigma/\omega\varepsilon)^2 \ll 1$ $(\sigma/\omega\varepsilon)^2 = 0.0044$	Conductor $(\sigma/\omega\varepsilon)^2 \gg 1$	Value	Unit
μ_0 Free-space permeability	$4\pi \times 10^{-7}$			1.25664×10^{-6}	H/m
μ_r Relative permeability				1	
μ Medium permeability	$\mu = \mu_0\mu_r$			1.25664×10^{-6}	H/m
ε_0 Free space permittivity	$10^{-9}/36\pi$			8.84194×10^{-12}	F/m
ε_r Relative permittivity				3	
ε Medium permittivity	$\varepsilon = \varepsilon_0\varepsilon_r$			2.65258×10^{-11}	F/m
σ Medium conductivity				0.01	$1/\Omega m$
σ_0 Free space conductivity				0.0106	$1/\Omega m$
E Electric field	V/d			0.97	V/m
H Magnetic field	E/Z			0.00446031	A/m
Z Wave impedance	$|E|/|H|$	$\mathrm{sqrt}(\mu/\varepsilon)$	$(1+j)\mathrm{sqrt}(\omega\mu/2\sigma)$	217.7	Ω
Z_0 Free space				377.0	Ω
v Phase velocity		$1/\mathrm{sqrt}(\mu\varepsilon)$	$2\pi\mathrm{sqrt}(2\omega/\mu\sigma)$	173205081	m/s
v_0 Free space	c			300000000	m/s
n Refractive index	c/v			1.73	
k Wave number		$\omega\mathrm{sqrt}(\mu\varepsilon)$	$\mathrm{sqrt}(\omega\mu\varepsilon/2)$		1/m
k_0 Free space		ω/c		18.85	1/m
ω Angular frequency	$2\pi f$			5654866776	rad/s
λ Wavelength	v/f	$2\pi/\omega\mathrm{sqrt}(\mu\varepsilon)$	$\mathrm{sqrt}(2/\omega\mu\sigma)$		m
λ_0 Free space	c/f	$2\pi c/\omega$		0.33	m
α Attenuation constant		$\sigma\mathrm{sqrt}(\mu/\varepsilon)/2$	$\mathrm{sqrt}(\omega\mu\sigma/2)$		$m-1$
α_0 Free space				2.00	
S Pointing vector	$EHA/2$			0.00217	W

Table 9.43   Medium assumptions

d	Distance to point		200	m
f	Frequency		9×10^8	Hz
P_t	Transmitted power		100	W
V	Voltage	sqrt($P_t Z$)	194.2	V
A	Receiving area		1.0	m^2

Table 9.44   Permittivity and conductivity of some dielectric materials

Material	Permittivity ε_r	Conductivity σ(S/m)
Air	1	0.0106
Ground	4 to 30	0.001 to 0.030
Fresh water	81	0.01
Seawater	81	5
Snow	1.2 to 1.5	
Ice	3.2	
Glass	3.8 to 8	
Wood	1.5 to 2.1	
Gypsum board	2.8	
Brick	4	
Concrete	4 to 6	
Marble	11.6	

The permeability of dielectric materials is quite constant for high RF frequencies and is considered equal to 1. Typical permittivity and conductivity ranges for common materials are listed in Table 9.44.

9.4.1.2.1   Wave Polarisation

Electromagnetic waves are created by electrical charges that pass through the antenna. These charges create an electrical field and a magnetic field perpendicular to it as illustrated in Figure 9.118.

A vertical wire antenna creates a vertical electrical field and a horizontal magnetic field. This wave is called vertically polarised. Conversely, a horizontal wire antenna creates a horizontal electrical field and a vertical magnetic field. This wave is called horizontally polarized. Figure 9.119 illustrates both cases.

The direction of the electric field in relation to the magnetic field can be determined using a practical method called 'the right hand rule' as illustrated in Figure 9.120.

The propagation of the magnetic wave can also be demonstrated in a space-time diagram as illustrated in Figure 9.121.

In commercial wireless communication systems, both the base station and the mobile antennas are usually vertically polarised. However, due to user movement, the mobile

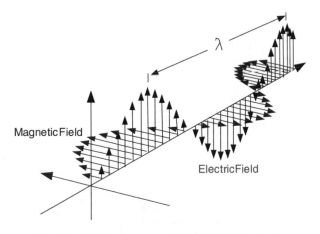
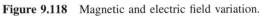

Magnetic Field

Electric Field

Figure 9.118 Magnetic and electric field variation.

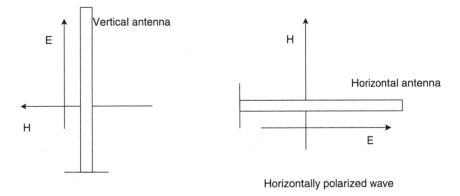

Vertical antenna

E

H

H

Horizontal antenna

E

Horizontally polarized wave

Figure 9.119 Wave polarisation.

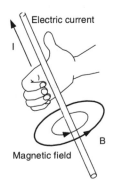

Electric current

I

B

Magnetic field

Figure 9.120 Right hand rule.

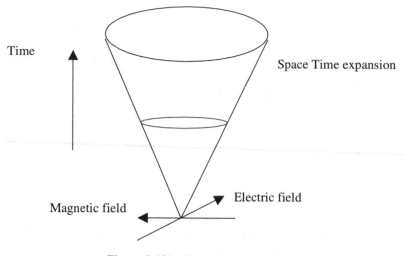

Figure 9.121   Space-time expansion.

antenna is always changing position. The incoming wave (from the base station) may also have its polarisation changed because it is composed of several multi-paths that have been reflected several times. Part of the polarisation loss is computed in the propagation loss, part is computed in the fading, and part is computed as human body loss, therefore, usually, there is not a specific polarisation loss.

9.4.1.2.2   Laws of Reflection and Refraction

To analyse the reflection mechanism this chapter represents the RF wave as rays. In Figure 9.122, the transmitted wave is represented as a ray that travels through medium 1 in a scattering plane and reaches a boundary for transition to medium 2. Part of the original wave is transmitted into medium 2 with a bend (refraction) and part is reflected. The frequency does not change from one medium to another but the phase velocity does.

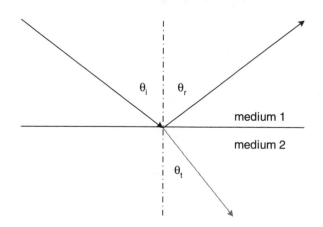

Figure 9.122   Reflection and refraction angles.

Pierre de Fermat (1601 1665) stated that electromagnetic waves follow the path of 'least time'. Willebrord Snell, based on Fermat's principle, derived, in 1621, two important laws governing reflection (eqn (9.38)) and refraction (eqn (9.39)). These laws are valid for a dielectric medium. All angles in the formulas are measured in relation to the normal of the boundary of the scattering plane

$$\theta_i = \theta_r \tag{9.38}$$

$$\sin \theta_i / \sin \theta_t = \text{sqrt}(\varepsilon_2 \mu_2 / \varepsilon_1 \mu_1) \tag{9.39}$$

These laws indicate that part of the energy is transmitted into medium 2 and part is reflected back into medium 1. The amount of energy reflected depends on the polarisation of the electric field in relation to the scattering plane.

The refracted ray bends towards the normal line when it goes from a lower to a higher refractive index medium and otherwise it bends away from the normal line. Figure 9.123 shows the relationship between incident and refracted angles for a ray going from the air into water.

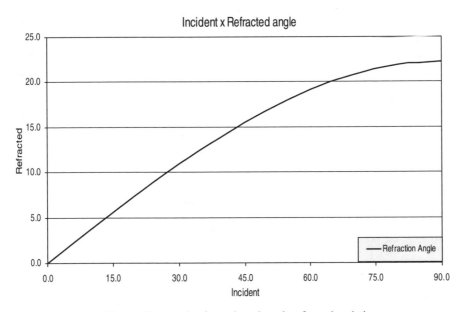

Figure 9.123 Incident and refracted angle going from the air into water.

The formulas in Table 9.45 are used to calculate the transmission and reflection coefficients. In the table, the symbol ∥ indicates that the electric field is parallel to the scattering plane, whereas ⊥ indicates that it is perpendicular. If the scattering plane is the ground, ∥ and ⊥ denote, respectively, vertical and horizontal polarisation.

Figure 9.124 shows these coefficients in a chart considering different angles. The calculations considered the propagation from free space into the ground. The signs of the coefficient depend on the original choices of field directions.

Table 9.45   Transmission and reflection formulas

$R_\parallel$	Reflection coefficient for parallel incidence	E_r/E_i	$(Z_2 \cos \theta_t - Z_1 \cos \theta_i)/Z_2 \cos \theta_t + Z_1 \cos \theta_i)$
$R_\perp$	Reflection coefficient for perpendicular incidence	E_r/E_i	$(Z_2 \cos \theta_i - Z_1 \cos \theta_t)/(Z_2 \cos \theta_i + Z_1 \cos \theta_t)$
$T_\parallel$	Transmission coefficient for parallel incidence	E_t/E_i	$2Z_2 \cos \theta_i/(Z_2 \cos \theta_t + Z_1 \cos \theta_i)$
$T_\perp$	Transmission coefficient for perpendicular incidence	E_t/E_i	$2Z_2 \cos \theta_i/(Z_2 \cos \theta_i + Z_1 \cos \theta_t)$

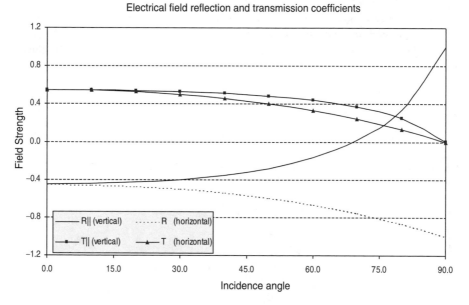

Figure 9.124   Electrical fields and transmission coefficients.

The area of the incident and refracted beams is different and the refracted power relationship must account for this difference. The incident and reflected beams, however, have the same area, because they do not change propagation media. The ratio of the incident and the refracted areas is given by

$$n_2 \cos \theta_t/n_1 \cos \theta i. \tag{9.40}$$

Equation (9.40) shows the calculation of the energy flux for each component of the scattered wave. Figure 9.125 shows the results.

For small incidence angles (ray perpendicular to the boundary) the bulk of the energy is refracted, whereas for large incident angles (ray grazing the boundary) the bulk of the energy is reflected. There is a transition point between these situations, called the Brewster angle, around which this transition occurs. The calculation of this point depends on the refractive

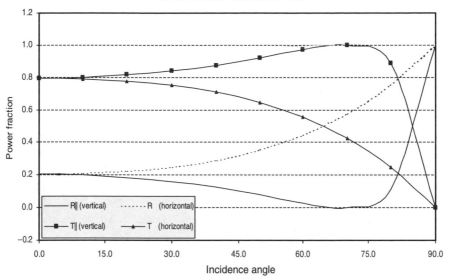

Figure 9.125 Reflected and transmitted power.

index of the mediums

$$\theta_B = \arctan(n_2/n_1) \tag{9.41}$$

9.4.1.3 Flat Earth Propagation

Propagation loss when measured on the earth surface differs from free space.

A simplified propagation model has been used by some authors to explain why the slope of the propagation loss is higher than the slope calculated for free space propagation.

The consideration in this simplified model is that the direct path is summed with a ground reflection to generate the received signal and the propagation happens over flat earth (plane). The reflected wave for a grasing angle (incidence angle close to 90°) is fully reflected and has its phase inverted for horizontal polarisation and the phase is maintained for vertical polarisation. This creates a series of nulls when the excess path is equal to λ for horizontal polarisation or $\lambda/2$ for vertical polarisation. With increased distance the nulls cannot occur anymore and the effect of the sum results in a steeper slope.

Figure 9.126 shows the two rays propagating from the tower to the subscriber.

$$d_1 = \text{sqrt}((h_t - h_r)^2 + d^2) \tag{9.42}$$

$$d_2 = \text{sqrt}((h_t + h_r)^2 + d^2) \tag{9.43}$$

$$d_2 - d_1 = d[(\text{sqrt}((h_t + h_r)/d) + 1)^2 - (\text{sqrt}((h_t + h_r)/d) + 1)^2] \tag{9.44}$$

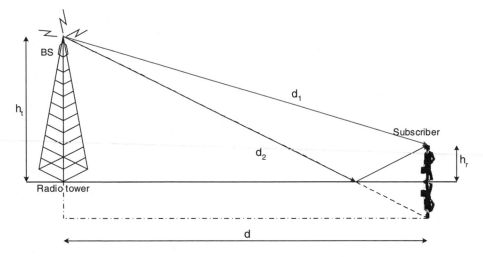

Figure 9.126   Flat earth propagation.

Approximating $(1 + x)^n \approx 1 + nx$ in eqn (9.44)

$$d_2 - d_1 = 2\,h_t h_r / d \qquad\qquad (9.45)$$

Tables 9.46 and 9.47 show an example of null distance calculation. The first table shows the variables considered and the second table shows the calculation results.

Figure 9.127 shows a chart representing the flat earth propagation gain. The calculations to build this chart considered a transmitting antenna height of 20 m, a receiver antenna height of 1 m and vertical polarisation.

Table 9.46   Conditions

f	850	MHz
λ	0.35	m
h_r	1	m
h_t	20	m

Table 9.47   Null distances

n	d	
	$2h_t h_r / n\lambda$	
1	113.33333	m
2	56.66667	m
3	37.77778	m
4	28.33333	m
5	22.66667	m

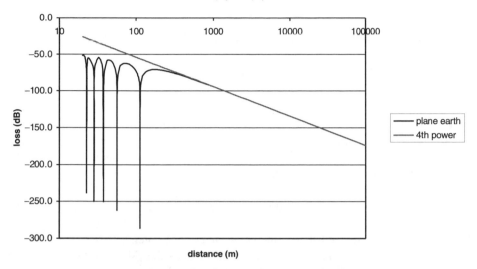

Figure 9.127 Flat Earth path loss.

The break distance at which the slope becomes constant can be calculated using eqn (9.46). In the example of Figure 9.127, this break distance is equal to 266.66 m

$$d_{\text{break}} = 4h_t h_r / \lambda \qquad (9.46)$$

It is also possible to calculate the path loss for when the slope becomes constant (eqn (9.47)).

$$\text{Loss} = (h_t h_r)^2 / d^4 \qquad (9.47)$$

Some RF propagation loss field measurements have a slope much larger than the one predicted by free space equations, and can be approximated by the flat earth propagation slope. This fact led many authors to adopt the flat earth propagation mode as a basis for their own propagation models. However it is difficult to explain how the reflection that this model is based on occurs in real life, as the majority of the paths do not have a direct path for the reflection to occur and consequently the calculations used in this model do not apply directly, but multi-paths and multiple phase inversions have to be considered. This is impractical to do deterministically as the number of paths can be in the hundreds and a statistical method is required. Anyway this method provides an insight into the propagation mechanism and is used by many propagation models as a starting point.

9.4.1.4 Refraction

The Earth shape can be approximated as being a sphere with a radius of $R_0 = 6370$ km. The earth diameter at the equator is 40 024 km. The transmission between two points on Earth

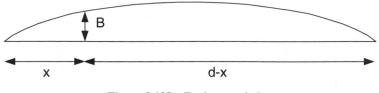

Figure 9.128 Earth curve bulge.

surface happens over the Earth sphere and the distance between these points makes its curvature more or less evident. When a geographical profile is drawn for transmit and receive points that are far apart, the earth bulge is more clearly visible as shown in Figure 9.128. This bulge can be calculated using eqn (9.48).

$$B = 1000 \times (d - x)/(2R_0) \tag{9.48}$$

Depending on the distance between the points and on the antenna heights, this bulge may become a line-of-sight obstruction. Equation (9.49) allows the calculation of the maximum distance that can be seen (line-of-sight) over flat earth, which is called the optical horizon

$$D = \text{sqrt}(2\,R_0 h) \tag{9.49}$$

where

D distance between antenna location and horizon or line-of-sight (LOS) point (km)
R_0 nominal earth radius (km)
h antenna height above sea level (km)

RF waves are curved when transmitted in a non-uniform medium, such as the air. The air density decreases with the elevation and this makes the waves bend down. This bending is called refraction. The refractive index of the atmosphere is approximately $n = 1.003$. The atmospheric refractivity expresses this variation in relation to the unity in part per million, therefore, $N \approx 300$. In reality this atmospheric refractivity varies with the environment temperature and water vapor pressures of the atmosphere, as shown in eqn (9.50).

$$N = 77.6(P + 4810\,e/T)/T \tag{9.50}$$

where

P atmospheric pressure in hPA (mb)
e water vapor pressure in hPA (mb)
T environment absolute temperature (K) $= 273 +$ temperature (°C)

The average mean refractivity gradient is the ratio of the delta of N to the delta of the heights, as in eqn (9.51).

$$G = \Delta N/\Delta h \tag{9.51}$$

According to eqn (9.50) it is necessary to determine the environment temperature and atmospheric and water vapor pressures at each of the desired heights. The heights chosen for this example were 0 and 1 km. The following variables were considered

$$h = 0\,\text{km} \quad P = 760\,\text{mb}, \quad e = 32.1\,\text{mb}, \text{ and } T = 300\text{K}$$
$$h = 1\,\text{km} \quad P = 751\,\text{mb}, \quad e = 21.4\,\text{mb}, \text{ and } T = 294\text{K}$$

Applying these variables to eqn (9.50), N is 329.6 for height 0 and 290.5 for height of 1 km. Using these values in eqn (9.51), the average mean refractivity gradient $G = \Delta N / \Delta h = -39.1$.

The average mean refractivity gradient can be calculated for any region in the world by determining the variables required in eqn (9.50). Table 9.48 shows the gradient calculated for some countries.

The refractivity has a nearly constant gradient between -40 and -48 N unit/km. The bending caused by this gradient causes the radio horizon or radio line-of-sight point to be farther away. Considering a larger earth radius in RF propagation calculations can compensate this. This increase in the earth radius is represented by the k factor.

$$k = R/R_0 = (1 + (\Delta N / 157\,\Delta h) - 1 \qquad (9.52)$$

Applying the data from the example in eqn (9.52), k is 1.33167. Equations (9.48) and (9.49) can be rewritten to consider the corrected Earth radius. The corrected bulge formula is given by eqn (9.53) and the radio horizon by eqn (9.54).

$$B = 1000 \times (d - x)/(2\,kR_0) \qquad (9.53)$$
$$D = \text{sqrt}(2\,kR_0 h) \qquad (9.54)$$

Table 9.48 Mean refractivity gradients

Average mean refractivity gradient	$G = \Delta N / \Delta h$	$G(N$ units/km)
USA	-7.32 exp $(0.005577\,N_s)$	-46.02
Germany	-9.3 exp $(0.004565\,N_s)$	-41.88
United Kingdom	-3.95 exp $(0.0072\,N_s)$	-42.40

Table 9.49 Typical K factors

k	Earth RF radius (km)	Earth RF perimeter (km)	Radio horizon for 30 m antenna height (km)
2	12740	80048	27.65
4/3	8493	53365	22.57
1	6370	40024	19.55
2/3	4247	26683	15.96

The k factor may vary between 2/3 and 2 depending on the area. The most commonly used value is 4/3. Table 9.49 shows typical k factors.

It is recommended that radio engineers calculate this value for the region involved in the network design using typical data. If they choose not to calculate a specific value, 4/3 should be used as the correction factor (k). This factor may vary with seasons; an average should be calculated between summer and winter using the busy hour as a reference.

9.4.1.5 Diffraction

The analysis of propagation in the previous sections has considered a flat earth model. However, in real life, obstacles can be present in the path that connects two communication points. The effect of these obstacles has to be understood and considered as a loss in the propagation path.

The calculation of obstruction effects is very complex. Therefore, several simplifications are done to derive the mathematical formulas. The main simplification is to consider all possible obstructions as screens or knife edges (sharp obstructions that block the propagation). This approach facilitates the development of formulas whose results can be compared to real life measurements and adjusted.

Christian Huygens (1629–1695) wrote 'Traite de la Lumiere', in 1678. In this paper he analysed the propagation of light, considering the wave front of a light wave as being formed by a series of spherical wavelets emanating at every point of the wave as illustrated in Figure 9.129.

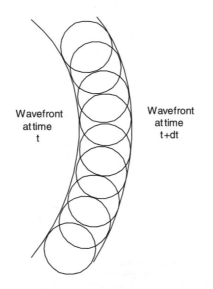

Wavefront at time t

Wavefront at time t+dt

Figure 9.129   Wave front.

This allowed Huygens to derive the laws of reflection and refraction. Augustin Fresnel (1788–1827) extended the Huygens principle by stating that the waveform itself is the sum of the superposition of all wavelets at any point. Gustav Kirchhoff (1824–1887) demonstrated how this principle can be deduced from the Maxwell equations.

9.4.1.5.1 Single Edge Diffraction

According to Huygens principle, the signal at the reception side is made of signals coming from all points of the wave front. Considering a transmitting point T and a receiving point R separated by a distance d as shown in Figure 9.130, the signal arriving at point R travels different distances and is summed out of phase. A signal that travels an additional $\lambda/2$ cancels the original signal.

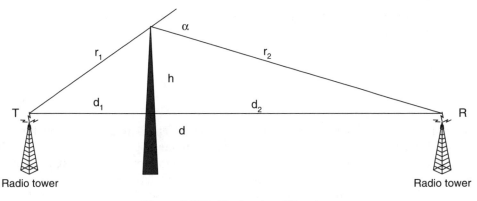

Figure 9.130   Single edge diffraction.

Fresnel theorised that these signals were coming from wavelets located on an ellipsoid that has its foci at T and R and for which the distance $r_1 + r_2$ is $n\lambda/2$ longer than the distance d. Figure 9.130 illustrate r_1, r_2, d_1 and d_2. Each ellipsoid defined by n establishes a Fresnel zone and each zone is numbered according to n. The radius of a Fresnel zone at any point can be calculated by eqn (9.55)

$$R_n = \mathrm{sqrt}(n\,\lambda(d_1 d_2/d) \tag{9.55}$$

The Fresnel–Huygens principle leads to complex equations solved using the Bessel function.[6] $F(v)$ defines the ratio of the diffraction loss to the free-space loss and is given by eqn (9.56)

$$|F(v)| = \frac{1}{2}\left(\frac{1}{2} + C^2(v) - C(v) + S^2(v) - S(v)\right) \tag{9.56}$$

$C(v)$ and $S(v)$—Fresnel cosine and sine integrals.

This function can be plotted against the Fresnel diffraction parameter (v) and can be approximated using simplified equations. Equation (9.57) shows the formula for calculating v. The parameter v was chosen to normalise the equation in relation to h, λ and d_1 and d_2.

$$v = -h\,\mathrm{sqrt}(2\,d/\lambda d_1 d_2) \tag{9.57}$$

---

[6] Bessel functions were derived to solve planetary motion equations derived by Kepler. The function of the first kind solves the Bessel differential equation $x^2 y'' + xy' + (x^2 - n^2)y = 0$.

where

h obstruction height above line that interconnects transmitter and receiver
d distance between transmitter and receiver
λ wavelength
d_1 distance from the transmitter to the obstacle over the line that interconnects transmitter
      and receiver
d_2 distance from the receiver to the obstacle over the line that interconnects transmitter
      and receiver.

The diffraction loss in dB is then given by

$$L(v) = 20 \log F(v) \tag{9.58}$$

Several other methods were proposed for segmented approximations of the Fresnel–Huygens diffraction loss. The most common methods are presented in Tables 9.50 and 9.51, respectively, the Lee (William Lee) and the Japan Radio Corporation (JRC) formulas. Another set of formulas was developed at CelPlan Technologies, Inc. and are the Korowajczuk formulas shown in Table 9.52. They are faster in terms of computational effort than the two other methods. Figure 9.131 compares these four diffraction methods.

Table 9.50   Lee diffraction loss formulas

	Lee
$v \leq -1$	0
$-1 \leq v \leq 0$	$20 \log (0.5 - 0.62\,v)$
$0 \leq v \leq 1$	$20 \log (0.5\mathrm{e}^{-0.95v})$
$1 \leq v \leq 2.4$	$20 \log (0.4 - \mathrm{sqrt}(0.1184 - (0.38 - 0.1\,v)^2))$
$v > 2.4$	$20 \log (0.225/v)$

Table 9.51   JRC diffraction loss formulas

	JRC
$-0.8 > v$	0
$-0.8 \leq v < 0$	$6.02 + 9\,v + 1.65v^2$
$0 \leq v < 2.4$	$6.02 + 9.11v - 1.27v^2$
$v > 2.4$	$13 + 20 \log v$

Table 9.52   Korowajczuk diffraction loss formulas

	Korowajczuk
$v \leq 0$	0
$0 < v \leq 4$	$0.0056\,v^6 - 0.0906\,v^5 + 0.5406\,v^4 - 1.339\,v^3 + 0.1008\,v^2 + 8.5679\,v + 6.1154$
$v > 4$	$20 \log (0.225/v)$

Diffraction loss

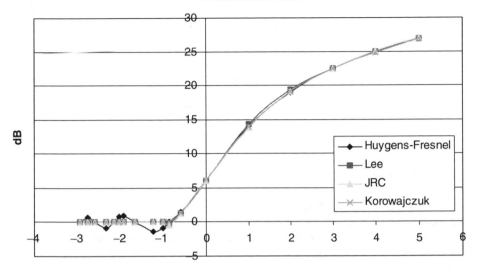

Figure 9.131   Comparison of diffraction loss methods.

Figure 9.131 shows that for LOS ($h = 0$ in Figure 9.130), the diffraction loss is 6 dB and for $h = -0.6 \, R_1$ (radius of the first Fresnel zone), i.e., v of -0.8485, the loss is 0 dB. Obstructions below 0.6 R_1 may even cause some gain, but it is quite reasonable to approximate these values to 0 dB. Therefore, the clearance to avoid obstruction loss should be $h = -0.6 \, R_1$.

These calculations apply to knife-edge obstructions; many obstructions, however, are not so sharp and the roundness of the obstruction should be considered as illustrated in Figure 9.132.

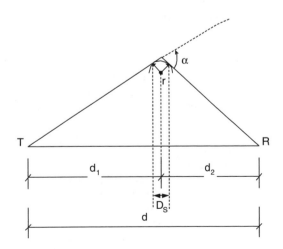

Figure 9.132   Rounded obstruction.

This roundness leads to an additional loss that can be expressed by eqn (9.59). In general, the literature suggests this loss to be multiplied by 0.65 if trees cover the obstacle. However, because of the difficulty in assessing the effective radius of the obstruction, the roundness correction is usually not applied

$$L_{add} = 11.7\,\text{sqrt}(\pi\,r/\lambda)\alpha \tag{9.59}$$

where

r Obstruction radius (m)
λ Wavelength (m)
α Diffraction angle (rad)

9.4.1.5.2   Multiple Edge Diffraction

The calculations developed by Fresnel apply to single obstructions. In practice, however, multiple obstructions are common. Several solutions have been proposed over time to consider them. All the proposed methods make an assumption of how the loss in one obstacle influences those due to the other obstacles. Table 9.53 summarises the main methods proposed.

Each of these methods in Table 9.53 is described and illustrated in the following sections. In all cases, 'D' represents the total diffraction loss. The calculation of individual peaks loss refers to the edges of the diffraction triangle as peak vertices (transmit vertices, receive vertices).

Table 9.53   Multiple edge diffraction models

Model	Year	Characteristic
Bullington	1947	Loss for a single equivalent obstacle is derived
		Origin is the transmitter
		Destination is the receiver
Epstein Peterson	1953	Loss between obstacles is accumulated
		Origin is the previous obstacle
		Destination is the next obstacle
		Millington correction
Japanese Atlas	1957	Loss between obstacles is accumulated
		Origin is the projection of the previous obstacle to origin line
		Destination is the next obstacle
Millington	1962	Correction for Epstein Peterson when two obstacles are very closely spaced
Deygout	1966	Loss of main obstacle is calculated first
		Secondary losses are added next
		Three peaks only
Causebrook	1971	Correction for Deygout when two obstacles are very closely spaced
Giovanelly (Vogler)	1981	Loss between obstacles is accumulated
		Origin is the previous obstacle
		Destination is the projection of the next obstacle on the destination line
Picquenard	1962	Loss is calculated for each obstruction separately
Korowajczuk	2003	Loss is calculated from the peak projection to the previous obstacle

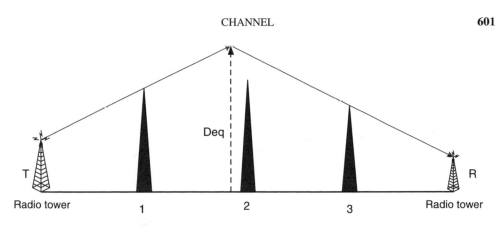

Figure 9.133   Bullington diffraction calculation method.

Bullington Method

This method replaces two obstacles by an equivalent knife-edge obstacle of D_{eq} height as shown in Figure 9.133. The calculation uses this single edge diffraction method described in Section 9.4.1.5.1

$$D = D_{eq}(T, R) \tag{9.60}$$

Epstein Peterson

This method treats the obstacles sequentially or in tandem; for each we define a source, an obstacle and a destination. The calculation starts from the transmitter, or from the previous obstacle. It then considers the obstacle and the destination is the next obstacle, or the receiver. This is repeated for subsequent obstacles until the receiver is reached. The total loss is the sum of the individual losses. Figure 9.134 shows the diagram for this method

$$D = D1(T, 2) + D2(1, 3) + D3(2, R) \tag{9.61}$$

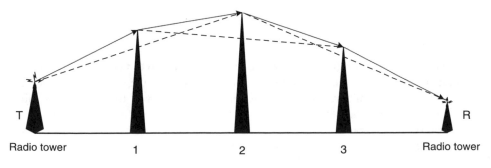

Figure 9.134   Epstein Peterson diffraction calculation method.

Millington

This method proposes a correction to the Epstein Peterson method. When two obstacles are closely spaced ($v > 1$ for both) an additional loss is calculated. This loss is expressed by eqn (9.62).

$$L_{\text{add}} = 20 \, \log(\text{cosec} \, a) \tag{9.62}$$

$$\text{cosec} \, a = \text{sqrt}((d_1 + d_2)(d_2 + d_3)/(d_2(d_1 + d_2 + d_3))) \tag{9.63}$$

Japanese Atlas

This method is similar to Epstein Peterson's method but the source is at the transmitter line wherever the line that connects the destination to the obstacles crosses with it. Figure 9.135 shows the diagram for this method.

$$D = D1(T, 2) + D2(T1, 3) + D3(T2, R) \tag{9.64}$$

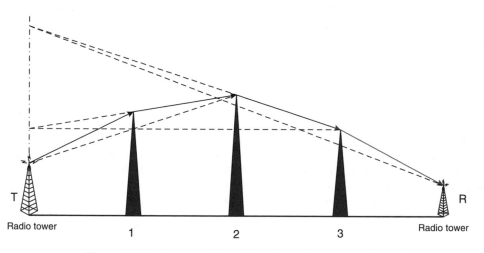

Figure 9.135   Japanese Atlas diffraction calculation method.

Deygout

In this method, the main obstruction between transmit and receive is determined and its loss is calculated. The path is then divided in two smaller parts with the selected obstacle acting as a new receiver/transmitter. The process is repeated for each of these smaller segments until all obstacles are exhausted. Figure 9.136 shows the diagram for this method.

This method is generally limited to the three main obstacles because it becomes too pessimistic when too many obstacles are considered.

$$D = D2(T, R) + D1(T, 2) + D3(2, R) \tag{9.65}$$

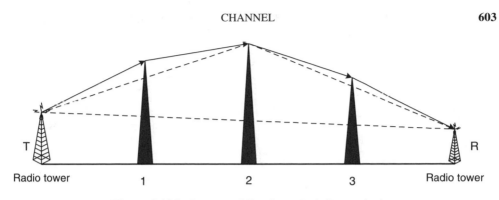

Figure 9.136   Deygout diffraction calculation method.

Causebrook

The Causebrook method proposes a correction to Deygout for situations when two obstacles are closely spaced ($v > 1$ for both). In these cases, an additional loss is calculated. This loss is calculated by eqn (9.66)

$$L_{add} = 20 \log(\operatorname{cosec}^2 a - v_2 \operatorname{cosec} a \cot a/v_1) \tag{9.66}$$

$$A = \operatorname{arc cosec} a = \operatorname{sqrt}(d_1 + d_2)(d_2 + d_3)/(d_2(d_1 + d_2 + d_3)) \tag{9.67}$$

Giovanelly (Vogler)

This method is similar to the Japanese Atlas method but the reception side is always moved to a line aligned to the original receiver. This method can be extended to paths with several obstructions. Figure 9.137 shows the diagram for this method

$$D = D1(T, R2) + D2(1, R3) + D3(2, R) \tag{9.68}$$

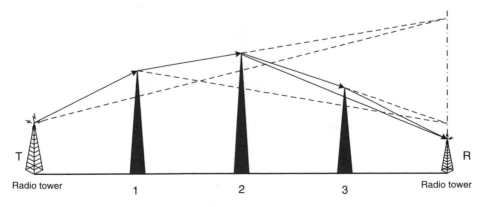

Figure 9.137   Giovanelli diffraction calculation method.

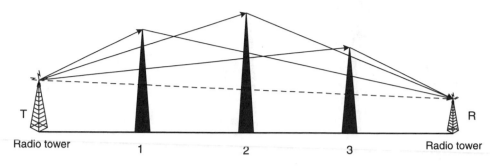

Figure 9.138   Picquenard diffraction calculation method.

Picquenard

In this method, each obstruction is considered separately between transmit and receive. After calculated the losses are added up to determine the total loss. This method becomes pessimistic with the increase in obstacles. Figure 9.138 shows a diagram representing this method

$$D = D1(T,R) + D2(T,R) + D3(T,R) \tag{9.69}$$

Korowajczuk

This method was developed for CelPlan technologies and assumes that the distance between peaks is large enough for the wave to recover from the effects of the previous peak, allowing the loss for each peak to be calculated using the line of the previous peak or transmitter as a reference. Figure 9.139 illustrates this method

$$D = D1(T1,R) + D2(11,R) + D3(21,R) \tag{9.70}$$

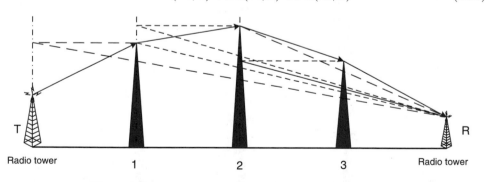

Figure 9.139   Korowajczuk diffraction calculation method.

The above obstruction calculations consider only the ground as an obstruction, ignoring the effects of morphology. This is reasonable for long distances in which the effect of the morphology is negligible. However, many times predictions are executed for very small distances (hundreds of meters). This implies that the morphology should be considered as an effective obstruction. The morphology height can not be simply added to the topography as it does not represent a knife-edge type obstruction.

9.4.1.5.3 Other Diffraction Calculation Methods

Other diffraction calculation methods were developed to calculate diffraction in urban and micro-cell environments. Several studies were performed using a half screen and physical optics (OP) principles.

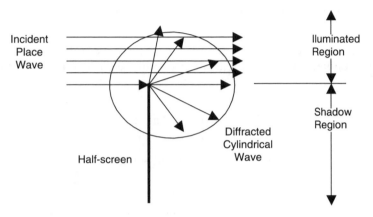

Figure 9.140 Physical optics diffraction.

The diffracted wave is considered as being formed by a direct plane wave, existent only in the illuminated region, and a diffracted cylindrical wave, existent in both regions as shown in Figure 9.140. The sum of both signals results in the received signal.

This theory is presented in the Geometrical Theory of Diffraction (GTD) as illustrated in Figure 9.141. A point is represented in the illuminated region at a distance ρ from the obstruction, which is defined by the distances x and y.

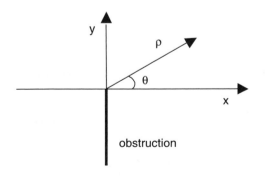

Figure 9.141 Geometric theory of diffraction (GTD).

Table 9.54   GTD parameters

f	Frequency			9.E + 08	Hz
λ	Wavelength	c/f	$2\pi c/\omega$	0.33	m
ω	Angular frequency	$2\pi f$		5654866776	rad/s
k	Wave number		ω/c	18.85	1/m
ρ	Distance from diffraction edge			16.67	m
θ	Angle from diffraction edge			0.175	rad
				10	°
x				16.4135	m
y	sqrt$(\rho^2 - x^2)$			2.8941	m

Table 9.54 shows the parameters used for a study case of the GTD shown in Figure 9.142. The diffraction loss, in this theory, is a function of the angle θ and the distance ρ as in eqn (9.71).

$$D(\theta) = -(1 + \cos\theta)/(\text{sqrt}(2\,\pi k)2\sin\theta) \tag{9.71}$$

$$L = -(1 + D(\theta)/\text{sqrt}(\rho)) \quad \text{illuminated region} \tag{9.72}$$

$$L = -D(\theta)/\text{sqrt}(\rho) \quad \text{shadow region} \tag{9.73}$$

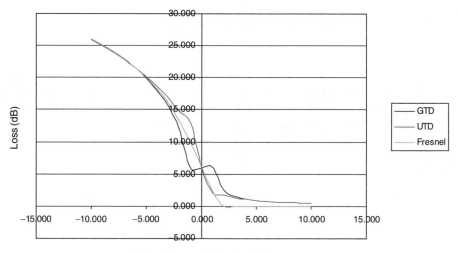

Diffraction Loss Algorithms

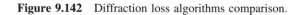

distance from obstruction (m)

Figure 9.142   Diffraction loss algorithms comparison.

Table 9.55   UTD formulas

$F(S)$	sqrt$(2\pi S)f(\xi)$
ξ	sqrt$(2S/\pi)$
$f(\xi)$	$(1 + 0.926\xi)/(2 + 1.792\xi + 3.104\xi^2)$
S	$k\rho\tan^2\theta/2$

The GTD, however, does not represent well the diffraction in the transition zone between the illuminated and the shadow regions. Therefore another theory, called Uniform Theory of Diffraction (UTD), was presented. The UTD adds an averaging function to the original GTD calculations, by averaging the transition between the illuminated and shadow regions.

Table 9.55 shows the formulas used to calculate UTD parameters.

$$D_T(\theta) = D(\theta)F(S) \tag{9.74}$$

$$L = -(1 + D_T(\theta)/\text{sqrt}(\rho)) \quad \text{illuminated region} \tag{9.75}$$

$$L = -D_T(\theta)/\text{sqrt}(\rho) \quad \text{shadow region} \tag{9.76}$$

Figure 9.142 was prepared for $x = 30$ m and y varying between $+10$ to -10 m.

A similar situation was calculated using the Huygens/Kirchhoff solution approximated by the Korowajczuk equation (named Fresnel in the graph) introduced above. The distances of Table 9.56 were considered in the calculations.

All solutions have an equal performance for large diffraction losses. GTD, however, presents a large discontinuity close to the origin. This discontinuity is partially corrected in UTD. The traditional theory applies quite well over the entire region.

9.4.1.5.4 Multiple Screens Diffraction

A theory equivalent to multiple edges obstruction was developed using multiple screens to represent morphology obstructions (mainly buildings). Both OP and GTD/UTD techniques are used in this theory. Figure 9.143 shows an example of a diagram used in the development of the theory.

The loss is divided in four parts when considering multiple screens diffraction.

- Free space loss due to the distance.

- Diffraction loss to the first building.

Table 9.56 Sample data

$d1$	1000	m
$d2$	30	m

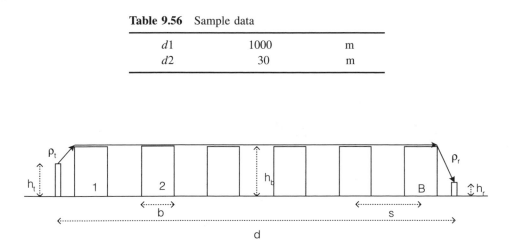

Figure 9.143 Multiple screens diffraction.

Table 9.57 Multiple screen diffraction formulas

	Loss		$h_t \leq h_b$	$h_t < h_b$
L_0	Free space	$-20\,\log(\lambda/4\pi d)$		
L_1	Diffraction loss to first building		$-10 \times \log(\lambda\rho_t/2\,\pi^2 \\ \times (h_b - h_t)^2)$	0
L_2	Diffraction loss over row of building		$20\,\log(B(B-1))$	$-20 \times \log(2.347(h_t - h_b) \\ \times \mathrm{sqrt}(s/\lambda)/d)^{0.9})$
L_3	Diffraction from last building to mobile		$-10 \times \log(\lambda\rho_t/2\pi^2 \\ \times (h_b - h_r)^2)$	$-10 \times \log(\lambda\rho_t/2\,\pi^2(h_b - h_r)^2)$

- Diffraction loss over the row of buildings.

- Diffraction from last building to mobile.

Each of these losses is calculated using a different formula. Table 9.57 shows the multiple screen diffraction formulas.

Figure 9.144 shows the loss calculated for different transmitter antenna heights when considering the conditions in Table 9.58.

The slopes are different for transmit antennas located below or above buildings. Figure 9.145 analyses the gain when the receive antenna height is changed. These values may change considerably with the user position in relation to the buildings.

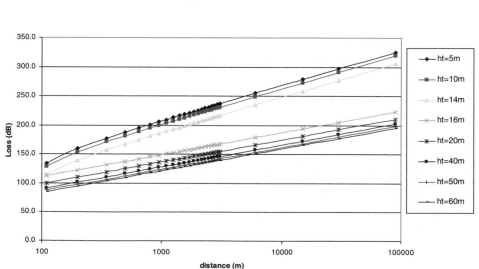

Figure 9.144 Multiple screen diffraction loss.

Table 9.58 Multi-screen diffraction example

f		900	MHz
λ		0.33	
ρ_t	$\mathrm{sqrt}((h_b - h_t)^2 + ((s - b)/2)^2)$	11.18	
ρ_r	$\mathrm{sqrt}((h_b - h_r)^2 + ((s - b)/2)^2)$	14.40	
b		20	m
s		30	m
h_r		1.5	m
H_b		15	m

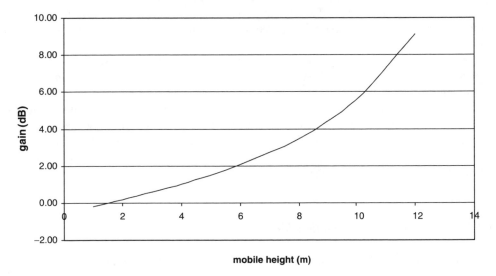

Figure 9.145 Mobile height influence.

9.4.2 RF Propagation Predictions

RF propagation prediction has been a challenge since the first radio transmissions were made. Over the years, several propagation models have been proposed to predict RF propagation, each model improving upon its predecessors.

Traditional models can be divided into two categories with similar characteristics: empirical and physical models. Empirical models are tailored to reproduce field measurements whereas physical models are based on diffraction theories. Both these types of models failed to address the diversity of situations found in real life and had to be adjusted for every prediction. Designers lost faith in propagation models and resorted to other techniques such as exhaustive measurement sample collections.

To overcome the limitation of traditional models, advanced models were developed trying to better match predictions to real life scenarios, while using the minimum amount of measurement samples required to reproduce RF performance. This section discusses some of these models, presenting their capabilities and limitations.

All prediction models have parameters that can be adjusted for a specific area. Traditional models require adjustments on a cell basis, whereas advanced models can be adjusted (calibrated) for scenarios that apply to groups of cells. This calibration process is complex and requires specialised tools.

RF predictions are made to a specific number of points defined by a grid that defines the prediction resolution. Typical resolutions are 5, 30 and 90 m although any resolution can be used. Each grid point is called a pixel as it will form a raster-like file. RF predictions do not require accurate prediction of every single pixel but they should be a valid statistical representation of the network performance. Exactly how good an RF prediction has to be is not an easy answer and it can make the difference between success and failure when designing or optimising a network.

RF path loss predictions give us an average of the path loss, but this average value may vary significantly because of fading (multi-path and shadowing). The analysis of the statistical distribution of fading effects, therefore, is important for the designer to calculate network availability.

Power control, handoff and resource management are important network strategies that impact on its performance. All these features should be considered when predicting network performance.

Users are not distributed uniformly throughout the network; they are usually located in different network environments. They may also use different services and have different mobile terminals. Therefore it is important to model all users. A convenient way of modelling users is by grouping them into service classes and then analysing each class separately and all of them together.

9.4.2.1   Traditional Propagation Models

Initially the goal of an RF designer was to provide coverage to large areas. Prediction models were therefore designed to predict very large areas. Several empirical models were developed based on measurements collected in the field. These models, however, were restricted by the computational capabilities of the time.

The rapid evolution of wireless communications imposed the requirement of predicting smaller areas. New approaches were then developed to cope with these requirements, leading to the development of physical models. These models however are also limited, but, this time, the limitations are the specific conditions for which the models were developed. The following sections describe and analyse traditional models, highlighting their advantages and disadvantages.

9.4.2.1.1   Empirical Models
Empirical models are based on field experiments in which results are tentatively reproduced using mathematical expressions. This analysis is limited to models employed in the UHF band and higher.

Table 9.59 Main empirical model dates

Egli	1957
Okumura–Hata	1968
JRC	1969
Lee	1982
Ibrahim–Parson	1983

Table 9.59 shows the dates when each of the main models was proposed. Figure 9.146 shows the diagram used to characterise the empirical model.

Table 9.60 shows the parameters used in this text to express the empirical models so all of them can be described within a similar frame. Units are standardised to allow the comparison of results.

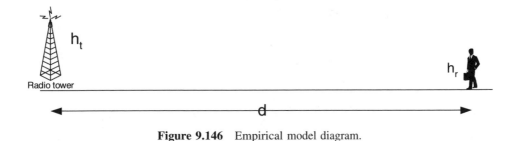

Figure 9.146 Empirical model diagram.

Table 9.60 Empirical models parameters

Parameter	Description	Unit	Range
f	Frequency	MHz	
λ	Wavelength	m	
h_t	Base antenna height	m	
h_{tef}	Effective base antenna height	m	
h_r	Mobile antenna height	m	
L	Percentage of area covered by buildings (500×500 m^2)	%	
H	Height difference between ht and hr	m	
U	Percentage of area covered by buildings with 4 or more floors (500×500 m^2)	%	
γ	Slope of the path loss	dB/decade	20 to 50
L_0	Loss up to 1 mile	dB	45 to 80
n	Frequency factor		2 to 3
D	Diffraction loss	dB	
d	Distance	km	

Table 9.61 Empirical models formulas

Models/factors	Constant	Base height	Mobile height	Frequency	Distance	Morphology loss/gain	Diffraction
Free space	32.44			$20 \log f$	$20 \log d$		Not applicable
Plane earth	120.0	$-20 \log h_t$	$-20 \log hr$		$40 \log d$		Not applicable
Egli	120.0	$-20 \log h_t$	$-20 \log hr$	$-32.04 + 20 \log f$	$+40 \log d$		Not applicable
Okumura-Hata urban large city $f \leq 200\,\text{MHz}$	69.55	$-13.82 \log h_t$	$8.29(\log 1.54 h_r)^2 - 1.1$	$26.16 \log f$	$(+44.9 - 6.55 \log h_t) \log d$		Not applicable
Okumura-Hata urban large city $f \geq 400\,\text{MHz}$	69.55	$-13.82 \log h_t$	$-(3.2(\log 11.75 h_r)^2 - 4.97)$	$26.16 \log f$	$(+44.9 - 6.55 \log h_t) \log d$		Not applicable
Okumura-Hata urban small/medium city	69.55	$-13.82 \log h_t$	$-(1.1 \log f - 0.7)h_r + (1.56 \log f - 0.8)$	$26.16 \log f$	$(+44.9 - 6.55 \log h_t) \log d$		Not applicable
Okomura-Hata suburban	69.55	$-13.82 \log h_t$	$-(1.1 \log f - 0.7)h_r + (1.56 \log f - 0.8)$	$26.16 \log f$	$(+44.9 - 6.55 \log h_t) \log d$	$-2[\log(f/28)]^2 - 5.4$	Not applicable
Okumura-Hata open area	69.55	$-13.82 \log h_t$	$-(1.1 \log f - 0.7)h_r + (1.56 \log f - 0.8)$	$26.16 \log f$	$(+44.9 - 6.55 \log h_t) \log d$	$-4.78(\log f)^2 + 18.33 \log f - 40.94$ 3	Not applicable
Okumura-Hata PCS metro	46.3	$-33.9 \log h_t$	$-(3.2(\log 11.75 h_r)^2 - 4.97)$	$13.82 \log f$	$(+44.9 - 6.55 \log h_t) \log d$		Not applicable
Okumura-Hata PCS medium city	46.3	$-33.9 \log h_t$	$-(1.1 \log f - 0.7)h_r + (1.56 \log f - 0.8)$	$13.82 \log f$	$(+44.9 - 6.55 \log h_t) \log d$	0	Not applicable
JRC	max(LF,LP)						D
Ibrahim-Parson empirical		$-20 \log(0.7 h_t)$	$-8 \log h_r$	$f/40 + 26 \log(f/40)$ $- 86 \log(f + 100)/156$	$[40 + 14.15$ $\log((f + 100)/$ $156)] \log 1000 d$	$0.256 L - 0.37 H$ $+ 0.094 U - 5.9$	D
Ibrahim-Parson semi-empirical		$-20 \log h_t$	$-20 \log h_r$	$20 + f/40$	$40 * \log 1000 d$	$0.18 L - 0.34 H$ $+ 0.094 U - 5.9$	D
Lee	$40 + L_0$	$-20 \log$ $(h_{\text{tef}}/30.5)$	$-10 \log(h_r/3)$	$10 n \log(f/900)$	$\gamma \log(d/1.5)$		D

Table 9.61 presents the formulas of path loss calculation for each of the empirical models in Table 9.59. All formulas can be divided into a set of basic parameters (groups):

- a constant value that adapts the units used;

- a frequency factor that adapts the loss with the frequency and is related to the antenna gain;

- a transmitter height parameter;

- a receiver height parameter;

- a distance related parameter;

- a morphology parameter; and

- a diffraction parameter.

The complete model formulation is the sum of all groups associated with it in Table 9.6.1. The free space propagation and plane Earth models are also shown in this table for comparison purposes. The formulas in each category are very similar and reflect only minor variations. Some models include diffraction into their formulas for obstructed paths, whereas others use knife-edge diffraction calculations as an additional loss. The first models proposed did not differentiate between different morphologies. These distinctions were only introduced in the most recent models. For all cases of empirical models only an average morphology type is considered. Table 9.62 shows the valid ranges of frequency, distance and mobile and BTS antennas applicable to each model.

Figures 9.147 and 9.148 display the path loss calculated by each model for the reference conditions presented in Table 9.62 at varying distances.

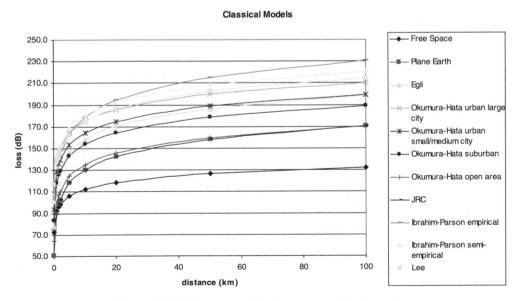

Figure 9.147   Comparative loss between empirical models.

Classical Models

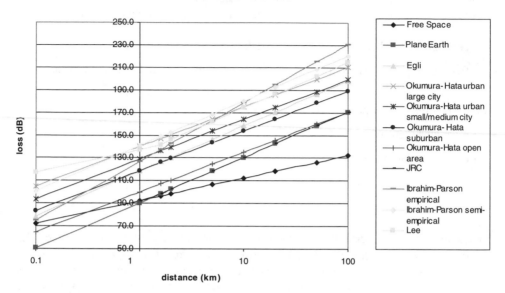

Figure 9.148　Comparative loss between empirical models (detail).

The path loss slope can be seen easier when using a logarithmic representation. This slope varies from 40 dB/decade to 50 dB/decade (excluding the free-space zone). This generic result may apply well for an average path but is, certainly, a very limited representation for different paths.

Table 9.62　Empirical models limits

Models	Frequency (MHz)	Mobile height	Distance (km)	BTS height (m)
Egli	$400 < f < 1500$	$1 \leq h_r \leq 10$ m	$1 < d < 20$	$30 < h_t < 200$
Okumura-Hata urban large city	$150 < f < 1500$	$1 \leq h_r \leq 10$ m	$1 < d < 20$	$30 < h_t < 200$
Okumura-Hata urban small/medium city	$150 < f < 1500$	$1 \leq h_r \leq 10$ m	$1 < d < 20$	$30 < h_t < 200$
Okumura-Hata suburban	$150 < f < 1500$	$1 \leq h_r \leq 10$ m	$1 < d < 20$	$30 < h_t < 200$
Okumura-Hata open area	$150 < f < 1500$	$1 \leq h_r \leq 10$ m	$1 < d < 20$	$30 < h_t < 200$
Okumura-Hata PCS metro	$1500 < f < 2000$	$1 \leq h_r \leq 10$ m	$1 < d < 20$	$30 < h_t < 200$
Okumura-Hata PCS medium city	$1500 < f < 2000$	$1 \leq h_r \leq 10$ m	$1 < d < 20$	$30 < h_t < 200$
JRC				
Ibrahim-Parson empirical	$150 < f < 1000$	$h_r < 3$ m	$d < 10$	$30 < h_t < 300$
Ibrahim-Parson semi-empirical	$150 < f < 1000$	$h_r < 3$ m	$d < 10$	$30 < h_t < 300$
Lee			$d > 1.5$	

The charts shows that the models have a large dispersion in the results, part of it explained by the representation of different morphological environments (urban, suburban, open area). The remaining difference, however, exists because the models were developed over measurements for a specific area. This proves that these models reflect average values and do not fit all areas.

Several variations of the models were developed allowing users to adjust parameters. These adjustable models are called general models and usually allow users to adjust parameters on a site-by-site basis. This implies a measurement collection for every site causing the prediction task to become a rough approximation.

This issue is not as crucial with models developed for applications in which the cells have radii of 10 or more km because in these situations averages can be used. When the cells shrink, however, morphology diversity starts playing an important role and varies significantly from one path to another. None of these models considered morphology heights and the relative position of the antennas in relation to the morphology could not be considered.

Prediction models predict only a single point inside a pixel, with size defined by the prediction resolution. The path loss inside the pixel, however, is not constant and varies even when fast fading is discounted. Okomura recognised this fact and published standard deviations for urban and suburban environments considering a pixel area with 2 km of radius. Figure 9.149 shows these curves.

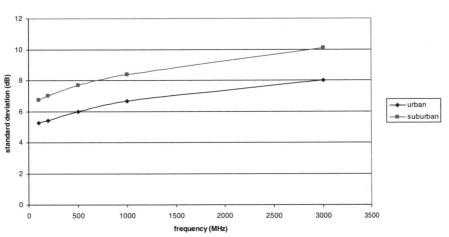

Figure 9.149 Okumura path loss variability with location inside a pixel.

9.4.2.1.2 Physical Models

The limitations of classical empirical models were clearly recognised mainly in dense environments where buildings are present. Several authors decided to apply physics to study the issue. They adopted simple and uniform configurations. The Huygens-Kirchhoff principle was resurrected and Optical Physics (OP) was used to study specific configurations. The multiple screen solution for the diffraction problem was used as a basis for the physical models theory.

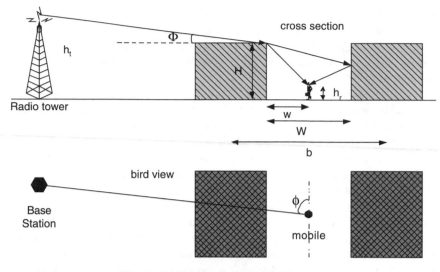

Figure 9.150 Physical models diagram.

Physical studies were restricted to flat terrain and uniform buildings, i.e., buildings with uniform height and similar spacing, situation common in some cities in Europe. Antenna heights were varied below and above the buildings height so insight was gain about the different situations and models were formulated to accommodate the practical results.

Figure 9.150 illustrates a reference scenario for physical models. Table 9.63 describes the parameters and the values assumed so we could compare the different models under the same frame.

Several authors studied this situation and proposed solutions to it. The General Theory of Diffraction (GTD) improved in some cases by the Uniform Theory of Diffraction (UTD) was applied to calculate the diffraction caused by the morphology.

Table 9.63 Physical models parameters

f	Frequency	1000	MHz
ht	Base antenna height	30	m
hr	Mobile antenna height	1	m
H	Building height	30	m
W	Street width	30	m
w	Distance to building	15	m
b	Distance between buildings	60	m
lr	Reflection coefficient	0.25	
Φ	Incidence angle on last building	0.0000	rad
	Incidence angle on last building	0.00	
φ	Incident angle with street axis	1.48	rad
	Incident angle with street axis	85.00	°
City type	Medium/suburban $= 1$, large $= 2$	1.00	

Table 9.64 Physical models formulas

Physical models	Conditions	Constant	Frequency	Distance	Receiver height	Transmitter height	Other
Ikegami		32.4 −5.8	$20 \log f$ $10 \log f$	$20 \log d$	$20 \log (H - h_r)$		$-10 \log (1 + 3/\text{lr}^2)$ $10 \log (\sin \Phi)$ $-10 \log W$
Bertoni		32.4	$20 \log f$	$20 \log d$	$5 \log ((H - h_r)^2 + (0.5 W)^2)$ $-20 \log (H - h_r)$		
Walfisch and Bertoni		11.82 89.55	$-10 \log f$ $21 \log f$	$38 \log d$ $-18 \log(1 - d^2/$ $17(ht - H)$	$5 \log [(b/2)2 + (H - h_r)^2]$ $20 \log \{ \text{tg}^{-1} [2(H - h_r)/b] \}$	$-18 \log(h_t - H)$	$-9 \log b$
Walfisch Ikegami (Cost 231)		32.4	$20 \log f$	$20 \log d$	$20 \log (H - h_r)$		$-10 \log W$
	$0 < \varphi < 35$ $35 \leq \varphi < 55$ $55 \leq \varphi \leq 90$	−16.90	$10 \log f$	L_{ori}	$20 \log (H - h_r)$	$-18 \log (1 + h_r - H)$	$-9 \log b$ $-10 + 0.3574 \varphi$ $2.5 + 0.075$ $(\varphi - 35)$ $4 - 0.114$ $(\varphi - 55)$
	$ht - H \geq 0$ $ht - H < 0,$ $d \geq 0.5$ $ht - H < 0,$ $d < 0.5$ $ht - H \geq 0$ $ht - H < 0$	54				$54 - 0.8(h_t - H)$ $54 - 0.8(h_t - H)$ $(d/500)$ $18 \log d$ $(18 - 15(h_t - H)/H)$ $\log d$	
	City type = 1 City type = 2		$(-4 + 0.7(f/$ $925 - 1)) \log f$ $(-4 + 1.5(f/$ $925 - 1)) \log f$				
Xia and Bertoni (GTD/UTD)		32.4	$20 \log f$	$20 \log d$ $Q(g)$	$10 \log (D^2(\theta)/$ $\pi k \cos \Phi_r)$		

Table 9.65 Walfish Ikegami and XIa Bertoni additional equations

D(θ)		$1/\theta - 1/(\theta + 2\pi)$
θ		$a \tan (H - h_r)/w$
r		$\text{sqrt}(w^2 + (H - h_r)^2)$
k (wave number)		$2\pi f/300$
Q(g)		$3.502g - 3.327g^2 + 0.962g^3$
g		$\text{sqrt}(b \cos \varphi/\lambda) * a \tan ((h_t - H)/(1000d))$
L_{ori}	$0 < \varphi < 35$	$-10 + 0.3674\,\varphi$
L_{ori}	$35 \leq \varphi < 55$	$2.5 + 0.075(\varphi - 35)$
L_{ori}	$55 \leq \varphi \leq 90$	$4 - 0.114(\varphi - 55)$

Physical models recognise the importance of the morphology height, of the relative height and position of the antennas in relation to the morphology, and of canyons (streets) when analysing the RF path loss.

The equations involved in the path loss calculation are very complex and can only be solved for the simplest situations making the applicability of these models very restricted.

Table 9.64 shows the equations used by the models to solve the reference situation presented in Figure 9.150. Table 9.65 shows the applicability limits of these models. These equations are grouped according to the main parameters that they describe (groups):

- A constant value to adjust the units;

- A frequency value usually connected to the antenna;

- A distance related value;

- A receiver height related value (differently from classical empirical models, these values are compared to the building height);

- A transmitter height related value (differently from classical empirical models, these values are compared to the building height); and

- A miscellaneous factor chosen by the author as important and related to the width of the canyon (street) and the angle of the canyon with the propagation direction.

The complete model formulation in the sum of all groups associated with it in Table 9.64. Three situations were analysed using these models: antenna above morphology, antenna at the morphology height, and antenna below morphology. Each of these situations is characterised by a specific building height and have their results presented as a chart. Not all models can predict all variations of building heights because some are limited to situations were the transmit antenna is above the buildings.

- Antenna above morphology (building height of 10 m)—results presented in Figures 9.151 and 9.152;

- Antenna at the morphology height (building height of 30 m)—results presented in Figures 9.153 and 9.154; and

- Antenna below morphology (building height of 40 m)—results presented in Figures 9.155 and 9.156.

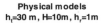

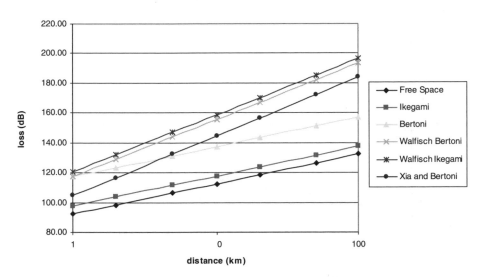

Figure 9.151 Physical models loss with transmitter below building height.

The path loss slopes can be easily seen in the logarithmic graphs. They vary from as low as 20 dB/decade to as much as 45 dB/decade. The path loss variation between the models is very high (more than 30 dB). This is partially explained because the models represent slightly different situations, but, even discounting this fact, the variation would still be large.

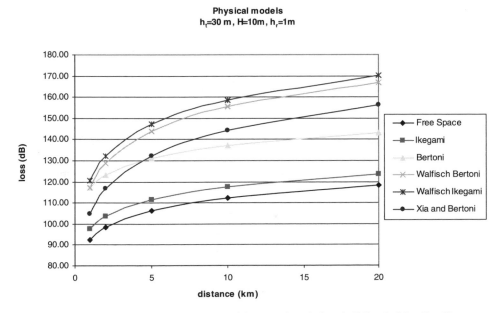

Figure 9.152 Physical models loss with transmitter below building height (detail).

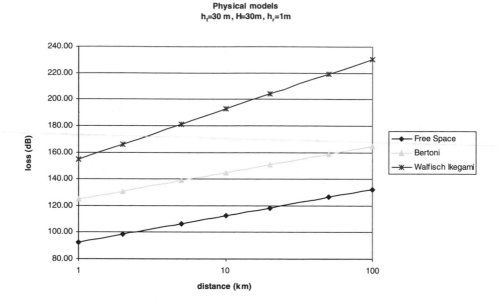

Figure 9.153   Building height variation with transmitter at building height.

Because these are physical models, there are no measurement samples that can be used to compare the predictions with the real world. However, even though these models provide valuable insight into the principles of RF propagation, it is certain that the applicability of physical models is limited because of the variations found in the field.

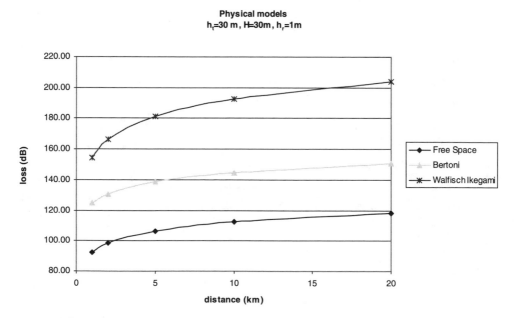

Figure 9.154   Physical models loss with transmitter at building height (detail).

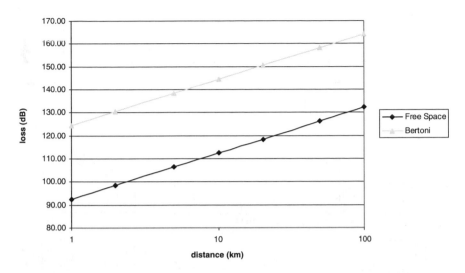

Figure 9.155 Physical models loss with transmitter above building height.

9.4.2.1.3 Microcell Models

Special models were developed to characterise microcells' path losses. They are empirical models that try to reproduce the behaviour observed in the field. Figures 9.157 and 9.158 illustrate a typical microcell scenario. Table 9.66 defines the parameters considered to calculate microcell path loss.

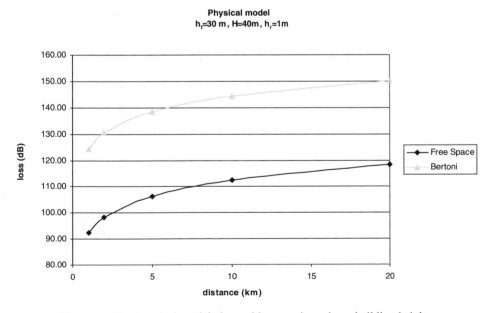

Figure 9.156 Physical models loss with transmitter above building height.

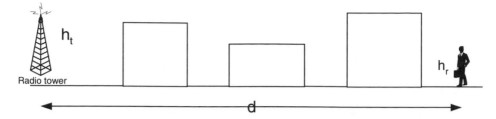

Figure 9.157 Microcell model diagram.

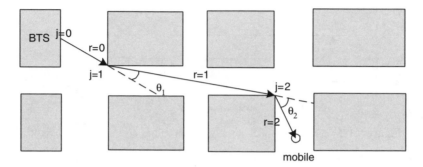

Figure 9.158 Microcell Berg model diagram.

Table 9.66 Microcell parameters

	Parameter	Formula	Value	Range
f	Frequency		1000	
λ	Wavelength	$300/f$	0.3 m	
h_t	Transmitter height		30	
h_r	Receiver height		1	
L_1	Free space loss at 1 m		32.44 dB	
n_1	Coefficient before Brewster point		2	2–2.3
n_2	Coefficient after Brewster point		6	3.3–13
d	Total path length		1000 m	
d_b	Brewster break point	$4\,h_t\,h_r/\lambda$	400 m	
θ	Angle with straight ray		60°	0°–90°

Table 9.67 shows the formulas of three microcell models commonly used: the single breakpoint, the double breakpoint and the Berg models. The single breakpoint model has two path loss slopes that intercept at the breakpoint, while the double breakpoint has three path loss slopes that intercept at the breakpoints. The Berg model is a simple ray tracing model that uses building information and looks for the smallest path loss path. A ray tracing models shoots rays at certain angle intervals (15° as an example) from the transmitter and calculates the path loss at the receiver considering diffractions and in some cases reflections. More sophisticated models analyse the rays on the horizontal planes and vertical planes simulating a 3D environment.

Table 9.67 Microcell formulas

Single break point	$L_1 + 10n_1 \log d$	dB	for $1 < d < d_b$
	$L_1 + 10n_1 \log d_b + 10n_2 \log (d/d_b)$	dB	$d > d_b$
Double break point	$40 + 25 \log d$	dB	$d < d_b/2$
	$40 + 25 \log d_b/2 + 40 \log(2d/d_b)$	dB	$d_b/2 <= d < 4d_b$
	$40 + 25 \log(d_b/2) + 40 \log 8 + 60 \log (d/4\ d_b)$	dB	$d >= 4\ d_b$
Berg (1995)	$32.44 + 20 \log f + 20 \log d_j$	dB	$r <= d_b$
	$32.44 + 20 \log f + 20 \log d_j + 20 \log d/d_b$	dB	$r > d_b$
	$d_j = k_j\ r_{j-1} + d_{j-1}$		
	$k_j = k_{j-1} + d_{j-1}\ q(\theta_{j-1})$		
	$q(\theta_j) = (0.5\theta_j/90)^{1.5}$		
	$k_0 = 1,\ d_0 = 0$		
	r_j path segment length		
	d_j equivalent path segment length		
	θ_j ray angle with straight line		

The Berg model divides the path between the transmitter and the receiver in segments as illustrated in Figure 9.158. Each segment has a real length r_j and an equivalent length d_j that includes the corner losses. The total loss is calculated as the plane earth loss over the equivalent length.

A simple one-slope model (single breakpoint) and a three-slope model (double breakpoint) are analysed in Figure 9.159 and 9.160 with the values presented in Table 9.66. The single breakpoint model has a path loss slope of 25 dB/decade, changing to 60 dB/decade. The three-slope model has a break point at half the Brewster distance (200 m) and at four times the Brewster distance (1600 m). The slopes in each segment are 25 dB/decade, 40 dB/decade and 60 dB/decade.

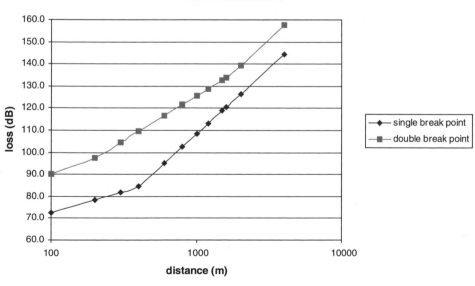

Figure 9.159 Microcell loss.

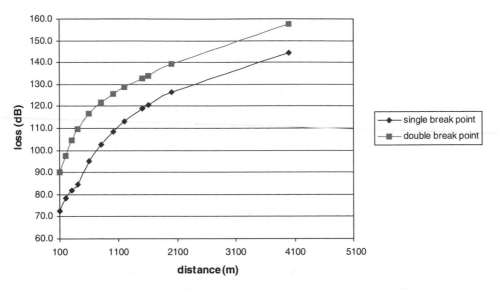

Figure 9.160 Microcell loss (detail).

These microcell models are simplistic and do not consider the complex scenarios encountered in real world. The Berg model produces better results if a good resolution is used but requires very detailed databases and the processing time is very large compared to any other method. The processing time increases prohibitively with the cell radius and the analysis is generally limited to 1 or 2 km radius.

9.4.2.1.4 Indoor Models
Indoor models are empirical models that match measured observations to predicted results. Two approaches are followed to perform indoor predictions:

- A single path is considered between the transmitter and the receiver and an empirical formula is applied to analyse it; or

- Multiple paths (ray tracing) are considered between the transmitter and the receiver and an empirical formula is applied to each of them. The result of the formulas is summed to determine the path loss.

Both solutions are approximations and produce similar results. Ray tracing, however, requires much more complex databases and longer processing time.

Figure 9.161 shows a typical scenario used as reference for indoor propagation models. The scenario represents a multi-room, five-story building. Table 9.68 defines the parameters used in these models.

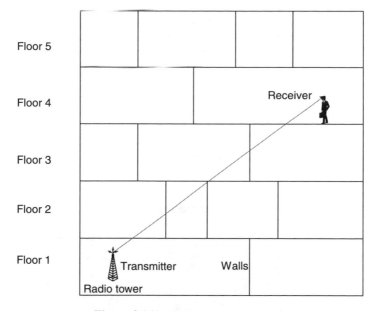

Figure 9.161 Indoor models diagram.

Table 9.68 Indoor models parameters

	Parameter	Unit	Range
f	Frequency	MHz	
λ	Wavelength	m	
L_1	Free space loss at 1 meter	dB	
L_c	Constant loss	dB	
γ	Path loss exponent		
f_i	Floors of type i		
nf_i	Number of type i floors		
L_{fi}	Loss per floor	dB	12–32 dB
w_j	Walls type j		
nw_j	Number of type j walls		
L_{wj}	Loss per wall	dB	1–20 dB
α	Attenuation factor	dB/m	0.23–0.62dB/m
$X\sigma$	Environmental clutter	dB	3–14 dB
d	Distance	m	

Table 9.69 Indoor models formulas

Log distance path loss model	$L_1 + 10\,\gamma \log{(d/1)} + X_\sigma$
Attenuation factor model	$L_1 + 20 \log{(d/1)} + \alpha d + \mathrm{sum}\,(n_{fi} * L_{fi})$
Keenan-Motley model	$L_1 + 10\,\gamma \log d + \mathrm{sum}\,(n_{fi} * L_{fi}) + \mathrm{sum}\,(n_{wj} * L_{wj})$
Multiple wall model	$L_{fs} + L_c + L_f n_f^E + \mathrm{sum}\,(n_{wj} * L_{wj})$

Table 9.69 presents the mathematical formulas of some indoor models used today. The log distance path loss model is very simple and applies only to single floors. The other models allow the calculation on a single and multiple floors and present similar results.
where

$$L_{fs} = -27.56 + 20 \log f + 20 \log d$$
$$E = (n_f + 2)/(n_f + 1) - 0.46$$

Figure 9.162 shows the results for the models presented in Table 9.69. All results are within a range of 15 dB from each other.

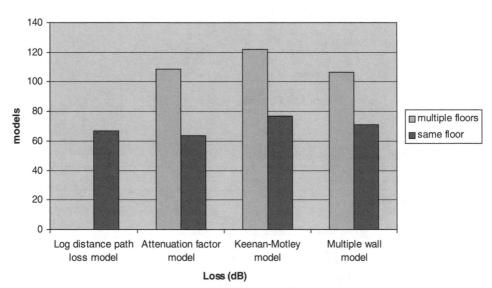

Figure 9.162   Indoor propagation models loss.

In indoor propagation models it is important to determine how much of the signal leaks from indoor into outdoor and vice-versa. A model called Cost 231 (1998) proposed methods to calculate this value for situations with and without line of sight. Figures 9.163 and 9.164 illustrate the diagrams considered when calculating loss between indoor and outdoor environments, respectively, for situations with LOS and NLOS.

Table 9.70 presents the formulas used to calculate the loss between indoor and outdoor environments according to Cost 231. A typical loss value is provided in the calculated values column.

These formulas are approximations and suffer from the same defficiencies as traditional models because they do not consider the real propagation to each building floor.

Generally there is no information about the buildings and the variety of buildings does not allow modelling all of them, so general values have to be derived that can be applied to a whole area (Table 9.71). The indoor losses are then approximated by an average loss and a

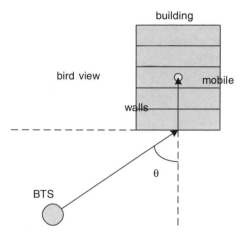

Figure 9.163 LOS Outdoor to indoor loss (through walls).

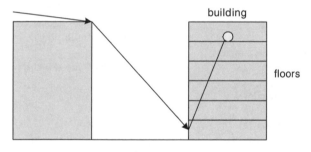

Figure 9.164 NLOS Outdoor to indoor loss with height gain.

standard deviation. CelPlan Technologies recommends the following values for indoor losses.

9.4.2.2 Advanced Propagation Models

CelPlan Technologies has developed models that address many of the weak points in empirical and physical models. These models can be considered as hybrid models and their main characteristics are described in the following sections.

9.4.2.2.1 Fractional Morphology

Classical models consider the role that morphology plays in RF propagation by changing some of the model parameters, but a single category is used for the whole area. Extensions of these models, such as the general model, consider a single morphology type per path or, in more recent implementations, a single morphology per receiver location.

Table 9.70   Cost231 indoor/outdoor loss calculation

Cost 231	Outdoor to indoor loss	Formulas and typical values	Calculated values	Units
L_{tLOS}	Overall loss	$L_f + L_e + L_g(1 - \cos\theta)^2 + \max(L_1, L_2)$	142	dB
L_{tNLOS}	Overall loss	$L_{out} + L_e + L_{ge} + \max(L_1, L_3) - G_{fh}$	129	dB
L_e	Path loss through external wall		7	dB
L_g	Additional wall loss, due to grasing incidence	20	20	dB
L_1	Loss per internal wall	$n_w L_i$	8	dB
L_2	Internal loss based on a specific attenuation and grasing angle	$\alpha(d_i - 2)(1 - \cos\theta)^2$	1.4	dB
L_3	Internal loss based on a specific attenuation	αd_i	18	dB
L_{out}	Path loss to building wall at 2 m height		122	dB
G_{fh}	Floor height gain	$n\,G_n$ or $h\,G_h$	21.6	
G_n	Gain per floor	1.5 to 2 (1st and 2nd floor)	1.5	dB/floor
		4 to 7 (3rd and 4th floor)	4	dB/floor
G_h	Gain per meter	1.1 to 1.6 (above 4th floor)	1.2	dB/m
h	Floor height above 2 meters			
L_{ge}	Excess penetration loss	4 for 900 MHz and 6 for 1800 MHz	4	dB
L_e or L_i	Wooden wall	4	4	dB
	Concrete with windows	7	7	dB
	Concrete no windows	20		dB
α	Specific attenuation	0.6	0.6	dB/m
n_w	Number of internal walls		2	
d_i	Internal distance		30	m
d_e	External distance		70	m
f	Frequency		1000	MHz
θ	Grazing angle		45	°
n	Number of floors		6	
h	Height		18	m

Table 9.71   Typical indoor loss values

Location	Average (dB)	Standard deviation (dB)
Window	6	4
Indoor plaster building	12	6
Indoor masonry building	16	8
Indoor enclosed (garage, elevator)	20	12

This implies that a parameter used to characterise the propagation path applies to a mix of morphologies and, because this mix varies significantly between paths, it cannot represent properly all the mixes. Only single morphology sites can be properly represented by the traditional methods. Every site has to be measured to adjust the predictions and, even so, only average values can be adjusted, presenting large variations for individual points.

In a single path between a transmitter and a receiver, many different morphology types are usually found as illustrated in the geographical profile in Figure 9.165.

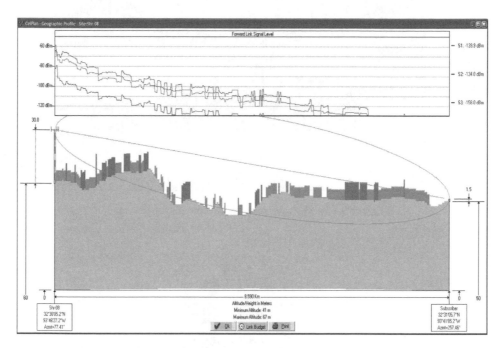

Figure 9.165 Terrain geographical profile showing the Fresnel zone.

Figure 9.166 shows the level predictions at different user heights over a geographical profile using the Lee model. Only terrain obstructions are considered here. Figure 9.167 provides the legend for the levels used in the profile. Colour saturation variations are used inside each range to illustrate level variations inside it.

CelPlan Technologies, aware of this issue, developed, in 1993, a fractional morphology method that considered the effect of different morphologies in the propagation path. In this method, propagation parameters are assigned to morphologies and can be re-used for new sites and new areas. This method can be applied to any of the classical propagation models using single or multiple slopes. Figure 9.168 illustrates the concept of fractional morphology.

Table 9.72 shows the fractional morphology parameters, assuming an area characterised by m different morphology types $(m_a, m_b, \ldots, m_m)$. Each type has two path slopes that characterise the loss due to the morphology according to the distance.

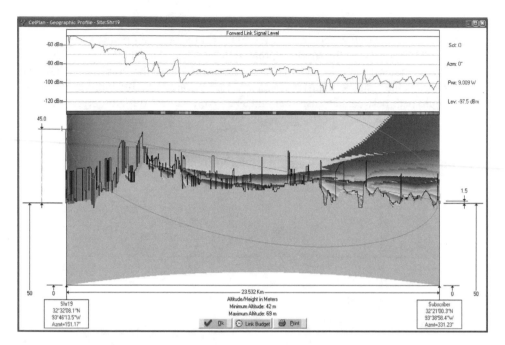

Figure 9.166   Terrain geographical profile for Lee model.

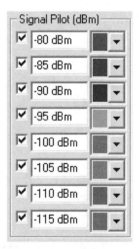

Figure 9.167   Legend for propagation loss profile.

In the example of Figure 9.168, $m_1, m_2, \ldots, m_{10}$ represent the morphology types. A start (d_s) and an end distance (d_e) are assigned to each of these types. The slope break distance (d_b) varies depending on the propagation model being used. This distance represents the point before which slope 1 is used, and after which slope 2 is used. The overall loss is given by eqns (9.77) and (9.78).

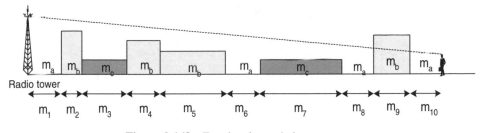

Figure 9.168 Fractional morphology concept.

Table 9.72 Fractional morphology parameters

	Slope 1	Slope 2
m_a	$S_{1\,ma}$	$S_{2\,ma}$
m_b	$S_{1\,mb}$	$S_{2\,mb}$
m_c	$S_{1\,mc}$	$S_{2\,mc}$
...	...	...
m_m	$S_{1\,mm}$	$S_{2\,mm}$

For $d <= d_b$, slope 1 (S_1) is used:

$$Loss_1 = \sum_{m_i=1}^{n} S_{1\,mi} \log(d_{smi}/d_{emi}) \tag{9.77}$$

For $d > d_b$, slope 2 (S_2) is used:

$$Loss_2 = \sum_{m_i=1}^{n} S_{2\,mi} \log(d_{smi}/d_{emi}) \tag{9.78}$$

CelPlan applied the fractional morphology method to Lee's model with great success. In this case, the first slope is defined by the signal strength at 1 mile and the second slope, by the loss in dB/decade over the distance as shown in Figure 9.169. These parameters can be re-used between sites and even between cities. Section 9.4.2.3 discusses the calibration of these parameters and its re-use depending on the area and databases.

9.4.2.2.2 Korowajczuk Model for Outdoor Propagation

This model was developed by Korowajczuk for *CelPlan Technologies* to overcome short-comings of the traditional models when predicting real life networks.

The use of fractional morphologies improved the predictions significantly but there were still many situations in which the outcome was not satisfactory:

- Antenna heights much higher than the morphology or smaller than it did not result in good predictions;

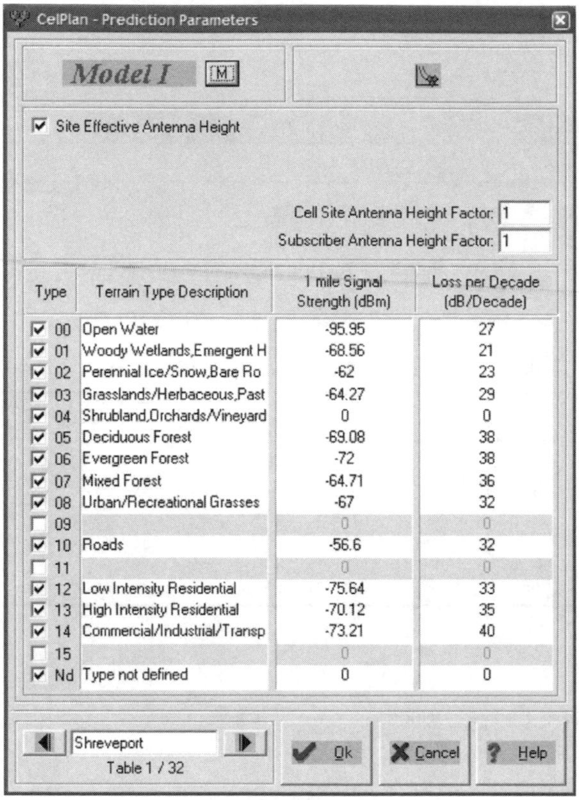

Figure 9.169 Fractional morphology parameters for Lee's model.

- The effect of canyons (streets) was not well represented;
- The dual slope dichotomy caused situations in which the loss was not the same when the transmitter and receiver were reversed;
- Small cells were not properly predicted.

A new approach was required to cope with these issues. The physical models address some of these issues but do not have the capability of predicting the variety of situations encountered in real life.

One of the problems diagnosed on the propagation models available is how the morphology is considered to affect the signal. RF propagation happens through RF waves. These waves are longitudinal waves as suggested in Figure 9.170.

Sound waves are also longitudinal as illustrated in Figure 9.171. RF waves vary the density of magnetic and electrical fields similarly to the way sound waves propagate varying the air pressure.

In two dimensions, this propagation can be represented by concentric circles or ellipses as shown in Figure 9.172.

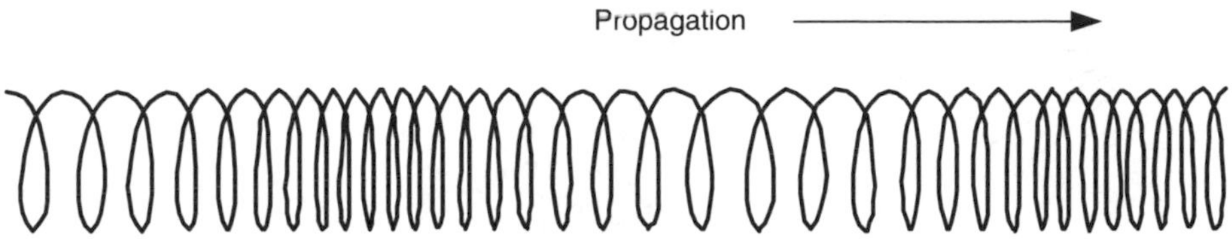

Propagation ————————→

Figure 9.170 Longitudinal wave.

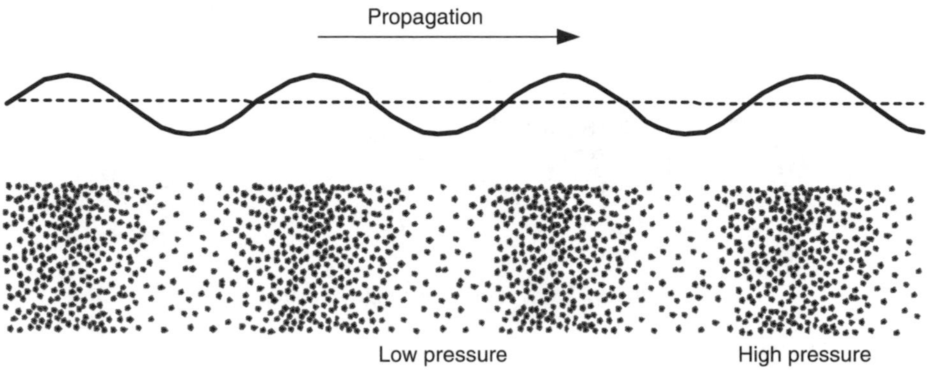

Propagation ————————→

Low pressure   High pressure

Figure 9.171 Sound motion through air molecules.

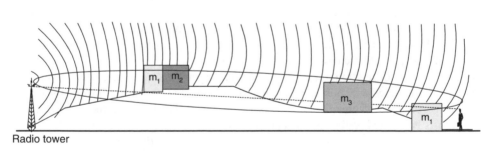

Radio tower

Figure 9.172 Wave propagation over morphology.

Figure 9.172 shows that the propagation in a real environment is obstructed and distorted by the morphology, indicating that the morphology height should be considered as an obstruction. The issue is that morphologies are not continuous and sometimes are not even compact (trees), so the knife-edge treatment for obstructions does not fully apply. Therefore, the Korowajczuk model proposes a morphing factor (m_m) that adapts the knife-edge loss to a morphology edge loss.

The Huygens-Kirchhoff theory says that, as long as 0.6 of the Fresnel zone is not obstructed, free-space propagation can be considered between transmit and receiver. This chapter refers to this zone as the inner part of the Fresnel zone as shown in Figure 9.173.

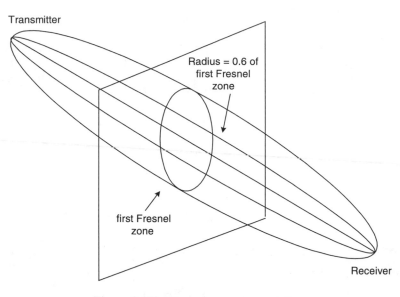

Figure 9.173   Fresnel zone representation.

All methods (empirical and physical) used to analyse the effect of multiple obstructions have limitations and should be used only within their limits. In our understanding, the best compromise to find diffraction losses is to use the Deygout or Korowajczuk methods limited to three peaks. The peaks should be determined using the morphology, but the losses should be calculated for the topography and morphology at each of the points. The three peaks are determined by choosing the three points (topography + morphology) that intrude more into the Fresnel zone, i.e., that cause the most obstruction.

To apply these diffraction methods when considering morphology, the transmitter is automatically assigned a clearance equivalent to the antenna height in all directions around it. If the morphology where the receiver is located is higher than the receiver height, the diffraction is calculated to the top of that morphology as shown in Figure 9.174.

The selected diffraction model, Deygout or Korowajczuk, is then applied to calculate the loss at each point for topography and morphology heights separately. The difference

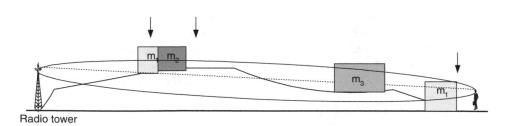

Figure 9.174   Diffraction considering terrain and morphology.

between both models is that, in the latter, the loss attributed to the morphology should be multiplied by the morphing factor of the corresponding morphology type as in eqn (9.79).

$$L_D = D_{1t} + m_m + D_{1m} + D_{2t} + m_m D_{2m} + D_{3t} + m_m D_{3m} \qquad (9.79)$$

After the diffraction loss is determined, the Korowajczuk propagation model requires the division of the propagation loss into four parts (Figure 9.175): initial distance (d_i), propagation zone (d_p), obstructed zone (d_m) and penetration zone (d_f).

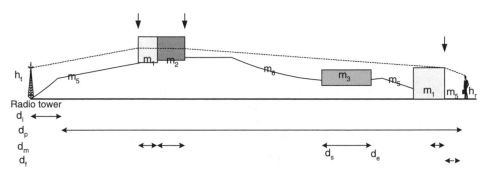

Figure 9.175   Propagation loss according to Korowajczuk model.

Initially, a path based on three diffractions is established and will be analysed for further losses analysis.

The initial distance is set as the height of the transmit antenna. The loss for this distance is always considered as free-space as in equation (9.80) where d_s indicates the start distance of an occurance and d_e the end distance of it.

$$L_i = 32.44 - 20 \log(f) - 20 \log(d_i) \qquad (9.80)$$

To the remaining distances for which the ellipsoid of 0.6 of the Fresnel zone does not touch the morphology (almost free space zone), a slope similar to free space is assigned. This slope varies for different areas and depends on local conditions. A slope S_p is assigned for each area as in equation (9.81).

$$L = \sum_{m_i=1}^{n} S_p \log(d_{smi}/d_{emi}) \qquad (9.81)$$

For the distances where the inner Fresnel ellipsoid touches the morphology (obstructed zone) a different slope is assigned to each morphology type as in eqn (9.82).

$$L_m = \sum_{m_i=1}^{n} S_{mi} \log(d_{smi}/d_{emi}) \qquad (9.82)$$

Finally, a penetration loss is assigned. This loss represents the combination of all losses/gains in the environment surrounding the receiver. If the receiver is higher than the morphology, the penetration loss represents all signals that are added to the main signal, e.g. signals reflected from other morphologies or from the ground. If the receiver is inside the morphology, the loss is applied proportionally to the height difference between receiver and top of morphology as in eqn (9.83) where p_m represents the penetration loss for the morphology and h_{morph} the height of morphology.

$$L_f = p_m \log(h_{\mathrm{morp}} - h_{\mathrm{r}}) \tag{9.83}$$

The final path loss is a sum of the diffraction loss and the loss calculated for each of the four parts in the propagation path as indicated in eqn (9.84).

$$L = L_{\mathrm{D}} + L_i + L_p + L_m + L_f \tag{9.84}$$

Even though this procedure is computationally intensive, it can be executed with today's computers in a short time. All slopes are empirically calculated from measurements. Typical values depend on the area and the database used.

A typical parameters table for the Korowajczuk model is shown in Figure 9.176.

This model allows the calculation of path loss for users at street level or at different floors of a building. Figures 9.177 and 9.178 show signal levels predicted for different user heights according to the legend of Figure 9.179. Colour saturation variations are used inside each

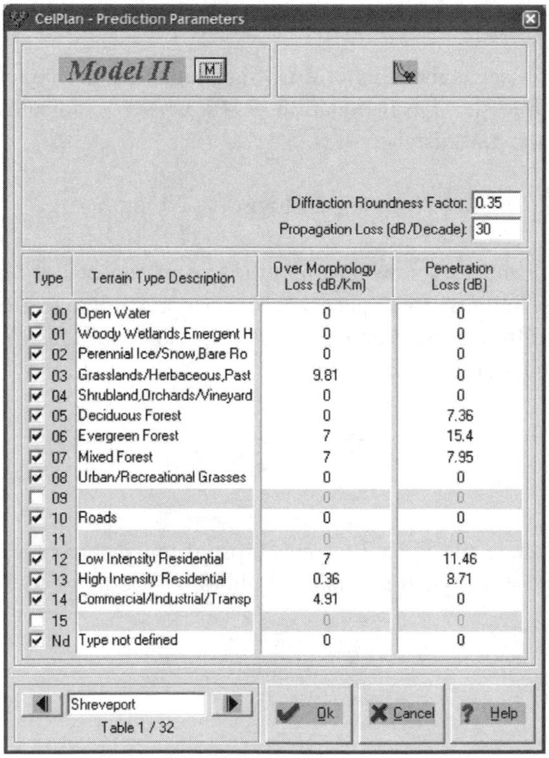

Figure 9.176 Korowajczuk model propagation parameters.

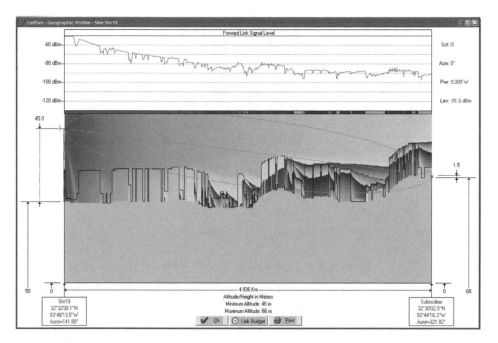

Figure 9.177 Korowajczuk model propagation loss profile (short distance).

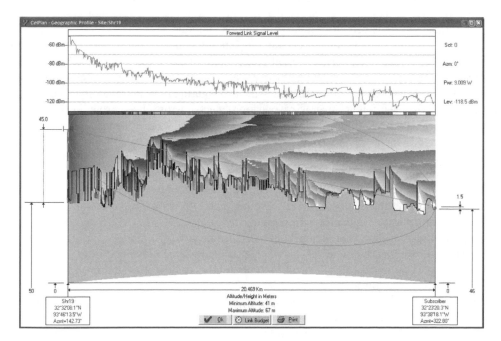

Figure 9.178 Korowajczuk model propagation loss profile (large distance).

Figure 9.179   Legend for propagation loss profile.

range to illustrate level variations inside it. The figures clearly show the diffraction loss caused by morphologies and the signal loss when penetrating into the morphologies.

9.4.2.2.3   CelPlan Microcell Model

Microcells are used in high-traffic areas and have small footprints by reducing power and placing the antennas at low heights. The morphology surrounding microcells is much higher than the cell transmitting antennas themselves and the canyons in which microcells are located are usually narrow. This means that the diffraction loss over the morphology is high and the transmission is primarily done through the canyons.

At the beginning of the propagation path, propagation happens mainly through and around the morphology obstructions. However, after a certain distance, the propagation mechanism becomes mixed: part of the signal goes over the morphology and part goes through and around the obstructions. At further distance, the propagation over morphology becomes predominant. This means that several propagation mechanisms play a role in a microcell propagation analysis, therefore multiple slopes are required to represent it.

The distances involved are small, which means that the Fresnel zone is also small, therefore the analysis of direct rays is a reasonable assumption. The propagation inside canyons (streets) and open spaces can be considered as free-space propagation.

One option is an analysis that uses ray-tracing techniques. However, these techniques require accurate databases that are rarely available and that never represent all morphology nuances, mainly when vegetation is involved.

CelPlan found a good correlation between the morphology composition of a direct ray connecting transmitter and receiver and the signal received as illustrated in Figure 9.180. A certain slope is assigned to each of the morphologies, representing the obstruction to the passage of the signal through or around it.

Figure 9.181 shows the different paths that the strongest signal can take. Direct paths are routed around street corners as illustrated in Figure 9.182.

To calculate the microcell propagation loss, initially the topography based diffraction loss is calculated using Deygout with up to three diffraction points. The microcell model

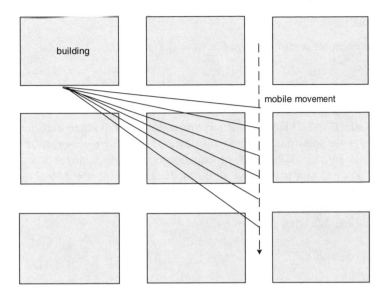

Figure 9.180 Microcell model diagram.

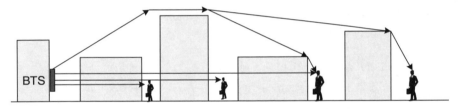

Figure 9.181 Microcell model diagram.

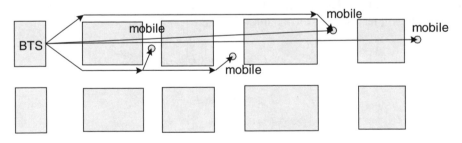

Figure 9.182 Microcell model diagram (bird view).

developed at CelPlan uses a fractional morphology multiple slope approach, with the following break distances for each slope:

- d = total path distance
- d_p = distance to last diffraction point

- $d_i =$ antenna height distance
- $d_2 =$ plane earth break distance $= 4\,h_t h_r (d - d_p) / \lambda \mathrm{sqrt}((h_t - h_r)^2 + (d - d_p)^2)$
- $d_1 = d_2/2$
- $d_3 = 2\,d_2$

The distances $d1$, $d2$ and $d3$ begin at the last diffraction point or at the transmitter, whereas d always begins at the transmitter. The significance of these distances can be understood in Figures 9.181 and 9.182 as the main propagation component changes from a path below rooftops to a mix and to a path above rooftops. Each segment has a slope attributed to it. These slopes are empirically calculated from measurements. The distances are calculated separately for each diffraction segment.

Let's assume the following path losses:

- $0 < d \le d_i : S_i = 20$
- $d_i < d \le d_1 : S_1$
- $d_1 < d \le d_2 : S_2$
- $d_2 < d \le d_3 : S_3$

Equation (9.85) shows the diffraction calculation. The overall loss for *CelPlan*'s microcell model is given by eqn (9.86).

$$D = D_1(T, R) + D_2(T, 1) + D_3(2, R) \tag{9.85}$$

$$L = 32.44 + 20\,\log f + D + L_i + L_{11} + L_{21} + L_{31} + L_{12} + L_{22} + L_{32}$$
$$+ L_{13} + L_{23} + L_{33} \tag{9.86}$$

Equations (9.87)–(9.90) show the calculation of each of the losses involved in eqn (9.86). In these equations n represents a diffraction segment.

$$L_i = 20\,\log(d_i) \tag{9.87}$$
$$L_{1n} = S_1\,\log(d_1) \tag{9.88}$$
$$L_{2n} = S_2\,\log(d_2) \tag{9.89}$$
$$L_{3n} = S_3\,\log(d_3) \tag{9.90}$$

Figure 9.183 shows the parameter configuration table for *CelPlan*'s microcell model.

9.4.2.3   *Model Calibration*

All models benefit from empirical calibration of its parameters. Models that use a single morphology per site or path are easily calibrated. Fractional morphology models are more difficult to calibrate and require automated methods to do it.

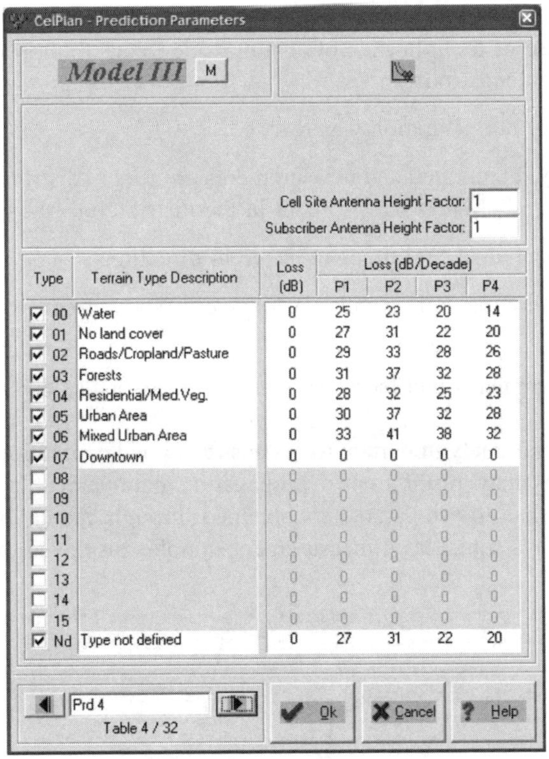

Figure 9.183 CelPlan's microcell model propagation parameters.

9.4.2.3.1 Measurements

Measurements used to calibrate a model should be very precise and be representative of the area. Bad measurements lead to bad results. The following is a check-list for the collection of measurements and calibration:

- Measurement coordinates are compatible with database coordinates;

- The measured site is properly characterised. Transmitter and antenna parameters were verified;

- The measured signal did not have interference and was assigned to the correct site (key off recommended when possible);

- Fast fading was filtered (average for every 40 λ);

- Collected samples are properly located, i.e., a signal measured on a street falls on this same morphology type in the GIS database;

- Samples collected in morphologies not represented on the GIS database (bridges, tunnels) were removed;

- Measurements in the same vicinity, (eg. inside a pixel of 30 m) were averaged;

- Samples collected close (less than three times tower height) to the transmitter were eliminated because of the influence of antenna side lobes. Samples should be collected within the main antenna lobe;

- Neither noise floor nor saturation was reached;

- The area is evenly represented and measurements are evenly distributed (averages can be done over small grids $(4\,\mathrm{m} \times 4\,\mathrm{m})$ to help in the distribution statistics);

- Streets were measured several times in different directions;

- After the first calibration, points with excessive errors were removed and the model was re-calibrated;

- Number of measurements and locations represent statistically the area.

This check-list is extremely important to make sure a good calibration is obtained. Failure to follow these steps may result in bad propagation parameters. It is better to assume standard parameters than to use parameters obtained through bad calibrations.

Figure 9.184 displays signal level measurement samples displayed over a map of streets.

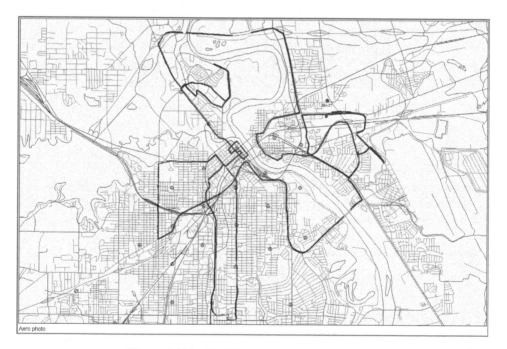

Figure 9.184 Signal level measurement samples.

Measured samples that may distort the calibration for being bad or non-representative samples must be eliminated. This elimination process is usually referred to as filtering. Some of the filtering techniques required are not trivial and computational resources such as *CelTools* are usually used to perform these tasks. Figure 9.185 shows some of the filtering dialog boxes available in *CelTools*.

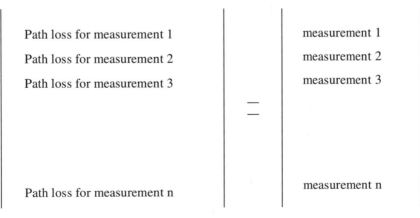

Figure 9.185 CelTools filtering dialog boxes.

A statistically valid number of measurements are required after filtering to obtain a good calibration. A reasonable number is 10 000 measurements remaining after all filtering operations are completed.

9.4.2.3.2 Parameter Extraction

Many mathematical algorithms can be used to calibrate propagation parameters with the objective of reducing the average error to zero, while, at the same time, minimising the standard deviation.

If the parameters suggested by the algorithm are out of range that means that there is something wrong with the calibration set up. Errors are usually found to be related with the database or measurements.

The first step in the calibration process is to establish a matrix that associates path loss equations with the measured samples as shown in Figure 9.186.

The unknowns in this matrix are the propagation parameters. The path loss equations are statistically solved to define these parameters. Parameters that are weakly represented should not be considered. These parameters can be obtained from other measurements or can be estimated by users.

$$
\begin{vmatrix}
\text{Path loss for measurement 1} \\
\text{Path loss for measurement 2} \\
\text{Path loss for measurement 3} \\
\\
\\
\text{Path loss for measurement } n
\end{vmatrix}
=
\begin{vmatrix}
\text{measurement 1} \\
\text{measurement 2} \\
\text{measurement 3} \\
\\
\\
\text{measurement } n
\end{vmatrix}
$$

Figure 9.186 Measurement path loss matrix.

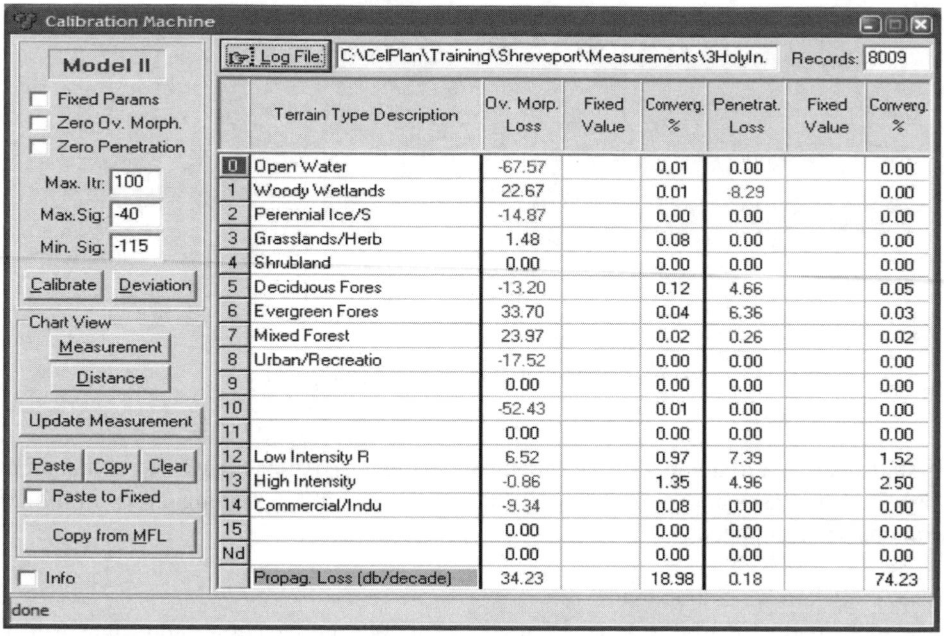

	Terrain Type Description	Ov. Morp. Loss	Fixed Value	Converg. %	Penetrat. Loss	Fixed Value	Converg. %
0	Open Water	-67.57		0.01	0.00		0.00
1	Woody Wetlands	22.67		0.01	-8.29		0.00
2	Perennial Ice/S	-14.87		0.00	0.00		0.00
3	Grasslands/Herb	1.48		0.08	0.00		0.00
4	Shrubland	0.00		0.00	0.00		0.00
5	Deciduous Fores	-13.20		0.12	4.66		0.05
6	Evergreen Fores	33.70		0.04	6.36		0.03
7	Mixed Forest	23.97		0.02	0.26		0.02
8	Urban/Recreatio	-17.52		0.00	0.00		0.00
9		0.00		0.00	0.00		0.00
10		-52.43		0.01	0.00		0.00
11		0.00		0.00	0.00		0.00
12	Low Intensity R	6.52		0.97	7.39		1.52
13	High Intensity	-0.86		1.35	4.96		2.50
14	Commercial/Indu	-9.34		0.08	0.00		0.00
15		0.00		0.00	0.00		0.00
Nd		0.00		0.00	0.00		0.00
	Propag. Loss (db/decade)	34.23		18.98	0.18		74.23

Figure 9.187 Propagation parameters calibration.

When using fractional morphology models, this matrix can be assembled using signal level information related to different sites because the intention is to calibrate morphologies and not sites.

Figure 9.187 shows the calibration dialog box used in CelTools for the Korowajczuk model. Celtools uses a proprietary algorithm (CelExtract) that provides a non constrained optimal result. Next, a simmulated annealing method is used to obtain results constrained to expected ranges. Both results can then be compared and the ranges can be fine tunned to provide reasonable and optimized values.

The list of propagation parameters associated with each of the morphology types is referred to as a parameter table. This table can be applied to sites/sectors on an individual basis. Average error and standard deviation are derived from it according to eqn (9.91). The statistics of these parameters are important to characterise the prediction error.

$$S = S + A + \sigma A(p) + \Sigma(\sigma\Sigma(p))(p) \tag{9.91}$$

where

A	Average of the average error
σA	Standard deviation of A
Σ	Average of the standard deviation
$\sigma\Sigma$	Standard deviation of Σ

An important verification of the quality of the model and its calibration is the re-usability of propagation parameters. The comparison of predicted results to measured values should result in consistent results. The average error should be less than 1 dB and the standard deviations should be in range of 4–8 dB.

9.4.2.3.3 Comparing Predictions and Measurements

The comparison of predictions and measurement samples is essential to determine the quality of the calibration. Figures 9.189 and 9.190 show three types of information: the measured samples, the predicted values corresponding to these samples and the error (or difference) between these two values. Figure 9.188 shows the legend for these figures.

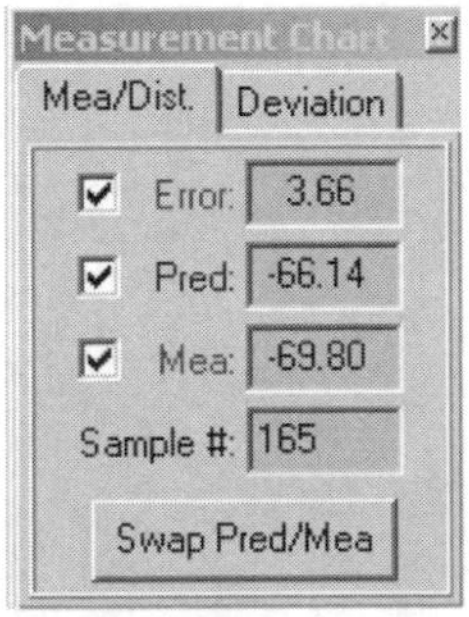

Figure 9.188 Legend.

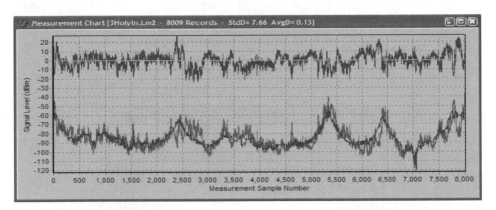

Figure 9.189 Comparing measurements and predictions by sequence.

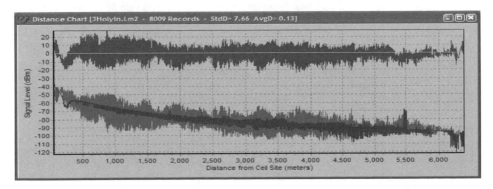

Figure 9.190 Comparing measurements and predictions by distance.

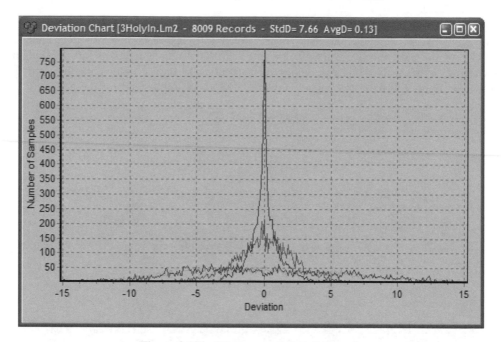

Figure 9.191 Measurement statistics graph.

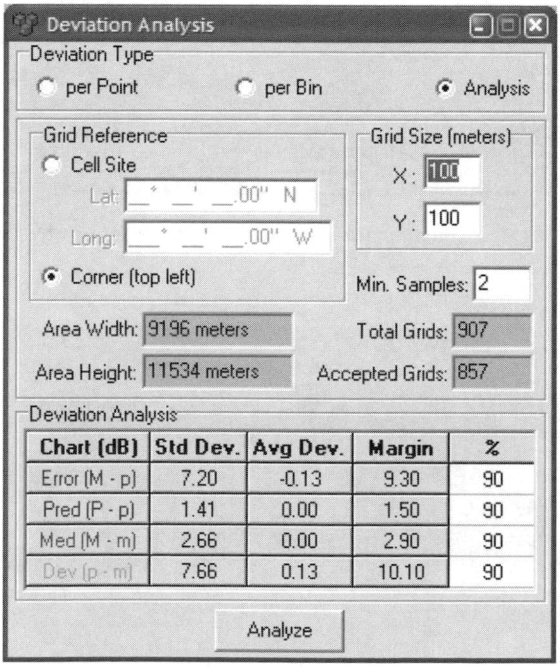

Figure 9.192 Measurement statistics.

Figure 9.189 shows these values in the order of collection of measured samples. Figure 9.190 re-arranges the samples in terms of the distance from the transmitter.

Figures 9.191 and 9.192 display a statistical analysis of the collected samples. In the table shown in Figure 9.192, 'p' stands for predicted values and 'm' for measured values. Small letters indicated the values themselves whereas capital letters indicate the average of values within the specified grid. For the purpose of this analysis predictions are done to the exact locations where the measurements were done. Only then the measurements and predictions are averaged within each pixel (grid).

Measurements present dispersion inside the grid characterising the slow fading effect. Prediction values have less dispersion in areas where the database is simplified, providing a representation closer to the real morphologies. The average difference is practically zero, meaning that the prediction, in average, is reproducing real life.

9.4.2.4 Model Comparison

To determine how much calibration is required, designers must know the model capabilities. The propagation model can be evaluated against measurements collected in the calibrated area.

The process of model comparison starts by the collection of measurement samples for typical sites. This data should be plotted over the GIS database to verify its integrity. Statistical analysis should be performed with the measured results, especially pixel standard deviation.

If the samples are considered statistically valid, the propagation parameters can be calibrated. Predictions should be executed using the calibrated parameters. These predictions can then be compared to the collected samples. This comparison should include average and standard deviation calculation.

Table 9.73 shows the comparison results between measurements and predictions obtained with different models when using a GIS database that combines 90 m, 30 m and street level morphology information. The table shows the comparison using default and calibrated parameters.

Models perform better when calibrated, but it would be very expensive and time consuming to calibrate each site in a market. Besides it is very difficult to calibrate non-existing (new) sites. Parameter re-usability is essential for a good model.

Average deviation is the parameter that mostly affects results because it shows a deviation that is not being compensated. Standard deviation is important but its imprecision is partially compensated when we analyse a large number of points statistically.

Table 9.73 Typical prediction performance for different models

Model	Same site				Other site			
	Default values		With calibration		Default values		With calibration	
	Average	Std	Average	Std	Average	Std	Average	Std
Okumura traditional	6	14	2	10	6	14	5	12
Okumura fractional	3	12	0	8	3	12	4	10
Lee fractional	2	10	0	7	2	10	3	8
Korowajczuk fractional	1.5	8	0	6	1.5	8	2	7

9.4.2.5 RF Signal Level Predictions

RF signal level predictions are calculated using the path loss values obtained for the propagation model. Measurements are usually taken on streets using an external antenna. When a propagation model is calibrated using these measurements, the loss it predicts also refers to external receivers. If the database does not have streets carved in it, the predictions reflect the signal in virtual embedded streets inside other morphology types. It is common for different people to do the calibration, the predictions and the analysis of results. The real meaning of the prediction is easily lost and the interpretation is wrongly done. This factor and the inadequacy of classical propagation models have led people to discredit RF prediction efficacy.

If the prediction constraints are understood, they can be adjusted to reflect other situations. If the prediction loss was calculated for streets, additional penetration losses can be added to represent indoor values, for example. Databases that have streets embedded into different types of morphologies should be treated carefully and it should be clearly identified if the propagation parameters represent the values obtained at virtual street level or they represent the values inside the morphology (outside streets).

Once we have understood what the prediction loss is representing in the different morphology types we have to understand what the value we see does represent. The loss values calculated are average with a statistical distribution associated to them. Some authors go to great lengths defining the loss at L (50,50) meaning that the loss value is assumed to be larger or smaller than the average 50% of the time on 50% of the area (the area being the resolution pixel).

The values calculated for loss can be analysed as a statistical distribution with a standard deviation. These values can be calculated from measurements or extrapolated from theoretical calculations. The reason these values are statistically distributed (dispersion) is the existence of slow and fast fading. The different types of fading are described in more detail in Section 9.4.2.7.1.

Slow fading (shadowing) depends on the morphology variation in the measured pixel and on neighbour pixels. It means that if several points are measured (and time filtered for fast fading) inside a pixel the values obtained will be different and they can be expressed by an average and a standard deviation. The values of the standard deviation represent the slow fading. It can be calculated by tools like CelTools that expresses the $M - m$ (pixel average—measured value inside the pixel) statistics for a specified pixel size over the area under study. This distribution is generally log normal and the standard deviation represents the slow fading.

Fast fading is caused by multi-path and Doppler effects. It can be measured for the different morphologies through a set of measurements averaged within distances of 20λ. The speed of the measurement equipment, for example, the speed of the car, should be maintained the same during collection of samples.

The prediction error does not increase signal dispersion and should not be considered when representing the signal.

The final distribution of loss values can be approximately represented by the statistical sum of the single distributions as in eqn (9.92).

$$\sigma_{tot} = \mathrm{sqrt}(\sigma^2 \text{slow fading} + \sigma^2 \text{fast fading (multipath)} + \sigma^2 \text{fast fading (doppler)}) \quad (9.92)$$

Not only fading affects predictions. The location of subscribers is also an important factor when performing prediction studies. Subscribers use their phones in different environments

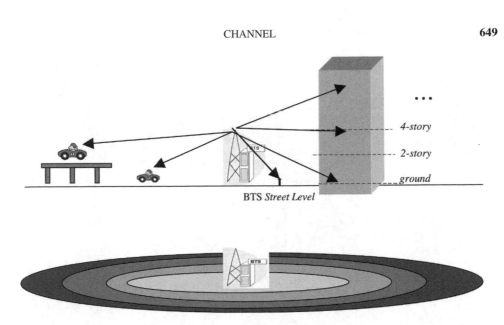

Figure 9.193 User locations.

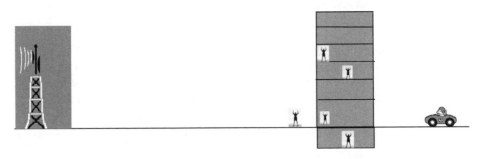

Figure 9.194 User distribution.

and this has to be somehow considered. A user on the 20th floor of a building gets a signal different than a user standing few meters away but outside the building (Figure 9.193).

User distributions vary from market to market and Figure 9.194 shows possible user locations while Table 9.74 provides some typical values for user distributions.

Table 9.74 Typical user distribution

Outdoor calls	15%
Indoor ground level calls close to window	10%
Indoor ground level calls internal	5%
Indoor elevated calls close to window	20%
Indoor elevated calls internal	10%
Encapsulated calls	5%
In car calls	35%

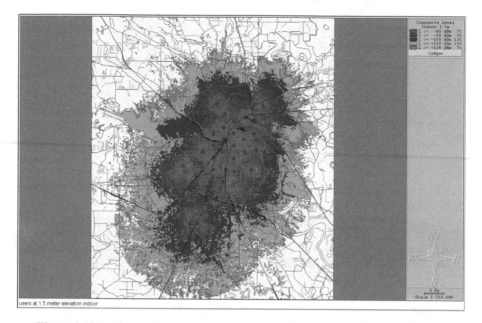

Figure 9.195 Composite signal level predictions for ground level (1.5 m) indoor.

User elevation significantly changes the path loss and cannot be corrected using a simple adjustment as obstructions may be cleared at different heights. Therefore path loss prediction must be calculated for different heights. Figures 9.195 and 9.196 show signal level predictions for different heights and different environments (indoor and outdoor).

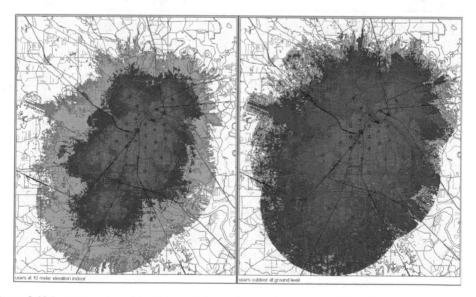

Figure 9.196 Different user height predictions: 10 m height indoor (left) and ground level outdoor (right).

9.4.2.6 Prediction Error

It is possible to calculate, or forecast, the prediction error with reasonable accuracy. There are, however, limits to the minimisation of this error and small improvements may have large costs. The problem then is to determine how good a prediction needs to be. The solution depends on what the predictions are going to be used for.

This analysis can be done by first formulating the statistics of the desired predictions application and then analysing what tolerance for errors it has. Typical applications for predictions are coverage and service availability (C/I, S/N, or E_b/N_0). An average error affects the accuracy of the results, while the standard deviation will influence only the precision, therefore a good prediction should have a low average error. The standard deviation will impact the precision by giving a larger uncertainty to the results. Prediction error analysis shows that the error distribution ressembles Gaussian as shown in Figure 9.191. The impact of the prediction error on the signal statistics can be evaluated using a Monte Carlo analysis by applying the error over the signal while keeping the probability distributions of both and calculating the composed distribution. This analysis applies to signals and interferers and should be done over the linear values. This will increase slightly the standard deviation of the signal.

For the majority of wireless network designs acceptable values for the average error lie in the range of -2 dB to $+2$ dB, while error standard deviations should be less then 6 dB (after removing slow and fast fading deviations). This does not mean that some predictions with larger deviations cannot be used, as it is difficult to get the same accuracy and precision over the entire network.

Section 9.4.3.4 provides more details about the signal and interference ratio issue.

9.4.2.7 Fading

Even when transmit signal levels are constant, received signal levels are not, because of a combination of mechanisms called fading. Fading can be divided into two types: fast (or multipath) fading, and slow (or shadow) fading. Each of these types is analysed in the following sections.

9.4.2.7.1 Multi-path Fading (Fast Fading)
This fading is caused by the reception of a signal coming from many different directions but transmitted from a single source as illustrated in Figure 9.197.

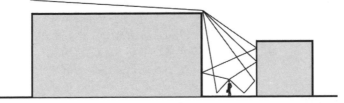

Figure 9.197 Multi-path fading.

Each of the waves received has a different strength and a different phase. It is impractical to deterministically calculate all the multi-path components, therefore a statistical approach is used.

Three situations are possible when statistically analysing fading:

- There is no multi-path and the signal is received in a line-of-sight (LOS) situation with only environment noise present (AWGN channel);

- There are only non-line-of-sight (NLOS) components and there is noise present (Rayleigh distribution);

- There is a LOS component, NLOS components and noise (Rice distribution).

In the LOS scenario, a LOS signal is combined with many NLOS signals. The ratio between the LOS and NLOS signal components determines the type of fading. When the LOS is pre-dominant, fading has a Gaussian distribution; when the NLOS is pre-dominant, fading has a Rayleigh distribution; when there is a balance, fading follows a Ricean distribution. There is not a distinctive line separating the distributions and all three of them can be derived from a Rice distribution just changing its parameters.

We will briefly analyse the main characteristics of each of these fading distributions.

AWGN Channel

When the signal reaches the receiver through a single path and noise is present in the environment, this channel is characterised as an Additive White Gaussian Noise (AWGN) channel.

$$s(t) = a(t) \cos \left[2 \pi f_c t + \theta(t) \right] \tag{9.93}$$

where

$a(t)$ envelope
f_c carrier
$\theta(t)$ phase

$$\text{rms} \cos(t) = a(t)/\text{sqrt}(2) \tag{9.94}$$
$$P_s = a^2(t)/2 \tag{9.95}$$

where

P_s mean power

$$y(t) = L u(t) + n(t) \tag{9.96}$$

where

$y(t)$ received signal
L path loss

$u(t)$ modulated signal
$n(t)$ noise, zero mean independent Gaussian process

$$P_n = \sigma_n^2 \tag{9.97}$$

where

σ_n noise standard deviation
P_n mean noise power.

Considering $u(t) = 1$, eqn (9.98) can be derived from eqns (9.93)–(9.97).

$$\gamma = \text{SNR} = L^2/2\,P_n = L^2/2\,\sigma_n^2 \tag{9.98}$$

For digital signals, the symbols energy (E_s) can be calculated by eqn (9.99).

$$E_s = L^2\,T/2 \tag{9.99}$$

where

L path loss
T fading duration

   Considering that the bandwidth can be expressed as eqn (9.100), the mean noise power (P_n) can be expressed in terms of the power spectral density (N_0) as in eqn (9.101).

$$B = 1/T \tag{9.100}$$
$$P_n = B\,N_0 = N_0/T \tag{9.101}$$

Considering that m is the number of bits per symbol. . . .

$$\gamma = E_s/N_0 \tag{9.102}$$
$$\gamma_b = \gamma/m = E_b/N_0 \tag{9.103}$$

   Figure 9.198 shows a Probability Density Function (PDF) of a Gaussian (normal) distribution, and Figure 9.199 shows the Cumulative Distribution Function (CDF), or Gaussian cumulative distribution. Sigma represents the standard deviation and the cumulative distribution shows what is the probability that the average value is exceeded by a number of standard deviations. Table 9.75 correlates the PDF to the cumulative distribution function for different number of standard deviations of a Gaussian distribution.

Rayleigh Distribution
In an NLOS situation, the sum of multiple independent signals assumes a normal (Gaussian) distribution. Equation (9.104) represents the calculation of the resulting signal (sum of

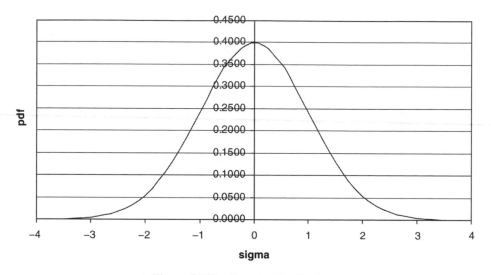

Figure 9.198   Gaussian distribution.

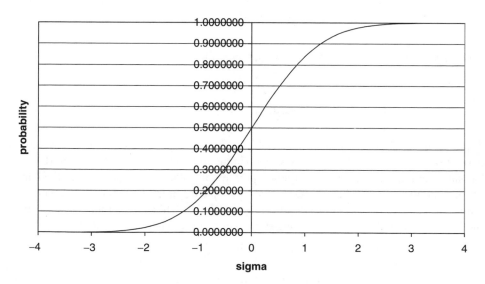

Figure 9.199   Gaussian cumulative distribution.

Table 9.75 Probability density and cumulative distribution of a Gaussian distribution

Std	PDF	CDF
−5	0.0000	0.0000003
−4.5	0.0000	0.0000034
−4	0.0001	0.0000317
−3.5	0.0009	0.0002326
−3	0.0044	0.0013499
−2.5	0.0175	0.0062097
−2	0.0540	0.0227501
−1.5	0.1295	0.0668072
−1	0.2420	0.1586553
−0.5	0.3521	0.3085375
0	0.3989	0.5000000
0.5	0.3521	0.6914625
1	0.2420	0.8413447
1.5	0.1295	0.9331928
2	0.0540	0.9772499
2.5	0.0175	0.9937903
3	0.0044	0.9986501
3.5	0.0009	0.9997674
4	0.0001	0.9999683
4.5	0.0000	0.9999966
5	0.0000	0.9999997

multiple independent signals).

$$z = x + i\,y \tag{9.104}$$

The probability density function of the module of the z vector is given by eqn (9.105).

$$r = \text{sqrt}(x^2 + y^2) \tag{9.105}$$

The Rayleigh distribution is then given by eqn (9.106).

$$p_{\text{Ray}}(r) = (r/\sigma^2)\exp(-r^2/2\,\sigma^2) \tag{9.106}$$

The Rayleigh channel presents the most severe fading as shown in Figures 9.200–9.202. All the graphs in these figures were calculated for $\sigma = 4$.

Figure 9.200 shows the Rayleigh distribution. The mean of this distribution is 1.2533σ, the median is 1.177σ, and the peak is 0.6065σ.

Figure 9.201 shows the Rayleigh Cumulative Distribution Function (CDF).

Figure 9.202 shows the error probability for different SNR.

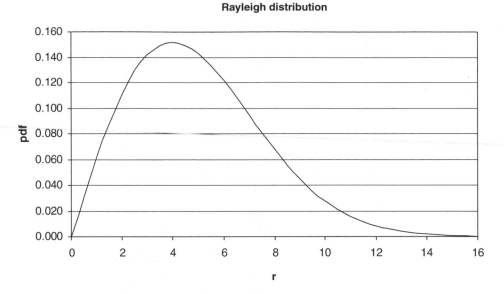

Figure 9.200   Rayleigh distribution.

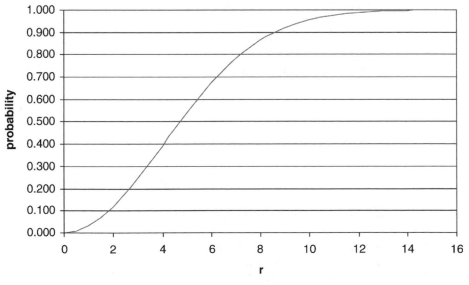

Figure 9.201   Rayleigh CDF.

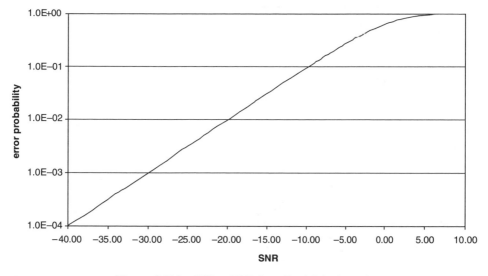

Figure 9.202   SNR × BER for a Rayleigh channel.

Rice Distribution (Nakagami-n)
In situation where there is balance between NLOS and LOS signal, the sum of multiple independent signals assumes a Ricean distribution. Equation (9.107) shows the calculation of the Rice k factor (defined by the signal-to-noise power ratio).

$$k = \text{LOS power}/\text{NLOS power} = s^2/2\,\sigma^2 \tag{9.107}$$

The probability distribution function (PDF) for a Ricean distribution is given by eqn (9.108), where I_0 is the modified Bessel function of the first kind and order zero.

$$p_{\text{Rice}}(r) = (r/\sigma^2)\exp(-r^2/2\sigma^2\exp(-k)I_0(r\;\text{sqrt}(2\,k)/\sigma) \tag{9.108}$$

Figure 9.203 shows the pdf of a Rice distribution for several values of k. The graph indicates that for a small k, the distribution approaches Rayleigh's distribution and, for a large k, it approaches the AWGN distribution.

Fading Statistics
The distributions above indicate the signal levels that can be expected because of fading but give no indication about the duration of the fading. These dynamic effects (fading variations with time) can be obtained by analysing the Doppler effect caused by the movement of the receiver. When the receiver moves toward the transmission source, the frequency apparently increases, when it moves away from the source, the frequency seems to decrease.

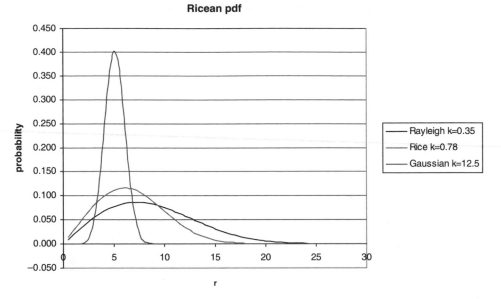

Figure 9.203 Rice distribution.

Equation (9.109) indicates how to calculate the frequency shift (f_m) based on the initial frequency (f_0), the receiver velocity (v) and the speed of light (c).

$$f_m = vf_0/c \tag{9.109}$$

Equation (9.110) defines the fade depth (ρ).

$$\rho = -20 \log (V/V_{\text{rms}}) \tag{9.110}$$

where

V signal voltage at fade
V_{rms} signal root-mean-squared voltage

The crossing rate (N_r) is given by eqn (9.111), based on the value of the fade depth.

$$Nr = f_m\rho \ \text{sqrt} \ (2\,\pi) \exp(-\rho^2) \tag{9.111}$$

The average duration of a fade (τ) is determined by the crossing rate and by the fade depth as indicated in eqn (9.112).

$$\tau = (1 - \exp(-\rho^2))/Nr \tag{9.112}$$

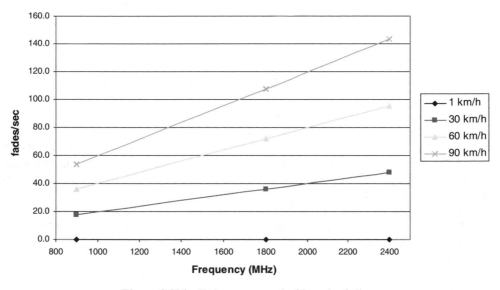

Figure 9.204   Fades per second of Doppler fading.

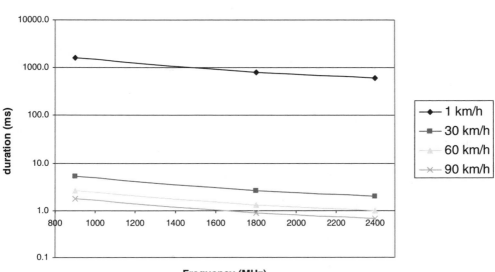

Figure 9.205   Average duration of Doppler fading.

Figure 9.204 shows the Doppler fading according to the receiver velocity. Figure 9.205 shows the average duration of the Doppler fading, which also varies with the receiver velocity.

Channel Fading Bandwidth
The channel fading bandwidth defines how fading is affecting the transmitted signal. The fading bandwidth can be narrowband (frequency non-selective) or wideband (frequency selective).

The fading bandwidth is defined by the ratio of the fading duration to the duration of the information. In narrowband fading, the multi-path components have small path length differences (Tc) when compared to the symbol duration (T_s). This type of fading results from multi-path originating in the vicinity of the receiver and, even though the difference may be of several wavelengths, the time arrival difference is a fraction of the symbol duration.

Wideband fading occurs when objects that are far away from the receiver cause the multi-path. The path length difference (T_c) is large in this case, in the order of the symbol duration (T_s) or larger. Wideband fading has a severe impact on the BER because it presents a higher error floor than narrowband fading.

The path length difference (T_c) is also called coherence time and expresses the period of time after which the correlation function of two samples of the channel response taken at the same frequency but different time intervals falls below a defined threshold.

Narrowband fading happens when:

$$T_c \ll T_s \qquad (9.113)$$

Wideband fading happens when:

$$T_c > T_s \qquad (9.114)$$

Figure 9.206 illustrates narrowband fading and Figure 9.207 illustrates wideband fading.

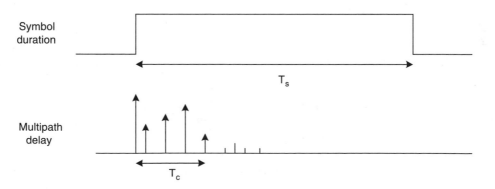

Figure 9.206   Narrowband fading.

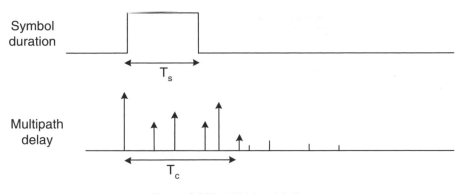

Figure 9.207 Wideband fading.

Overcoming Multi-path Fading

Several precautions can be taken to minimise fading, including data rate reduction, cell size reduction, diversity, equalisation and use of directional antennas.

The reduction of the data rate increases the bit duration and consequently minimises the occurrence of wideband fading. The cell size reduction avoids long multi-paths because small cells use low power. The diversity technique minimises multi-path by tapping on more than one source of signal. The equalisation transforms the wideband fading channel into narrowband by applying adaptive filters. The use of directional antennas reduces the multi-path by offering less gain to undesired directions therefore improving co-channel interference.

9.4.2.7.2   Shadow Fading (Slow Fading)

After the fast (or multi-path) fading is filtered by averaging the signal over an interval larger than the largest propagation time difference, the signal calculated is said to be valid for 50% of the time. This means that, statistically, for any measured value the signal is 50% of the time above it and 50% of the time below it.

A small movement of the receiver may result in a different value, because of variations on the path, such as different obstructions or stronger reflections entering into play. It is impractical to account for all these variations therefore it is necessary to integrate the signal over an area. The first studies in this area analysed concentric circles. Okumura implemented the analysis area as 2 km square. Today resolutions of 30 m or even 3m are used.

The variation inside this analysis area is called shadow fading or slow fading and is characterised by local variability σ_s. This value is the standard deviation of the time-averaged signal inside the area of analysis.

Previous studies of shadow fading calculated this local variability over the whole area and over the points close to the edge of the area. The edge shadowing only applied to the concentric circles analysis. The circle edges were considered to be the worst areas. The area shadowing should be used when we divide the area in pixels, as it is done in today's predictions. The value of shadow fading varies significantly with the resolution chosen because the local variability diminishes with the resolution size.

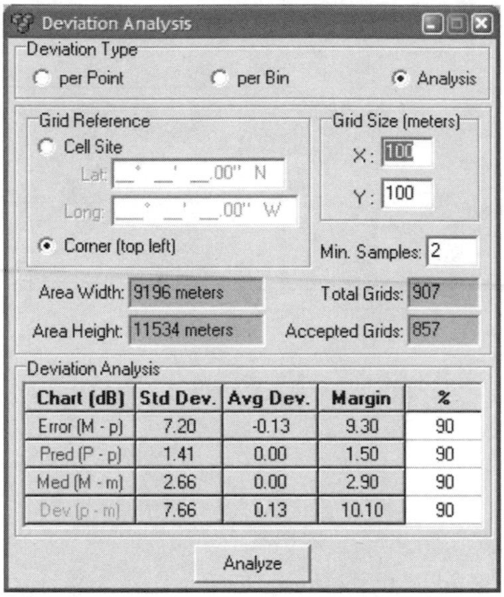

Figure 9.208   Shadow fading (CelTools dialog box).

After calculating the multi-path and shadow fading, it is possible to predict a minimum signal value with the desired probability of time and area. This probability is expressed as a pair of percentages, for example, 90% of the time and 90% of the area is a common goal. This example can also be indicated as (90, 90).

The local variability parameter can be obtained by examining measurements made inside several resolution grids, as shown in Figure 9.208. A typical value for a grid of 100 m × 100 m is 2.9 dB, whereas for a grid of 30 m × 30 m it is 2 dB.

9.4.2.7.3   Diversity

Diversity is a way of mitigating multi-path fading by providing the transmitted signal to the receiver through more than one channel. Each channel represents a diversity branch. Figure 9.209 depicts a typical situation.

There are several techniques to provide diversity. All of them have to provide low fading cross-correlation between the channels, i.e, the diversity paths should have a length difference larger than ¼ of a wavelength. The diversity channels have to use similar power levels so one of them can replace the one that is in fading.

Diversity schemes require additional hardware therefore more than two channels are seldom used. The diversity benefit is usually presented as a gain in dB, but the benefit is not the gain itself but the reduction of fading effects. Diversity does not improve the performance when fading is not present. The most common diversity techniques are space, angle, time and frequency diversity, which are described in the following topics.

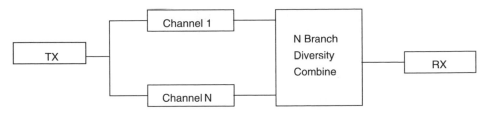

Figure 9.209   Diversity diagram.

There are several types of diversity that can be implemented and they are presented below.

Space Diversity
This diversity is implemented by assembling two antennas with separation enough so that the multi-path signals created from them have lengths with a difference of more than ¼ of a wavelength from each other as in Figure 9.210.

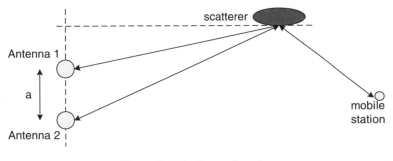

Figure 9.210   Space diversity.

The difference of path lengths depends on the scatterer (reflective element) position. Figures 9.211 and 9.212 show the effect of space diversity for antennas separation of 1 wavelength (1 λ) and 40 wavelengths (40 λ) on a study executed for a simplified case of space diversity. The graphs map the diversity for different scatterer angles (θ) and distances.

The graphs show that space diversity is only effective for scatterers that are more than 30° away from the mobile position. This significantly reduces the efficacy of space diversity. The diversity gain should be preferably measured in the field to provide reliable results. As a rule of the thumb it is considered that 60% of the fading loss is reduced when using two antennas.

Angle Diversity
Another diversity technique is the use of angle diversity. This method consists on receiving signals from different directions. Both antennas receive the direct signal but each of them receives the multi-path signals from a different side. This type of diversity

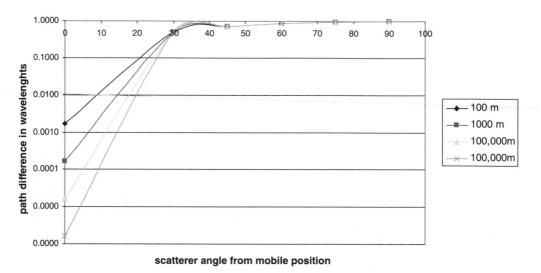

Figure 9.211   Space diversity for 1 λ.

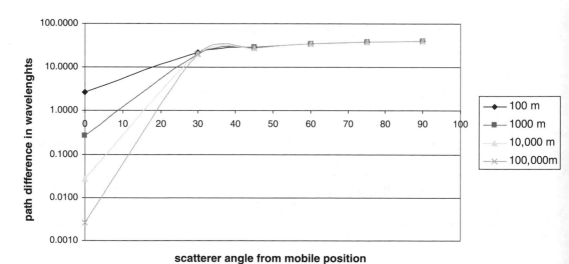

Figure 9.212   Space diversity for 40 λ.

requires uniform morphology around the mobile in order for the multi-path signals to be evenly distributed (as in downtown areas).

Time Diversity
In the time diversity implementation, the same signal is re-transmitted more than once. Therefore, this technique requires additional bandwidth and is not commonly used. The interleaving process is a kind of time diversity.

Frequency Diversity
In frequency diversity the signal is divided in segments and each segment is sent through a channel using a different frequency. This technique spreads eventual errors and improves the efficacy of error correction. Orthogonal Frequency Division Multiplexing (OFDM) is a technique that uses this kind of diversity.

Several combining methods can be used to combine the multiple diversity paths. Some method increases the degree of processing required but in general present similar results. This section briefly illustrates the most common methods, which include selection, switching, equal gain and maximum ratio combining.

Selection
In the selection combining method, all branches have the instantaneous SNR measured and the branch that has the best ratio is connected to the receiver.

Switching
In switching combining, the receiver is connected to one of the branches. When the SNR drops below a pre-defined level the receiver is switched to the other branch.

Equal Gain
In equal gain combining, all branches are phase adjusted and then combined and sent to the receiver.

Maximal Ratio
In the maximal ratio combining method, all branches are phase adjusted and weight combined according to their SNR before being sent to the receiver.

9.4.2.8   Equalisers

Diversity mitigates the impact of multi-path by choosing the best source of information at any instant. This solution works for narrowband fading but does not work for wideband fading where the delays are of the order of the bit period being received. Equalisers provide the solution for wideband by cancelling or extracting the largely delayed multi-path, improving signal detection.

The equaliser samples the received sequence at symbol intervals and then processes these values to reduce inter-symbol interference (ISI). There are several types of equalisers. This section briefly describes the four most common types: linear, adaptive, non-linear and rake receivers.

Linear Equalisers

A linear equaliser, as in Figure 9.213, adds time-shifted copies of the signal samples with different coefficients to optimise signal detection. These shifts can have the duration of the symbol interval or only a fraction of it.

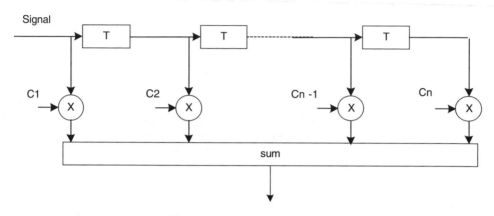

Figure 9.213 Linear equaliser.

The zero-forcing equaliser objective is to cancel the ISI at the sampling points, whereas the least mean square equaliser aims to minimise the total interference (including noise) at the sampling points.

Adaptive Equalisers

The wireless channel response varies over time and also does the ideal delay. Different algorithms were developed to optimise these delays, including convergent and blind algorithms.

Convergent algorithms require a known training sequence that is sent together with the data to allow them to adjust the delays by minimizing the error of the detected training sequence. Blind algorithms do not require training sequences but are less efficient than convergent algorithms.

Non-Linear Equalisers

Maximum Likelihood Sequence Estimators (MLSE) are a type of non-linear equalisers that analyse the statistics of the disturbance over a certain number of symbols and by applying those statistics to all possible sequences that can be received, find the one that is most likely to have been transmitted. This requires extensive calculations and limits the usage of the method.

Viterbi developed an algorithm that simplifies this computation by building a Trellis diagram that computes the maximum likelihood sequence. The results are similar to the maximum ratio diversity combining for narrowband signals. The maximum improvement achievable is the elimination of the impact of signal fading.

Rake Receivers
The Viterbi algorithm is widely used in wireless communications in general, in CDMA, however, it is replaced by another method that explores the technology capability of extracting shifted copies of the signal from a mix of signals. This is possible because of the orthogonality presented by PN sequences to their shifted copies.

In the rake receiver method, several receivers are connected in parallel, each with a shifted copy of the spreading sequence. This arrangement is called a Rake receiver (named after the Garden Rake). Figure 9.214 illustrates this arrangement.

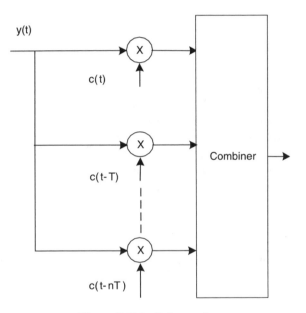

Figure 9.214   Rake receiver.

Each of the rake receiver branches is called a finger. The number of fingers is limited and the best delays have to be chosen. The output of each finger is combined in a maximum ratio combiner.

Rake receivers have an excellent performance to minimise wideband fading.

9.4.3   Signal, Noise and Interference as seen by the Receive stage

The receive stage gets a mix of signals, noise and interference from the wireless channel. Although some aspects of these elements were already analysed, we need to evaluate their effect on the receiver stage and how this element combine. The goal is to extract the signal sent by the transmit stage.

9.4.3.1 Signal

The desired signal is, in reality, a composition of signals arriving through different paths added to signal variations caused by the terminal mobility and network power control. The composite signal presents a statistical variation that is formed by the aggregation of all these effects. Each effect causes a statistical variation of the signal. Some of the variations may tend to specific distributions but they can be approximated to a normal distribution because there are many independent sources responsible for the signal statistics.

The main sources of signal variation are:

- Narrowband fading;

- Wideband fading;

- Shadow fading; and

- Power control dispersion.

An average power value can be associated with the signal and a standard deviation applies to each of the sources of signal variation. These deviations may vary geographically and can have different values for each prediction point.

The calculations should use linear units and the results should be converted to log units (dB). Equation (9.115) shows the formula to calculate signal standard deviation (σ). In this equation, σ_i indicates the standard deviation of the different components.

$$\sigma = \text{sqrt}(\text{sum}(\sigma_i^2)) \qquad (9.115)$$

Table 9.76 shows a typical calculation of the signal composition. The signal may be present only during part of the time because traffic may be smaller than the channel capacity.

9.4.3.2 Noise

Noise is any unwanted power received at the receiver. It is intrinsic to the random motion of charged carriers in matter due to its temperature and is also generated by man-made sources.

Table 9.76 Signal composition

	Signal	Unit	σ	Unit	Distribution
Signal	−60	dBm			
Signal narrowband multi-path fading			5	dB	Gaussian
Signal wideband multi-path fading			4	dB	Gaussian
Signal shadow fading			3	dB	Gaussian
Power control dispersion			1	dB	Gaussian
Combined signal	−60.0	dBm	6.7	dB	Gaussian

Thermal Notice

The thermal noise can be calculated based on quantum mechanics as in eqn (9.116).

$$N_0 = hf[(1/(e^{hf//k_bT} - 1)) + 1], \text{in W} \quad (9.116)$$

where

f channel frequency in Hz
h planck's constant $= 6.6261755 * 10^{-34}$ Js
k_b boltzmann's constant $= 1.380662 * 10^{-23}$ J/K
T absolute temperature in K
N_0 noise per Hertz

For frequencies below 10 GHz this formula can be simplified to eqn (9.117).

$$N_0 = k_bT, \text{in W/Hz} \quad (9.117)$$

The total noise power in a bandwidth B (in Hz) is calculated below as well as its standard deviation

$$N = k_bTB, \text{in W} \quad (9.118)$$

$$\sigma = \text{sqrt}(k_bTB), \text{in W} \quad (9.119)$$

An equipment noise can be expressed in terms of an equivalent noise temperature as indicated in eqn (9.120). This is the temperature that will produce a thermal noise power equivalent to the equipment generated noise power.

$$T_e = N_e * T_0 / N_0 \quad (9.120)$$

Isotropic Background Noise

The isotropic background noise is present in the whole universe and is the remnant form of the Big Bang[7]. This component is very small when compared to other noises present in the environment. The value of this noise is given by:

$$T = 2.726 \text{ K}$$

where T_0 ambient noise temperature N_e equipment noise power T_e equipment noise temperature

Galactic Noise

The galactic noise has a bathtub[8] characteristic and has it minimum value between 1 and 10 GHz. This component can be calculated by eqn (9.121).

$$T_g = 7 * 10^{26} / f^3 \quad (9.121)$$

---

[7] Dominant scientific theory about the Universe origin.
[8] Bathtub curves have a high initial value, a low and stable intermediate value and a high end value. They are typically used to represent components reliability.

Table 9.77 Noise composition

t	Temperature		17	°C
T	Absolute temperature		290	K
T_0	Ambient noise temperature		290	K
		$h_f[(1/(e^{h_f/kbT}-1))+1]$	4.00422E $-$ 21	W/Hz
			-173.97	dBm/Hz
T_b	Background noise	2.2726	2.2726	K
			3.1676E $-$ 23	W/Hz
			-194.99	dBm/Hz
T_g	Galactic noise	$7 \times 10^{26}/f^3$	0.9602	K
			1.35578E $-$ 23	W/Hz
			-198.68	dBm/Hz
N_i	Total input noise		293.23	K
			4.049E $-$ 21	W/Hz
			-173.93	dBm/Hz

Assuming the average ambient temperature at 17°C, Table 9.77 indicates the calculation off the total input noise from the atmosphere.

This noise density (expressed in W/Hz) can be extended to the whole bandwidth using eqn (9.122).

$$N = N_i + 10 \log B \tag{9.122}$$

For CDMA B = 1 228 800 Hz

$N = -83.1$ dBm
$\sigma = 26.5$ dB

Man Made Noise
This is typically an impulsive type of noise (that is usually generated by things such as car ignition systems and electric utilities. Parsons [5] presented an interesting study in Britain that showed that this noise diminishes with the increase of frequency and that, in urban centres, there are a few hundreds crossings per second with a level 25 dB above the thermal noise with pulse durations in the order of 10 μs.

9.4.3.2.1 Noise in Cascading Stages
Receive equipment is constituted by several stages of circuits and each one has a different noise characteristic. When several stages are in cascade, such as in a typical receiver, the noise contribution of each stage should be added according to eqn (9.123).

$$T_e = T_1 + T_2/G_1 + T_3/G_1G_2 + T_4/G_1G_2G_3 \cdots \tag{9.123}$$

In this equation, T_n is the noise contribution of each stage (in temperature units) and G_n is the gain of each stage. From the formula it is clear that the noise in the initial stages is very important to the total result. Figure 9.215 shows the different cascading stages for noise calculation.

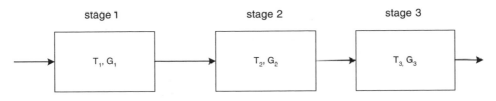

Figure 9.215 Noise in cascading stages.

A noise ratio (NR) can be defined to represent the relative noise contribution of each stage as indicated in eqn (9.124). When this ratio is expressed in dB as in eqns (9.125) and (9.126), it becomes the noise figure (NF). Equation (9.127) calculates the value of the noise in temperature units from the noise figure. A range of value is shown in Table 9.78.

$$NR = (T_e + T_0)/T_0 \tag{9.124}$$

$$NF = 10 \log NR \tag{9.125}$$

$$NF = 10 * \log((T_e + T_0)/T_0) \tag{9.126}$$

$$T_e = T_0 * 10^{NF/10} - T_0 \tag{9.127}$$

The noise figure can be directly added to the input noise to calculate the noise offered to the receiver stage.

The cascading formula becomes the Friis' noise formula shown in eqn (9.128). This calculation has to be done with linear units and then be transformed to logarithmic units.

$$NR_t = NR_1 + (NR_2 - 1)/G_1 + (NR_3 - 1)/G_1G_2 + (NR_4 - 1)/G_1G_2G_3 + \cdots \tag{9.128}$$

Tables 9.79 and 9.80 show, respectively, examples of system noise figure calculation for ground mounted and tower mounted LNAs. The tower mounted LNA provides an improvement of 0.6 dB in this case.

Table 9.78 Noise figure to noise temperature relation

T_e (K)	NF (dB)	NF (dB)	T_e (K)
0	0	0	0.0
1	0.01	0.01	0.7
2	0.03	0.02	1.3
3	0.04	0.05	3.4
5	0.07	0.1	6.8
10	0.15	0.2	13.7
20	0.29	0.5	35.4
30	0.43	1	75.1
50	0.69	2	169.6
100	1.29	5	627.1
200	2.28	10	2610.0

Table 9.79 Noise parameters—ground mounted LNA

	Ground mounted LNA		
T_1	Antenna noise temperature	35	K
	Antenna gain	10	
T_2	Cable noise temperature	290	K
	Cable gain	0.5	
T_3	LNA noise temperature	169	K
	LNA gain	1000	
T_4	Receiver noise temperature	2000	K
T_e	Equivalent temperature	98.2	K
NF	Noise figure	1.3	dB

Table 9.80 Noise parameters—tower mounted LNA

	Tower mounted LNA		
T_1	Antenna noise temperature	35	K
	Antenna gain	10	
T_2	LNA noise temperature	169	K
	LNA gain	1000	
T_3	Cable noise temperature	290	K
	Cable gain	0.5	
T_4	Receiver noise temperature	2000	K
T_e	Equivalent temperature	52.3	K
NF	Noise figure	0.7	dB

9.4.3.3 Interference

Interference is characterised by signals coming from the same or neighbouring systems. The interference can be in the same channel as the signal (co-channel interference) or on a neighbour channel (adjacent channel interference).

Same Channel Interference
This interference is also called co-channel interference and may come from similar or different sources. If the impairment is only on part of the signal band, the following correction should apply to the interfering power:

$$W = 10 \log(b_i/B_i) \qquad (9.129)$$

where

W correction factor (dB)
b_i bandwidth being interfered
B_i total interferer bandwidth

Other Channel Interference

This interference is also known as adjacent channel interference and is filtered at the source, at the destination, or at both ends. The filter loss at the centre of the band should be discounted from the signal level received.

The power amplifier amplifies a signal before being transmitted. This amplifier has to have a large dynamic range and usually it is not fully linear. Therefore the actual signal is distorted and harmonics are generated. The inter-modulation of these harmonics with the carrier may create out-of-band signals.

The receiver uses filters to remove out-of-band signals. However, any practical filter presents a progressive loss to frequencies that are just outside the pass band. After one channel width, the filter usually has attenuated the signal significantly and whatever is left can be ignored, if not multiple adjacent channels have to be considered.

Filters are generally specified by an out-of-band attenuation, and this single value is used to calculate the out-of-band signals, like is done to calculate adjacent channel attenuation. In practice, the interference can be largely spread and the use of filter masks (with progressive attenuations) will provide more precise results.

The interference is subject to the same variations as the signal, described in Section 9.4.3.1. However one additional variation must be added to the interference component. This additional variation is related to the position of the interferer in its own cell. The position of the interferer determines its power control, which impacts on the amount of interference. This variation is statistically calculated considering the possible location of interferers according to their traffic distribution.

This effect is illustrated in Figure 9.216. In this figure E_r stands for Interferer and E_d for interfered. The interference coming from E_r depends on the position of the user in its cell because of power control. In a CDMA system, all active users have to be simultaneously considered. This can be done by determining the number of simultaneous users by simulation and then applying the ratio of users to the required powers of all possible user positions.

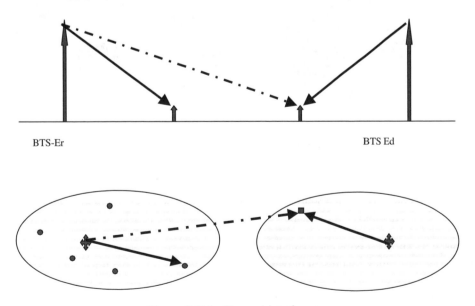

BTS-Er   BTS Ed

Figure 9.216 Forward interference.

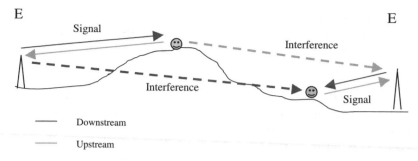

Figure 9.217   Comparing forward and reverse interference.

Downstream (forward link) and upstream (reverse link) interferences can be different as shown in Figure 9.217. As the communication is bi-directional both links have to be considered. Figure 9.218 illustrates the interference calculation for the reverse link.

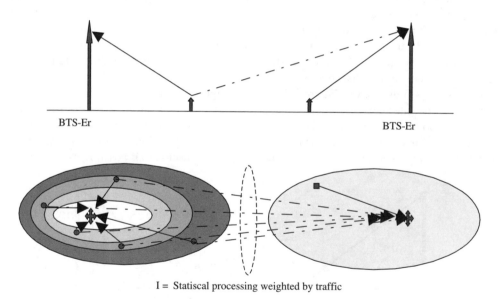

Figure 9.218   Reverse interference.

More than one interference source (different cells) may be present and their influence can be considered separately or in conjunction. In the latter case, noise is also summed to the interference.

Table 9.81 shows a combined calculation of interference.

Each interferer may be present only during part of the period as the traffic may be smaller than its channel capacity. We will identify the channel occupancy by O_{in}. When several interferers are combined the occupancy factor should be used to weight the interference. A composite occupancy will be then obtained.

Table 9.81 Interference budget

	Weight	Signal		σ		Distribution
Thermal noise	1	−83.1	dBm	6.9	dB	Gaussian
Interference 1	0.6	−65	dBm	10	dB	Gaussian
Interference 2	1	−70	dBm	5	dB	Gaussian
Interference 3	1	−75	dBm	6	dB	Gaussian
Combined noise + Interference		−64.9	dBm	9.7	dB	Gaussian

9.4.3.4 Signal-to-Noise and Interference Ratio (SNIR)

This SNI ratio is a very important value because it determines the raw Bit Error Rate (BER) and, consequently, the quality of the recovered signal. The SNIR can be calculated as a ratio of the average power values of S and $(I + N)$, but, in this situation, the ratio would reflect the quality of the signal only for 50% of the time. To calculate the signal quality over any period of time the distribution of S and I and its standard deviation has to be calculated first.

There is not a unique or exact distribution for the SNIR and significant variations can be found from one location to another. The signal power has a predominantly lognormal distribution (meaning that the values in dB have a Gaussian distribution) due to shadowing (long-term fading). Additionally the signal envelope (peak-to-peak amplitude) has a Ricean distribution due to multi-path (fast fading). This Ricean distribution can be characterised as a Rayleigh distribution if multi-path without line-of-sight is present or a Suzuki distribution if line-of-sight is present. Each pixel in the analysis region will have a variation of these distributions and an individual analysis becomes impractical. Not knowing the exact signal power distribution poses a problem, but in the majority of the cases it will be a uni-modal distribution (single peak) and in our opinion the best compromise is to represent it by a Gaussian distribution.

The interference on the other side comes in general from different independent sources and can be even more certainly be represented by a Gaussian distribution by applying the central limit theorem. This theorem states that N independent random variables with a mean and finite variance will have a limiting cumulative distribution function that approaches a Gaussian distribution.

The easiest approximation we can make is to assume the signal and the interference power as lognormal so their values in dB will have a Gaussian distribution and consider that the SNIR has also a Gaussian distribution as the sum of two independent Gaussian random variables is also a Gaussian distribution.

In the case that the signal and interference power are considered lognormal and consequently their distributions expressed in dB have a Gaussian distribution, we can apply the following equations.

$$S(\text{dB}) = G(S_{med}(\text{dB}), \sigma_{S,\text{dB}}) \tag{9.130}$$

If the interference pre-dominates over the receiver thermal noise, the SNIR becomes SIR, which can be expressed in dB by:

$$SIR(\text{dB}) = G(S_{med}(\text{dB}), \sigma_{S,\text{dB}}) - G(I_{med}(\text{dB}), \sigma_{i,\text{dB})}. \tag{9.131}$$

The sum of two independent Gaussian variables is also a Gaussian variable.

$$aG(m_1, \sigma_1) + bG(m_2, \sigma_2) = G\left(am_1 + bm_2, \sqrt{a^2\sigma_1^2 + b^2\sigma_2^2}\right) \qquad (9.132)$$

It follows that SIR in dB is Gaussian and is given by the formula:

$$SIR(\text{dB}) = G\left(S_{med}(\text{dB}) - I_{med}(\text{dB}), \sqrt{\sigma_{S,\text{dB}}^2 + \sigma_{I,\text{dB}}^2}\right) \qquad (9.133)$$

An example of this calculation is shown in Table 9.82.

The SNIR for a guaranteed availability of 90% can be calculated by the Gaussian cdf as 15.1 dB.

Sometimes two interferers do not have the same bandwidth, in which case the power density can be used to adjust the amount of interference perceived. Table 9.83 shows an example of power density calculation.

Table 9.82   Ratio of two Gaussian distributions

	$S_{med}(\text{dB})$	$\sigma_{S,\text{dB}}$
S = Signal	−60	6.7
N + I = Interference + Noise	−64.9	9.7
SNIR	4.9	11.79

Table 9.83   Example of power density calculation

B	Interfering channel bandwidth	1 228 800	KHz
B_i	Interfered channel bandwidth	200	KHz
b_i	Overlap bandwidth	100	KHz
	$10 * \log b_i/B_i$	−3.01	dB

Table 9.84   SNIR calculation

	Weight	Signal		σ		Distribution
Signal	1	−60	dBm			
Signal narrowband multi-path fading	1			5	dB	Gaussian
Signal wideband multi-path fading	1			4	dB	Gaussian
Signal shadow fading	1			3	dB	Gaussian
Power control dispersion	1			1	dB	Gaussian
S = combined signal		−60.0	dBm	6.7	dB	Gaussian
Thermal noise	1	−83.1	dBm	6.9	dB	Gaussian
Interference 1	0.6	−65	dBm	10	dB	Gaussian
Interference 2	1	−70	dBm	5	dB	Gaussian
Interference 3	1	−75	dBm	6	dB	Gaussian
$(N + I)$ = Combined noise + Interference		−64.9	dBm	9.7	dB	Gaussian
SNIR = $S/(N + I)$		4.9	dB	11.79	dB	Gaussian

The SNIR ratio can be calculated considering only one interferer or all interferers. The signal-to-interferer calculation is used to express relationships in matrix form so the possible relations can be analysed separately. The ratio of the signal to all interferers is used for performance calculations. Table 9.84 shows an example of this calculation.

The network requires a certain SNIR to provide a desired level of quality. In this example we will assume a SNIR of 0 dB. Based on the standard deviation calculated above (11.79 dB) it is possible to calculate the probability that the ratio will fall below the required value and this probability will express the network outage (O). In this case the outage is 33.57%.

It is also possible to calculate the SNIR required for any desired outage from the same cdf. If we assume an outage of 10% the SNIR should be 15 dB and consequently the existing SNIR should be improved by 10.1 dB.

The availability (A) represents the percentage of time in which the SNIR is better than a specified limit value. The availability (A) is the complement of the outage (O) as in eqn (9.134).

$$A = 1 - O \tag{9.134}$$

The availability can be calculated on a pixel basis, on a cell basis or on a network basis. As each pixel has traffic associated to it, this traffic can be multiplied by the pixel availability to determine the available traffic for a given quality level. This is a powerful concept as it captures a performance that includes time variations. Figures 9.219–9.221 show that the network can be analysed for different quality levels.

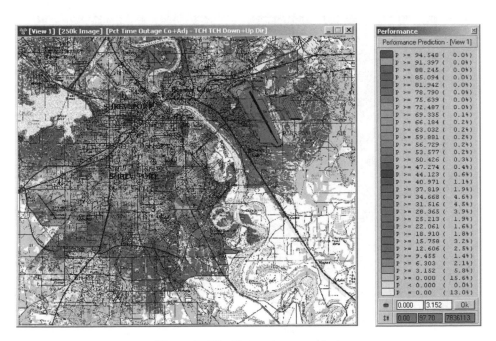

Figure 9.219 Time outage per pixel.

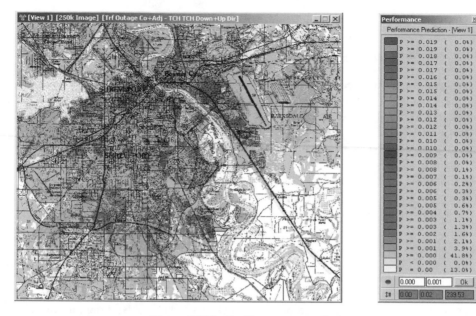

Figure 9.220   Traffic outage per pixel.

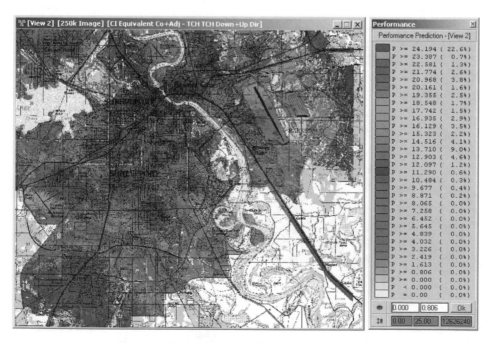

Figure 9.221   SNIR for 50% of the time per pixel.

9.5 LINK MANAGEMENT

As illustrated in Figure 9.1, a wireless communication link is composed of a transmit stage, a wireless communications channel and a receive stage. A commercial communications system requires that hundreds and in some cases thousands of links operate simultaneously. Links interfere with each other and to increase system capacity the disturbance caused by each link should be kept to a minimum. This is the function of the link management: to operate each link so it will interfere as little as possible with other links. The link management function is not a centralised function so the only information it has about the other links is the interference the link it is supervising suffers from the others. The link management behaves as a good citizen by restricting its environmental pollution to the bare minimum to tolerate the impairments it is receiving and still provide the service required from its own link. The tools the link management has at its disposal are handoff, power control and, lately, resource management. Each one of these tools is described below.

9.5.1 Handoff

Handoff is included in the link management category because it is responsible for the transfer of a mobile from one cell to another to preserve the continuity of its own call and minimise interference to other calls. There are three types of handoff: idle, hard or soft.

The idle-handoff happens when the mobile is not in a call but is looking for the best signal. The hard handoff happens when, during a call, a new frequency has to be assigned to the mobile. The circuitry has to leave the existing frequency and tune to the new one (a process known as break before make).

The soft-handoff happens when, during a call, a new cell is in a condition to provide service to the mobile. The circuitry then supports one or more cells simultaneously connected to the mobile (a process known as make before brake). This handoff is further classified in:

- Soft-handoff—between two sectors in different BTS;
- Softer handoff—between two sectors of the same BTS; and
- Soft softer handoff—between multiple cells and multiple sectors within one of the cells.

The mobile itself makes the idle-handoff decision. It periodically measures the available signals and chooses the strongest one.

The hard-handoff happens when there is a need to change carriers. In CDMA, hard-handoff can happen between CDMA carriers or between an AMPS carrier and a CMDA carrier. Handoff from a CDMA carrier to an AMPS carrier was not implemented because of commercial reasons.

Soft-handoff is only implemented in CDMA systems. This type of handoff is requirement of the technology because all cells use the same frequency. The fact of sharing the same frequency among all cells reduces the network capacity but some of this reduction is recovered through a diversity scheme that is provided by soft-handoff gain.

The decision of handoff is always based on the signal level received from possible handoff candidates. A mobile has four rake receivers from which one is used to constantly scan the pilots. The mobile listens to the BTS with the strongest pilot while monitoring other pilots to check their levels.

In the reverse link all neighbour stations receive the mobile signal and its level is adjusted for the one it receives the strongest pilot. It can happen that a second or a third station are being strongly interfered by this mobile and the decision of putting it in soft-handoff will provide some additional diversity gain and consequently will allow the mobile power reduction by the same amount. It should be noted that the diversity gain here is not a theoretical value but it is a result of performance measurements (FER). The decision to perform soft-handoff by a mobile is determined by the intensity of the received pilot and is defined by a T_{add} parameter (generally specified as -12 dB for IS-95). The decision of adding a mobile to a soft-handoff connection requires that a channel element is assigned to it and consequently reducing the available resources and limiting the capacity of the system. It is important that the soft-handoff be cancelled as soon as possible and a T_{drop} parameter is specified (generally as -15 dB in IS-95) so the soft-handoff connection is released when the pilot drops below the specified level.

In the forward link only one site transmits to the mobile at a time, but when the decision is made to go to soft-handoff a second site transmits. The mobile will then benefit from the diversity gain and the transmit level of the first site can be dropped a little.

On one aspect the soft-handoff increases the network capacity as the mobile power in the reverse link is reduced due to the diversity gain, but an additional transmission is added in the forward link and resources (channel elements) are used in the network. It is difficult to predict the net impact of the soft-handoff and simulation tools like the one used in the CelPlanner tool proved to be invaluable to analyse the best settings to the different parameters. In real life system soft handoff reduces the system capacity by depleting its resources and should be limited to the minimum required to transition the calls from one sector to the other.

Table 9.85 lists the steps followed to set up a typical call.

Table 9.85   Typical call set up procedure

Steps	Message	Channel	Direction
1	Call origination	Access	Reverse
2	Mobile station order	Paging	Forward
3	Channel assignment	Paging	Forward
4	Base station acknowledge	Traffic	Forward
5	Pilot strength measure	Traffic	Reverse
6	Handoff direction	Traffic	Forward
7	Handoff completion	Traffic	Reverse
8	Service connect	Traffic	Forward
9	Voice & signalling	Traffic	Forward
10	Voice & signalling	Traffic	Reverse
11	Release order	Traffic	Forward
12	Release order	Traffic	Reverse
13	Sync channel message	Sync	Forward

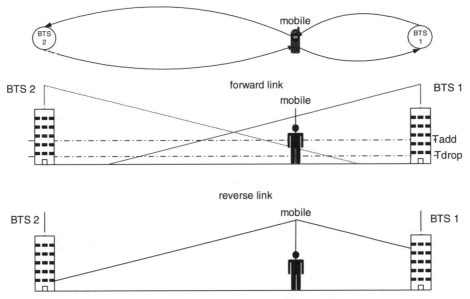

Figure 9.222 Mobile starts call with BTS 1.

The example illustrated in Figure 9.222 starts with a mobile connected to a TRX in BTS 1. The mobile has BTS1 listed in its active pilot set but it is constantly searching for other pilots that may be above Tadd to include in its candidate pilot set. The mobile can request a connection to any of these candidate pilots. The MSC sends a list of neighbours of the active pilots, which is stored by the mobile as the neighbour pilot set. The remaining pilots are stored in a remaining pilot set list.

The BTS signals are defined by their PN offsets. A list of the cell neighbours is sent to the mobile by the BTS. The mobile classifies these neighbours in three sets:

- Active: BTS signals that are in soft-handoff

- Candidates: all signals that may be in soft-handoff but do not have hardware available

- Neighbours: remaining signals from the list received by the mobile

- Remaining: all other possible signals, not present in the previous lists, derived by dividing the pilot offset by a PILOT_INC parameter sent by the BTS

The mobile scans the active and candidate list constantly and for each complete scan of the list it scans one of the neighbours. This takes approximately 1 s. When all the neighbours in the list were scanned the mobile scans one of the BTS signals in the remaining list. The complete scanning process of the neighbours can take several seconds (typically 4) and the scanning of all pilots in the remaining set can take as long as 120 s.

Although the call is placed using only BTS 1, BTS 2 is also involved. It receives the signals from the mobile as the mobile transmission is omni directional. This signal adds to

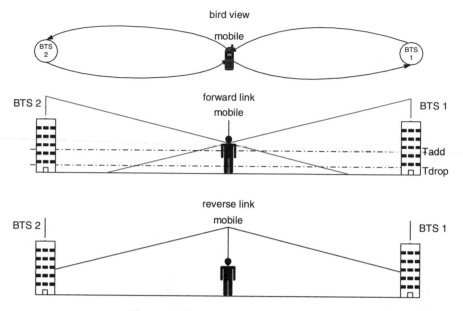

Figure 9.223 Mobile is in soft handoff.

the noise floor. The mobile monitors BTS2 pilot and when it reaches a Tadd level it will request a soft-handoff.

In the example of Figure 9.223, the mobile detects another pilot above the T_{add} threshold and sends a pilot strength measurement report (step 5 in Table 9.85) requesting a soft-handoff to this pilot. The MSC checks network availability, authorises the mobile to start soft-handoff, and assigns a traffic channel for it (step 6). The network availability is determined by hardware (typically 20 channels per BTS) or power availability.

The mobile assigns a traffic rake receiver to this traffic channel and confirms soft-handoff (step 7). The MSC confirms service connection. Now, the information is carried through two paths in both directions. A diversity combiner can improve the reception by minimising narrowband fading effects.

Both BTSs have equivalent signals and the traffic capacity is reduced on both sides. There is an area between the BTSs in which this effect occurs. Even though it reduces capacity, this effect should be minimised but not eliminated, allowing handoff to happen smoothly, affecting the traffic capacity as little as possible.

When the mobile moves away from BTS1, its E_c/I_0 drops below T_{drop} and BTS1 is excluded from the active set. This is informed to the MSC through a pilot strength measure report and the MSC sends a new neighbour list. Figure 9.224 illustrates this situation.

9.5.1.1.1   Neighbourhood

The neighbourhood of each cell is defined by the operator and its descriptive parameters are stored in the MSC, which sends them to the mobiles when required to allow them to search for the strongest signals.

A CDMA neighbour is defined as a cell that may affect the network E_c/I_0 when receiving or sending a signal from/to a mobile. This happens, for example, when a mobile clears an

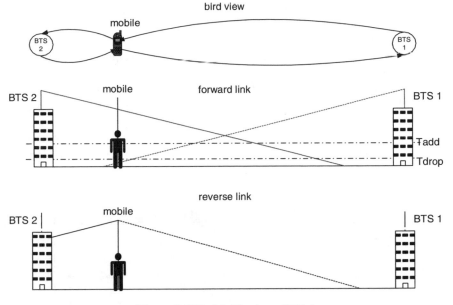

Figure 9.224   Mobile drops BTS 1.

obstacle. Because there is a scanning time involved, if the BTS is in the remaining set it takes a long time for the mobile to detect the interference issue. Adding too many cells to the neighbour list also does not help because it also increases the scanning time.

The determination of a neighbour list is not an easy task and requires the use of sophisticated tools. Topological neighbourhood is not good enough because it does not consider

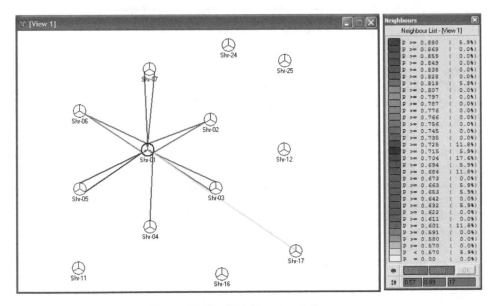

Figure 9.225   Neighbourhood diagram.

signal strength and user mobility. To determine if a site should be included in the neighbour list of another site, it is necessary to identify the impact of a user arriving at a certain pixel without being a neighbour. If the impact is small, it does not need to be included in the list, otherwise it must be included. This impact should be traffic weighted and totalised over all pixels for each site.

Figure 9.225 shows traffic weighted BTS sector neighbourhood relationships. The strength of neighbourhood expressed in affected traffic is indicated in the legend next to it.

Figure 9.226 shows, in text format, the neighbour relationships automatically created by *CelTools*. This illustration makes clear that the definition of neighbours is a decision of the operator. The list can be edited to change the weight of relationships or to include/exclude relationships.

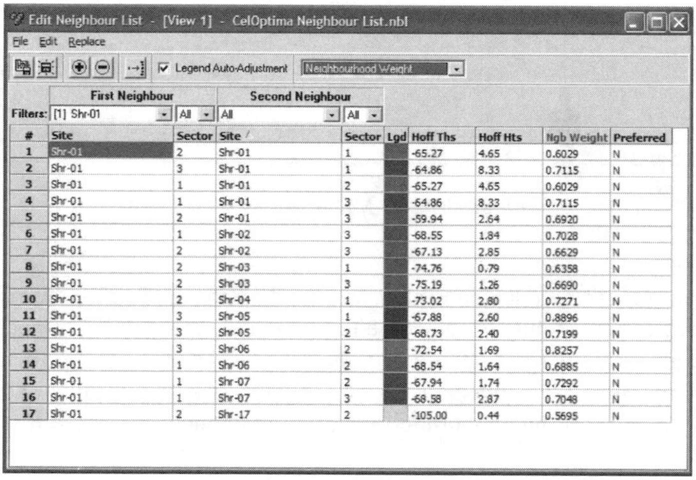

#	Site	Sector	Site	Sector	Lgd	Hoff Ths	Hoff Hts	Ngb Weight	Preferred
1	Shr-01	2	Shr-01	1		-65.27	4.65	0.6029	N
2	Shr-01	3	Shr-01	1		-64.86	8.33	0.7115	N
3	Shr-01	1	Shr-01	2		-65.27	4.65	0.6029	N
4	Shr-01	1	Shr-01	3		-64.86	8.33	0.7115	N
5	Shr-01	2	Shr-01	3		-59.94	2.64	0.6920	N
6	Shr-01	1	Shr-02	3		-68.55	1.84	0.7028	N
7	Shr-01	2	Shr-02	3		-67.13	2.85	0.6629	N
8	Shr-01	2	Shr-03	1		-74.76	0.79	0.6358	N
9	Shr-01	2	Shr-03	3		-75.19	1.26	0.6690	N
10	Shr-01	2	Shr-04	1		-73.02	2.80	0.7271	N
11	Shr-01	3	Shr-05	1		-67.88	2.60	0.8896	N
12	Shr-01	3	Shr-05	2		-68.73	2.40	0.7199	N
13	Shr-01	3	Shr-06	2		-72.54	1.69	0.8257	N
14	Shr-01	1	Shr-06	2		-68.54	1.64	0.6885	N
15	Shr-01	1	Shr-07	2		-67.94	1.74	0.7292	N
16	Shr-01	1	Shr-07	3		-68.58	2.87	0.7048	N
17	Shr-01	2	Shr-17	2		-105.00	0.44	0.5695	N

Figure 9.226 Neighbourhood table.

Figure 9.227 shows soft-handoff areas calculated by CelPlanner as no handoff, softer, soft between 2 sectors, soft softer and soft between 3 sectors.

9.5.1.1.2 Handoff Optimisation

Handoff in CDMA is a balancing act because it consumes network resources but at the same time provides diversity gain, minimising the effect of fading. The resources of a CDMA network are represented by the BTS power, BTS RF channels (processing hardware) and mobile power.

In the forward link, the designer should limit the coverage area of a BTS by setting its pilot power. This is an interactive process because the required receive power is a function of the network load. To break the deadlock, the designer estimates a uniform margin to compensate the load effect. The pilot coverage should provide coverage over the entire service area (outdoor and indoor). This automatically provides a pilot coverage overlap between cells. This overlap should be kept to a minimum but without disrupting the coverage in the service area.

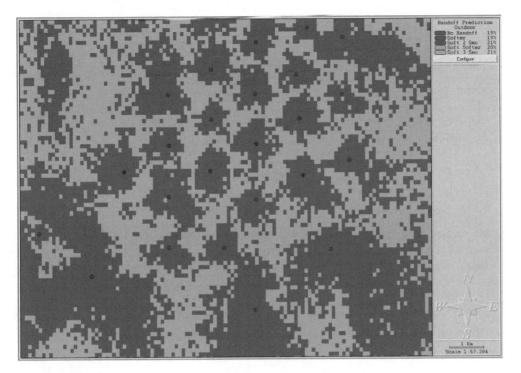

Figure 9.227 Soft handoff.

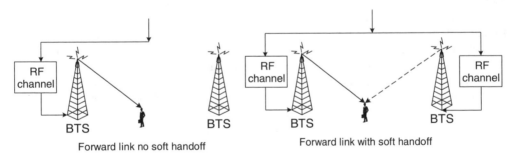

Figure 9.228 Forward link soft-handoff.

After the pilot coverage is defined, the designer must define which call should be considered for handoff. In the forward link (Figure 9.228), when a call is placed in soft-handoff, the overall network noise increases. In the reverse link (Figure 9.229), there is no impact to the network capacity when a call is placed in handoff, because the interference is already there (with or without handoff), but the call quality can be improved because of soft-handoff gain. Network resources are reduced each time a call is placed in soft-handoff.

The designer should consider the trade-offs according to its network limitations. If a network area is limited in resources, the soft-handoff should be reduced; if it is first limited

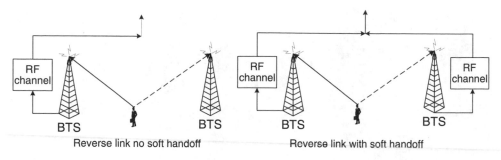

Figure 9.229   Reverse link soft-handoff.

by the reverse power, the soft-handoff area should be increased; and if it is limited by the forward power both alternatives should be tried to choose the one that gives best results.

Calls placed in handoff can be adjusted by the handoff threshold parameters Tadd and T drop. These parameters represent the E_c/I_0 ratios that determine if a call should be placed or dropped from handoff.

Sending signal power above the level required for a proper detection little improves the signal quality but produces unduly interference and diminishes system capacity.

9.5.2   Power Control

Power control is used to dynamically adapt the transmit signal power level to the receiver requirements. The capacity of CDMA systems is directly related to interference and power. Power control not only improves these two factors but also brings the additional benefit of saving battery power of the mobile unit, consequently increasing talk time.

Without power control, mobiles close to the BTS would receive a much stronger signal than far away mobiles. Because both mobiles use the same frequency, the distant ones would not be able to obtain the required SNIR. This effect is called the near-far interference. This section briefly explains the implementation of power control in CDMA. Chapter 7 provides a detailed description of the power control procedure.

Power control can be implemented in the downstream, in the upstream or in both directions. The implementation can differ in each direction depending on network require-ments. The power control target in the forward link (downstream) is to make all mobiles connected to a given BTS to arrive with the same power level, with just sufficient margin to be detected with the required quality. In the reverse link (upstream), the idea is to make the received signal be just strong enough to be detected with quality. The quality objectives are defined by the technology and include BER, FER, capacity, dropped calls and coverage.

The power control goal is to keep a certain network parameter, such as signal level or SNIR, within a pre-established range. This range can be a target level or a window, and the adjustment is done in steps within pre-defined limits.

Power control can be performed in two ways: as an open loop or as a closed loop. In the open loop, power control decisions are taken locally based on pre-established parameters. In the closed loop, feedback information is sent from the network and the loop is continuously adjusted.

9.5.2.1.1 Forward Link Power Control in IS-95

The network designer assigns a nominal power level to the forward link, which can vary by ±6 dB. This means that mobiles close to the BTS transmit initially with power levels 6 dB lower than the nominal power whereas mobiles far from the BTS transmit with power levels 6 dB higher than the nominal power. This creates an extremely bad SNIR for the close mobiles requiring a dynamic power control to adjust the power level based on the overall mobiles distribution.

The strategy adopted in the forward link of IS-95 systems is to transmit using the lowest possible power, just enough to reach the required quality at each mobile. In this strategy, the BTS drops the power in steps of 0.5 dB at every 20 ms, relying in a feedback from the mobile, which determines the quality of the received signal.

The quality evaluation happens at the mobile receiver, which analyses the FER measured for every frame (20 ms). This information is sent through messages using the 'dim and burst' (vocoder rate is reduced to send signalling bits) or 'blank and burst' (vocoder bits are replaced by signalling bits) techniques. The message is received by the BTS, which then controls the channel gain. Figure 9.230 shows schematics of the forward power control.

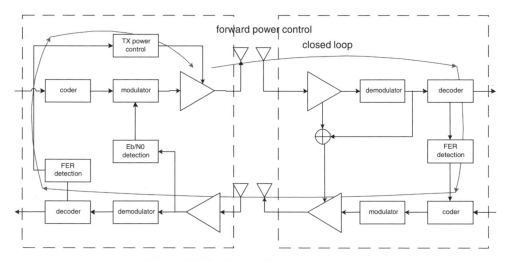

Figure 9.230 Forward link power control.

The forward link power control is able to compensate for the average path loss, including shadow slow fading. Multi-path fading effects, however, cannot be compensated. Signals coming from different BTS channels will be received by the mobile spread by several dB additionally to the fading. This is tolerable in the forward link because of the coherent detection and corresponding processing gain.

9.5.2.1.2 Reverse Link Power Control in IS-95

The reverse link power control is initially based on the pilot power received from the BTS. Because this power is fixed (no power control is applied to it), it gives a good indication of the path loss. This value is averaged and provides what is called an open loop strategy. Figure 9.231 shows the reverse power control strategy.

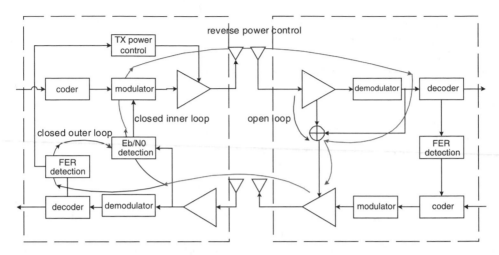

Figure 9.231   Reverse link power control.

The pilot transmitted in the forward link uses a different frequency than the signal transmitted in the reverse link. These frequencies are sufficiently apart to have independent multi-path (fast) fading, therefore multi-path fading information from the forward link is useless to the reverse link and should be averaged leaving only the shadow fading information. A value of 20 ms for the integration time is enough to obtain this effect.

The open loop power control value is given by eqn (9.135). The constant value is standardised in IS-95 systems as 73 dB for the cellular band and 76 dB for the PCS band. The open loop dynamic range is limited to 40 dB from the mobile nominal power.

$$\text{Power mobile (dBm)} = \text{Constant (dB)} - \text{Received power (dBm)} \qquad (9.135)$$

Because the reverse link uses noncoherent detection, it requires less dispersion between the signal levels received from different sectors. This means that additional control is required to reduce the variation between the different signals and to compensate at least part of the multi-path fading.

The quality evaluation is done at the BTS receiver by analysing the FER at every frame (20 ms). This process constitutes the closed outer loop of the power control. This period is too long to provide the required protection in the reverse link, requiring an additional analysis at the E_b/N_0 level. The required E_b/N_0 is calculated based on the measured FER, using an inner loop to adjust the power levels. In this inner loop, power control indications are sent to the mobile at 800 Hz (1.25 ms) requesting an increase or reduction of the power level by 1 dB with a dynamic range (variation range) of 24 dB over the open loop level and limited to the mobile power of +23 dBm.

The accuracy of the reverse power control provides a variation between the different signal levels that goes from 1 to 3 dB depending on the mobile speed.

9.5.2.1.3   Power Control in cdma2000

cdma2000 presents advanced power control features. Fast power control is used both in the forward and reverse links in a way similar to the reverse link power control of IS-95 systems.

However, cdma2000 implements steps of 0.5 and 0.25 dB for power changes. Additionally, the required E_b/N_0 was added to the open loop equation because in these systems, different rates require different E_b/N_0 values.

This type of power control is not effective when packet-switching (data transmissions) is used because the packets are too short and the interval between packets can be too long losing the adjustment of the information. cdma2000 uses the slotted Aloha principle for packet data transmission. This feature is used to provide the required power control by progressively increasing the random access burst after an unsuccessful access attempt.

9.5.3 Resource Management

IS-95 CDMA networks use handoff management and power control as resource management tools as its main traffic is voice. cdma2000 networks were designed to handle voice as well as data. Data services can be divided in: conversational, streaming, interactive and background. Each of there categories has different data rate and delay requirements.

Conversational data service has to have a low delay (<400 ms) and an even lower delay variation. The data traffic is more or less constant over the call period and BER requirements are not very stringent. This service covers mainly voice applications.

Streaming data service has to have low delay and low delay of variation, adapt to large variations in data rate and has quite stringent BER requirements. This service covers mainly video or audio streaming.

Interactive data service has to have low round trip delay, but the delay of variations are not important and BER requirements are stringent but can be reduced by the use of ARQ based application protocols. This service covers web browsing, interactive games and location based services.

Background data services can support large delays and delay variations and the BER requirements are stringent but can be reduced by ARQ application based protocols. This service covers SMS, e-mail notification, e-mail download and file transfer.

This variability of requirements allows accommodating congestion periods by managing the data rate and the delay of the different services. The function of the resource management is to schedule the different services according to the network availability maximising its overall throughput.

Voice calls are well understood and basically a logical channel is available for the whole duration of a call.

Data calls are different and we will describe them next. Each data user is assigned a low data rate connection (fundamental channel of up to 9.6 kbps) during the duration of the call to maintain the call, provide infrequent signaling frames and aid in the data transmission. Burst of data are transferred at high rates through supplemental channels of up to 153.6 kbps each based on current interference background, mobile RF conditions, amount of data to be sent and history of the data session. The supplemental channels are dynamically allocated as required and are limited in number.

The supplemental channels are generally allocated for a short period of time and do not perform soft-handoff. An equivalent of a hard-handoff is implemented if the anchor cell changes the next time the supplemental channel has to be allocated.

Voice calls traffic can be analysed using Erlang B models based on Poisson distribution, but data calls have long tailed distributions and cannot be modelled mathematically yet. A reasonable approximation though is to use Erlang B and Erlang C approximations. Due to the different traffic characteristics of each service the geographical distribution is expressed in users and the rates are calculated at simulation time.

The data rate available generally drops from the centre of a cell to its edges and the reduction of the data rate as the mobile moves away from the centre of the cell (worsening of the RF conditions) is a consequence of the resource management that keeps the data transfer but reduces the data rate.

The bandwidth is then shared between different users and in the forward direction the burst allocation is triggered when data gets backlogged on the network side, while in the reverse direction data builds up at the mobile which requests a supplemental channel through a message.

Complex simulations are required to predict service when a mix of voice and data is being offered. Each service class has to be specified in terms of service offered, user terminal used, environment and user geographical distribution to configure the demand for each one. Attention should be paid to the user distribution according to calls made outdoor, indoor on the ground floor and on other building floors.

Once the demand is defined the simulation can be done. To perform statistically valid simulations static and dynamic conditions have to be considered. CelPlanner implements a Monte Carlo simulation in which static and dynamic considerations are made. This procedure is described in more detail in chapters 10 and 11.

9.6   RECEIVE STAGE

The last part of the diagram shown in Figure 9.1 is the receive stage. In this stage, the signal is received from the air–interface and the original information is recovered.

9.6.1   Receive Antenna

The receive antenna can capture the energy of a certain area of the propagation sphere. Section 9.3.6 provides the description of the transmit antenna, which is also valid for the receive antenna. Many times the antenna shares transmit and receive functions.

Receive antennas are usually used in diversity configurations at base stations to mitigate narrowband fading as explained in Section 9.4.2.7.3. Some vendors are currently studying diversity configurations at the mobile station, however there is not a practical solution for it yet.

9.6.2   Reception (RX) Line

The signals received by the antenna have very low energy; therefore any additional noise added by the receiving circuitry is critical. The connection issues analysed on the transmit side, in Section 9.3.5, also apply to the receive side. Figure 9.232 shows the several stages involved in the receive process.

The reception and transmission antennas can be independent from each other or shared. The same applies to the cable, jumpers and connectors.

Figure 9.232 RX losses.

The filter is very important because it limits the receive band to avoid saturating the first amplifier stage. When transmit and receive share the same cable this filter is part of the duplexer. The duplexer separates transmit and receive paths. Section 9.3.5.1 describes the duplexer in detail.

The first active stage is the low noise amplifier (LNA), described in Section 9.6.2.1. After the LNA, the multi-coupler distributes the signal to all the channels in the equipment. The multi-coupler can be a multi-stage circuit, from which the first stage is usually integrated with the LNA, thus they are analysed together in Section 9.6.2.1. These losses include the cable and connector losses. It includes also the splitter losses as the signals have to be send to several receivers.

9.6.2.1 Low Noise Amplifier (LNA)

A low noise amplifier (LNA) is the first active stage used to amplify the signal. In the cascading stages example analysed above the first stage is the most important and practically defines the system noise. For this reason the LNA should be as close as possible to the antenna.

The typical LNA characteristics are the following:

- Noise figure: 3 dB

- IP3: +27 dBm

- Return loss: 20 dB

- Impedance: 50

The filter characteristics are important as it will limit the power that reaches the LNA and will avoid overloading the input. Figures 9.233 and 9.234 show filter responses respectively for Bands *A* and *B* used in the USA.

Cryogenically cooled elements explore the properties of High Temperature Superconductors (HTS) discovered in 1986 that can be cooled by liquid nitrogen (77 K) instead of Helium (4 K). The most common materials are YBCO (yttrium compounds – 90 K), BSCCO (bismuth compounds – 110 K) and TBCCO (thallium compounds – 127 K).

The low noise amplifier (LNA) is the first active stage used to amplify the signal. In the cascading stages example analysed in Section 9.4.3.2.1, the first stage is the most important and practically defines the system noise.

Figure 9.235 shows a scenario with a ground-mounted amplifier. Tables 9.86 and 9.87 show, respectively, the link budget for ground-mounted amplifiers at 900 and 1800 MHz.

Band A filter

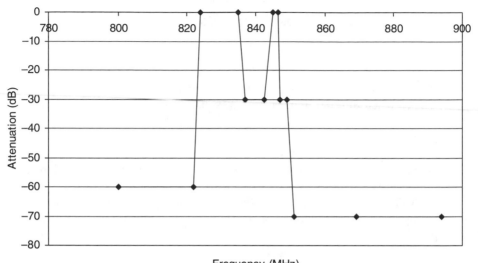

Frequency (MHz)

Figure 9.233 Filter response band A.

Figure 9.236 shows a scenario with a tower-mounted amplifier. Tables 9.88 and 9.89 show, respectively, the link budget for ground-mounted amplifiers at 900 and 1800 MHz.

In this examples, the noise reduction provided by the tower-mounted amplifiers is approximately 4 dB.

Band B filter

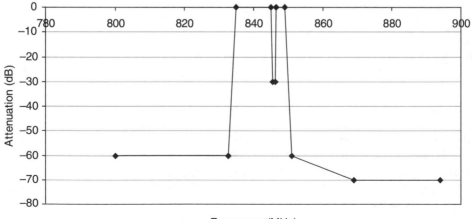

Frequency (MHz)

Figure 9.234 Filter response band *B*.

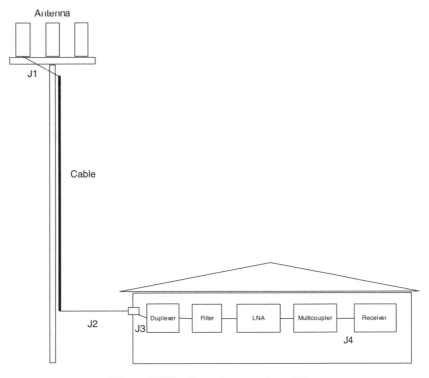

Figure 9.235   Ground-mounted amplifier.

9.6.3   Receiver

The minimum signal level that can be detected by a receiver is defined as the level that produces a minimum SNIR for analogue systems or a minimum bit error rate (BER) for digital systems.

Table 9.86   Ground-mounted amplifier budget (900 MHz)

			Cable loss (dB/100 m)	Connector loss (dB)	Length (m)	Gain (dB)	Gain	NF (dB)	F
J1	3/4		8.00	0.135	4	−0.59	0.87	0.87	1.22
C	1 1/2		0.825	0.045	30	−0.34	0.93	0.93	1.24
J2	3/4		8.00	0.135	4	−0.59	0.87	0.87	1.22
J3	3/4		8.00	0.135	4	−0.59	0.87	0.87	1.22
Filter	Ground mount	300 K				−1	0.79	0.79	1.20
LNA	Ground mount	300 K				12	15.85	3.00	2.00
Multi-coupler	Ground mount	300 K				−3	0.50	3.00	2.00
J4	3/4		8.00	0.135	6	−0.75	0.84	0.75	1.19
Receiver	Ground mount	300 K						7	5.01
Total noise figure								11.32	

Table 9.87 Ground-mounted amplifier budget (1800 MHz)

		Temperature	Cable loss (dB/100 m)	Connector loss (dB)	Length (m)	Gain (dB)	Gain	NF (dB)	F
J1	3/4		12.40	0.27	4	−1.04	0.79	0.79	1.20
C	1 1/2		1.22	0.09	30	−0.55	0.88	0.88	1.23
J2	3/4		12.40	0.27	4	−1.04	0.79	0.79	1.20
J3	3/4		12.40	0.27	4	−1.04	0.79	0.79	1.20
Filter	Ground mount	300 K				−1	0.79	0.79	1.20
LNA	Ground mount	300 K				12	15.85	3.00	2.00
Multi-coupler	Ground mount	300 K				−3	0.50	3.00	2.00
J4	3/4		12.40	0.27	6	−1.28	0.74	1.28	1.34
Receiver	Ground mount	300 K						7	5.01
Total noise figure								12.53	

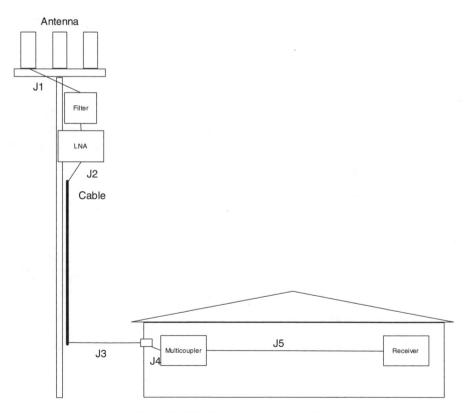

Figure 9.236 Tower-mounted amplifier.

Table 9.88 Tower-mounted amplifier budget (900 MHz)

	Temperature	Cable loss (dB/100 m)	Connector loss (dB)	Length (m)	Gain (dB)	Gain	NF (dB)	F
J1	3/4	8.00	0.135	4	−0.59	0.87	0.87	1.22
Filter	Ground mount	127 K			−0.25	0.94	0.94	1.24
LNA	Ground mount	127 K			12	15.85	0.4	1.10
J2	3/4	8.00	0.135	4	−0.59	0.87	0.87	1.22
C	1 1/2	0.825	0.045	30	−0.34	0.93	0.93	1.24
J3	3/4	8.00	0.135	4	−0.59	0.87	0.87	1.22
J4	3/4	8.00	0.135	4	−0.59	0.87	0.87	1.22
Multi-coupler	Ground mount	300 K			−3	0.50	3.00	2.00
J5	1/4	20.00	0.135	4	−1.07	0.78	0.78	1.20
Receiver	Ground mount	300 K					7	5.01
Total noise figure							7.39	

The receiver has to distinguish between the signal it receives as input and the noise generated internally, thus each receiver may have a different sensitivity level. To determine this level, a very low signal is injected at the receiver input and it is slowly raised while the signal-to-noise ratio of the output is monitored. The sensitivity point is when there is a jump in the S/N ratio. This point is usually known as 12 dB point.

Table 9.89 Tower-mounted amplifier budget (1800 MHz)

	Temperature	Cable loss (dB/100 m)	Connector loss (dB)	Length (m)	Gain (dB)	Gain	NF (dB)	F
J1	3/4	12.40	0.27	4	−1.04	0.79	0.79	1.20
Filter	Ground mount	127 K			−0.25	0.94	0.94	1.24
LNA	Ground mount	127 K			12	15.85	0.4	1.10
J2	3/4	12.40	0.27	4	−1.04	0.79	0.79	1.20
C	1 1/2	1.22	0.09	30	−0.55	0.88	0.88	1.23
J3	3/4	12.40	0.27	4	−1.04	0.79	0.79	1.20
J4	3/4	12.40	0.27	4	−1.04	0.79	0.79	1.20
Multi-coupler	Ground mount	300 K			−3	0.50	3.00	2.00
J5	1/4	30.00	0.27	4	−1.74	0.67	0.67	1.17
Receiver	Ground mount	300 K					7	5.01
Total noise figure							8.43	

The input noise level should not be amplified above the receiver sensitivity level otherwise it reduces the receiver's dynamic range and increases system noise.

An important characteristic of the receiver is the dynamic range, defined by the sensitivity level and the 3 dB point (the output presents 3 dB distortion component). Signals that pass the 3 dB point impact the cell performance significantly because the distortion affects all channels. In CDMA, the power control should take care of this issue but mobiles that fail power control cause large disruptions to the system.

Another cause of receiver saturation are strong out-of-band signals emanating from nearby antennas. In this cases sharp filters should be added to minimise external interference.

9.6.4   Demodulator and Decoder

The demodulator demodulates the signal back to the base band, detecting, from the received signal, the information that was most likely transmitted. The decoder maps the signal received to possible states to detect the transmitted information. Several processes are involved in this task and the demodulation and decoding process are intermingled.

In the forward link, the mobile receives signals from several base stations, demodulates them to the base band and searches for the strongest signal by comparing the composed signal with different PN offsets. This task allows the mobile to extract the base station information by combining it with the PN sequence. During this process the mobile can also align the Walsh code to the information, to extract the channel information. The information is then de-interleaved and the encoder corrects the errors by mapping the data stream to the most probable values.

In the reverse link, the base station receives signals from many mobiles and has to extract the information of each one. The alignment task is simplified because the mobile sends information within a specified delay after receiving the data from the base station; therefore the base station knows approximately when to look for it. In IS-95, there is no pilot in the reverse link and the task of exactly aligning the PN sequence is harder because it uses a non-coherent detection process. In IS-2000, the pilot is available and the alignment can be more easily achieved. The mobile long PN sequence is used to extract the specific user data. The desired data is extracted by the PN sequence and sent to a comparator that decides which of the Walsh codes was sent and, consequently, retrieves the data stream. The information is then de-interleaved and the encoder corrects the errors by mapping the data stream to the most probable values.

The whole process is very complex because perfect orthogonality cannot be assured and fading affects drastically the ratio of signal-to-noise, and interference. The designer needs to know the performance of the technology in different environments to properly design the network. In many cases, this has to be done in before the network is deployed, therefore it is very important for the designers to understand all the trade-offs and what to expect from the technology. Mathematical representation of the whole network is very complex but each part can be modelled properly, so simulators are used to calculate the required performance.

Voice circuits are generally designed for a FER of 1%, while data circuits are designed for a FER of 3–5% requirements for the forward and reverse links in terms of required E_b/N_0 for different FERs. This information is important for the designer because it provides the

RSQI range

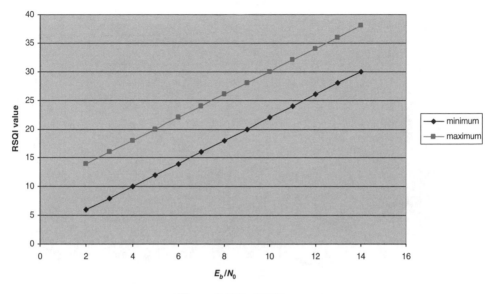

Figure 9.237   RSQI range.

sensitivity of how the performance varies with different factors. These minimum requirements are presented as graphs to be easily observed.

The mobile measures the received signal E_b/N_0 and expresses it in Receive Signal Quality Indicator (RSQI) units. These units are mapped to ranges of E_b/N_0 as shown in Figure 9.237.

9.6.4.1   Forward Link

This section presents the forward link minimum requirements, first for an AWGN channel and then for fading channels at different speeds and number of paths. These requirements vary for the fundamental or supplemental channels and if convolutional or turbo encoding is used. The cases shown herein are based on 3 GPP standards.

AWGN
Figures 9.238–9.242 show the minimum requirements for an AWGN channel for the most common radio configurations and rates available in cdma2000. These figures allow the derivation of relationships between different rates and coding techniques. The curves are valid for the whole frequency range, from cellular to PCS.

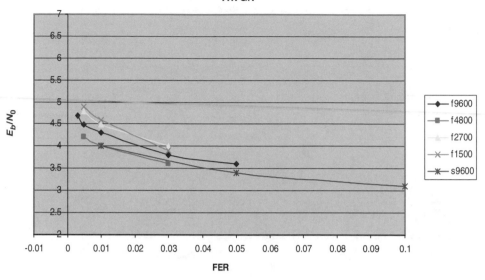

Figure 9.238   RC1 fundamental AWGN.

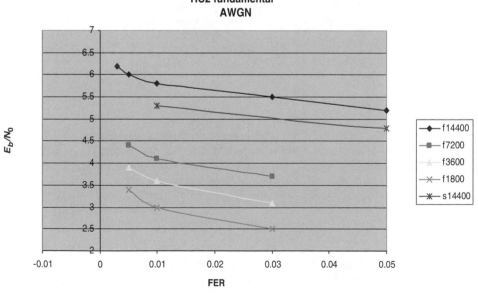

Figure 9.239   RC2 fundamental AWGN.

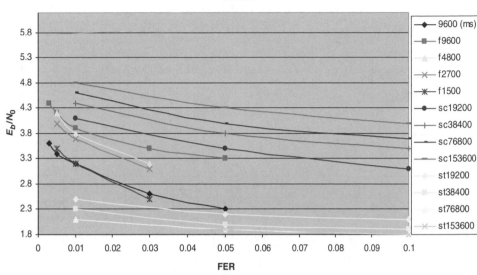

Figure 9.240   RC3 fundamental, supplemental, convolutional and turbo AWGN.

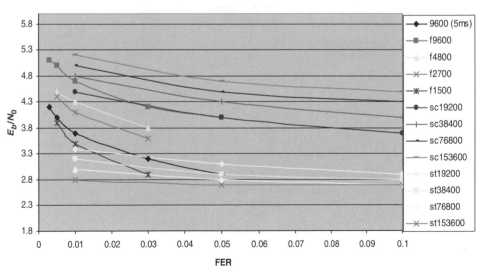

Figure 9.241   RC4 fundamental AWGN.

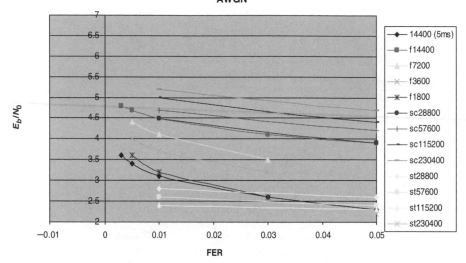

Figure 9.242   RC5 fundamental AWGN.

Fading Environment

Figures 9.243–9.255 show the minimum requirements in fading channels for the most common radio configurations and rates available in cdma2000. These figures allow the derivation of relationships between different rates and coding techniques. The curves vary with the frequency because of fading. The figures are grouped in a cellular range (800–1000 MHz) and a PCS range (1800–2200 MHz).

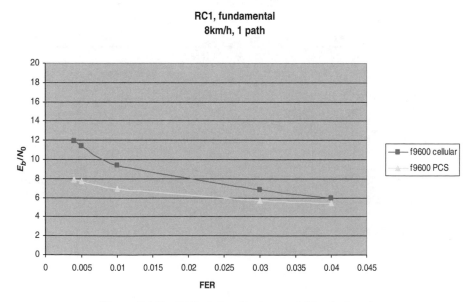

Figure 9.243   RC1, fading fundamental 8 km/h, 1 path.

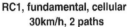

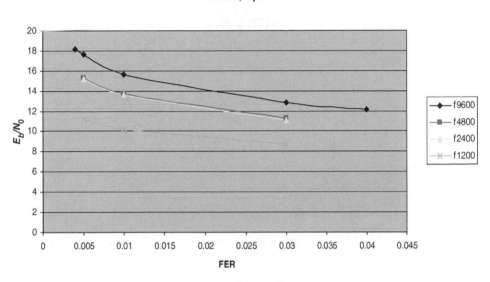

Figure 9.244   RC1, fading, fundamental, cellular, 30 km/h, 2 paths.

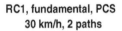

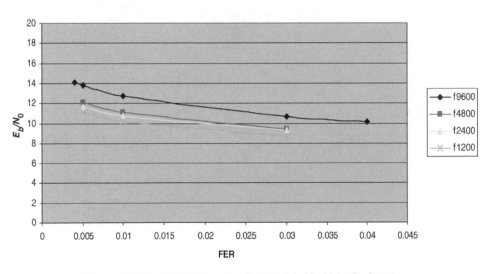

Figure 9.245   RC1, fading, fundamental, PCS, 30 km/h, 2 paths.

RC2, fading, 100 km/h, 3 paths

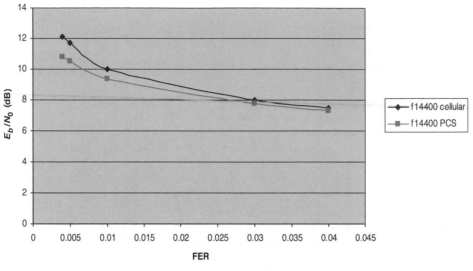

Figure 9.246   RC2, fading 100 km/h, 3 paths.

RC2, fading, 0 km/h, 2 paths, cellular

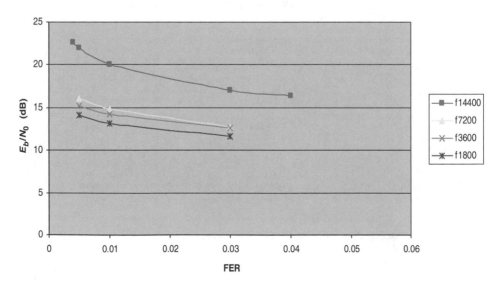

Figure 9.247   RC2, fading 0 km/h, 2 paths, cellular.

RC2, fading, 3 km/h, 1 path, cellular

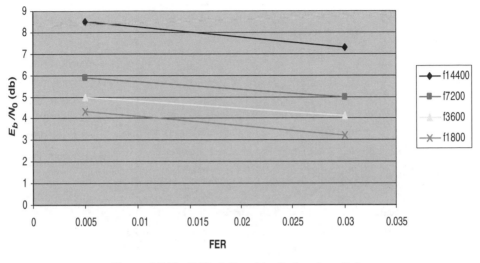

Figure 9.248   RC2, fading, 3 km/h, 1 path, cellular.

RC2, fading, 3 km/h, 1 path, PCS

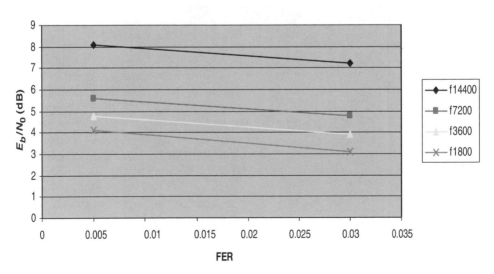

Figure 9.249   RC2, fading, 3 km/h, 1 path, PCS.

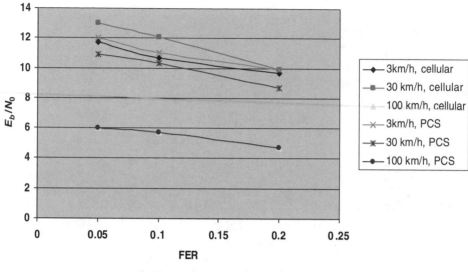

Figure 9.250   RC3, fundamental, fading.

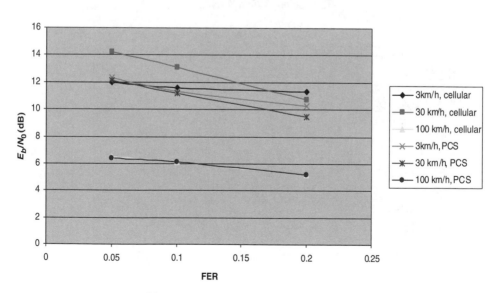

Figure 9.251   RC5, fundamental, fading.

RC3, fading, 8 km/h, 1 path, cellular

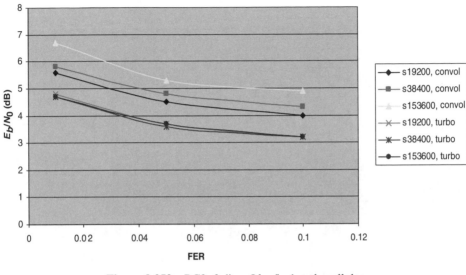

Figure 9.252 RC3, fading, 8 km/h, 1 path, cellular.

RC3, fading, 8 km/h, 1 path PCS

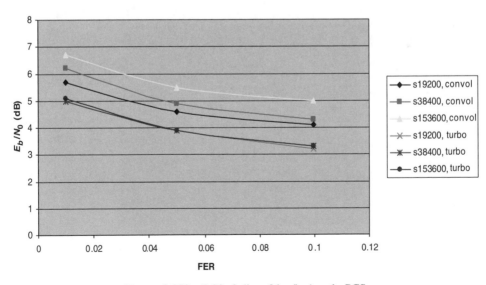

Figure 9.253 RC3, fading, 8 km/h, 1 path, PCS.

RC5, fading, 8 km/h, 1 path, cellular

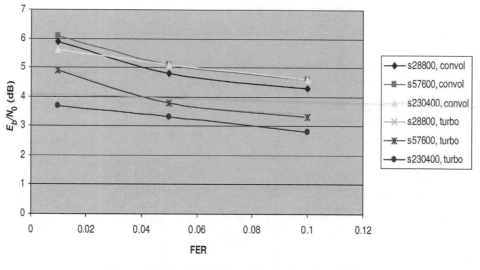

Figure 9.254   RC5, fading, 8 km/h, 1 path, cellular.

RC5, fading, 8 km/h, 1 path, Pcs

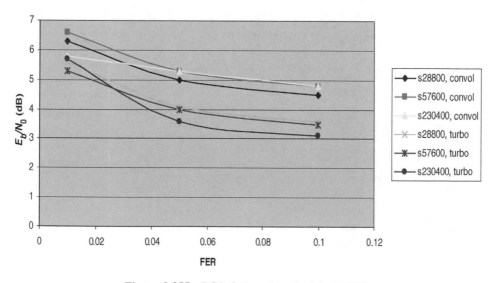

Figure 9.255   RC5, fading, 8 km/h, 1 path, PCS.

Transmit Diversity Schemes

The forward link is the limiting link in cdma2000. One of the reasons for this is the fact that it does not benefit from the diversity and low noise reception provided by the base station in the reverse link. However, transmit diversity can be used to improve the forward link. Two methods are proposed in their standard: Orthogonal Transmit Diversity (OTD) and Space Time Coding (STC).

OTD is an open loop method in which interleave symbols are split into two symbol streams (even/odd), which are transmitted by two different antennas using different Walsh codes as shown in Figure 9.256. The Walsh code length is doubled; therefore the number of available codes is not reduced.

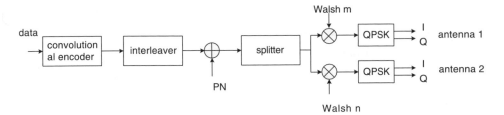

Figure 9.256   Orthogonal transmit diversity (OTD).

The OTD method can be further improved by sending both symbol streams through both antennas as in the STC diversity. This is achieved by combining the even and odd symbol streams with each Walsh code and it's inverse as shown in Figure 9.257 (also see Figures 9.258–9.266).

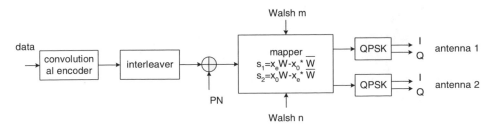

Figure 9.257   Space time coding (STC).

RC3, fading, OTD, 9600

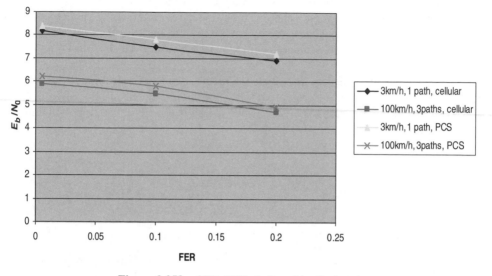

Figure 9.258   OTD, RC1, fading, 8 km/h, 1 path.

RC3, fading, STC, 9600

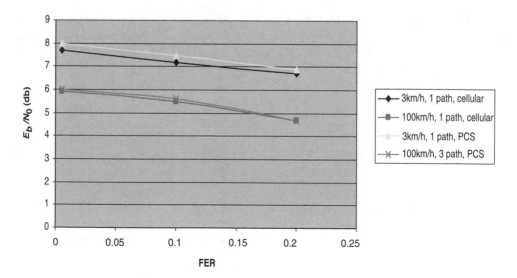

Figure 9.259   STC, RC1, fading, 8 km/h, 1 path.

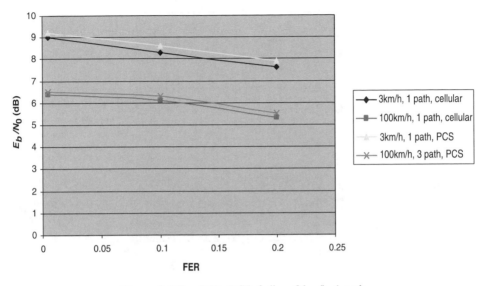

Figure 9.260   OTD, RC5, fading, 8 km/h, 1 path.

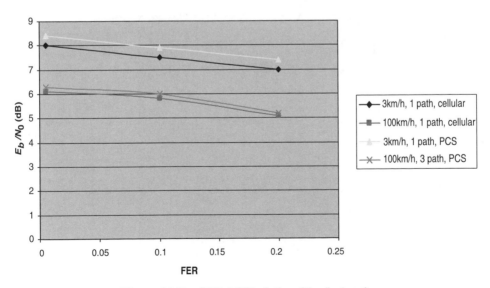

Figure 9.261   STC, RC51, fading, 8 km/h, 1 path.

RC3, fading, 3km/h, 1 path, OTD, 38400

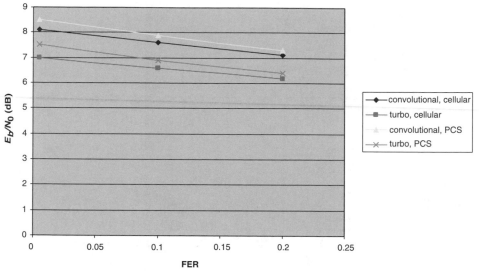

Figure 9.262   OTD, RC3, supplemental, fading, 8 km/h, 1 path.

RC3, fading, 3km/h, 1 path, STC, 38400

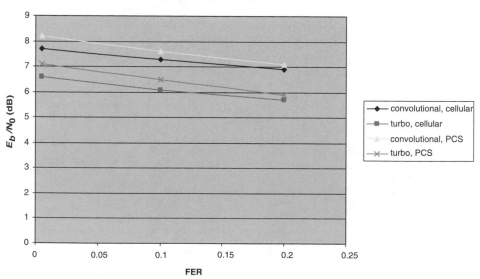

Figure 9.263   STC, RC3, supplemental, fading, 8 km/h, 1 path.

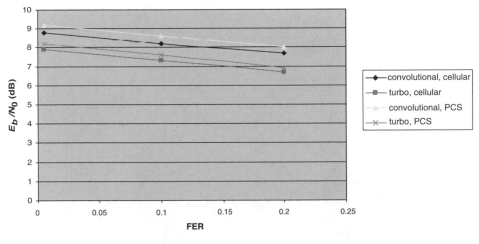

Figure 9.264   OTD, RC5, fundamental 8 km/h, 1 path.

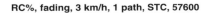

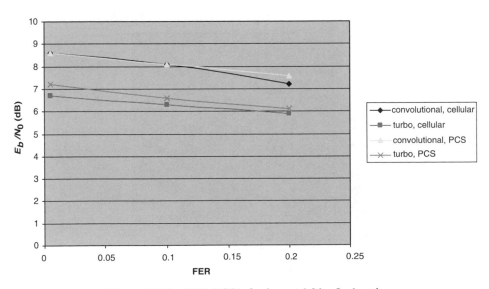

Figure 9.265   STC, RC51, fundamental 8 km/h, 1 path.

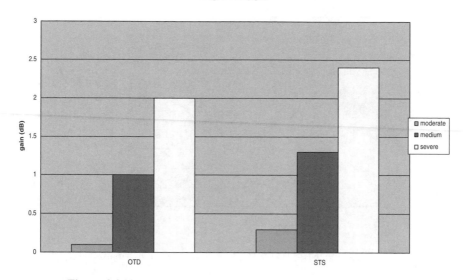

Figure 9.266   Average diversity gain (limited to the fading value).

9.6.4.1.1   Forward Link Budget Assumptions

Table 9.90 presents values that can be used in the link assignment and link performance tasks. These values were derived based on the minimum requirements presented in the graphs of Section 9.6.4.1. Exact values should be discussed with the equipment vendors because there may be variations depending on the exact implementation of each feature. Figure 9.267 to 9.269 present a summary of the forward link budget requirements.

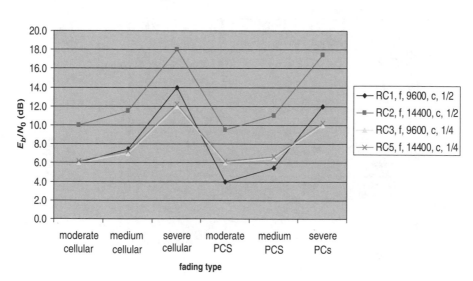

Figure 9.267   Fundamental.

Table 9.90 Forward link budget assumptions

Suggested E_b/N_0 design values for forward link cdma2000

E_b/N_0 (dB) for a FER of 1%

Radio configuration	Channel	Data rate (kbps)	Coding	Rate	Cellular			PCS			FER 1 % to 3 % (dB)
					Moderate fading cellular	Medium fading cellular	Severe fading cellular	Moderate fading cellular	Medium fading PCS	Severe fading PCS	
RC1	Fundamental	9600	Convolutional	1/2	6.0	7.5	14.0	4.0	5.5	12.0	−3.0
RC2	Fundamental	14 400	Convolutional	1/2	10.0	11.5	18.0	9.5	11.0	17.5	−1.0
RC3	Fundamental	9600	Convolutional	1/4	6.0	7.0	12.0	6.0	6.5	10.0	−2.0
	Supplemental	9600	Convolutional	1/4	6.0	7.0	12.0	6.0	6.5	10.5	−1.5
	Supplemental	19 200	Convolutional	1/4	6.2	7.2	12.2	6.2	6.7	10.7	−1.5
	Supplemental	38 400	Convolutional	1/4	6.4	7.4	12.4	6.4	6.9	11.0	−1.5
	Supplemental	76 800	Convolutional	1/4	6.7	7.7	12.8	6.7	7.2	11.3	−1.5
	Supplemental	153 600	Convolutional	1/4	7.0	8.0	13.2	7.0	7.5	11.8	−1.0
	Supplemental	9600	Turbo	1/4	5.0	6.0	11.0	5.0	5.5	9.5	−1.0
	Supplemental	19 200	Turbo	1/4	5.2	6.2	11.2	5.2	5.7	9.7	−1.0
	Supplemental	38 400	Turbo	1/4	5.3	6.3	11.3	5.3	5.8	9.9	−1.0
	Supplemental	76 800	Turbo	1/4	5.4	6.4	11.5	5.4	5.9	10.0	−1.0
	Supplemental	153 600	Turbo	1/4	5.4	6.4	11.6	5.4	5.9	10.2	−1.0
RC5	Fundamental	14 400	Convolutional	1/4	6.2	7.2	12.2	6.2	6.7	10.2	−0.6
	Supplemental	14 400	Convolutional	1/4	6.2	7.2	12.2	6.2	6.7	10.7	−0.6
	Supplemental	28 800	Convolutional	1/4	6.4	7.4	12.4	6.4	6.9	10.9	−0.6
	Supplemental	57 600	Convolutional	1/4	6.6	7.6	12.6	6.6	7.1	11.2	−0.6
	Supplemental	115 200	Convolutional	1/4	6.9	7.9	13.0	6.9	7.4	11.5	−0.6
	Supplemental	230 400	Convolutional	1/4	7.2	8.2	13.4	7.2	7.7	12.0	−0.6
	Supplemental	14 400	Turbo	1/4	5.2	6.2	11.2	5.2	5.7	9.7	−0.5
	Supplemental	28 800	Turbo	1/4	5.2	6.2	11.2	5.2	5.7	9.7	−0.5
	Supplemental	57 600	Turbo	1/4	5.2	6.2	11.2	5.2	5.7	9.8	−0.5
	Supplemental	115 200	Turbo	1/4	5.4	6.4	11.5	5.4	5.9	10.0	−0.5
	Supplemental	230 400	Turbo	1/4	5.6	6.6	11.8	5.6	6.1	10.4	−0.5

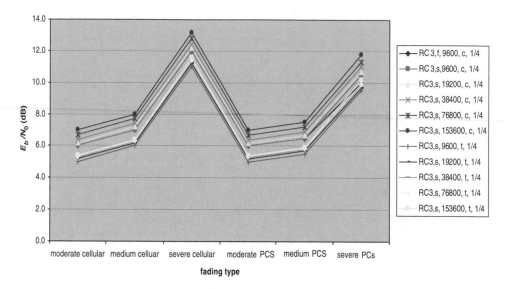

Figure 9.268   RC3.

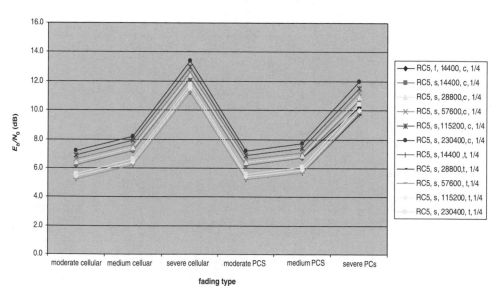

Figure 9.269   RC5.

9.6.4.2 Reverse Link

This section presents the reverse link minimum requirements, first for an AWGN channel and then for fading channels at different speeds and number of paths. These requirements vary for the fundamental or supplemental channels and if convolutional or turbo encoding is used. The cases shown herein are based on 3 GPP standards.

AWGN

Figures 9.270–9.273 show the minimum requirements in an AWGN channel for the most common radio configurations and rates available in cdma2000. These figures allow the derivation of relationships between different rates and coding techniques. The curves are valid for the whole frequency range, from cellular to PCS.

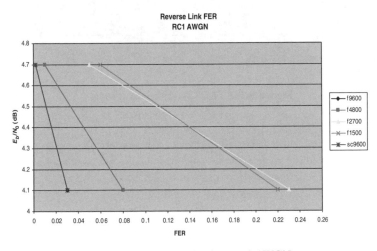

Figure 9.270 RC1 fundamental AWGN.

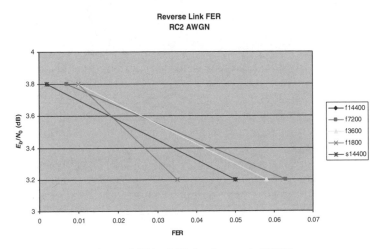

Figure 9.271 RC2 fundamental AWGN.

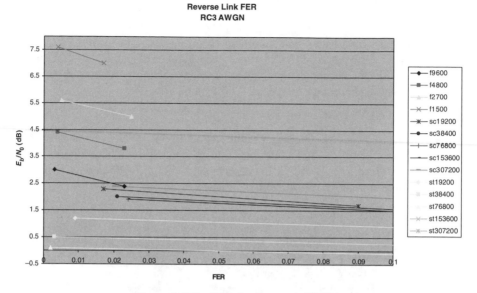

Figure 9.272   RC3 fundamental AWGN.

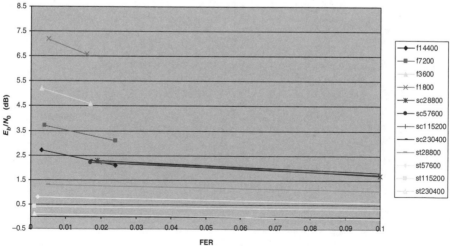

Figure 9.273   RC4 fundamental AWGN.

9.6.4.2.1 Fading Environment

Figures 9.274–9.289 show the minimum requirements in fading channels for the most common radio configurations and rates available in cdma2000. These figures allow the derivation of relationships between different rates and coding techniques. The curves vary with the frequency because of fading. The figures are grouped in a cellular range (800–1000 MHz) and a PCS range (1800–2200 MHz).

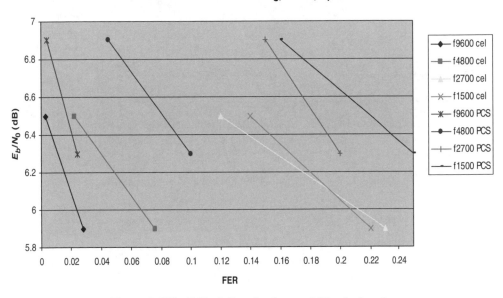

Figure 9.274 RC1, fading, fundamental 8 km/h, 2 paths.

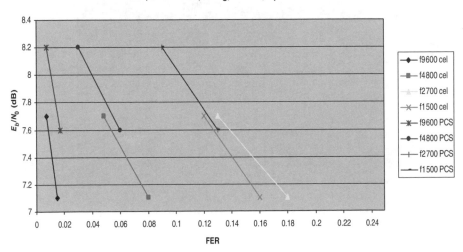

Figure 9.275 RC1, fading, fundamental 30 km/h, 1 path.

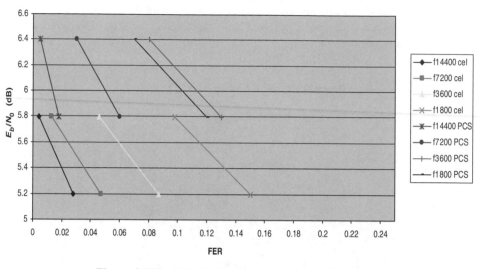

Figure 9.276   RC2, fading, fundamental 8 km/h, 2 paths.

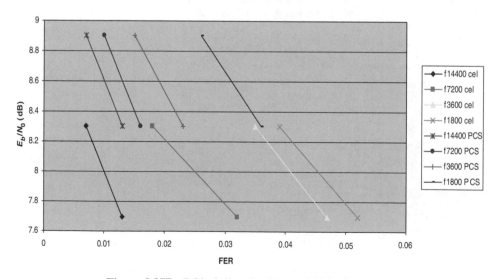

Figure 9.277   RC2, fading, fundamental 30 km/h, 1 path.

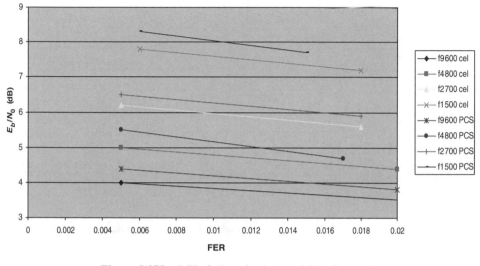

Figure 9.278 RC3, fading, fundamental 3 km/h, 1 path.

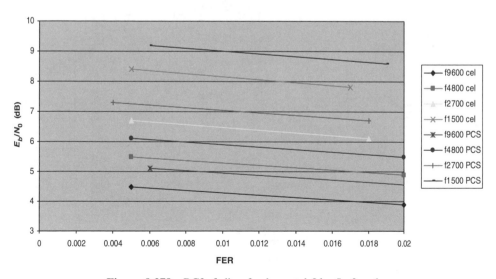

Figure 9.279 RC3, fading fundamental 8 km/h, 2 paths.

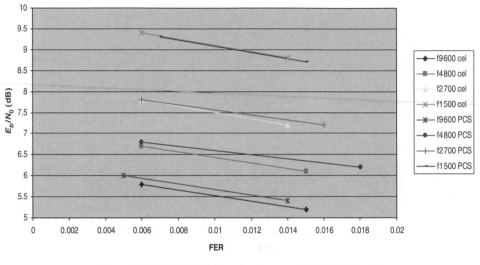

Figure 9.280   RC3, fading, fundamental 30 km/h, 1 path.

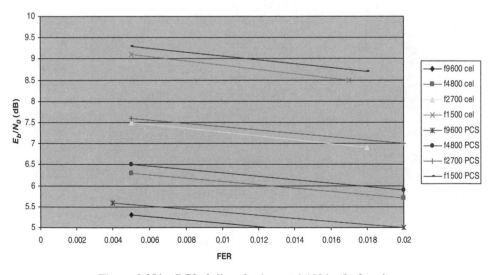

Figure 9.281   RC3, fading, fundamental 100 km/h, 3 paths.

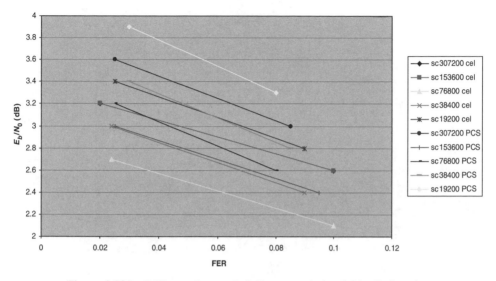

Figure 9.282 RC3, supplemental, fading, convolutional 8 km/h, 2 paths.

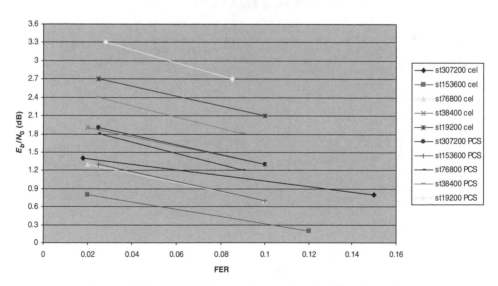

Figure 9.283 RC3, fading, supplemental, turbo, 8 km/h, 2 paths.

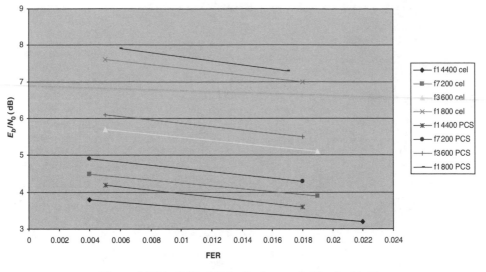

Figure 9.284   RC4, fading, fundamental, 3 km/h, 1 path.

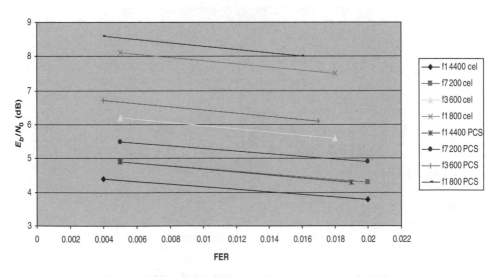

Figure 9.285   RC4, fading, fundamental, 8 km/h, 2 paths.

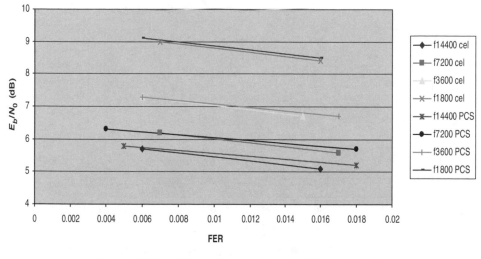

Figure 9.286   RC4, fading, fundamental 30 km/h, 1 path.

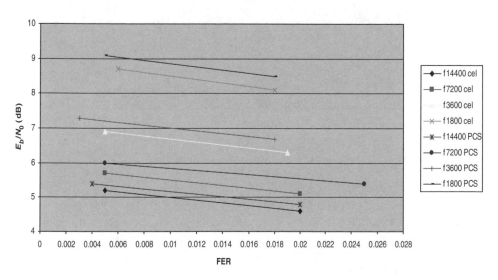

Figure 9.287   RC4, fading, fundamental 100 km/h, 3 paths.

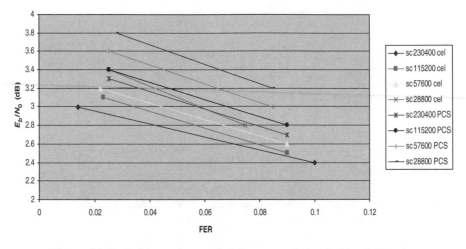

Figure 9.288   RC4, supplemental, fading, convolutional, 8 km/h, 2 paths.

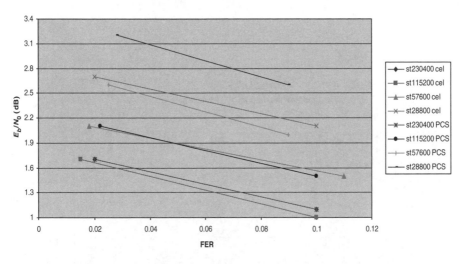

Figure 9.289   RC4, fading, supplemental, turbo, 8 km/h, 2 paths.

9.6.4.2.2   Reverse Link Budget Assumptions

Table 9.91 presents figures that can be used in the link Figures (9.290–9.293) assignment and link performance tasks. These values were derived based on the minimum requirements presented in the graphs of Section 9.6.4.2. Exact values should be discussed with the equipment vendors because there may be variations depending on the exact implementation of each feature. Figures 9.290 to 9.293 present a summary of the forward link budget requirements.

Table 9.91 Reverse link budget assumptions

Suggested E_b/N_0 design values for Reverse Link cdma2000

E_b/N_0 (dB) for a FER of 1%

Radio configuration	Channel	Data rate (kbps)	Coding	Rate	Cellular			PCS			FER 1 % to 3 %
					Moderate fading cellular	Medium fading cellular	Severe fading PCS	Moderate fading PCS	Medium fading PCS	Severe fading (dB)	
RC1	Fundamental	9600	Convolutional	1/3	6.2	7.5	8.5	6.7	8	9	−0.5
RC2	Fundamental	14 400	Convolutional	1/2	5.4	8	9	5.9	8.6	9.6	−0.5
RC3	Fundamental	9600	Convolutional	1/4	4.2	5.5	6.2	4.6	5.8	6.5	−0.3
	Supplemental	19 200	Convolutional	1/4	2.3	3.6	4.3	2.9	4.1	4.7	−0.3
	Supplemental	38 400	Convolutional	1/4	1.9	3.2	3.9	2.4	3.6	4.2	−0.3
	Supplemental	76 800	Convolutional	1/4	1.55	2.85	3.55	2.25	3.45	4.05	−0.3
	Supplemental	153 600	Convolutional	1/4	2	3.3	4	2	3.2	3.8	−0.3
	Supplemental	307 200	Convolutional	1/2	2	3.3	4	2.7	3.9	4.5	−0.3
RC3	Supplemental	19 200	Turbo	1/4	1.6	2.9	3.6	2.3	3.5	4.1	−0.2
	Supplemental	38 400	Turbo	1/4	0.7	2	2.7	1.4	2.6	3.2	−0.2
	Supplemental	76 800	Turbo	1/4	0.2	1.5	2.2	0.8	2	2.6	−0.2
	Supplemental	153 600	Turbo	1/4	−0.4	0.9	1.6	0.2	1.4	2	−0.2
	Supplemental	307 200	Turbo	1/2	0.2	1.5	2.2	0.9	2.1	2.7	−0.2
RC4	Fundamental	14 400	Convolutional	1/4	4	5.5	6.2	4.5	5.7	6.4	−0.3
	Supplemental	28 800	Convolutional	1/4	2.1	3.6	4.3	2.9	4.1	4.8	−0.3
	Supplemental	57 600	Convolutional	1/4	1.85	3.35	4.05	2.6	3.8	4.5	−0.3
	Supplemental	115 200	Convolutional	1/4	1.8	3.3	4	2.4	3.6	4.3	−0.3
	Supplemental	230 400	Convolutional	1/4	1.6	3.1	3.8	2.3	3.5	4.2	−0.3
RC5	Supplemental	28 800	Turbo	1/4	1.3	2.8	3.5	2.3	3.5	4.2	−0.2
	Supplemental	57 600	Turbo	1/4	0.7	2.2	2.9	1.6	2.8	3.5	−0.2
	Supplemental	115 200	Turbo	1/4	0.3	1.8	2.5	1.15	2.35	3.05	−0.2
	Supplemental	230 400	Turbo	1/4	0.4	1.9	2.6	0.6	1.8	2.5	−0.2

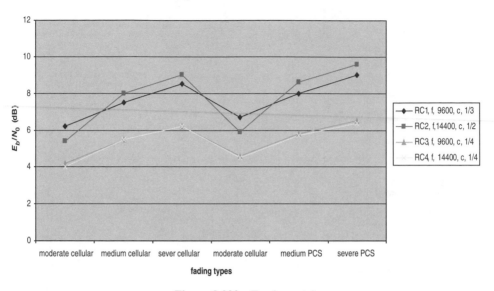

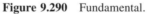

Figure 9.290   Fundamental.

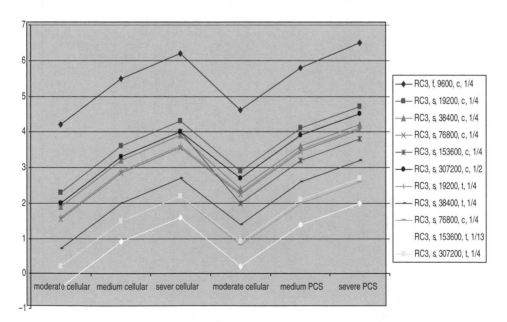

Figure 9.291   RC1.

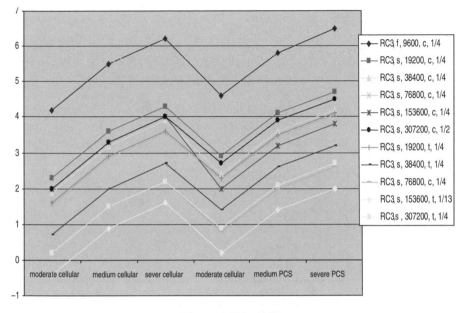

Figure 9.292 RC3.

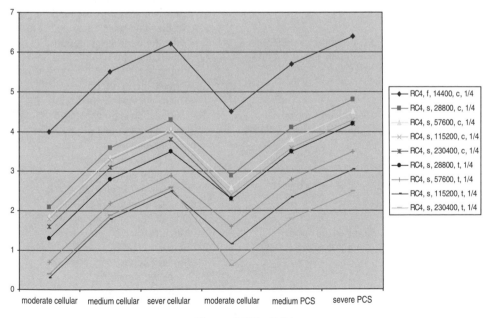

Figure 9.293 RC4.

9.6.5   Source Decoding

The information retrieved may contain errors after the decoding process. If these errors are within the tolerable margin, the information can be accepted. However, if they are below the desired quality level, the information cannot be used and the service is considered as not available.

The prediction software should calculate the expected quality of the information received and help to best configure (optimise) the network parameters to maximise throughput within the quality constraints. One of the difficulties found in this task is the analysis of the impact of residual errors on the information quality.

The SNR results in a certain raw Bit Error Rate (BER) after the information goes through the demodulator. The coding process removes part of these errors, determining the final

Table 9.92   Mean opinion score

Method	Mean opinion score	Grade
Absolute category rating (ACR)	Mean listening-quality opinion score (MOS)	5 - Excellent 4 - Good 3 - Fair 2 - Poor 1 - Bad
	Mean listening-effort opinion score (MOSLE)	5 - Complete relaxation possible 4 - Attention necessary 3 - Moderate effort required 2 - Considerable effort required 1 - No meaning understood
	Mean loudness-preference opinion score (MOSLP)	5 - Much louder then preferred 4 - Louder then preferred 3 - Preferred 2 - Quieter then preferred 1 - Much quieter then preferred
Degradation category rating (DCR)	Degradation mean opinion score (DMOS)	5 - Degradation is imperceptible 4 - Degradation is just perceptible but not annoying 3 - Degradation is perceptible but slightly annoying 2 - Degradation is annoying but not objectable 1 - Degradation is annoying and objectable
Comparison category rating (CCR)	Comparison mean opinion score (CMOS)	3 - Much better 2 - Better 1 - Slightly better 0 - about the same -1 - Slightly worse -2 - Worse -3 - Much worse

BER. The information is then coded in frames. However, frames may present errors, which are accounted for in a quality criterion known as Frame Error Rate (FER). Frames can be further corrected or repeated using Automated Repeat Request (ARQ) algorithms to replace damaged frames.

The quality of the voice, however, is still a very important parameter and is usually assessed by means of subjective tests. Table 9.92 shows some of the subjective methods commonly used for this assessment.

Packet voice presents additional challenge because the delay becomes an important factor in the quality perceived. Several methods have been proposed to assess this quality. One of the most commonly used is the R factor. This method represents the measure of voice quality due to packeting effects. Results of this method are expressed from 1 to 100.

Equation (9.136) shows how to calculate the R factor based on the impairment associated with the mouth-to-ear delay (d), expressed in terms, of the path (I_d) and I_f

$$R = 93.2 - I_d - I_f \tag{9.136}$$

I_d is calculated based on d and H

$$I_d = 0.024\,d + 0.11(d - 177.3)H(d - 177.3) \tag{9.137}$$

$$H(x) = 0 \quad \text{for} \quad x < 0 \tag{9.138}$$

$$H(x) = 1 \quad \text{for} \quad x \geq 0 \tag{9.139}$$

Table 9.93 R-factor

R factor	Quality of voice (user satisfaction)	MOS
$90 < R < 100$	Best	4.34 to 4.5
$80 < R < 90$	High	4.03 to 4.34
$70 < R < 80$	Medium	3.6 to 4.03
$60 < R < 70$	Low	3.1 to 3.6
$50 < R < 60$	Poor	2.58 to 3.11

Table 9.94 Vocoder delay

Vocoder	Rate (kbps)	MOS	Delay (ms)
PCM (G711)	64	4.1	0.75
ADPCM (G726)	32	3.8	1
LD-CELP (G728)	16	3.6	2.5
CS-CELP (G729)	8	3.7	10
G 723.1	5.3	3.9	30

The equipment impairment factor (I_f) is given by eqn (9.140). In this equation 'e' indicates the total packet loss probability, expressed between 0 and 1.

$$I_f = 30 \, \mathrm{Ln}(1 + 15\,e)H(0.04 - e) + 19 \, \mathrm{Ln}(1 + 70\,e)H(e - 0.04) \qquad (9.140)$$

Table 9.93 shows the mapping of the R-factor into the MOS system.

Table 9.94 shows delays and MOS provided by typical vocoders. Landline toll quality MOS is considered as 4.2.

BIBLIOGRAPHY AND REFERENCES

1. Jhong Sam Lee and Leonard E. Miller, CDMA Systems Engineering Handbook, 1998, Artech house, Inc.
2. Simon R. Saunders, Antennas and Propagation for Wireless Communication Systems, 1999, John Wiley & Sons Ltd.
3. Henry L. Bertoni, Radio propagation for Modern Wireless Systems, 2000, Prentice Hall, Inc.
4. Kazimierz Siwiak, Radio Propagation and Antennas for Personal Communications, 1995, Artech House, Inc.
5. David Parson, The Mobile Radio Propagation Channel, 1996, John Wiley & Sons Ltd.
6. Manuel F. Cátedra and Jesús Pérez-Arriaga, Cell Planning for Wireless Communications, 1999, Artech House, Inc.
7. José M. Hernando and F. Pérez-Fontan, Introduction to Mobile Communications Engineering, 1999, Artech House, Inc.
8. Saleh Faruque, Cellular Mobile Systems Engineering, 1996, Artech House, Inc.
9. Roger L. Freeman, Radio System Design for Telecommunications, 1997, John Wiley & Sons Ltd.
10. John Doble, Introduction to radio Propagation for Fixed and Mobile Communications, 1996, Artech House, Inc.
11. Trevor Manning, Microwave Radio Transmission Design Guide, 1999, Artech House, Inc.
12. Samuel C. Yang, CDMA RF System Engineering, 1998, Artech House, Inc.
13. John B. Groe and Lawrence E. Larson, CDMA Mobile Radio Design, 2000, Artech House, Inc.
14. Andrew J. Viterbi, CDMA Principles of Spread Spectrum Communications, 1995, Addison-Wesley Publishing Company.
15. Theodore S. Rappaport, Wireless Communications, 2002, 1996, Prentice Hall, Inc.
16. Bernard Sklar, Digital Communications, 2001, Prentice Hall, Inc.
17. John G. Proakis and Masoud Salehi, Communication System Engineering, 2002, Prentice Hall, Inc.
18. M.R. Karim and Mohsen Sarraf, WCDMA and cdma2000, 2002, McGraw-Hill.
19. Tommy Oberg, Modulation Detection and Coding, 2001, John Wiley & Sons Ltd.
20. George M. Calhoun, Third Generation Wireless Systems, 2003, Artech house, Inc.
21. Marvin K. Simon and Mohamed-Slim Alouini, Digital Communications over Fading Channels, 2000, John Willey & Sons Ltd.
22. Robert M. Gagliardi, Introduction to Communications Engineering, 1988, John Wiley & Sons Ltd.
23. Claude E. Shannon and Warren Weaver, The Mathematical Theory of Communications, 1949, 1998, University of Illinois.
24. Claude E. Shannon, A Mathematical Theory of Communications, Bell System Technical Journal, vol. 27, pp 379–423, 625–656, July, October, 1948.
25. Kaveh Phlavan and Allen H. Levesque, Wireless Information Networks, 1995, John Wiley & Sons Ltd.

26. 3GPP2- 'Recommended Minimum Performance Standards for cdma2000 Spread Spectrum Mobile Stations' Release B, Version 1, December 13, 2002.

27. 3GPP2- 'Recommended Minimum Performance Standards for cdma2000 Spread Spectrum Base Stations' Release B, version 1, December 13, 2002.

28. Lawrence Harte, Morris Hoenig, Daniel McLaughlin, Roman K. Kta, CDMA IS-95 for Cellular and PCS1999, McGraw-Hill.

29. Steve Lee, Spread Spectrum CDMA IS-95 and IS-2000 for RF Communicaions, 2002, McGraw-Hill.

30. Harry Holma, Antti Toskala, WCDMA for UMTS Radio Access for Third Generation Mobile Communications, 2001, John Willey & Sons Ltd.

31. Vijay K. Garg, Joseph E. Wilkes, Principles & Applications of GSM, 1999, Prentice-Hall, Inc.

32. Tero Ojanperä, Ramjee Prasad, WCDMA: Towards IP Mobility and Mobile Internet, 2000, Artech house, Inc.

33. Simon Haykin, Digital Communications, 1988, John Willey & Sons Ltd.

34. Vijay K. Garg, Is-95 CDMA and cdma2000, 1999, Prentice-Hall, Inc.

35. Marvin K. Simon, Mohamed-Slim Alouini, Digital Communications over Fading Channels, 2000, John Willey & Sons Ltd.

36. Hazysztof Wesolowski, Mobile Communication Systems, 2002, John Willey & Sons Ltd.

37. Saleh Faruque, Cellular Mobile System Engineering, 1996, Artech house, Inc.

38. Husni Hammuda, Cellular Mobile Radio Systems, 1997, John Willey & Sons Ltd.

39. Michael Daoud Yacoub, Foundations of Mobile Radio Engineering, 1993, CRC Press.

40. Michael Daoud Yacoub, Wireless Technology Protocols, Standards, and Techniques, 2001, CRC Press.

41. Robert M. Gagliardi, Introduction to Communications Engineering, 1988, John Willey & Sons Ltd.

42. Reinaldo Perez, Wireless Communications Design Handbook, 1998, Academic Press.

43. Cost Action 231, Digital mobile radio towards future generation systems-Final report, 1999, European Comission.

44. John G. Proakis, Digital Communications, 2001, McGraw-Hill.

45. Bernard Sklar, Digital Communications Fundamentals and Applications, 2001, Prentice-Hall, Inc.

46. John P. Castro, The UMTS Netwrok and Radio Access Technology Air Interface Techniques for Future Mobile Systems, 2001, John Willey & Sons Ltd. Kaveh Pahlavan, Allen H. Levesque, Wireless Information Networks, 1995, John Willey & Sons, Inc.

47. M. R. Karim, Mohsen Sarraf, W-CDMA and cdma2000 for 3G Mobile Networks, 2002, McGraw-Hill.

48. William C. Y. Lee, Mobile Cellular Tellecommunications Analog and Digital Systems, 1989, McGraw-Hill.

49. J. S. Milton, Jesse C. Arnold, Introduction to Probability and Statistics-Principles and Applications for Engineering and the Computing Sciences, 1995, McGraw-Hill.

50. Alberto Leon-Garcia, Probability and Random Processes for Electrical Engineering, Addison-Wesley Publishing Company.

51. Roger L. Freeman, Radio System Design for Telecommunications, 1997, John Willey & Sons, Inc.

10

Network Design

LEONHARD KOROWAJCZUK

In this chapter network design steps are analysed. The databases required are explained with practical examples shown. The concepts of user classes and traffic for today's wireless networks are introduced as well as link budget issues. Several network predictions are introduced. Wireless network simulation is introduced and techniques for network enhancement, optimisation and performance evaluation are presented.

10.1 NETWORK DESIGN FLOW

Network design is an iterative, complex task that has to reconcile marketing goals and technical capabilities with the budget available. Marketing goals are established based on service offering and expected demand. This leads to a business case that provides for CAPEX (Capital Expenditures) and OPEX (Operational Expenditures). The expected revenues should cover these expenditures and financial costs while still providing profit to shareholders.

Figure 10.1 shows the network design flowchart, indicating the main iteration loops faced by designers during network planning and optimisation. Each of the tasks involved in the network design flow are briefly described in this section.

10.1.1 Service and Market Definition

When designing a network, marketing goals determine services to be offered, the quality parameters and expected demand, which include growth over time and geographical distribution of users. It is common, however, for the marketing department not to provide the geographical component, which is then left as an additional task for network designers.

10.1.2 Demand Characterisation

The demand for wireless communications has and continues to grow drastically, especially because of the new services being offered, requiring spectrum usage to be stretched to the

Designing CDMA 2000 Systems L. Korowajczuk et al.
© 2004 John Wiley & Sons, Ltd ISBN: 0-470-85399-9

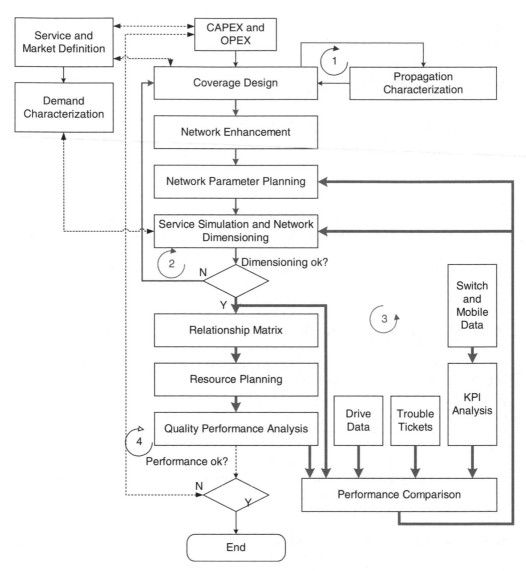

Figure 10.1 Network design flowchart.

limit to cope with it. The estimation of the geographical distribution of this demand is key for proper network design and full spectrum utilisation.

Traditionally, demand is characterised by the traffic expressed in Erlangs, which is appropriate for voice systems where all users conform to a similar behaviour. New services, however, can not be appropriately represented in this way; data demand must be expressed according to the service traffic distribution. Because users of a data system are characterised by a combination of service, environment and terminal, the demand of these systems should be determined based on the different types of users.

10.1.3 CAPEX and OPEX

The Capital Expenditures (CAPEX) budget defines the amount of equipment that can be deployed in the network, whereas the Operational Expenditures (OPEX) budget determines the maintenance effort expenditures, including network reconfigurations.

10.1.4 Coverage Design and Propagation Characterisation

One of the first tasks of the network designer is to define the number of BTSs required to meet desired traffic and coverage specifications. The traffic demand allows an initial approximation of the number of BTSs required based on the maximum traffic that can be carried by each BTS. The coverage, however, depends on propagation and terrain characteristics and on the traffic distribution. Therefore designers usually make use of propagation prediction tools to help on assessing these factors.

Prediction tools need to be adjusted (calibrated) to reflect actual propagation behaviour. Continuous Wave (CW) measurements provide the data required to characterise the propagation parameters required by propagation models. After calibrated, the tools can be used to generate coverage predictions that can be compared to real life measurements. Further adjustments can be done until the designer is satisfied with the predicted RF coverage. This iterative process is the calibration loop and is represented in Figure 10.1 as loop number 1.

The BTSs placement is a highly iterative process because of zoning, airport restrictions, tower availability and economical issues. It is common that network designers propose several possible locations for each BTS. In this case, the final location is only defined after a field inspection and lease negotiations.

10.1.5 Network Enhancement

The overlap of cell site coverage area is a requirement for good network designs but, at the same time, it is a burden to the system. Not enough overlap results in dropped calls because of lack of handoff areas, whereas too much overlap results in interference and, consequently, capacity reduction. It is essential to enhance cell overlaps to provide just enough handoff area while keeping interference to a minimum.

Enhancement tools such as CelEnhancer can analyse different network configurations and select the one providing the best overall results in terms of traffic and interference. Tools like this are able to suggest the best configuration settings to each of the network sites, including antenna type (out of a pre-defined set), tilt, azimuth, power and height.

10.1.6 Network Parameter Planning

The complexity level of wireless networks, especially after the addition of data services, requires the definition of the desired quality level for each of the services offered. The quality criteria usually include the definition of the required SNIR level (or equivalent parameter). This is where the link budget is defined.

10.1.7   Service Simulation and Network Dimensioning

The CDMA network performance depends on the mix of services and their geographical distribution, consequently varying over time (significant variations can be observed within the busy hour), because both these factors are dynamic. Therefore, the only way to analyse the performance is to assess these parameters statistically. Monte Carlo simulation is one of the possible techniques for this analysis. In this simulation, calls are randomly placed according to the estimated traffic density, representing a snapshot of the network. The operational convergence is then calculated for each snapshot. Section 10.8 explains this technique in more detail. The statistics of several snapshots represent the network performance and provide the average value of parameters such as

- average BTS traffic channel power and its standard deviation;
- average BTS noise rise (or load) and its standard deviation;
- average BTS channel usage and its standard deviation;
- average mobile power.

10.1.8   Dimensioning

The second loop presented in Figure 10.1 is related to traffic dimensioning. This loop is required when traffic requirements are not met, and may imply addition of new sites or change in existing site locations.

10.1.9   Relationship Matrix

After the layout of sites is ready, the next step in designing the network is to create a resource plan, which varies depending on the technology being used. To help on resources assignment, a relationship matrix is built, establishing the relation among all sectors in the network. This relation represents how bad it would be to assign the same resource for each pair of sectors in the network. The matrix is ideally traffic weighted to prioritise planning of areas that generate more revenue to the operator.

10.1.10   Resource Planning

Not only the relationship matrix is required for resource planning, but also handoff levels, neighbourhood and SNIR levels. The calculation of all these parameters is not a simple task and usually requires the use of optimisation tools such as CelOptima.

10.1.11   Quality Performance Analysis

At this point the network design is complete and the next step is to assess quality performance. A new traffic simulation is executed and statistical parameters are applied to calculate network outage at different levels of quality. This quality can be mapped to bad call quality and to dropped calls, indicating system outage.

10.1.12 Performance Comparison

The predicted performance can be compared to measured data for a fine calibration of predictions. Drive tests and trouble tickets (user complaints) are compared to the predictions and discrepancies are analysed and corrected.

Statistics provided by the MSC (Mobile Switching Center) can also be used to evaluate network performance. These statistics are usually referred to as KPI (Key Performance Indicators) and provide information about network accessibility (congestion), call retention (dropped calls) and call quality (frame error rates).

The comparison of predictions to KPI can be a powerful maintenance tool that can pinpoint hardware issues, allowing the forecast of the result of network changes before implementing them, minimising the OPEX.

10.1.13 Performance

The final loop in Figure 10.1 is related to performance. If the desired network performance is not achieved, designers should adjust the initial CAPEX and OPEX requirements and re-start the design process. The use of a good prediction tool is essential in these situations because it allows designers to study multiple what-if scenarios before implementing them, consequently saving deployment time and substantial amounts of money.

10.2 DATABASES

To design or analyse a system it is essential to properly characterise the network and the environment where it resides. This information is stored in multiple databases, which can be divided into several categories, including GIS (Geographic Information System), network and demographic databases.

10.2.1 GIS Database

Geographic information system databases store the characteristics of the environment where the network is located. This information consists mainly of terrain and land cover data. Sections 10.2.1.2 describes these databases in more detail.

10.2.1.1 Coordinate Systems and Map Projections

Geodesy is the science that measures and represents the earth, studying its size, shape and gravity field. These gravitational measurements allow the determination of an element called geoid, which represents the real shape of the earth, that is, the shape that most closely approximates the mean sea level all around the planet. The geoid is used as a reference for heights, therefore essential for determining terrain height. Figure 10.2 shows the variations of the earth geoid.

The geoid can be approximated as an ellipsoid. Figure 10.3 illustrates the earth (topographic) surface (thick solid line), the geoid (thin solid line) and the ellipsoid (dotted line).

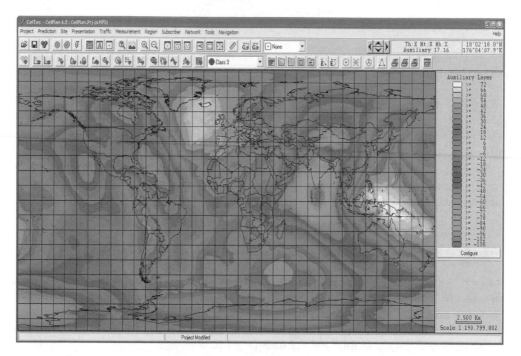

Figure 10.2 WGS-84[1] geoid height grid.

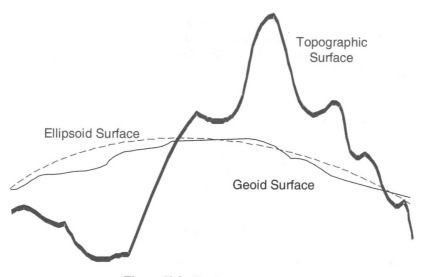

Figure 10.3 Earth representations.

[1]See Table 10.1 for the configuration of World Geodetic System (WGS) 84.

Table 10.1 Main ellipsoids

Ellipsoid	Semi-major axis (a)	1/flattening
Everest	6 377 276.345	300.801 700 000
Airy 1830	6 377 563.396	299.324 964 600
Bessel 1830	6 377 397.155	299.152 812 800
Bessel 1841	6 377 397.155	299.152 812 800
Clarke 1866	6 378 206.400	294.978 698 200
Clarke 1880	6 378 249.145	293.465 000 000
International 1928	6 378 388.000	297.000 000 000
Krassovsky 1940	6 378 245.000	298.300 000 000
Hough 1956	6 378 270.000	297.000 000 000
WGS 1960	6 378 165.000	298.300 000 000
Fisher 1960	6 378 166.000	298.300 000 000
WGS 1966 (c)	6 378 145.000	298.250 000 000
GRS 1967 (Geodetic Reference System)	6 378 160.000	298.247 167 427
Fisher 1968	6 378 150.000	298.300 000 000
South American 1969	6 378 160.000	298.250 000 000
WGS 1972	6 378 135.000	298.260 000 000
GRS 1975	6 378 140.000	298.257 000 000
GRS 1980	6 378 137.000	298.257 222 101
WGS 1984	6 378 137.000	298.257 223 563

The ellipsoid representing the geoid is aligned with the earth rotational axis. An ellipsoid is defined by its semi-major axis and by a polar flattening factor, which is given by eqn (10.1).

$$f = (a - b)/a \tag{10.1}$$

Where

a = semi-major axis

b = semi-minor axis

Several different values have been proposed for these parameters as measurement techniques evolved. Table 10.1 lists some of the ellipsoids proposed.

The rotation of the ellipsoid around its north-south axis forms the virtual surface of the earth that is used as a reference for the location of any point on the planet surface as illustrated in Figure 10.4. In 1637 René Descartes proposed a co-ordinate system based on parallels and meridians, later named as the Cartesian system. This system divides the Earth in graticules (spherical grid of co-ordinate lines over the Earth surface).

These grids are formed by parallels that are perpendicular to the north-south axis and by meridians that cross at the north and south poles.

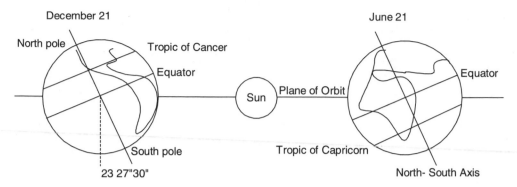

Figure 10.4   Earth axis relative to the orbit plane.

The main parallel is the Equator, located on the surface generated by the ellipsoid at the tip of the ellipsoid major axis. The Equator is considered as 0° and all other parallels are labelled according to their angle to the north (up to 90°) or to the south as illustrated in Figure 10.6. There are two methods to measure this angle, each of them used by different co-ordinate systems.

The geodetic method measures the angle that the normal to the parallel does with the equatorial plane, while the geocentric measures the angle that the line that connects the parallel to the centre of the earth with the equatorial plane. This is illustrated in Figure 10.5.

Surface lines that connect the north and south poles of the ellipsoid represent the meridians. The prime meridian was defined in 1884 as the line that crosses a specific marker in the Royal Observatory in Greenwich (close to London), England. This meridian is

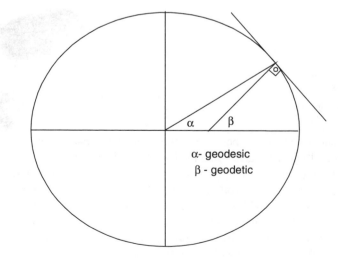

Figure 10.5   Geodesic and geodetic methods.

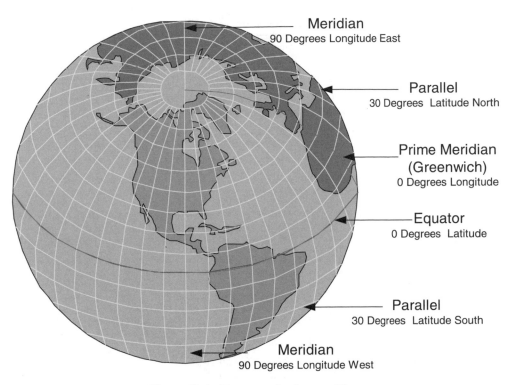

Figure 10.6 Equator and prime meridian.

known as the Greenwich Meridian and is considered as 0°. All other meridians are labelled according to the angle they make with the prime meridian as illustrated in Figure 10.6.

After the ellipsoid is defined, a reference point has to be selected to act as the origin of the co-ordinates. Until very recently, all cartographic work was done using triangulation on the ground. Cartographers found out, however, that relative measurements from a local origin are more practical and defined a local datum (geodetic). More recent systems use the earth centre of mass as the reference point.

An example of a local datum is NAD27, which uses ellipsoid Clarke 1866 with the reference at a mark in Meades Ranch in Kansas, USA. The WGS 84 (World Geodetic System), which is the co-ordinate system used by the GPS (Geographic Position System) in its raw data, uses the earth centre of mass as a reference point.

The representation of the earth on a plane surface brings serious difficulties and cartographers have proposed many different methods of doing it. The transformation of the earth ellipsoid into a plane surface is done using projection techniques or mathematical manipulation of graticules.

Cartographic representations can be classified according to the projection surface, surface position, origin and type of projection, and properties envisaged for the end product.

Table 10.2 lists the main projection surfaces used. Figure 10.7 illustrates the azimuthal, cylindrical, and conical projections, whereas Figure 10.8 shows, respectively, a tri-dimensional view of the cylindrical and conic projections.

Table 10.2 Main projection surfaces

	Projection Surface
Azimuthal or planar	Projected on a plane surface
Cylindrical	Projected on a cylinder
Conic	Projected on a cone
Pseudo cylindrical, elliptic or sinusoidal	Calculated from cylindrical

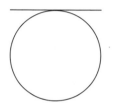 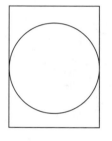 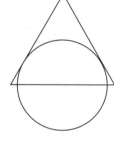

azimuthal or planar   cylindrical   conical

Figure 10.7 Main projection surfaces.

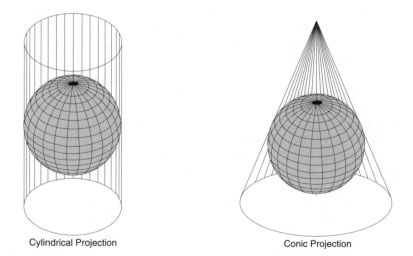

Cylindrical Projection   Conic Projection

Figure 10.8 Tri-dimensional view of cylindrical and conical projections.

Table 10.3 Surface positions

Surface position
Normal
Transverse
Oblique

The surface position can be tangent or secant to the earth ellipsoid. Table 10.3 shows the main surface positions, illustrated in Figure 10.9. Figure 10.10 shows a tri-dimensional view of the transverse cylindrical projection.

Table 10.4 list the possible origin and types of projection, also illustrated in Figure 10.11.

Projections are also classified according to some of their properties as listed in Table 10.5. There are a large number of cartographic representations; Table 10.6 shows some of the

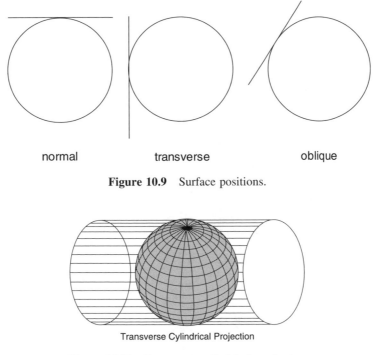

| normal | transverse | oblique |

Figure 10.9 Surface positions.

Transverse Cylindrical Projection

Figure 10.10 Transverse cylindrical projection.

Table 10.4 Origin and type of projection

Origin and type of projection
Orthographic
Stereographic
Vertical perspective
Gnomonic

Table 10.5   Properties of main projections

	Properties
Equidistant	Preserves distances
Equal area	Preserves relative areas
Conformal	Scale is the same in any direction
	Meridians and parallels cross at right angles
	Shape is preserved locally
Orthophanic	Right appearing

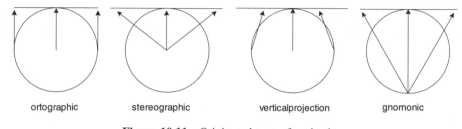

| ortographic | stereographic | verticalprojection | gnomonic |

Figure 10.11   Origin and type of projection.

Table 10.6   Cartographic representations

Cylindrical	Cylindrical equal area	Behrmann cylindrical equal area
		Gall's stereographic cylindrical
	Peters	
	Mercator	
	Miller cylindrical	
	Oblique mercator	
	Transverse mercator	British National Grid
		Universal Transverse Mercator
Pseudo-cylindrical	Mollweide	
	Eckert	Eckert IV equal area
		Eckert VI equal area
	Robinson	
	Sinusoidal equal area	
	Albers equal area conic	
Conic projections	Equidistant conic	
	Lambert conformal conic	
	Poly-conic	
Azimuthal	Azimuthal equidistant	
	Lambert azimuthal equal area	
	Oblique aspect orthographic projection	
	North polar stereographic	
Miscellaneous	Un-projected	
	Robinson	
	Winkel Tripel	

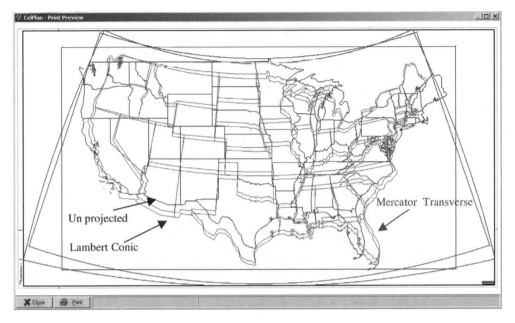

Figure 10.12 USA map with different projections.

representations commonly used. Each option represents a certain part of the globe better; therefore the representation is chosen or created by local geographical institutes.

Cartographic representations used in mobile communications apply only to limited areas (some hundreds of km in size). Figure 10.12 shows the USA map using different projections. The representation of a line has to be carefully done, especially when using two distant points to determine this line. In a sphere, the smallest distance between two points is defined by a great circle (segment of the ellipsoid that connects two points), which does not coincide with straight lines in some projections. Sections 10.2.1.1.1 and 10.2.1.1.2 describe in more detail the two most commonly used cartographic representations for designing networks.

10.2.1.1.1 Universal Transverse Mercator (UTM) Representation
This is not a real projection but instead a mathematical representation of distances in a two-dimensional way, measured from a local meridian and the equator. Figure 10.13 shows the UTM representation of the earth.

UTM coordinates are represented through three parameters: zone, northing and easting. The graticules of the earth surface are divided in 60 parts of 6°, called zones (meridians); 18 parts of 8°, called bands (parallels) (C to V, no I or O); and one 12° band (band X). The central meridian of the zone is used as an east-west reference whereas the equator is used as a north-south reference. Figure 10.14 illustrates a zone from the UTM system.

Coordinates in the Universal Transverse of Mercator system are expressed by the distance of a point in relation to the equator towards the northern hemisphere, or in relation to the parallel located 10 000 km south towards the southern hemisphere, and by the distance of the point in relation to an imaginary line located 500 km west from a reference meridian. This is illustrated in Figure 10.15.

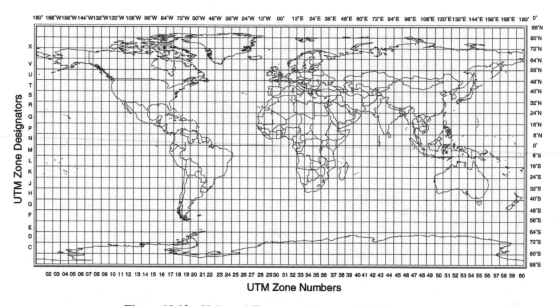

Figure 10.13 Universal Transverse Mercator (UTM) system.

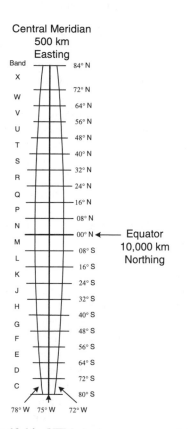

Figure 10.14 UTM single-zone representation.

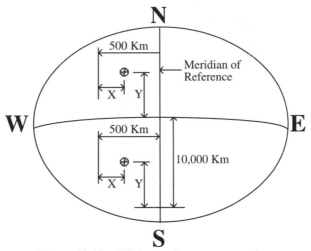

Figure 10.15 UTM co-ordinate representation.

10.2.1.1.2 Unprojected Representation

In this representation, graticules (pixels) are represented by their centre or corner co-ordinates. Each graticule is defined by a rectangle. The vertical size of the rectangle is constant anywhere in the globe and its representation is perfect. The horizontal size varies with latitude and, to make its representation feasible, the mean size of the area of interest is used to represent all graticules. This introduces a progressive error to the north and south of the area.

The size of the graticule represents the database resolution. Each graticule has information associated to it, generating a raster database as graticules are listed in sequence from side to side and top to bottom. Figure 10.16 shows an unprojected world representation.

10.2.1.2 Terrain Database–Topography

The terrain elevation above the reference ellipsoid is given in topographical maps. These maps show the ground elevation through equal-height contour curves, which form a vector representation of the terrain.

It took nations decades to chart their countries and, today, most of the world topography has been mapped with resolutions as good as 10–50 m spacing contour curves. Some of these maps are quite old but as terrain heights take a long time to change, their use is still acceptable.

The first challenge to create the topography database is to rasterise the information in the topographic maps. The interpolation process, which calculates the height of points between the contour curves, is extremely complex and it is common to find gross interpolation errors in the databases. Therefore it is recommended to inspect databases and correct eventual abnormalities before using them for designing networks. Some database formats, such as CelGeo (see www.celplan.com) store the contour and raster information simultaneously, making it easier to recover the original information.

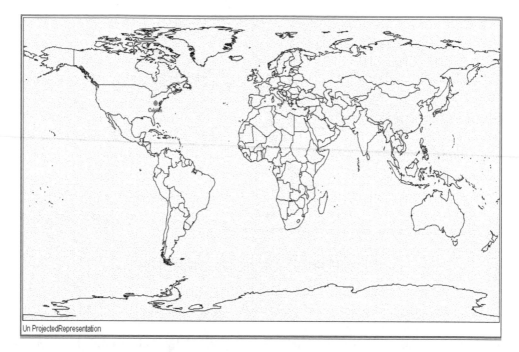

Figure 10.16   Unprojected representation of the world.

Figure 10.17 shows the topography in raster and in contour format. The raster format can be derived from the contours though sophisticated interpolation algorithms.

Because of the 1 m vertical resolution, coastal areas may get confused with the water, thus many databases store an additional indication, such as the use of $+0$ and -0, to distinguish between ground and water.

Figure 10.18 shows the unprojected representation of topography.

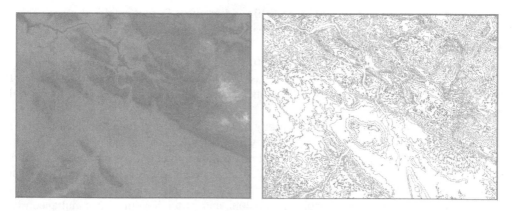

Figure 10.17   Topography database in CelGeo format – interpolated (left) and level contours (right).

Figure 10.18   Unprojected representation of topography.

10.2.1.3   Land Cover/Land Use Database – Morphology

The terrain is not only represented by the altitude (height) but also by the morphology formations over it. These formations can be divided in categories such as water (in its many forms), natural vegetation and man made constructions. A morphology database shows the presence of each of these categories over a map.

Morphological information has many applications, including RF studies. Unfortunately, maps were very seldom created with RF in mind; therefore the morphological categories selected are not appropriate for this type of application. A solution is to use maps with abundant categories and group them according to the morphological behaviour with respect to RF waves.

The morphology height plays a big role in RF propagation. Good databases have this parameter expressed on a pixel-by-pixel basis together with the morphology type. Morphology and topography pixels should be independent from each other because morphology and topography vary differently.

The number of morphologies to be used when designing a network depends on the characteristics of the area to be studied. It is difficult to choose a set of morphologies applicable to all areas; thus there is always a certain regionalism in the description of morphology types.

Maps represent morphology in clusters, using a certain resolution. This chapter classifies these clusters in mixed types and pure types. Table 10.7 shows a suggestion of pure

Table 10.7 Pure morphology types

			Pure morphology types	Height (m)	Resolution (m)
1	Water	1	Salt water	0	30
		2	Sweet water	0	30
		3	Snow	0 to 3	30
		4	Ice	0 to 1	30
		5	Swamp	0	30
2	Ground cover	6	Bare ground	0	30
		7	Sand	0	30
		8	Grass	0	30
		9	Bush	0–2	30
3	Vegetation	10	Sparse shrubs	0–2	30
		11	Medium shrubs	1–2	30
		12	Dense shrubs	2–2	30
4	Trees	13	Deciduous	2–30	12–18
		14	Perennial	2–30	12–18
		15	Conifer	2–30	12–18
		16	Tropical	2–40	12–18
		17	Palm	2–12	12–18
5	Building rooftop	18	Wood/plastic	3–12	3–6
		19	Masonry	3–512	3–6
		20	Steel	3–512	3–6
6	Canyons	21	Street	0	1
		22	Road	0	1

morphological types that can be used in RF studies. The table presents 22 types but, usually, less than 16 types are used for a single area. The resolution column in the table indicates the horizontal resolution of the pixel. The vertical resolution is indicated as height.

Figures 10.19–10.21 show, respectively, examples of low-resolution (200 m), median-resolution (30 m) and high-resolution morphology (up to 1 m). Street canyons are very important when analysing RF behaviour but they imply an increase in the database resolution, consequently increasing size and cost. Figure 10.22 shows an example of a morphology database with streets data.

Databases with a resolution worse than 1 m required to imprint roads and street canyons can be over-sampled and 'carved' to show roads and streets with pre-defined widths. This solution adds canyons to the database with a fraction of the cost. The size of the database is still large because of the new resolution (1 m); but, because of the over-sampling, the information is highly redundant, therefore the databases are highly compressible. The compression helps in reducing the space required to store the data but it requires software that can de-compress the information in real time. CelZip is a *CelPlanner* proprietary algorithm that uncompressed in real time only the required parts of the database, making the process fast and transparent to users.

RF does not distinguish between the height and width of an obstacle, it will 'go around' the smallest one; therefore the height used in RF databases should not be the actual height

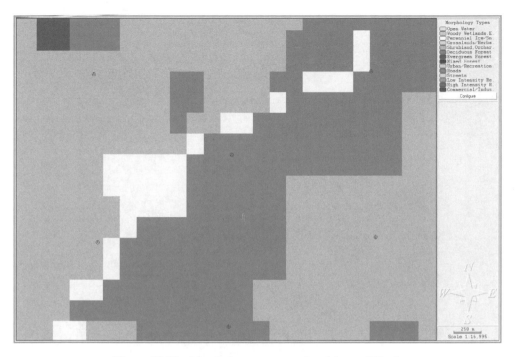

Figure 10.19 Morphology representation (cluster 200 m).

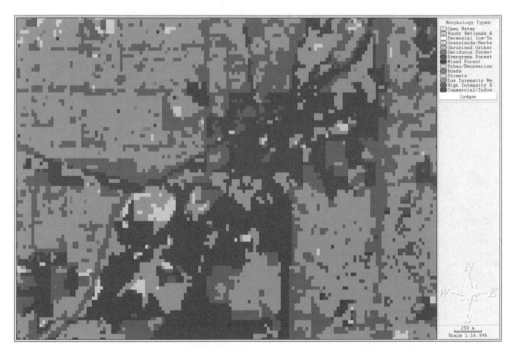

Figure 10.20 Morphology representation (cluster 30 m).

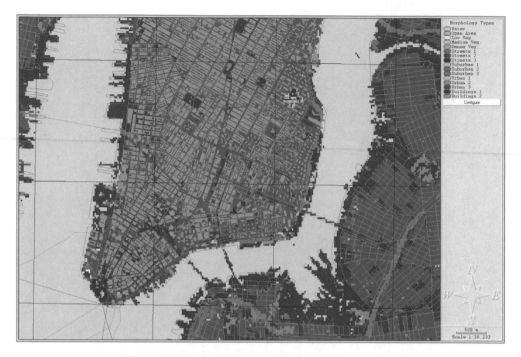

Figure 10.21   Morphology representation (buildings).

Figure 10.22   Morphology representation (streets).

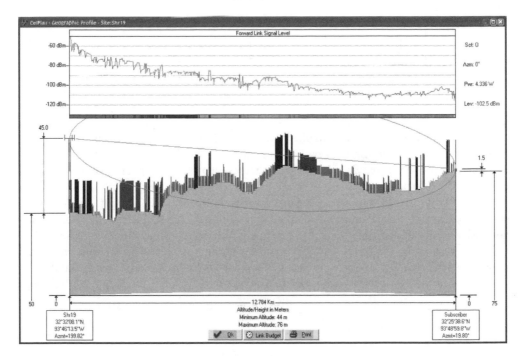

Figure 10.23 Building profile.

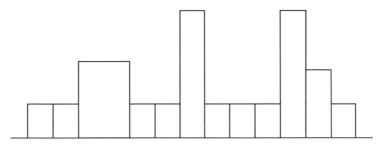

Figure 10.24 Building profile diagram.

Figure 10.25 RF corrected building profile.

Figure 10.26   Building profile approximation for RF studies.

of the obstacle but instead the smallest measure, that is either the height or the width. Figure 10.23 shows a profile with building heights.

Figures 10.24–10.26 show different height representations of the same set of buildings. Figure 10.24 shows the actual height of the buildings whereas Figure 10.25 shows the ideal RF height. Figure 10.26 shows a reasonable approximation of the RF height, where a single height representing the average of all buildings is used. These illustrations show a two-dimensional visualisation but the same concept applies for tri-dimensional analyses.

Lower resolution databases can be used for certain applications. In this case, each morphological type represents only the predominant type in the area, as in Table 10.8. The height of these morphological types follows the same rules as the height for buildings.

Table 10.8   Mixed morphology types

	Mixed morphology types	Height (m)	Resolution (m)
1	Green field	0–2	90
2	Trees cover	2–30	90
3	Rural	2–4	60
4	Sub-urban	3–8	30
5	Urban	6–18	30
6	Dense urban	12–60	30
7	Skyscrapers cluster	40–400	30

10.2.1.4   Imagery and Landmarks Database

The use of graticules provides a good way of storing and manipulating topography and morphology data but does not provide good visual references to designers. Two categories of additional information layers, raster and vector, can be displayed as auxiliary visual references.

The raster layer consists of geo-referenced maps and photos. The vector layer represents several different types of landmarks that can be displayed simultaneously on screen without cluttering the background. Roads, streets, rivers, boundaries, transmission lines and many others can be displayed in vector format.

Figures 10.27 and 10.28 show, respectively, a street map and an aerial photo being used as background. Both these backgrounds are raster representations. Figure 10.29 shows landmarks (streets, roads and rivers) in a vector representation.

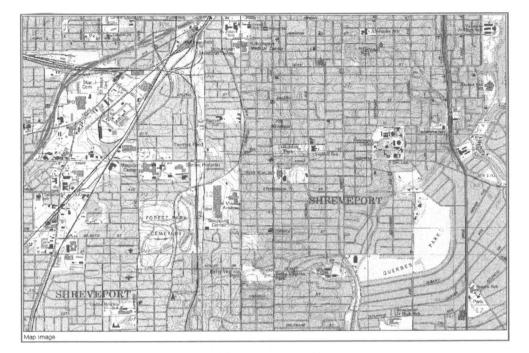

Figure 10.27 Street map as background.

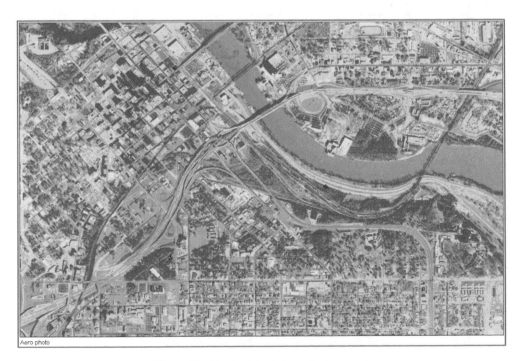

Figure 10.28 Aerial photo as background.

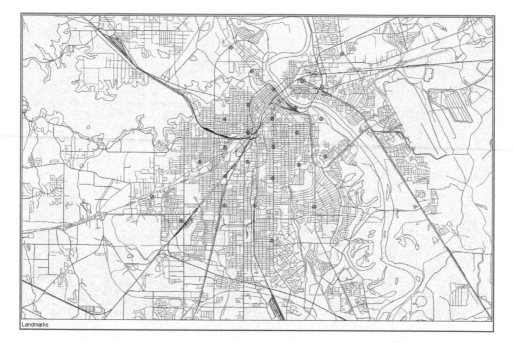

Figure 10.29   Landmarks.

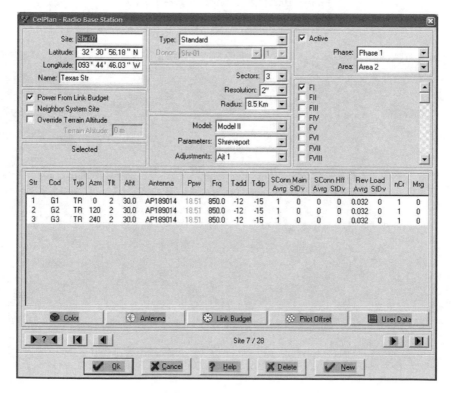

Figure 10.30   Typical RBS database entry.

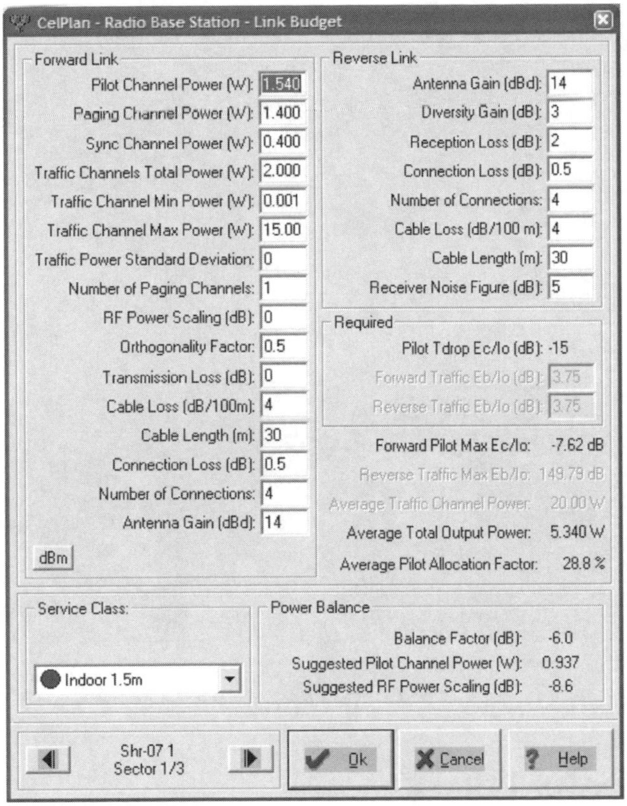

Figure 10.31 Typical BTS database entry.

10.2.2 Network Databases

Network databases store information regarding equipment configuration and system characteristics. The network information can be divided in smaller databases. Figures 10.30 and 10.31 show, respectively, a typical database entry for a radio base station (BTS):

- antenna;

- resources (e.g. frequency channels, SAT, DCC and PN offset);

- system parameters;

- prediction parameters;

- user service classes;

- radio base stations (site data, including location, status, prediction, sector data and user data);

- sector data (including equipment, electrical parameters, capacity parameters, link budget, neighbours and resources).

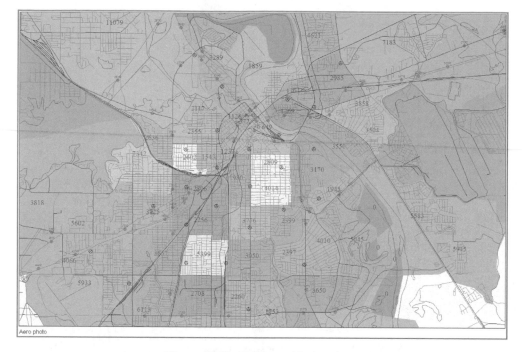

Figure 10.32 Demographical regions.

10.3 REGIONS DATABASE

To define the requirements of a network it is necessary to specify characteristics that are geographically distributed. Region files are used for this purpose. Each region file is composed of one or multiple geographically defined regions, with or without attributes associated to them.

A region is constituted of one or more geographically defined closed polygons. Usually regions do not overlap, allowing a dichotomy between them.

A set of regions defines the attributes of a certain area. Typical region attributes are population, households, businesses, income or any other parameter expressed in geographical terms. Figure 10.32 shows demographical regions with population as attribute. Sections 10.5.1 and 11.2.3 show the use of regions to distribute traffic geographically.

10.4 SERVICE CLASSES

Operators offer different services to different users. The network analysis has to consider this mix because it profoundly affects the overall performance. Each component of this mix is called a service class.

The service class configuration includes the service being used, the terminal type and the environment where the communication is happening. Each service class has its own traffic distribution associated to it. Figures 10.33 and 10.34 show a typical service class config-uration, respectively, for 3 G (1XRTT) and (EVDV) networks.

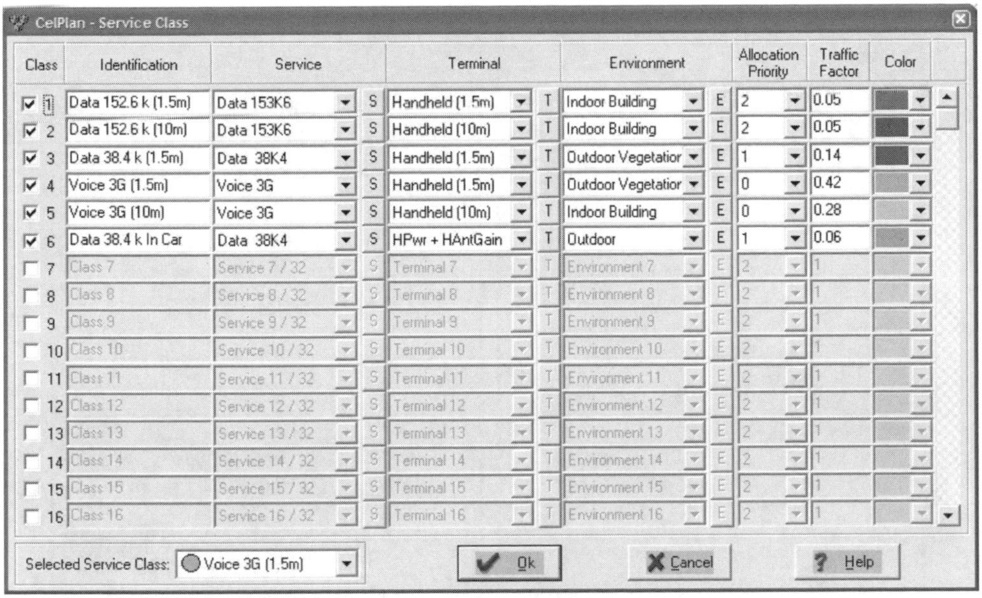

Class	Identification	Service		Terminal		Environment		Allocation Priority		Traffic Factor	Color
☑ 1	Data 152.6 k (1.5m)	Data 153K6	S	Handheld (1.5m)	T	Indoor Building	E	2		0.05	
☑ 2	Data 152.6 k (10m)	Data 153K6	S	Handheld (10m)	T	Indoor Building	E	2		0.05	
☑ 3	Data 38.4 k (1.5m)	Data 38K4	S	Handheld (1.5m)	T	Outdoor Vegetation	E	1		0.14	
☑ 4	Voice 3G (1.5m)	Voice 3G	S	Handheld (1.5m)	T	Outdoor Vegetation	E	0		0.42	
☑ 5	Voice 3G (10m)	Voice 3G	S	Handheld (10m)	T	Indoor Building	E	0		0.28	
☑ 6	Data 38.4 k In Car	Data 38K4	S	HPwr + HAntGain	T	Outdoor	E	1		0.06	
☐ 7	Class 7	Service 7 / 32	S	Terminal 7	T	Environment 7	E	2		1	
☐ 8	Class 8	Service 8 / 32	S	Terminal 8	T	Environment 8	E	2		1	
☐ 9	Class 9	Service 9 / 32	S	Terminal 9	T	Environment 9	E	2		1	
☐ 10	Class 10	Service 10 / 32	S	Terminal 10	T	Environment 10	E	2		1	
☐ 11	Class 11	Service 11 / 32	S	Terminal 11	T	Environment 11	E	2		1	
☐ 12	Class 12	Service 12 / 32	S	Terminal 12	T	Environment 12	E	2		1	
☐ 13	Class 13	Service 13 / 32	S	Terminal 13	T	Environment 13	E	2		1	
☐ 14	Class 14	Service 14 / 32	S	Terminal 14	T	Environment 14	E	2		1	
☐ 15	Class 15	Service 15 / 32	S	Terminal 15	T	Environment 15	E	2		1	
☐ 16	Class 16	Service 16 / 32	S	Terminal 16	T	Environment 16	E	2		1	

Selected Service Class: ◉ Voice 3G (1.5m)   ✔ Ok   ✗ Cancel   ? Help

Figure 10.33 Service class definition for 3 G.

Class	Identification	Service		User Terminal		Environment		Traffic		Factor	Color
☑ 1	VoiceInGSM	Voice	S	GSM	U	Indoor	E	Traffic I	D	0.2	
☑ 2	VoiceOutGSM	Voice	S	GSM	U	Outdoor	E	Traffic I	D	0.3	
☑ 3	VoiceInCDMA	Voice	S	CDMA	U	Indoor	E	Traffic I	D	0.2	
☑ 4	VoiceOutCDMA	Voice	S	CDMA	U	Outdoor	E	Traffic I	D	0.3	
☑ 5	EmailInCDMA	Email	S	CDMA2000	U	Indoor	E	Traffic IV	D	0.25	
☑ 6	EmailOutCDMA	Email	S	CDMA2000	U	Outdoor	E	Traffic IV	D	0.09	
☑ 7	EmailInGPRS	Email	S	GPRS	U	Indoor	E	Traffic IV	D	0.25	
☑ 8	EmailOutGPRS	Email	S	GPRS	U	Outdoor	E	Traffic IV	D	0.08	
☑ 9	EmailInEDGE	Email	S	EDGE	U	Indoor	E	Traffic IV	D	0.25	
☑ 10	EmailOutEDGE	Email	S	EDGE	U	Outdoor	E	Traffic IV	D	0.08	
☑ 11	VoiceInCDMA2000	Voice	S	CDMA2000	U	Indoor	E	Traffic II	D	0.4	
☑ 12	VoiceOutCDMA2000	Voice	S	CDMA2000	U	Outdoor	E	Traffic II	D	0.6	
☑ 13	WebnEDGE	Web	S	GPRS	U	Indoor	E	Traffic III	D	0.25	
☑ 14	WebOutEDGE	Web	S	EDGE	U	Outdoor	E	Traffic III	D	0.08	
☑ 15	WebInGPRS	Web	S	EDGE	U	Indoor	E	Traffic III	D	0.25	
☑ 16	WebOutGPRS	Web	S	EDGE	U	Outdoor	E	Traffic III	D	0.08	

Selected User Class: ◉ VoiceOutCDMA   ✔ Ok   ✗ Cancel   ? Help

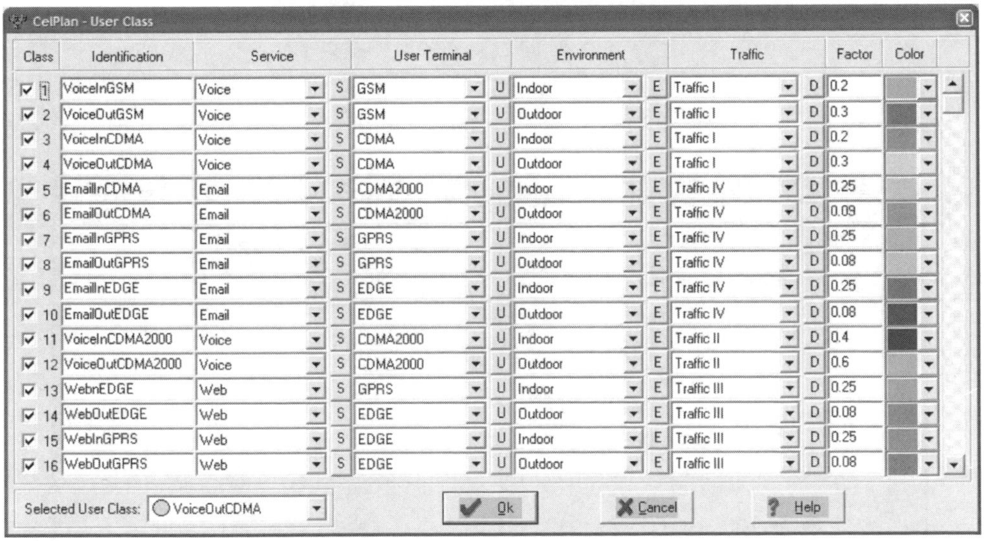

Figure 10.34 Service class definition for 3 G.

10.4.1 Service Type

The definition of services includes the configuration of the required data rate and target quality. Figures 10.35–10.37 show typical configurations for voice, low-speed data and high-speed data for 2.5G systems. Figure 10.38 shows the configuration of voice and video service classes in 3G networks.

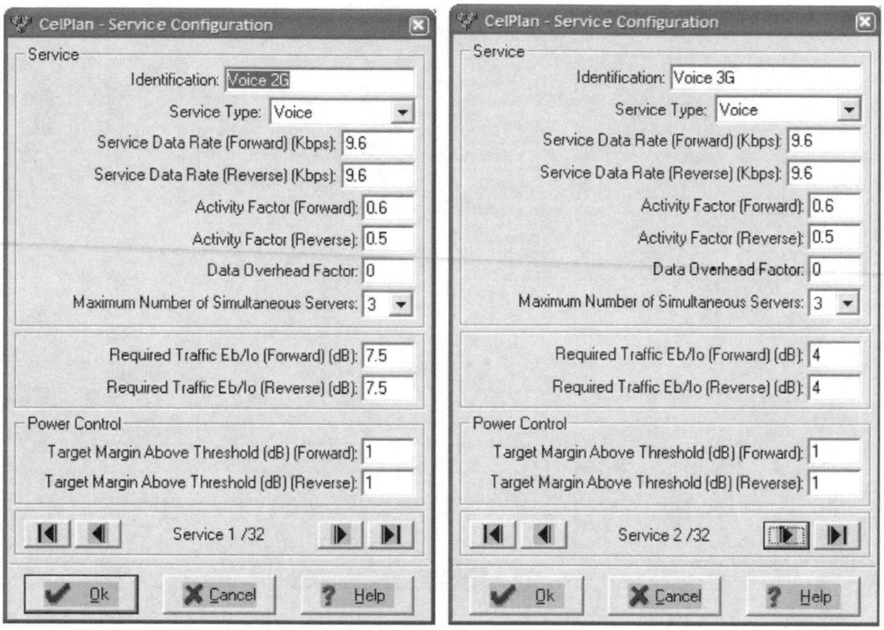

Figure 10.35 Voice service classes.

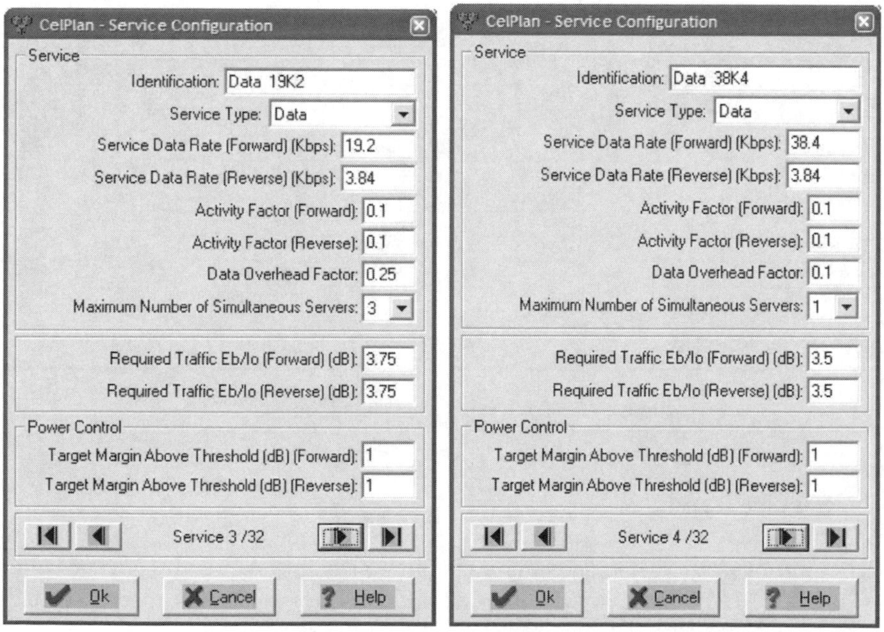

Figure 10.36 Low-speed data service classes.

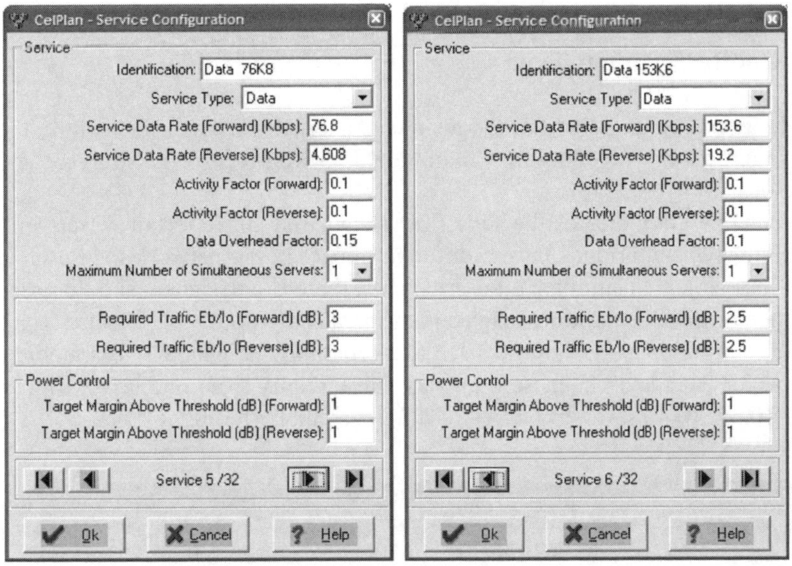

Figure 10.37 High-speed data service classes.

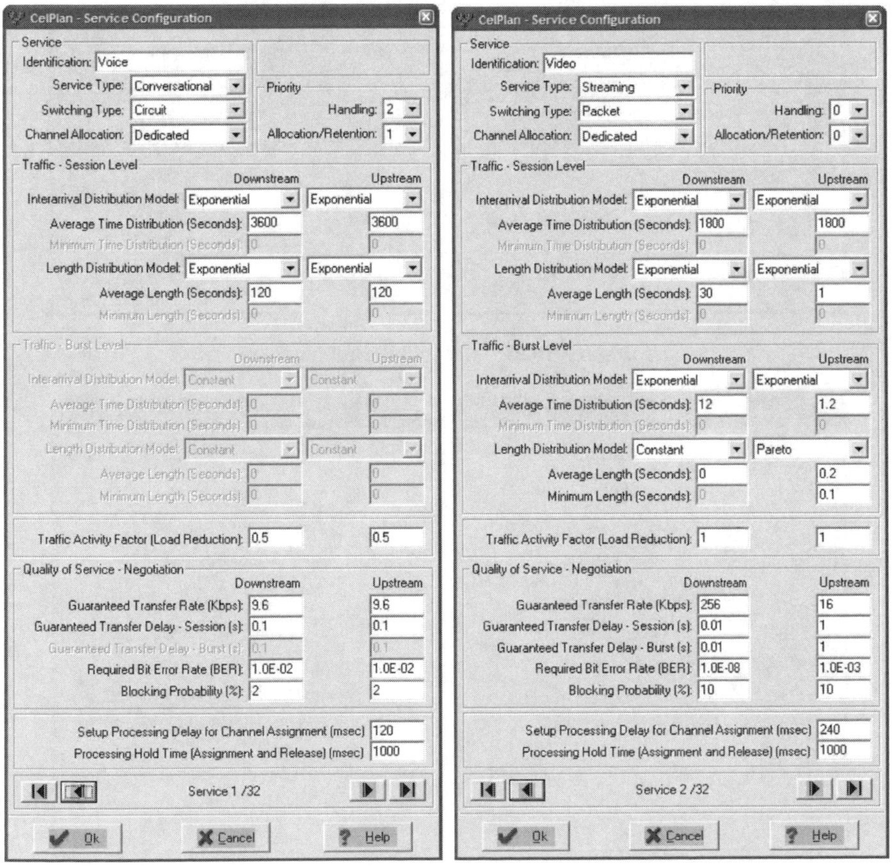

Left window:

CelPlan - Service Configuration

Service
Identification: Data 76K8
Service Type: Data
Service Data Rate (Forward) (Kbps): 76.8
Service Data Rate (Reverse) (Kbps): 4.608
Activity Factor (Forward): 0.1
Activity Factor (Reverse): 0.1
Data Overhead Factor: 0.15
Maximum Number of Simultaneous Servers: 1

Required Traffic Eb/Io (Forward) (dB): 3
Required Traffic Eb/Io (Reverse) (dB): 3

Power Control
Target Margin Above Threshold (dB) (Forward): 1
Target Margin Above Threshold (dB) (Reverse): 1

Service 5 /32

Ok   Cancel   Help

Right window:

CelPlan - Service Configuration

Service
Identification: Data 153K6
Service Type: Data
Service Data Rate (Forward) (Kbps): 153.6
Service Data Rate (Reverse) (Kbps): 19.2
Activity Factor (Forward): 0.1
Activity Factor (Reverse): 0.1
Data Overhead Factor: 0.1
Maximum Number of Simultaneous Servers: 1

Required Traffic Eb/Io (Forward) (dB): 2.5
Required Traffic Eb/Io (Reverse) (dB): 2.5

Power Control
Target Margin Above Threshold (dB) (Forward): 1
Target Margin Above Threshold (dB) (Reverse): 1

Service 6 /32

Ok   Cancel   Help

Figure 10.38 3 G services configuration.

Left window:

CelPlan - Service Configuration

Service
Identification: Voice
Service Type: Conversational
Switching Type: Circuit
Channel Allocation: Dedicated

Priority
Handling: 2
Allocation/Retention: 1

Traffic - Session Level

	Downstream	Upstream
Interarrival Distribution Model:	Exponential	Exponential
Average Time Distribution (Seconds):	3600	3600
Minimum Time Distribution (Seconds):	0	0
Length Distribution Model:	Exponential	Exponential
Average Length (Seconds):	120	120
Minimum Length (Seconds):	0	0

Traffic - Burst Level

	Downstream	Upstream
Interarrival Distribution Model:	Constant	Constant
Average Time Distribution (Seconds):	0	0
Minimum Time Distribution (Seconds):	0	0
Length Distribution Model:	Constant	Constant
Average Length (Seconds):	0	0
Minimum Length (Seconds):	0	0

Traffic Activity Factor (Load Reduction): 0.5 / 0.5

Quality of Service - Negotiation

	Downstream	Upstream
Guaranteed Transfer Rate (Kbps):	9.6	9.6
Guaranteed Transfer Delay - Session (s):	0.1	0.1
Guaranteed Transfer Delay - Burst (s):	0.1	0.1
Required Bit Error Rate (BER):	1.0E-02	1.0E-02
Blocking Probability (%):	2	2

Setup Processing Delay for Channel Assignment (msec): 120
Processing Hold Time (Assignment and Release) (msec): 1000

Service 1 /32

Ok   Cancel   Help

Right window:

CelPlan - Service Configuration

Service
Identification: Video
Service Type: Streaming
Switching Type: Packet
Channel Allocation: Dedicated

Priority
Handling: 0
Allocation/Retention: 0

Traffic - Session Level

	Downstream	Upstream
Interarrival Distribution Model:	Exponential	Exponential
Average Time Distribution (Seconds):	1800	1800
Minimum Time Distribution (Seconds):	0	0
Length Distribution Model:	Exponential	Exponential
Average Length (Seconds):	30	1
Minimum Length (Seconds):	0	0

Traffic - Burst Level

	Downstream	Upstream
Interarrival Distribution Model:	Exponential	Exponential
Average Time Distribution (Seconds):	12	1.2
Minimum Time Distribution (Seconds):	0	0
Length Distribution Model:	Constant	Pareto
Average Length (Seconds):	0	0.2
Minimum Length (Seconds):	0	0.1

Traffic Activity Factor (Load Reduction): 1 / 1

Quality of Service - Negotiation

	Downstream	Upstream
Guaranteed Transfer Rate (Kbps):	256	16
Guaranteed Transfer Delay - Session (s):	0.01	1
Guaranteed Transfer Delay - Burst (s):	0.01	1
Required Bit Error Rate (BER):	1.0E-08	1.0E-03
Blocking Probability (%):	10	10

Setup Processing Delay for Channel Assignment (msec): 240
Processing Hold Time (Assignment and Release) (msec): 1000

Service 2 /32

Ok   Cancel   Help

10.4.2   Mobile Terminal Type

The mobile terminal type identifies the main characteristics of the user radio in the forward and reverse directions, including maximum output power, receive noise figure and antenna characteristics.

The antenna height indicates the subscriber's elevation above terrain. Users in buildings, elevated highways and bridges have a different impact in the network than users at ground level. It is impractical to simulate all user heights; therefore designers should group users in layers. Depending on the building heights in the area analysed, 2–8 layers are typically used. The choice of layers must be carefully tied to morphology heights. RF loss predictions must be executed for each layer because they vary substantially from one layer to the next.

Figures 10.39–10.41 show examples of different mobile terminal types.

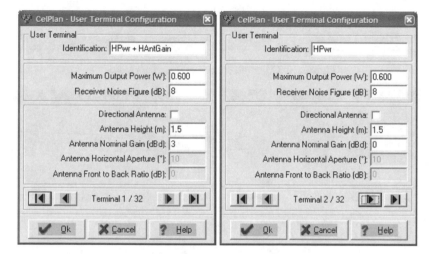

Figure 10.39   Car mobile terminals.

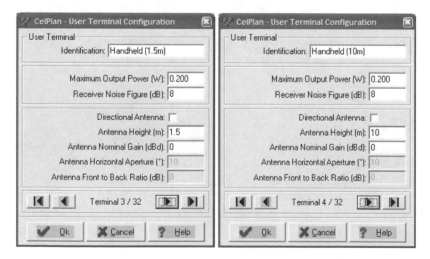

Figure 10.40   Handhelds at different user elevations.

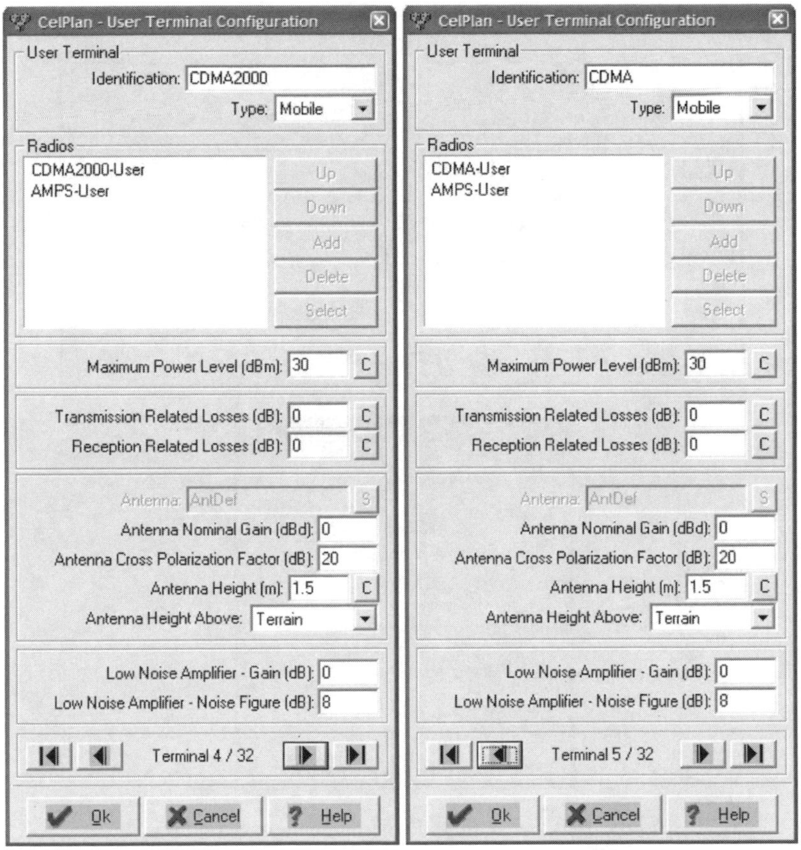

Figure 10.41 User terminal configurations for 3G systems.

10.4.3 Environment

The environment configuration encompasses the mobile terminal antenna position in relation to the user and RF characteristics in the area around the user, as shown in Figures 10.42 and 10.43. The main aspects covered by this configuration are the following:

- human body attenuation – the antenna position in relation to the user may cause an attenuation because of the closeness to the human body;

- penetration attenuation – an indoor antenna may suffer additional loss because of its surroundings (e.g. close to a window or secluded);

- fading margin – short term and long term fading variations can be considered as a fixed value, a distribution or geographically defined by regions;

- orthogonality indicates loss of orthogonality (between codes in CDMA) with distance;

- cross polarisation applies to fixed deployments where cross polarised antennas are commonly used;

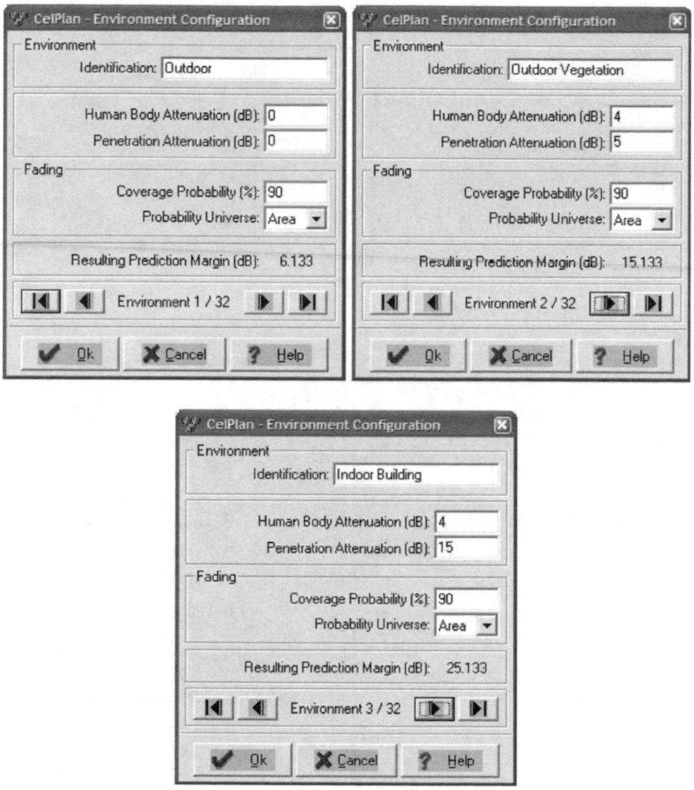

Figure 10.42 Simplified environment configurations.

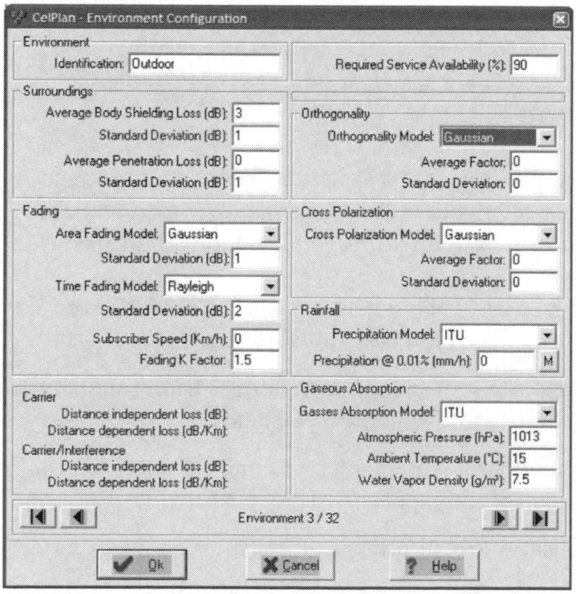

Figure 10.43 Detailed environment configuration.

- rainfall – significant only for systems operating on frequencies above 10 GHz;

- gaseous absorption applies only to systems operating on frequencies above 10 GHz.

10.5 USER DISTRIBUTION

After service classes are defined, users should be geographically identified and quantified. The user distribution can be generated based on demographical sources, knowledge of the area, switch data or a mix of this information.

10.5.1 Demographics

Figures 10.44–10.46 show examples of demographical distributions: vehicles per km on roads, residential population per tract and commercial buildings area. Although each demographic layer may be expressed in different units, it can be normalised to users or to simultaneous active users.

The demographical information is usually coarse but it can be re-spread within regions using weights to determine the distribution of users among the different morphological types. Figure 10.47 shows an example of a table for weight assignment. In this table, a weight of zero indicates that no users are present in the associated morphology. The weights indicate how many more users can be found in one morphology type than in another. For example, according to Figure 10.47, a low intensity residential area would have only half of the users that a high intensity residential area of the same size would have.

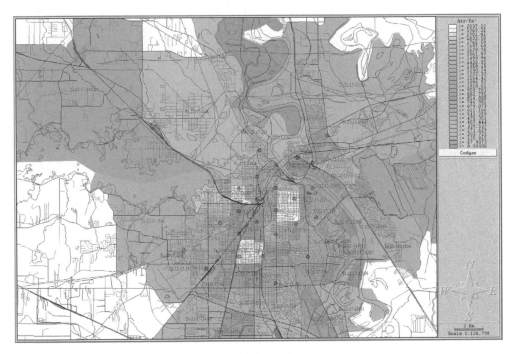

Figure 10.44 Residential population demographics.

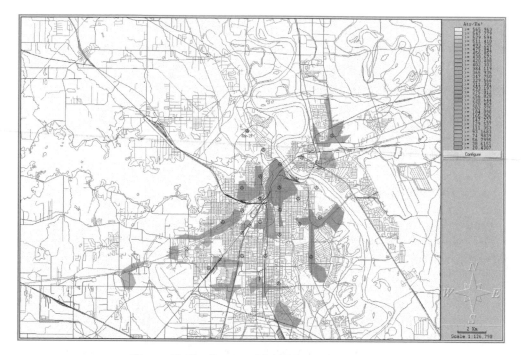

Figure 10.45   Commercial buildings demographics.

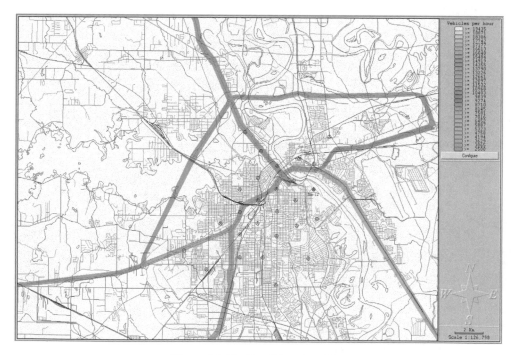

Figure 10.46   Road demographics.

Terrain Type	Factor
Open Water	0
Woody Wetlands,Emergent	0
Perennial Ice/Snow,Bare Ro	0
Grasslands/Herbaceous,Past	0
Shrubland,Orchards/Vineyard	0
Deciduous Forest	0
Evergreen Forest	5
Mixed Forest	5
Urban/Recreational Grasses	5
	0
Roads	100
	0
Low Intensity Residential	40
High Intensity Residential	80
Commercial/Industrial/Transp	100
	0
Type not defined	0

Figure 10.47   Relative demographics distribution between morphologies.

The concept of users and traffic is thoroughly analysed in Chapter 11; therefore this chapter limits the explanations to the basic concepts.

The geographical traffic distribution obtained from the demographics is stored in demographical traffic layers. Figures 10.48–10.50 show traffic distributions derived, respectively, from residential, commercial and vehicular demographics. The advantage of this approach is that it allows modelling different configurations and not just the peak hour.

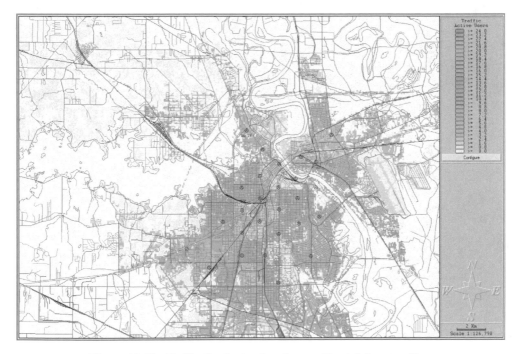

Figure 10.48   Traffic distribution based on residential demographics.

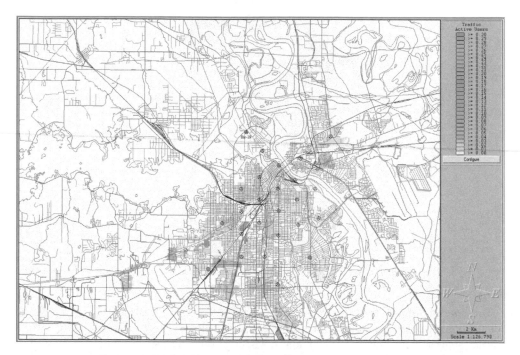

Figure 10.49   Traffic distribution based on commercial buildings.

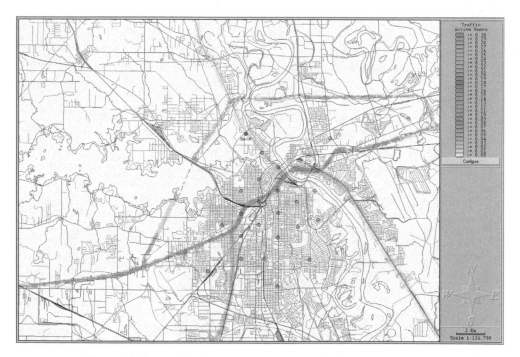

Figure 10.50   Traffic distribution based on vehicles per hour on roads.

Normalized traffic distribution

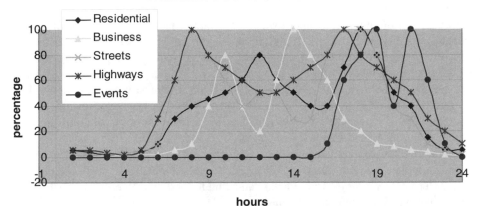

Figure 10.51 Traffic variations during a day for different demographics.

Figure 10.51 shows hourly variations of the traffic per demographic layer. Geographical traffic patterns can be generated for different hours by multiplying each layer by a factor and then combining all layers. Figure 10.52 shows the result of this combination.

Traditionally, network traffic intensity is represented in Erlangs.[2] An average traffic intensity value and a probability distribution (Rayleigh for voice) are assigned to each type

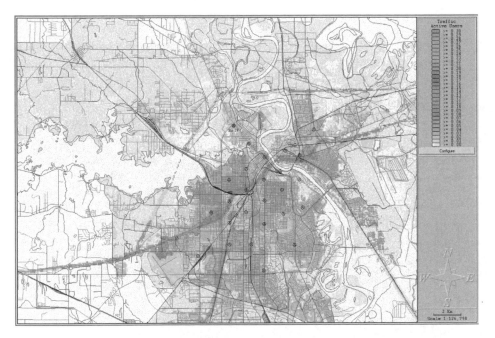

Figure 10.52 Multi-layer composed traffic distribution.

[2]The Erlang represents the percentage of occupancy of a circuit.

of user. This information is enough to calculate blocking probabilities and to dimension circuit requirements.

With the addition of services other than voice, a mix of packet and circuit switching must be considered, making the traffic representation more complex because different services present different distributions and can not be easily mixed in a single number. Besides, some services require real-time transmission, whereas others can be delayed, using an adaptive transmission rate. The best way to express traffic intensity for these networks is to determine the number of active users for each service. The service usage is then characterised by average rates and statistical distributions. These distributions are similar to Rayleigh but have distributions with longer or shorter tails.

After the number of simultaneous users of each class is calculated, it can be applied to statistical simulations. This number is presented geographically as traffic demand maps (layer). Alternatively, a single traffic layer can be multiplied by different factors to represent the participation of different service classes.

10.5.2 Switch Traffic Data

The switch can measure the combined traffic for the peak hour but the use of this information is limited because it includes a mix of services on a sector basis. Specialised tools can re-distribute this traffic according to the pilot best server and morphologies in the area to provide a more meaningful data. Figure 10.53 shows the pilot best server footprint. Figure 10.54 shows the sector traffic distributed across the area using the morphology

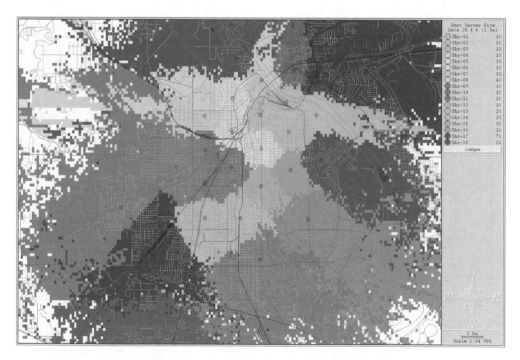

Figure 10.53 Pilot best server plot used for traffic distribution.

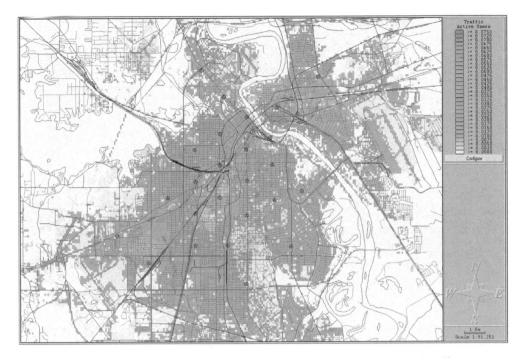

Figure 10.54 Traffic distribution based on the pilot best server and switch reported traffic per sector.

weights defined in Figure 10.47. Because the pilot coverage is traffic dependent, some iteration may be required to reach convergence between the pilot and traffic footprints.

The results in Figure 10.54 were obtained using CelPlanner, and match field distributions well enough to produce valid simulations. When multiple services are offered, the overall traffic can be split between them.

10.5.3   Multiple Traffic Layers

Each service class should have a traffic layer defining its user distribution as shown in Figure 10.55. The user distribution should represent users that are outdoor, inside vehicles, and indoor.

Outdoor users represent people walking on the street, in parks, and open areas in general, usually at ground level. In-vehicle users are also usually at ground level, with two important exceptions that should be modelled in certain cities: bridges and elevated highways. Indoor users are located inside buildings, which require further considerations as they can be at different elevations. It is very important to consider the height of the user in relation to the ground because of its impact on the signal to interference ratio.

User distribution between the different categories varies between cities, but the relative number of in-building calls is increasing. More than fifty percent of the calls today happen above ground level thus it is essential to consider this during design and planning phases.

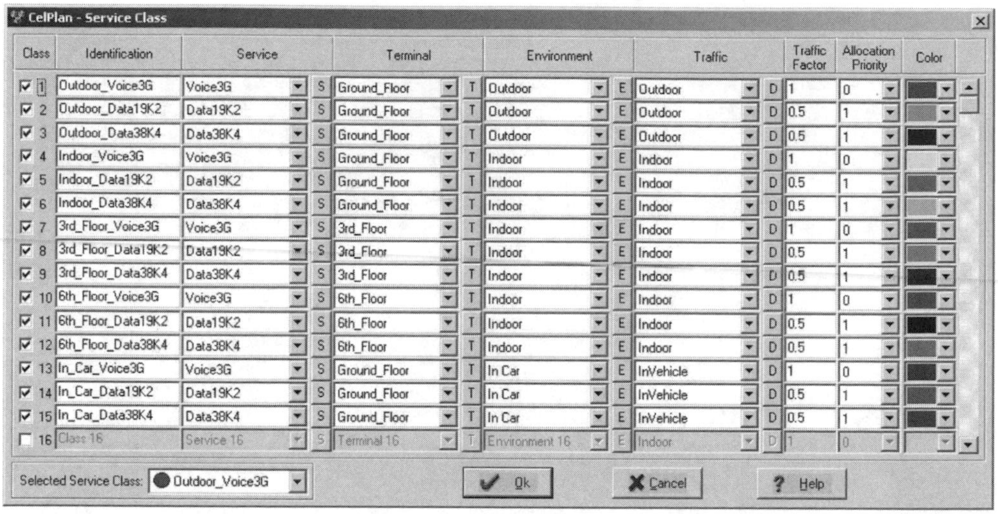

Figure 10.55 Multiple Service Classes with multiple traffic layers.

Table 10.9 presents typical user distribution among sample categories on a typical metropolitan city in the USA.

Figures 10.56 and 10.57, respectively, display the traffic distribution for outdoor and indoor. The indoor traffic is divided among different building floors through the use of multiplying factors.

In the example of Figure 10.55, three heights are used to represent buildings with up to eight floors: ground level, third floor (which also includes second and forth floors), and sixth floor (representing fifth to eight floor). It is not necessary to model all possible elevations and three to five different levels suffice for the majority of the cases. The traffic of several heights

Table 10.9 Network Call Distribution in a large metropolitan area

Network Calls		Height (m)	Average penetration loss (dB) best	Average penetration loss (dB) worst	Percentage (%)
Outdoor calls		1.5	3	4	15.00
In-vehicle calls at ground level		1	5	7	30.00
In-vehicle calls on elevated highways		6	5	7	5.00
Indoor calls	Ground floor	1.5	12	25	25.00
Indoor calls	Second floor	4.5	12	25	12.00
Indoor calls	Fourth floor	13.5	12	25	6.00
Indoor calls	Eighth floor	25.5	12	25	3.00
Indoor calls	16th floor	49.5	12	25	1.50
Indoor calls	32nd floor	97.5	12	25	0.70
Indoor calls	64th floor	193.5	12	25	0.20
Indoor calls	128th floor	385.5	12	25	0.05
Encapsulated calls		1.5	25	35	1.55

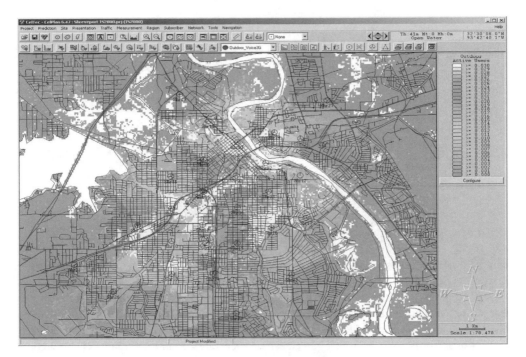

Figure 10.56 Outdoor traffic layer.

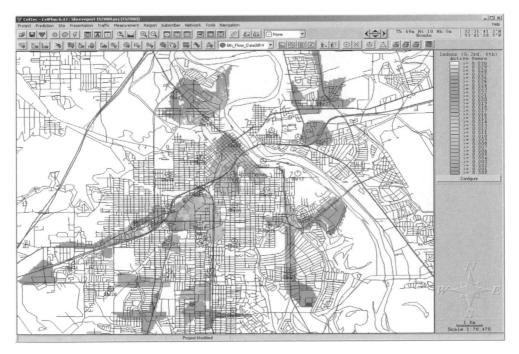

Figure 10.57 Indoor traffic layer for floors 1 to 8.

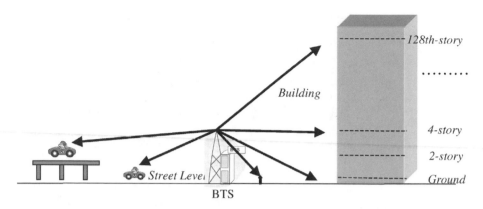

Figure 10.58   Influence of users at multiple heights over the network.

can be considered as concentrated in one of the representative heights. Figure 10.58 represents the impact of calls at different heights.

A good prediction tool should actually execute RF signal predictions for multiple heights, instead of only applying height correction factors, which can be misleading. For example, obstacles can be cleared depending on the height, resulting in completely different predictions. Tools such as *CelPlanner* scan service classes and calculate RF predictions for all specified heights. Each height is analysed independently for transmission and diffractions losses.

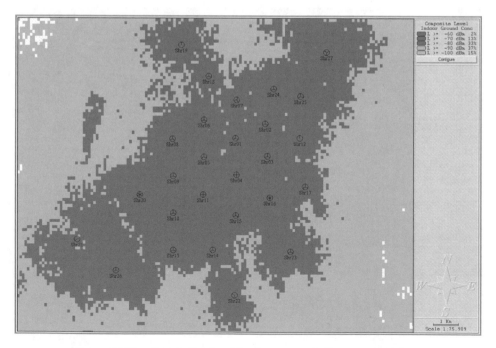

Figure 10.59   Composite signal level coverage indoor at ground floor.

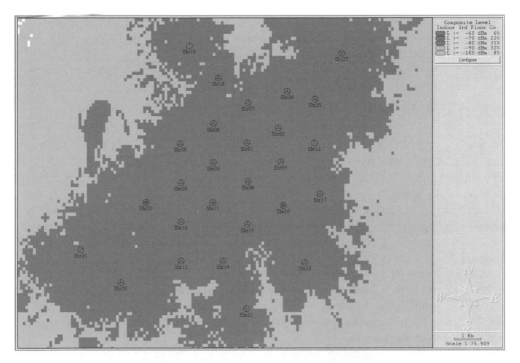

Figure 10.60 Composite signal level coverage indoor at third floor.

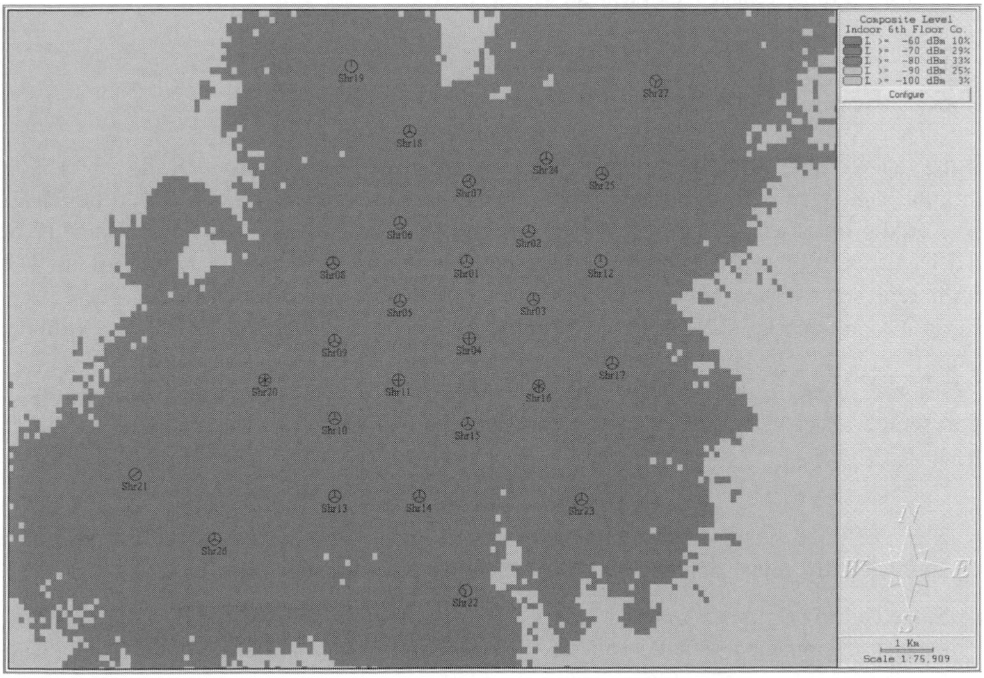

Figure 10.61 Composite signal level coverage indoor at sixth floor.

Figures 10.59–10.61 show the composite signal level for ground, third and sixth floors, respectively. The coverage increases significantly with the height, with the results being mainly influenced by obstructions and antenna pattern.

10.6   LINK BUDGET

Once user classes are defined and traffic is distributed across the area, it is time to define the sites location and BTS parameters. This task requires careful evaluation of the network performance considering equipment constraints and network design goals. This is achieved by establishing a link budget.

The objective of the link budget is to establish the main link parameters and analyse their influence on the site coverage. The link budget provides initial information for site spacing and equipment settings, being used as a reference for troubleshooting and multiple scenario analyses, but without replacing a call placement simulation of the network.

A different link budget applies to the reverse and forward links, especially in CDMA based systems. Service is only provided when the requirements of both links are met; therefore it is important to adjust links to provide similar coverage. The reverse link is expected to be the limiting one because of the restricted power at mobile terminals; however, this is not always the case.

If the forward link is preponderant (larger average then the reverse link), some mobiles will receive enough signal strength to establish calls but as there is insufficient signal strength in the reverse link, they will fail, frustrating users. Unbalanced links such as these, result also in unnecessary excessive interference.

10.6.1   Reverse Link Budget for Voice

In the reverse link, all mobiles adjust their transmit power level to arrive at the sector BTS with the same receive level because all of them require the same E_b/N_0. The user in focus is interfered primarily by other users in the same sector, in neighbour sectors of the same BTS, and by users in other BTSs close to the sector border, as illustrated in Figure 10.62. The figure represents a three-sectored BTS (sectors α, β, and γ) and the active users. These users transmit to its serving sector and adjust their transmit power to arrive at the receiver with the same level as all others.

One user close to the border of the sector is analysed. The main interfering users are represented as connections to that sector. The main interferers can be divided in three groups:

- Same cell and sector – All active users of this sector are interferers and because they automatically adjust their power, all of them have the same level.

- Same cell in neighbour sector – Only users close to the border are significant interferers, because the antenna pattern provides some isolation. The main issue in this category is that the power control is adjusted according to requirements of another sector, which can lead to interference.

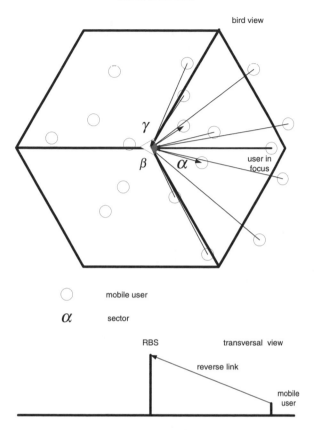

Figure 10.62 Reverse link diagram.

- Other cell – Mainly users close to the border interfere, although terrain variations and buildings may cause distant users to interfere. Once again the main issue in this category is that the power control is adjusted according to requirements of a neighbour cell.

In the reverse link, the interference that arrives at the BTS comes from different locations (user positions) and is subject to different path loss.

The majority of the interference is expected to come from same cell users, which arrive at the BTS with the same level. The voice activity factor represents the reduction of this interference while users are not speaking. Other cell users are attenuated by the antenna pattern and distance, although users located close to the cell border still present a strong interference possibility.

An accurate representation can only be done by distributing traffic and performing a statistical analysis through call placement simulation. To simplify things in the link budget analysis this chapter considers interference coming from other cells as a fraction of the own cell interference, referring to it as inner to outer cell interference ratio.

Without users there is only noise arriving at the BTS, and then the first user has to adjust its signal to -12.31 dB below the noise level to provide the required E_b/N_0. As more signals arrive their signals have to be beefed up so they maintain the desired E_b/N_0.

Table 10.10   Reverse link interference margin and load for IS-95 voice 14 400 bps

Users	S/N (dB)	Interference margin (dB)	I_t/N (dB)	$I_t/(I_t + N)$ (load)
0	−12.47	0.00		
1	−12.31	0.16		0.00
2	−12.15	0.32	−12.15	0.04
3	−11.98	0.49	−8.97	0.07
4	−11.81	0.66	−7.04	0.11
5	−11.62	0.84	−5.60	0.15
6	−11.43	1.03	−4.44	0.18
7	−11.24	1.23	−3.45	0.22
8	−11.03	1.44	−2.58	0.26
9	−10.81	1.66	−1.78	0.29
10	−10.58	1.89	−1.03	0.33
11	−10.33	2.13	−0.33	0.37
12	−10.08	2.39	0.34	0.40
13	−9.80	2.67	0.99	0.44
14	−9.51	2.96	1.63	0.48
15	−9.20	3.27	2.27	0.51
16	−8.86	3.61	2.90	0.55
17	−8.49	3.98	3.55	0.59
18	−8.09	4.38	4.22	0.62
19	−7.65	4.82	4.91	0.66
20	−7.16	5.31	5.63	0.69
21	−6.60	5.87	6.41	0.73
22	−5.97	6.50	7.26	0.77
23	−5.22	7.25	8.20	0.80
24	−4.32	8.15	9.30	0.84
25	−3.18	9.28	10.62	0.88
26	−1.64	10.83	12.34	0.91
27	0.78	13.24	14.93	0.95
28	6.71	19.18	21.02	0.99

The load factor represents load imposed to the network by the interference generated by all users. The load factor is defined as in eqn. (10.2).

$$\text{Load factor} = I_t/(I_t + N) \tag{10.2}$$

Where

I_t- Total interference from other users

N- Total noise in the bandwidth

Tables 10.10–10.12 associate the number of users to the margin required to overcome interference, presenting the load factor, respectively, for IS-95 voice at 14 400 kbps, IS-95 voice at 9600 kbps, and IS-2000 voice using 1XRTT. The tables are graphically represented by Figures 10.63–10.65.

Additional external interference can also be added to the existing noise level, therefore any interference close to the noise threshold should be considered in the calculations. The interference is present over the whole bandwidth and, in general, if it has a smaller bandwidth, it can be prorated over the entire bandwidth, reducing its effective value.

Table 10.11 Reverse link interference margin and load for IS-95 voice 9600 bps

Users	S/N (dB)	Interference margin (dB)	I_t/N (dB)	$I_t/(I_t + N)$ (load)
0	−14.18	0.00		
1	−14.07	0.10		0.00
2	−13.96	0.21	−13.96	0.02
3	−13.85	0.32	−10.84	0.05
4	−13.74	0.43	−8.97	0.07
5	−13.63	0.55	−7.61	0.10
6	−13.51	0.67	−6.52	0.12
7	−13.39	0.79	−5.60	0.15
8	−13.26	0.92	−4.81	0.17
9	−13.13	1.05	−4.10	0.20
10	−13.00	1.18	−3.45	0.22
11	−12.86	1.32	−2.86	0.24
12	−12.72	1.46	−2.30	0.27
13	−12.57	1.61	−1.78	0.29
14	−12.42	1.76	−1.28	0.32
15	−12.26	1.92	−0.80	0.34
16	−12.09	2.08	−0.33	0.37
17	−11.92	2.25	0.12	0.39
18	−11.75	2.43	0.56	0.41
19	−11.56	2.61	0.99	0.44
20	−11.37	2.81	1.42	0.46
21	−11.17	3.01	1.84	0.49
22	−10.96	3.22	2.27	0.51
23	−10.73	3.44	2.69	0.54
24	−10.50	3.68	3.12	0.56
25	−10.25	3.93	3.55	0.59
26	−9.99	4.19	3.99	0.61
27	−9.71	4.47	4.44	0.63
28	−9.41	4.77	4.91	0.66
29	−9.09	5.09	5.38	0.68
30	−8.74	5.44	5.88	0.71
31	−8.36	5.81	6.41	0.73
32	−7.95	6.23	6.96	0.76
33	−7.49	6.68	7.56	0.78
34	−6.98	7.20	8.20	0.80
35	−6.40	7.77	8.91	0.83
36	−5.73	8.44	9.71	0.85
37	−4.94	9.23	10.62	0.88
38	−3.98	10.20	11.70	0.90
39	−2.73	11.44	13.06	0.93
40	−0.98	13.19	14.93	0.95
41	2.00	16.17	18.02	0.98
42	20.63	34.80	36.75	1.00

Table 10.12 Reverse link interference margin and load for 1XRTT voice 9600 bps

Users	S/N (dB)	Interference margin (dB)	I_t/N (dB)	$I_t/(I_t+N)$ (load)
0	−17.12	0.00		−0.01
1	−17.07	0.05		0.00
2	−17.02	0.11	−17.02	0.01
3	−16.96	0.16	−13.95	0.02
4	−16.91	0.21	−12.14	0.04
5	−16.85	0.27	−10.83	0.05
6	−16.80	0.33	−9.81	0.06
7	−16.74	0.38	−8.96	0.07
8	−16.68	0.44	−8.23	0.09
9	−16.63	0.50	−7.59	0.10
10	−16.57	0.56	−7.02	0.11
11	−16.51	0.62	−6.51	0.12
12	−16.45	0.68	−6.03	0.13
13	−16.38	0.74	−5.59	0.15
14	−16.32	0.80	−5.18	0.16
15	−16.26	0.87	−4.80	0.17
16	−16.19	0.93	−4.43	0.18
17	−16.13	1.00	−4.09	0.20
18	−16.06	1.06	−3.76	0.21
19	−15.99	1.13	−3.44	0.22
20	−15.92	1.20	−3.14	0.23
21	−15.86	1.27	−2.84	0.24
22	−15.78	1.34	−2.56	0.26
23	−15.71	1.41	−2.29	0.27
24	−15.64	1.49	−2.02	0.28
25	−15.56	1.56	−1.76	0.29
26	−15.49	1.64	−1.51	0.31
27	−15.41	1.71	−1.26	0.32
28	−15.33	1.79	−1.02	0.33
29	−15.25	1.87	−0.78	0.34
30	−15.17	1.95	−0.55	0.35
31	−15.09	2.04	−0.32	0.37
32	−15.00	2.12	−0.09	0.38
33	−14.92	2.21	0.13	0.39
34	−14.83	2.29	0.36	0.40
35	−14.74	2.38	0.57	0.42
36	−14.65	2.48	0.79	0.43
37	−14.55	2.57	1.01	0.44
38	−14.46	2.67	1.22	0.45
39	−14.36	2.76	1.44	0.46
40	−14.26	2.86	1.65	0.48
41	−14.16	2.97	1.86	0.49
42	−14.05	3.07	2.07	0.50
43	−13.95	3.18	2.29	0.51
44	−13.83	3.29	2.50	0.53
45	−13.72	3.40	2.71	0.54
46	−13.61	3.52	2.93	0.55
47	−13.49	3.64	3.14	0.56

Table 10.12   (*Continued*)

Users	S/N (dB)	Interference margin (dB)	I_t / N (dB)	$I_t/(I_t + N)$ (load)
48	−13.36	3.76	3.36	0.57
49	−13.24	3.89	3.58	0.59
50	−13.11	4.02	3.80	0.60
51	−12.97	4.15	4.02	0.61
52	−12.83	4.29	4.24	0.62
53	−12.69	4.44	4.47	0.64
54	−12.54	4.58	4.70	0.65
55	−12.39	4.74	4.94	0.66
56	−12.23	4.90	5.17	0.67
57	−12.06	5.06	5.42	0.68
58	−11.89	5.23	5.67	0.70
59	−11.71	5.41	5.92	0.71
60	−11.53	5.60	6.18	0.72
61	−11.33	5.79	6.45	0.73
62	−11.13	5.99	6.72	0.75
63	−10.92	6.21	7.01	0.76
64	−10.69	6.43	7.30	0.77
65	−10.46	6.67	7.61	0.78
66	−10.20	6.92	7.92	0.79
67	−9.94	7.19	8.26	0.81
68	−9.66	7.47	8.61	0.82
69	−9.35	7.77	8.97	0.83
70	−9.03	8.10	9.36	0.84
71	−8.67	8.45	9.78	0.86
72	−8.29	8.83	10.22	0.87
73	−7.87	9.25	10.70	0.88
74	−7.40	9.72	11.23	0.89
75	−6.88	10.24	11.81	0.90
76	−6.29	10.84	12.46	0.92
77	−5.60	11.52	13.21	0.93
78	−4.79	12.34	14.08	0.94
79	−3.78	13.35	15.14	0.95
80	−2.47	14.66	16.51	0.97
81	−0.58	16.55	18.46	0.98
82	2.85	19.97	21.93	0.99

The reverse link budget in Table 10.13 was prepared for IS-2000 1XRTT and IS-95 voice only networks using vocoders at 14 400 and 9600 bps. Other traditional technologies (AMPS, TDMA, and GSM) were added in the table for comparison.

Link budget items do not apply equally to all technologies and adjustments had to be made to represent them in this table. The comment column provides formulas based on the letters in the reference (Ref) column[3].

---

[3]The majority of the calculations are done using logarithmic values (dB), but some have to be done using linear values. This requires that some parameters be presented in both forms, linear and logarithmic values use the same reference but linear values have a 1 added as a suffix.

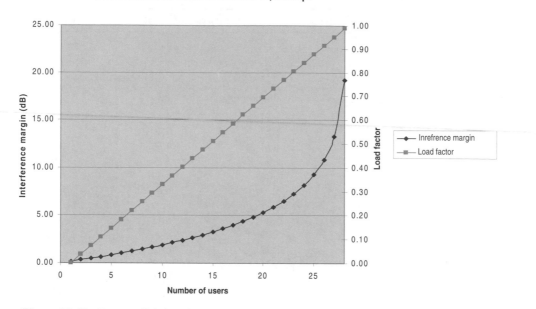

Figure 10.63 Reverse link interference margin and load factor for IS-95 voice at 14 400 kbps.

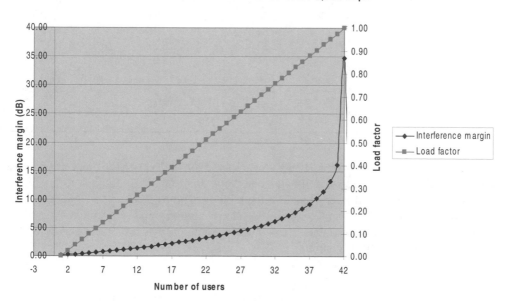

Figure 10.64 Reverse link interference margin and load factor for IS-95 voice at 9600.

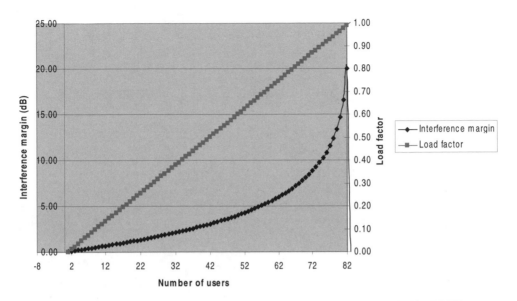

Figure 10.65   Reverse link interference margin and load factor for IS-2000 1XRTT.

To calculate the reverse link budget, the transmitted EIRP has to be calculated first. Then the minimum signal power required at the BTS antenna that allows extraction of the signal with the desired quality is calculated. The difference between these two values corresponds to the nominal path loss for fifty percent probability of coverage and without additional impairments.

The operational margin calculation includes all impairments taken at this probability. A compounded margin spread that provides for the desired certainty in service coverage is considered (in this case estimated at ninety percent of the time and ninety percent of the area). This margin spread has to be estimated based on measurements and user experience.

10.6.1.1   Mobile Transmit Power

The maximum transmitted power per traffic channel is a characteristic of the mobile but is also limited by standards. The power output of the mobile PA is attenuated by the duplexer circuitry and amplified by the antenna. The effective isotropic power (EIRP) is determined based on these elements.

10.6.1.2   Receiver Sensitivity at BTS Antenna

Thermal noise has to be calculated over the effective bandwidth and added to the receiver noise figure as explained in Section 9.4.3.2.1.

Table 10.13 Reverse Link Budget for voice

Ref. Reverse link (voice)	AMPS	TDMA	GSM	IS-95	IS-95	1XRTT	Unit	Comment
a1 Data rate (Hz for AMPS)	3600	9600	14 400	14 400	9600	9600	bps	Input
a Data rate	36	40	42	42	40	40	dB	Conversion
Mobile transmit power								
b Maximum transmitted power per channel	27	27	27	21	21	21	dBm	Input
c Transmit losses	1	1	1	1	1	1	dB	Input
d Transmit antenna gain	2	2	2	2	2	2	dBi	Input
e Transmit EIRP per channel	28	28	28	22	22	22	dBm	$b + d - c$
Receiver sensitivity at BTS antenna								
f Thermal noise density	-174	-174	-174	-174	-174	-174	dBm/Hz	Input
g1 Channel bandwidth	12 000	12 000	160 000	1 228 800	1 228 800	1 228 800	Hz	Input
g Channel bandwidth	40.8	40.8	52.0	60.9	60.9	60.9	dB	Conversion
h Bandwidth noise	-133.2	-133.2	-122.0	-113.1	-113.1	-113.1	dBm	$v + g$
i BTS receiver noise figure	5	5	5	5	5	5	dB	Input
j Total noise power	-128.21	-128.21	-116.96	-108.11	-108.11	-108.11	dBm	$h + i$
j1 Total noise power	1.51×10^{-16}	1.51×10^{-16}	2.01×10^{-15}	1.55×10^{-14}	1.55×10^{-14}	1.55×10^{-14}	W	Conversion
k External interference	-130.00	-130.00	-119.00	-110.00	-110.00	-110.00	dBm	Input
k1 External interference	1.00×10^{-16}	1.00×10^{-16}	1.26×10^{-15}	1.00×10^{-14}	1.00×10^{-14}	1.00×10^{-14}	W	Conversion
l Total noise/external interference power	2.51×10^{-16}	2.51×10^{-16}	3.27×10^{-15}	2.55×10^{-14}	2.55×10^{-14}	2.55×10^{-14}	W	$j1 + k1$
l Total noise/external interference power	-126.00	-126.00	-114.85	-105.94	-105.94	-105.94	dBm	Conversion
m Processing gain	0.00	0.00	0.00	19.31	21.07	21.07	dB	Input
n Required E_b/N_t (CDMA) or S/N (PCS)	17	17	12	7	7	5.5	dB	$g - a$
o Receiver sensitivity	-109.00	-109.00	-102.85	-118.25	-120.01	-121.51	dBm	$l + n - m$
p BTS cable and connector loss	2.00	2.00	2.00	2.00	2.00	2.00	dB	Input
q Receive antenna gain	12	12	12	12	12	12	dBi	Input
r Receiver sensitivity at BTS antenna	-119.00	-119.00	-112.85	-128.25	-130.01	-131.51	dBm	$o + p + q$

		(1)	(2)	(3)	(4)	(5)	Units	
s	*Operational margins*							
t	User body loss	2	2	2	2	2	dB	Input
u	Building/vehicle penetration	0	0	0	0	0	dB	Input
v	Fast fading margin	4	4	4	4	4	dB	Input
w	Slow fading margin	3	3	3	3	3	dB	Input
x	Power control error	0	0	1	1	0.5	dB	Input
y	Handoff gain	0	0	2.5	2.5	2.5	dB	Input
z	Diversity gain	2	2	2	2	2	dB	Input
aa	Margin spread (90% time and/or area)	3	3	3	3	3	dB	Input
ab	Total operational margin	10	10	8.5	8.5	8	dB	$t+u+v+w+x+aa-y-z$
	Interference margins							
ac	Own and other cell users interference	2	2	3.27	3.68	5.23	dB	50% load for IS-95 14.4, 55% load for IS-95 9.6, 70% load for 1XRTT
ad	Margin spread (90% time and/or area)	0.5	0.5	1.5	2.00	3.00	dB	Input
ae	Total interference margin	2.5	2.5	4.77	5.68	8.23	dB	$ac+ad$
af	Maximum mean path loss	134.5	128.4	137.0	137.8	137.3	dB	$e-r-ab-ae$
	Average cell radius (Reverse Link)							
ag	800/900 MHz	7.25	5.46	8.12	8.45	8.23	km	
ah	1800/1900 MHz	5.50	4.14	6.16	6.41	6.25	km	
ai	2400 MHz	4.90	3.69	5.49	5.71	5.57	km	

Interference coming from uncorrelated sources (external to the network) must be added to the noise, because it impacts the detection threshold. This factor is frequently ignored and predictions are blamed for lack of accuracy. For the calculations in this link budget, interference was considered equal for all technologies at -170.79 dBm/Hz.

The required E_b/N_t is obtained using simulators. Chapter 9 showed a simplified version of such simulations in Section 9.3.2.3 and extensive data in Section 9.6.4. For CDMA systems, processing gain must also be considered. For AMPS, TDMA and GSM, the required S/N ratio is used. The more robust coding and detection schemes used in 1XRTT are reflected in the smaller value of the required E_b/N_t.

Finally, the antenna gain should be added to calculate the minimum signal level required at the antenna (receiver sensitivity at BTS antenna).

10.6.1.3   Operational Margins

The path loss varies from location to location and these variations should be considered as a margin, which can be

- Environmental (comprising user body shielding and vehicle or building attenuation);

- Or Fading (comprising slow fading due to environment variations and fast fading due to multipath).

Some technological solutions can provide a gain by diminishing the effects of these impairments. Slow and fast fading effects can be compensated by diversity and soft handoff gain as long as these gains are not larger than the fading itself. Usually they are between fifty and eighty percent of the impairment value in dB.

Link budget must also account for equipment imperfection. For example, power control is not perfectly executed requiring the addition of a margin in the calculations. 1XRTT systems, which have an improved power control scheme, show the effect of this margin in the link budget.

All values considered in the link budget calculations are mean values and supposedly have a combined distribution. For practical purposes, the calculations in this chapter assume this distribution to be Gaussian with a certain standard deviation. This deviation cannot be accurately calculated and has to be estimated from empirical data and experience.

Once the standard deviation is estimated, the margin spread can be calculated for any specified statistical certainty (ninety percent being the commonly used margin). The majority of the parameters considered in this margin are time sensitive. Some, however, are area sensitive (such as slow fading) and some (such as the margin) are the combination of both. This is an acceptable simplification due to the subjectivity of the deviation estimation. More accurate values can be obtained by applying two margins, one for time-variant parameters and another for area-variant parameters. The calculations in this chapter assumed a margin of 5 dB for CDMA networks.

10.6.1.4   Interference Margins

A margin was added to represent the interference coming from the serving cell and neighbouring cells. This interference is based on the load factor. The higher the load factor, the more users the network can handle, but larger the interference. More interference implies on a smaller cell size, which reduces the number of subscribers. Therefore designers should carefully choose the best objective for the network. For IS-95 networks, practical load factors vary between 0.5 and 0.7 (15–20 users), whereas for 1XRTT, values of 0.5–0.8 (42–60 users) are acceptable. A margin spread is also applied to the interference margins.

10.6.1.5 Maximum Path Loss and Average Cell Size

After calculation of the interference margin, the maximum path loss with mean probability can then be derived for the reverse link. The comparison between technologies should be carefully done as each technology supports a different number of users. Among CDMA technologies path losses are similar, but a much larger load was considered for 1XRTT, resulting in a larger user capacity. In this example, IS-95 at 14 400 bps supports 15 users, IS-95 at 9600 bps supports 24 users and 1XRTT at 9600 bps supports 58 users. The improved performance of 1XRTT relies on the coding and detection improvements allied to a better power control. The path loss allows estimation of the cell size. In this example, a single 50 dB/decade slope was used.

10.6.2 Forward Link Budget for Voice

In the forward link, the BTS sector transmits the pilot, paging, and sync channel signals at constant power, adjusting the traffic channel to the requirements of each user. Besides their desired signal, users also receive other traffic channel signals from their serving sector and signals from neighbour cells. The worst situation, in principle, is sustained by users close to the border of neighbouring cells, as illustrated in Figure 10.66.

Three neighbouring sectored BTSs are represented (sectors α, β, and γ). Each sector transmits signals to all its users, adjusting the power of the traffic channel signal through power control to meet user requirements. In principle the signal strength increases with the distance to the user. The example in Figure 10.66 analyses a user close to the border. The main interfering users are represented connected to their sectors. The main interferers can be divided in two groups:

- Same cell and sector – Larger the number of users in the cell, the more interference is received by each user. Distant users may interfere more than closer ones and indoor users (which require higher power). This makes calculations more difficult because the amount of interference depends on the user distribution within the sector. As same sector channels use the same PN offsets, the distinction between them is done through the Walsh code; therefore the orthogonallity plays an important role in this scenario. If the signals were kept fully orthogonal they could be recovered independently of their relative powers; in real life, however, the received signal has a main component and a myriad of multipath signals, some with large delays. If the delay becomes significant the Walsh codes do not perform as well and errors are introduced. The orthogonallity factor reflects the ratio of the main component to the sum of all multipath components. It is considered then that only this fraction of the total transmitted power interferes with the desired signal.

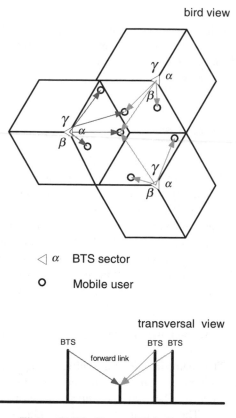

Figure 10.66   Forward link diagram.

- Other cell – Users of neighbour cells are prone to interfere on users close to the cell borders. The amount of interference depends on their distribution. This interference is different from the one coming from the same cell because, in this case, the cells use different PN offsets. This interference has to be considered as noise and in this case a ratio is specified to express the average relation of the other cell signal to the same sector signal at the edge of the sector. It must be noted that the same sector power to be considered here is the one multiplied by the orthogonallity factor.

In the reverse link, the signal comes from the BTS and is received at the MS antenna, also affected by interference generated in the sector itself and coming from neighbour sectors. In principle the required E_b/N_0 should be different for each of the interference types, but this would complicate the analysis considering that this data is not readily available from simulators, therefore an average value is used.

Additional external interference can also be added to the existing noise level, therefore any interference close to the noise threshold should be considered in the calculations. The interference is present over the whole bandwidth and, in general, if it has a smaller bandwidth, it can be prorated over the entire bandwidth, reducing its effective value.

Table 10.14 presents forward link budget calculations for IS-2000 1XRTT and IS-95 systems using vocoders at 14 400 and 9600 bps. Other traditional technologies (AMPS, TDMA, and GSM) are also presented for comparison.

Link budget items do not apply equally to all technologies and adjustments had to be made to represent them in this table. The comment column provides formulas based on the letters in the reference (Ref) column[4].

In the forward link budget, the EIRP transmitted by the BTS is initially calculated. Then the minimum signal power required at the mobile (MS) antenna to allow extraction of the signal with the desired quality is calculated. The difference between these two values corresponds to the nominal path loss for fifty percent probability, without additional impairments. The operational margin calculation includes all the impairments taken at a fifty percent probability. A compounded margin spread is then considered to provide for the desired certainty in service coverage (in this case estimated at ninety percent of the time and ninety percent of the area). This margin spread has to be estimated based on measurements and user experience.

10.6.2.1 BTS Transmit Power

The BTS sector output power is provided by a highly linear broadband amplifier and is generally limited between 16 and 25 W, depending on the equipment generation. This transmit power includes pilot, paging, sync, and traffic channels. The pilot, paging, and sync channels should provide coverage balanced to the traffic channel with the biggest footprint.

The pilot power is usually expressed as a percentage of the total power available with fifteen percent being the typical figure. The paging channel is expressed as a percentage of the pilot power because it has a lower data rate (35% is the usual value). The synch channel is also expressed as a percentage of the pilot power, with a typical figure of 10%.

The rest of the power is available for the traffic channels. Each active mobile station (MS) gets one or more traffic channels depending on its soft handoff status, which defines the required number of traffic channels. In CDMA, the channel is only used if there is information being transmitted. When the user is not speaking the channel is off, reducing the average power required per traffic channel. The mean voice activity factor represents this effect. The typical value for this factor is 0.479.

After applying the activity factor, cable losses and antenna gain are considered and the traffic channel EIRP when the user is speaking at full vocoder rate can be calculated. The total BTS EIRP can then be calculated by applying the losses and the antenna gain to the total available power.

10.6.2.2 Receiver Sensitivity at MS Antenna

Thermal noise and interference coming from uncorrelated sources must be accounted for because of their impact on the detection threshold. The thermal noise is calculated over the effective bandwidth and added to the receiver noise figure as explained in Section 9.4.3.2.

---

[4]Link budget items do not apply equally to all technologies and adjustments had to be made to represent them in this table. The comment column provides formulas based on the letters in the reference (Ref) column.

Table 10.14 Forward Link Budget for voice

Ref.	Forward link (voice)	AMPS	TDMA	GSM	IS-95	IS-95	1XRTT	Unit	Comment
a1	Data rate (Hz for AMPS)	3600	9600	14 400	14 400	9600	9600	bps	Input
a	Data rate	35.6	39.8	41.6	41.6	39.8	39.8	dB	Conversion
	Transmit power calculation								
b	BTS Nominal available power	55.8	55.8	49.8	42.0	43.0	44.0	dBm	Input
b1	BTS Nominal available power	384.0	384.0	96.0	16.0	20.0	25.0	W	Conversion
c	Pilot channel power (CDMA only)	0.0	0.0	0.0	33.8	34.8	35.7	dBm	$B + 10 \log (0.15)$
c1	Pilot channel power (CDMA only)	0.0	0.0	0.0	2.4	3.0	3.8	W	Conversion
d	Sync channel power (CDMA only)	0.0	0.0	0.0	23.8	24.8	25.7	dBm	$C + 10 \log (0.1)$
d1	Sync channel power (CDMA only)	0.0	0.0	0.0	0.2	0.3	0.4	W	Conversion
e	Paging channel power (CDMA only)	0.0	0.0	0.0	29.2	30.2	31.2	dBm	$C + 10 \log (0.35)$
e1	Paging channel power (CDMA only)	0.0	0.0	0.0	0.8	1.1	1.3	W	Conversion
f1	Power available for the traffic channel	384.0	384.0	96.0	12.5	15.7	19.6	W	$b1c1 - d1 - e1$
f	Power available for the traffic channel	55.8	55.8	49.8	41.0	41.9	42.9	dBm	Conversion
g	Number of mobiles per sector	24.0	72.0	48.0	20.0	20.0	35.0		Input
h1	Number of simultaneous traffic channels per sector	24.0	24.0	6.0	20.0	20.0	35.0		Four mobiles per TDMA carrier, eight mobiles per GSM carrier
h	Number of simultaneous traffic channels per sector	13.8	13.8	7.8	13.0	13.0	15.4	dB	Conversion
i1	Soft handoff factor (CDMA only)	1.0	1.0	1.0	1.9	1.9	1.8		Input
i	Soft handoff factor (CDMA only)	0.0	0.0	0.0	2.8	2.8	2.6	dB	Conversion
j1	Total number of active channels	24.0	24.0	6.0	38.0	38.0	63.0		$h1 * i1$
j	Total number of active channels	13.8	13.8	7.8	15.8	15.8	18.0	dB	Conversion

Symbol	Description							Unit	Relation
k	Average traffic channel power per user	42.0	42.0	42.0	25.2	26.1	24.9	dBm	$f - j$
k1	Average traffic channel power per user	16.0	16.0	16.0	0.3	0.4	0.3	W	Conversion
l1	Mean voice activity factor (CDMA only)	1.0	1.0	1.0	0.479	0.479	0.479		Input
l	Mean voice activity factor (CDMA only)	0.0	0.0	0.0	−3.2	−3.2	−3.2	dB	Conversion
m	Peak traffic channel power per user	42.0	42.0	42.0	28.4	29.3	28.1	dBm	$k - l$
n	Cell site cable and connector loss	3.0	3.0	3.0	3.0	3.0	3.0	dB	Input
o	Cell site transmit antenna gain	12.0	12.0	12.0	12.0	12.0	12.0	dB	Input
o1	Cell site transmit antenna gain	15.8	15.8	15.8	15.8	15.8	15.8	dBi	Conversion
p	Traffic channel EIRP per user at full rate	51.0	51.0	51.0	37.4	38.3	37.1	dBm	$m + o - n$
p1	Traffic channel EIRP per user at full rate	127.1	127.1	127.1	5.5	6.8	5.1	W	Conversion
q	Total BTS EIRP	64.8	64.8	58.8	51.0	52.0	53.0	dBm	$b + m - n$
q1	Total BTS EIRP	3050.2	3050.2	762.6	127.1	158.9	198.6	W	Conversion
Receive sensitivity at MS antenna									
r	Thermal noise density	−174.0	−174.0	−174.0	−174.0	−174.0	−174.0	dBm/Hz	Input
s1	Channel bandwidth	12 000	12 000	160 000	1 228.800	1 228.800	1 228.800		Input
s	Channel bandwidth	40.8	40.8	52.0	60.9	60.9	60.9	dB	Conversion
t	Bandwidth noise	−133.2	−133.2	−122.0	−113.1	−113.1	−113.1	dBm	$r + s$
u	MS receiver noise figure	10.1	10.1	10.1	10.1	10.1	10.1	dB	Input
v	Total noise power	−123.1	−123.1	−111.9	−103.0	−103.0	−103.0	dBm	$t + u$
v1	Total noise power	4.9×10^{-16}	4.9×10^{-16}	6.5×10^{-15}	5.0×10^{-14}	5.0×10^{-14}	5.0×10^{-14}	W	Conversion
w	External interference	−109.9	−109.9	−109.9	−109.9	−109.9	−109.9	dBm	Input
w1	External interference	1.0×10^{-14}	1.0×10^{-14}	1.0×10^{-14}	1.0×10^{-14}	1.0×10^{-14}	1.0×10^{-14}	W	Conversion
x1	Total noise/external interference power	1.1×10^{-14}	1.1×10^{-14}	1.7×10^{-14}	6.0×10^{-14}	6.0×10^{-14}	6.0×10^{-14}	W	$v1 + w1$
x	Total noise/external interference power	−103.0	−103.0	−103.0	−103.0	−103.0	−103.0	dBm	Conversion

Table 10.14 (Continued)

Ref.	Forward link (voice)	AMPS	TDMA	GSM	IS-95	IS-95	1XRTT	Unit	Comment
y	Processing gain (CDMA only)	0.0	0.0	0.0	19.3	21.1	21.1	dB	$s - a1$
z	Required E_b/N_t (CDMA) or S/N (PCS)	17.0	17.0	12.0	7.5	7.5	4.0	dB	Input
aa	Mobile receive antenna gain	2.0	2.0	2.0	2.0	2.0	2.0	dBi	Input
ab	Receiver Sensitivity at MS antenna	−88.0	−88.0	−93.0	−116.8	−118.6	−122.1	dBm	$x + z - y - aa$
	Operational margins								
ac	User body loss margin	2.0	2.0	2.0	2.0	2.0	2.0	dB	Input
ad	Building/vehicle penetration	0.0	0.0	0.0	0.0	0.0	0.0	dB	Input
ae	Fast fading margin	6.0	6.0	6.0	6.0	6.0	6.0	dB	Input
af	Slow fading margin	5.0	5.0	5.0	5.0	5.0	5.0	dB	Input
ag	Power control error	0.0	0.0	0.0	2.0	2.0	1.0	dB	Input
ah	Handoff gain	0.0	0.0	0.0	2.5	2.5	2.5	dB	Input
ai	Diversity gain	0.0	0.0	0.0	0.0	0.0	0.0	dB	Input
aj	Margin spread (90% time and/or area)	4.0	4.0	4.0	5.0	5.0	5.0	dB	Input
ak	Total operational margin	17.0	17.0	17.0	17.5	17.5	16.5	dB	$ac + ad + ae +$ $af + ag+$ $aj - ah - ai$
	Interference margins								
al1	Other users orthogonality factor (CDMA)	1.0	1.0	1.0	0.16	0.16	0.16		Input
al	Other users orthogonality factor (CDMA)	0.0	0.0	0.0	−8.0	−8.0	−8.0	dB	Conversion
am1	Other users interference for E_b/N_t calculations (CDMA)				19.5	24.3	30.9	W	$al1*(q1 - p1)$
am	Other users interference for E_b/N_t calculations (CDMA)				4.3	4.4	4.5	dBm	Conversion
an1	Ratio of mean other sector to same sector power at cell edge	0.01	0.01	0.02	0.4	0.4	0.4		Input
an	Ratio of mean other sector to same sector power at cell edge	−20.00	−20.00	−16.99	−3.98	−3.98	−3.98	dB	Conversion

								Units	
ao	Other cell interference power	44.8	44.8	41.8	47.1	48.0	49.0	dBm	$An + q$
ao1	Other cell interference power	30.5	30.5	15.3	50.8	63.5	79.4	W	Conversion
ap1	Total interference	30.5	30.5	15.3	70.3	87.9	110.4	W	ao1 + am1
ap	Total interference	44.8	44.8	41.8	48.5	49.4	50.4	dBm	Conversion
aq	Interference spread	4.0	4.0	4.0	5.0	5.0	5.0	dB	Input
ar	Total interference margin	−2.2	−2.2	−5.2	16.1	16.1	18.3	dB	$ap + aq - p$
	Maximum mean path loss								
as	Maximum mean propagation loss for forward link	124.2	124.2	132.2	120.6	123.3	124.4	dB	$p - ab - ak - ar$
at	Active user capacity per sector in the forward link	24.0	72.0	48.0	20.0	20.0	35.0		g
au	Maximum mean propagation loss for reverse link	134.5	134.5	128.4	137.0	137.8	137.3	dB	From reverse link
	Average cell size (Forward and Reverse Links)								
av	800/900 MHz	4.52	4.52	5.46	3.82	4.33	4.55	km	
aw	1800/1900 MHz	3.43	3.43	4.14	2.90	3.28	3.45	km	
ax	2400 MHz	3.05	3.05	3.69	2.58	2.93	3.07	km	

The interference (external to the network) is added to the noise. These factors are frequently ignored and the predictions are blamed for lack of accuracy. In this example, the interference was considered the same for all technologies at −170.79 dBm/Hz.

The required E_b/N_t is obtained using simulators. Section 9.3.2.3, in Chapter 9 showed a simplified version of such simulations and extensive data was provided in Section 9.6.4. For CDMA systems, processing gain must also be considered. For AMPS, TDMA and GSM, the required S/N ratio is used. The more robust coding and detection schemes used in 1XRTT are reflected in the value of the required E_b/N_t.

Finally, the antenna gain should be added to calculate the minimum signal level required at the antenna (receiver sensitivity at MS antenna).

10.6.2.3   Operational Margins

The path loss varies from location to location and these variations should be considered as a margin, which can be

- Environmental (comprising user body shielding and vehicle or building attenuation);
- Or Fading (comprising slow fading due to environment variations and fast fading due to multipath).

Some technological solutions can provide an apparent gain by diminishing the effects of these impairments. Slow and fast fading effects can be compensated by diversity and soft handoff gain as long as these gains are not larger than fading. Usually they are between fifty and eighty percent of the impairment value in dB.

Additionally equipment imperfections should be added and this is the case of the power control. This control is not perfect and a margin should be added due to this. The power control is improved in 1XRTT and this is reflected in the link budget.

All values considered in the link budget calculations are mean values and supposedly have a combined distribution. For practical purposes, the calculations in this chapter assume this distribution to be Gaussian with a certain standard deviation. This deviation cannot be accurately calculated and has to be estimated from empirical data and experience.

Once the standard deviation is estimated, the margin spread can be calculated for any specified statistical certainty (ninety percent being the commonly used margin). The majority of the parameters considered in this margin are time sensitive. Some, however, are area sensitive (such as slow fading) and some (such as the margin) are the combination of both. This is an acceptable simplification due to the subjectivity of the deviation estimation. More accurate values can be obtained by applying two margins, one for time-variant parameters and another for area-variant parameters. The calculations in this chapter assumed a margin of 5 dB for CDMA networks.

10.6.2.4   Interference Margins

A margin was added to represent the interference coming from the serving cell and neighbouring cells. For the first case (interference within the cell), the orthogonallity factor is applied. Typical values for this factor range from 0.1 to 0.2; this example uses 0.16 (−8 dB).

The interference is also based on the load factor. The higher the load factor, the more users the network can handle, and larger is the interference. More interference implies on a smaller cell size, which reduces the number of subscribers. Therefore designers should carefully choose the best objective for the network. For IS-95 networks, practical load factors vary between 0.5 and 0.7 (15–20 users), whereas for 1XRTT, values of 0.5–0.8 (42–60 users) are acceptable. A margin spread is also applied to the interference margins.

10.6.2.5   Maximum Path Loss and Average Cell Size

After calculation of the interference margin, the maximum path loss with mean probability can then be derived for the forward link. The comparison between technologies should be carefully done as each technology supports a different number of users. Among CDMA technologies path losses are similar, but a much larger load was considered for 1XRTT, resulting in a larger user capacity. In this example, IS-95 at 14 400 bps considered 20 users (13 dB), IS-95 at 9600 bps considered 20 users (13 dB), and 1XRTT at 9600 bps considered 35 users (15.4 dB). The improved performance of 1XRTT relies on the coding and detection improvements allied to a better power control. The path loss allows estimation of the cell size. In this example, a single 50 dB/decade slope was used.

10.6.3   Link Balance

The reverse and the forward links should be balanced to achieve optimum performance as there are only disadvantages to have more coverage in one link than in the other, as this excessive signal will generate only more interference but not carry any traffic.

Technologies such as AMPS, TDMA, and GSM are clearly limited by the reverse link and their path loss is relatively independent of the number of active users. However, the more users are present in the network, the larger the interference margin required for co-channel and adjacent channel interference. CDMA systems' analysis is more complex because the path loss depends largely on the number of active users.

The next step in the analysis is to equalise (according to the more restrictive value) the number of active users in the forward and reverse links and then adjust the link budget parameters to equalise the path loss for both links.

In the examples in items 10.6.1 and 10.6.2, the IS-95 network (14 400 bps vocoder), the IS-95 (9600 vocoder) and 1XRTT (9600 vocoder) are limited by the forward link. These considerations are true for this sample link budget but they may vary for different vendor equipments.

Prediction software can provide very detailed link budget information for any point in the network. Figure 10.67 shows the link budget analysis of the CelPlanner tool.

10.6.4   3G Voice and Data Networks

The term 3G is used to describe the generation of mobile services that provides enhanced voice quality and high-speed Internet and multimedia services.

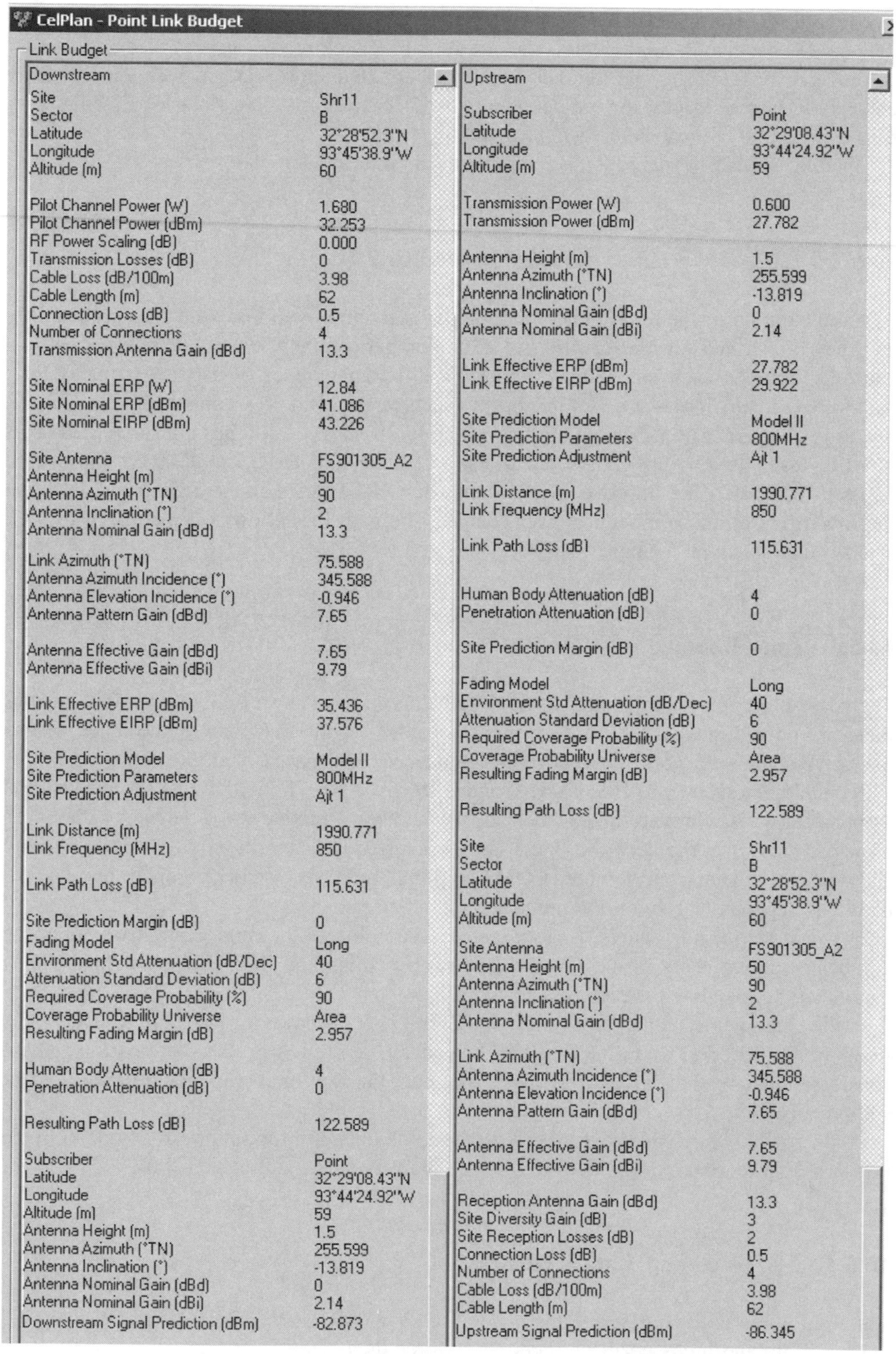

CelPlan - Point Link Budget

Link Budget

Downstream		Upstream	
Site	Shr11	Subscriber	Point
Sector	B	Latitude	32°29'08.43"N
Latitude	32°28'52.3"N	Longitude	93°44'24.92"W
Longitude	93°45'38.9"W	Altitude (m)	59
Altitude (m)	60		
		Transmission Power (W)	0.600
Pilot Channel Power (W)	1.680	Transmission Power (dBm)	27.782
Pilot Channel Power (dBm)	32.253		
RF Power Scaling (dB)	0.000	Antenna Height (m)	1.5
Transmission Losses (dB)	0	Antenna Azimuth (°TN)	255.599
Cable Loss (dB/100m)	3.98	Antenna Inclination (°)	-13.819
Cable Length (m)	62	Antenna Nominal Gain (dBd)	0
Connection Loss (dB)	0.5	Antenna Nominal Gain (dBi)	2.14
Number of Connections	4		
Transmission Antenna Gain (dBd)	13.3	Link Effective ERP (dBm)	27.782
		Link Effective EIRP (dBm)	29.922
Site Nominal ERP (W)	12.84		
Site Nominal ERP (dBm)	41.086	Site Prediction Model	Model II
Site Nominal EIRP (dBm)	43.226	Site Prediction Parameters	800MHz
		Site Prediction Adjustment	Ajt 1
Site Antenna	FS901305_A2		
Antenna Height (m)	50	Link Distance (m)	1990.771
Antenna Azimuth (°TN)	90	Link Frequency (MHz)	850
Antenna Inclination (°)	2		
Antenna Nominal Gain (dBd)	13.3	Link Path Loss (dB)	115.631
Link Azimuth (°TN)	75.588		
Antenna Azimuth Incidence (°)	345.588		
Antenna Elevation Incidence (°)	-0.946	Human Body Attenuation (dB)	4
Antenna Pattern Gain (dBd)	7.65	Penetration Attenuation (dB)	0
Antenna Effective Gain (dBd)	7.65	Site Prediction Margin (dB)	0
Antenna Effective Gain (dBi)	9.79		
		Fading Model	Long
Link Effective ERP (dBm)	35.436	Environment Std Attenuation (dB/Dec)	40
Link Effective EIRP (dBm)	37.576	Attenuation Standard Deviation (dB)	6
		Required Coverage Probability (%)	90
Site Prediction Model	Model II	Coverage Probability Universe	Area
Site Prediction Parameters	800MHz	Resulting Fading Margin (dB)	2.957
Site Prediction Adjustment	Ajt 1		
		Resulting Path Loss (dB)	122.589
Link Distance (m)	1990.771		
Link Frequency (MHz)	850	Site	Shr11
		Sector	B
Link Path Loss (dB)	115.631	Latitude	32°28'52.3"N
		Longitude	93°45'38.9"W
Site Prediction Margin (dB)	0	Altitude (m)	60
Fading Model	Long		
Environment Std Attenuation (dB/Dec)	40	Site Antenna	FS901305_A2
Attenuation Standard Deviation (dB)	6	Antenna Height (m)	50
Required Coverage Probability (%)	90	Antenna Azimuth (°TN)	90
Coverage Probability Universe	Area	Antenna Inclination (°)	2
Resulting Fading Margin (dB)	2.957	Antenna Nominal Gain (dBd)	13.3
Human Body Attenuation (dB)	4	Link Azimuth (°TN)	75.588
Penetration Attenuation (dB)	0	Antenna Azimuth Incidence (°)	345.588
		Antenna Elevation Incidence (°)	-0.946
Resulting Path Loss (dB)	122.589	Antenna Pattern Gain (dBd)	7.65
Subscriber	Point	Antenna Effective Gain (dBd)	7.65
Latitude	32°29'08.43"N	Antenna Effective Gain (dBi)	9.79
Longitude	93°44'24.92"W		
Altitude (m)	59	Reception Antenna Gain (dBd)	13.3
Antenna Height (m)	1.5	Site Diversity Gain (dB)	3
Antenna Azimuth (°TN)	255.599	Site Reception Losses (dB)	2
Antenna Inclination (°)	-13.819	Connection Loss (dB)	0.5
Antenna Nominal Gain (dBd)	0	Number of Connections	4
Antenna Nominal Gain (dBi)	2.14	Cable Loss (dB/100m)	3.98
Downstream Signal Prediction (dBm)	-82.873	Cable Length (m)	62
		Upstream Signal Prediction (dBm)	-86.345

Figure 10.67 Link budget detail for a network point.

The ITU (International Telecommunications Union) defined 3G networks as IMT-2000 (International Mobile Telecommunications for year 2000) and approved in 1999 the ITU-R M.1457 Recommendation that defines five terrestrial radio interfaces for 3G networks.

These interfaces are as follows.

- IMT-2000 CDMA direct spread, specified as UMTS (Universal Mobile Telecommunication System) using WCDMA (Wideband CDMA) as radio access technology.

- IMT-2000 CDMA multi-carrier, specified as CDMA2000 1X and 1XEV.

- IMT-2000 CDMA TDD, specified as UTRA TDD (Universal Terrestrial Radio Access with Time Division Duplex) using TD-SCDMA (Time Division Synchronous CDMA).

- IMT-2000 TDMA, specified as UWC-136 (Universal Wireless Communication-136) and EDGE (Enhanced Data Rates for Global Evolution).

- IMT-2000 FDMA/TDMA, specified as DECT (Digital Enhanced Cordless Technology).

CDMA2000 is one of these technologies that has been successfully deployed worldwide. It represents a family of technologies that include CDMA2000 1X and CDMA2000 1XEV. CDMA2000 is defined by IS-2000 recommendations. They define the RTT (Radio Transmission Technology) physical layer interface, the MAC (Medium Access Control) interface, the LAC (Signalling Link Access Control) interface and the Layer 3 (Upper Layer) interface.

CDMA2000 1X doubles cdmaOne (IS-95C) voice capacity and supports packet data speeds of 153 kbps (release)) and 307 kbps (release 1) over a single 1.25 MHz carrier. This is achieved by the EVRC vocoder, faster power control, lower coder rates (1/4 rate) and transmit diversity, allied to coherent reverse link as explained in the previous chapters. It provides backward compatibility to IS-95. SK Telecom and LG Telecom did the first commercial deployment of this technology in October 2000.

CDMA2000 1XEV stands for single carrier (1X) evolution (EV) and includes 1XEVDO and 1XEVDV.

CDMA2000 1XEVDO (single carrier evolution data only) was recognized as an IMT-2000 standard at the 2001 Stockholm Conference. It delivers peak data speeds of 2.4 Mbps and supports applications as video conferencing and MP3 (MPEG-1 audio layer3) transfers. This technology offers an always-on access to the users, so they can be permanently connected to the Internet. This technology uses the same carrier bandwidth as 1X, but requires a separate carrier to be operated. SK Telecom did the first commercial deployment in January 2002.

CDMA2000 1XEVDV (single carrier evolution data and voice) was submitted to the ITU in July 2002. It is compatible with CDMA2000 1X and will provide integrated voice and high-speed data packet services up to 3.09 Mbps.

10.6.4.1   3G Voice and Data technologies

10.6.4.1.1   CDMA2000 1X

This technology mixes voice and packet data by establishing separate sessions for each. It is also known as CDMA 1XRTT.

Voice is a real-time application and requires a permanent circuit to be assigned to a call. Voice calls cannot be stored, therefore there is no delay associated with them. To transmit voice, regular Erlang B tables are used to calculate the number of required circuits for a specific blockage.

Data applications, however, do not usually require real-time transmission and, therefore, can use packet switching. Because data can be stored until a circuit becomes available, Erlang C tables are used to calculate the number of circuits required for a specific delay.

Circuit-switched data uses voice-like channels with a very similar link budget. The difference is that it requires lower E_b/N_t ratios because it can usually operate with higher FER/PER (Frame Error Rate/Packet Error Rate). Higher FER/PER rates are tolerated in non-real-time data because of the possibility of using ARQ (automatic retransmission request) when errors occur. The equivalent to a voice call in circuit-switched data is a data message or a set of messages for which the circuit is allocated.

Packet-switched data of several users can use a common channel. This task however requires an additional addressing overhead that has to be considered into the data rate. It also requires a lower E_b/N_t ratio than voice because of the same no real-time characteristic.

In CDMA2000 1X, one or more supplemental data channels are available to send packet-switched data. The user data is stored and sent using the regular traffic channel (fundamental traffic channel) at 9600 bps. When the stored rate exceeds a specific threshold, a supplemental channel is requested. The same supplemental channel can be assigned to many users that will share it. An additional supplemental channel is only assigned if the first one reaches capacity.

In a data call the fundamental channel (FCH) is established first and is kept during the whole session. If the rate exceeds the FCH capacity a supplemental channel (SCH) is allocated. If more capacity is required, a higher capacity SCH is allocated. The FCH is kept connected for a dormancy time after the transmission is finished. However, the PPP (Point-to-Point Protocol) connection is maintained throughout the transition. When the supplemental channel is used the FCH rate is reduced to 1200 bps.

Figure 10.68 shows a typical web session in which a user clicks a mouse and receives a page back. Only the forward link is shown. When data is received for a certain user a FCH is established and the data is stored to a buffer and sent through the FCH. When the buffer reaches a pre-established threshold, access to a supplemental channel is requested and data is sent through it. Varying speeds are assigned as required by the buffer level or due to network capability. When data stops for a period superior to the dormancy timer the FCH is disconnected.

10.6.4.1.2   CDMA2000 1XEVDO

EVDO technology is based on an initial Qualcomm's HDR (High Data Rate) specification. It requires a dedicated carrier and special terminals. The technology is compatible in terms of bandwidth with IS-95 and 1XRTT and is totally packet oriented for non-real-time services, mainly Internet.

EVDO uses CDMA spreading in the forward link, but provides access to one single user at a time through a TDM structure with slots of 1.667 ms each (600 slots per second). Each slot corresponds to 2048 chips). Slots are assigned dynamically for the transmission of user packets.

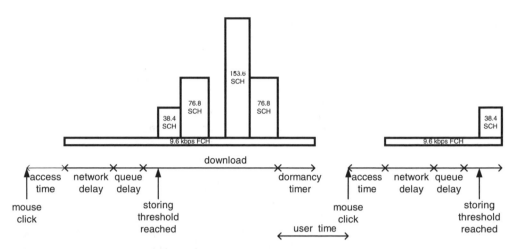

Figure 10.68 Supplemental channel allocation for data (web application).

The forward-link packets vary from 1024 to 4028 bits and EVDO assigns 1 to 16 slots to transmit a single user packet, and with this the forward link supports data rates from 38 400 to 2 457 600 bps.

The reverse-link packets vary from 256 to 4096 bits and it always assigns 16 slots per packet resulting in rates that vary from 9600 to 153 600 bps.

10.6.4.1.2.1 Forward Link

The forward link multiplexes in each slot (2048 chips) the user information according to the following structure.

1. Pilot – 192 bits.

2. Mac, composed of: RA (Reverse Activity Channel), DRC (RDClock Channel), and RPC (Reverse Link Power Control) – 256 bits.

3. Data – 1600 bits.

The forward link transmits always at full power (no power control). As it should provide the same footprint as IS-95 and 1XRTT its maximum power per traffic channel should be smaller than the one used in 1XRTT, as there will be only one user transmitting at a time from each BTS sector (called access network or AN). Table 10.15 provides the main characteristics of this link.

The link adaptation is done by transmitting at the maximum data rate achievable for each user (called access terminal or AT) based on the value of E_c/N_0 at the reception of the forward link. This measurement is done over a pilot that is send synchronously from all AN. The structure supports trip delay differences of up to 15 km.

The data rate determination is done by the AT and informed to the AN. The AT can choose from a standardized set of 9 data rates (12 combinations of coding and modulation). See Table 10.15.

The AN will always follow this determination as it is this data rate that will be expected by the AT.

Table 10.15   EVDO Forward Link characteristics

Packet size (bit)	Number of slots	Data rate (bps)	Processing gain	Data effective processing gain	Modulation	Code rate	Effective code rate
1024	16	38 400.00	32.0	13.80	QPSK	1/5	1/48
1024	8	76 800.00	16.0	10.79	QPSK	1/5	1/24
1024	4.00	153 600.00	8.0	7.78	QPSK	1/5	1/12
1024	2.00	307 200.00	4.0	4.77	QPSK	1/5	1/6
1024	1	614 400.00	2.0	1.76	QPSK	1/3	1/3
2048	4.00	307 200.00	4.0	4.86	QPSK	1/3	8/49
2048	2	614 400.00	2.0	1.85	QPSK	1/3	16/49
2048	1	1 228 800	1.0	−1.25	QPSK	1/3	2/3
3072	2	921 600	1.3	0.09	8PSK	1/3	16/49
3072	1	1 843 200	0.7	−3.01	8PSK	1/3	2/3
4096	2	1 228 800	1.0	−1.16	16QAM	1/3	1/3
4096	1	2 457 600	0.5	−4.26	16QAM	1/3	2/3

There is no soft handoff in the forward link but the connection is always done with the best server, although congestion considerations should be also taken into account.

10.6.4.1.2.2   Reverse Link

The reverse link is also TDM. Only one user transmits at a time. Power control is based on the strength of the received signal. It uses only one modulation BPSK and there are four basic channels per user, identified by different Walsh codes that are transmitted simultaneously.

1. Pilot + RRI (Reverse Rate Indicator) – 112.5 bps

2. DRC (Data Rate Control) – 2.4 kbps

3. ACK (Acknowledge) – 1.2 kbps

4. Data (see Table 10.16)

Packet size varies from 256 to 4096 bits and 16 slots are always assigned to a packet. The Data channel can vary from 9.6 to 153.6 kbps. The user should transmit at the highest

Table 10.16   EVDO Reverse-Link characteristics.

Packet size (bit)	Data rate (bps)	Processing gain	Repetition factor	Modulation	Code rate	Symbols
256	9600	128	8	BPSK	1/4	1024
512	19 200	64	4	BPSK	1/4	2048
1024	38 400	32	2	BPSK	1/4	4096
2048	76 800	16	1	BPSK	1/4	8192
4096	153 600	8	1	BPSK	1/2	8192

possible speed allowed by its available power. The reverse link supports soft and softer handoff. Table 10.16 shows the main characteristics of the reverse link.

10.6.4.1.3   CDMA2000 1XEVDV

CDMA2000 1XEVDV is compatible with 1XRTT and EVDO terminals, allowing the transmission of voice and packet data, including video conferencing.

10.6.4.2   Voice and Data Service Classes

Similar to voice, data services have to be defined. Figure 10.69 shows configuration screens used for web and E-mail services.

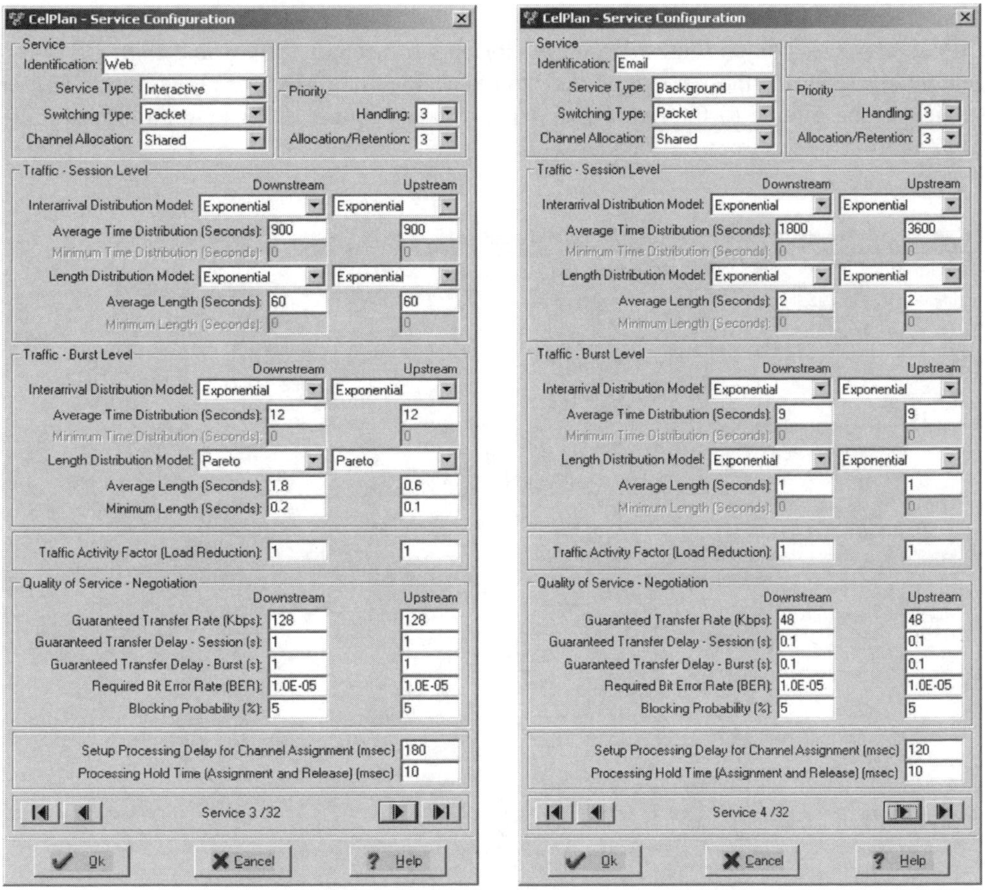

Figure 10.69   Examples of data service configurations.

10.6.4.3   Radio Configurations

CDMA2000 provides many implementation options, which have to be carefully defined to allow proper data allocation. The following values represent data rates obtained with RC3 configuration considering different error targets:

- 9.6 kbps with 1% FER

- 19.2 kbps with 2% FER

- 38.4 kbps with 3% FER

- 76.8 kbps with 5% FER

- 153.6 kbps with 10% FER

The supplemental channel does not execute soft handoff because it may be operating only on short fractions of time.

Real life networks present a mix of equipments of second (IS-95) and third (IS-2000) generations requiring these configurations to be specified in a way that allows different user services to be offered to the radios that support them. Figure 10.70 shows a typical radio configuration screen from CelPlanner tool.

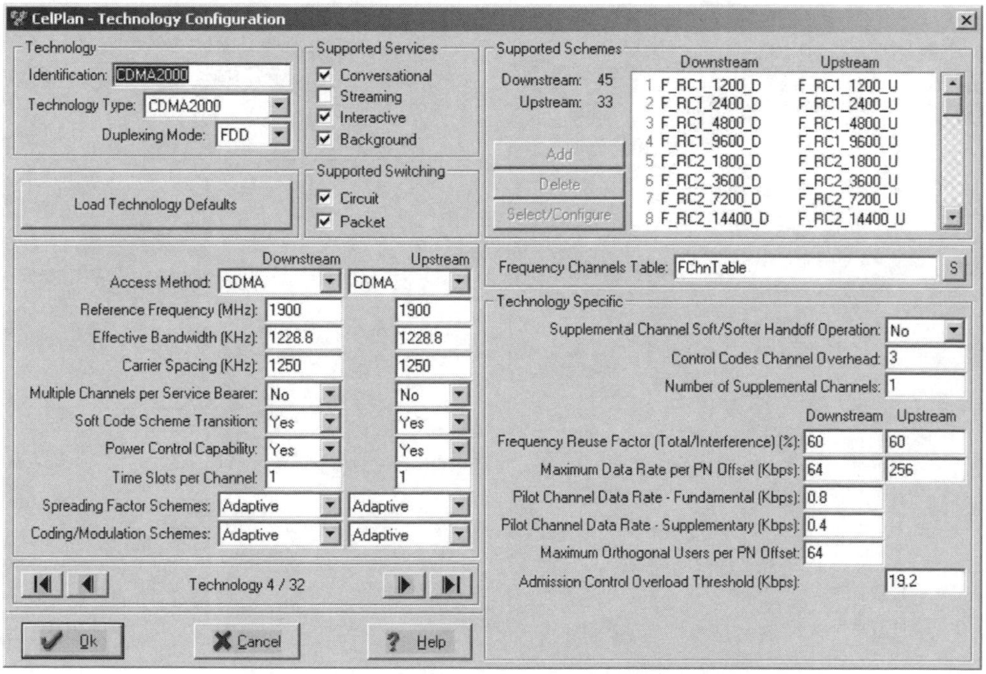

Figure 10.70   Examples of radio configuration for CDMA 2000.

A network can be equipped with IS-95 carriers that support only legacy radio terminals carrying voice and with 1XRTT carriers that provide support for voice (including legacy terminals) and data (new 3G terminals). A third type of carrier may also be available to carry EVDO traffic with support for data only.

10.6.4.4 Data Traffic Layers

To simulate the system, data traffic layers need to be generated for each of the service classes defined. In existing networks, data traffic is always associated to voice users; therefore data traffic layers can be derived from the voice layers just by applying a traffic multiplier.

Data traffic calculation is very complex because data does not follow traditional distributions such as Poisson. A more complete coverage is given in Chapter 11. The best way to perform this analysis is to use simulators that can calculate, for a network population, the number of simultaneous users of each data rate within the specified delay.

10.6.4.5 Data Link Budget

The difficulty to simulate data traffic makes the analysis of data networks extremely complex, mainly when considering simultaneous voice and multiple data services. Another complicating factor is that the adaptive network characteristics and consequent scheduling issues must also be considered. The only way to tackle this additional problem is to use tools that provide call placement simulation.

Because this link budget analysis is specially focused in the extreme cases, a large simplification can be done in this simulation. In this example, the link budget is determined for services at specific rates. This gives the maximum path loss for these rates and consequently the expected maximum coverage area. The real area would, in fact, be smaller due to the mix of services and can only be determined after performing call placement simulation.

The E_b/N_t requirements for each technology have to be established from characterisation measurements on the equipments or estimated by simulating the technology components. Values compiled in Table 10.17 for 1XRTT were derived from data presented in Section 9.6.4.

Values of E_b/N_t for EVDO are presented in Table 10.18 and were derived from 1XRTT values applying corrections for the technology variations between 1XRTT and EVDO. In EVDO the error rate is expressed in PER (Packet Error Rate).

10.6.4.6 Reverse Link Budget for 1XRTT

Table 10.19 presents the reverse link budget for IS2000 1XRTT data considering several data speeds. It is similar to the voice link budget. The maximum mean path loss diminishes with the data rate consequently reducing the cell footprint for each data rate. Section 9.6.4 presents E_b/N_t requirements for several data rates and equipment configurations.

The comment column provides formulas based on the letters in the reference (Ref.) column.[5]

---

[5]The majority of the calculations are done using logarithmic values (dB), but some have to be done using linear values. This requires that some parameters be presented in both forms, linear and logarithmic values use the same reference but linear values have a 1 added as a suffix.

Table 10.17 E_b/N_t values for 1XRTT

1XRTT (Medium fading)

Data rate (bps)	Configuration	Type	Coder	Coder rate	Modulation	Cellular Fundamental FER (%)	Cellular Fundamental E_b/N_t (dB)	Cellular Supplemental FER (%)	Cellular Supplemental E_b/N_t (dB)	PCS Fundamental FER (%)	PCS Fundamental E_b/N_t (dB)	PCS Supplemental FER (%)	PCS Supplemental E_b/N_t (dB)
					Forward link								
9600	RC2	Voice	Convolutional	1/2	QPSK	1	7.5			1	5.5		
9600	RC3	Data	Turbo	1/4	QPSK	3	5.0			3	4.5		
19 200	RC3	Data	Turbo	1/4	QPSK			3	5.2			3	4.7
38 400	RC3	Data	Turbo	1/4	QPSK			3	5.3			3	4.8
76 800	RC3	Data	Turbo	1/4	QPSK			3	5.4			3	4.9
153 600	RC3	Data	Turbo	1/4	QPSK			3	5.4			3	4.9
307 200	RC6	Data	Turbo	1/4	QPSK			5	4.4			5	4.0
614 400	RC7	Data	Turbo	1/4	QPSK			10	3.6			10	3.0
1 036 800	RC9	Data	Turbo	1/2	QPSK			10	4.2			10	3.6
					Reverse link								
9600	RC1	Voice	Convolutional	1/2	QPSK	1	7.5			1	8		
9600	RC3	Data	Turbo	1/4	QPSK	3	4.0			3	3.8		
19 200	RC3	Data	Turbo	1/4	QPSK			3	2.7			3	3.3
38 400	RC3	Data	Turbo	1/4	QPSK			3	1.8			3	2.4
76 800	RC3	Data	Turbo	1/4	QPSK			3	1.3			3	1.8
153 600	RC3	Data	Turbo	1/4	QPSK			3	0.7			3	1.2
307 200	RC3	Data	Turbo	1/2	QPSK			5	1.2			5	1.9
614 400	RC5	Data	Turbo	1/3	QPSK			10	−1.0			10	−0.4
1 036 800	RC6	Data	Turbo	1/2	QPSK			10	−1.0			10	−0.4

Table 10.18 E_b/N_t values for EVDO

Data rate (bit/s)	Coder	Coder rate	Packet/s	Modulation	Packet size (bit)	Slots	EVDO data effective processing gain (dB)	Cellular traffic PER (%)	Cellular traffic E_b/N_0 (dB)	Cellular traffic S/N (dB)	PCS traffic PER (%)	PCS traffic E_b/N_0 (dB)	PCS traffic S/N (dB)
							Forward link						
38 400	Turbo	1/5	37.5	QPSK	1024	16	13.8	3	6.1	−8.90	3	5.6	−9.40
76 800	Turbo	1/5	75	QPSK	1024	8	10.8	3	6.2	−5.79	3	5.7	−6.29
153 600	Turbo	1/5	150	QPSK	1024	4	7.8	3	6.2	−2.78	3	5.7	−3.28
307 200	Turbo	1/5	300	QPSK	1024	2	4.8	5	5.2	−0.77	5	5.2	−0.77
614 400	Turbo	1/3	600	QPSK	1024	1	1.8	5	6.2	3.24	5	6.2	3.24
307 200	Turbo	1/3	150	QPSK	2048	4	4.9	10	5.4	−0.66	5	5.4	−0.66
614 400	Turbo	1/3	300	QPSK	2048	2	1.9	10	5.4	2.35	5	5.4	2.35
1 228 800	Turbo	1/3	600	QPSK	2048	1	−1.2	10	5.7	5.69	5	5.4	5.39
921 600	Turbo	1/3	300	8PSK	3072	2	0.1	10	8.0	6.80	10	7.7	6.50
1 843 200	Turbo	1/3	600	8PSK	3072	1	−3.0	10	11.7	13.51	10	11.4	13.21
1 228 800	Turbo	1/3	300	16QAM	4096	2	−1.2	10	13.6	13.60	10	13.3	13.30
2 457 600	Turbo	1/3	600	16QAM	4096	1	−4.3	10	17.3	20.31	10	17.1	20.11
							Reverse Link						
9600	Turbo	1/4	37.5	BPSK	256	8	21.1	3	5	−16.07	3	5.6	−15.47
19 200	Turbo	1/4	37.5	BPSK	512	4	18.1	3	5.2	−12.86	3	5.8	−12.26
38 400	Turbo	1/4	37.5	BPSK	1024	2	15.1	3	5.3	−9.75	3	5.9	−9.15
76 800	Turbo	1/4	37.5	BPSK	2048	1	12.0	3	5.4	−6.64	3	6	−6.04
153 600	Turbo	1/2	37.5	BPSK	4096	1	9.0	3	6.6	−2.43	3	7.2	−1.83

Table 10.19 Reverse link budget for 1XRTT

Ref.	Reverse link	1XRTT					Unit	Comment
a1	Data rate	9600	19 200	38 400	76 800	153 600	bps	Input
a	Data rate	40	43	46	49	52	dB	Conversion
	Mobile transmit power							
b	Maximum transmitted power per channel	21	21	21	21	21	dBm	Input
c	Transmit losses	1	1	1	1	1	dB	Input
d	Transmit antenna gain	2	2	2	2	2	dBi	Input
e	Transmit EIRP per channel	22	22	22	22	22	dBm	$b + d - c$
	Receiver Sensitivity at BTS antenna							
f	Thermal noise density	-174	-174	-174	-174	-174	dBm/Hz	Input
g1	Channel bandwidth	1 228 800	1 228 800	1 228 800	1 228 800	1 228 800	Hz	Input
g	Channel bandwidth	60.9	60.9	60.9	60.9	60.9	dB	Conversion
h	Bandwidth noise	-113.11	-113.11	-113.11	-113.11	-113.11	dBm	$f + g$
i	BTS receiver noise figure	5	5	5	5	5	dB	
j	Total noise power	-118.11	-118.11	-118.11	-118.11	-118.11	dBm	$h - i$
j1	Total noise power	1.55×10^{-15}	1.55×10^{-15}	1.55×10^{-15}	1.55×10^{-15}	1.55×10^{-15}	W	Conversion
k	External interference	-109.9	-109.9	-109.9	-109.9	-109.9	dBm	Input
k1	External interference	1.02×10^{-14}	1.02×10^{-14}	1.02×10^{-14}	1.02×10^{-14}	1.02×10^{-14}	W	Conversion
l1	Total noise/external interference	1.18×10^{-14}	1.18×10^{-14}	1.18×10^{-14}	1.18×10^{-14}	1.18×10^{-14}	W	$j1 + k1$
l	Total noise/external interference power	-109.28	-109.28	-109.28	-109.28	-109.28	dBm	Conversion
m	Processing gain	21.07	18.06	15.05	12.04	9.03	dB	$g - a$
n	Required E_b/N_t (PCS)	4	3.4	2.6	1.8	1	dB	Input
o	Receiver sensitivity	-126.36	-123.95	-121.74	-119.53	-117.32	dBm	$l + n - m$
p	Cable and connector losses	2	2	2	2	2	dB	Input
q	Receive antenna gain	12	12	12	12	12	dBi	Input
r	Receiver sensitivity at BTS antenna	-136.36	-133.95	-131.74	-129.53	-127.32	dBm	$o + p - q$

	Operational margins							
s	User body loss	2	2	2	2	2	dB	Input
t	Building/vehicle penetration	0	0	0	0	0	dB	Input
u	Fast fading margin	4	4	4	4	4	dB	Input
v	Slow fading margin	3	3	3	3	3	dB	Input
w	Power control error	0.5	0.5	0.5	0.5	0.5	dB	Input
x	Handoff gain	2.5	2.5	2.5	2.5	2.5	dB	Input
y	Diversity gain	2	2	2	2	2	dB	Input
z	Margin spread (90% time and/or area)	3	3	3	3	3	dB	Input
aa	Total operational margin	8	8	8	8	8	dB	$s+t+u+v+$ $w-z-x-y$
	Interference margins							
ab	Own and other cell users interference	5.23	5.23	5.23	5.23	5.23	dB	70% load for 1XRTT
ac	Margin spread (90% time and/or area)	3	3	3	3	3	dB	Input
ad	Total interference margin	8.23	8.23	8.23	8.23	8.23	dB	$ab+ac$
ae	*Maximum mean path loss*	142.13	139.72	137.51	135.30	133.09	dB	$e-r-aa-ad$
	Average cell radius (Reverse link)							
af	800/900 MHz	10.29	9.21	8.32	7.51	6.79	km	50 dB/decade
ag	1800/1900 MHz	7.81	6.99	6.31	5.70	5.15	km	50 dB/decade
ah	2400 MHz	6.96	6.23	5.62	5.08	4.59	km	50 dB/decade

Table 10.20 Forward link budget for 1XRTT

Ref.	Forward link		1XRTT			Unit	Comment
a1	Fundamental channel rate	9600	9600	9600	9600	bps	Input
a	Fundamental channel rate	39.8	39.8	39.8	39.8	dB	Conversion
b1	Supplemental channel rate	19 200	38 400	76 800	153 600	bps	Input
b	Supplemental channel rate	42.8	45.8	48.9	51.9	dB	Conversion
	Transmit Power Calculation						
c	BTS Nominal available power	44	44	44	44	dBm	Input
c1	BTS Nominal available power	25.12	25.12	25.12	25.12	W	Conversion
d	Pilot channel power	35.76	35.76	35.76	35.76	dBm	$c + 10 \log (0.15)$
d1	Pilot channel power	3.77	3.77	3.77	3.77	W	Conversion
e	Sync channel power	25.76	25.76	25.76	25.76	dBm	$d + 10 \log (0.10)$
e1	Sync channel power	0.38	0.38	0.38	0.38	W	Conversion
f	Paging channel power	31.20	31.20	31.20	31.20	dBm	$d + 10 \log (0.35)$
f1	Paging channel power	1.32	1.32	1.32	1.32	W	Conversion
g1	Power available for the traffic channels	19.66	19.66	19.66	19.66	W	$c1 - d1 - e1 - f1$
g	Power available for the traffic channels	42.93	42.93	42.93	42.93	dBm	Conversion
h1	Channel bandwidth	1 228 800	1 228 800	1 228 800	1 228 800	Hz	Input
h	Channel bandwidth	60.9	60.9	60.9	60.9	dB	Conversion
i	Required FCH E_b/N_T (PCS)	2.5	2.5	2.5	4.2	dB	Input
j	Processing gain for FCH	21.1	21.1	21.1	21.1	dB	$h - a$
k1	Number of FCH per sector	50	50	50	50		Input
k	Number of FCH per sector	17.0	17.0	17.0	17.0	dB	Conversion
l1	Soft handoff factor for FCH	1.75	1.75	1.75	1.75		Input
l	Soft handoff factor for FCH	2.4	2.4	2.4	2.4	dB	Conversion
m1	Total number of active FCH	87.5	87.5	87.5	87.5		$k1*l1$
m	Total number of active FCH	19.4	19.4	19.4	19.4	dB	Conversion
n1	Mean voice activity factor for FCH	0.125	0.125	0.125	0.125		Input
n	Mean voice activity factor for FCH	−9.0	−9.0	−9.0	−9.0	dB	Conversion
o	Proportionality factor between average FCH and SCH power level	1.5	9.6	5.5	10.6	dB	$r + j - s - i$
o1	Proportionality factor between average FCH and SCH power level	1.42	9.03	3.59	11.40	dB	Conversion

		0.20	0.17	0.12	0.11		
p1	Average traffic channel power per FCH channel	0.20	0.17	0.12	0.11	W	$n1*g1/(m1*n1+t1*w1*o1)$
p	Average traffic channel power per FCH channel	23.0	22.3	20.9	20.4	dBm	Conversion
q	Peak traffic channel power per FCH channel	32.0	31.3	29.9	29.4	dBm	$p - n$
r	Required SCH E_b/N_t	1	2	3	2.7	dB	Input
s	Processing gain for SCH	18.1	15.1	12.0	9.0	dB	$H - b$
t1	Number of SCH per sector	1	1	1	1		Input
t	Number of SCH per sector	0	0	0	0	dB	Conversion
u1	Soft handoff factor for SCH	1	1	1	1		Input
u	Soft handoff factor for SCH	0	0	0	0	dB	Conversion
v1	Total number of active SCH channels	1	1	1	1		Input
v	Total number of active SCH channels	0	0	0	0	dB	Conversion
w1	Mean activity factor for SCH	1	1	1	1		Input
w	Mean activity factor for SCH	0	0	0	0	dB	Conversion
x	Average traffic channel power per SCH channel	33.5	36.9	39.5	40.0	dBm	$p + w - n + o$
y	Peak traffic channel power per SCH	33.5	36.9	39.5	40.0	dBm	$x - w$
z	Cell site cable loss	3	3	3	3	dB	Input
aa	Cell site transmit antenna gain	12	12	12	12	dBi	Input
ab	FCH EIRP per user at full rate	41.0	40.3	38.9	38.4	dBm	$q + aa - z$
ab1	FCH EIRP per user at full rate	12.6	10.7	7.8	7.0	W	Conversion
ac	SCH EIRP per user at full rate	42.5	45.9	48.5	49.0	dBm	$y + aa - z$
ad	Total BTS EIRP	53	53	53	53	dBm	$c + aa - z$
ad1	Total BTS EIRP	199.5	199.5	199.5	199.5	W	Conversion
	Receive sensitivity at MS antenna						
ae	Thermal noise density	-174	-174	-174	-174	dBm/Hz	Input
af	Bandwidth noise	-113.1	-113.1	-113.1	-113.1	dBm	$ae + h$
ag	MS receiver noise figure	10.1	10.1	10.1	10.1	dB	Input
ah	Total noise power	-103.0	-103	-103	-103	dBm	$af + ag$
ah1	Total noise power	5.01×10^{-14}	5.01×10^{-14}	5.01×10^{-14}	5.01×10^{-14}	W	Conversion
ai	External interference	-109.9	-109.9	-109.9	-109.9	dBm	Input
ai1	External interference	1.02×10^{-14}	1.02×10^{-14}	1.02×10^{-14}	1.02×10^{-14}	W	Conversion
aj1	Total noise/external interference power	6.03×10^{-14}	6.04×10^{-14}	6.04×10^{-14}	6.04×10^{-14}	W	$ah1 + ai1$
aj	Total noise/external interference power	-102.2	-102.2	-102.2	-103	dBm	Conversion

Table 10.20 (Continued)

Ref.	Forward link	1XRTT				Unit	Comment
ak	Mobile receive antenna gain	2	2	2	2	dBi	Input
al	Receiver sensitivity for FCH at MS antenna	-122.8	-122.8	-122.8	-121.9	dBm	$aj + l - j - ak$
am	Receiver sensitivity for SCH at MS antenna	-121.3	-117.2	-113.2	-111.3	dBm	$aj + r - s - ak$
	Operational margins						
an	User body loss margin	2	2	2	2	dB	Input
ao	Building/vehicle penetration	0	0	0	0	dB	Input
ap	Fast fading margin	6	6	6	6	dB	Input
aq	Slow fading margin	5	5	5	5	dB	Input
ar	Power control error	1	1	1	1	dB	Input
as	Handoff gain	2.5	2.5	2.5	2.5	dB	Input
at	Diversity gain	0	0	0	0	dB	Input
au	Margin spread (90% time and/or area)	5	5	5	5	dB	Input
av	Total operational margin	16.5	16.5	16.5	16.5	dB	$an + ao + ap + aq$ $+ ar + au - as - at$
	Interference margins						
aw1	Other users orthogonality factor	0.16	0.16	0.16	0.16		Input
aw	Other users orthogonality factor	-8.0	-8.0	-8.0	-8.0	dB	Conversion
ax1	Other users interference for E_b/N_t calculations	29.90	30.20	30.67	30.81	W	$aw1\,(ad1 - ab1)$
ax	Other users interference for E_b/N_t calculations	44.8	44.8	44.9	44.9	dBm	Conversion
ay1	Ratio of mean other sector to same sector power at cell edge	0.4	0.4	0.4	0.4		Input
ay	Ratio of mean other sector to same sector power at cell edge	-4.0	-4.0	-4.0	-4.0	dB	Conversion
az	Other cell interference power	49.0	49.0	49.0	49.0	dBm	$ay + ad$
az1	Other cell interference power	79.8	79.8	79.8	79.8	W	Conversion
ba1	Total interference	109.7	110.0	110.5	110.6	W	$az1 + ax1$
ba	Total interference	50.4	50.4	50.4	50.4	dBm	Conversion
bb	Interference spread	5	5	5	5	dB	Input
bc	Total Interference margin for FCH	14.4	15.1	16.5	17.0	dB	$ba + bb - ab$
bd	Total Interference margin for SCH	12.9	9.6	6.9	6.4	dB	$ba + bb - ac$

	Propagation loss					
be	Maximum mean propagation loss for FCH	131.4	126.0	119.2	116.3	dB	$ab - \min(al, am) - av - bc$
bf	Maximum mean propagation loss for SCH	134.4	137.1	138.3	137.4	dB	$Ac - am - av - bd$
bg	Maximum mean propagation loss for forward link	131.4	126.0	119.2	116.3	dB	$\min(be, bf)$
bh	Maximum mean propagation loss for reverse link	142.13	139.72	137.51	135.30	dB	From reverse link
	Average cell radius (FW and RV link)						
bi	800/900 MHz	6.28	4.89	3.57	3.13	km	50dB/decade
bj	1800/1900 MHz	4.76	3.71	2.71	2.37	km	50 dB/decade
bk	2400 MHz	4.24	3.30	2.42	2.12	km	50 dB/decade

10.6.4.7   Forward Link Budget for 1XRTT

Table 10.20 shows the forward link budget for IS2000 1XRTT networks considering several data speeds, using a fundamental channel (FCH) and a supplemental channel (SCH). This link budget is different from the voice link budget because two simultaneous channels are considered: FCH and SCH. The main difference is that FCH and SCH have different rates and E_b/N_t requirements. The available transmit traffic power is divided between the various FCH and the SCH to allow all the channels to maintain the required signal to noise ratio. Chapter 9 presents E_b/N_t requirements for several data rates and equipment configurations.

The limiting link depends on the number of data users specified and on the number of supplemental channels required to transmit its data.

The comment column provides formulas based on the letters in the reference (Ref.) column.[6]

10.6.4.8   Reverse Link Budget for EVDO

The reverse data link for EVDO is very similar to 1XRTT and is presented in Table 10.21. The comment column provides formulas based on the letters in the reference (Ref.) column.[7]

10.6.4.9   Forward Link Budget EVDO

The forward data link for EVDO is significantly different to the others as it does not have power control and changing the data rate does the link adaptation (Table 10.22). Additionally, only one user is transmitted at a time. The comment column provides formulas based on the letters in the reference (Ref.) column.[8]

10.6.4.10   Link Budget for EVDV

EVDV must be compatible with 1XRTT and EVDO. Considering the link budgets of both technologies at the same time we can derive the link budget for EVDV.

---

[6]The majority of the calculations are done using logarithmic values (dB), but some have to be done using linear values. This requires that some parameters be presented in both forms, linear and logarithmic values use the same reference but linear values have a 1 added as a suffix.
[7]The majority of the calculations are done using logarithmic values (dB), but some have to be done using linear values. This requires that some parameters be presented in both forms, linear and logarithmic values use the same reference but linear values have a 1 added as a suffix.
[8]The majority of the calculations are done using logarithmic values (dB), but some have to be done using linear values. This requires that some parameters be presented in both forms, linear and logarithmic values use the same reference but linear values have a 1 added as a suffix.

Table 10.21 Reverse link budget for EVDO

Ref.	Reverse link	EVDO					Unit	Comment
a1	Data rate	9600	19 200	38 400	76 800	153 600	bps	$b*600/c$
a	Data rate	40	43	46	49	52	dB	Conversion
b	Packet size	256	512	1024	2048	4096	bit	Input
c	Slots	16	16	16	16	16		Input
	Mobile transmit power							
d	Maximum transmitted power per channel	21	21	21	21	21	dBm	Input
e	Transmit losses	1	1	1	1	1	dB	Input
f	Transmit antenna gain	2	2	2	2	2	dBi	Input
g	Transmit EIRP per channel	22	22	22	22	22	dBm	$d+f-e$
	Receiver sensitivity at BTS antenna							
h	Thermal noise density	−174	−174	−174	−174	−174	dBm/Hz	Input
i1	Channel bandwidth	1 228 800	1 228 800	1 228 800	1 228 800	1 228 800	Hz	Input
i	Channel bandwidth	60.9	60.9	60.9	60.9	60.9	dB	Conversion
j	Bandwidth noise	−113.11	−113.11	−113.11	−113.11	−113.11	dBm	$h+i$
k	BTS receiver noise figure	5	5	5	5	5	dB	
l	Total noise power	−118.11	−118.11	−118.11	−118.11	−118.11	dBm	$j-k$
l1	Total noise power	1.55×10^{-15}	1.55×10^{-15}	1.55×10^{-15}	1.55×10^{-15}	1.55×10^{-15}	W	Conversion
m	External interference	−109.9	−109.9	−109.9	−109.9	−109.9	dBm	Input
m1	External interference	1.02×10^{-14}	1.02×10^{-14}	1.02×10^{-14}	1.02×10^{-14}	1.02×10^{-14}	W	Conversion
n1	Total noise/external interference	1.18×10^{-14}	1.18×10^{-14}	1.18×10^{-14}	1.18×10^{-14}	1.18×10^{-14}	W	$l1+m1$
n	Total noise/external interference power	−109.28	−109.28	−109.28	−109.28	−109.28	dBm	Conversion
o	Processing gain	21.07	18.06	15.05	12.04	9.03	dB	$i-a$
p	Required E_b/N_t (PCS)	5.6	5.8	5.9	6	7.2	dB	Input
q	Receiver sensitivity	−124.76	−121.55	−118.44	−115.33	−111.12	dBm	$n+p-o$
r	Cable and connector losses	2	2	2	2	2	dB	Input
s	Receive antenna gain	12	12	12	12	12	dBi	Input
t	Receiver sensitivity at BTS antenna	−134.76	−131.55	−128.44	−125.33	−121.12	dBm	$q+r-s$

Table 10.21 (Continued)

Ref. Reverse link		EVDO			Unit	Comment
Operational margins						
u	User body loss	2	2	2	dB	Input
v	Building/vehicle penetration	0	0	0	dB	Input
w	Fast fading margin	4	4	4	dB	Input
x	Slow fading margin	3	3	3	dB	Input
y	Power control error	0.5	0.5	0.5	dB	Input
z	Handoff gain	2.5	2.5	2.5	dB	Input
aa	Diversity gain	2	2	2	dB	Input
ab	Margin spread (90% time and/or area)	3	3	3	dB	Input
ac	Total operational margin	8	8	8	dB	$u + v + w + w + x +$ $y + ac - aa$ $- ab$
	Interference margins					
ad1	Ratio of mean other sector to same sector power at cell edge	0.9	0.9	0.9		Input
ad	Ratio of mean other sector to same sector power at cell edge	−0.46	−0.46	−0.46	dB	Conversion
ae	Other cell interference power	21.54	21.54	21.54	dBm	$g + ad$
af	Margin spread (90% time and/or area)	3	3	3	dB	Input
ag	Total interference margin	3.46	3.46	3.46	dB	$g - ae + af$
ah	*Maximum mean path loss*	145.30	142.09	138.98	135.87	131.66	dB	$g - t - ae - ag$
	Average cell radius (Reverse link)					
ai	800/900 MHz	11.91	10.27	8.90	7.72	6.36	km	50 dB/decade
aj	1800/1900 MHz	9.04	7.79	6.75	5.85	4.82	km	50 dB/decade
ak	2400 MHz	8.05	6.95	6.02	5.22	4.30	km	50 dB/decade

Table 10.22 Forward link budget for EVDO

Ref.	Forward link	EVDO												Unit	Comment
a1	Data rate	38 400	76 800	153 600	307 200	307 200	614 400	614 400	921 600	1 228 800	1 228 800	1 843 200	2 457 600	bps	$z*600/d$
a	Data rate	45.8	48.9	51.9	54.9	54.9	57.9	57.9	59.6	60.9	60.9	62.7	63.9	dB	Conversion
b	Modulation	QPSK	QPSK	QPSK	QPSK	QPSK	QPSK	QPSK	8PSK	QPSK	16QAM	8PSK	16QAM		Input
c	Packet size	1024	1024	1024	1024	2048	1024	2048	3072	2048	4096	3072	4096		Input
d	Number of slots	16	8	4	2	4	1	2	2	1	2	1	1		Input
	Transmit power calculation														
e	BTS Nominal available power	45	45	45	45	45	45	45	45	45	45	45	45	dBm	Input
f1	Channel bandwidth	1 228 800	1 228 800	1 228 800	1 228 800	1 228 800	1 228 800	1 228 800	1 228 800	1 228 800	1 228 800	1 228 800	1 228 800	Hz	Input
f	Channel bandwidth	60.9	60.9	60.9	60.9	60.9	60.9	60.9	60.9	60.9	60.9	60.9	60.9	dB	Conversion
g	Processing gain for data	15.1	12.0	9.0	6.0	6.0	3.0	3.0	1.2	0.0	0.0	-1.8	-3.0	dB	$f - a$
h	Diversity soft handoff factor	0.5	0.5	0.5	0.5	0.5	0.5	0.5	0.5	0.5	0.5	0.5	0.5	dB	Input
i	Cell site cable loss	3	3	3	3	3	3	3	3	3	3	3	3	dB	Input
j	Cell site transmit antenna gain	12	12	12	12	12	12	12	12	12	12	12	12	dBi	Input
k	Total BTS EIRP	54.5	54.5	54.5	54.5	54.5	54.5	54.5	54.5	54.5	54.5	54.5	54.5	dBm	$e - i + j + h$
	Receive sensitivity at MS antenna														
l	Thermal noise density	-174	-174	-174	-174	-174	-174	-174	-174	-174	-174	-174	-166	dBm/Hz	Input
m	Bandwidth noise	-113.1	-113.1	-113.1	-113.1	-113.1	-113.1	-113.1	-113.1	-113.1	-113.1	-113.1	-105.1	dBm	$l - j$
n	MS receiver noise figure	10.1	10.1	10.1	10.1	10.1	10.1	10.1	10.1	10.1	10.1	10.1	10.1	dB	Input
o	Total noise power	-103.0	-103.0	-103.0	-103.0	-103.0	-103.0	-103.0	-103.0	-103.0	-103.0	-103.0	-95.0	dBm	$m + n$
o1	Total noise power	5.0×10^{-14}	5.0×10^{-14}	5.0×10^{-14}	5.0×10^{-14}	5.0×10^{-14}	5.0×10^{-14}	5.0×10^{-14}	5.0×10^{-14}	5.0×10^{-14}	5.0×10^{-14}	5.0×10^{-14}	3.2×10^{-13}	W	Conversion
p	External interference	-109.9	-109.9	-109.9	-109.9	-109.9	-109.9	-109.9	-109.9	-109.9	-109.9	-109.9	-109.9	dBm	Input
p1	External interference	1.0×10^{-14}	1.0×10^{-14}	1.0×10^{-14}	1.0×10^{-14}	1.0×10^{-14}	1.0×10^{-14}	1.0×10^{-14}	1.0×10^{-14}	1.0×10^{-14}	1.0×10^{-14}	1.0×10^{-14}	1.0×10^{-14}	W	Conversion

Table 10.22 (*Continued*)

Ref.	Forward link	EVDO										Unit	Comment
q1	Total noise/external interference power	6.0×10^{-14}	6.0×10^{-14}	6.0×10^{-14}	6.0×10^{-14}	6.0×10^{-14}	6.0×10^{-14}	6.0×10^{-14}	6.0×10^{-14}	6.0×10^{-14}	3.3×10^{-13}	W	$p1 + o1$
q	Total noise/external interference power	−102.2	−102.2	−102.2	−102.2	−102.2	−102.2	−102.2	−102.2	−102.2	−94.9	dBm	Conversion
r	Required E_b/N_T (PCS)	5.6	5.8	5.2	5.4	6.2	7.7	5.4	13.3	11.4	17.1	dB	Input
s	Mobile receive antenna gain	2	2	2	2	2	2	2	2	2	2	dBi	Input
t	Receiver sensitivity for FCH at MS antenna	−113.6	−110.5	−107.4	−105.0	−104.8	−101.8	−98.8	−90.9	−91.0	−76.8	dBm	$q + r - g - s$
	Operational margins												
u	User body loss margin	2	2	2	2	2	2	2	2	2	2	dB	Input
v	Building/vehicle penetration	0	0	0	0	0	0	0	0	0	0	dB	Input
w	Fast fading margin	6	6	6	6	6	6	6	6	6	6	dB	Input
x	Slow fading margin	5	5	5	5	5	5	5	5	5	5	dB	Input
y	Power control error	1	1	1	1	1	1	1	1	1	1	dB	Input
z	Handoff gain	2.5	2.5	2.5	2.5	2.5	2.5	2.5	2.5	2.5	2.5	dB	Input
aa	Diversity gain	0	0	0	0	0	0	0	0	0	0	dB	Input
ab	Margin spread (90% time and/or area)	5	5	5	5	5	5	5	5	5	5	dB	Input
ac	Total operational margin	16.5	16.5	16.5	16.5	16.5	16.5	16.5	16.5	16.5	16.5	dB	$u + v + w + x + y + z - aa - ab$
	Interference margins												
ad1	Ratio of mean other sector to same sector power at cell edge	0.9	0.9	0.9	0.9	0.9	0.9	0.9	0.9	0.9	0.9		Input
ad	Ratio of mean other sector to same sector power at cell edge	−0.5	−0.5	−0.5	−0.5	−0.5	−0.5	−0.5	−0.5	−0.5	−0.5	dB	Conversion
ae	Other cell interference power	54.0	54.0	54.0	54.0	54.0	54.0	54.0	54.0	54.0	54.0	dBm	$k + ad$
af	Interference spread	5	5	5	5	5	5	5	5	5	5	dB	Input
ag	Total Interference margin	4.5	4.5	4.5	4.5	4.5	4.5	4.5	4.5	4.5	4.5	dB	$ae + af - k$

														Units	$k - t - ac - ag$
	Propagation loss														
ah	Maximum mean propagation loss for forward link	147.1	144.0	140.9	138.5	138.3	134.5	135.3	131.2	132.3	124.4	124.5	110.2	dB	
ai	Maximum mean propagation loss for reverse link (153 600 bps)	131.7	131.7	131.7	131.7		131.7	131.7	131.7	131.7	131.7	131.7	131.7	dB	From reverse link
	Average cell radius (FW and RV link)														
aj	800/900 MHz	6.36	6.36	6.36	6.36	6.36	6.36	6.36	6.22	6.36	4.54	4.57	2.37	km	50 dB/decade
ak	1800/1900 MHz	4.82	4.82	4.82	4.82	4.82	4.82	4.82	4.72	4.82	3.44	3.47	1.80	km	50 dB/decade
al	2400 MHz	4.30	4.30	4.30	4.30	4.30	4.30	4.30	4.21	4.30	3.07	3.09	1.60	km	50 dB/decade

10.7   SIGNAL STRENGTH PREDICTIONS

After the network and equipment parameters are set through the link budget they can be applied to network sites. Each site's path loss and received signal level can then be calculated.

Figure 10.71 shows the signal strength in dBm for a single BTS (composite of three sectors) whereas Figure 10.72 shows the composite signal strength in dBm for the whole network.

These individual signal strength predictions are used to provide path loss information for call placement simulations and other composite predictions.

10.8   CALL PLACEMENT SIMULATION

Network parameters are greatly influenced by user distribution and corresponding traffic. This is especially true for adaptive networks, such as CDMA2000 with its adaptive data rates. Network parameters change dynamically over time and the only way to evaluate them properly is to perform a statistical analysis over time. For this evaluation it is necessary to simulate the calls, adjust network parameters, and analyse the results. This procedure is called Call Placement Simulation.

A CDMA network must be well represented in its main characteristics for the simulation to provide results similar to reality. A poor representation still provides results, but they may

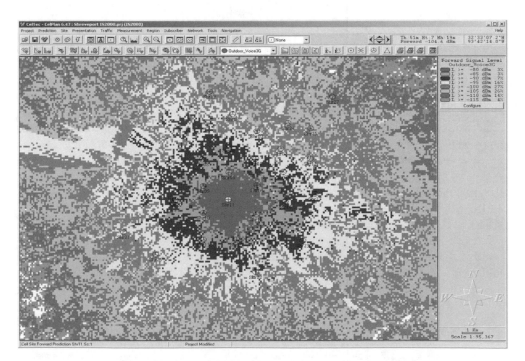

Figure 10.71   Individual BTS received signal level by MS.

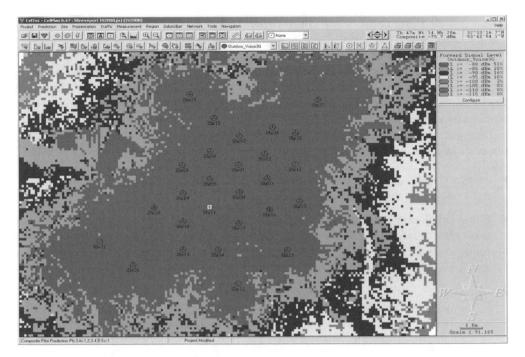

Figure 10.72   Network composite MS received signal level.

be useless. Important aspects in the network representation are

- Services provided should have their characteristics defined and be adequately grouped

- Users distribution and traffic characteristics should be geographically defined

- In-building user distribution should be performed for several representative heights, with a traffic distribution associated to each height

The most common mistake is to assume that all users are at ground floor, when, in reality, the majority of the calls come from buildings.

The main parameters to be determined during the simulation are the power levels required at the BTSs and at MS locations. These levels have to be adjusted to provide enough E_b/N_0 at the BTS and MS receivers.

A limitation of CDMA networks is when a BTS or MS runs out of power and cannot provide enough signal strength to maintain the desired quality of service. This power information can be analysed from the receiver side, being represented by the load factor or noise rise as shown in eqns (10.3) and (10.4).

$$\text{Load factor} = I_t/(I_t + N) \tag{10.3}$$

$$\text{Noise Rise} = (I_t + N)/N \tag{10.4}$$

Where
  I_t- is the total interference received from all sectors
  N- is the total noise including the equipment noise figure

10.8.1   Call Placement Simulation Types

Call placement simulation can be performed to analyse performance or to dimension the network. Each of these aspects requires different procedures during the simulation. In the performance case, network resources are pre-defined and the goal is to determine blockage and delays. In the dimensioning case required resources are dimensioned for a certain blockage and delay.

There are several ways of implementing call placement simulation; the most common are static analysis, static simulation, dynamic analysis, dynamic simulation, and performance analysis.

- Static analysis – network parameters are pre-defined by the user as design goals and the performance of the network is then assessed.

- Static simulation – network parameters are derived from a series of network snapshots and defined statistically. The snapshots are based on the traffic distribution of different service classes. Not all static simulations are the same; some consider traffic fluctuations and fading variations whereas others do not.

- Dynamic simulation – After the static simulation is performed and the main parameters of the network are defined, an additional simulation can be performed covering a short time period, considering arrival and termination of voice calls and data sessions. This procedure allows the calculation of blockage, data delays, and storage queues.

- Dynamic analysis – This analysis can be performed over the static or dynamic simulation results. It consists on a series of snap shots that analyse the mobile movement in the network. It is ideal to analyse handoff issues and code conflicts.

- Performance analysis – This simulation considers static and dynamic simulation results. A call is placed on all network locations and the outcome is statistically analysed to calculate the success probabilities.

Call placement simulation is a very extensive topic and could be a subject for a whole book. This chapter covers only the main aspects.

Call placement analysis uses independent snapshots of the network call distribution. The traffic offered to the network is characterised by traffic distribution maps and by the simultaneous number of calls. The Monte Carlo technique is used to distribute calls across the service area based on the traffic probability of each pixel.

Each snapshot represents the network at a different moment in time. Calls are assigned to cells considering forward and reverse links configuration, power control, and soft handoff. Due to the adaptive characteristic of CDMA systems and to the different data rates used, each snapshot has to go through multiples iterations to reach convergence. The statistics of the snapshots are calculated in terms of noise level received and power transmitted in the traffic channels.

The Monte Carlo simulation process is repeated for several snapshots before calculating the statistics. The results obtained from the simulation are the average and the standard deviation of

- Noise level at the input of each cell;

- Traffic power level transmitted from each cell; and

- Blocking for each service class.

Figure 10.73 shows an example of call placement and the iterative convergence loops (left) for one snapshot.

The top part of the figure shows the simultaneous calls offered to the system from one snapshot to the next. For each snapshot there are several iterations on the forward and reverse link until equilibrium (stable noise and power values) are reached.

The central and bottom parts of the figure show, respectively, the convergence of the forward and reverse links during the iterations to reach equilibrium. These links are calculated together because the results of one influence the other. The dotted lines show the limits within which the equilibrium is considered satisfactory. To reach this equilibrium a specific number of iterations must fall within this range before the iterations stop.

Figure 10.74 shows calls placed after several snapshots indicating blocked calls and the result by class of service.

The statistics obtained from the simulations above can be used to perform network availability calculations, with the network loaded with all service classes, verifying what percentage of the traffic of each class of service is within pre-established limits (desired quality).

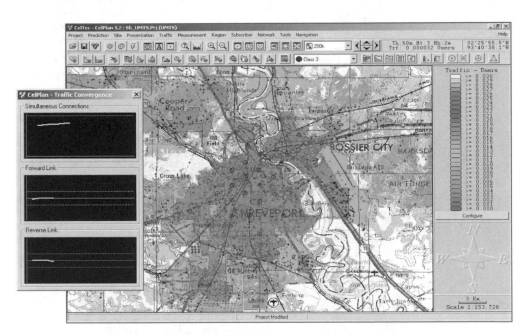

Figure 10.73   Call placement simulation.

10.9   COMPOSITE PREDICTIONS

After the design area is characterised, cell coverage prediction studies can be performed. The prediction process is lengthy because the profile to each pixel has to be considered. The prediction time increases at least quadratically with the prediction radius. Therefore predictions should store enough information to allow fast recalculations when site

NETWORK DESIGN

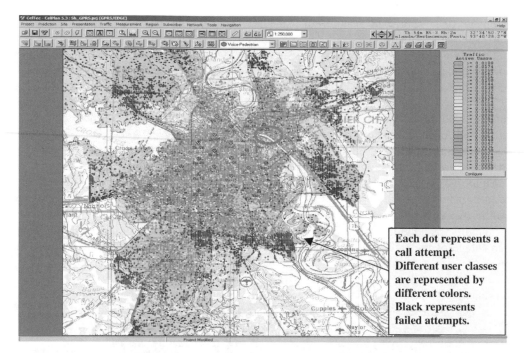

Figure 10.74 Call placement results.

Figure 10.75 Single site signal level prediction (dBm)—voice user, ground floor, outdoor.

Figure 10.76 Single site signal level prediction (dBm)—voice user, ground floor, indoor.

parameters are modified, requiring prediction files to store the path loss for different heights and the angle to the first and last obstructions.

This solution allows immediate recalculation of prediction studies when changes in antenna type, tilt, power, and azimuth are performed. Changes in antenna height should still require regeneration of the predictions. Even though some models provide a rough estimate for MS height changes, in the majority of the cases the results are not acceptable. The ideal is to generate predictions that account for the different heights specified in the service classes as stated above.

Figures 10.75–10.77 show the coverage of a single BTS in dBm for voice users located, respectively, outdoor, indoor at ground floor, and indoor at the sixth floor. These figures clearly indicate that the coverage shape changes with the height and that the linear correction done by some propagation models is not enough to reflect this.

In CDMA technology systems, single cell predictions can also present calculations for E_c/I_0, forward traffic, and reverse traffic. Figures 10.78–10.80 show these predictions for voice users at ground level.

The prediction radius, that is, the area for which the prediction was calculated, should be enough to allow precise interference calculations. Factors such as terrain and BTS height affect the prediction radius definition. Microcell radii from 2 to 10 km are common whereas outer cells usually have radii varying from 20 to 80 km. Good prediction tools usually offer built-in automatic prediction radius calculation.

After single cell predictions are done, designers may execute composite predictions. The result of these predictions is tied to traffic distribution when working with CDMA or other adaptive technologies. Therefore, before executing composite studies, designers must perform traffic simulations to calculate cell load parameters. For CDMA systems, the parameters required are cell load and cell power. More details about this simulation are given in Section 10.8 and in Chapter 11.

After obtaining the required cell parameters, the composite predictions can be calculated. These predictions are useful for system analysis, allowing designers to better understand the network. The following composite predictions are usually generated:

- Composite Pilot Channel in dBm – strongest pilot level in dBm at each pixel;

- Composite Pilot Channel in E_c/I_0 – strongest value of E_c/I_0 at each pixel;

- Composite Forward Traffic – identifies pixels that have coverage in the forward traffic channel;

- Composite Reverse Traffic – identifies pixels that have coverage in the reverse traffic channel;

- Pilot Best Server – identifies which pilot is the best server at each pixel. This prediction can be used to display any network parameter attributed to the best server pilot;

- Pilot Delta – displays the delta between the first and the second strongest pilots;

- Pilot Offset Conflict – provides the pilot offset conflict at each pixel;

- Forward Traffic Probability – provides the percentage of forward traffic that complies with the quality requirements;

- Reverse Traffic Probability – provides the percentage of reverse traffic that complies with the quality requirements;

- Forward/Reverse Service – provides an indication of how well the forward and reverse links are balanced;

- Forward Reuse Factor – provides I_{same}/I_t for the forward link. I_{same} is the interference coming from the same sector, while I_t is the interference from all sectors;

- Forward Noise Rise – provides $(I_t + N)/N$ for the forward link, where N is the total noise;

- Forward Load Factor – provides the ratio of $I_t/(I_t + N)$ for the forward link;

- Number of Servers – provides the number of candidate pilots at each pixel;

- Service Classes – overlaid coverage of each service class;

- Handoff – identifies the areas in soft, softer, and soft-softer handoff.

Figures 10.81–10.96 present examples of each of these predictions considering voice users, outdoor, at ground level.

Figures 10.97–10.100 compare the number of servers composite prediction for different services (voice, 19.2 kbit/s, 38.4 kbit/s, and 153.6 kbit/s) considering users at ground floor in an indoor environment.

Figures 10.101–10.103 compare the number of servers for voice users respectively located in outdoor, indoor ground level, and indoor sixth floor environments.

Figure 10.77   Single site signal level prediction (dBm)—voice user, sixth floor, indoor.

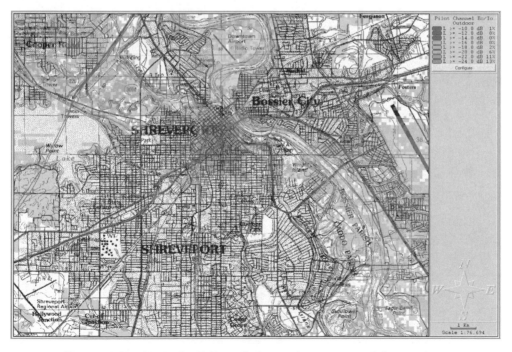

Figure 10.78   Single site E_c/I_0 prediction—voice user, ground floor, outdoor.

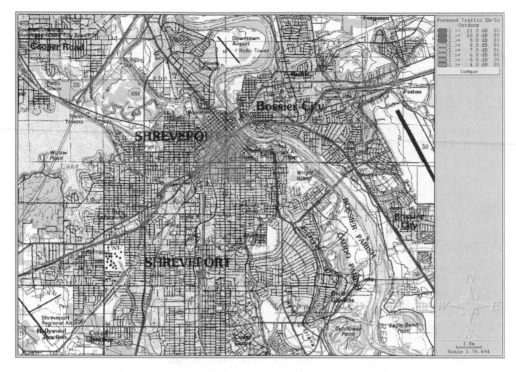

Figure 10.79   Single site forward traffic prediction—voice user, ground floor, outdoor.

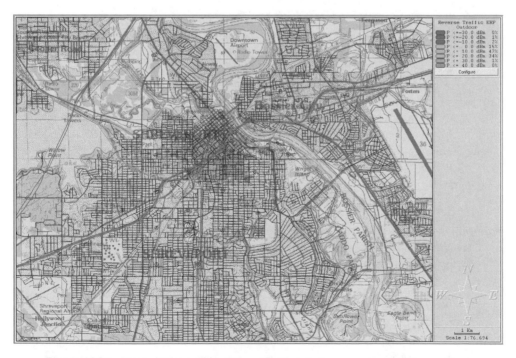

Figure 10.80   Single site reverse traffic prediction—voice user, ground floor, outdoor.

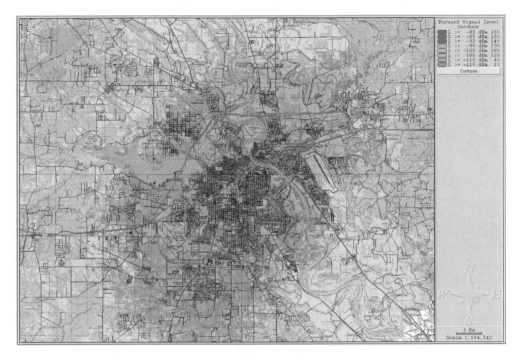

Figure 10.81 Composite pilot (dBm) prediction—voice user, ground floor, outdoor.

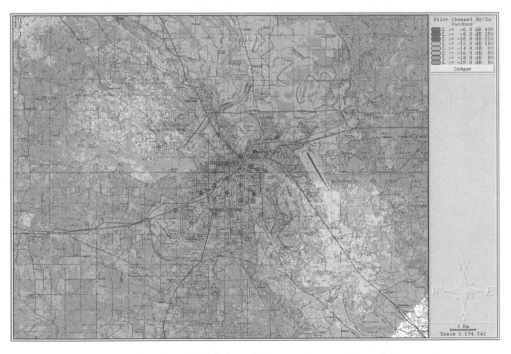

Figure 10.82 Pilot channel (E_c/I_0) prediction—voice user, ground floor, outdoor.

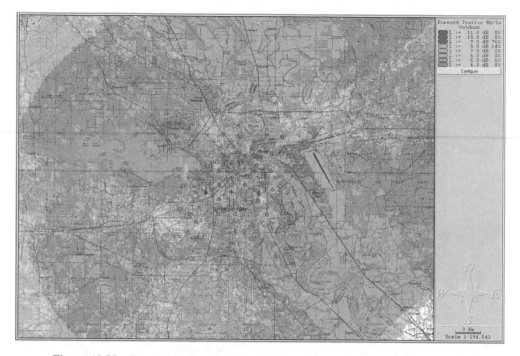

Figure 10.83 Forward traffic (E_b/I_0) prediction—voice user, ground floor, outdoor.

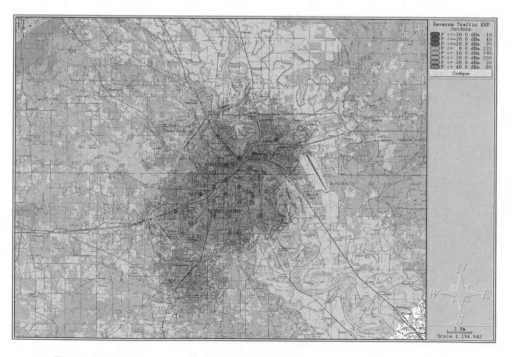

Figure 10.84 Reverse traffic ERP prediction—voice user, ground floor, outdoor.

Figure 10.85 Best server site prediction—voice user, ground floor, outdoor.

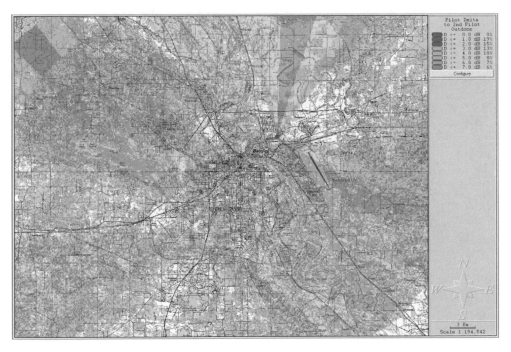

Figure 10.86 Pilot delta to second pilot prediction—voice user, ground floor, outdoor.

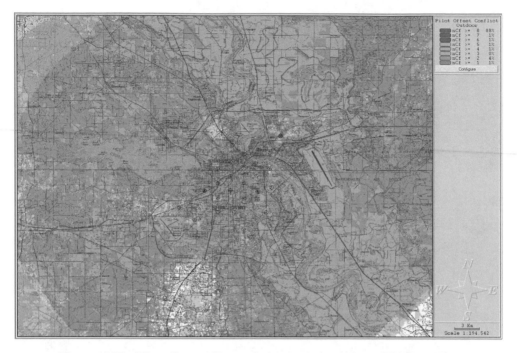

Figure 10.87   Pilot offset conflict prediction—voice user, ground floor, outdoor.

Figure 10.88   Forward probability prediction—voice user, ground floor, outdoor.

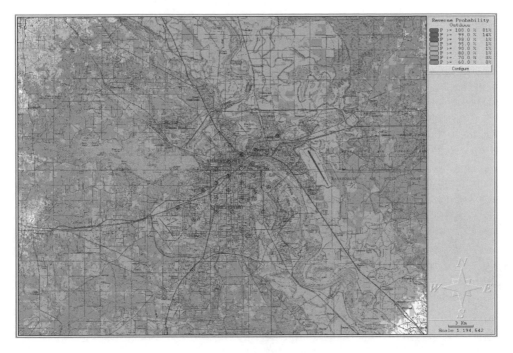

Figure 10.89 Reverse probability prediction—voice user, ground floor, outdoor.

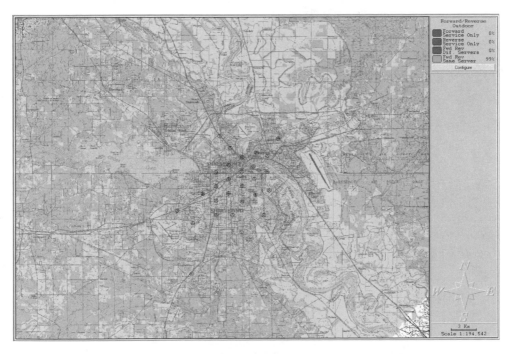

Figure 10.90 Forward/reverse service prediction—voice user, ground floor, outdoor.

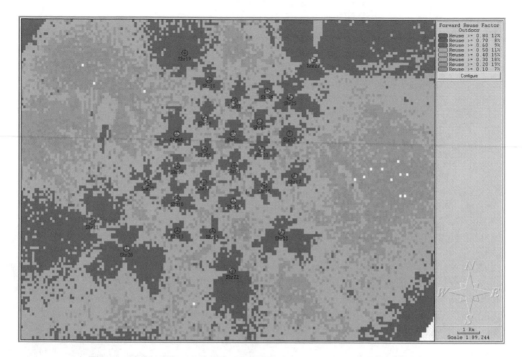

Figure 10.91 Forward reuse factor voice user, ground floor, outdoor.

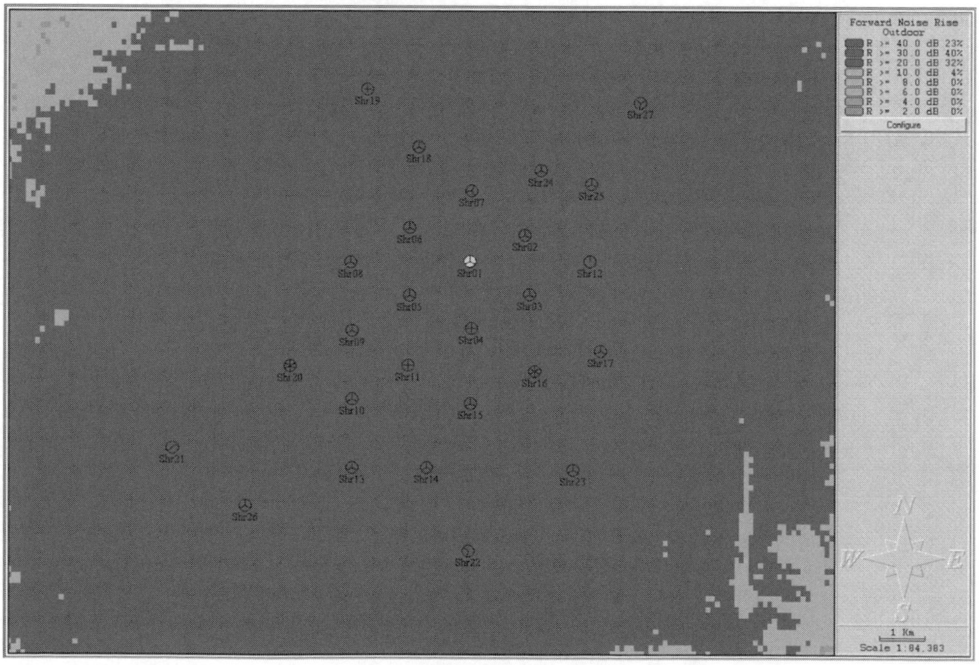

Figure 10.92 Forward noise rise prediction—voice user, ground floor, outdoor.

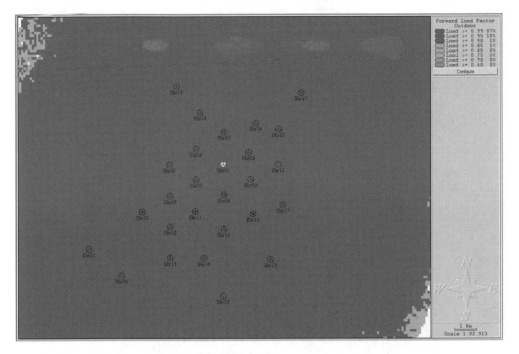

Figure 10.93   Forward load factor prediction—voice user, ground floor, outdoor.

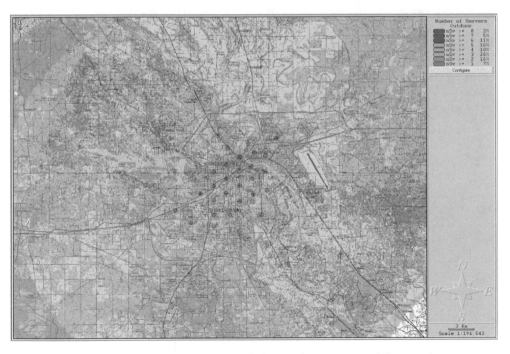

Figure 10.94   Number of servers prediction—voice user, ground floor, outdoor.

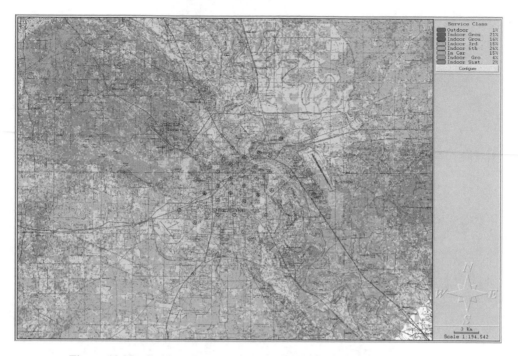

Figure 10.95   Service classes prediction—voice user, ground floor, outdoor.

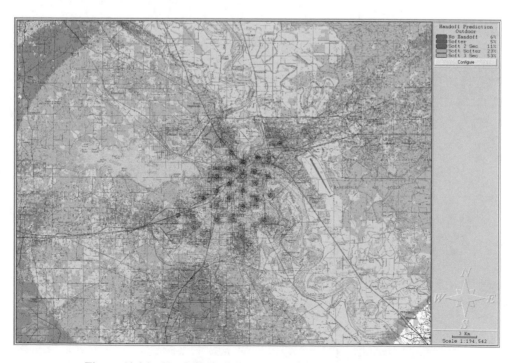

Figure 10.96   Handoff prediction—voice user, ground floor, outdoor.

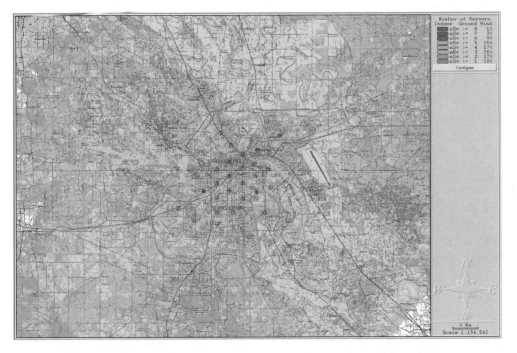

Figure 10.97 Number of servers prediction—voice user, ground floor, indoor.

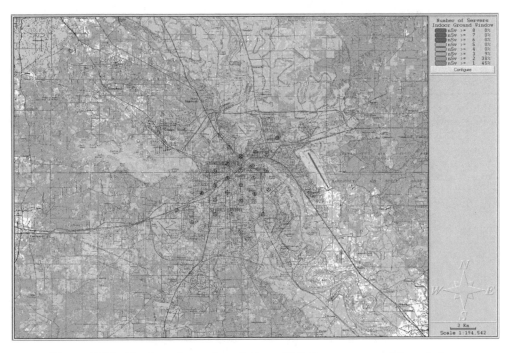

Figure 10.98 Number of servers prediction—19.2 kbps user, ground floor, indoor.

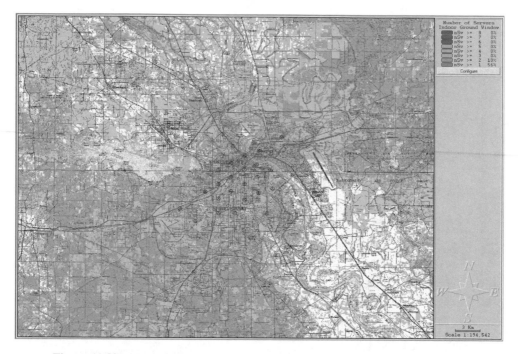

Figure 10.99   Number of servers prediction—38.4 kbps user, ground floor, indoor.

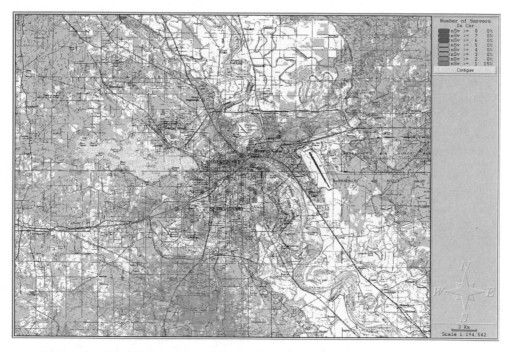

Figure 10.100   Number of servers prediction—153.6 kbps user, ground floor, indoor.

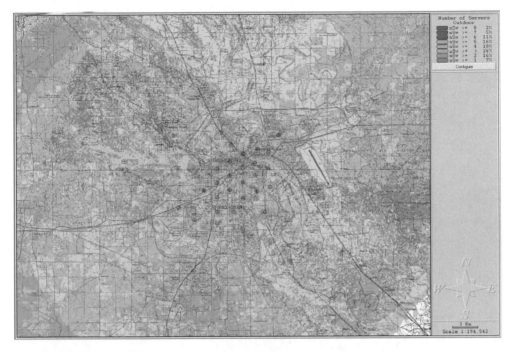

Figure 10.101 Number of servers prediction—voice user, ground floor, outdoor.

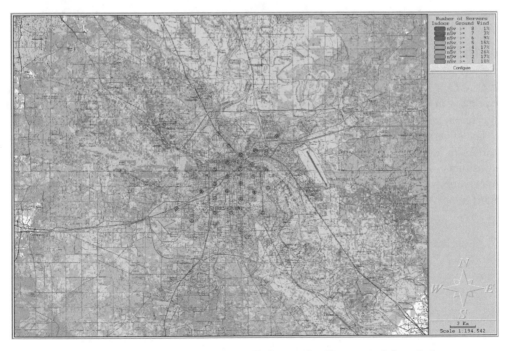

Figure 10.102 Number of servers prediction—voice user, ground floor, indoor.

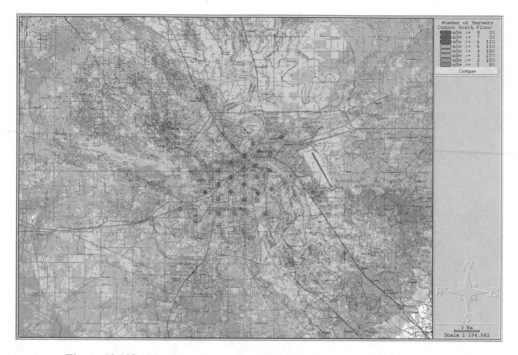

Figure 10.103   Number of servers prediction—voice user, sixth floor, indoor.

10.10   NETWORK ENHANCEMENT

In CDMA systems, network capacity is limited by the received noise level and by the power available in the cells and mobiles. As the required transmitted power depends on the interference level, it is important to minimise interference as much as possible.

The coverage of a cell should be adjusted to provide enough service to handle the offered traffic, minimising the disruption caused by interference to neighbouring cells. This adjustment is the goal of network enhancement. Figures 10.104 and 10.105 show an example of coverage being reduced without disrupting service, but diminishing interference.

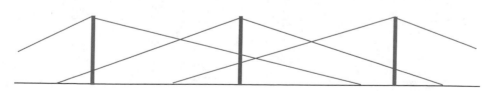

Figure 10.104   Network enhancement (before).

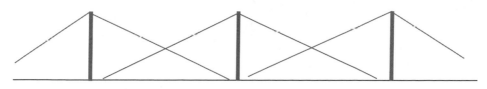

Figure 10.105   Network enhancement (after).

Different type of adjustments can be done to change the shape of a cell footprint. The main parameters that can be modified for this purpose are the following:

- Power adjustment;

- Antenna tilt;

- Antenna type change;

- Antenna azimuth change; and

- Antenna height.

The enhancement process is extremely complex because of the adaptive nature of CDMA and of the large number of cells involved in a real system. In this situation tools such as *CelEnhancer*, which can perform automatic enhancement based on user-configurable parameters, are highly recommended. Network enhancement in CDMA is, in terms, equivalent to frequency planning in TDMA, having a large impact in the overall network capacity and performance.

10.11   NETWORK OPTIMISATION

Even though CDMA does not require frequency planning, it still needs a sensible distribution of network resources such as neighbour lists and PN offset. Tools like CelOptima are used for this task.

Neighbourhood planning is performed building a matrix that records the relationship between sectors on a pixel by pixel basis and accumulates the percentage of traffic that is affected based on the SNIR (E_b/N_0) distribution. Figure 10.106 illustrates an example of neighbourhood relationships. PN offset planning is performed considering the mobile distribution and the distances between sites.

10.12   NETWORK PERFORMANCE

The prediction of the outcome of a network design or modification is very important to judge between different scenarios and to decide on capital expenditures. The network availability for a desired quality is the best way of verifying its performance. This availability can be expressed in terms of time availability or traffic availability.

Figures 10.107–10.108 show the geographical representation, respectively, of time and traffic availabilities for a sample market, obtained using CelPerformance. Availability can

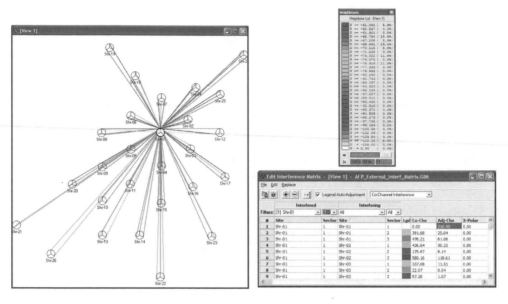

Figure 10.106   Neighbourhood planning.

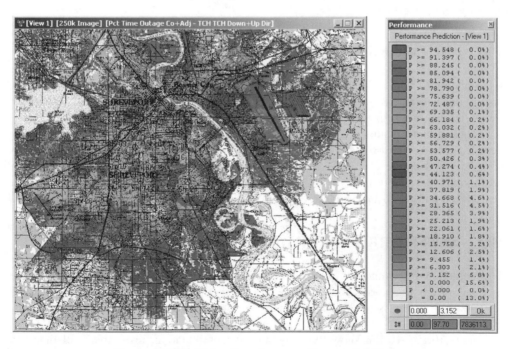

Figure 10.107   Network performance (time availability).

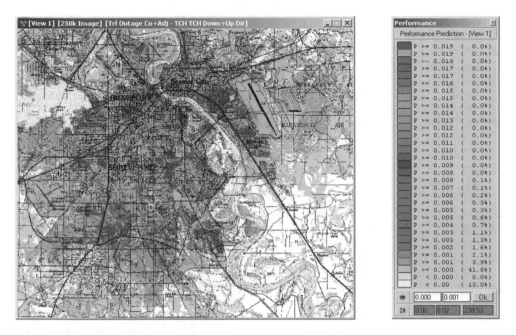

Figure 10.108   Network performance (traffic availability).

also be calculated on a sector or network basis, expressed as an overall value. This value can reflect the network performance for a single service class or for a combination of all classes.

10.13   CONCLUSION

The fast evolution of wireless networks frustrated designers because traditional methods do not provide the required results anymore. The solution adopted by many to solve this problem was the introduction of a trend based on extensive measurements to adjust each sector of the network in a continuous loop. This approach is a practical solution for the lack of design methods.

Network design, however, has evolved significantly in the last years, bringing better models and techniques. Because not all designers are familiar with them, this chapter covered the fundamentals required for a proper design.

Any planning technique requires a good representation of the network; a wireless design is not an exception to this rule. After the initial representation of a network is complete, it is easy to maintain it. The careful design of a network can provide substantial improvements and savings to operators.

BIBLIOGRAPHY AND REFERENCES

1. Jhong Sam Lee and Leonard E. Miller, CDMA Systems Engineering Handbook, 1998, Artech House Ltd.
2. Simon R. Saunders, Antennas and Propagation for Wireless Communication Systems, 1999, Wiley, New York.

3. Henry L. Bertoni, Radio propagation for Modern Wireless Systems, 2000, Prentice Hall, Englewood Cliffs, NJ.
4. Kazimierz Siwiak, Radio Propagation and Antennas for Personal Communications, 1995, Artech House, Inc.
5. David Parson, The Mobile Radio Propagation Channel, 1996, Wiley, New York.
6. Manuel F. Cátedra and Jesús Pérez-Arriaga, Cell Planning for Wireless Communications, 1999, Artech House, Inc.
7. José M. Hernando and F. Pérez-Fontan, Introduction to Mobile Communications Engineering, 1999, Artech House, Inc.
8. Saleh Faruque, Cellular Mobile Systems Engineering, 1996, Artech House, Inc.
9. Roger L. Freeman, Radio System Design for Telecommunications, 1997, Wiley, New York.
10. John Doble, Introduction to radio Propagation for Fixed and Mobile Communications, 1996, Artech House, Inc.
11. Trevor Manning, Microwave Radio Transmission Design Guide, 1999, Artech House, Inc.
12. Samuel C. Yang, CDMA RF System Engineering, 1998, Artech House, Inc.
13. John B. Groe and Lawrence E. Larson, CDMA Mobile Radio Design, 2000, Artech House, Inc.
14. Andrew J. Viterbi, CDMA Principles of Spread Spectrum Communications, 1995, Addison-Wesley Publishing Company, London.
15. Theodore S. Rappaport, Wireless Communications, 2002, 1996, Prentice Hall, Englewood Cliffs, NJ.
16. Bernard Sklar, Digital Communications, 2001, Prentice Hall, Englewood Cliffs, NJ.
17. John G. Proakis and Masoud Salehi, Communication System Engineering, 2002, Prentice Hall, Englewood Cliffs, NJ.
18. M.R. Karim and Mohsen Sarraf, WCDMA and CDMA 2000, 2002, McGraw-Hill, New York.
19. Tommy Oberg, Modulation Detection and Coding, 2001, Wiley, New York.
20. George M. Calhoun, Third Generation Wireless Systems, 2003, Artech house, Inc.
21. Marvin K. Simon and Mohamed-Slim Alouini, Digital Communications over Fading Channels, 2000, Wiley, New York.
22. Robert M. Gagliardi, Introduction to Communications Engineering, 1988, Wiley, New York.
23. Claude E. Shannon and Warren Weaver, The Mathematical Theory of Communications, 1949, 1998, University of Illinois, USA.
24. Kaveh Phlavan and Allen H. Levesque, Wireless Information Networks, 1995, Wiley, New York.

11

Traffic Dimensioning

LEILA ZURBA RIBEIRO AND LUIZ A. DASILVA

This chapter discusses the process of traffic dimensioning for multimedia (voice, image, video and data) systems. While most of the aspects discussed herein apply to 3G systems in general, the main focus is on IS2000 networks.

11.1 INTRODUCTION

The process of dimensioning a wireless system consists of the appropriate allocation of resources that satisfy certain system performance requirements given knowledge of the expected traffic demand. Based on this definition, three main building blocks are discussed in this chapter.

- Demand characterisation, which consists in describing, as accurately as possible, the traffic offered to a given network, appropriately considering multiple demographic input variables, different traffic patterns expected for different user profiles, and the time dependent relationships among these patterns.

- A method to estimate system response to the traffic offered. This task is typically done by simulating the system for a given configuration. This chapter discusses simulation methods that appropriately consider the demand characterisation mentioned above in a statistical manner.

- Techniques to characterise the initial configuration and to tune the system to maximise capacity based on simulation results. This process is typically iterative and requires successive rounds of modifications to the original network configuration and corresponding simulation execution.

These three main building blocks correspond to phases of system dimensioning typically performed in a recursive way, together with the network design, until the process converges to an optimised and well-dimensioned network. Figure 11.1 illustrates this relationship, highlighting in grey the three blocks listed previously.

Designing CDMA 2000 Systems   L. Korowajczuk et al.
© 2004 John Wiley & Sons, Ltd   ISBN: 0-470-85399-9

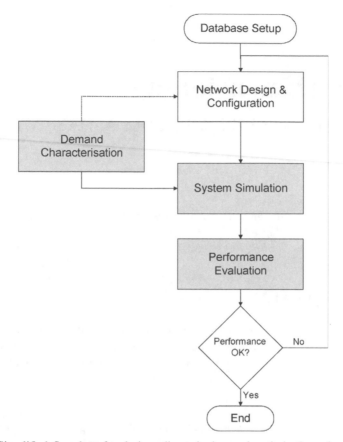

Figure 11.1 Simplified flowchart for design, dimensioning and optimisation of wireless systems. This, chapter discusses the shaded blocks.

In Figure 11.1, results from demand characterisation, together with the current network configuration, are used as inputs in the process of simulating the operation of the system. Statistics collected in the simulation allow system designers to decide whether the performance obtained is acceptable or if further tuning of the current design is needed.

Some of the results obtained from demand characterisation, mainly those related to the definition of user profiles and services supported by the system, are also used as inputs in the process of network configuration. Those inputs are necessary for the definition of initial link budgets and target coverage thresholds, as discussed in chapters 9 (Design Fundamentals) and 10 (Network Design).

11.2   DEMAND CHARACTERISATION

This section is based on Ref. [1].

A framework is presented to model traffic offered to a 3G system providing voice and data services, focusing on modelling this traffic as an input for capacity and performance analysis

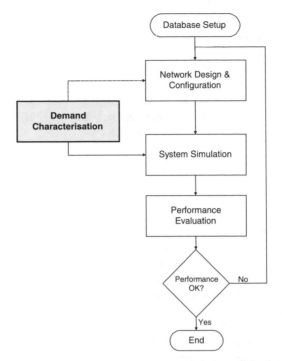

Figure 11.2 Traffic dimensioning process–demand characterisation procedure.

using simulation methods. This discussion corresponds to the "demand characterisation" block highlighted in the block diagram in Figure 11.2.

The section starts with a discussion of the processes of service and market identification, which lead to the definition of multiple user profiles to be supported by the system. A detailed description of the concept of user profile is presented, describing parameters relevant to the specification of services, user terminal and environmental conditions to be supported in the system. It also presents a review of traffic models applicable to the different types of services expected in 3G systems. These models include traditional assumptions used for circuit-switched voice traffic and more complex models that adopt self-similarity to characterise the expected burstiness of packet-switched data traffic in a wide range of time scales.

The section presents a methodology to define geographical traffic distribution of users for different user profiles, based on available demographic data and on existing 2G network traffic reports from network switch reports (for instance, as available from existing 2G systems) and other system statistics available.

11.2.1 Market and Product Definition

To define a business plan and therefore justify implementation, operation and expansion of third-generation networks, operators must have a clear view of the product to be offered and its potential market.

Product specification is a key step in the definition of the user classes to be supported by the system. The different types of applications and services offered to the user influence user behaviour and Quality-of-Service (QoS) requirements. The knowledge of these services' characteristics and requirements allows the adoption of adequate pricing policies.

Besides product definition from the users' point of view, another important step in the dimensioning of a system is the process of technology and implementation decisions. This process determines system equipment characteristics and capabilities, which are key to defining how services and QoS will be provided.

Market forecast, allied to business case analysis, helps to determine the scope and extension of the target market, contributing to a more solid business model and maximising return on investment.

11.2.1.1 User Profiles

When dimensioning the system, we must be aware of the different types of user supported. Users can be distinguished by a combination of attributes that define their service class, propagation environment and equipment characteristics. The following paragraphs explain the user profile concept in detail [1].

In multimedia systems, users can generate one or more simultaneous flows that present different characteristics regarding their traffic nature, volume and QoS requirements. These characteristics define the service parameters. Considering the multiple possibilities of user capabilities and applications expected in mobile broadband systems, these flows may be grouped in categories that present common characteristics, referred herein as service classes. Examples of service classes include typical traffic behaviours expected in Internet access, voice and video conferencing.

Further, the type of terminal used by a subscriber has distinct capabilities and limitations that have a direct impact on the type of application and QoS supported by it. Differences in antenna gain, maximum transmission power and receiver implementation (lower noise figures) constitute the main factors that impact link margins allowed for a given user from the terminal point of view. Terminals likely to be actively present in any 3G system include plain voice terminals (such as legacy 2G phones), handheld voice-data terminals (such as Personal Digital Assistants (PDAs)), laptops, vehicular phones and vehicular computers.

Different mobility characteristics and the context in which the mobile is inserted define its environment. These factors also impact the propagation channel experienced, imposing different interference and fading effects that depend on user location, user speed, line of sight, etc. All these factors influence link availability, which, in turn, affects achievable data rates and transmission quality experienced by users. User environment characterisation includes factors such as mobile speed, forward link orthogonality factor, diffraction and penetration losses and whether the user terminal is experiencing line-of-sight (LOS) coverage or receiving only through indirect paths. Examples among the main categories of user environment are indoor, pedestrian and vehicular environments.

In summary, performance requirements and capabilities depend on factors and parameters from all three groups that describe a given user traffic flow: service class, terminal and environment. This is clear when considering that the same type of application may be able to achieve different QoS levels depending on the terminal being used and on the propagation environment. For instance, Internet access from a laptop within a building may achieve higher data rates than Internet access by a pedestrian carrying a palmtop, due to differences in the transceiver equipment and fading conditions. To further explore the influence and inter-dependence among these three groups of characteristics, consider the impact of different environments on some typical service characteristics and terminal types, as in Ref. [1].

- *Indoor users*: Attenuation from building walls constitutes a major propagation impairment, impacting link strength and limiting coverage area and bandwidth. Indoor signal boosters and micro-cells may be used to compensate for wall attenuations. Typical applications for this type of user are Internet access and connection to office LANs using laptops.

- *Pedestrian users*: Typical terminal types include handheld phones and palmtops (PDAs). Attenuation due to building walls is not expected, although other propagation impairments typical of urban areas such as reflections and building diffractions are common. Link margins may be tighter as a consequence of lower transmission power limits, terminal and antenna sizes and receiver implementations. Voice, image, Internet access and small video applications are the primary applications for these users.

- *Vehicular users*: A fast-moving subscriber unit means that the effects of rapidly changing fading environment may be severe. Expected applications include voice, video conferencing, Internet access and navigation systems. Typical terminal types are on-board car computers, hands-free terminals and handset phones.

A complete user profile is described by grouping together the characteristics that identify a user in terms of service class, terminal and environment. The concept of user profile [1], which takes the Cartesian product of these three dimensions, is adopted for the process of demand characterisation. Therefore, the parameters that specify a user service class (application, performance requirements and traffic characteristics), propagation environment (mobility, fading channel, penetration losses) and terminal type (antenna, transmission power limits, technology capabilities, receiver implementation) define the user profile. Examples of user-profile categories may include, for instance, indoor-laptop-video conference and pedestrian-palmtop-web browsing.

Figure 11.3 illustrates this concept.

The concept suggested for user profile in Ref. [1] is applicable to different multimedia wireless systems. In the particular case of 3G networks, such as IS2000, examples of parameters that may be defined in each group include the following.

Service class definition includes

- service type (voice, video conference, web, etc.);

- service category (conversational, streaming, etc.);

- traffic model description (e.g. session duration, packet lengths and call inter-arrival times);

- required data rates (kbps);

- type of switching (circuit/packet);

- activity factors (forward or reverse link);

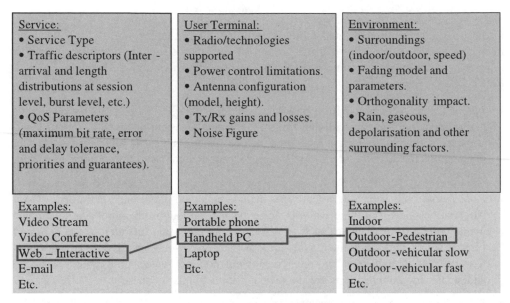

Figure 11.3 Concept of user profile as a combination of service, user terminal and environment configurations. The user profile in the example (connected boxes) is 'web on handheld PC at a slow moving vehicle'.

- required quality metrics such as bit error rate (BER), frame error rate (FER) or block error rate (BLER), both on the forward and reverse links.

Environment definition includes

- mobile speed and location;
- channel model;
- body shielding loss;
- in-building losses;
- orthogonality factor;
- fading model.

User terminal definition includes

- radio and technology parameters;
- radio configurations supported;
- power settings and limits;
- power control loop;
- available number of rake fingers.

Table 11.1 Example of multiple user classes and performance requirements

	User profile	1	2	3	4
	Service name	Video conference	Voice	Streaming video	Web browsing
	Service class group	Conversational	Conversational	Streaming	Interactive
	Type of switching	Circuit	Circuit	Packet	Packet
	Maximum bit rate down (kbps)	64	13	144	1000
	Maximum bit rate up (kbps)	64	13	0	500
Service class	Session arrival distribution	Exponential	Exponential	Exponential	Exponential
	Burst arrival distribution	Exponential	Exponential	Exponential	Exponential
	Burst length distribution	Pareto	Exponential	Pareto	Pareto
	Required BER - down	1.0×10^6	1.00×10^3	1.00E-06	1.00E-08
	Required BER - up	1.0×10^6	1.00×10^3	–	1.00E-08
	Required E_b/N_o - down (dB)	8	7	12	1
	Required E_b/N_o - up (dB)	8	7	–	1
	Tolerated delay (ms)	100	100	200	–
Propagation environment	Propagation scenario	Indoor slow	Outdoor fast	Outdoor Slow	Indoor slow
	Terminal speed (km)/h	0	120	120	3
Terminal type	User terminal	Laptop	Mobile phone	Palmtop	Laptop

Table 11.1 shows an example of multiple user profiles.

For the service class description, particularly for variables that refer to the traffic model, different parameters must be considered depending on the type of application, including the definition of whether the service is circuit or packet switched. The next section presents a review of traffic theory discussing the main models applicable to voice and data traffic.

11.2.2 Traffic Modelling

This section presents a brief review of traffic theory, discussing models available for voice and data traffic, as well as the most recent fundamental developments in this area.

In networks that support voice and data traffic (hybrid traffic), the applicability of the Erlang theory to network dimensioning has been greatly challenged in the recent years. The dispersion of hold times has increased in magnitude, and networks dimensioned for average holding times in the order of minutes are confronted now with subscribers connected for hours. For circuit-switched networks, there is a real inefficiency in dealing with bursty data communications, as resources allocated for one subscriber can not be shared with others even if they are not effectively being used. To avoid such inefficiencies in a medium that is already starved for resources, new technologies and standards came aboard to support broadband wireless access with packet-switched services. For these networks, pure Erlang theory does not hold any more, and other modelling techniques must be considered.

This section describes ways to characterise the broadband traffic source, and discuss some traffic models used to correlate the source description with the required network capacity and desired quality of service. It starts with a review of voice traffic modelling (Erlang analysis) and then presents techniques to model data traffic sources.

11.2.2.1 Voice Traffic and Poisson Models

For the past decades, traffic modelling and dimensioning of voice networks has relied on the work developed by Erlang in 1917 [2]. The Erlang theory models traffic using two parameters: the call arrival rate (in calls/time unit) and the average call holding time (in time unit/call). The product of these two parameters is a metric of traffic load, which is a dimension-less quantity, but using a unit traditionally referred to as 'Erlangs' (Erl) in the telephony universe. For further insight into the concept of Erlang, one can think from the point of view of the circuit usage. A traffic load given in Erlangs corresponds to the usage of a certain resource over a certain time scale. For instance, if the resource is used 80% of the time, it is carrying 0.8 Erlangs.

To calculate the total traffic offered to one cell, one needs to consider the forecast (or switch-measured) values for the parameters described above as an average over the number of subscribers in the cell coverage area. Therefore, the following relations hold

$$\text{Traffic per subscriber (Erl)} = \text{Call arrival rate per subscriber}$$
$$\times \text{ Average holding time per call} \tag{11.1}$$
$$\text{Total traffic offered (Erl)} = \text{Traffic per subscriber}$$
$$\times \text{ Number of subscribers in service area} \tag{11.2}$$

In circuit-switched voice applications, call arrivals are modelled as a Poisson stochastic process, with inter-arrival times following an exponential distribution. The system is assumed to have a finite number of channels and, as a consequence, to support a maximum number of simultaneous calls or connections. Because offered traffic may exceed system capacity, it is possible that the system fails to assign channels when users attempt to make a call. This type of failure is referred to as *blocking*. For a certain traffic load A, in Erlangs, it is possible to calculate the number of simultaneous channels needed (N), constrained by the maximal blocking probability that is acceptable in the system. For system dimensioning purposes, well-known mathematical relationships that relate those three variables [2] are used.

There are two ways to define call attempt failure, depending on how the system deals with calls that arrive when all circuits are busy. Most traditional fixed telephony systems and many cellular systems, including Advanced Mobile Phone Service (AMPS) and IS-136, immediately block incoming calls when all channels are busy. Users immediately receive an indication of failure; it is up to users to retry the call. This approach is usually referred to as *blocked calls cleared*. It is modelled by the Erlang B formula, which provides the probability that all servers are busy in an M/M/*c*/*c*/ queuing system.[1] An alternative approach, used in

---

[1] Here we use standard queuing theory notation. M/M/c/c refers to a queuing system where arrivals occur according to a Poisson random process (the first M, for Markovian), service times follow an exponential distribution (the second M, also for Markovian), there are *c* servers and a maximum of *c* users in the system (in other words, no queue).

GSM, for instance, consists of placing arriving calls into a queue, in the hope that channels will soon become available and the calls will be served. If the waiting time for a call that is placed in such a queue exceeds a certain pre-defined limit, the call attempt is considered failed and users are free to try again. This mechanism is known as *blocked calls delayed* and the figure of merit is the probability of the waiting time exceeding some pre-established threshold. This probability is modeled by the Erlang C formula, obtained from an M/M/*c* queueing system [3]. Both formulas, eqns (11.3) and (11.4), allow direct mathematical evaluation of the blocking probability as a function of the number of channels and offered traffic. However, they require numerical methods for situations when one desires to know the required number of channels, N, to support a given offered traffic, A, subject to a given blocking probability. The latter, in fact, is the typical dimensioning problem. For CDMA systems, the same formulas apply, with the exception that the definition of whether a server is available is based on link quality level, which varies with the system load.

Erlang B model: probability (no server available)

$$P_{blocking} = \frac{\dfrac{A^N}{N!}}{\sum_{k=0}^{N} \dfrac{A^k}{k!}} \tag{11.3}$$

Erlang C model: probability (waiting time > T)

$$P[W > T] = \frac{A^N}{A^N + N!\left(1 - \dfrac{A}{N}\right)\sum_{k=0}^{N-1}\dfrac{A^k}{k!}} e^{(N-A)T/\tau} \tag{11.4}$$

where

T = maximum allowed delay before attempt is considered as failed
W = waiting time
τ = average call holding time
A = total traffic supported by the system
N = number of channels

For system dimensioning purposes, it is usually assumed that user behaviour parameters (call arrival rate and holding time) have normalised standard deviation of small order.

11.2.2.2 Data Traffic Modelling

Broadband data networks employ packet switching, as opposed to the circuit-switched architecture of voice telephony networks. In packet-switched networks, data exchanges can occur in a connectionless (or datagram-oriented) manner, where every information packet contains enough information to enable its correct delivery, or in a connection oriented (virtual-circuit switched) manner, which include an initial negotiation of a virtual path from source to destination (connection setup) followed by data transfer.

11.2.2.2.1   Traffic Descriptors
For data communications, typical parameters used to characterise a traffic source are the peak and sustained (long-term average) data rates as well as some measure of traffic burstiness. These variables are explained next.

- *Average data rate*: Amount of data that the source generates over a certain time interval, usually expressed in bits per second or packets per second. The measured interval is often a large time scale, such as hours or days.

- *Peak data rate*: Maximum instantaneous data rate that the source generates during typical time intervals.

- *Burstiness*: Describes how infrequently a source sends traffic. A simple equation to define burstiness is given in Ref. [4] as

$$\text{Burstiness} = \frac{\text{Peak Data Rate}}{\text{Average Data Rate}} \tag{11.5}$$

If a source sends data at a constant rate, it is considered as not bursty (Burstiness = 1). It is considered very bursty if it allows long periods of 'silence' mixed with periods of transmission close to peak rate (Burstiness $\gg$ 1).

The definition above is just a simplistic measure of burstiness. More complex approaches suggest measuring burstiness based on its sustainability across multiple time scales [5].

The diversity of services offered in broadband connections leads to a large dispersion of burstiness, holding time and peak data rates.

11.2.2.2.2   QoS Requirements
As discussed in Section 11.2.2.1, QoS in voice networks is reasonably well characterised through the use of one single metric, the probability of call blocking. For data networks, the definition of QoS involves a larger set of metrics, some more important than others depending on the service class and type of application. These metrics may include [6,7]

- packet loss rate,

- throughput,

- availability (flow blocking probability),

- delay,

- delay variability (jitter).

In multimedia services, such as those supported by third-generation wireless systems, different types of applications have different characteristics and performance requirements. The larger set of possible applications is grouped into four main categories of service classes, according to the UMTS definition, as follows [8].

- *Conversational class*: includes voice, video conference, video games, etc. The most important performance requirement for this class is consistency in time relations, including both low delay and low jitter requirements. This includes preservation of the source data rate. Data integrity (loss rate) is not as critical.

- *Streaming class*: includes streaming audio and/or video, for instance, to support Video On Demand (VOD). It requires preservation of time relations to low jitter effect but is not as critical as the conversational class with regard to low delay requirements. Data integrity (loss rate) is not critical.

- *Interactive class*: includes web browsing, database retrieval and remote LAN access. An important parameter is the round-trip delay, which characterises the request-response time. Data integrity (low loss rate) is very important for these applications.

- *Background class*: Applications include non-real-time background download of E-mails, file transfers, etc. There are no strong restrictions with regards to time relations or delay, but data integrity is critical.

These categories can be further grouped in two main sets: real-time traffic, which includes classes where time and rate variation characteristics must be preserved (conversational and streaming); and elastic or non-real-time traffic, less sensitive to time variations but for which data integrity is more critical (interactive and background).

The issues involved in incorporating QoS mechanisms into mobile environments have been receiving considerable attention from the industry as well as the research community [4,9,10]. Renegotiation of performance guarantees during handoff, differentiated allocation of resources through power control and prioritised access to the medium are some of the mechanisms through which QoS may be incorporated into 3G systems.

11.2.2.2.3 Traffic Modelling Assumptions

Due to the layered nature of data networks, most commonly modelled by the Open Systems Interconnection (OSI) architecture, there are many levels at which one may describe traffic behaviour. Assuming that we are looking for a model capable of describing connectionless networks, it would be natural to choose the packet (network) layer as our focus for analysis. However, even though the definition of connection-less packet-switched networks permits packet independence, in practice, it is reasonable to consider that packets sent from the same instance of an application to the same end destination are indeed very correlated [11]. Based on this idea, it is useful to define, for the purpose of resource allocation, the concept of flow as a sequence of packets exchanged between a source/destination host pair as part of a given application.

In QoS-based systems, such as the model assumed for third-generation networks, both real-time and elastic classes can be supported with some sort of admission control mechanism, where flows are considered much like 'connections' at their setup phase, with the network deciding on whether to refuse or accept new flow requests after estimating the performance impact caused on already established flows.

The problem of modelling the flow arrival process and flow duration for broadband multimedia applications is a complex one. For simplicity, we divide these modelling approaches into two major groups. The first and more traditional group describes data traffic using Markovian models. The second group uses a more recent approach, based on the self-similar nature observed from multimedia data sources.

Markovian models, which have the Poisson-related model as their main sub-set, constitute the fundamental pillar of existing queuing theory. The Erlang theory, where calls are modelled as having Poisson arrivals, is an example of such an approach. Others in this group model packet arrivals, for instance, using Poisson-related models such as Poisson-batch or Markov-modulated Poisson processes. Further description of such models are available in Ref. [12].

The second group comprises a more recent set of models that rely on the observation that bursty traffic patterns generated by data sources and variable bit-rate applications tend to exhibit certain degrees of correlation between flow arrivals and show long-term dependence in time (self-similar traffic) [6]. The seminal paper on the study of self-similar traffic was published in 1993 and further expanded in 1994 [13]. It shattered the basic assumptions of queuing analysis using Poisson models. Based on massive traffic measurements on an Ethernet network, this work, and many others that came after it, shows that data traffic displays structural similarities across a wide range of time scales.

11.2.2.2.4   Self-Similarity

Self-similarity is a property associated with 'fractals', which are objects whose appearance is unchanged regardless of the scale at which they are viewed [14]. A self-similar stochastic process looks or behaves similarly when viewed at different degrees of magnification or different scales in one dimension (time or space).

A mathematical definition of such behaviour would show that these processes present approximately the same statistics as the magnified (or aggregated) versions of the same process. In other words, when compressing a discrete-time stationary process by a factor m, the mean, variance and correlation are preserved. This suggests that the burstiness would be preserved at different time scales. A weaker condition is asymptotical self-similarity, which states that auto-correlation functions keep the same form as the auto-correlation of the non-compressed process. An interesting consequence of this property is that as the compressing factor tends to infinity, the correlation does not tend to zero, as in stochastic models previously used for packet data. This characteristic is referred to as 'long-range dependence'.

Self-similarity and long-range dependence are not exactly the same concept and do not necessarily imply each other, but, in the context of traffic dimensioning, these terms are often used interchangeably, as well as 'scale-invariant' burstiness.

It is often observed that file sizes, connection durations or, more broadly, the duration of on/off periods exhibit a heavy-tailed distribution.

Although the observation of self-similar behaviour disproved the traditional assumption of exponential distribution for flow durations, the flow arrival process for many types of data applications may still be assumed as following Poisson processes for each individual source [15]. In this case, the candidate model to describe the system is an M/G/∞ queuing model that assumes Poisson arrivals with generic service time following a heavy-tailed distribution. However, this model assumes independence between the sources and this assumption is not realistic for all types of data applications. Particularly, World Wide Web (WWW) connection arrivals have been shown to present self-similar behaviour [16], which is explained by the correlation between multiple flows, including both the correlation among flows generated by the same source and the correlation among flows coming from different sources but governed by congestion-control mechanisms such as those implemented by TCP.

As mentioned before, the description of multimedia traffic as self-similar has brought much debate over traditional assumptions used for decades in traffic theory. Much research has been done on the subject as a result of those initial findings. Some of those topics are discussed next [16].

One of the research topics on self-similarity refers to measurement-based traffic modelling (source modelling), where different types of data sources are measured with the intention of quantifying their self-similar characteristics. The degree of self-similarity of a traffic pattern can be expressed by the Hurst parameter, which ranges from 0.5 (non-self-similar) to 1 (completely self-similar). Examples of sources that have already been investigated and demonstrated to exhibit self-similar behavior include [16] Local Area Network (LAN), Wide Area Network (WAN), Internet Protocol (IP), File Transfer Protocol (FTP), copper, fiber optic and WWW WAN traffic.

Another line of recent research focuses on physical modelling of traffic sources in an attempt to understand physical causes of self-similarity in network traffic. The idea is that explaining the causes of such effects may help selecting among models that fit equally well and give additional insight into the model description. Two main causality possibilities are considered [16]. In single source causality, the arrival pattern of a single data source, e.g. the variability found at multiple time scales on a Variable-Bit-Rate (VBR) video stream (Moving Picture Experts Group (MPEG)), could be explained by the variability in time duration between two successive scene changes. Structural causality is attributed to the heavy-tailed distribution of files or object sizes. If end hosts exchange files whose sizes are heavy tailed, then the resulting network traffic at multiplexing points in the network layer is self-similar [16]. An ON/OFF model [17] is often employed to model this behaviour, typically using heavy-tailed distributions for both active (ON) and idle (OFF) periods.

Researchers have also been focusing on the impact of self-similar sources on queuing behaviour. An important point to consider is that infinite buffer systems with non-self-similar traffic input (short-range dependent) generate an 'exponentially decreasing' queue length distribution. Self-similar traffic input (long-term dependent), however, generates slower-than-exponential (or sub-exponentially) decreasing queue length distribution, sometimes with polynomial decreasing behaviour. This implies that increasing buffering yields to little improvement in the packet-loss rate, at the cost of the queuing delay penalty it imposes on the system. Therefore, there are proposals advocating small buffer capacity/large bandwidth resource provisioning strategies [16].

Traffic control research explores techniques to improve resource allocation efficiency, based on some knowledge of the traffic nature. For instance, if it is expected, based on the assumption of 'heavy tailedness', that long connections will last even longer, then techniques to shape traffic or dedicate more efficient resources to those flows could be implemented.

While research currently focuses more on self-similar traffic modelling, the recognition of limitations in the previously used Poisson-based approach does not necessarily void its usefulness. For instance, Poisson assumptions may provide useful bounds on performance metrics of interest for the system. The particular property of these traditional methods, which makes them attractive, is the simple mathematical formulation for the quality of service in terms of offered traffic and available network capacity.

For traffic simulation purposes, typical models for reproducing self-similar traffic include superposition of many ON/OFF sources with sojourn times following Pareto or other heavy-tailed distributions.

The Pareto probability distribution function is given by

$$F(x) = 1 - \left(\frac{k}{x}\right)^{\alpha}, \, x > k, \, \alpha > 0 \qquad (11.6)$$

where

k is the minimum value that the random variable X can take
α is a parameter for the Pareto function

In the process of generating data traffic, α typically assumes values between 1.2 and 1.5 (the smaller the value, the more self-similar the traffic). The Pareto random variable has infinite variance for $\alpha \leq 2$.

Comparing Poisson and self-similar traffic models. When multiple self-similar sources are aggregated, there is still burstiness in the resulting traffic (clustering). This results in higher buffering requirements than Poisson traffic, where traffic smoothes when aggregated.

In Poisson processes, samples of the process that are far from each other, in time, have little correlation between them (auto-covariance), whereas in self-similar processes auto-covariance decays much more slowly.

Self-similar processes are expected to demand more queuing (buffering) capacity as well as more servers to handle the same amount of traffic.

11.2.2.2.5 Multiple Time Scales

One way to describe broadband traffic is to model its different time scales with appropriate stochastic processes used for each level. Table 11.2 shows a simplified summary of the most relevant time scales for the resource allocation problem adapted from Refs. [18–21].

The following paragraphs provide a brief description of each of these scales as well as possible models for each of them.

Packet/burst scale. This scale models arrivals of packets or groups of packets from multiple channels that will be multiplexed using a specific queue model. One phenomenon

Table 11.2 Traffic modelling: scales in time

	Stochastic phenomenon	Stochastic models	Traffic impairments
Packet/burst scale	Statistical multiplexing of packets or groups of packets from different sources	Various queue and traffic models	Packet loss, delay, jitter
Flow/session scale	Admission of connections	Various population models	Blocking
Human activity scale	Correlated human activity giving daily, weekly and yearly cycles	Statistical inference of cycles and correlations	Blocking
Planning scale	Forecasting of traffic demand after planning lead time	Numerous forecasting techniques	Blocking

of interest is the packet grouping that occurs due to packet/framing performed at upper layers or due to packet rate variations existing in each flow because of application demand variation, such as real-time applications on video transmissions.

The main parameter to be modeled at this time scale is the queue buffer size, which should be chosen based on the desired transmission quality, typically expressed by the probability of packet loss and/or end-to-end latency.

Flow/session scale. The flow scale layer is the initial scale for circuit-switched systems, because no packet or burst scale exists. In old narrowband telephony systems, a flow is the equivalent of a call, and failure is described as a refused call attempt. In an ATM data network, the flow scale is described as the level at which a connection setup (virtual circuit) is established. At the more general level, a flow can be defined as [22–24] the unidirectional succession of packets relating to one instance of an application. Packets belonging to a flow have the same identifiers (source and destination addresses and port numbers) and occur with a maximum separation of a few seconds.

In the case of circuit-switched networks, such as traditional voice telecommunications networks, the basic model to look at for predicting the number of simultaneous users in the system is the $M/G/\infty$ queue. In this scenario, the system is modelled as following a Poisson process for arrivals (i.e. flow setup attempts), a generic service time distribution and an unlimited number of servers. For practical systems, the number of servers is physically limited and the system quality is measured as the probability of the number of servers in use being larger than the number supported by the system.

In more general packet-switched data networks, the typical variable of interest in dimensioning a system is the probability of the total effective bandwidth required being greater than the channel capacity, given the statistical description of individual flow behaviour. An interesting observation is that the circuit-switched case is usually a simplification of the general case, assuming that all flows are at constant bit rate and the total link capacity is expressed as the total number of supported flows at that fixed rate.

Human activity cycle scale. At the human activity cycle scale, we should look at the hourly variations of parameters used as input at the flow level. Parameters such as the number of flow setup attempts, holding times and mean and peak data rates used in the dimensioning dependencies described at the flow level have local averages that vary hourly, weekly, seasonally, etc. For system dimensioning, one needs to consider the strong correlations in human traffic patterns, to predict when traffic demand will reach its peaks and planning the capacity for these situations. A typical example of this time scale is the concept of the 'busiest hour', used in typical voice telephony systems.

A wireless scenario must consider human cycle variations, which include not only time but also strong spatial fluctuations. In this case, dynamic dimensioning techniques, such as dynamic channel allocation, where the spectrum distribution among cells and sectors can be changed to accommodate the user demand concentration at different times of the day, can be used.

This chapter discusses a suggestion of a practical approach in considering the human activity cycle scale for broadband networks.

Planning scale. At the planning scale, the main aspect considered is system growth, which depends on variables such as progression of market share, market penetration and long-term changes in user behaviour.

The basic trade-offs in designing a system to accommodate long-term growth are the risk of over-investing in a network and having it under-utilised if too much margin is allowed, versus the risk of sub-dimensioning the network and losing revenue and market share. The appropriate demand characterisation for optimal system dimensioning needs to consider variables such as latency between the times a requirement for growth is perceived and the expansion can be implemented, and costs involved in system expansion as compared to planning ahead for a large system. In the particular case of wireless networks, system deployment allows comparatively fast response time, due to the cellular architecture that allows operators to 'drop' sites in busy areas without really building expensive infrastructure to get to these places by land. However, resources are usually wasted when cell splitting techniques have to be used to address traffic problems compared to the design that would have been done if the same traffic had been considered at the initial deployment. This happens because cell splitting usually requires de-activating and moving sites or keeping under-utilised cells in some areas. Some additional practical aspects of the planning scale are further discussed later in this chapter.

11.2.2.2.6   Traffic and Link Asymmetry

Implicit in user-profile characterisation is the fact that data demand is usually asymmetric, i.e. traffic is typically denser in one direction (typically, base station to mobile) than in the other (mobile to base station). Many multimedia services offered in broadband wireless systems are heavily un-balanced towards downstream. Examples include video and audio streaming, web browsing and simulcast. For these applications, there is usually little interaction and transmission from the user end. Other applications like real-time video conferencing and voice and data transfer present similar proportions between downstream and upstream traffic volumes, coming closer to symmetry. There are still a few applications, mostly related to telemetry, that present heavier upstream traffic.

The process of traffic modelling described in the previous sections could be initially thought of as two separate problems: the separate demand characterisation of downstream and upstream user behaviour. However, these two problems are strongly correlated and the probability distribution of a subscriber sending data is, in many ways, related to that of the user receiving data, significantly depending on the type of application. The correct characterisation of user asymmetry and its evolution as more and more broadband applications are available is a challenging traffic-engineering problem and has strong consequences on the way a system is dimensioned.

While traffic on multimedia applications is typically heavier on the downlink, the wireless mobile coverage is typically limited on the uplink due to antenna size and battery restrictions at the mobile unit. This asymmetry towards a weaker uplink may partially counter-balance traffic asymmetry.

In wireless systems there are many techniques to deal with asymmetry. They may be grouped based on the system's duplexing mode, i.e. whether it uses frequency-division duplexing (FDD) or time-duplexing division (TDD).

In FDD systems, separate slices of the spectrum are allocated for the uplink and downlink. Support for asymmetry may include differentiated spectrum allocations for the downstream and upstream and differentiated coding/modulation techniques, which may be achieved in a static or adaptive mode (link adaptation). Differentiated spectrum allocation is typically used in fixed-wireless systems (MMDS, LMDS), where technologies do not follow strict standards and operators have flexibility to allocate the available spectrum. For mobile systems (cellular, PCS and 3G) spectrum allocation is typically standardised and symmetric (paired channels) and, as a consequence, coding/modulation techniques are the most common techniques used to compensate for traffic asymmetry.

When using different coding and/or modulation schemes in each direction, stronger schemes are used on the upstream to compensate for lower signal strength requirements at the mobile. On the downlink, more spectral-efficient schemes may be used to provide for heavier traffic demand, at the expense of higher power requirements that may be satisfied by the base station's stronger power capabilities. Although link adaptation works nicely to compensate un-balances, depending on the intensity of traffic asymmetry, one side of the spectrum may still end up being used less efficiently than the other.

In TDD systems, the same slice of spectrum is allocated for both downlink and uplink, but different time slots are used in each direction. To support asymmetry, the system can use un-even distribution of time-slots between downstream and upstream. This option provides more flexibility to the system in terms of accommodating future changes in the user asymmetry proportions, as asymmetry may be adapted on a dynamic basis as traffic behaviour changes, with the objective of balancing spectrum efficiency in both directions.

As mentioned before, although traffic is typically heavier on the downlink (i.e. larger throughput requirements), signal strength is typically limited on the uplink due to battery restrictions of the mobile unit. At the same time, for most 3G systems (with the exception of TDD), spectrum allocation is still symmetric (paired channels). Thus, the final balance of spectrum efficiency is typically asymmetric, and spectrum in one direction may be used less efficiently than in the other.

11.2.3 Geographical Traffic Characterisation

Accurate traffic analysis requires a detailed characterisation of the different user profiles. This includes modelling temporal parameters that describe those profiles, as well as the spatial distribution of such users. This section presents and explains the most relevant parameters and procedures used to define geographic traffic distribution. These parameters should lead to an appropriate description of how users of different profiles are spread over the market area.

To achieve such spatial description, demographics information is evaluated with regard to their implications to generated traffic. Network designers should do this analysis carefully in collaboration with the market forecast group, using tools that allow the individualisation of different profiles, with parameters describing their differentiated behaviour and geographical distribution. The traffic simulation should consider a multi-layered traffic description, as explained in more detail in this section.

The dimensioning process must also look into forecasts of how the market for mobile services will evolve over near and medium terms.

11.2.3.1 Creating User Distributions

Demand characterisation deals with the creation and description of geographically referenced databases that describe densities of users belonging to each user profile.

The following are among the main requirements of demand characterisation:

- to reflect different geographical distributions that are possible for different user profiles;
- to reflect distinguished downlink/uplink asymmetries per user profile;

- to correctly model the fact that different geographical distributions achieve their busiest hour at different times of the day.

Based on these requirements, the methodology proposed herein follows the framework suggested in Ref. [1], which consists of the following main steps:

- input multiple layers of demographic data;

- explore possibilities on how to map those data into a set of multiple user layers;

- use additional existing data from legacy networks (2G system) when available for the same market.

The following sections describe and illustrate each of these steps, based on Ref. [1].

11.2.3.2   Residential Database

The residential database is one of the most important layers to be considered in forecasting traffic demand. Typically given by census data in the United States, this Geographical Information System (GIS) database is described by region files, which include a collection of polygons corresponding to census tracts or census lots, smaller sub-division of tracts. Each of these polygons has relevant information about the geographical area within its limits. Typical attributes are

- residential population;

- number of households;

- ethnicity, gender and age distribution of residents.

Figure 11.4 shows an example of this type of database.

This type of database constitutes one of the layers of information to be used in the demand forecast. User profiles and their correlation with some proportionality factors to the attributes available for the geographical areas must be forecast. For instance, typical mobile forecasts assume that the market penetration is a factor applied directly over the population in each polygon, whereas fixed wireless applications may apply the proportionality over the household attribute.

Besides census data, other commercially available databases classify areas with other attributes that may be of interest to the operator such as average income, average number of years in school and level of familiarity with 'high-tech' consumer items. This allows another level of refinement where sub-layers can be used. For instance, differentiated values of market penetration and average bandwidth could be assigned to different regions depending on the attribute ranges defined for household income in those polygons.

11.2.3.3   Business Database

Business databases describe the location of all businesses available, or in a certain market, and may include many parameters of interest in estimating traffic to be offered for a 3G

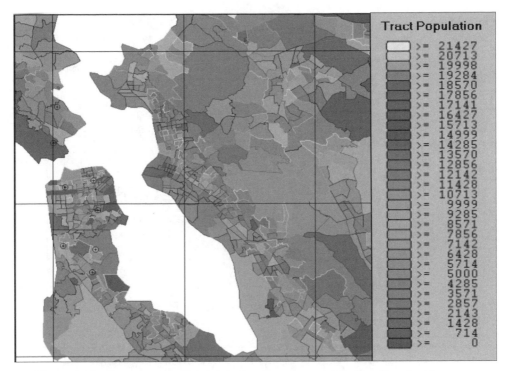

Figure 11.4 Example of residential database (census tracts) for San Francisco bay area.

system. This type of database may come from research institutions, chambers of commerce, GIS retailers, government databases, etc. As a consequence of this diversity of sources, database formats are not presented in any standard form. Nevertheless, they are typically given as geo-referenced databases (GIS), which include a collection of polygons or a collection of points, geographically distributed, with some or all of the attributes listed below:

- number of employees at the location;
- average business revenue;
- average monthly expenditures on telephone/data services;
- available office area in square feet;
- street address;
- type of business.

Figure 11.5 shows an example of a business database for a dense urban area.

A solution to forecast traffic distribution based on this type of database is to evaluate market penetration data per type of business and estimate the range of required data rate depending on database attributes such as the number of employees or monthly telephone bill expenditure. For mobile broadband forecasts, one could look at the market penetration

Block Code	Street Address	Area (1000sq.ft)	Height (ft)	Owner	Business Group
260021	46 Wall Street	493	312	46 Wall St. LLC	D9
270001	53 Wall Street	206	130	55 Wall Company NC	O1
270009	59 Wall Street	390	399	63 Wall Inc.	O3
270017	67 Wall Street	303	292	Gesellschaft Fuer MM	O4
310011	86 Wall Street	650	370	Eight-Five Wall Etc.	O4
330011	96 Wall Street	493	312	Chemical Bank	O3
330022	99 Wall Street	91	335	PA Building Co	O3
350010	107 Wall Street	990	296	Citicorp	O4

Figure 11.5   Example of business database in Manhattan area.

applied to some sort of combination of the number of businesses and number of employees at a location, and estimate ranges of required bandwidth depending on business classification (e.g. a high-tech office is more likely to have 'high-tech employees' than a food store).

11.2.3.4   Road Traffic

This type of database contains locations of roads, streets and highways and provides attributes that describe the density of vehicles on these roads. Other attributes of relevance in these databases may include road width and number of vehicles that transit in sections of the road per day, or per hour, for different hours of the day.

When estimating data or voice traffic volume based on this type of database, one must account for the typical busiest hours. It may be difficult to map this relationship between 'cars per day' and 'subscriber density at busiest hour', and some assumptions may be necessary in the process. Nevertheless, this type of information allows some level of forecast for areas that are usually neglected in residential or business geographical databases.

Different methods can be used to obtain data traffic forecast from vehicle traffic statistics for each road or road section. In some situations, database attributes may be given in terms of the 'rush hour' or as a distribution of cars per hour for different times of the day. This type of information could be directly used to estimate vehicular traffic distribution during different times of the day on each road or road section. For smaller roads and streets, some assumptions need to be made to convert car traffic information on a 'per day' basis into 'busiest hour' data, as well as the estimation of the average car speed for each road section.

The procedure described next highlights the main steps that can be used in this task.

1. Obtain the length of the road section.

2. Estimate what percentage of the overall number of cars per day occurs during the busiest hour (normalisation reference point). Apply that factor to the total number of cars in a day, to estimate 'total cars in the busiest hour' for the road section in question.

3. Combine the number of cars, road length and mean car speed to obtain a 'snapshot' of how many cars are present in that section of the road during the busiest hour.

4. Apply market penetration factor to estimate how many of these cars are potential users to the system. This gives the traffic offered at that section at the busiest hour.

5. Spread traffic offered evenly over that section.

Using one of the multiple forecast method possibilities, one traffic layer distribution must be created for each applicable user profile coming from the road database.

Figure 11.6 shows a sample of a road database.

11.2.3.5 Morphology Weighting

As in the previous sections, typical GIS demographic databases are described as sets of polygons, such as those presented in residential, business and roads databases. When mapping those attributes into geographically distributed subscribers for each profile, there are multiple approaches on how to handle these distributions within the limits of each polygon.

The homogenous approach is the simplest one, and consists of uniformly distributing a polygon attribute within the geographical area limited by the polygon. Figure 11.7 illustrates an example of such approach.

In the example in Figure 11.7, a census tract with population attribute equal to 2530 habitants is used to distribute subscribers within the same polygon limits. A 10% penetration factor is estimated for the area, resulting in a total of 253 subscribers for that polygon area. In the homogeneous spreading approach, these 253 subscribers are evenly spread within the census tract polygon area.

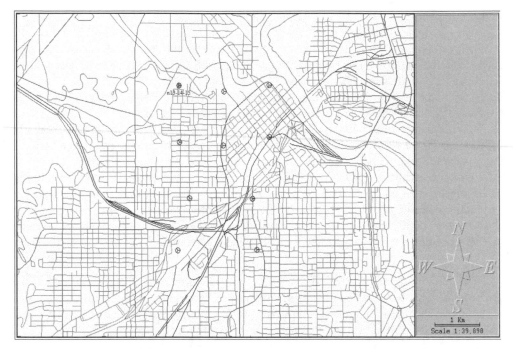

Figure 11.6   Example of road database in Shreveport, LA.

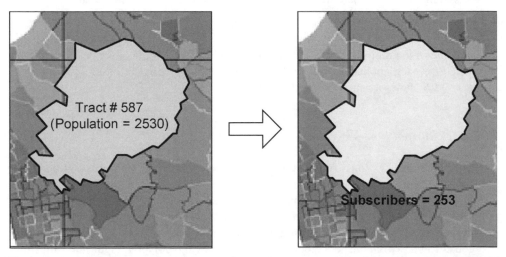

Figure 11.7   Example of homogenous spreading within demographic polygon given by a residential census tract (market penetration applied = 10%).

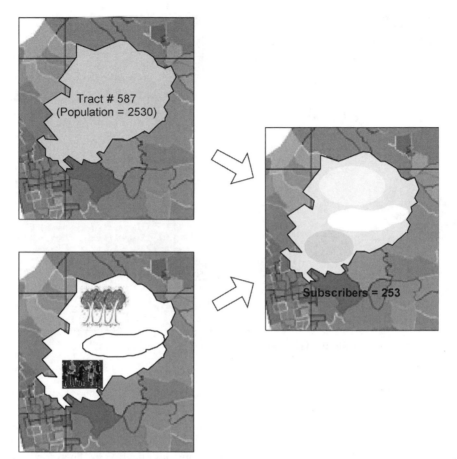

Figure 11.8 Example of morphology-weighted spreading within demographic polygon given by a residential census tract. Demographic and morphology database information are combined to provide more detail (granularity) in the subscriber distribution (market penetration applied = 10%).

A more refined approach is possible if information regarding land usage (morphology) is available. Morphology databases classify terrain as to what type of land cover exists at each location. Figure 11.8 shows a simplified example. Suppose that the same census tract discussed in the previous example covers an area where there are distinct land usage classifications. While most of the area is classified as medium density sub-urban area, part of the region is covered by forest, by a lake and a small area to the south presents a higher density urban concentration. In that case, it makes sense to distribute the same 253 subscribers in a morphology dependent manner, where we assume, for instance, that no subscribers will be talking from the lake, and higher densities will occur in the urban areas compared to the sub-urban area or to the forest in the north (lowest concentration).

Examples of land cover types may range from low-detail classifications such as urban, sub-urban and rural areas, to finer morphology databases that may show each individual building and/or construction type in a given area. Figure 11.9 shows an example of a medium-resolution morphology database. In this example, the database granularity goes to

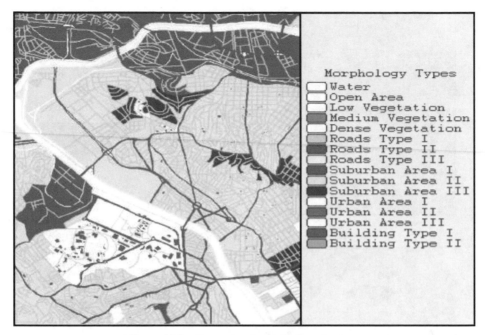

Figure 11.9   Example of morphology database.

the level of blocks, but not individual buildings. This type of classification is usually referred to as *canopy*.

When geographically distributing traffic within a polygon from a demographic database, the different spreading weights (ω_m) should account for the relative probability that a user within that polygon will be located in each morphology classification m. Therefore, the user density per pixel σ_m inside the polygon area is the same for all pixels of morphology m, and differs from pixels of different morphologies, with proportionality following that of morphology weights

$$\sigma_m = \omega_m \frac{Subscribers_{Polygon}}{\sum_m (\omega_m Npixels_m)} \qquad (11.7)$$

In the equation

- σ_m is the density of users per pixel of morphology type m in a certain polygon,

- ω_m is the weight assigned for morphology type m,

- $Subscribers_{Polygon}$ is the total number of subscribers to be spread within the polygon area,

- $Npixels_m$ is the number of pixels within the polygon that have morphology type m.

For instance, suppose a census polygon is distinguished in two different morphology types: urban and suburban areas, and that 200 users are to be distributed within that polygon (Table 11.3). Table 11.1 shows additional information for this example.

Table 11.3 Numerical example for traffic spreading weighted by morphology

N users in polygon	200	*Users*
Polygon area	10	Km^2
Pixel size (resolution)	100 × 100	m
Pixel area	0.01	Km^2
N pixels	1000	

Information per morphology

Type	1-Urban	2-Suburban
Pixels per type m	300	700
Morphology weight	5	2

In the previous example, $\sigma_1 = 0.34$ and $\sigma_2 = 0.14$, giving a total of 103.4 users distributed among the 300 pixels of morphology type 1 (urban area) and 96.6 users distributed among the 700 pixels of morphology type 2 (suburban area).

Table 11.4 illustrates multiple examples of morphology weight distributions that could be applied to the morphology database in Figure 11.9.

11.2.3.6 Event-Related Demographics

In addition to the residential, business and road traffic databases described in the previous sections, event-related information can also be considered. This may include special treatment for locations such as stadiums, theatres and show arenas with regards to routine or temporary gathering of people due to festivals, athletic games and other events.

Table 11.4 Example of morphology weights for traffic distribution using the morphology classification provided in Figure 11.9

Morphology types	Type 1 city Urban market	Type 2 city Sub-urban market	Type 1 city Rural market
Water	0	0	0
Open area	5	5	5
Low vegetation	2	2	2
Dense vegetation	1	1	1
Roads - type I	10	10	10
Roads - type II	12	12	12
Roads - type III	15	15	15
Suburban area I	40	40	40
Suburban area II	60	60	60
Suburban area III	80	80	80
Urban area I	100	100	100
Urban area II	120	120	120
Urban area III	150	150	150
Building block–type I	200	200	200
Building block–type II	250	250	250

11.2.3.7   Market Penetration Forecast

In estimating the number of subscribers expected from a set of demographic databases, the network designer must consider the expected penetration of wireless services in the region of interest, as well as its growth over the near and medium terms. The task of mapping those relationships between demographic layer attributes and expected traffic demand for the applicable user profiles will rely on metrics such as the following.

- *Market penetration*: determines what percentage of the population will be potential consumers of the type of product being offered. Typically, market penetration is described on a per-year basis, but it may be specified at smaller intervals near launching.

- *Market share factor*: considers the multiple operators expected to offer products in this category, and how market is expected to be split among the competitors. As with market penetration, market share is typically described on a per-year basis, but it may be specified at smaller intervals near launching.

- *Demographic growth*: projected growth of the population in the region of interest.

Clearly, the network operator has a direct influence on some of these factors, as the growth of its customer base (which reflects on both market penetration and market share) will depend on the provided service quality as well as the aggressiveness of marketing and pricing strategies.

On the other hand, service quality and return on investment depend on a good system dimensioning that should accurately consider the estimated demand.

Therefore, at the same time that the operator can 'forecast' market penetration and market share, it can also influence these factors and make this forecast become a reality.

11.2.3.8   Combining Multiple Traffic Layers: Time Dependency

Once all user layers described in the previous sections are characterised, multiple sets must be combined, one coming from each demographic layer. This consists of grouping together layers that come from different demographic sources but correspond to the same user profile. From a modelling point of view, this combination does not consist of a simple sum. First, each user profile must be considered separately and be appropriately weighted. On this weighting, one must consider time dependence dynamics, as weights will likely vary with time of the day and day of the week.

Using the factors described in the previous section (market penetration, market share, etc.), suppose that, for each layer, we have obtained the traffic distribution during the busiest hour. However, the busiest hour for each layer does not necessarily correspond to that of the other layers, as in Figure 11.10.

It is not reasonable, therefore, to combine all layers at their peak hours, since it would lead to over-dimensioning the network. Time dependency weighting must be applied to correctly assess demand at any given time.

For instance, considering the time weights suggested in Figure 11.10, suppose the system behaviour is being evaluated at 14:00 hours. The business layer should be considered with 100% weight, but the residential layer would be scaled at around 50% of its peak value.

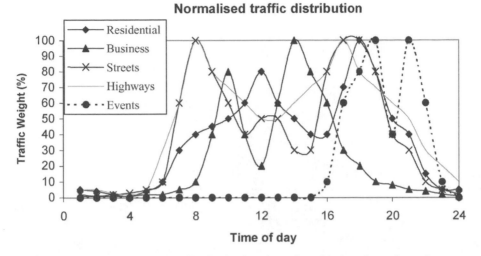

Figure 11.10 Example of traffic distribution dynamics with time for a given class.

Looking at the combined traffic grid at this time, the highest bandwidth demand is most likely to be found on cells located in business areas, and fewer channels needed on residential cells. However, looking at the 19:00 hours time frame, the heaviest demand for bandwidth is more likely to come from residential areas.

Because we are looking for ways to dimension channels on a per sector basis, not only on a system basis, we can not ignore the spatial distribution of the busiest hour over different layers. The required bandwidth per sector is the maximum demand by looking at all time periods. To allow all possibilities one would need to create combined class grids for all hours of the day. For simplicity, however, usually only two or three hours are selected. A typical system analysis looks into hours where most relevant layers have their peaks, such as residential and business in the previous example.

Figure 11.11 illustrates a traffic grid created as a combination of multiple input layers (residential, business, etc.) for a certain application type and time of day. Small boxes of fixed size (grid resolution) represent this grid, each box with a particular set of traffic attribute values. Attributes can be expressed in terms of user density, throughput demand density, etc.

Figure 11.11 displays the effective traffic demand load offered for a given user profile. One of these layers should be created corresponding to each user profile defined for a given system. One set of such traffic grids constitutes a demand characterisation scenario, as in Figure 11.12.

Figure 11.13 shows an example of a demand characterisation scenario.

The results of this process are, therefore, geographical distributions of subscribers, each distribution corresponding to a different user profile. This information, as discussed next, is used as input in traffic simulation algorithms, which consider, together with each geographical layer, traffic models associated to the corresponding user profile. This allows simulations to reproduce the system response to the traffic offer, based on the actual number of active users at the system at any given time.

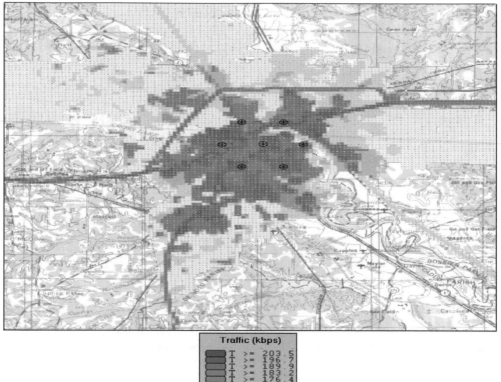

Figure 11.11 Example of one-layer traffic grid for a specific service class and time of the day.

11.2.3.9 *Additional Input Variables for Existing Systems*

Besides the methodology described in the previous section for demand characterisation based on demographic databases, other valuable information may be used when the network being dimensioned evolves from an existing wireless system.

In cases where third-generation systems grow from existing PCS and cellular networks, there is previous (and valuable) information known about traffic in the system. This situation is expected to happen in most cases of third-generation systems where legacy network

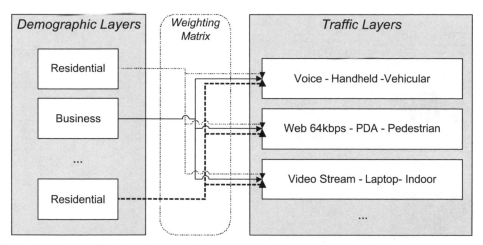

Figure 11.12 Mapping from demographic layers to traffic layers for a given weighting scenario. The resulting set of traffic layers correspond to one demand characterisation scenario.

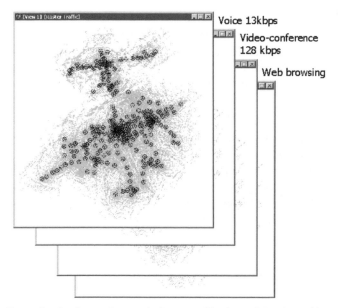

Figure 11.13 Example of a demand scenario for a specific time of the day, with multiple traffic grids (one grid for each service class).

operators upgrade their networks to satisfy user demand for data communications. The most important information is the following.

- *Switch traffic data (Erlangs/cell)*: Switch traffic data (Erlangs/cell) statistics describe existing voice traffic distribution per cell. This information is expected to show strong correlation with the data traffic distribution for the same system when services are

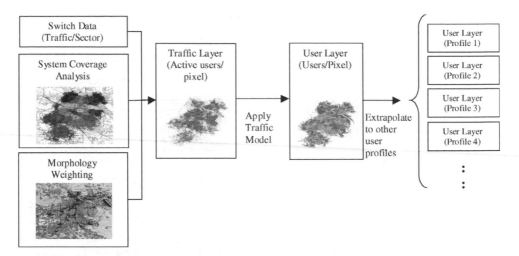

Figure 11.14   Using existing switch data to estimate traffic distribution.

available. As an alternative to the traffic grid creation method described in the previous section, one option would be to create traffic grids based on the information of traffic per sector, spreading the traffic associated to each sector under its best-server coverage area. Factors can be applied to map Erlangs into number of users and use extrapolations to factor the voice-user distributions into data-user distributions for different profiles. Figure 11.14 illustrates this process.

- *Switch handoff data (inter-cell dynamics)*: Switch handoff statistics (inter-cell dynamics) are a valuable information to allow appropriate dimensioning of the demand overhead caused by handoff, where traffic may be served by sectors that are not the best server at the location, or may be taking resources from more than one sector at a certain time (soft handoff). Handoff data for existing networks also allows appropriate dimensioning of the neighbourhood relationships and their relative importance.

11.3   TRAFFIC SIMULATION

The previous section described methods to estimate the traffic offered to the system. This included the discussion of traffic models applicable for different service categories (voice, real-time data, non-real-time data, etc.), followed by the description of methods to characterise the demand expected in a given market. This characterisation process uses demographics and existing network data as inputs, and results in geographical user distributions (user layers), with possibly different distributions for each user profile.

The next step in the process of system dimensioning is the system operation simulation, keeping track of its response to the traffic offer, with collection of the appropriate statistics.

The simulation utilises performance metrics that should be used to evaluate whether enough resources have been allocated to each site, and to decide how the system should be changed to achieve the desired performance levels.

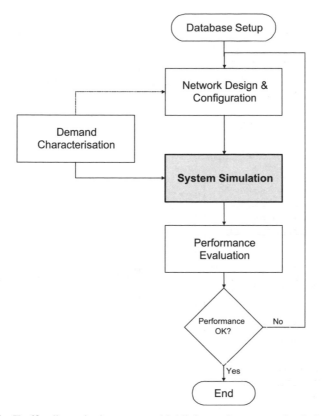

Figure 11.15 Traffic dimensioning process–highlight on the system simulation procedure.

Figure 11.15 highlights the simulation block and how it interacts with other tasks in the process of system dimensioning.

11.3.1   Traffic Simulation Input

After demand characterisation is performed, resulting in the creation of multiple traffic layers, it is used as input in the traffic simulation analysis, together with the full description of the system design, considering cell locations and coverage areas.

The simulation process also needs to consider as inputs the type of technology selected for the system, access methods, modulation schemes, etc., which also need to be defined as input parameters for the traffic simulation.

Figure 11.16 illustrates the main input blocks for traffic simulation.

11.3.1.1   Multi-Layer Geographical Traffic Distribution

As mentioned in Section 11.2, the simulation should be performed for different times of the day. Each simulation scenario consists of multiple geographic user distributions, one for

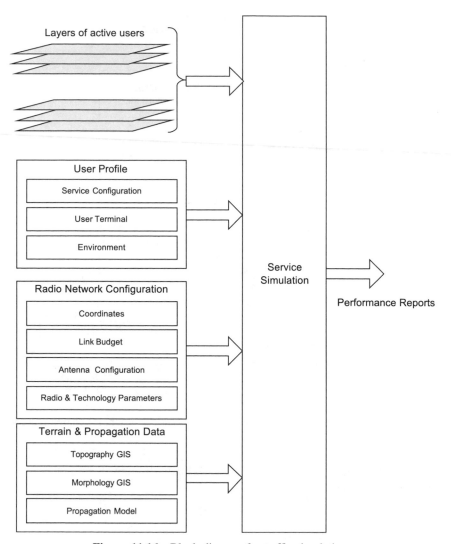

Figure 11.16   Block diagram for traffic simulation.

each user profile. Figure 11.17 illustrates the dynamics of creating multiple scenarios of demand characterisation to simulate different times of the day. Each simulation scenario represented at the right side of the figure is the result of a different set of weights used to map demographic data into traffic for different times of the day or different days of the week.

Although each traffic scenario is estimated as a single set of layers, each layer results in a different number of active subscribers on the downlink and uplink, because all parameters used in defining the traffic distribution are dependent on the link direction, and data traffic is modeled as asymmetric for most applications.

From each simulation scenario, resource-limited sectors are identified, and network capacity improvements techniques may be applied. The process of system dimensioning,

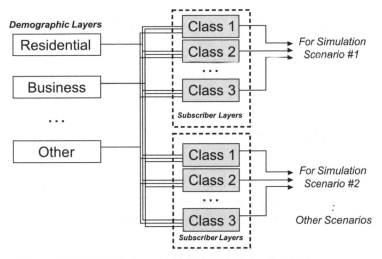

Figure 11.17 Block diagram of the demand characterisation process.

therefore, includes the simulation of multiple scenarios, where the resource allocation solution adopted needs to satisfy QoS requirements in all scenarios.

11.3.1.2 User Profile and Traffic Models

As explained in Section 11.2, each traffic layer is associated with the description of a user profile, which is defined by a set of service configuration, user terminal and environment parameters. Figure 11.18 summarises these elements.

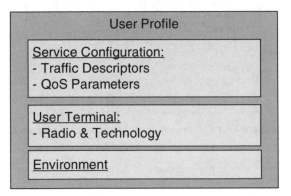

Figure 11.18 User-profile configuration elements.

The appropriate description of the traffic model associated to each user profile, together with QoS parameters that define its quality level targets and possible guarantees required for each class, allows the correct quantification of the number of users active in the system at each moment, as well as the corresponding bandwidth requirements. This quantification may be achieved in different ways, depending on whether the simulation is performed in a static (snapshots) or dynamic way, as discussed later in this chapter.

The description of user terminal capabilities allows the correlation of mobiles associated to each class with the sectors installed in the system, determining which sectors may serve each mobile, and which capabilities are available.

The environment information contains data such as speed, location type, fading, etc., which allows the estimation of impairments in signal coverage, and as a consequence, the impact of these impairments in the service provided for each user and for the system load.

11.3.1.3 Radio Network Configuration

The radio network configuration describes the radio access network, including its topology, base station configurations and capabilities.

For the simulation to correctly capture the way the network responds to the traffic offered, it has to consider its topology, antenna configuration, link budget and technology specific parameters.

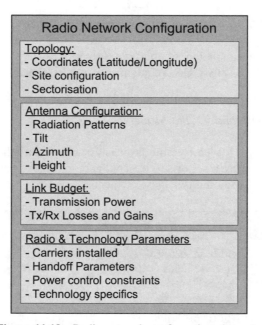

Figure 11.19　Radio network configuration elements.

Figure 11.19 illustrates these main groups of input information.

The topology information should include cell coordinates (latitude and longitude), their configurations, such as cell type (tower, monopole, building, indoor, etc.), and sectorisation data (e.g. number of sectors).

Antenna configuration parameters describe, for each installed sector, information such as antenna model, radiation patterns, tilt, azimuth, diversity configurations, installed height above ground, etc.

Link-budget parameters include information such as gains and losses at transmission and reception, noise figure, diversity gains, etc. They also include transmission values such as power configuration for control channels, transmission power limitations, etc.

In the radio and technology specific group, we include parameters that are specific to the technology being used for the particular system under analysis, and its implementation possibilities. It includes information such as sector capacity (installed carriers), frequency plan, PN offset plan and radio resource management parameters such as handoff thresholds, power control targets and window limits, etc.

A more detailed description of some technology specific parameters that should be included in the radio network description for IS2000 networks is provided next.

11.3.1.3.1 Technology Parameters and Equipment Description

In a wireless network supporting IS2000 technology, base-station sectors and user terminals may support multiple radios. These radios fully or partially support the technology characteristics and may include different implementation options.

Some technology parameters that may be configured on a per-radio basis on an IS2000 system are presented next.

Example Technology Parameters

- System's evolutionary stage (e.g. IS-95, IS-95B, 1×RTT, 3×RTT, 1×EV-DO, 1×EV-DV).

- Channel bandwidth (single carrier, multiple carrier).

- Spreading rates (chip rate)
  1. spreading rate 1 = 1 228 800 chips per second (over 1.25 MHz bandwidth).
  2. spreading rate 3 = 3 686 400 chips per second (over 3.75 MHz bandwidth).

- Supported radio configurations, in the physical layer, determine coding and data rates supported

  Downlink

  1. RC1–RC2 on IS-95B
  2. RC1–RC5 on 1×RTT.
  3. RC6–RC9 on 3×RTT.

  Uplink

  1. RC1–RC2 on IS-95B;
  2. RC1–RC4 on 1×RTT;
  3. RC5–RC6 on 3×RTT.

- Spreading factors supported

  1. Transmission possibilities for each radio configuration, depending on Walsh code combinations allowed.

  Forward

  (a) RC2: 64 codes available (rate = 1/4) if using 9.6 kbps. Higher rates use groups of codes (Walsh tree).
  (b) RC4: 128 codes available (rate = 1/2) if using 9.6 kbps. Higher rates use groups of codes (Walsh tree).

Reverse

(a) Multiple channels per mobile at the same time.

(b) For Supplemental channel 1: R-SCH 1, spreading factor 2, 4, 8 or 16 (e.g. rates: 9.6, 19.2, 38.4 or 76.8 kbps) at maximum rate, or 1/2 rate.

(c) For Supplemental channel 2: R-SCH 2, where maximum is half the rate used by R-SCH 1.

2. Quasi-orthogonal functions

- Overhead channels implemented

Forward

1. Minimum: F-Pilot, F-Sync, Paging.

2. Optionals: F-QPCH (Quick Paging Channels), F-BCH (Broadcast Channels), F-CPCCH. (Common Power Control Channels), F-CACH (Common Assignment Channels), F-CCCH (Common Control Channels).

Reverse

1. Minimum: R-Pilot, R-ACH (Access Channel).

2. Optionals: R-EACH (Enhanced Access), R-CCCH (Common Control Channel)

(a) User Channels Implemented

Forward: (i) Minimum: F-FCH (Fundamental Channel), F-SCH (Supplemental Channel); (ii) Optional: F-DCCH (Dedicated Control Channel).

Reverse: (i) Minimum: R-FCH (Fundamental Channel), F-SCH (Supplemental Channel); (ii) Optional: F-DCCH (Dedicated Control Channel).

(b) Technology enhancements supported:

(i) Forward link Orthogonal Transmit Diversity (OTD); (ii) Quick Paging Channel; (iii) Auxiliary pilots for beam forming and smart antennas.

(c) Data rate capabilities.

(d) QoS management capabilities.

(e) Performance curves (BER/BLER vs CIR) per supported scheme.

Example radio configuration parameters

- Transmission specifications.

- Power settings.

- Power control loop.

- Reception.

- Filter parameters (noise figure, sensitivity, bandwidth, etc.).

- Implementation performance curves.

- Transmission schemes supported.

- Coding schemes.

Example parameters for supported schemes

- Supported services.

- Coding type (convolutional, turbo).

- Coding rates.

- Puncturing rates.

- Frame overhead.

- Coding gain curves.

- Additional overheads.

- Spreading factors.

- Spectral efficiency.

- Performance curves.

11.3.1.4   Terrain and Propagation Data

After sector locations are defined through the sector database and user locations are defined for a given simulation instant or snapshot, propagation analyses must take place to evaluate path loss between transmission and reception between these elements. Path loss analysis allows appropriate calculation of transmission powers needed on power controlled channels, as well as calculation of noise rise at each point. Figure 11.20 shows a block diagram of the terrain and propagation data components.

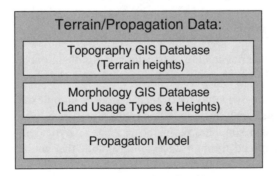

Figure 11.20   Terrain and propagation data.

The information about terrain and morphology profile over the propagation path for each MS-BS pair is used, together with the selected propagation model, for estimating the path loss between these points. This information also allows further calculations on the service analysis process.

11.3.2   Reproducing Traffic Offered to the System

After demand characterisation results are obtained and the inputs for traffic simulation (user layers) are defined for each simulation scenario, the next step is to reproduce traffic behaviour offered to the system for the simulation process to properly capture its response. For this, the full description of the traffic model associated with each profile should be considered.

In simulating the system operation in response to the traffic, two main types of simulation may be used: static and dynamic. This section revises the main concepts involved in system-level simulations and the methods used for simulating mobile networks today.

11.3.2.1   Static Simulation

The main characteristic of static simulations is that they do not model the time evolution of the simulated system, i.e. there is no correlation between consecutive simulation times.

The following list provides a summary of the main characteristics of static simulation [25].

- Outage is defined as signal-to-interference ratio falling below a given threshold.

- Handoff levels and signal and interference levels impact the analysis.

- Fading is simulated for each call at each instant of time (snapshot) independently, but its impact is evaluated instantaneously through probability (e.g. if that fading is randomly found to be below the acceptable threshold for a call, then the call is in outage).

As additional points, we highlight the following limitations of the static simulation method.

- Handoff thresholds impact QoS; however, handoff timers and hysteresis can not be perfectly modelled.

- Static simulations do not capture time correlations of signal due to fading and mobility.

- These simulations can not capture burst admission probabilities and performance during bursts.

- Static simulations may widely vary in their degree of complexity. They may range from methods as simple as analytical modelling to more statistically representative approaches such as Monte Carlo simulations. In the case of analytical modelling, coverage and capacity are determined through analysis assuming that mobile densities and fading models are analytically (although statistically) tractable. This chapter does not discuss this option, because it is not considered to be a 'true' simulation method.

- For further discussion of static simulation methods, we divide them in two main categories, statistical pixel simulation and Monte Carlo simulation.

11.3.2.1.1   Statistical Pixel Simulation
This approach is referred to, in Ref. [26], as 'statistical snap-shot' analysis. That paper (Ref. [26]) proposes type of analysis as a solution to evaluate performance of second-generation

CDMA systems (single-class voice users) based on the traffic (mean number of active users) assumed per pixel on a traffic layer.

Instead of simulating calls from different combinations of user locations, this method assumes that calls are generated from all pixels and then it scales the results according to traffic values.

For each pixel, it obtains BS-MS and MS powers (more than one BS if in soft handoff) that would be transmitted on the downlink and uplink, respectively, for a call at that pixel and then scales those powers according to the traffic declared for the pixel. For instance, if, at a given pixel, the traffic is 0.2 active users (Erlangs), the power computed for the pixel would be 20% of that required for an actual call.

The method is still iterative in the sense that the results have to converge, after power control adjustments are applied to powers on each iteration, based on the total received interference at downlink and uplink. Although not mentioned in Ref. [26], a natural metric to test convergence is the sum of all powers transmitted and received at each sector.

11.3.2.1.2 Monte Carlo Simulation (or Monte Carlo Snapshots)
In this case, multiple instants of the system operation (snapshots) are simulated in the system, each of them with a different set of mobile locations. The number of users active in the snapshots should reflect an expected distribution of active users in the network area.

For each snapshot, multiple calls are generated and offered to the system at randomly distributed locations and with density proportional to the expected geographical traffic density.

For each call, the relevant BS and MS transmitted powers are obtained, and power control algorithms are emulated to adjust these powers to achieve target signal-to-interference or carrier-to-interference (C/I) ratios.

Using powers calculated by the step described previously, new C/I ratios are obtained, and an iterative process may be necessary between these steps until powers converge, similarly to the method described in the statistical pixel simulation method.

One fundamental difference between the statistical pixel simulation and the Monte Carlo simulation methods is that, in the pixel simulations, outages tend always to happen on the last pixels tested for each sector, when load is already close to capacity limits, while on the latter, calls in outage are statistically averaged over many iterations. One way to overcome this limitation, in the statistical pixel simulation, and to avoid biased outage maps is to evaluate pixels in random order, but, in this case, we are gradually transforming the pixel simulation into a Monte Carlo simulation, as multiple random scenarios would have to be evaluated for statistical significance.

One of the most severe limitations of static simulations is the difference in the concept of outage. While in static simulations outage is defined as the event of the C/I ratio falling below a certain threshold, in the case of dynamic simulations (discussed next) the concept of outage accounts for the amount of time a certain signal-to-interference ratio stays below that threshold. One possibility to overcome that particular limitation is to introduce this broader outage concept in static simulations as well, through the use of minimum duration outage probabilities, as discussed next.

For static simulations, fading effects are typically considered through the use of random variables and associated Cumulative Distribution Functions (CDF). Typically, those curves express the probability of a given signal level or signal-to-interference ratio falling below a specified threshold, given the probability distribution parameters. Typical distributions used in this context are log-normal distribution (for shadow fading) and Rayleigh distribution (for non-line-of-sight multipath fading).

The use of the level crossings theory [27] allows one to obtain different types of curves that express the probability of the signal level being above (or below) a certain threshold level for a period longer than a specified limit, given the signal's mean level and other relevant distribution parameters. If these types of curves are used instead of typical signal strength distributions, they can enhance the results achieved with Monte Carlo static simulations when compared to typical static simulation, getting results closer to those of dynamic simulations. In case this type of approach is to be considered, different distributions should be used for different environments and mobile speeds, where each distribution represents the probability that a signal stays below a certain value for a time longer than a certain limit at a given environment and mobile speed [27].

11.3.2.2 Dynamic Simulation

In dynamic simulations, there is a time correlation between consecutive simulation times, i.e. events trigger other events. The most important consequence of this approach is that it is possible to model feedback loops that are present in many radio link control and radio resource management algorithms. Additionally, but not as importantly, dynamic simulations allow modelling user mobility along pre-defined user trajectories.

Dynamic simulators may be grouped in two categories.

- *Event driven*: Events simulated in the system are the triggers of new events or tasks. The simulation time jumps from event to event.

- *Time driven*: Simulation time steps are fixed. Time steps are typically selected according to the network frame structure.

A summary highlighting the main advantages of dynamic simulations is given next [25].

- Outage is defined as a signal-to-interference ratio falling below a given threshold for a given duration of time. This is the concept of 'minimum duration outage', also presented in Ref. [27].

- Fading and outage durations are functions of fading environment and mobile speed.

- Handoff timers and handoff delays impact QoS.

- Captures performance at call level.

However, one of the disadvantages of this type of simulation is that it is more complex to implement and, most importantly, may require impractically long execution times before statistically representative results are obtained.

11.3.3 Simulating System Operation

The simulation must reproduce key system behaviours to capture their effects. These behaviours include

- signal coverage;

- target noise-rise used for admission and load control;

- link adaptation;

- multiple radio types and technologies;

- link asymmetry;

- power control, handoff and soft handoff analysis;

- interference budget (for CDMA systems).

Once all blocks have been defined, system simulation can be performed. The use of the framework described in previous sections allows the verification, for each active user, of what would be the system resources (physical channels, power, etc.) needed to satisfy the required QoS metrics. For that purpose, the simulator performs iterative calculations for system load and power control algorithm convergence. Figure 11.21 shows a block diagram summarising the information flow for the system simulation, illustrating the main functions performed by the simulation module, as well as the main iterative functions performed in that algorithm.

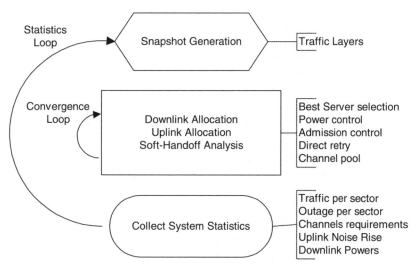

Figure 11.21   Iterative algorithm in 3G system simulation.

As a consequence of allocating the appropriate resources for each active user (within the limits of availability of the radio access network), it is possible to calculate the overall interference being generated by the sectors at each user location (downlink noise rise) and the overall interference generated by the users at each sector location (uplink noise rise). The updated noise-rise calculation creates new conditions for power control algorithms, which create new figures of noise rise. This iterative process goes alternately on downlink and uplink until convergence is reached in both directions.

In addition to convergence for the offered load, some other radio resource management functions performed by the simulation algorithm are listed below.

- Admission control: Monitor load and refuse service if a specified load limit has been achieved. Service may also be refused due to limits on number of codes, total throughput, etc., depending on the criteria defined.

- Load control: To give preferential treatment to real-time traffic, the system may temporarily restrict rates available to active data services.

- Bandwidth scheduling: The system can schedule packets belonging to different service classes according to pre-established policies.

- Handover: Based on pilot E_c/I_o levels and user class capabilities to support soft handoff, the same user may be served by more than one sector. Handover gains are calculated based on the difference of signal levels available at each pixel.

The main outputs of the simulation process are sector parameters that depend on the traffic load, namely the noise rise per sector and the power transmitted by traffic channels. Besides, statistics may be collected with regards to many variables including throughput, number of channels/codes used, main/handoff traffic allocation, outage, etc.

11.3.3.1 Interference Convergence

The convergence loop illustrated in Figure 11.21 refers to the convergence of interference levels (directly related to load and noise rise) at the system on both downlink and uplink. Appendix A at the end of this chapter provides a formal description of the concept of load and noise rise, and how they can be estimated in real and simulated systems. The main steps to achieve interference convergence are described next.

- Adjust transmitted powers on downlink and uplink based on link information and power control algorithm.

- Calculate overall interference being generated by the sectors at each user location (downlink noise rise).

- Calculate overall interference generated by users at each sector location (uplink noise rise).

- The updated noise rise calculation creates new conditions for power control algorithms, which, in turn, create new figures of noise rise.

The steps above represent an iterative process that goes on alternately on downlink and uplink until convergence is reached in both directions.

11.3.3.2 Radio Resource Management

In IS2000 systems, variable data rates are supported through the use of different spreading factors. To maximise capacity and at the same time to provide QoS guarantees for the services allocated, Radio Resource Management (RRM) needs to be in place. The following are the main tasks related to RRM, which should be simulated to reproduce system behaviour in response to the traffic offered.

- *Power control*: Used in the downlink and uplink to maintain transmitting powers at the minimum requirements that satisfy the targets expected for each class (eventually updated by an outer loop). Implemented on both subscriber and base-station ends of the air interface.

- *Handover*: To manage mobility, the system needs to support soft handoff between neighbouring cells.

- *Load control*: Based on received power or on throughput, load control determines how the system monitors the levels of interference to keep them under acceptable planned ranges and how it manages to bring the load back to acceptable levels when these ranges are exceeded.

- *Admission control*: Based on estimates of the increase of total interference power on the uplink or on total throughput being supported, the network estimates the noise rise (or throughput rise) that would result from accepting the requested Radio Access Bearers (RABs) and decides whether to grant access and with which QoS parameters. Admission control is performed both on downlink and uplink.

- *Packet scheduling functions*: Scheduling should support non-real-time transmissions over packet data channels. The system decides on allowed bit rates and schedules transmissions to achieve optimum resource utilisation while satisfying the required QoS parameters.

RRM functions can be implemented based on hard blocking or soft blocking. In either case, the system stops accepting new connections at some point to avoid sacrificing existing connections and planned coverage. Typically, soft blocking mechanisms allow more efficient system utilization.

Although the subject of radio resource management is fascinating and that has been widely explored in recent research [28–40], it is not within the scope of this chapter to compare different methods or strategies for radio resource management, such as whether it is better to use call admission control or bandwidth reservation. Instead, the focus of this discussion is on the importance of presenting the correct inputs to a system simulator so that, given a certain RRM strategy, the correct system capacity can be estimated and appropriate dimensioning of the system is possible.

Based on many published works on the subject [29, 31–37, 41, 42], it is observed that link load is widely used as the metric of choice to indicate a system's ability to accept more users and, therefore, it is used as the main criteria in algorithms for load control, admission control and packet scheduling. Moreover, it is the most widely used system metric of capacity, and, therefore, its target operating point is typically a design parameter. Appendix A presents a formal description on the concept of load and noise rise, and how they can be estimated in real and simulated systems.

For a given resource management strategy and a given system target load, the total system capacity, or throughput, still depends on factors such as user geographical distribution and service mix. These variables are typically overlooked and it is common to find important conclusions being drawn from homogeneous traffic distributions and arbitrarily determined service mixes. Appendix B shows a brief illustration of the impact of different user mixes on the total capacity achieved by the system.

Congestion control is the method used to limit the amount of traffic accepted in a system in response to over-load conditions in the network. Particularly in 3G systems, where the quality of service of active users must be guaranteed within certain thresholds, the system cannot afford to accept more connections in detriment of the existing ones, or it has to take some actions if the existing load grows beyond certain limits. Therefore, sector loads are kept under strict control, through the use of admission control and load control techniques. Algorithms used for that purpose are not completely standard, and different vendors and implementations may adopt different congestion control algorithms.

In *admission control* techniques, the system typically tries to estimate the amount of load increase that would result from a service request (RAB) being accepted. It then accepts or rejects the RABs depending on pre-specified acceptance thresholds.

In *load control*, the system keeps monitoring (or estimating) the current load in each sector. Typical methods of estimating load in WCDMA systems are through the measurement of total received power in the uplink and total transmitted power on the downlink.

When a problem is flagged, i.e. when load has reached a certain threshold (plus margins allowed for stability), the system takes some action to prevent or solve over-load problems. Those actions may include the following.

- Reduction of traffic volume in non-real-time traffic classes, through the use of packet scheduling algorithms. These reductions consider classes and priorities negotiated for these bearers, respecting the minimum guaranteed rates.

- Reduction of data rates of real-time traffic classes such as voice or data, within the limits of acceptable quality of service.

- Temporary blocking, or reduction, in power control commands, which ultimately also reduces quality for existing RABs.

- Forced handoffs to sectors with lower loads.

While it is not part of network planning and optimisation to decide which algorithms to use, typical parameters to be optimised with regards to congestion control are the target values for maximum load, which may be different for downlink and uplink, and may also vary for admission control and load control functions. Besides, target margins must be configured to give the system some room to exceed, by small amounts, the load values that it is trying to keep on average.

In optimising the parameters for load control targets, the effects of non-uniform traffic density must be considered once again, as the choice of target loads, which may be configurable per sector or for the whole system, depends on the traffic distribution and on ratios of inter-cell to intra-cell interference that vary according to location and sector geometry.

11.3.4   The Dimensioning Loop

The appropriate simulation of the system for a given demand characterisation allows the performance analysis of the network, i.e. allows correct identification of the percentage of outage expected at the system due to the radio network limitations.

As soon as these limitations are identified, on a per sector basis, the next step in the design is to increase resources in bottleneck sectors, based on the feedback provided by the simulation algorithm. After that, new simulations are required to make sure that the new performance figures are within acceptable levels. This procedure of course is also an iterative process, where the feedback from the simulation algorithm, namely achieved QoS results, serve as input for modifications in the radio network. Figure 11.22 illustrates this process.

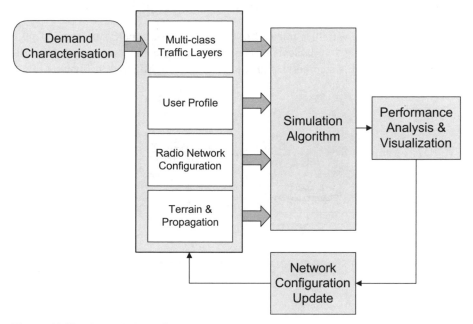

Figure 11.22   Overall block diagram of dimensioning process proposed in this framework.

In dimensioning traffic for wireless systems, the main resource for allocation is the spectrum, i.e. operators need to estimate and maximise the capacity that can be handled per carrier installed for a given network design. Assuming a system layout with given site locations and sector configurations and using the methodology previously described to estimate the demand on that network, the most important dimensioning problem is to determine the amount of resources that need to be allocated at each sector to satisfy the demand while providing the required quality of service. If demand and QoS are to be satisfied, in the event that the number of resources needed exceeds the spectrum availability (dictated by license agreements), a re-design of the network with the addition of new cells and cell splitting is considered. If the system design is for a new network layout, this process allows more flexibility and many iterations may be executed to optimise site location and sector configuration.

Although QoS must be achieved from end to end in the 3G networks, we are particularly concerned in this section with the bottleneck of the air interface, which is strongly limited by spectrum availability and hardware implementation, among other things. After the radio network is appropriately dimensioned, e.g. the appropriate number of required carriers and sector locations has been obtained, the final simulation of the system provides the relevant figures of throughput per sector and its statistical distribution. These figures are used for the core network dimensioning, which is not within the scope of this discussion.

11.4   PERFORMANCE ANALYSIS

The global description of the dimensioning process discussed at the beginning of this chapter introduced the process of performance evaluation as the method by which network behaviour is judged, indicating to designers whether the required QoS metrics are satisfied or if design modifications are needed.

This section discusses variables relevant in evaluating system performance, and suggests methods to visualise these variables in ways that facilitate the decision making during network optimisation. This step corresponds to the block highlighted in Figure 11.23.

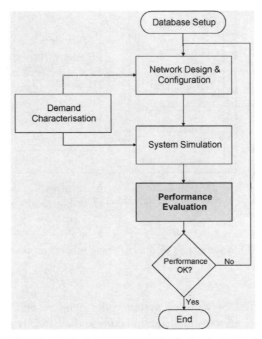

Figure 11.23   Traffic dimensioning process – highlight on the performance analysis.

Once a simulation is performed for a certain scenario, a large volume of information is available for post-processing. One example of the information available from simulation results is shown in Figure 11.24, which contains a summary report from CelPlanner IS2000 simulation and network planning module. Besides the summary display, more detailed reports may be exported as text files, which may be opened as a spreadsheet for post-processing. An example of such reports available for IS2000 simulation results is shown in Figure 11.25.

These reports contain important information that allows the assessment of system capacity in response to the traffic offered, as well as system diagnosis that identifies the main reasons why traffic could not be handled in each scenario.

Out of the many output variables available in the report from CelPlanner simulation tool, we highlight here the ones that are most relevant to system dimensioning. Unless otherwise indicated, the variables described below are available on a per-sector, per-user profile basis, as well as in overall results for the whole system. When computed over multiple snapshots, all variables are presented in terms of their mean and standard deviation. The following metrics are considered.

CelPlan - Traffic / Load Simulation

Site	Str	Simultaneous Connections						Forward Traffic Channel Power (W)		Reverse Traffic Load Factor		Suggested Number of Carriers
		Main		Handoff		Total						
		Average	StdDev	Average	StdDev	Average	StdDev	Average	StdDev	Average	StdDev	
n01-02-03	01	47.00	0.00	8.00	0.00	55.00	0.00	7.98	0.00	0.583	0.000	1
	02	19.00	0.00	19.00	0.00	38.00	0.00	5.91	0.00	0.393	0.000	1
	03	26.00	0.00	12.00	0.00	38.00	0.00	4.73	0.00	0.393	0.000	1
n04-05-06	04	25.00	0.00	13.00	0.00	38.00	0.00	5.04	0.00	0.393	0.000	1
	05	33.00	0.00	15.00	0.00	48.00	0.00	6.82	0.00	0.496	0.000	1
	06	20.00	0.00	16.00	0.00	36.00	0.00	4.92	0.00	0.355	0.000	1
n07-08-09	07	24.00	0.00	12.00	0.00	36.00	0.00	4.61	0.00	0.396	0.000	1
	08	34.00	0.00	16.00	0.00	50.00	0.00	7.99	0.00	0.503	0.000	1
	09	36.00	0.00	9.00	0.00	45.00	0.00	6.21	0.00	0.513	0.000	1
n10-11-12	10	45.00	0.00	6.00	0.00	51.00	0.00	6.80	0.00	0.572	0.000	1
	11	14.00	0.00	13.00	0.00	27.00	0.00	3.63	0.00	0.297	0.000	1
	12	39.00	0.00	11.00	0.00	50.00	0.00	7.24	0.00	0.505	0.000	1
n13-14-15	13	21.00	0.00	22.00	0.00	43.00	0.00	6.82	0.00	0.402	0.000	1
	14	31.00	0.00	19.00	0.00	50.00	0.00	8.00	0.00	0.492	0.000	1
	15	42.00	0.00	3.00	0.00	45.00	0.00	5.46	0.00	0.519	0.000	1
n16-17-18	16	36.00	0.00	13.00	0.00	49.00	0.00	7.79	0.00	0.505	0.000	1

Simulated Traffic:	1140.0 Active Users	
Carried Traffic:	1361.0 Active Users	119.4 %
Main Traffic:	977.0 Active Users	85.7 %
Handoff Traffic:	384.0 Active Users	33.7 %
Not Carried Traffic:	163.0 Active Users	14.3 %
Pilot channel coverage:	0.0 Active Users	0.0 %
Directed to other carriers:	0.0 Active Users	0.0 %
Throughput per carrier limit:	0.0 Active Users	0.0 %
Forward traffic channel power limit:	120.0 Active Users	10.5 %
Forward sector total power limit:	43.0 Active Users	3.8 %
Forward load factor limit:	0.0 Active Users	0.0 %
Reverse mobile power limit:	0.0 Active Users	0.0 %
Reverse load factor limit:	0.0 Active Users	0.0 %

Service Class: ○ All Service Classes ▼
Carrier: ○ First Carrier ▼

Accept Simultaneous Connections
Accept Traffic Channel Power
Accept Reverse Traffic Load Factor
Accept Suggested Number of Carriers

✓ Ok   ? Help   🖨 Print   T Export

Figure 11.24 Example of simulation results summary from CelPlanner IS2000 simulation tool.

Traffic Simulation Report

Simulated Traffic 1139.978 Active Users
Iterations 1
System Blocking 2 %

Site	Sector	First Carrier Carried Traffic															First Carrier Not Carried					
		FCSCMain (Conn)		FCSCHff (Conn)		SCTotal (Conn)		FWDRTot (Kbps)		RVDRTot (Kbps)		FWPower (W)		RVLoad (Factor)		NoiseRise (dB)	OtherCarriers (Conn)		DataRate (Conn)		FWPower (Conn)
		Avrg	StDv	Avrg	StDv	Avrg	StDv	Avrg	StDv	Avrg	StDv	Avrg	StDv	Avrg	StDv		Avrg	StDv	Avrg	StDv	Avrg
Class Data 8k Portable																					
n01-02-03	1	10	0	4	0	14	0	112	0	112	0	2.051	0				0		0	0	5
...																					
n28-29-20	30	17	0	4	0	21	0	168	0	168	0	3.351	0				0		0	0	12
Sum		469	0	87	0	556	0	4448	0	4448	0	83.52	0					0	0	0.00	201.00
Class Data 64k Palmtop																					
n01-02-03	1	3	0	0	0	3	0	192	0	192	0	2.186	0				0		0	0	4
...																					
n28-29-20	30	5	0	0	0	5	0	320	0	320	0	3.468	0				0		0	0	4
Sum		92	0	13	0	105	0	6720	0	6720	0	71.49	0					0	0	0.00	90.00
Class Data 144k Laptop																					
n01-02-03	1	3	0	0	0	3	0	432	0	432	0	3.763	0				0		0	0	3
...																					
n28-29-20	30	1	0	0	0	1	0	144	0	144	0	1.156	0				0		0	0	4
Sum		56	0	3	0	59	0	8496	0	8496	0	72.58	0					0	0	0.00	74.00
All Service Classes																					
n01-02-03	1	16	0	4	0	20	0	736	0	736	0	8	0	0.58	0	3.770995	0	0	0	0	12
...																					
n28-29-20	30	23	0	4	0	27	0	632	0	632	0	7.975	0	0.586	0	3.834159	0	0	0	0	20
Sum		617	0	103	0	720	0	19664	0	19664	0	227.6	0					0	0	0.00	365.00

First Carrier
Carried traffic - Main traffic	617 Active Users	54.1 %
Carried traffic - Handoff traffic	103 Active Users	9 %
Not carried traffic - Pilot channel coverage	0 Active Users	0 %
Not carried traffic - Directed to other carriers	0 Active Users	0 %
Not carried traffic - Throughput per carrier limit	0 Active Users	0 %
Not carried traffic - Forward traffic channel power li	365 Active Users	32 %
Not carried traffic - Forward sector total power limit	158 Active Users	13.9 %
Not carried traffic - Forward load factor limit	0 Active Users	0 %
Not carried traffic - Reverse mobile power limit	0 Active Users	0 %
Not carried traffic - Reverse load factor limit	0 Active Users	0 %

All Suggested Carriers
Carried traffic - Main traffic	1140 Active Users	100 %
Carried traffic - Handoff traffic	951 Active Users	83.4 %
Not carried traffic - Pilot channel coverage	0 Active Users	0 %

Figure 11.25 Example of simulation results report from CelPlanner IS2000 simulation tool.

- Number of accepted connections (service bearers).

- Number of overhead connections (in soft handoff).

- Overall throughput.

- Total power transmitted per sector on traffic channels (in addition to common channels, which have fixed and pre-defined powers, input variables for the simulation).

- Uplink load factor and noise rise.

- Number of rejected connections, grouped by rejection reasons, which include the following.

  Rejection due to lack of coverage. Because this means that no sector was able to provide coverage for this service request, this statistic is not grouped on a per-sector basis, but only as a global result for the system.

  Rejection due to limit on downlink power allowed per RAB.

  Rejection due to limit on total downlink power available at sector.

  Rejection due to limit on uplink power available at mobile station.

  Rejection due to load control on downlink.

  Rejection due to load control on uplink.

  Rejection due to hard blocking based on maximum channel count or data rate allowed per sector.

After simulation is executed for a given scenario, a large set of additional analyses may be performed based on the key results obtained from simulation. These analyses include the geographical display (prediction) of performance parameters, such as the achievable signal-to-interference ratios for the pilot and traffic channels, soft handoff overheads, etc. These predictions must be calculated after the traffic simulation phase is completed, because they depend on results such as the total power transmitted and received per sector, which varies depending on the traffic load. Figures 11.26–11.40 show examples of the most relevant predictions available from CelPlanner IS2000 simulation and network planning tool.

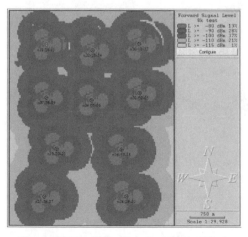

Figure 11.26   Pilot signal strength (dBm).

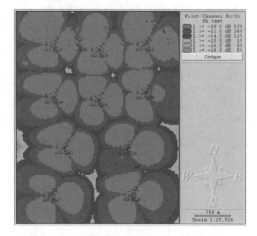

Figure 11.27   Pilot channel E_c/I_o (dB).

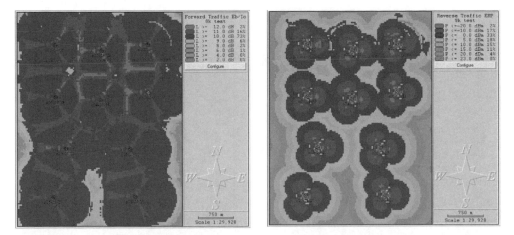

Figure 11.28   Downlink traffic channel E_b/I_o (*dB*).   Figure 11.29   Uplink required MS power (*dBm*).

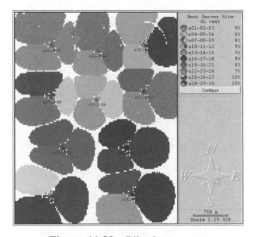

Figure 11.30   Pilot best server.

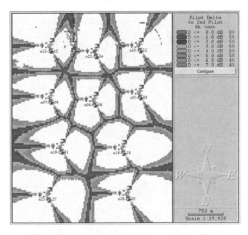

Figure 11.31   Pilot delta to 2nd pilot.

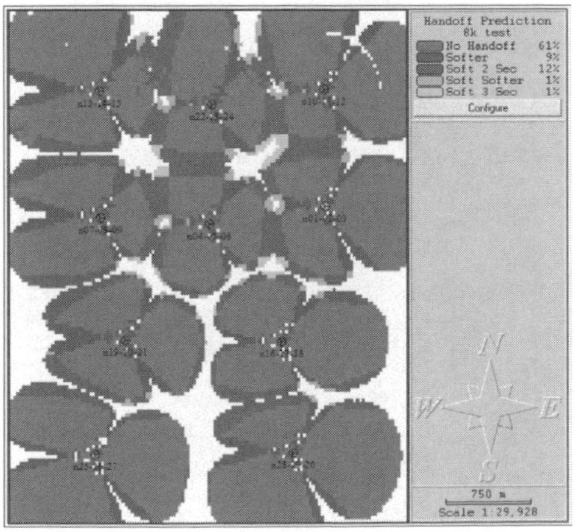

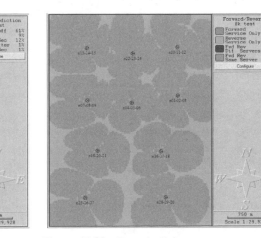

Figure 11.32   Handoff areas.

Figure 11.33   Down/uplink coverage comparison.

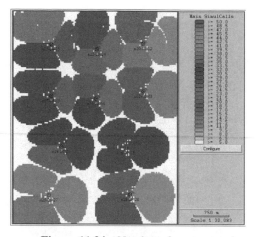

Figure 11.34　Number of users served.

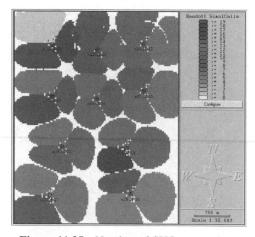

Figure 11.35　Number of SHO connections.

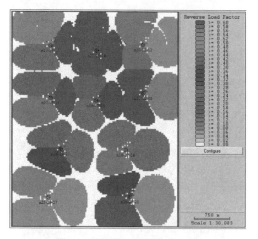

Figure 11.36　Uplink load factor.

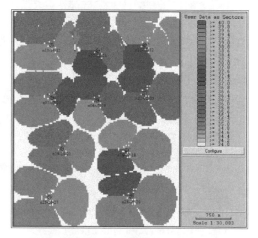

Figure 11.37　Total BS Tx power (dBm).

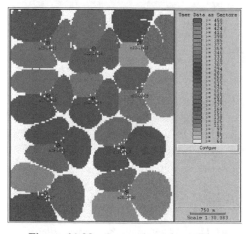

Figure 11.38　Sector throughput (kbps).

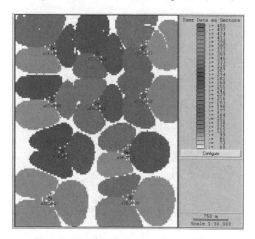

Figure 11.39　Sector total rate (kbps).

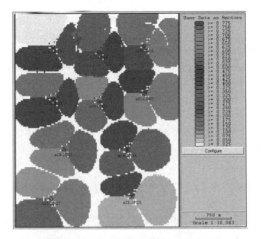

Figure 11.40   Sector SHO overhead.

11.5   SUMMARY

This chapter discussed the process of traffic dimensioning for multimedia wireless systems, particularly for third-generation IS2000 systems.

We divided the process of dimensioning a wireless system into three major blocks: demand characterisation, system simulation and performance analysis. The previous sections illustrated and described each block.

The process of demand characterisation consists of describing the traffic to be offered to a given network, considering multiple demographic input variables, different traffic patterns expected for different user profiles and time dependence relationships among these patterns.

Demographic input variables such as residential, business and road databases can be used to estimate subscriber distributions for different user profiles. To add granularity, morphology weighting may be used inside demographic database elements when spreading traffic. Market forecasting includes many unknowns but may also be influenced by market strategy. After multiple layers of user distributions are created for each profile from different sources, they may be combined to represent a given simulation scenario. The layer combination is time dependent, and multiple scenarios must be simulated to ensure proper capture of the system's different bottleneck times.

In the system simulation process, some of the inputs come from the results of the demand characterisation process. These include the multi-layer geographical traffic distribution, user profiles and their associated traffic models. Other inputs to the traffic simulation are terrain and propagation data and radio network configuration, which should include the network topology with site locations, sector configurations, antenna parameters, link budget and radio specifications.

Two approaches for traffic simulation were discussed, the static and dynamic simulations. Whereas the static simulation (e.g. snapshots, Monte Carlo analysis) presents some limitations in reproducing system operation, it may be the best choice in some occasions, because dynamic simulations often require impractical simulation periods before statistically representative results are achieved.

The system simulation should reproduce the system operation by considering radio resource management functions such as load and admission control, handoff, power control, etc. An interference convergence loop is needed to obtain reliable levels for received powers on the forward and reverse links, and, therefore, correctly characterise load levels in each direction.

To allow performance evaluation, statistics are collected during the system simulation process. At the end of the process, reports and geographical displays may be generated to allow designers to decide on whether network configuration updates are needed to achieve the desired QoS metrics.

APPENDIX A - LINK LOAD AND NOISE RISE

An important step for verification of the system capacity in the system-level simulation is to include the same approach used in the real system for admission and congestion control. This includes the verification of whether the system is exceeding "target loading factors." In this sense, the loading factor in a CDMA system is one of the most important design parameters. Maximizing capacity achievable for a given loading factor is a design objective.

The load or loading factor expresses the ratio of received interference divided by the sum of received interference plus thermal noise [43]

$$\ell = \frac{I}{I + N} = \frac{I_0}{I_0 + N_0} \tag{11.8}$$

The load factor is always smaller than 1 and tends to 1 as interference grows, i.e. as the system gets more loaded.

The noise rise indicates how much the interference raises the noise floor. It is therefore always greater than 1, and typically expressed in dB. The noise rise is equivalent to

$$N_{rise} = \frac{I + N}{N} = \frac{I_0 + N_0}{N_0} \qquad N_{rise}\ \text{dB} = 10\log(N_{rise}) \tag{11.9}$$

Therefore, the noise rise is related to the loading factor by [6]

$$N_{rise} = \frac{1}{1 - \ell} \qquad N_{rise}\ \text{dB} = -10\log(1 - \ell) \tag{11.10}$$

or

$$\ell = \frac{N_{rise} - 1}{N_{rise}} \tag{11.11}$$

Downstream Load

The load of a sector on the downstream is calculated considering the load contribution of each user, which varies with the user location. It is given by

$$\ell_{Dw} = \sum_{j=1}^{N} \left\{ \nu_j \frac{\left(\frac{E_b}{N_0}\right)_j}{\frac{W}{R_j}} \left[(1 - \alpha_j) + \left(\frac{o}{s}\right)_j\right] \right\} \tag{11.12}$$

In the equation, W is the chip rate, R_j is the data rate for user j and $(E_b/N_o)_j$ is the required signal energy per bit divided by noise plus interference spectral density to achieve a certain quality of service for user j when connected at data rate R_j. ν_j is the activity factor for user j at the physical layer, α_j is the orthogonality factor in the downlink and $(o/s)_j$ represents other-cell-to-same-cell interference ratio as experienced by user j at each specific location. The sum is calculated for all N users connected to the sector.

Equation (11.12) is a theoretical formula that expresses the overall interference received at the mobile location by evaluating the overall intra-cell interference and using a proportionality factor to estimate the additional fraction coming from other cells. From that expression, the total received interference at one location depends on the combination of all services and data rates supported at a given moment.

Estimating Downstream Load (for Practical Purposes)

Equation (11.12) can not be used to estimate the load when the intention is to use the load factor as a criterion for admission control for real systems or simulations. For this, we would have to assume or estimate the orthogonality factor at each location or use an average value for that variable. For the same reason, the ratio $(o/s)_j$, which also varies per location, would have to be measured by the mobile or estimated through simulations. Therefore, one way to estimate the downstream load is

$$\ell_{Dw} \cong \sum_{j=1}^{N} \left\{ \nu_j \frac{\left(\frac{E_b}{N_0}\right)_j}{\frac{W}{R_j}} [(1 - \bar{\alpha}) + \bar{i}] \right\} \tag{11.13}$$

In this equation, $\bar{\alpha}$ is the mean (estimated or assumed) orthogonality factor for all locations served by the sector and $\bar{i}$ is the average of $(o/s)_j$, i.e. it represents average other-cell-to-same-cell interference ratio.

Another (more practical) way to estimate the downlink load factor in real systems or simulations is based on the total throughput of the sector, and is given by

$$\ell_{Dw} = \frac{\sum_{j=1}^{N} R_j}{R_{max}} \tag{11.14}$$

In this equation, R_j is the bit rate of the jth user and R_{max} is the maximum achievable throughput for the cell.

One additional option of load estimation uses the total sector transmit power, as shown next

$$\ell_{Dw} = \frac{P_{total}}{P_{max}} \tag{11.15}$$

where P_{total} is the total base-station power being transmitted and P_{max} is the maximum base-station transmission power allowed.

Upstream Load

The formula used to estimate the upstream loading factor for a multi-user multi-rate WCDMA system can be written as [43]

$$\ell_{up} = \left(1 + \frac{o}{s}\right) \sum_{j=1}^{N} \left[\frac{1}{1 + \frac{W}{\left(\frac{E_b}{N_0}\right)_j R_j \nu_j}} \right] \tag{11.16}$$

In the equation, W is the chip rate, R_j is the data rate for user j and $(E_b/N_o)_j$ is the required signal energy per bit divided by noise plus interference spectral density to achieve a certain quality of service for user j when connected at data rate R_j. ν_j is the activity factor for user j at the physical layer and o/s represents other cell to same cell interference ratio as seen by the sector receiver.

Estimating Upstream Load (for Practical Purposes)

When the intention is to use the load factor as a criterion for admission control for real systems or simulations, there are different approaches for estimating the downlink load factor.

Because all parameters relevant to the uplink load calculation may be directly estimated or measured by the system (real or simulated), one could use eqn (11.16). Another way to estimate upstream load is through direct application of the definition of loading, given by eqn (11.8). This means that the uplink load may be calculated directly by measuring the total power received at a sector, together with an assumed value for the noise floor. That relationship is rearranged next

$$\ell_{up} = 1 - \frac{N_{floor}}{P_{Rx\_Total}} \tag{11.17}$$

In this equation, P_{Rx_Total} is the total measured power (noise plus interference) at the reception of a sector, and N_{floor} is an estimate of the thermal noise floor, typically calculated as

$$N_{floor} = KTBN_f \tag{11.18}$$

K is the Boltzman constant (1.38×10^{-23}), T is the ambient temperature in Kelvin, B is the receiver bandwidth in Hertz and N_f is the receiver's noise figure. Instead of using eqn (11.18), N_{floor} may also be measured in real systems as the received power at the unloaded sector (no active users in the system).

Finally, for simulation purposes, the uplink load is typically estimated by computing the sum of all received powers at a sector from all users in the system (not only the users connected to that sector). This option is shown next

$$\ell_{up} = 1 - \frac{N_{floor}}{N_{floor} + \sum_{k}^{all} P_{Rx_k}} \tag{11.19}$$

In eqn (11.19), P_{Rxk} is the power received from the kth user and is summed for all users in the system.

APPENDIX B - CAPACITY DEPENDENCE ON USER MIX

Inspecting eqns (11.13) and (11.16), we notice that different scenarios of user mix (proportion of users in different classes) would result in approximately constant total throughput values for a given loading factor if the required $(E_b/N_o)_j$ thresholds are all set to the same value.

To demonstrate that observation, the total data rate supported by one channel at one sector in a multi-user multi-rate scenario can be defined as

$$R_{Total} = \sum_{All\_cell\_users} R_j \nu_j \tag{11.20}$$

Assuming $(E_b/N_o)_j$ is the same for all users $(j = 1, \ldots, N)$, $(E_b/N_o)_j$ in eqns (11.13) and (11.16) can be replaced as a constant ρ where

$$\rho = \left(\frac{E_b}{N_0}\right)_{required} \tag{11.21}$$

where E_b/N_o is assumed constant for all classes.

Therefore, the downlink load given in eqn (11.13) can be rewritten as

$$\ell_{Dw} \cong [(1 - \bar{\alpha}) + \bar{i}] \frac{\rho}{W} \sum_{All\_cell\_users} R_j \nu_j$$

$$\ell_{Dw} \cong [(1 - \bar{\alpha}) + \bar{i}] \frac{\rho}{W} R_{Total} \tag{11.22}$$

For the uplink load, given in eqn (11.16), we may consider that, in practice, the right side of the denominator is much greater than 1, which is a reasonable assumption [43]. Therefore, eqn (11.16) could be rewritten as

$$\ell_{Up} \cong (1 + i) \frac{\rho}{W} \sum_{All\_cell\_users} R_j \nu_j$$

$$\ell_{Up} \cong (1 + i) \frac{\rho}{W} R_{Total} \tag{11.23}$$

Considering that the total loading factor is used as a criterion for admission control, it determines the capacity of the cell as a design parameter. Therefore, it is valid to say that, if the $(E_b/N_o)_j$ requirement for all classes was homogeneous, the total throughput capacity of a cell to achieve a desired loading factor would not depend on the particular mix of service classes. However, in the more realistic scenario where different classes have different threshold sets for operation in terms of E_b/N_o required, the total load is given by a weighted sum of the different data rates times the required thresholds. This means that different total throughput per cell can be achieved for the same loading factor and network layout depending on the user mix that defines the demand offered to the system. This observation emphasises once more the importance of appropriately describing the traffic demand offered to the system.

BIBLIOGRAPHY AND REFERENCES

1. Ribeiro, L.Z. and DaSilva, L.A., A Framework for the Dimensioning of Broadband Mobile Networks Supporting Wireless Internet Services, IEEE Wireless Communications, **9**, 6–13, 2002.
2. Kleinrock, L., *Queueing systems Volume 1: Theory.* New York: Wiley, 1975.
3. Leon-Garcia, A., *Probability and Random Processes for Electrical Engineering*, 2nd ed. Reading, MA: Addison-Wesley, 1994.
4. Spohn, D.L., *Data Network Design.* New York, N.Y.: McGraw-Hill, 1997.
5. Willinger, W. and Paxson, V., Where Mathematics Meets the Internet, Notices of the AMS, September 1998.
6. Sahinoglu, Z. and Tekinay, S., On Multimedia Networks: Self-Similar Traffic and Network Performance, *IEEE Communications Magazine*, **37**, 48–52, 1999.
7. Lee, J.S. and Miller, L.E., *CDMA Systems Engineering Handbook.* Boston, MA: Artech House, 1998.
8. 3GPP TS 23.107 Version 5.3.0 Release 5, *QoS Concept and Architecture*, 3rd Generation Partnership Project Technical Specification, 2002-01.
9. Jabbari, B., Teletraffic Aspects of Evolving and Next-Generation Wireless Communications Networks, IEEE Personal Communications, **3**, 4–9, 1996.
10. Oliphant, M.W., The Mobile Phone Meets the Internet, IEEE Spectrum, **36**, 20–28, 1999.
11. Peterson, L.L. and Davie, B.S., *Computer Networks: a Systems Approach*, 2nd ed. San Francisco: Morgan Kaufmann, 2000.
12. Heffes, H. and Lucantoni, D.M., A Markov Modulated Characterization of Packetized Voice and Data Traffic and Related Statistical Multiplexor Performance, IEEE Journal on Selected Areas in Communications, **4**, 856–868, 1986.
13. Leland, W.E., Taqqu, M.S., Willinger, W. and Wilson, D.V., On the Self-similar Nature of Ethernet Traffic (Extended Version), IEEE/ACM Transactions on Networking, **2**, 1–5, 1994.
14. Crovella, M.E. and Bestravros, A., Self-Similarity in World Wide Web Traffic – Evidence and Possible Causes, IEEE/ACM Transactions on Networking, **5**, 835–845, 1997.
15. Roberts, J.W., Traffic Theory and the Internet, IEEE Communications Magazine, **39**, 94–99, 2001.
16. *Self Similar Network Traffic and Performance Evaluation*, edited by Park, K. and Willinger, W., New York: Wiley, 2000.
17. Willinger, W. *et al.*, Self-Similarity through High-Variability: Statistical Analysis of Ethernet LAN Traffic at the Source Level, Proceedings ACM SIGCOMM '95, 100–113, 1995.
18. Sexton, M. and Reid, A., *Broadband Networking.* Norwood, MA: Artech House, 1997.
19. 3GPP TS 23.101 Version 3.1.0 Release 1999, *General UMTS Architecture*, 3rd Generation Partnership Project, Technical Specification, 2001-12.
20. 3GPP TS 23.107 Version 3.7.0 Release 1999, *QoS Concept and Architecture*, 3rd Generation Partnership Project, Technical Specification, 2002-01.
21. 3GPP TS 23.107 Version 4.3.0 Release 4, *QoS Concept and Architecture*, 3rd Generation Partnership Project, Technical Specification, 2002-01.
22. *Broadband Network Teletraffic: Final Report of Action COST 242,* edited by Roberts, J., Mocci, U. and Virtamo, J. Berlin: Springer, 1996.
23. Laiho, J., Wacker, A. and Novosad, T., *Radio Network Planning and Optimization for UMTS*, West Sussex, England: Wiley, 2002.
24. Yacoub, M.D., *Wireless Technology: Protocols, Standards, and Techniques.* Boca Raton, CA: CRC Press, 2001.
25. Kumar, S. and Nanda, S., High Data-Rate Packet Communications for Cellular Networks Using CDMA: Algorithms and Performance, IEEE Journal on Selected Areas in Communications, **17**, 472–492, 1999.

26. Dziong, Z., Krishnan, M., Kumar, S. and Nanda, S., Statistical 'Snap-Shot' for Multic-Cell CDMA System Capacity Analysis, IEEE VTC, 1234–1237, 1999.

27. Mandayam, N., Chen, P. and Holtzman, J., Minimum Duration Outage for Cellular Systems: A Level Crossing Analysis, IEEE Vehicular Technology Conference, VTC 1996 46th, **2**, 879–883, 1996.

28. Dixit, S., Guo, Y. and Antoniou, Z., Resource Management and Quality of Service in Third-Generation Wireless Networks, IEEE Communications Magazine, **39**, 125–133, 2001.

29. Jorguseski, L., Farserotu, J. and Prasad, R., Radio Resource Allocation in Third-Generation Mobile Communication Systems, IEEE Communications Magazine, **39**, 117–123, 2001.

30. Dimitriou, N., Tafazolli, R. and Sfikas, G., Quality of Service for Multimedia CDMA, IEEE Communications Magazine, **38**, 88–94, 2000.

31. Kim, K., Han, Y., Yim, C.H. and Jeong, K.S., A Call Admission Algorithm with Optimal Power Allocation for Multiple Class Traffic in CDMA Systems, IEEE Vehicular Technology Conference, VTC'2000 52nd, **6**, 2666–2671.

32. Guo, Y. and Chaskar, H., A Framework for Quality of Service Differentiation on 3G CDMA Air Interface, IEEE Wireless Communications and Networking Conference, WCNC'2000, **3**, 975–979, 2000.

33. Guo, Y. and Aazhang, B., Resource Allocation and Capacity in Wireless CDMA Networks Using Adaptive Power Control and Antenna Array Multiuser Receiver, IEEE Symposium on Computers and Communications, ISCC'2000 5th, 723–730, 2000.

34. Guo, Y. and Aazhang, B., Call Admission Control in Multi-Class Traffic CDMA Cellular System using Multiuser Antenna Array Receiver, IEEE Vehicular Technology Conference, VTC'2000-Spring, Tokyo, **1**, 365–369, 2000.

35. Guo, Y. and Aazhang, B., Capacity of Multi-Class Traffic CDMA System with Multiuser Receiver, IEEE Wireless Communications and Networking Conference, WCNC'1999, **3**, 975–979, 1999.

36. Koodli, R. and Puuskari, M., Supporting Packet-Data QoS in Next-Generation Cellular Networks, IEEE Communications Magazine, **39**, 180–187, 2001.

37. Imbeni, D. and Karlsson, M., Quality of Service Management for Mixed Services in WCDMA, IEEE Vehicular Technology Conference, VTC'2000-Fall, Boston, **2**, 565–572, 2000.

38. Jiang, J. and Lai, T.H., An Efficient Approach to Support QoS and Bandwidth Efficiency in High-Speed Mobile Networks, IEEE International Conference on Communication, ICC 2000, New Orleans, **2**, 980–984, 2000.

39. Jiang, J. and Lai, T.H., Call Admission Control vs Bandwidth Reservation: Reducing Handoff Call Dropping Rate and Providing Bandwidth Efficiency in Mobile Networks, International Conference on Parallel Processing, Toronto, August 2000, pp. 581–588.

40. Jiang, J. and Lai, T.H., Bandwidth Management Providing Guaranteed Call Dropping Rates for Multimedia Mobile Networks, IEEE International Conference on Multimedia and Expo, ICME 2000, New York, **2**, 963–966, 2000.

41. Sipilä K., *et al.*, Estimation of Capacity and Required Transmission Power of WCDMA Downlink Based on a Downlink Pole Equation, IEEE Vehicular Technology Conference, VTC'2000-Spring, Tokyo, **2**, 1002–1005, 2000.

42. Laiho-Steffens, J., Wacker, A. and Aikio, P., The Impact of the Radio Network Planning and Site Configuration on the WCDMA Network Capacity and Quality of Service, IEEE Vehicular Technology Conference, VTC'2000-Spring, Tokyo, **2**, 1006–1010, 2000.

43. *WCDMA for UMTS – Radio Access for Third Generation Mobile Communications*, edited by Holma, H. and Toskala, A., New York: Wiley, 2000.

Acronyms

1G	First Generation
1XRTT	Single Carrier Radio Transmission Technology
1XTREME	Branded term for evolution of CDMA-1X systems
2G	Second Generation
3G	Third Generation
3GPP2	Third Generation Partnership Project
3XRTT	Multiple Carriers Radio Transmission Technology
A/D	Analogue to Digital
AAA	Authentication, Authorisation, and Accounting
AC	Authentication Centre
ACF	Auto-correlation Function
ACMP	Access Channel MAC Protocol
ACMP	Access Channel MAC Protocol
ACPAC	Access Channel MAC Layer Packet Authentication Code
ACTS	Advanced Communications Technologies and Services
ALMP	Air Link Management Protocol
AM	Amplitude Modulation
AMC	Adaptive Modulation and Coding
AMP	Address Management Protocol
AMP	Address Management Protocol
AMPS	Advanced Mobile Phone Service
AM-SSB	Single-Side Band Amplitude Modulation
ANSI	American National Standards Institute
AP	Authentication Protocol
APiCh	Auxiliary Pilot Channel
ARIB	Association of Radio Industries and Businesses
ARQ	Automatic Repeat and Request
ATDPiCh	Auxiliary Transmit Diversity Pilot Channel
ATI	Access Terminal Identifier
AUC	Authentication Centre
AWGN	Additive White Gaussian Noise
BAM	Basic Access Mode
BATI	Broadcast Access Terminal Identity
BCCh	Broadcast Control Channel
BCD	Binary Coded Decimal
BER	Bit Error Rate

Designing CDMA 2000 Systems   L. Korowajczuk et al.
© 2004 John Wiley & Sons, Ltd   ISBN: 0-470-85399-9

BLER	Block Error Rate
BPSK	Binary Phase Shift Keying
BRO	Bit-Reversal Order Interleaving
BS	Base Station
BSC	Base Station Controller
BTS	Base Transceiver Station
C/I	Carrier signal to Interference Ratio
CACh	Common Assignment Channel
CACh	Common Assignment Channel
CAVE	Cellular Algorithms for Validation and Encryption
CCF	Cross-correlation Function
CCITT	Consultative Committee on International Telegraphy and Telephony
CCMP	Control Channel MAC Protocol
CCMP	Control Channel MAC Protocol
CCS BRO	Complex Cyclic-Shift Bit-Reversal Order Interleaving
CDCP	Call Data Collection Point
CDG	CDMA Development Group
CDGP	Call Data Generation Point
CDIS	Call Data Information Source
CDMA	Code Division Multiple Access
cdmaOne	Branded term for IS-95A and IS95B CDMA systems
CDRP	Call Data Rating Point
CELP	Code-Excited Linear Predictive
CF	Collection Function
CODEC	Coder/Decoder
CODIT	Code Division Testbed
CoS	Class of Service
CPCCh	Common Power Control Channel
CRC	Cyclic Redundancy Check
CRDB	Coordinate Routing Database
CSC	Customer Service Centre
CSP	Connected State Protocol
CTIA	Cellular Telecommunications and Internet Association
CWTS	China Wireless Telecommunication Standard
D-AMPS	Digital AMPS (IS-54)
DBR	Data Burst Randomiser
DC	Direct Current
DCCh	Dedicated Control Channel
DCE	Data Circuit Equipment
DEMUX	Demultiplexer
DF	Delivery Function
DH	Diffie-Hellman
DLL	Delay Lock Loop
DQPSK	Differential Quadrature Phase Shift Keying
DRC	Data Rate Control
DS	Direct Sequence
DSP	Digital Signal Processing

DS-SS	Direct Sequence Spread Spectrum
DTMF	Dual Tone Multi Frequency
EACh	Enhanced Access Channel
EDGE	Enhanced Data GSM Environment
EHDM	Extended Handoff Direction Message
EIA	Electronic Industries Alliance
EIB	Erasure Indicator Bit
EIR	Equipment Identity Register
EP	Encryption Protocol
EPSMM	Extended Strength Measurement Message
ERP	Effective Radiated Power
ESME	Emergency Service Message Entity
ESN	Electronic Serial Number
ESNE	Emergency Service Network Entity
ETSI	European Telecommunications Standards Institute
EVDO	Evolution to Data Only
EVDV	Evolution to Data and Voice
EVRC	Enhanced Variable Rate Coder
FAPiCh	Forward Auxiliary Pilot Channel
FB BRO	Forward-Backwards Bit-Reversal Order Interleaving
FBCCh	Forward Broadcast Control Channel
FCC	Federal Communications Commission
FCCCh	Forward Common Control Channel
FCP	Flow Control Protocol
FCPCCh	Forward Common Power Control Channel
FCS	Frame Check Sequence
FDCCh	Forward Dedicated Control Channel
FDD	Frequency Division Duplex
FDMA	Frequency Division Multiple Access
FEC	Forward Error Correction
FER	Frame Error Rate
FFCh	Forward Fundamental Channel
FFH	Fast Frequency Hopping
FH	Frequency Hopping
FM	Frequency Modulation
FM	Format Mode
FMA1	FRAMES Multiple Access Based on TDMA
FMA2	FRAMES Multiple Access Based on CDMA
FPC	Forward Power Control
FPCh	Forward Paging Channel
FPDCCh	Forward Packet Data Control Channel
FPDCh	Forward Packet Data Channel
FPiCh	Forward Pilot Channel
FQPCh	Forward Quick Paging Channel
FRAMES	Future Radio Wideband Multiple Access System
FSCCh	Forward Supplemental Code Channel
FSCh	Forward Supplemental Channel

FTCh	Forward Traffic Channel
FTCMP	Forward Traffic Channel MAC Protocol
FTCMP	Forward Traffic Channel MAC Protocol
FTDPiCh	Forward Transmit Diversity Pilot Channel
FTP	File Transfer Protocol
GHDM	General Handoff Direction Message
GIS	Geographical Information System
GPM	General Page Message
GPRS	General Packet Radio Service
GPS	Global Positioning System
GSM	Global System for Mobile Communications
GSM-MAP	GSM- Mobile Application Part
HA	Home Agent
HCM	Handoff Completion Message
HDR	High Data Rate
HLR	Home Location Register
HRPD	High Rate Packet Data
HTTP	Hypertext Transfer Protocol
IAP	Intercept Access Point
IDP	Idle State Protocol
IEEE	Institute of Electrical and Electronics Engineers
IMSI	International Mobile Subscriber Identifier
IMT2000	International Mobile Telecommunications 2000
IP	Internet Protocol
IS-95	International Standard 95
ISDN	Integrated Service Digital Network
ISP	Initialisation State Protocol
ITAR	Internal Traffic and Arms Regulation
ITAR	International Traffic and Arms Regulation
IWF	Inter-working Function
KEP	Key Exchange Protocol
KTF	Korea Telecom Freetel
LA	Link Adaptation
LAC	Layer Access Control
LAN	Local Area Network
LC	Long Code
LFSR	Linear Feedback Shift Register
LMDS	Local Multipoint Distribution System
LOS	Line of Sight
LPDE	Local Position Determining Entity
LSB	Least Significant Bit
LUP	Location Update Protocol
MAC	Media Access Control
MAHO	Mobile Assisted Handoff
MATI	Multicast Access Terminal Identity
MC	Multi-carrier
MCC	Mobile Country Code

MC CDMA	Multi-carrier CDMA
MCS	Modulation and Coding Scheme
ME	Mobile Equipment
MIN	Mobile Identification Number
MLS	Maximal Length Sequences
MM	Mixed Mobile
MMDS	Multichannel Multipoint Distribution Service
MNC	Mobile Network Code
MPC	Mobile Position Centre
MPEG	Moving Picture Experts Group
MS	Mobile Station
MSB	Most Significant Bit
MSC	Mobile Switching Centre
MSGR	Modular Shift-Register Generator
MSIN	Mobile Station Identification Number
MT	Mobile Terminal
MUD	Multi-User Detection
MUX	Multiplexer
MWNE	Managed Wireless Network Entity
N-AMPS	Narrowband AMPS
NE	Network Entity
NID	Network Identification
NLUM	Neighbour List Update Message
NMS	Nordic Mobile System
NMSI	National Mobile Station Identity
NPDB	Number Portability Database
OMP	Overhead Messages Protocol
OOK	On-Off Keying
O-QPSK	Offset Quadrature Phase Shift Keying
OSF	Operations Systems Function
OSI	Open Systems Interconnection
OTAF	Over-the-Air Service Provisioning Function
OTD	Orthogonal Transmit Diversity
PACA	Priority Access and Channel Assignment
PCB	Power Control Bit
PCF	Packet Control Function
PCG	Power Control Group
PCM	Pulse Code Modulation
PCP	Packet Consolidation Protocol
PCS	Personal Communication Systems
PCSCh	Power Control Sub-Channel
PDA	Personal Digital Assistant
PDC	Personal Digital Communications
PDE	Position Determining Entity
PDN	Packet Data Network
PDSN	Packet Data Serving Node
PDU	Protocol Data Unit

PER	Packet Error Rate
PLMN	Public Land Mobile Network
PMRM	Power Measurement Report Message
PN	Pseudo-noise
PNLC	Long PN Sequence
PS	Pilot Strength
PSK	Phase Shift Keying
PSMM	Pilot Strength Measurement Message
PSMMM	Pilot Strength Measurement Mini-Message
PSPDN	Packet-Switched Public Digital Network
PSTN	Public Switched Telephone Network
QAM	Quadrature Amplitude Modulation
QCELP	Qualcomm's Code-Excited Linear Predictive
QIB	Quality Indicator Bit
QOF	Quasi-Orthogonal Function
QoS	Quality of Service
QPCh	Quick Paging Channel
QPSK	Quadrature Phase Shift Keying
R	Reserved
RA	Reverse Link Activity
RAB	Reverse Activity Bit
RACE	Research of Advanced Communication Technologies in Europe
RACh	Reverse Access Channel
RackCh	Reverse Acknowledgement Channel
RAM	Reservation Access Mode
RAN	Radio Access Network
RATI	Random Access Terminal Identity
RBS	Radio Base Station
RC	Radio Configuration
RCCCh	Reverse Common Control Channel
RCELP	Relaxed Code-Excited Linear Predictive
RCQICh	Reverse Channel Quality Indicator Channel
RDCCh	Reverse Dedicated Control Channel
RF	Radio Frequency
RFCh	Reverse Fundamental Channel
RLP	Radio Link Protocol
RLP	Radio Link Protocol
RPC	Reverse Link Power Control
RPC	Reverse Power Control
RPCSCh	Reverse Power Control Sub-Channel
RPiCh	Reverse Pilot Channel
RRI	Reverse Link Data Rate Index
RRI	Reverse Rate Indicator
RRM	Radio Resource Management
RS	Rate Set
RSCCh	Reverse Supplemental Code Channel
RSCh	Reverse Supplemental Channel

RTCh	Reverse Traffic Channel
RTCMP	Reverse Traffic Channel MAC Protocol
RUP	Route Update Protocol
SCh	Supplemental Channel
SCI	Synchronised Capsules Indicator
SCM	Station Class Mark
SCP	Service Control Point
SCP	Session Configuration Protocol
SFH	Slow Frequency Hopping
SID	System Identification
SKT	South Korean Telecom
SLP	Signalling Link Protocol
SMCS	Spreading Factor and Modulation Coding Scheme
SME	Short Message Entity
SMP	Session Management Protocol
SMP	Session Management Protocol
SMS	Short Messaging Service
SN	Service Node
SNR	Signal to Noise Ratio
SOM	Start of Message
SP	Security Protocol
SR	Spreading Rate
SS	Spread Spectrum
SSD	Secret Shared Data
SSDT	Site Selection Diversity Transmission
SSDUP	SSD Update Process
SSRG	Simple Shift-Register Generator
STS	Space Time Spreading
SyncCh	Synchronisation Channel
T1P1	Standards Committee Telecommunications 1 - Technical Subcommittee P1
TA	Terminal Adapter
TCP	Transmission Control Protocol
TD	Transmit Diversity
TDC	Transmission Duty Cycle
TDD	Time Division Duplex
TDMA	Time Division Multiple Access
TDPiCh	Transmit Diversity Pilot Channel
TE	Terminal Equipment
TH	Time Hopping
TH-CDMA	Time Hopping CDMA
TIA	Telecommunications Industry Association
TM	Traffic Mode
TT	Traffic Type
TTA	Telecommunication Technology Association
TTC	Telecommunication Technology Committee
UATI	Unicast Access Terminal Identity
UCRP	Unique Challenge-Response Procedure

UHDM	Universal Handoff Direction Message
UIM	User Identity Module
UM	Radio Interface
UMTS	Universal Mobile Telecommunications System
UPR	User Performance Requirements
UTC	Universal Coordinated Time
UZID	User Zone Identification
VLR	Visitor Location Register
VMS	Voice Message System
VOD	Video on Demand
VoIP	Voice over IP
VPN	Virtual Private Network
WAN	Wide Area Network
WCDMA	Wideband CDMA
WNE	Wireless Network Entity
WWW	World Wide Web

Index

1xEVDO (*See* EVDO)
1xEVDV (*See* EVDV)
1xRTT 19, 20, 42, 45
1XTREME 21
2.5G 18
2G 15, 16, 18
3G 18, 21, 25, 26, 29, 34, 35, 46, 50
3GPP 29
3GPP2 21, 22, 29, 34, 45, 49, 352, 467
3xRTT 19, 20, 42
64-ary encoding 448

A_Key 281, 285, 289
Access Channel 187, 206, 207, 215
Access
  attempt 240, 242, 254–8, 260–2, 278, 281, 285,
    286, 289, 339
  handoff 291, 316–8, 320–1, 323–1, 333–40
  parameters message 249, 254, 257, 258, 260,
    272, 274, 280
  priority access 248, 249, 262, 271
  probe 255–8, 260, 261, 294, 296–300, 302,
    303, 313, 338, 339
  State 239–43, 246–9, 252–4, 260–7, 270–2,
    274, 279
  system access procedure 254
ACF 56, 58, 59
ACK 45
ACMP 361, 362, 367, 371, 390, 395, 406, 407
Acquisition
  pilot channel acquisition 242, 243
  sync channel acquisition 242, 243, 246
Active mode 267
Active set 318, 320, 325–8, 330, 338, 339
Adaptive modulation and coding 353, 357
Admission control 853, 885–7, 895, 897
ALMP 360, 362, 367, 370, 383
Always-on 347, 348
AMC 43, 353, 357
AMP 360, 367, 369, 376, 381

AMPS 15–7, 19, 29
Analog Combiner 491
ANSI-J-STD-008 17
Antenna
  azimuth 572, 576, 581
  cross-polarized antenna 575
  dipole antenna 573–5, 577
  directive antenna 577, 581
  elevation 572, 581, 596
  far field 569
  horn antenna 576, 579
  isotropic antenna 569, 572, 575, 582, 586
  near field 568, 569
  omni antenna 572, 573, 575, 576
  polarization 581
  tilt 581, 582
AP 361
Application
  application layer 381
  default packet application 377, 379, 380
  default signaling application 377
  signaling application 359, 377, 378, 380
Architecture 16, 17, 21, 22, 44, 357
ARQ 353, 467
Attempt
  request 252, 254, 255, 257, 258, 260, 262,
    264–7, 271, 272, 278, 294, 296, 300, 301,
    311, 324, 337
  response 254–7, 260, 261, 281, 282, 288, 290,
    294, 295, 300, 339
Authentication 239, 261, 262, 266, 271, 275, 276,
    280–2, 285, 286, 288–90, 352, 354, 358,
    360, 367, 385–8
AWGN 479, 544, 661, 729, 730

Background 27, 853
Bandwidth 1–4, 9, 10, 471, 473–6, 479, 506, 513,
    514, 518, 553, 558, 578, 658, 662, 667, 673, 692
Bandwidth scheduling 834
Base Station Controller 480, 484

Designing CDMA 2000 Systems   L. Korowajczuk et al.
© 2004 John Wiley & Sons, Ltd   ISBN: 0-470-85399-9